T0177139

Nanostructures and Nanotechnology

Focusing on the fundamental principles of nanoscience and nanotechnology, this carefully developed textbook will equip students with a deep understanding of the nanoscale.

- Each new topic is introduced with a concise summary of the relevant physical principles, emphasising universal commonalities between seemingly disparate areas, and encouraging students to develop an intuitive understanding of this diverse area of study.
- Accessible introductions to condensed matter physics and materials systems provide students from a broad range of scientific disciplines with all the necessary background.
- Theoretical concepts are linked to real-world applications, allowing students to connect theory and practice.
- Chapters are packed with engaging color illustrations and problems to help students develop and retain their understanding, and are accompanied by suggestions for additional reading.

Containing enough material for a one- or two-semester course, this is an excellent resource for senior undergraduate and graduate students with backgrounds in physics, chemistry, materials science, and electrical engineering.

Douglas Natelson is a Professor of Physics and Astronomy at Rice University, where he has taught courses on nanoscale science and technology for fifteen years. He is a Fellow of the APS and AAAS and blogs at nanoscale.blogspot.co.uk.

Nanostructures and Nanotechnology

DOUGLAS NATELSON

Rice University, Houston

CAMBRIDGE
UNIVERSITY PRESS

CAMBRIDGE
UNIVERSITY PRESS

University Printing House, Cambridge CB2 8BS, United Kingdom

Cambridge University Press is part of the University of Cambridge.

It furthers the University's mission by disseminating knowledge in the pursuit of education, learning and research at the highest international levels of excellence.

www.cambridge.org
Information on this title: www.cambridge.org/9780521877008

© Cambridge University Press 2015

First published 2015

Printed in the United Kingdom by Bell and Bain Ltd

A catalogue record for this publication is available from the British Library

Library of Congress Cataloguing in Publication data
Natelson, Douglas, 1970–
Nanostructures and nanotechnology / Douglas Natelson, Rice University, Houston.
pages   cm
Includes bibliographical references and index.
ISBN 978-0-521-87700-8 (Hardback)
1. Nanostructured materials.   I. Title.
TA418.9.N35N36489 2015
620$'$.5–dc23   2014044739

ISBN 978-0-521-87700-8 Hardback

Contents

Preface

This book is intended to provide a physical foundation for students interested in nanoscale science and technology. Developed while teaching a two-course graduate sequence on the topic, this book is my attempt to lay out the physical underpinnings of this incredibly broad topic while striking a balance between depth and approachability.

When I set out to develop and teach these courses, I found that most books on this subject were very specialized (for example, dealing only with nanoscale electronics), more focused on research rather then pedagogy (collections of review articles rather than an actual textbook), or not sufficiently technical (more like a series of *Scientific American* articles rather than a quantitative approach). I have tried to get to the physical basis of nanoscale science, the origins of the fascinating properties of materials at previously inaccessible size scales. A common thread through much of the material is the breakdown of the simplifying approximations that we have made in developing our physical models of macroscopic systems. I've also tried to indicate the underlying connections between some superficially disparate topics (e.g., band theory, coupled mechanical oscillators, and plasmons). Hopefully this approach allows students to develop an intuition for, and the ability to reason critically about, the nanoscale world. By focusing on the fundamentals rather than the latest research results (though those are mentioned when appropriate), I also hope that this text will stand the test of time, rather than appearing dated as soon as it is published. Of course, during the writing of this book, a number of other texts more or less in a similar or complementary spirit have appeared. These include *Introduction to Nanoscale Science and Technology*, edited by M. Di Ventra, S. Evoy, and J. R. Heflin, Jr. (Springer, 2004); and *Introduction to Nanoscience* by S. Lindsay (Oxford, 2009).

When teaching this material as a course or course sequence, I recommend supplementing the exercises with short-answer questions based on readings from the current literature. I had reasonable success assigning midterm and final papers. Whether they want careers in academia or industry, students need to become facile at writing, both short (one paragraph) responses to conceptual questions or questions about readings, and longer (5–10 page) essays that demonstrate analysis and critical thinking.

Many of the topics in this book deserve much more extensive treatment than what I have been able to provide. To compensate for the limitations necessitated by finite space, I have tried to give ample suggestions for further reading, including book-length treatments and review articles. Some areas, while extremely interesting, I decided were "too physicsy" for the intended broader audience. This is the reason for my extremely limited mention of both nanoscale superconductivity and the integer and fractional quantum Hall effects. I have similarly steered clear of quantum computing, a discipline certainly connected to nanoscale science and nanoelectronics, but just as certainly a distinct field. Likewise, some topics (e.g., the physical chemistry of catalysis at nanoparticle surfaces; the molecular biology of many biological motors) are far enough removed from my own expertise that I could not possibly treat them adequately. If you feel that your favorite nano topic gets short shrift, *mea culpa*. I have also done my best to cite explicitly in the text and in the "Suggested Reading" the many books and reference works that I consulted during the writing of this book. Truly, I would never have been able to put this together without the hard work of many authors before me. Any omissions or mistakes are my responsibility.

This book would never have been possible without the support and encouragement of many people and groups over the last several years. I would particularly like to acknowledge the Alfred P. Sloan Foundation, for their original sponsorship of a professional masters program that spurred the development of this course material. I also owe a debt to the National Science Foundation (specifically awards DMR-0347253, DMR-0855607, and DMR-1305879), whose educational mandate dovetailed perfectly with the opportunity to create these courses and this book. I hope this work has the educational broader impact that NSF is meant to encourage.

My colleagues within Rice University have been nothing but supportive, especially my former department chairman Professor Barry Dunning, my former dean Professor Kathy Matthews, my current chairman Professor Tom Killian, and my faculty colleagues within the Physics and Astronomy Department. I thank all the students, both in my research group and in the courses, that have helped me formulate my thinking about these subjects through their rigorous questions and insightful conversations. They are too numerous to mention, but I'm very appreciative of their insights. My father, Michael Natelson, and my faculty colleague Professor Rui-Rui Du deserve special gratitude for their time spent reading the manuscript. Finally, special thanks to my wife and sons, whose love and support have helped keep me sane during this whole process.

PART I

Introduction and overview 1

Nanotechnology has seized both the public and scientific communities' imaginations, and it's not hard to see why. From K. Eric Drexler's and Ray Kurzweil's visions of self-reproducing engineered nanomachines building macroscale structures out of single crystal diamond and swimming through our capillaries repairing damaged cells, to talk of building an elevator directly to geosynchronous orbit, the promise of nanoscale science and engineering has been described to an enthusiastic public. Billions of dollars of research funding have been directed into this area, and further billions of dollars are already being spent on commercial products that are self-described as examples of nanotechnology. Estimates of the global economic impact of nanotechnology in the next ten years exceed $1 trillion, and are rising.

How seriously should we take these exciting visions? Is nanotechnology a "disruptive technology", a distinct advance that will reshape the world? What are the *real* potential impacts of the ability to engineer the structure and composition of materials on the nanometer scale? What are the limitations imposed by Nature (that is, physics and chemistry) on what is possible? What physical principles become relevant at small scales that on the one hand set limitations, but on the other provide new opportunities? What scientific questions remain to be answered at the nanometer scale, and why? How can people manipulate and characterize materials on these scales?

Hopefully this book will help you answer these questions, or at least give you a good idea of what to consider when trying. Even though the nanometer scale is very different from our everyday experience, it is possible to develop an intuition about how matter will behave at that extremely small level.

1.1 What is nanotechnology?

Nanotechnology is an extraordinarily broad term. According to the US government,[1] nanotechnology is "Research and technology development at the atomic, molecular or macromolecular levels, in the length scale of approximately 1 – 100 nanometer range, to provide a fundamental understanding of phenomena and materials at the nanoscale and to create and use structures, devices and systems that have novel properties and functions because of their small and/or intermediate size." That's quite a mouthful.

---

[1] See www.nano.gov.

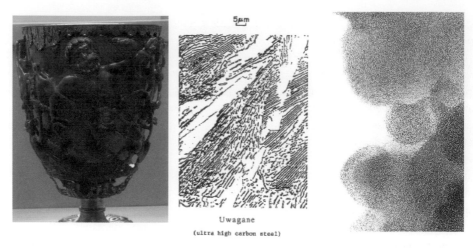

Uwagane
(ultra high carbon steel)

Fig. 1.1   Nanotechnology in history. Left: the Lycurgus cup, a Roman goblet made using glass containing colloidal gold nanoparticles (British Museum [1]). Center: an electron micrograph of micro- and nanostructured steel, *uwagane*, of the type used to produce Japanese *katana* swords. From [2]. Right: carbon black nanoparticles used to improve the stiffness and strength of rubber (Electrochemistry Encyclopedia [3]).

Colloquially, nanotechnology has come to encompass just about any technology that involves controllably engineering structures with at least some critical dimension less than 100 nm in extent. By that definition, we have been living in the Age of Nanotechnology for hundreds of years (see Fig. 1.1). Colored Roman glass and some medieval stained glasses owe their remarkable optical properties to embedded colloidal metal particles on that scale. Damascene steel and Japanese *katana* sword steel are made by elaborate processes that end up producing nanostructured materials with impressive mechanical properties. Metal films for coating other metals or glass (mirrors) have long been produced with thicknesses less than 100 nm. Carbon black, the sooty stuff used as an additive to enhance the mechanical properties of rubber, contains particles as small as a few nanometers.

What distinguishes current efforts from these past achievements are understanding and control. In the past few decades, there have been tremendous advances in the chemistry and physics of materials relevant at the nanometer scale. Simultaneously and symbiotically, new tools have been developed that allow people to "see" what is happening at that scale with unprecedented precision. Computational prowess has also grown astronomically over the same time period, enabling calculations that formerly would have been impractical. The net result of all this is that people now have the ability to design, engineer, and construct a vast array of structures on the nanometer scale to take advantage of the unique properties of matter when tailored on those dimensions.

One reason it is difficult to define nanotechnology is that much nano research takes place at the boundaries between traditional disciplines. For example, consider a single recent innovation: the use of semiconductor nanocrystals as markers for particular biological processes. The unique optical properties of the nanocrystals result directly from physics, specifically the impact of finite crystal size on the allowed electronic states in the material. The nanomaterial itself is made using a batch synthesis process in solution at moderate temperatures in a typical chemistry laboratory fume hood. Surface chemistry is exploited to coat the nanocrystals with a surfactant layer that prevents them from aggregating and falling out of solution. The nanocrystals are characterized using standard materials science tools such as transmission electron microscopy (TEM) and x-ray photoemission spectroscopy (XPS). Biochemistry expertise is required to get the nanocrystals to bind appropriately in cells, and biological microscopy techniques are needed to do the actual sensing of processes in living systems. In this respect nanotechnology is truly an interdisiplinary endeavor.

1.2 Sizes of things

It helps to have a "big picture" view of the very small so we can see what's likely to be relevant. Just how small *is* the nanometer scale? Figure 1.2 is an example of the typical answer to this question, trying to bridge the realm of everyday experience with the nanoscale. Rather than just looking at the relative sizes of things, let's consider the figure in terms of the physics relevant in the different size regimes.

At the comparatively macroscopic scale is a human hair, on the order of 100 μm in diameter. From the physics perspective, this object and its interactions with the world are

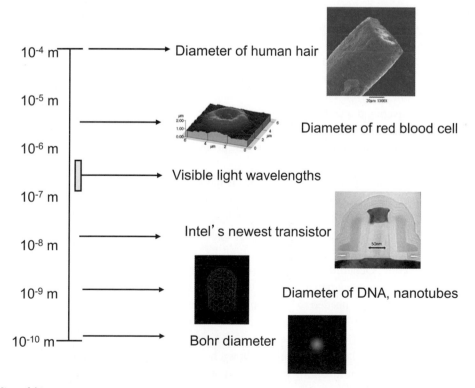

Sizes of things.   Fig. 1.2

"big". It contains a large number, $\sim 10^{21}$, of atoms, bound together at the molecular level into macromolecules called proteins, primarily of a single type, keratin.

To describe the mechanical properties of the hair, no one would ever consider solving the quantum mechanical or even classical equations of motion for all the consituent atoms; that would be computationally untenable, and manifestly unwise. Fortunately, it is also unnecessary from a practical standpoint. The atoms, locked in the protein molecules, interact with each other strongly enough and in the right ways to take on certain *emergent properties* that describe the mechanical response of the whole ensemble. It makes sense to talk about an average mass density, and an elastic modulus that relates the elongation of the hair to the average internal forces in the hair – related to the Hooke's Law spring constant you think of from first-year physics. If the hair is pulled tight and plucked, the intelligent way to think about the collective motion of all the atoms is in the form of sound waves with wavelengths spanning many thousands of atoms.

Similarly, no one would ever consider describing the optical properties of the hair starting from the optical transitions known from gas phase spectroscopy of the component atoms. Instead, the ensemble of molecules collectively has optical properties that can be succinctly understood as some (complex) index of refraction.

These collective properties and their emergence in "large" systems are the intellectual core of *condensed matter* physics. Even when the underlying rules governing the properties and interactions of individual objects are simple, the ensemble can show collective properties that are profound, surprising, and extremely rich. As Nobel laureate physicist Phil Anderson said in his now famous 1972 Science paper, "More is *different*".

Moving down the ladder, we come to a single red blood cell a few microns in diameter. It's not too inaccurate to describe this cell as a bag made from a lipid bilayer membrane, filled with a bunch of liquid. Even though the membrane is only a couple of molecules thick, it is quite extended and has some features of macroscopic objects, like an elastic modulus within the layer. The mostly-water inside the cell is also pretty fairly described by "continuum" averaged quantities, like a mass density, a viscosity that describes dissipation within the fluid, an index of refraction, ionic concentrations, etc. Except in some very reductionist sense, quantum mechanics is not necessary to think about in understanding these properties.

We now pass through the wavelengths of visible light. While electromagnetic radiation comes in quanta called photons, it is still often useful to consider the continuum (huge number of photons) limit of thinking about light as propagating waves of electric and magnetic fields. There is nothing particularly special about this length scale, except that it corresponds to photons with energies of about 1–2 eV that our eyes are photochemically adept at detecting. Generally it is challenging to resolve (via scattered light) objects separated by distances much smaller than the wavelength of light being used for illumination. In short, other tools are needed to "see" objects smaller than this.

The next item down is a cross-sectional image of a silicon metal-oxide-semiconductor field-effect transistor (MOSFET), the basis for much of modern electronics. We will discuss these in detail later. The example shown here is state-of-the-art as of 2006, with a critical spacing between incoming (source) and outgoing (drain) electrodes of 50 nm, and a crucial insulating layer only 1.2 nm thick. In 2006 there were $\sim 10^8$ of these in a typical high end microprocessor. Clearly the semiconductor industry has been working at the nanoscale for some time! Interestingly, even with dimensions such as these, the material properties are still well described as "bulk-like"; the operating characteristics of the MOSFET are not terribly different from those of appropriately scaled much larger devices.

This would no longer be the case, however, if the Si were structured on the 1–2 nm scale, the next stop on the road to the atomic regime. As indicated in the figure, this is approximately the diameter of an individual single-walled carbon nanotube, or the double helix of deoxyribonucleic acid (DNA). At this size scale it is often no longer appropriate to use continuum quantities to understand the mechanical, electronic, and optical properties of materials, and quantum mechanical details begin to matter. A significant portion of this book will examine why this is so, and the implications.

Finally, at the bottom end of the size scale for neutral, stable matter, we are left with individual atoms. For hydrogen, the Bohr diameter is approximately 0.1 nm. As should be familiar from chemistry and physics classes back to high school, the electronic and optical properties of single atoms are completely dominated by quantum mechanics.

If we consider the surface of the hair on this kind of distance scale, it is extremely lumpy and inhomogeneous. If the hair is exposed to ambient atmosphere, it is undoubtedly coated with a physisorbed layer of water, as well as some amount of small molecular weight hydrocarbon compounds. Further, its surface is in constant motion, with adsorbed contaminants reshuffling themselves, bonds vibrating at terahertz frequencies, ions and solvated electrons swapping charge over a broad distribution of time scales, gas molecules from the air impinging at a rate high enough that each atom on the surface is hit on average once a nanosecond. At the single nanometer level, apparently clean surfaces are often dirty, and apparently quiescent equilibrium is alive with activity.

Somewhere between the macroscale and the atomic scale, we have passed from the classical, bulk world of Newton's laws, continuum elasticity theory, and fluid mechanics. Instead we have entered the realm of quantum mechanics, interfacial effects, local fluctuations and deviations from equilibrium, and the breakdown of continuum approximations. This transition happens because as a system's size, L, is reduced, length scales relevant for particular physical processes shift from being negligibly small compared to L to being comparable to or much larger than L. In the next section, we will see just how this works in a bit more detail, and introduce several of these physically motivated lengths, using a particular example familiar, at least crudely, from everyday experience.

1.3 Important length scales: breaking a wire

What happens when a metal wire breaks? As a concrete example, let's consider a gold wire with a diameter $d = 25$ μm. The choice of gold already simplifies our thought experiment: gold is ductile, so we won't have to worry about crack nucleation and propagation, and bulk gold is chemically inert under ambient conditions, so we needn't be concerned about oxidation. To the naked eye, nothing particularly interesting or surprising occurs. When a tensile force is applied to the ends of a wire of length L, the wire deforms, first elastically (so that when the force is removed the wire springs back to its original length), then plastically (irreversibly). Under continued pulling, a "neck" forms somewhere along the wire, where the plastic deformation reduces the local diameter. The deformation speeds up, the neck shrinks further, and eventually the wire breaks completely, popping into two separate pieces. If one were to measure the electrical conductance through the wire every 0.1 s by applying a known voltage, V, across it and measuring the resulting current, I, one would see that the conductance, $G \equiv dI/dV$ remains fairly constant, perhaps decreasing slightly before dropping to zero after the wire breaks.

1.3.1 Length scales in play

Examining the wire and the breaking process on smaller length scales and more finely in time is very revealing, and serves as a guide to some of the physics that we will see later. First, consider the microstructure of the wire before the breaking process. The wire is polycrystalline, composed of many individual Au grains, typically 20–40 nm in size. Each grain consists of Au atoms stacked in a close-packed arrangement called face-centered cubic (FCC), with atom centers separated by a *lattice parameter*, $a \approx 0.4$ nm. Generally the grains are randomly oriented, so that the surface of the wire consists of a collection of different crystallographic faces.

The detailed crystal structure is determined in part by the electronic interactions between the atoms. It turns out not to be too bad an approximation to think of the metal as consisting of a lattice of FCC-stacked ion cores, with each ion having a positive charge, surrounded by an electronic "fluid" with a density of around 5.9×10^{22} electrons per cm^3. Even though the electrons relevant to electronic conduction in the wire are negatively charged and therefore interact repulsively, we can often get away with ignoring their interactions. One reason for this is that, by rearranging themselves, the electrons can *screen* excess charge on a distance scale of roughly the *Thomas–Fermi screening length*, $r_{\text{TF}} \approx 0.4$ nm in this system.

One can think of these electrons "semiclassically",[2] and ask how far a typical electron travels before scattering – that is, before participating in some process that reorients its direction of propagation. Processes that do not change the energy of the electron are called elastic, while those in which the electron gains or loses energy to some other degree of freedom in the material are inelastic. The typical distance traveled by an electron before such a scattering process is called the *mean free path*, ℓ. As we shall see later, any break in the periodicity of the stacking of the atoms (vacancies, grain boundaries, surfaces) can cause elastic scattering, In the absence of inelastic scattering, the grain size mentioned above is a reasonable estimate of ℓ_{e}, the elastic mean free path.

The quantum mechanical nature of the conduction electrons means that they have wave-like properties. The effective wavelength of the electrons relevant for conduction, the *Fermi wavelength*, is around $\lambda_{\text{F}} \approx 0.5$ nm in bulk Au.[3] Like all wave phenomena, the electronic waves have some phase. Waves that are in phase at a given position interfere constructively, while those that are out of phase by π radians interfere destructively. These interference

---

[2] We'll learn about the details of this in Chapter 2.

[3] As we shall see, the similar magnitudes of a, r_{TF}, and λ_{F} are not coincidental.

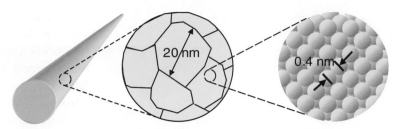

A gold wire that seems roughly homogeneous is really polycrystalline with a grain size of 20 nm, and each grain is a face-centered cubic crystal with atomic separation of 0.4 nm.

Fig. 1.3

effects, which have no classical analog, are generally undetectable on macroscopic length scales because of *decoherence*. Inelastic interactions with environmental degrees of freedom effectively randomize the phases of the electronic waves on a distance scale, L_ϕ, the *coherence length*. At room temperature in Au, $L_\phi \sim 1-2$ nm.

Some of these same inelastic interactions, including electron–electron and electron–vibrational scattering, are what establish thermal equilibrium for the electrons. There is therefore a related distance scale, L_i, the inelastic length, over which a nonthermal distribution of electrons equilibrates to the temperature of the system as a whole. At room temperature it's not too bad to assume $L_i \sim L_\phi$, though this can change at low temperatures.

Finally, for completeness recall that electrons possess intrinsic angular momentum called "spin". In the absence of magnetic correlations like ferromagnetism, the spins point in no particular direction. One can ask, if a particular electron is prepared with its spin aligned along some known direction, how far does the electron propagate before some process effectively randomizes the direction of the electron's spin. In gold, there is strong spin–orbit scattering; that is, from the point of view of the moving electron, the electric field from the gold ion cores looks like an effective magnetic field that causes the spin to precess. The *spin–orbit scattering length* in this material is around $L_{SO} \approx 30$ nm. On distances short compared to L_{SO}, it's not bad to think about the spin of an electron as a well defined quantity, while on longer distance scales, the electron trades angular momentum back and forth with the lattice.

So, the stage is set. Before the breaking begins, the hierarchy of lengths is:

$$L \gg d \gg \ell_e \sim L_{SO} > \ell_i \sim L_\phi \sim L_i > \lambda_F \sim r_{TF} \sim a. \qquad (1.1)$$

There are actually two more length scales relevant during the breaking process, as shown in Fig. 1.4: δ, defined here as the rough diameter of the most constricted part of the "neck", and λ, the rough length of that region. Before the breaking process, it doesn't make sense to think about δ and λ, since there is no constriction.

1.3.2 The breaking process up close

Start pulling on the ends of the wire. The tensile force is distributed across the wire cross-section. Pull hard enough, and somewhere within the wire, grains start to change

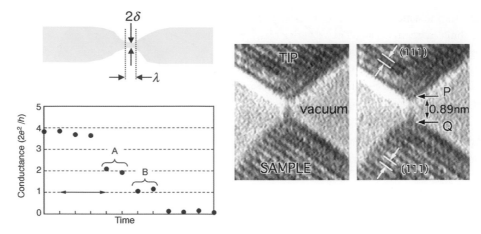

Breaking a gold wire. Top left: definitions of λ and δ, the characteristic length scales of the necking constriction. Bottom left: conductance as a function of time as a gold wire is pulled apart. Right: *in-situ* transmission electron micrographs of the end of the breaking process, showing the formation of a one-dimensional chain of Au atoms just before complete breaking. Adapted from [4].

irreversibly. Atoms at the boundaries between grains are less constrained than those within grains, so they are more free to move in response to local forces. Defects within the grains (vacancies, interruptions in the fcc stacking called dislocations) can also move, propagating to grain boundaries under the applied forces. Once the cross-section somewhere along the wire is reduced below $\pi d^2/4$, the local tensile *stress* increases, since the same applied force can only be spread over a smaller area.

Interesting things start to happen as δ and λ shrink, and trade places in the hierarchy of Eq. (1.1) with the intrinsic scales of different physical processes. The first of these crossovers occurs when λ (on the order of δ for such a ductile material) and δ become smaller than a typical grain size, and therefore smaller than ℓ_e. If a slight voltage bias is applied between the two ends of the wire, one finds that the conductance of the whole system is now limited by the constriction. Electrons passing through the narrow neck are more likely to scatter elastically from the boundaries of the neck than within the neck itself. Similarly, since $\lambda < L_{SO}$, the spin of the electrons remains approximately conserved when traversing the neck.

When $\delta, \lambda \to \ell_i$, another crossover occurs. Now the constriction is so short and narrow that electrons that traverse it do so without any significant scattering taking place in the constriction itself. This is the *ballistic* or *point contact* regime, and raises some interesting questions. The neck now strongly limits the total conduction through the wire, but does so even though electrons pass through that region ballistically. What is the origin of the constriction's resistance? As we shall see in Chapter 5, a ballistic constriction's resistance in this situation is an example of a *contact resistance*, rooted in the difficulty of getting electrons to enter the ballistic region in the first place. If there is no scattering within the constriction, yet the constriction has some resistance R_c, where is the power dissipated from the simple $I^2 R_c$ Joule heating? We shall find that the dissipation takes place where there is inelastic scattering, typically a distance L_i away from the constriction itself in the bulk wire material. This implies that when a voltage is applied to the wire, the electrons in the constriction are not in thermal equilibrium; this complicates the detailed description of such junctions. Indeed, once $\lambda < L_\phi$, properly treating conduction through such a constriction requires consideration of both nonequilibrium effects and quantum interference corrections to the conduction.

The real excitement comes when $\delta \rightarrow \lambda_F, a$, when the diameter of the constriction at its narrowest is only a couple of lattice spacings and (effective) electron wavelengths. The wave character of the electrons becomes of paramount importance. The number of electronic "modes" that span from one side of the constriction to the other is a handful, and depends on the detailed atomic configuration at the junction between the two wire pieces. Measurements of the conductance while passing through this regime reveal discrete steps in G as the pulling continues, corresponding to particular arrangements of the junction atoms. The exact G values vary from junction to junction, but a histogram reveals that the conductance in Au often changes by integer multiples of $G_0 \equiv 2e^2/h$, where e is the charge of the electron (-1.602×10^{-19} C), and h is Planck's constant (6.63×10^{-34} Js). This is *conductance quantization*, and is an example of electronic conduction dominated entirely by quantum mechanics that can be seen at room temperature on a benchtop!

Unsurprisingly, in this limit of a few-atom constriction the details of the particular chemistry of the wire material become very important, and it is no longer sensible to describe the mechanical properties of the neck in terms of the elastic characteristics of bulk Au. Experiments have found that the final stages of neck deformation in breaking Au wires lead to remarkable changes in neck structure. The Au atoms take on certain particular configurations very different from the bulk fcc arrangement, including helical stacking and extended chains only one atom in diameter! In configurations like these, the chemical reactivity, thermodynamic properties, and even the detailed electronic structure of the metal can change; every atom is at a surface, with different chemical bond arrangements than in the bulk.

Finally, one critical Au–Au bond in the constriction breaks, and the wire becomes two separate pieces. The last steps of the breaking process take tens of microseconds, and are followed by complicated reconstructions of the wire surfaces as the applied forces on the freshly exposed wire ends drop precipitously to zero. As this simple example shows, even something as conceptually simple as pulling a wire into two pieces reveals a wealth of physical processes, including rich quantum mechanical effects, once critical system scales pass into the nanometer regime.

1.4 The structure of this book

This text is the outgrowth of teaching a popular two-semester graduate course sequence at Rice University. Students in these classes tended to be first-year graduate students or senior undergraduates, with backgrounds including physics, physical chemistry, electrical engineering, and materials science. This text is intended for a similar audience, though it is written more from a physics point of view largely because of the background of the author. A familiarity with quantum mechanics and statistical physics ideas is very helpful. Many students in the courses had never taken a solid state course of any kind, but they had seen the Schrödinger equation before.

There are two main challenges in approaching such a broad audience in a rapidly moving field: establishing a common foundation of knowledge from which to proceed, and formulating (tractable!) exercises and problems that convey fundamental concepts and ideas. I have tried to do both, at sufficient depth for students to develop an appreciation for and an intuition about nanoscale science, though by necessity individual subtopics (*e.g.* photonics, microfluidics) cannot be covered at a level similar to that of a focused graduate course. In each chapter I try to provide references to other more specialized works – review articles, textbooks, web resources – to enable readers to pursue their particular interests in more detail.

Exercises are a mixture of calculations, derivations, and short response writing. As anyone who has studied quantum mechanics, statistical mechanics, or solid state physics can attest, the number of exactly solvable homework problems is limited. As a result many of the exercises here are comparatively simple, and are meant to stimulate critical thought about major concepts rather than turn students into theorists capable of calculational *tours de force*. The short answer questions, while challenging to grade, have proven very worthwhile. Students need to be able to read critically and write concisely, and the more practice at this the better.

The structure of this book is straightforward. I assume reasonable familiarity with undergraduate level quantum mechanics from either the physics or physical chemistry perspectives. A few particularly important ideas (time independent and time dependent perturbation theory; Fermi's Golden Rule) are reviewed in an appendix. Some background in solid state physics is not essential, but wouldn't hurt.

Without an understanding of why macroscopic materials have the properties they do, it's very hard to appreciate what is different about nanostructured materials. Therefore, Chapter 2 is a lengthy review of solid state physics, with particular emphasis on properties and processes that are likely to be affected by nanostructuring. This chapter is not meant to be an exhaustive treatment or a substitute for a two-semester solid state course, but should give readers of various backgrounds an overview of the concepts and the beginnings of an intuition about solids at the nanoscale. This will include a comparison and contrasting of, for lack of a better description, the physicist's and chemist's approaches to the solid state.

Following this discussion, Chapter 3 briefly examines several of the most common material systems relevant for nanotechnology. In addition to common bulk inorganic materials such as silicon and gallium arsenide, specific attention is given to nanostructured semiconductors, complex and strongly correlated oxides, magnetic semiconductors, and fullerene nanomaterials. A more complete discussion of magnetic materials is left for Chapter 7.

Chapter 4 provides a very abbreviated overview of various techniques commonly used for the fabrication and characterization of materials down to the nanometer scale. Fabrication methods include top–down lithographic patterning, material deposition and removal approaches, and more bottom–up, self-assembly techniques. Characterization methods start from traditional solid state bulk approaches and extend toward the inherently nanoscale, such as scanned probe microscopy. A complete, in-depth treatment of this topic is deserving of its own book. Here the idea is to emphasize the physical bases of the techniques, including their advantages and limitations.

Once readers know "big" physics, Chapter 5 takes a look at specific physical effects relevant on the nanoscale. This includes a discussion of quantum corrections to classical electronic conduction, such as conductance quantization mentioned above, as well as tunneling and quantum confinement. Some thermodynamic consequences of nanostructuring will be included as well.

Chapter 6 begins Part II of the book. With the overview out of the way, chapters will now turn to specific subtopics. Chapter 6 focuses on nanoelectronics, and opens with a more specific physics overview of conventional semiconductor electronic devices, especially the metal-oxide-semiconductor field-effect transistor (MOSFET). This is the longest chapter, largely because this is the nanoarena that has seen the most (physics) research as well as the most direct technological impact. After considering the basic parameters of the MOSFET, technological trends, and physical limitations to transistor scaling, the chapter turns to potential successor technologies. Topics presented in some detail include unconventional MOSFET designs, nanotube and nanowire transistors, single-electron devices, and molecular electronics. Organic semiconductors are also presented briefly, since nanoscale physics and chemistry issues are critical to their development, even though most devices incorporating such materials are not nanostructures. The underlying physical principles behind these approaches are emphasized, as are the pitfalls of large-scale adoption of these as functional technologies.

Similarly, Chapter 7 begins with an overview of magnetism and magnetic materials, to set the stage for an introduction to the science of magnetic data storage and magneto-electronics. The physics underlying current technologies such as giant magnetoresistance (GMR) sensors and magnetic random access memory (MRAM) is introduced. The text also discusses recent developments in novel materials and the concepts of "spintronics", in which the spin degree of freedom of electrons is considered for manipulation and information processing.

The beginning of Chapter 8 reviews the fundamentals of electromagnetic radiation and its interactions with matter, to get readers all on the same page. Refraction and reflection from interfaces between media are covered, followed by a discussion of (scalar) diffraction theory. These concepts enable discussions of photonic bandgap systems and near-field optical phenomena. Fundamentals of lasers are introduced, and semiconductor lasers are described along with some of the elements of modern optical communications. Recent developments in nanophotonics are discussed, including an introduction to plasmon-based effects and their use in surface-enhanced spectroscopies.

Following this look at photonics, Chapter 9 begins with an overview of continuum mechanics and elasticity, including concepts like areal moments of inertia that explain why I-beams are stiff. After introducing mechanical oscillators based on beams and torsion members, actual implementations of micromechanical systems are discussed. We then turn to the physical limits imposed on device performance as mechanical systems are reduced to the nanometer scale, including the origins of damping, the ultimate limits of nanomechanical properties, the mechanisms behind friction, and the prospect of true *quantum* mechanics.

Chapter 10 opens with an introduction to dimensional analysis and dimensionless products, a subject dear to mechanical and chemical engineers and often sorely neglected in the education of physicists and chemists. This is followed by an introduction to basic fluid mechanics, with an emphasis on laminar flow and the underlying assumptions, such as the continuum approximation and the no-slip boundary condition. We then examine fluids containing ions, and introduce the basic concepts (Debye–Hückel screening, charge double layers) required to understand electro-osmosis, electrophoresis, and dielectrophoresis. Putting these pieces together, microfluidics is discussed, including the challenge of mixing. This chapter concludes with nanofluidics issues, such as the breakdown of continuum density and viscosity in nanoconfined geometries, and the failure of the no-slip condition at the nanoscale.

The fluidic discussion naturally leads into Chapter 11's introduction to the tools of biology as applied to nanotechnological problems. As with a number of the topics above, entire courses could readily be taught around this subject, so this discussion is limited to the basic concepts.

Finally, Chapter 12 looks at global energy production as an example of an issue to which nanostructures and nanotechnology are likely to be relevant. There I also write a bit about dangers and limitations of nanoscale science, and I conclude with remarks that try to put the material of the preceding chapters into perspective.

I have tried to connect the fundamentals of nanoscale science with real world applications – this was an explicit goal of the original courses, and one that I wanted to preserve. Nanoscale science is not pursued purely out of curiosity, but in part because the ability to control materials at the nanometer scale has tremendous technological potential. Nanotechnology is, in many ways, *already here*, and it's not the science fiction concept of self-reproducing nanobots repairing our cells or constructing rocket engines out of single-crystal diamond. This book won't make anyone an expert on all facets of nanotechnology, but after reading it I hope you will have good intuition and critical insights regarding the underlying science, and you will know where to look to learn more. Good luck!

Solid state physics in a nutshell

Solid state physics, or condensed matter as it is now more commonly known, is the underpinning of the vast majority of modern technology. This (long) chapter will examine key elements of the solid state physics of bulk systems. Understanding the origins of the properties of macroscale materials leads to insights about *why* nanoscale materials are different. This is not intended to be a complete treatment of modern solid state physics – that would take hundreds of pages alone! At the end of the chapter I list a few of the many excellent books on this vast subject for those interested in learning more.

As shall become clear, the quantum mechanical character of electrons is essential to understanding matter in the solid state, even though the more exotic quantum effects evade everyday experience. However, describing realistic solids is, in principle, incredibly complicated. Restricting the discussion to relatively low energy scales, in a cubic centimeter of copper are $\sim 10^{23}$ nuclei, surrounded by $\sim 10^{24}$ electrons. Exactly solving the detailed quantum many-body problem for this system is clearly a practical impossibility. This makes the case for the other key component of approaches to understanding the solid state: statistical physics. Solving classical equations of motion for every gas molecule in your room is just as impractical as the solid state problem. Fortunately, statistical mechanics has been developed, allowing us to consider *distributions* of particle velocities. Similar approaches are routinely applied in the solid state, along with related concepts like a density of states.

Note that all of the electrons and nuclei are electrically charged and therefore interact with one another. These interactions are not necessarily weak! Strongly interacting many-body problems are exceedingly hard to solve, but the observation that atoms remain a useful way of thinking about solid materials hints at a way out of this dilemma. Natural separations of energy scale allow approximations to simplify the solid state problem into something tractable. For example, the strongest electron–nuclei electrostatic interactions are very energetic (hundreds of eV), and are well treated by dividing the electrons into "core" electrons, tightly bound to the nuclei and essentially the same as in isolated atoms, and "valence" electrons, involved in the bonding between atoms.

This interatomic bonding is responsible for the detailed spatial arrangement of the atoms, and the emergence of bulk effects like the elastic rigidity of solids. These structural properties bring up another helpful separation of scales: the difference in time scales between the mechanical dynamics of the atoms and the response of the valence electrons. This is already familiar from quantum mechanics and physical chemistry courses as the Born–Oppenheimer approximation: to solve the electronic problem of the valence electrons, it is generally useful to consider the background atoms to be *fixed* in space, providing a static potential for the electrons. We will begin with this assumption as well, and will correct for it later when necessary, or at least point out when its validity is questionable.

The simplest approach to dealing with the remaining interacting electrons is to neglect the interactions entirely. This converts an n-body interacting problem into a 1-body problem that must be solved n times. Called the independent electron approximation, this underlies the basic approaches to electronic structure of solids. Later we will see why one can often get away with this. Interaction corrections can be included at a later point. Some of those interaction effects can be neatly encapsulated in particular terms. For example, "exchange" effects that lead to magnetic correlations between electrons really originate from the interelectron Coulomb interaction, and in atomic systems this complicated process may be summarized succinctly via Hund's Rules, familiar from chemistry. Magnetic effects will be discussed in more detail later, in Chapter 7.

There are two different directions from which to approach even the noninteracting idealization of materials. One possibility is to start by ignoring even the atomic potentials, except to assume some overall uniform confinement of the electronic system. This is called the *free electron* approach, and it is the initial direction followed in most elementary solid state physics books (e.g. Kittel [5], Ashcroft and Mermin [6]). As we shall see in Section 2.1, even this extreme simplification contains many features that will survive in more complete treatments, and is pedagogically useful for introducing key concepts. Treatments evolving from the free electron picture typically assume that the most sensible basis for describing electronic wavefunctions in solids is to think in terms of single-particle states that are *extended* in space, spanning many atoms.

The other major approach to the noninteracting problem is to start with atoms and bonding between atoms as basic ingredients, as we discuss in Section 2.3. This is the direction often emphasized in chemistry curricula, and is also that taken in some solid state physics books that emphasize the calculation of realistic electronic structures (e.g. Harrison's texts [7, 8]). This molecular language reveals complementary concepts compared to the free electron approach, such as the underlying chemical basis for particular features in crystal and band structure. This approach has its roots in the idea that a good basis set for describing the electrons in real materials is the atomic orbitals (and closely related single-particle states based on linear combinations of those orbitals).

For understanding why nanoscale materials differ from bulk, each approach has something to offer. We begin by coming from the free electron direction.

2.1 Free electrons

As we know from quantum mechanics courses, the single-particle energy eigenstates of a particular system are given by solving the Schrödinger equation,

$$i\hbar \frac{\partial \psi}{\partial t} = H\psi, \tag{2.1}$$

where ψ is the wavefunction, $\hbar$ is Planck's constant divided by 2π, and H is the Hamiltonian of the system. Assuming a time-independent Hamiltonian, we can separate out the time dependence, and are left with the time-independent Schrödinger equation, given in the spatial representation as:

$$-\frac{\hbar^2}{2m}\nabla^2\psi + V(\mathbf{r})\psi = E\psi, \tag{2.2}$$

here m is the mass of the particle, $V(\mathbf{r})$ is the spatially varying potential, and E is the energy. Solutions to Eq. (2.2) must satisfy appropriate boundary conditions for the system being described. A given energy eigenstate then has a time evolution given by $\psi e^{-iEt/\hbar}$.

For truly free particles, we can set $V = 0$, and the solutions to Eq. (2.2) can be written as plane waves,

$$\psi \sim A e^{i\mathbf{k}\cdot\mathbf{r}} + B e^{-i\mathbf{k}\cdot\mathbf{r}}, \tag{2.3}$$

where A and B are complex amplitudes. Throwing in the time dependence, we find that these are traveling waves with wavevector $\mathbf{k}$ and frequency $\omega = E/\hbar$. Recall that the wavevector tells us about the spatial periodicity of the wavefunction: $|\mathbf{k}| \equiv k = 2\pi/\lambda$, where λ is the wavelength. If we apply the momentum operator to these states, we find the familiar result that $\mathbf{p} = \hbar\mathbf{k}$.

Plugging the plane-wave solutions into Eq. (2.2) gives us the relationship between E and $\mathbf{k}$,

$$E(\mathbf{k}) = \frac{\hbar^2 k^2}{2m}. \tag{2.4}$$

Equation (2.4) is an example of a *dispersion relation*. Figure 2.1 illustrates this parabolic relationship for the 1d case. Note that there are no restrictions here on the allowed values of $\mathbf{k}$ (and therefore E).

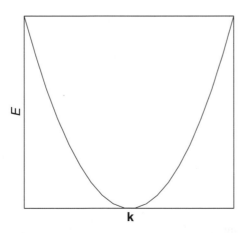

Dispersion relation for free particles in one dimension.

Fig. 2.1

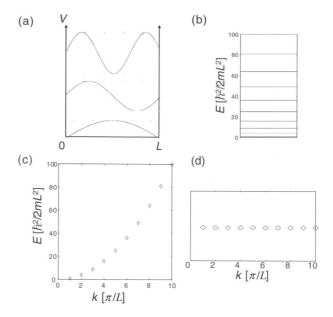

Fig. 2.2   (a) Single-particle states for the lowest three energy levels of the 1d particle-in-a-box. (b) Energy level diagram for the same system. (c) Energy as a function of wavevector for the particle-in-a-box, showing the discretization of allowed k values due to boundary conditions. (d) A k-space plot, showing the allowed values of k compatible with boundary conditions.

2.1.1 Particle-in-a-box

Now consider what happens when we restrict the particles to a limited domain. Let's work in one dimension,[1] and choose the "infinite square well" or "particle-in-a-box" potential,

$$V = 0, \quad 0 < x < L,$$
$$\infty, \quad x \le 0; x \ge L, \tag{2.5}$$

where L is the size of the box. In the interior of the box, $V = 0$ as before, so the allowed wavefunctions must still look like (superpositions of) plane waves. However, ψ must vanish at the walls ($x = 0, L$) because of the infinite potential there. The answer is familiar from introductory quantum mechanics. The allowed states are

$$\psi_n(x) = \sqrt{\frac{2}{L}} \sin \frac{\pi}{L} nx, \tag{2.6}$$

where $n = 1, 2, 3, \ldots$ and we have included a normalization factor for the wavefunction. The solutions are standing waves, as shown in Fig. 2.2, corresponding to allowed k values of $n\pi/L$. With this particular definition of k we have restricted ourselves to positive k values. Since the wavefunction vanishes outside the box, we can find the energies of the

---

[1] The word "dimension" is used in many different ways, often sloppily, when one talks about nanostructures. A *strictly* one dimensional spatial problem has only a single spatial coordinate, here chosen to be x for convenience.

allowed states readily using Eq. (2.4):

$$E(k) = \frac{\hbar^2 k^2}{2m} = \frac{\hbar^2 \pi^2 n^2}{2mL^2}.$$

$$(2.7)$$

Even from this 1d example, we can draw some important facts.

- **Finite sample size drastically restricts the allowed momenta and energy levels.** The energy spectrum is discrete rather than continuous.
- The allowed values of k are uniformly spaced in "k-space" by an amount π/L.
- The system size, L, determines the spacing of allowed wavevectors and single-particle energies, with a smaller system giving larger spacings.

What about an analogous two-dimensional box of side length L, still with infinitely high walls? Choosing rectangular coordinates for convenience, the solution may be written as the product of two 1d solutions:

$$\psi_{n_x,n_y}(x,y) = \frac{2}{L}\left(\sin\frac{\pi}{L}n_x x\right)\left(\sin\frac{\pi}{L}n_y y\right),$$

$$k_x \equiv \frac{n_x \pi}{L}, \quad n_x = 1, 2, 3, \ldots,$$

$$k_y \equiv \frac{n_y \pi}{L}, \quad n_y = 1, 2, 3, \ldots. \qquad (2.8)$$

Figure 2.3 shows the ground state and lowest excited states for this system. Now we are considering a two-dimensional k-space, with allowed **k** values distributed uniformly, with one allowed spatial state per "area" $(\pi/L)^2$. The energy spectrum again reflects this discreteness:

$$E(\mathbf{k}) = \frac{\hbar^2 k^2}{2m} = \frac{\hbar^2}{2m}(k_x^2 + k_y^2) = \frac{\hbar^2 \pi^2}{2mL^2}(n_x^2 + n_y^2). \qquad (2.9)$$

Figure 2.3 better shows what is happening through three common diagrams. Figure 2.3b shows the 2d free particle dispersion relation, $E(\mathbf{k})$, and how the imposition of system boundaries discretizes both the wavevector and energy spectrum. Figure 2.3(c) is an energy level diagram for the 2d square well, with each symbol representing one particular single particle spatial state. Figure 2.3(d) is a plot of the 2d k-space, showing clearly the uniform spacing of allowed **k** values.

The 3d generalization is clear:

$$\psi_{n_x,n_y,n_z}(\mathbf{r}) = \left(\frac{2}{L}\right)^{3/2}\left(\sin\frac{\pi}{L}n_x x\right)\left(\sin\frac{\pi}{L}n_y y\right)\left(\sin\frac{\pi}{L}n_z z\right),$$

$$k_x \equiv \frac{n_x \pi}{L}, \quad n_x = 1, 2, 3, \ldots,$$

$$k_y \equiv \frac{n_y \pi}{L}, \quad n_y = 1, 2, 3, \ldots,$$

$$k_z \equiv \frac{n_z \pi}{L}, \quad n_z = 1, 2, 3, \ldots, \text{ and} \qquad (2.10)$$

$$E(\mathbf{k}) = \frac{\hbar^2 k^2}{2m} = \frac{\hbar^2}{2m}(k_x^2 + k_y^2 + k_z^2) = \frac{\hbar^2 \pi^2}{2mL^2}(n_x^2 + n_y^2 + n_z^2). \qquad (2.11)$$

Now allowed **k** states are distributed uniformly in a 3d k-space, one per k-space "volume" $(\pi/L)^3$.

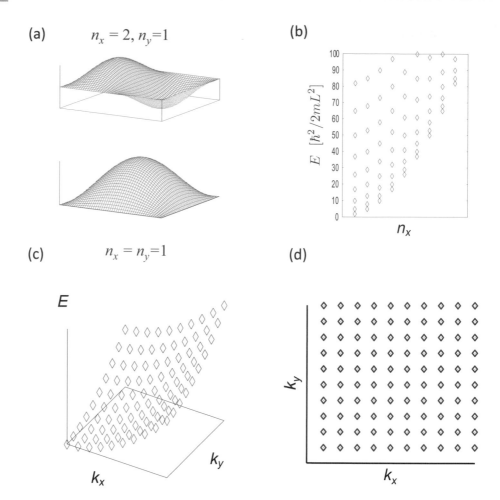

Fig. 2.3   (a) Single-particle states for the ground state and one excited state of the 2d particle-in-a-box. (b) Energy level diagram for the same system. (c) Energy as a function of wavevector for the 2d particle-in-a-box, showing the discretization of allowed k values due to boundary conditions. (d) A k-space plot, showing the allowed values of k compatible with boundary conditions.

2.1.2 Periodic boundary conditions

Just how important was our particular choice of boundary conditions? After all, we know that the "infinitely high walls" approximation is unphysical – otherwise processes like photoemission, field emission, and thermionic emission, all of which result in electrons leaving a bulk material, would be impossible. *Periodic boundary conditions* are another popular choice, particularly when describing large bulk materials. Imposing periodic boundary conditions is intuitively appealing because it is a way of capturing some finite size effects while avoiding potentials with singularities and drastic changes in the single-particle wavefunctions.

Mathematically, periodic boundary conditions are, for the 1d case above:

$$\psi(x) = \psi(x + L),$$
$$\psi'(x) = \psi'(x + L), \tag{2.12}$$

where the prime designates a derivative with respect to x. Solving the 1d problem for this case, we find that the solutions look familiar:

$$\psi_n(x) = \sqrt{\frac{2}{L}} \cos \frac{2\pi}{L} nx, \tag{2.13}$$

and now $n = 0, \pm 1, \pm 2, \ldots$. This implies that k now takes on both positive and negative values, $k = 2\pi n/L$. The dispersion relation, which comes from the Schrödinger equation rather than the boundary conditions, is unchanged, but the different allowed k values change the spectrum:

$$E(k) = \frac{\hbar^2 k^2}{2m} = \frac{\hbar^2}{2m} \left(\frac{2\pi}{L} \right)^2 n^2. \tag{2.14}$$

The generalization to higher dimensions is straightforward.

So, with the change of boundary conditions, the details of the allowed single-particle spatial states have changed, as has the spectrum. However, the main points mentioned in Section 2.1.1 remain correct. As we shall see below, the *statistical* properties of the spectra for the various dimensionalities are also insensitive to the boundary conditions for the free electron case, a fact that will remain true in more sophisticated treatments of bulk materials.

2.1.3　Filling up the box

Once we have found a set of single-particle energy eigenstates, either in the free electron approximation or some other noninteracting limit, we need to consider how such states would be occupied if we put N electrons into the system. Since the electrons are non-interacting in our initial model, we don't have to worry about the eigenstates changing with N. Indeed, we should be able to write the many-particle states as some kind of linear combination of products of the single-particle states. That is, if particle 1 is in a single-particle state, $|\phi_a\rangle$, and particle 2 is in single-particle state $|\phi_b\rangle$, then we can write

$$|\Psi(1, 2)\rangle = |\phi_a\rangle_1 |\phi_b\rangle_2. \tag{2.15}$$

(From here on we will use the Dirac bra–ket notation for quantum states. If this is unfamiliar to you, consult, e.g., [9].)

Life is, however, more complicated than this, and one cannot neatly say that each particle is in a particular single-particle state. Recall that electrons are *fermions*, and therefore are subject to the Pauli exclusion principle: no two identical fermions may have identical quantum numbers. This follows naturally from the Fermi–Dirac statistics that give fermions their name. For an N-fermion system, the many-particle wavefunction state must be *antisymmetric* under the exchange of any two (identical) fermions. For our particular example, we would have to write

$$|\Psi(1,2)\rangle = \frac{1}{\sqrt{2}} \left(|\phi_a\rangle_1 |\phi_b\rangle_2 - |\phi_b\rangle_1 |\phi_a\rangle_2 \right). \qquad (2.16)$$

Recall, too, that each electron has *spin*, an internal angular momentum degree of freedom (with a corresponding magnetic moment). To accomodate this we just need to say that our states $|\phi_a\rangle$ and $|\phi_b\rangle$ include spin – that is, they are a product of a spatial wavefunction and a spin wavefunction.

The situation gets very complicated for large numbers of particles. One way of guaranteeing this antisymmetry is to write the many-particle wavefunction as a *Slater determinant* – the determinant of a matrix such as this one:

$$\Psi(1,2) = \frac{1}{\sqrt{2}} \begin{vmatrix} |\phi_a\rangle_1 & |\phi_b\rangle_1 \\ |\phi_a\rangle_2 & |\phi_b\rangle_2 \end{vmatrix}. \qquad (2.17)$$

Here row i corresponds to particle i, while each column enumerates a particular single-particle state. The prefactor is the normalization, and would be $1/\sqrt{N!}$ if N is the number of particles and single-particle states. Recall that the specific single-particle states we're considering are not unique; one could write linear superpositions of those states that are equally valid solutions to the Schrödinger equation. The most general way of writing the many-particle state is as a linear combination of such Slater determinants.

As students of chemistry are well aware, we can often get away with neglecting this and instead using the *aufbau* process. Rather than write the electronic ground state of carbon as some linear combination of Slater determinants, we instead write it as $1s^2\, 2s^2\, 2p^2$. This is short-hand for saying that the $1s$ spatial orbital is doubly occupied – that is, with one electron in the spin state $|\uparrow\rangle$ and another in $|\downarrow\rangle$. Similarly, in common usage we would say that the same is true for the $2s$ and $2p$ single-particle spatial states. This approach is extremely useful for counting up occupancy of the allowed single-particle spatial states of electrons, and is much less cumbersome than constantly worrying about writing the many-body state in an explicitly totally antisymmetric form.

When worrying about the many-electron ground state, then, of particles in a box, having the allowed spectrum allows us to make rapid progress. We consider dropping electrons into the box, filling up each spatial single-particle state with two electrons of opposite spin, starting with the lowest (kinetic) energy states first. We then continue until all the electrons are "used up". We can then talk sensibly about the energy of the highest occupied state (when what we really mean in the full gory treatment is the highest energy single-particle state that must be included in relevant Slater determinants).

This highest occupied state is special, in the sense that all single-particle states of lower energy are full, and all of higher energy are empty. We will see in the next chapter and beyond that this has important consequences. In brief, often we are interested in inelastic processes that take electrons and scatter them into new states. For example, if we apply an electric field to a metal, we expect to drive a current. In this free electron picture, that's equivalent to the electrons redistributing themselves in k-space so that the total momentum of the electrons is no longer zero. To do this while obeying the Pauli principle, the electrons have to make transitions from filled single-particle states into empty ones. The lowest energy processes that do this involve taking an electron near the highest occupied state, and putting it into an empty state just above the highest occupied state. This is why we usually care a lot about how the single-particle states are distributed in energy and momentum, and how they're occupied.

In the particle-in-a-box example it is clear that an electron in that highest occupied state has a spatial wavefunction with more nodes (and therefore more "wiggles", more kinetic energy, and a shorter effective wavelength) than one in the single-particle spatial ground state. The more electrons there are in the box, the higher up the energy "ladder" that highest energy electron is. Counting up states properly to assess this energy scale is important for a number of reasons. In the next section we see that there are better ways of doing this, and that nanoscale confinement can strongly affect this procedure.

2.1.4 Density of states

The fact that single-particle states[2] are evenly distributed in k-space is convenient, and makes statistical treatments of those states (counting them; averaging some quantity over them) relatively easy. For large systems of many particles, it would be inefficient to count up the states individually, as a summation:

$$\text{number of states} = \sum_{\text{allowed } \mathbf{k} \text{ values}} 1, \qquad (2.18)$$

particularly if there are 10^{23} states involved. When the number of particles, N, is large, it makes sense to count the single-particle states by converting the sum to an integral, and to introduce the concept of a *density of states*, acting as if the single-particle states are continuously distributed in k-space. In this way of thinking about electronic structure, **many of the special electronic and optical properties of nanostructures result from the fact that in small structures one cannot neglect the discreteness of the allowed single-particle states**.

Consider the ground state of the many-particle system in this limit. For periodic boundary conditions in 1d, we know that the allowed single-particle states are spaced by $2\pi/L$ in k-space, and that (in the language of the previous section) each such state can hold two electrons of opposite spins. We could say that the density of states in k-space is $2 \times L/2\pi$. We can then write:

$$N = 2 \times \int_{-k_F}^{k_F} dk/(2\pi/L). \qquad (2.19)$$

Here, k_F is the magnitude of the highest occupied wavevector, and is called the *Fermi wavevector*. All the single-particle states with $|k| \leq k_F$ are full, and those with $|k| > k_F$ are empty. This is shown in Fig. 2.4(a). The Fermi wavevector has an associated *Fermi momentum*, $p_F \equiv \hbar k_F$.

We can do the same thing in 2d and 3d systems, again capitalizing on the even distribution of allowed states in k-space. In 2d, this gives

---

[2] Remember, this phrase is short-hand for "single-particle spatial states that are energy eigenstates of the single-particle Hamiltonian, and are compatible with chosen boundary conditions".

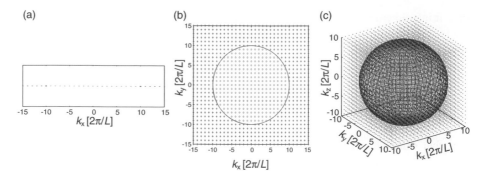

Fig. 2.4 Fermi gas at $T = 0$ in k-space, with filled single-particle states indicated by green points, empty single-particle states by red points, and the Fermi wavevector magnitude indicated in blue, for (a) 1d, (b) 2d, and (c) 3d.

$$N = 2 \times \int_0^{k_F} \frac{2\pi k \, \mathrm{d}k}{(2\pi/L)^2}$$
$$= \frac{L^2}{2\pi} k_F^2, \tag{2.20}$$

where we integrate over a *disk* in 2d k-space, as in Fig. 2.4(b). The disk of radius k_F contains all the filled single-particle states and is called the *Fermi disk*. In 3d,

$$N = 2 \times \int_0^{k_F} \frac{4\pi k^2 \mathrm{d}k}{(2\pi/L)^3}$$
$$= \frac{L^3}{3\pi^2} k_F^3. \tag{2.21}$$

Now the filled single-particle states form the *Fermi sphere* in 3d k-space. The spherical boundary between the filled and empty states is called the *Fermi surface*, as shown in Fig. 2.4(c). One can rearrange Eqs. (2.19–2.21) to write k_F in terms of the d-dimensional particle density, $n_d \equiv N/L^d$.

Very often it is more convenient to use the dispersion relation, Eq. (2.4), to rewrite the state-counting integrals in terms of the energy. The density of states as a function of energy, $\nu_d(E)$, is the number of single-particle states per unit energy per d-dimensional volume. In the ground state we can then count up to the *Fermi energy* or Fermi level, the energy of the highest occupied single-particle state,

$$E_F \equiv \frac{\hbar^2 k_F^2}{2m}. \tag{2.22}$$

Sometimes this written in terms of a *Fermi temperature*, $T_F \equiv E_F/k_B$.

Simply doing the change of variables, we find[3]

$$n_d = \int_0^{E_F} \nu_d(E) \mathrm{d}E \tag{2.23}$$

---

[3] When we wrote our dispersion relation in Eq. (2.4), we tacitly set the zero of energy to correspond to a particle with zero wavevector (and therefore zero momentum).

with

$$\nu_{1d}(E) = \frac{1}{2\pi}\left(\frac{2m}{\hbar^2}\right)^{1/2} E^{-1/2},$$

$$\nu_{2d}(E) = \frac{m}{\pi\hbar^2},$$

$$\nu_{3d}(E) = \frac{1}{2\pi^2}\left(\frac{2m}{\hbar^2}\right)^{3/2} E^{1/2}. \tag{2.24}$$

Notice that **dimensionality strongly affects the energy dependence of the density of states**. For 3d systems, as one considers allowed single-particle states of increasingly higher energies, one finds that the number of those states per unit energy per unit volume *increases*. That is, there are more single-particle states available at a higher energy than at lower energies. Conversely, for a 1d system of fixed length, as one tends toward higher energies the allowed single-particle states are more sparse. This shouldn't be surprising at all – it's obvious from Eq. (2.7). The 2d case is intermediate between the two, and is quite special.

Interestingly, one can redo this whole state counting analysis starting from hard-wall boundary conditions rather than periodic boundary conditions. In that case one would start with Eq. (2.19), restrict the discussion to positive values of k, but realize that the allowed k states compatible with hard walls are spaced by π/L rather than $2\pi/L$. In the end, the densities of states in Eq. (2.24) would be *exactly* the same.

Given a dimensionality, a number of electrons, and a characteristic system size, we can use Eqs. (2.19–2.21) to find $k_F(n_d)$. Then we can use Eq. (2.22) to find the Fermi energy, and we can use Eq. (2.24) to evaluate $\nu_d(E_F)$, the density of states at the Fermi energy. This is often an extremely useful quantity; as we discuss later, the rates of many processes are proportional to $\nu_d(E_F)$.

Let's look at the limits of this whole idea of a continuous, smooth density of states. Consider a 1 cm^3 block of a simple alkali metal like sodium, for which this free, noninteracting electron picture is a good first approximation. Counting just the 3s valence electrons (the rest end up being bound to the ion cores), we find $n_{3d} = 2.65 \times 10^{22}$ cm^{-3}. Using the free electron mass, $m = 9.1 \times 10^{-31}$ kg, we find that $E_F = 5.2 \times 10^{-19}$ J $= 3.2$ eV. In units of temperature, that's $E_F/k_B = 38{,}000$ K (!). Since the typical thermal energy scale at room temperature, $k_B T \approx 0.025$ eV, is far lower than E_F, it is not bad to approximate this metal as being close to its ground state. In the language of quantum statistical mechanics, an electron gas with $k_B T \ll E_F$ is said to be *degenerate*.

From E_F, we can find Δ, the typical *single-particle level spacing* at the Fermi level:

$$\Delta = \frac{1}{\nu_d(E_F)L^d}. \tag{2.25}$$

For the example in the previous paragraph, $\nu_{3d}(E_F) = 7.6 \times 10^{46}$ J^{-1}m^{-3}, and therefore $\Delta = 1.3 \times 10^{-41}$ J for our 1 cm^3 cube. This is an exceedingly small energy spacing, vastly smaller than $k_B T$ for any temperature achievable in the laboratory.

However, the situation becomes very different when we consider a metal nanoparticle! Suppose instead that we have (3 nm)^3 of sodium. The electron density is unchanged, and therefore so are E_F and $\nu_{3d}(E_F)$. With the vastly smaller volume, however, we find

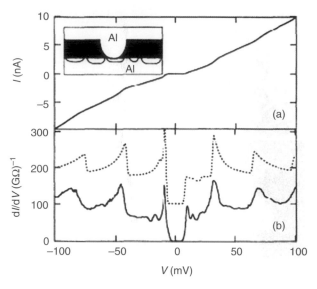

Fig. 2.5 Current–voltage characteristics of a single aluminum nanoparticle connected via tunnel junctions to "source" and "drain" electrodes (inset). The current I has a step-like features, and $\partial I / \partial V$ shows a corresponding peak structure, due to the discrete spectrum of the nanoparticle. Figure adapted from [10].

$\Delta = 4.9 \times 10^{-22}$ J, equivalent to 3 meV or 35 K. This is much more significant, and is actually measurable, since it is relatively simple these days to reach temperatures down to 50 mK. When worrying about the electronic properties (such as conduction) of a such a nanoscale metal grain, the individual electronic levels can dominate the response. This is a situation that simply doesn't arise in macroscopic systems.

Figure 2.5 shows an experimental realization of this. Using a clever fabrication technique, a structure is made consisting of a tiny aluminum nanoparticle sandwiched between thicker aluminum pads to allow electrical measurements. Aluminum oxide insulating spacer layers are thick enough that the nanoparticle can be considered "isolated" from the leads, but thin enough that some electrons can tunnel through, producing a measurable current. As the voltage between the big aluminum pads is ramped, the current, I, exhibits steps as each discrete single-particle level begins to contribute to conduction. This is more clearly seen as peaks in the conductance, $\partial I / \partial V$. In this structure the measured energy spacing between single-particle states has two components: Δ, and an additional "charging energy" due to the repulsion of the electrons already on the grain (an interaction effect that we've been neglecting so far). We will learn much more about such structures in Chapter 6.

A second example is shown in Fig. 2.6. In this case the investigators have assembled a 20 atom long Au chain on a NiAl substrate using a scanning tunneling microscope (STM; we'll learn more about those in Chapter 4). The measured current between the tip and the sample is proportional to the number of available single-particle states at the tunneling electron's energy and position (a local density of states). This allows the spatial mapping of the single-particle states, as shown in the figure. The experimentally inferred 1d single-particle states at different energies are qualitatively what one would expect for a 1d particle-in-a-box system; more energetic single-particle states have more "wiggles".

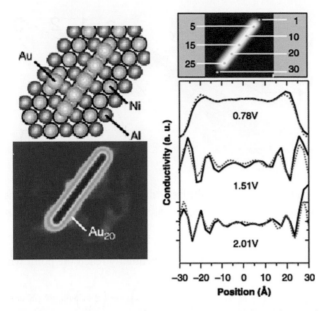

Fig. 2.6

A 1d Au chain assembled using a scanning tunneling microscope. By mapping the tunneling conduction as a function of bias voltage, it is possible to map the specific single-particle electronic states into which electrons are tunneling. These states are qualitatively similar to what would be expected for a 1d particle-in-a-box. Adapted from [11].

2.1.5 Dimensional crossovers – electronic structure

The relationship between the Fermi wavelength and the characteristic size of the object in question establishes the effective electronic dimensionality of a nanoscale system in its ground state. Armed with our expressions for λ_F and a fixed system size, we can imagine tuning the electronic density, and figuring out when the effective electronic dimensionality changes. Consider a particle-in-a-box situation again, but this time the length, L, width, w, and thickness, t, of the box are very different: $L \gg w \gg t$, with L oriented along the x direction, etc. We can write down the allowed single-particle electronic states readily, as before, but this time Eq. (2.11) looks more complicated:

$$E(\mathbf{k}) = \frac{\hbar^2}{2m}(k_x^2 + k_y^2 + k_z^2) = \frac{\hbar^2 \pi^2}{2m}\left(\frac{n_x^2}{L^2} + \frac{n_y^2}{w^2} + \frac{n_z^2}{t^2}\right). \qquad (2.26)$$

Filling these states from the bottom up (dialing up the Fermi level), it is clear (see Fig. 2.7) that initially the only states that are relevant are those with $n_y = n_z = 1$. Because of the hierarchy of the box's characteristic sizes, the various n_x states are much lower in energy than those with some excitation along the transverse (y and z) directions. Recasting this in the language of momentum, the lowest energy states only have (non-zero-point) momentum along the x axis – the system is effectively 1d. Another way of saying this is that only the "lowest 1d subband is occupied".

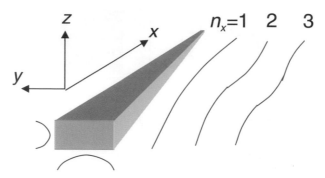

Fig. 2.7   A cartoon of the single-particle states of a 1d box that is extended much more in the x direction than in the transverse directions. Energetically, subsequent n_x states are much closer in energy than subsequent n_y and n_z states.

If we keep adding electrons to the box, we eventually reach the point where

$$\left(\frac{\pi}{L}n_x\right)^2 + \left(\frac{\pi}{w}\right)^2 \approx \left(\frac{\pi}{L}\right)^2 + \left(2\frac{\pi}{w}\right)^2, \tag{2.27}$$

while n_z remains 1. That is, it becomes energetically favorable for the system to start occupying the states with $n_y = 2$. If we consider further increasing the Fermi level such that n_x and n_y for the highest occupied single-particle state are both typically much greater than 1, but n_z remains 1, then the system is effectively two-dimensional. Finally, if we continue increasing E_F by dropping more and more particles into the box, we reach a point where all three principal quantum numbers of the highest occupied state are nonzero, then the system is effectively in the 3d limit.

We can recast this in the language of λ_F very succinctly. If λ_F is small compared to one transverse size of the system, then the system is effectively 1d. If λ_F is small compared to two transverse sizes of the system, then the 2d approach is reasonable. If λ_F is smaller than all physical dimensions of the system, then the electronic structure is effectively 3d. This remains a reasonable way to think about electronic dimensionality even in real materials. This is what is meant by dimensionality with respect to quantum confinement. Colloquially, this language is often generalized to include the case when λ_F is *comparable* to all the transverse dimensions of a system. This situation, relevant for a small molecule, metal, or semiconductor structure in which the wavefunction of the highest occupied single-particle state has only a few nodes within the system, is called the "0d" limit.

One can also consider what this implies for $\nu(E)$, the density of states for an anisotropically shaped system, shown in Fig. 2.8. At low E, the density of states should look like the 1d case of Eq. (2.24), with a peak as $E \to 0$ and decreasing as $E^{-1/2}$ as E is increased. As E is raised to the point where a new 1d subband is accessible, another peak appears in ν, on top of the original 1d behavior. At energies large compared to this transverse confinement energy scale, we can squint a little bit regarding the spikes in ν as each new 1d subband comes into play, and overall $\nu(E)$ begins to approach a roughly constant value. At still larger energies, in excess of the z subband scale, the steps that appear in $\nu(E)$ as each 2d subband is added look small, and overall $\nu(E)$ increases like $E^{1/2}$.

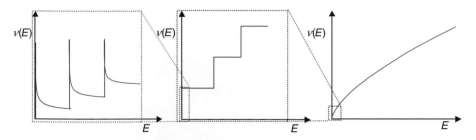

Fig. 2.8

The density of states, $\nu(E)$, for a 3d anisotropic hard-walled box, as in Eq. (2.26). At low energies ν looks like the 1d case, with $\nu(E)$ falling like $E^{-1/2}$ within each 1d subband, and new 1d subbands becoming relevant as criteria like those in Eq. (2.27) are met. At higher energies, $\nu(E)$ looks like the 2d case, since many x and y degrees of freedom matter before each z subband becomes relevant. Finally, at energies such that the electronic wavelength is far shorter than all three dimensions of the box, $\nu(E)$ scales with energy like $E^{1/2}$.

What this discussion tells us is that electronic dimensionality depends not just on the actual size of the system, but also its electronic population. For relatively low electronic densities (we'll see just how low shortly), it is not difficult for the Fermi wavelength to be on a scale at which structures may be engineered (e.g., 100 nm). Figure 2.9 shows an excellent example of this, in the form of a quantum point contact (QPC). Here investigators began with a two-dimensional electron gas (2deg) formed at the interface between GaAs and an AlGaAs layer. This material is specially grown via molecular beam epitaxy, a technique described in Chapter 4. Because of the energetics of the GaAs/AlGaAs interface, the free electrons are confined in the z direction by an approximately triangular potential, and remain in the first z subband. That is, the only momentum the electrons possess along z is zero-point momentum from their confinement, and therefore their motion is truly two-dimensional.

Using metal electrodes patterned and deposited on the free surface of the semiconductor (approximately 100 nm above the 2deg), it is possible to "gate" the 2deg using electrostatic potentials. If the top gate electrodes are biased to a negative potential relative to the 2deg, the electron gas may be depleted completely beneath the electrodes. As a result, two closely spaced electrodes, as drawn, may be used to constrain the 2deg *laterally*. As the interelectrode region is pinched down to the point where $w \rightarrow \lambda_F$, the quantized transverse momentum states become apparent, and the quantum point contact acts increasingly like a one-dimensional electronic system in the contact region. Figure 2.9(a) displays a theoretical calculation of this situation, while Fig. 2.9(b) shows real data taken using a scanned probe technique, demonstrating the quantization of transverse momentum through the contact region.

2.1.6 Fermi's Golden Rule

This restriction of allowed single-particle electronic states due to confinement and reduced dimensionality has profound consequences in nanoscale systems. *Rates* of many processes of physical interest are strongly affected by the size and form of the density of states,

(a)

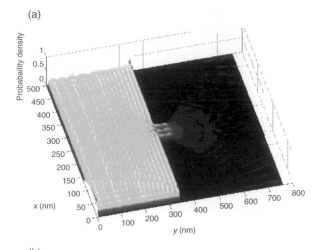

(b)

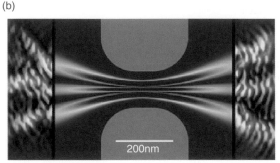

Fig. 2.9   (a) A calculation of electronic transport through a quantum point contact in a 2d electron system, with the constriction narrow enough to support only three transverse modes. From D. K. Ferry, Arizona State University [12]. (b) Data on such a device, obtained via scanned probe microscopy. The middle region is not imaged due to gate electrodes (light gray). From [13].

$v(E)$. This is most readily seen in the form of "Fermi's Golden Rule", a standard result of time-dependent perturbation theory in quantum mechanics courses.[4]

Consider a general time-dependent single-particle state, $|\Psi(t)\rangle$. The time-independent part of the Hamiltonian is designated H_0, and the time-independent Schrödinger equation may be solved to find a complete set of eigenstates, $|\psi_j^0\rangle$. Since those are a complete set of eigenstates, at any moment in time we can write $|\Psi(t)\rangle$ as an expansion based on the time evolution of those eigenstates:

$$|\Psi(t)\rangle = \sum_j c_j(t) e^{-iE_j^0 t/\hbar} |\psi_j^0\rangle, \qquad (2.28)$$

where E_j^0 is the energy eigenvalue for the jth eigenstate of H_0. The expansion coefficients let us interpret $|c_j(t)|^2$ as the instantaneous probability of finding the system in the state $|\psi_j^0\rangle$.

---

[4] A derivation of Fermi's Golden Rule is sketched out in the appendix.

The problem we care about is then to add in the time-varying part of the Hamiltonian, and solve for $P_{ja}(t)$, the probability of being in $|\psi_j^0\rangle$ at time t, if the particle started out in state $|\psi_a^0\rangle$ at $t = 0$. Rather than solving with a generic time-dependent potential, the common approach is to imagine writing $V(t)$ as a superposition of harmonic components, and solve for the single harmonic component case. This is physically relevant, for example, for the case of an atom illuminated with monochromatic electromagnetic radiation, when the local electric field at the position of the particle varies harmonically in time.

Fermi's Golden Rule is the result of solving this problem, and states:

$$\frac{1}{\tau_{ja}} \equiv \frac{\mathrm{d}}{\mathrm{d}t} P_{ja}(t) = \frac{2\pi}{\hbar} |V_{0,ja}|^2 v_d(E_j^0) L^d, \qquad (2.29)$$

where $V_{0,ja} \equiv \langle \psi_j^0 | V | \psi_a^0 \rangle$. Here the harmonic potential is $V \exp(i\omega t)$, with the caveat that $|E_j^0 - E_a^0| = \hbar\omega$; that is, *energy is conserved*, with the perturbation making up for the energy difference between the initial and final states. Note the appearance of $v_d(E_j^0) L^d$, the number of states available per unit energy at the final energy, in the system, approximated here as a d-dimensional box of side length L.

Equation (2.29) has a number of implications and caveats. Recall that conventionally the unperturbed single-particle states are chosen to be orthonormal; that is, $\langle \psi_i^0 | \psi_j^0 \rangle = \delta_{ij}$, where δ_{ij} is the Kronecker delta (1 if $i = j$; 0 otherwise). The only way to have a nonzero transition rate, τ_{ja}^{-1}, is for the time-dependent perturbation to mix the initial and final states. This requirement, that $|V_{0,ja}|^2 \neq 0$, is key to selection rules such as those governing transitions between different atomic states under optical illumination. The derivation of Eq. (2.29) also assumes that the transition rate is sufficiently slow that the initial state population never gets depleted. We already knew this; the Golden Rule is a perturbative expression, and this provides a heuristic sense of what a weak perturbation is.

Most relevant for our discussion, if the *number of available single-particle states* at the final energy is zero, then the rate of the process in question is zero. This basic concept, that *nanoscale confinement and reduced dimensionality can strongly restrict the phase space available for transitions*, is a recurring theme in nanostructures and nanotechnology, from ballistic conduction in nanotubes (Chapter 6) to photonic crystals and cavities (Chapter 8).

Figure 2.10 shows an example of the former case. Nanotubes extending from a large bundle of "fluff" conductively attached to a metal tip are lowered gradually into contact with a mercury bath as the tip–bath electronic conductance is measured. Remarkably, once the tips of the tubes make contact with the liquid metal, the conductance is *independent* of the length of the contact region, over distances of hundreds of nanometers. This is convincing evidence for ballistic conduction in the nanotubes over these scales, much longer than the typical mean free path for scattering in bulk metals. One key is the unique 1d electronic properties of the nanotubes, which restrict scattering to be either forward or back. The result of this lack of small angle scattering is a dramatic deviation from the common sense experience of resistances that scale with length.

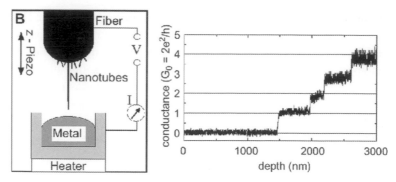

Fig. 2.10 An experiment strongly suggesting ballistic conduction in carbon nanotubes at room temperature over hundreds of nanometers. Once the lowest nanotube attached to the conducting tip makes contact with the liquid mercury bath, conduction is independent of immersed length until a second tube makes contact. Adapted from [14].

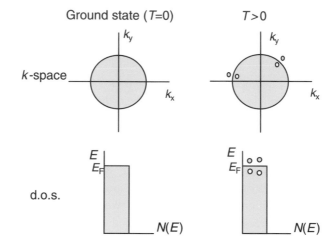

Fig. 2.11 Left: the ground state of a 2d electronic gas, showing the filled single-particle states (light blue), with energies $\leq E_F$ and $k \leq k_F$. Right: particle/hole excitations expected when $T > 0$. The typical deviation from E_F of the excited particle or hole is on the order of $k_B T$.

2.1.7 Excitations and finite temperature

Before we move on from the ideal noninteracting Fermi gas picture of electrons, we need to consider the effects of finite temperature. We have discussed the ground state of this many-particle model extensively. Now we ask: what are the low energy elementary excitations of the many-particle system?

The answer is shown in Figure 2.11. The excitations of the ideal noninteracting Fermi gas are *particles* and *holes*. Because of the Pauli principle, the only way to create an excitation above the ground state is to take an electron from inside the Fermi surface, and promote it to an empty state outside the Fermi surface. The result is a particle excitation with an energy $> E_F$, and the *absence* of an electron from some single-particle state with energy $< E_F$. This lack of an electron, a vacancy within the Fermi sea, can be thought of as a particle of its own, described by its own set of quantum numbers (e.g., charge $+e$, some particular momentum).

When considering the effects of temperature on the many-particle system, a statistical treatment is essential. It is necessary to find the population distribution of the electrons. For a given single-particle state, what is the probability that the state is occupied? Qualitatively this is easy. A typical thermal energy scale is $k_B T$. Single-particle states far below E_F must remain fully occupied, since thermal energy is insufficient to promote those particles up to empty states. So, most of the particle and hole excitations occupy single-particle states within $k_B T$ of E_F.

The probability of finding a given single-particle state with energy E and some spin occupied is given by the *Fermi–Dirac distribution*,

$$f(E, \mu, T) = \frac{1}{\exp\left(\frac{E - \mu}{k_B T}\right) + 1}. \tag{2.30}$$

Here μ is the *chemical potential*. When $T \ll E_F/k_B$, it is reasonable to approximate $\mu \to E_F$.

Figure 2.12(a) shows the Fermi–Dirac distribution at various temperatures. As $T \to 0$ the distribution approaches a step function, and the many-particle system is degenerate. As $T \to \infty$, the FD distribution smears out and looks more like the classical Maxwell–Boltzmann distribution at high energies. As mentioned in Section 2.1.4, using the free electron mass, m_0, and a fixed density of $n_{3d} \approx 8.47 \times 10^{22}$ electrons per cm^3 for copper,

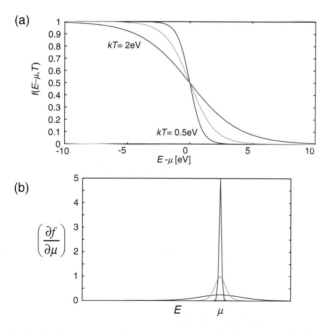

(a) The Fermi–Dirac distribution function, plotted at three temperatures ($k_B T = 0.5, 1$, and 2 eV). At $T = 0$, the function becomes a sharp step down at $E = \mu$. (b) The thermal broadening function, defined in Eq. (2.32). At $T = 0$, this becomes a Dirac delta function centered on $E = \mu$.

Fig. 2.12

we find that this typical metal has $E_F/k_B = 82\,000$ K. So, it is reasonable to say that electrons in copper are degenerate at room temperature. This implies that the Fermi sphere in k-space is sharply bounded, because of the step-like distribution function. By contrast, in the 2d conducting channel of a typical Si field-effect transistor with $n_{2d} \sim 10^{13}$ electrons per cm^2, using an effective mass (more on this concept below!) of 0.33 m_0, we find $E_F/k_B \approx 840$ K. Electrons in room temperature transistors are *not* very degenerate, and their 2d Fermi surface in k-space is correspondingly "blurred"; a significant fraction of all the electrons are thermally excited above the Fermi energy.

To see the effect of this "blurring", it is useful to think about computing average quantities while counting single-particle states by their energies. This involves integrals of the form:

$$\int_0^\infty [\nu_d(E)f(E,T,\mu)]\,g(E)\mathrm{d}E, \tag{2.31}$$

where $g(E)$ is some function of energy that we want to average. Here $\nu_d(E)\mathrm{d}E$ tells us about the number of single-particle states with energies between E and $E + \mathrm{d}E$, and f gives the occupancy of such a state. As shown in Figure 2.12(b), only the occupancy of single-particle states within roughly $k_B T$ of the Fermi surface is affected.

Similarly, often we care about computing the *difference* between two FD distributions (e.g., one with energy E and one with energy $E+eV$, where V is some applied voltage). This difference may be related to the derivative of the FD distribution if the energy difference between the two FD functions is small, and this allows us to quantify the thermal smearing of the Fermi surface. The function

$$\left(\frac{\partial f}{\partial \mu}\right) = -\left(\frac{\partial f}{\partial E}\right) = \frac{1}{4k_B T}\cosh^{-2}\left(\frac{E-\mu}{2k_B T}\right) \tag{2.32}$$

is often called the thermal broadening function and is plotted in Fig. 2.12(b). This function says how rapidly the occupancy of the single-particle states changes with energy, and is (unsurprisingly) peaked around $E = \mu$. It is useful to note that

$$\lim_{T\to 0}\left[\frac{1}{4k_B T}\cosh^{-2}\left(\frac{E-\mu}{2k_B T}\right)\right] = \delta(E-\mu). \tag{2.33}$$

The Dirac delta function reiterates the point that the distribution becomes sharp in the degenerate limit.

2.2 Nearly free electrons

Now that we have thoroughly examined the "free" noninteracting electron gas, we will introduce the lattice potential. As you might have guessed, many features of the no-lattice treatment survive (e.g., single-particle states labeled by a quantum number $\mathbf{k}$; a filled Fermi sea bounded by a Fermi surface in k-space; excitations above the many-particle ground state that look much like particles and holes). However, the spatial arrangement of the atoms, which in a crystal results in a periodic lattice potential, has profound consequences, leading to what is commonly called *band structure*. The single-particle states acquire an additional label, m, the band index. The relationship between a single-particle state's energy and k is no longer the simple parabola seen above. Indeed, $E(\mathbf{k})$ generally depends on the direction of $\mathbf{k}$, and there are whole domains of E for which there are no allowed single-particle states compatible with the lattice potential and boundary conditions. That is, there are bands of allowed single-particle state energies, and gaps where there are no allowed states. Below we see how this comes about naturally. Interestingly, these general features are relatively robust, and survive even in systems that do not possess a long-range, spatially periodic arrangement of constituent atoms.

2.2.1 The reciprocal lattice

In the discussion of plane waves, we saw that the wavevector $\mathbf{k}$ is directly related to the spatial periodicity of the wave. In particular, $\mathbf{k} \cdot \mathbf{r}$ is how much phase the plane wave accumulates in propagating along the vector $\mathbf{r}$. The natural way to describe the spatial periodicity of the atomic arrangement is also to work in k-space.

In real space, one can describe the ordered arrangement of the atoms in a crystalline solid using a *lattice* and a *basis*. A lattice is a spatially periodic array of points, and a basis is an atom or group of atoms positioned at each lattice site. Each lattice point in real space can be written as an integer linear combination of a set of *primitive lattice vectors*:

$$\mathbf{R} = n_1 \mathbf{a}_1 + n_2 \mathbf{a}_2 + n_3 \mathbf{a}_3, \tag{2.34}$$

where n_1, n_2, n_3 are integers. The $\mathbf{a}_i$ vectors have a magnitude of one lattice spacing in the ith direction, and are generally not required to be orthogonal. The statement that the lattice potential is periodic in real space with the periodicity of the lattice may then be written:

$$V(\mathbf{r} + \mathbf{R}) = V(\mathbf{r}) \tag{2.35}$$

where $\mathbf{R}$ is any vector of the type in Eq. (2.34).

One can picture a little volume in real space near the origin bounded by the three $\mathbf{a}$ vectors; see Fig. 2.13(a) for the 2d version. This is one kind of *unit cell*. An entire extended crystal can be built by stacking unit cells. Another common kind of unit cell can be found by starting at the origin, drawing vectors to neighboring lattice points, and bisecting the vectors with planes normal to those vectors, as in Fig. 2.13(b). Locations within that boundary are closer to the origin point than to any other lattice point; this is called a *Wigner–Seitz* unit cell.

The real space periodicity implied by the $\mathbf{a}_i$ can be translated immediately into special points in k-space. Define

$$\mathbf{b}_1 = 2\pi \frac{\mathbf{a}_2 \times \mathbf{a}_3}{\mathbf{a}_1 \cdot (\mathbf{a}_2 \times \mathbf{a}_3)},$$

$$\mathbf{b}_2 = 2\pi \frac{\mathbf{a}_3 \times \mathbf{a}_1}{\mathbf{a}_2 \cdot (\mathbf{a}_3 \times \mathbf{a}_1)},$$

$$\mathbf{b}_3 = 2\pi \frac{\mathbf{a}_1 \times \mathbf{a}_2}{\mathbf{a}_3 \cdot (\mathbf{a}_1 \times \mathbf{a}_2)}. \tag{2.36}$$

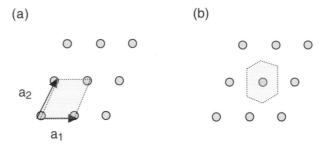

Fig. 2.13 (a) A few lattice points in a 2d lattice, showing the primitive lattice vectors $\mathbf{a}_i$. The shaded region is a unit cell. By repeated tiling of unit cells, the entire lattice may be reproduced. (b) The same lattice, showing a Wigner–Seitz unit cell. All locations within the shaded region are closer to the central lattice point than to any other lattice point.

While the $\mathbf{a}_i$ have units of length, the $\mathbf{b}_i$ have units of $(\text{length})^{-1}$. They are normalized such that

$$\mathbf{b}_i \cdot \mathbf{a}_j = \delta_{ij}. \tag{2.37}$$

The $\mathbf{b}_i$ are called *reciprocal primitive lattice vectors*, and they are a set of basis vectors for k-space. Just as the $\mathbf{a}_i$s define a real-space lattice given by the possible values of $\mathbf{R}$, the $\mathbf{b}_i$s define a set of points in k-space:

$$\mathbf{G} = n_1\mathbf{b}_1 + n_2\mathbf{b}_2 + n_3\mathbf{b}_3. \tag{2.38}$$

Each point $\mathbf{G}$ corresponds to a wavevector with a spatial period related (as a harmonic) to a real-space periodicity of the lattice. Indeed, any real-space function with the periodicity of the lattice (such as the potential felt by an electron due to the stacked atoms) may be written as a Fourier series:

$$f(\mathbf{r} + \mathbf{R}) = f(\mathbf{r}) = \sum_{\mathbf{G}} f_{\mathbf{G}} \exp(i\mathbf{G} \cdot \mathbf{r}). \tag{2.39}$$

This is proven in many solid-state books.

Just as in real space, one can use the $\mathbf{b}$ vectors to define a unit cell in k-space. A Wigner–Seitz unit cell in k-space centered on the origin is called the *first Brillouin zone*. These Brillouin zones and their boundaries are important in the weak periodic potential treatment, as will be seen shortly.

2.2.2 The reciprocal lattice and diffraction

Diffraction is the coherent scattering of some incident radiation (e.g., electrons, photons, neutrons) by a spatially periodic system. The reciprocal lattice, containing information about all the spatial periodicities of the real-space lattice, provides a great way to think about diffraction. Here is a brief summary of the salient points.

Consider some radiation described by a wavevector $\mathbf{k}$ incident on a lattice, as shown in Fig. 2.14. The radiation interacts with the lattice, and is scattered into some outgoing wavevector, $\mathbf{k}'$. Assuming that the interaction was elastic, $|\mathbf{k}| = |\mathbf{k}'| = k$.

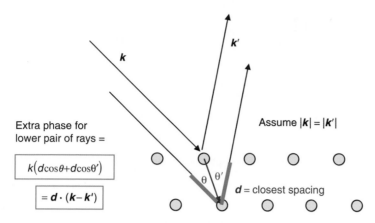

Radiation incident on a lattice, showing how the lower ray acquires extra phase during its propagation relative to the upper ray. A diffraction maximum occurs when rays impinging on subsequent layers interfere constructively.   **Fig. 2.14**

"Bright" features in diffraction patterns result from the *constructive* interference of waves diffracted off different planes of the lattice. Define the vector **d** that goes from one plane of interest to another, as shown in the figure. Compare waves scattered off the first layer with those scattered off the layer **d** away. How much extra phase did the longer-traveling waves accumulate? From the geometry the extra phase is $kd \cos \theta + k'd \cos \theta' = \mathbf{d} \cdot (\mathbf{k} - \mathbf{k}')$. To get constructive interference between the waves scattered from the initial layer and those scattered from the deeper layer, this extra phase must be an integer multiple of 2π. In other words, the extra path length traveled by the wave scattered from the deeper layer must be an integer number of wavelengths. This is called the *Bragg condition*.

Recall that since **d** connects two real-space lattice points, it is an example of a real-space lattice vector **R**, built out of integer multiples of the $\mathbf{a}_i$. Constructive interference then means

$$(\mathbf{k} - \mathbf{k}') \cdot \mathbf{R} = 2\pi j \rightarrow (\mathbf{k} - \mathbf{k}') = \mathbf{G}. \tag{2.40}$$

The difference between incident and outgoing **k** must be a reciprocal lattice vector to have constructive interference and therefore a diffraction maximum. This is called the *Laue condition*, and is a generalization of the Bragg condition above.

Knowing the incident **k** of some radiation such as x-rays, and measuring the pattern of diffracted intensity, one can infer the **G** reciprocal lattice vectors through which the radiation is scattered. From the set of **G**s, the spatial structure of the real-space lattice may be inferred. We will discuss this further in Chapter 4.

When discussing diffraction and the relationship between k-space and the real-space lattice, it is useful to know the terminology shown in Fig. 2.15. Consider some particular plane of atoms in real space. Determine the intercepts of that plane along the crystallographic axes (the directions of the $\mathbf{a}_i$), in units of $|\mathbf{a}_i|$. Take the *reciprocals* of those intercepts. By historical convention, multiply through to write the set of numbers as integers rather than fractions, and represent a negative value by an overline rather than a negative sign

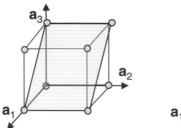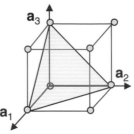

Fig. 2.15   Definitions of crystallographic Miller indices. Real-space atomic planes shown are the (110) plane (at left) and the (111) plane (at right).

(e.g., $(00\overline{1})$ rather than $(00\ \text{-}1))$. Depending on the particular $\mathbf{a}_i$ set, the result is either a triplet, (hkl), or a quadruplet, $(hjkl)$, of integers called *Miller indices*.

Besides a convenient means of labeling real-space lattice planes, of what use are Miller indices? Given a set of indices (hkl), construct a reciprocal lattice vector $\mathbf{G}_{hkl} = h\mathbf{b}_1 + k\mathbf{b}_2 + l\mathbf{b}_3$. The resulting $\mathbf{G}_{hkl}$ points along a direction normal to the real-space (hkl) lattice plane. Furthermore, the real-space spacing between (hkl) lattice planes is $2\pi/|\mathbf{G}_{hkl}|$.

2.2.3 Common lattices

There are fourteen different lattice types in three dimensions, commonly categorized by their symmetry properties. Here we only mention the simplest and most relevant. The most common lattices are summarized in Table 2.1, and are shown in Fig. 2.16. A more complete discussion of lattices and bases is found in Chapter 1 of Kittel [5] or Chapter 4 of Ashcroft and Mermin [6]. A number of resources are available online for animated viewing of crystal structures.

The simplest conceptual lattice is *simple cubic*. The simple cubic lattice has a reciprocal lattice in k-space that is also simple cubic. No element has this as its actual crystal structure, in large part because its packing efficiency is poor. Furthermore, a one atom basis would require 90 degree bond angles, not generally favored with standard chemical hybridization for chemical bonding. Ionic crystals (e.g., NaCl) have related structures, as if one put Na^+ and Cl^- ions on alternating sites. However, strictly speaking these are really described as a two-atom basis on our next common lattice type.

The face-centered cubic (FCC) lattice structure is very common, particularly among transition metals (Al, Cu, Ni, Sr, Rh, Pd, Ag). The FCC lattice with a two-atom basis is also the underlying structure of the ionic salts mentioned above. The reciprocal lattice in k-space is actually body-centered cubic (BCC). Similarly, the the BCC real-space lattice, common for the alkali metals (Li, Na, K, Rb, Cs) and others (W, V, Cr, Fe, Nb, Mo), has a reciprocal lattice that is FCC. Both BCC and FCC lattices have, for a single atom basis, higher packing efficiencies than simple cubic.

A fourth common lattice structure is the simple hexagonal lattice, formed by the vertical stacking of two-dimensional triangular nets. With the appropriate basis, one can use this lattice to build up the *hexagonal close-packed* (HCP) crystal structure. The HCP structure

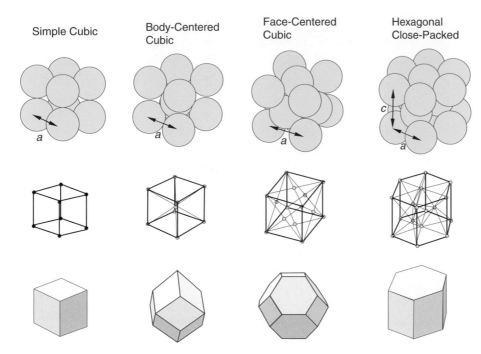

Simple Cubic | Body-Centered Cubic | Face-Centered Cubic | Hexagonal Close-Packed

Fig. 2.16

Conventional unit cells for four common lattices. Top row: the lattices shown assuming a single-atom basis (one hard-sphere atom per lattice site). Rigorously, the atoms at the boundaries of the unit cells are shared by neighboring cells. Middle row: the unit cells shown as points, with nearest-neighbor links indicated in red. For the HCP lattice, interior atoms are shown in darker gray. Bottom row: the shapes of the first Brillouin zones in k-space for these common lattices.

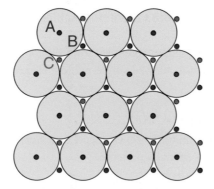

Fig. 2.17

Stacking scheme for the HCP and FCC lattices. Sites are labeled on a single close-packed layer of atoms. Stacking alternate subsequent layers AB/AB/AB would correspond to extending an HCP lattice along the c-axis (001) direction. Stacking subsequent layers ABC/ABC/ABC would correspond to extending the FCC lattice along the (111) direction.

and the FCC lattice with single-atom basis are both "close-packed" structures. One can imagine forming both by stacking successive triangular arrays of hard spheres. Labeling possible stacking sites as shown in Fig. 2.17, an ABC/ABC/ABC arrangement corresponds to FCC, while an AB/AB/AB arrangement would be HCP.

	Simple cubic	BCC	FCC	HCP
Table 2.1 Parameters of common lattice types, assuming a single atom basis and conventional unit cells, for lattice parameter a.				
Unit cell volume	a^3	a^3	a^3	$3\sqrt{2}a^3$
Lattice points/cell	1	2	4	6
Coordination (nearest neighbors)	6	8	12	12
Nearest neighbor distance	a	$\sqrt{3}a/2$	$a/\sqrt{2}$	a
Packing fraction	$\pi/6$	$\sqrt{3}\pi/8$	$\pi/(3\sqrt{2})$	$\pi/(3\sqrt{2})$

Working from these basic lattices one can account for the bulk structure of most technologically relevant materials. Specific examples will be discussed in Chapter 3. Many more complicated structures are possible, of course, such as the complex oxides (perovskites) relevant for high temperature superconductivity and collosal magnetoresistance phenomena. Even materials known for thousands of years (e.g., Fe_3O_4, known as magnetite or loadstone) can have relatively large and complicated unit cells.

The precise chemical nature of interatomic bonds is critical to determining what materials have what lattice structures. Related to this is the concept of *coordination number*, the number of nearest neighbors that an atom has in a crystalline solid. This is directly relevant when considering nanomaterials, since atoms at interfaces are "undercoordinated" and can exhibit chemical and electronic properties different from "fully coordinated" atoms in the bulk. This will be discussed further in Chapter 5.

2.2.4 Bloch's Theorem and Bloch waves

The spatial periodicity of the lattice potential (from the ion cores) seen by the (valence) electrons profoundly affects the allowed single-particle electronic states. We can define a translation operator by

$$T_{\mathbf{R}}f(\mathbf{r}) = f(\mathbf{r} + \mathbf{R}). \tag{2.41}$$

Suppose $\psi(\mathbf{r})$ is some solution to the time-independent Schrödinger equation, $H\psi(\mathbf{r}) = E\psi(\mathbf{r})$. Since the potential is spatially periodic,

$$V(\mathbf{r} + \mathbf{R}) = V(\mathbf{r}), \tag{2.42}$$

we see that H and $T_{\mathbf{R}}$ commute. That means that ψ can simultaneously be an eigenstate of $T_{\mathbf{R}}$:

$$T_{\mathbf{R}}\psi(\mathbf{r}) = c(\mathbf{R})\psi(\mathbf{r}), \tag{2.43}$$

where $c(\mathbf{R})$ is the eigenvalue of the translation operator. We know that two translations have to commute:

$$c(\mathbf{R} + \mathbf{R}') = c(\mathbf{R})c(\mathbf{R}') = c(\mathbf{R}')c(\mathbf{R}). \tag{2.44}$$

We also know that translation has to preserve the normalization of the wavefunction. These imply that $c(\mathbf{R})$ is some unit-magnitude complex number, with a phase set up so that translations add. We can write:

$$c(\mathbf{R}) = \exp(i\mathbf{k} \cdot \mathbf{R}), k \qquad (2.45)$$

where $\mathbf{k} = x_1\mathbf{b}_1 + x_2\mathbf{b}_2 + x_3\mathbf{b}_3$. That is, $\mathbf{k}$ is some vector in k-space (and thus can be written as a linear combination of the primitive reciprocal lattice vectors). Exactly which values of x_i are allowed will be set by our choice of boundary conditions on the wavefunction.

We now see that

$$T_\mathbf{R}\psi(\mathbf{r}) = \psi(\mathbf{r} + \mathbf{R}) = \exp(i\mathbf{k} \cdot \mathbf{R})\psi(\mathbf{r}). \qquad (2.46)$$

This is Bloch's Theorem, and with it we can see what constraints must be obeyed by single-particle states in the periodic potential.

Define a function $u_\mathbf{k}(\mathbf{r}) \equiv \exp(-i\mathbf{k} \cdot \mathbf{r})\psi(\mathbf{r})$. This function turns out to have the same periodicity as the lattice! You can see this using Bloch's Theorem:

$$
\begin{aligned}
u_\mathbf{k}(\mathbf{r} + \mathbf{R}) &= \exp(-i\mathbf{k} \cdot (\mathbf{r} + \mathbf{R}))\psi(\mathbf{r} + \mathbf{R}) \\
&= \exp(-i\mathbf{k} \cdot \mathbf{r})\psi(\mathbf{r}) \\
&= u_\mathbf{k}(\mathbf{r}). \qquad (2.47)
\end{aligned}
$$

Turning this around, if $u_\mathbf{k}(\mathbf{r})$ is a function that is strictly periodic in real space with the periodicity of the lattice (that is, $u_\mathbf{k}(\mathbf{r} + \mathbf{R}) = u_\mathbf{k}(\mathbf{r})$), then a single-particle wavefunction that automatically satisfies Bloch's Theorem is

$$\psi(\mathbf{r}) = \exp(i\mathbf{k} \cdot \mathbf{r})u_\mathbf{k}(\mathbf{r}). \qquad (2.48)$$

This is an important result. Single-particle eigenstates of periodic potentials look like plane waves modulated by functions that are strictly periodic with the same periodicity as the lattice. Two examples of such states are shown in Fig. 2.18. Note that the Bloch states are still labeled by a quantum number $\mathbf{k}$, though this no longer corresponds directly with momentum, as in the free particle case. Note, too, that the Bloch Theorem holds even for potentials that are not weak. These Bloch states are said to be *extended*. That is, they extend throughout the lattice in real space. In contrast, a single-particle electronic state such as a deeply-bound core electronic orbital of one of the lattice atoms is said to be *localized*.

To learn more about the allowed single-particle states, we have to impose boundary conditions. The usual Born–von Karman boundary conditions are to assume that the wavefunctions are *periodic* in the $\mathbf{a}_i$ direction after some number of lattice sites N_i,

$$\psi(\mathbf{r} + N_i\mathbf{a}_i) = \psi(\mathbf{r}). \qquad (2.49)$$

Bloch's theorem then implies that

$$\exp(iN_i\mathbf{k} \cdot \mathbf{a}_i) = \exp(2\pi iN_ix_i) = 1, \qquad (2.50)$$

where we have plugged in the expansion for $\mathbf{k}$ and used the orthogonality relationship between the **a**s and **b**s. This last equality is what specifies the allowed values of $\mathbf{k}$. For that

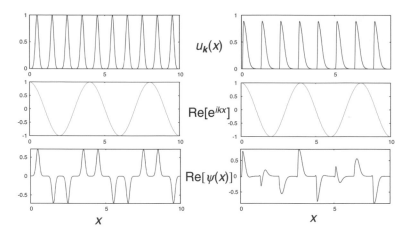

Fig. 2.18 Two examples of Bloch functions. At top, two functions $u_k(x)$ that are periodic in x with period 1 lattice spacing. In the middle, the real part of a plane wave with a different periodicity. At bottom, the product of the two.

equality to hold,

$$\rightarrow x_i = \frac{j_i}{N_i}, \quad j = 1, 2, 3, \ldots, N_i. \tag{2.51}$$

The allowed values of $\mathbf{k}$ are then

$$\mathbf{k} = \sum_{i=1}^{3} \frac{j_i}{N_i} \mathbf{b}_i. \tag{2.52}$$

So, for a lattice with $N = N_1 N_2 N_3$ sites, there are N allowed values of $\mathbf{k}$ that are consistent with periodic boundary conditions. Notice that the allowed volume in k-space for each single-particle spatial state is $(2\pi/L)^d$, just as in the free electron gas with periodic boundary conditions. Indeed, one can show that the first Brillouin zone in k-space contains precisely N allowed values of $\mathbf{k}$ that satisfy the boundary conditions. Unlike the free particle case, for each of these $\mathbf{k}$ values, there are multiple solutions to the single-particle Schrödinger equation with different energy eigenvalues. These are numbered with a *band index*, usually labeled n.

What about values of $\mathbf{k}$ that are outside the first Brillouin zone? It turns out that all the allowed $\mathbf{k}$ points outside the first zone correspond *identically* to points inside the first zone. To see this, consider adding some reciprocal lattice vector $\mathbf{G}$ to $\mathbf{k}$:

$$\begin{aligned} \psi_{n,\mathbf{k}+\mathbf{G}}(\mathbf{r}) &= \exp(i(\mathbf{k} + \mathbf{G}) \cdot \mathbf{r}) u_{n,\mathbf{k}+\mathbf{G}}(\mathbf{r}) \\ &= \exp(i\mathbf{k} \cdot \mathbf{r})[u_{n,\mathbf{k}+\mathbf{G}}(\mathbf{r}) \exp(i\mathbf{G} \cdot \mathbf{r})] \\ &= \exp(i\mathbf{k} \cdot \mathbf{r})\tilde{u}(\mathbf{r}) \\ &= \psi_{n',\mathbf{k}}(\mathbf{r}). \end{aligned} \tag{2.53}$$

Any single-particle state compatible with the periodic boundary conditions with $\mathbf{k}$ outside the first Brillouin zone corresponds exactly to a state inside the first zone, but with a different band index! This means that, if we want, we can limit $\mathbf{k}$ entirely to inside the first

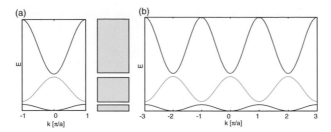

Schematic of energy as a function of k for allowed single-particle states in a 1d periodic potential. (a) The reduced zone scheme, showing the different bands of allowed energy. A complete set of single-particle states compatible with periodic boundary conditions may be specified by k values entirely within this first Brillouin zone. (b) The extended zone scheme, in which the periodic nature of $E(k)$ outside the first zone is apparent.   **Fig. 2.19**

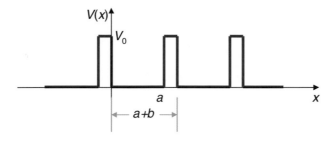

The periodic 1d potential of the Krönig–Penney model.   **Fig. 2.20**

zone and describe all the states. This is the *reduced zone* approach, shown in Fig. 2.19(a). Alternatively, we can allow **k** to range over all k-space and label the states such that

$$E_n(\mathbf{k} + \mathbf{G}) = E_n(\mathbf{k}). \tag{2.54}$$

That is, the energy is a periodic function in k-space. For each n, the energies of all the single-particle states labeled by the allowed **k** values are said to fall in a *band*. This is the *extended zone* scheme, as in Fig. 2.19(b). The range in energy allowed within a single band is called the *band width*.

2.2.5 The Krönig–Penney model

From Fig. 2.19 it is clear that there are bands of allowed single-particle state energies, separated by gaps, for which no single-particle states of the Bloch form can be found that are compatible with the boundary conditions. The Krönig–Penney model is a specific 1d toy model that exhibits these sorts of features. Consider finding the states that solve the 1d periodic potential shown in Fig. 2.20. We will work through this in some detail, to see how these features come about.

There are two regimes of interest: one with $V_0 > E > 0$, and one with $E > V_0$, where we have set the zero of energy at the bottom of the potential wells. Let's look at the first case. Inside the wells, the wavefunction has to look like plane waves, while in the classically forbidden regions between the wells, the wavefunction must be exponentially decaying or growing:

$$\psi(x) = Ae^{i\alpha x} + Be^{-i\alpha x},$$

$$\alpha \equiv \left[\frac{2mE}{\hbar^2}\right]^{1/2},$$

$$a > x > 0, \tag{2.55}$$

and

$$\psi(x) = Ce^{\beta x} + De^{-\beta x},$$

$$\beta \equiv \left[\frac{2m}{\hbar^2}(V_0 - E)\right]^{1/2},$$

$$-b < x < 0. \tag{2.56}$$

Knowing Bloch's Theorem, we can rewrite $\psi(x) = u_k(x)\exp(ikx)$, where

$$u_k(x) = Ae^{i(\alpha-k)x} + Be^{-i(\alpha+k)x}, \quad a > x > 0,$$
$$u_k(x) = Ce^{i(\beta-ik)x} + De^{-i(\beta+ik)x}, \quad -b < x < 0. \tag{2.57}$$

From the conditions that the wavefunction and its first derivative (and hence $u_k(x)$ and $du_k(x)/dx$) must be continuous at the edges of the potential wells, and $u_k(x)$ must be periodic in x with period $(a + b)$, we arrive at four equations and four unknowns:

$$A + B = C + D,$$
$$i(\alpha - k)A - i(\alpha + k)B = (\beta - ik)C - (\beta + ik)D,$$
$$Ae^{i(\alpha-k)a} + Be^{-i(\alpha+k)a} = Ce^{(\beta-ik)(-b)} + De^{-(\beta+ik)(-b)},$$
$$i(\alpha - k)Ae^{i(\alpha-k)a} - i(\alpha + k)Be^{i(\alpha+k)a} = (\beta - ik)Ce^{(\beta-ik)(-b)} - (\beta + ik)De^{-(\beta+ik)(-b)}. \tag{2.58}$$

The resulting condition that must be satisfied for solutions to exist is:

$$\cos(\alpha a)\cosh(\beta b) + \frac{\beta^2 - \alpha^2}{2\alpha\beta}\sin(\alpha a)\sinh(\beta b) = \cos(k(a + b)). \tag{2.59}$$

If one follows the same procedure for the *other* energy regime $E > V_0$, β is purely imaginary, so we can define $\beta \equiv i\gamma$, and find:

$$\cos(\alpha a)\cos(\beta b) - \frac{\alpha^2 + \gamma^2}{2\alpha\gamma}\sin(\alpha a)\sin(\gamma b) = \cos(k(a + b)). \tag{2.60}$$

Both of these may be rewritten as $F(E) = \cos(k(a + b))$. Clearly, if $|F(E)| > 1$, then this relationship cannot be satisfied with *real* values of k, meaning that no single-particle states exist with such energies. The result is *bands* of allowed states with real k, separated by *gaps* with no states, as we expected.

One useful limiting case of the Krönig–Penney model is the δ-function potential case, where $b \to 0$ and $V_0 \to \infty$ such that $\frac{1}{2}\beta ab = p$ remains constant. Then Eq. (2.59) becomes

$$\cos(ka) = \cos(\alpha a) + (p\sin(\alpha a))/(\alpha a). \tag{2.61}$$

If $p \to 0$, the potential barriers are completely ineffective, and we get back planewave solutions with no restrictions on k. Conversely, if $p \to \infty$, solutions only exist when $\sin(\alpha a) = 0$, which ends up being equivalent to the problem of (entirely decoupled) particles in boxes of width a with infinitely high walls. We shall return to this case when we discuss the effects of surfaces.

2.2.6 General weak periodic potentials

While the Krönig–Penney model is highly illustrative, we can also consider the general problem of some periodic potential. This is treated in detail in many solid-state texts, and is summarized below. As mentioned above, because of its periodicity we write the lattice potential $V(\mathbf{r})$ as a Fourier composition using only the reciprocal lattice vectors, $\mathbf{G}$:

$$V(\mathbf{r}) = \sum_{\mathbf{G}} V_{\mathbf{G}} \exp(i\mathbf{G} \cdot \mathbf{r}). \tag{2.62}$$

The single-particle wavefunctions $\psi_{\mathbf{k}}(\mathbf{r})$ may also be written as Fourier compositions, but since they do not share the lattice's periodicity, all values of the wavevector must be used that are compatible with the boundary conditions. We have already found the discrete allowed values of $\mathbf{k}$ back in Eq. (2.52).

$$\psi_{\mathbf{k}}(\mathbf{r}) = \sum_{\mathbf{k}} c_{\mathbf{k}} \exp(i\mathbf{k} \cdot \mathbf{r}). \tag{2.63}$$

Here, $V_{\mathbf{G}}$ and $c_{\mathbf{k}}$ are the coefficients for the potential and wavefunction, respectively. Note that $\psi_{\mathbf{k}}(\mathbf{r})$ must still satisfy the Bloch relation.

Using these expansions, the time-independent Schrödinger equation may be written term-by-term using these coefficients:

$$\left[\frac{\hbar^2 k^2}{2m} - E \right] c_{\mathbf{k}} + \sum_{\mathbf{G}} V_{\mathbf{G}} c_{\mathbf{k}-\mathbf{G}} = 0. \tag{2.64}$$

So far, this is exact. For the free electron case, the Vs are zero, and the solutions are plane waves with wavevectors $\mathbf{k}$ such that the energy eigenvalues are

$$E_{\mathbf{k}}^0 = \frac{\hbar^2 k^2}{2m}. \tag{2.65}$$

Looking at Eq. (2.64), we see that the single-particle state $\psi_{\mathbf{k}}(\mathbf{r})$ for some $\mathbf{k}$ is built out of plane waves differing from $\mathbf{k}$ only by a reciprocal lattice vector. That means that we can rewrite Eq. (2.63):

$$\psi_{\mathbf{k}}(\mathbf{r}) = \sum_{\mathbf{G}} c_{\mathbf{k}-\mathbf{G}} \exp(i(\mathbf{k} - \mathbf{G}) \cdot \mathbf{r}). \tag{2.66}$$

This can be factored into a form that explicitly satisfies the Bloch condition:

$$\psi_{\mathbf{k}}(\mathbf{r}) = \left(\sum_{\mathbf{G}} c_{\mathbf{k}-\mathbf{G}} \exp(-i\mathbf{G} \cdot \mathbf{r}) \right) \exp(i\mathbf{k} \cdot \mathbf{r}). \tag{2.67}$$

The first term is a Fourier series only in $\mathbf{G}$, and is therefore an example of a function $u_{\mathbf{k}}(\mathbf{r})$ in a Bloch wave.

Perturbation theory is very illustrative here, and is discussed at length in several books (e.g., [15, 6, 5]). The first-order correction to the single-particle energies vanishes. To second order the energy looks like

$$E_{\mathbf{k}} \approx E_{\mathbf{k}}^0 + \sum_{\mathbf{G}} \frac{|V_{\mathbf{G}}|^2}{E_{\mathbf{k}}^0 - E_{\mathbf{k}-\mathbf{G}}}. \tag{2.68}$$

This assumes nondegeneracy; that is, $E_{\mathbf{k}}^0 \neq E_{\mathbf{k}-\mathbf{G}}$. When this is applicable, the energy shift due to the lattice is second order in V.

Clearly exciting things happen when $\mathbf{k}$ is chosen to be near where

$$E_{\mathbf{k}}^0 = E_{\mathbf{k}-\mathbf{G}} \tag{2.69}$$

for some $\mathbf{G}$. This happens when

$$\mathbf{k} = \frac{1}{2}\mathbf{G}. \tag{2.70}$$

This is precisely at the boundary of a Brillouin zone in k-space. In this case, degenerate perturbation theory must be applied, and the result near this value of $\mathbf{k}$ is

$$E_{\mathbf{k}} = \frac{1}{2}(E_{\mathbf{k}}^0 + E_{\mathbf{k}-\mathbf{G}}^0) \pm \left[\left(\frac{E_{\mathbf{k}}^0 - E_{\mathbf{k}-\mathbf{G}}}{2} \right)^2 + |V_{\mathbf{G}}|^2 \right]^{1/2}. \tag{2.71}$$

Right *at* that point,

$$E_{\mathbf{k}} = E_{\mathbf{k}}^0 \pm |V_{\mathbf{G}}|. \tag{2.72}$$

The allowed energies for a single-electron state with $\mathbf{k}$ at a value that is half of a reciprocal lattice vector $\mathbf{G}$ are *split* around the unperturbed (free particle) value. The size of this band gap is proportional to the $\mathbf{G}$th component of the lattice potential.

The energetic condition, Eq. (2.69), leads to a way of thinking about where this gap comes from. The restriction that $|\mathbf{k}| = |\mathbf{k} - \mathbf{G}|$ is equivalent to the Bragg condition for a planewave of wavevector $\mathbf{k}$. Because there is a commensurate spatial periodicity between the electronic wave and the underlying lattice potential, the lattice potential strongly perturbs the energy of that state from the no-lattice case.

2.2.7 Summary and discussion of nearly free electrons

So far we have considered the single-particle states of electrons confined to a finite (but very large) periodic crystal. The main properties of this system that we've seen so far are:

- the spatial periodicity of the real-space lattice also defines a reciprocal lattice in k-space, with particular reciprocal lattice vectors $\mathbf{G}$;
- allowed single-electron states in a periodic potential are Bloch states, and may be written as a plane wave modulated by a function with the spatial periodicity of the lattice; Bloch states are extended states;
- these Bloch functions are labeled by a wave vector $\mathbf{k}$;

- the boundary conditions imply that only certain $\mathbf{k}$ values are allowed, with $(2\pi/L)^d$ of k-space volume in d dimensions associated with each allowed $\mathbf{k}$ point;
- when $\mathbf{k}$ is near $\mathbf{G}/2$, the energy of the Bloch state is strongly modified by the lattice potential, opening up gaps in the allowed single-particle state energies.

The details of the allowed single-particle states (e.g., how the energy depends on $\mathbf{k}$ for the various bands) are called the *band structure* of a material. Unsurprisingly, the band structure depends on just what the lattice and basis are. The resulting material properties depend critically on how the atoms are arranged in real space, how the atoms are bonded together, and how many atoms per unit cell are "free". We will discuss this further in the next chapter, when we consider what happens when we "fill up" those single-particle states.

Here are a couple of other points, stated without proof.

- Just as one can think of a plane wave carrying momentum $\hbar\mathbf{k}$, one can think of a Bloch wave as carrying *crystal momentum* $\hbar\mathbf{k}$. In the language of physics and symmetry principles, momentum is conserved for free particles in a potential-free space because that space is translationally invariant for arbitrary displacements. In an infinite lattice, translational invariance only holds for certain discrete translations (by lattice vectors $\mathbf{R}$). The consequence of this is that crystal momentum is only conserved to within a reciprocal lattice vector, $\hbar\mathbf{G}$. That is, when considering transitions between electronic states within the infinite lattice, energy must be strictly conserved, but crystal momentum can change by $\hbar\mathbf{G}$. That would correspond in some sense to the extended electronic states transferring momentum to the crystal as a whole.
- Bloch states are eigenstates in the presence of the lattice potential, and are constructed to be extended states. If one considers a propagating wave packet built out of Bloch states, deviations from perfect periodicity will lead to scattering – that is, transitions from states of one $\mathbf{k}$ to another, at the same energy. We will explicitly consider the effects of defects, surfaces, and interfaces in a later chapter.

We have spent quite a bit of time on the Bloch picture of nearly free electrons in a periodic potential, and seen that a number of the properties of the "free" electron gas model survive, though some are strongly modified. Interestingly, many of the Bloch picture's essential features still persist even when the periodicity requirement is relaxed. For example, bands of allowed single-particle state energies separated by gaps exist even in substitutionally disordered alloys. Extended states with wave-like properties exist even in certain liquids (e.g., mercury). To appreciate a bit more why this is so, in the next section we consider a more chemical approach to electronic structure, building up materials from constituent atoms.

2.3 Chemical approaches to electronic structure

Rather than treating solids as boxes full of electrons, with the lattice as a perturbation, the complementary approach to electronic structure begins with individual atoms. Rather than assuming that the basic single-particle states with which to work are plane waves, this approach assumes that atomic orbitals are the right starting point. Knowing the properties of atomic systems, one can consider building up molecules, and then eventually larger systems. We briefly look at Hückel theory, an approach commonly taught in undergraduate physical chemistry courses as an illustrative way of thinking about more complicated molecules. We then move on to a quick look at "tight-binding", including the Bloch condition in treating extended periodic systems. Our conclusions at the end of the day will be similar to those above: there are bands of allowed single-particle state energies, separated by gaps. What those bands mean and how those bands are filled are really the subject of the next chapter.

2.3.1 Born–Oppenheimer

Before we tackle electronic states of molecules, it's important to mention another approximation underlying nearly everything we're going to do: the *Born–Oppenheimer approximation* (BOA). The BOA treats the atoms as if the nuclear and electronic wavefunctions are completely separable, and ignores direct nuclear–nuclear interactions.

Instead of worrying about the nuclear positions being influenced by the electronic states, we treat the nuclear positions $\mathbf{R}_i$ as parameters. We can solve the electronic problem as a function of $\mathbf{R}_i$, and then say that the nuclei move in a potential created by the electrons. This is reasonable for many cases because nuclei are very heavy compared to electrons, and the timescale for electronic rearrangements is often much faster than that for significant nuclear motion. In what follows, consider the BOA to be in effect. (A side note: people doing research on femtosecond chemistry and reaction mechanisms involving excited complexes often need to take a more sophisticated approach than the BOA.) Completing the Born–Oppenheimer approximation, the resulting electronic states and how they are occupied are then used to compute an effective potential for the nuclei. The nuclear motions are then solved self-consistently with the electronic system.

2.3.2 How atoms are bound together

Rather than idealizing bulk solids as infinite periodic potentials, let's take a "bottom-up" approach and consider how atoms are actually bound together to form molecules. This is where chemistry comes into the picture, and this has been discussed in several books [16, 17, 8].

There are several types of bonding that can occur; which actually takes place depends on the details of the atoms in question.

- *Van der Waals bonding.* This doesn't involve any significant change to the electronic wavefunctions; atom A and atom B remain intact and interact via fluctuating electric dipole forces. This weak attractive interaction is most responsible for bonding in crystals of small molecules (e.g., C_{60}, pentacene, anthracene, naphthalene). While ordinarily small, under certain circumstances the aggregate binding energy from Van der Waals forces can be significant. For example, the van der Waals binding energy between adjacent carbon nanotubes can be as large as 1 eV per micron of tube length.

- *Covalent bonding.* We are familiar with the concept of covalent bonding from high school chemistry, and we shall discuss it further below. In brief, one treats the atoms as consisting of ion cores (the inner electrons plus the nucleus) and outer "valence" electrons. It is useful to describe the single-particle states as being made up of linear combinations of atomic orbital (LCAO) states, with each atomic orbital centered on a particular atom. The LCAO states are called molecular orbitals, and to some degree the valence electrons are delocalized over more than one ion core. When considering bonding between two identical atoms, each atom receives equal weight in the linear combination. Bonding between different atoms gets more complicated due to the differences in chemical properties of the atoms. Pauling introduced [18] the concept of *electronegativity* to describe quantitatively how lopsided the LCAO needs to be to give a good description of the energies of bond formation. Bonding between atoms with large differences in electronegativities tends to be more polar; that is, the filled molecular orbital has more probability amplitude for the electron to be found on the more electronegative ion core. Defining "covalency" and "ionicity" of bonding in a quantitative way can be complicated, but their qualitative meaning is clear. In large molecules there tends to be clustering of energy levels with intervening gaps in energy containing no allowed states. There is a *highest occupied molecular orbital (HOMO)* and a *lowest unoccupied molecular orbital (LUMO)*. Depending on the relative sign of the coefficients used in the linear combination, one can speak of "bonding" orbitals, which tend to have significant electronic probability density between atoms A and B, and "antibonding" orbitals, which have a deficit of electronic density between the atoms. The typical energy of a covalent bond is on the order of one or two electron volts.

- *Ionic bonding.* This is the limit of extreme asymmetry between the different atoms in the LCAO approach. For essentially Coulomb energy reasons, the situation is well approximated by saying that atom A donates a valence electron that is accepted by atom B. Atom A is positively charged and its remaining electrons are tightly bound to the ion core; atom B is negatively charged, and all the electrons are tightly bound. Atoms A and B "stick" to each other by electrostatics, but the wavefunction overlap between their electrons is minimal.

- *Metallic bonding.* When looking at just our two atoms A and B, remember that they're part of a larger, more extended solid. Therefore thinking of a molecular orbital bonding A and B together is only part of the picture. Qualitatively, because of overlap of that molecular orbital with those of other groups of atoms, the single-particle state described by that orbital has a finite lifetime, and therefore a width in energy. If that overlap is strong, the width in energy can be larger than the bonding/antibonding energy splitting. When that happens, the system is metallic, and is best described by delocalized molecular orbitals that extend over many atomic spacings.

- *Hydrogen bonding.* This is included for completeness, and is related to the polar bond concept introduced above. Bonds involving hydrogen can be very polar because of hydrogen's comparatively low electronegativity. As a result, the hydrogen atom often has a fair fraction of a proton's worth of excess positive charge, and the bond can have a significant dipole moment. That excess positive charge can be attracted to "lone pairs" and the negatively charged parts of other molecules. The effective bond energy can be on the order of 0.1 eV, and is significant in determining the structure of many biological molecules (proteins, DNA) as well as the local structure of water.

Since Van der Waals and hydrogen bonding do not involve significant changes in the electronic distributions between the bonded entities, they aren't directly relevant in trying to understand the electronic structure of solids. They can be very important, however, at the nanoscale, and we will return to them later.

2.3.3 Basic Hückel theory

Start by thinking about two hydrogen atoms very far from one another. Each atom has an occupied $1s$ orbital, and a number of unoccupied higher orbitals (p, d, etc.). If the atoms are moved sufficiently close that wavefunction overlap can be significant (a separation comparable to a couple of Bohr radii), an accurate set of energy eigenstates requires hybrid orbitals.

That is, do a perturbation theory calculation accounting for the Coulomb interaction between the electrons in this geometry. Instead of the ground state energy being minimized by two single-electron states ($1s$) of identical energy, we find two states (σ, σ^*) that differ in energy. The energy of two electrons (with antisymmetrized spins) in the σ state is lower than that of each electron in a $1s$ state. The σ and σ^* states are bonding and antibonding molecular orbitals.

Now think about combinations of many-electron atoms. It's reasonable to think about an *ion core* containing the nucleus and electrons that are firmly stuck in *localized* states around that nucleus, and *valence* electrons, which are more loosely bound to the ion core and can in principal overlap with their neighbors. We can then consider using the valence orbitals of those electrons as a basis for constructing molecular orbitals.

In 1931 Hückel formulated [19] an approximation procedure based on this idea of LCAO in an attempt to calculate the electronic levels of conjugated hydrocarbons. Rather than doing a full perturbation theory calculation, the idea is to use a *variational method*: a trial wavefunction for the molecule is *assumed* to be an LCAO. The coefficients of the individual orbitals are found by minimizing the energy expectation value with respect to changes in the coefficients, using a parametrized model for the Hamiltonian. **Note**: this method should be viewed as largely qualitative, giving some intuitive sense of level spacings, ionization energies, and molecular orbital shapes. This is not how electronic structure calculations are done anymore, as discussed below.

Let's do an example to make this concrete. Consider the allyl radical consisting of three carbons as shown in Fig. 2.21. Suppose that the Hamiltonian for the whole radical is H. The wavefunction $|i\rangle$ is the atomic p_z-orbital for the carbon at position i. Our trial wavefunction is assumed to be

$$|\Psi\rangle = c_1|1\rangle + c_2|2\rangle + c_3|3\rangle, \tag{2.73}$$

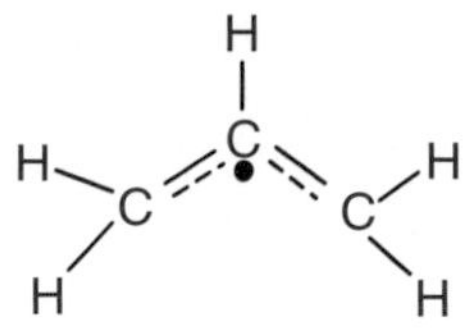

Fig. 2.21

An allyl radical. As drawn, the unpaired electron is in a molecular orbital that spans the three carbons and is constructed from a linear combination of carbon p_z atomic orbitals.

where the c_i are the (real) coefficients in question. We can then write the energy expectation value,

$$\langle E \rangle = \frac{\langle \Psi | H | \Psi \rangle}{\langle \Psi | \Psi \rangle} = \frac{(c_1 \langle 1 | + c_2 \langle 2 | + c_3 \langle 3 |)H(c_1 | 1 \rangle + c_2 | 2 \rangle + c_3 | 3 \rangle)}{(c_1 \langle 1 | + c_2 \langle 2 | + c_3 \langle 3 |)(c_1 | 1 \rangle + c_2 | 2 \rangle + c_3 | 3 \rangle)}. \tag{2.74}$$

Let's deal with the denominator first. Since the $|i\rangle$ correspond to orbitals around different atoms, it is a common approximation to assume

$$\langle i | j \rangle = \delta_{ij}. \tag{2.75}$$

Clearly this is not necessarily accurate if the two atoms are in close proximity, but since we're doing a perturbative calculation, we won't really care to first order. With this assumption, the denominator is just $c_1^2 + c_2^2 + c_3^2$.

The numerator is more complicated, with lots of cross-terms. A simplifying approximation is to worry only about interactions between nearest neighbor atoms. Define

$$\alpha = \langle i | H | i \rangle,$$
$$\beta = \langle i | H | j \rangle, \quad i \neq j \tag{2.76}$$

for i, j nearest neighbors. Other matrix elements are approximated as zero. The idea is that α and β can be compiled phenomenologically for different pairs of atoms. Rewriting,

$$\langle E \rangle \approx \frac{c_1^2 \alpha + 2c_1 c_2 \beta + c_2^2 \alpha + 2c_2 c_3 \beta + c_3^2 \alpha}{c_1^2 + c_2^2 + c_3^2}. \tag{2.77}$$

We've parametrized the interactions here. Note, too, that nowhere have we included the antisymmetrization of the total wavefunction – the Pauli principle is thrown in ad-hoc at the end. Electronegativity has also not been included, since we have been considering only bonding between carbon atoms. Harrison [8] has a good description of extending this picture to polar bonds. See below for more caveats.

We now consider varying each of the c_i, and for each variation we set

$$\frac{\partial \langle E \rangle}{\partial c_i} = 0. \tag{2.78}$$

(Extremizing $\langle E \rangle$ subject to the normalization constraint is equivalent to solving the time-independent Schrödinger equation.) This gives us a system of equations that, in our example, looks like

$$\begin{pmatrix} X & 1 & 0 \\ 1 & X & 1 \\ 0 & 1 & X \end{pmatrix} \begin{pmatrix} c_1 \\ c_2 \\ c_3 \end{pmatrix} = 0, \tag{2.79}$$

where

$$X \equiv \frac{\alpha - \langle E \rangle}{\beta}. \tag{2.80}$$

The matrix is called the Hückel matrix. For there to be a solution to this equation, the determinant of the Hückel matrix must equal zero. Notice that for a molecule with N atoms this results in an Nth order polynomial equation, and N resulting values for X (and thus $\langle E \rangle$). The remaining constraint that we must remember is the normalization condition for $|\Psi\rangle$.

For any given X that is a solution, we can solve for the c_i and construct a trial wavefunction called a Hückel molecular orbital. One can then look at the molecular energy levels, count valence electrons, and figure out the occupancy of the orbitals.

2.3.4 Caveats

This approach is *not* what's currently used by people calculating electronic structures today. It omits many things, most notably the antisymmetry condition on the total wavefunction. It also parametrizes any electron–electron interaction effects into α and β terms, so that we end up with an effective independent electron method.

Considerably more sophisticated methods exist that incorporate the Pauli principle, and will be discussed below. Similarly, parametrizing the matrix elements above has now largely been replaced by empirical interaction terms or direct first-principles calculations. Using modern approaches like those discussed in Section 2.4, far more quantitatively accurate calculations may be done than is possible with simple Hückel theory.

The Hückel approach does get qualitative trends correct, though.

- If Hückel is performed on large molecules, one finds that the predicted energy levels cluster into groups. The number of such groups goes like the number of orbitals included in the calculation.
- As overlap between adjacent atoms is increased, the width in energy of those groups of levels increases.
- As the number of atoms in the molecule increases toward infinity (the bulk limit), the level clusters evolve into *bands* of allowed energies separated by regions of energy with no allowed states.

2.3.5 A tight-binding picture

One related electronic structure approach is the "tight-binding" picture, reminiscent of the molecular orbital case described above. We're interested in approximating the ground state of the many-body system of interacting electrons plus the potential due to the ion cores.

In tight-binding we consider our solution to be built out of superpositions of localized single-electron wave functions centered on the positions of the atoms in the solid. That is, if the atoms in the solid are at positions $\mathbf{r}_n$, we try to write the ground state wavefunction as:

$$\psi(\mathbf{r}) = \sum_{\mathbf{R}} a(\mathbf{R})\phi(\mathbf{r} - \mathbf{R}), \tag{2.81}$$

where $\phi(\mathbf{r})$ is a localized atomic orbital wave function centered on the origin. The periodic nature of the lattice potential in a crystal enters through restrictions on the $a(\mathbf{R})$.

For a more complete description of the tight-binding method, refer to many solid state books, such as Ashcroft and Mermin [6]. For now, we only state the results and make a few qualitative observations. Suppose that the full potential energy at location $\mathbf{r}$ is given by the sum of two pieces:

$$V(\mathbf{r}) = V_{\text{atomic}}(\mathbf{r}) + \delta V(\mathbf{r}), \tag{2.82}$$

where the first term is the potential due to the atom located at $\mathbf{r}$, and δV is the potential energy at that site due to all the other atoms in the solid. Suppose further that the overlap between orbitals on adjacent atoms is small. Consider a very simple case involving only a single atomic orbital for each atom $\phi(\mathbf{r})$ (for example, the $1s$ orbital). Then we can define some parameters:

$$\alpha(\mathbf{R}) \equiv \langle \phi(\mathbf{r}) | \phi(\mathbf{r} - \mathbf{R}) \rangle, \tag{2.83}$$

$$\beta \equiv -\langle \phi(\mathbf{r}) | \delta V | \phi(\mathbf{r}) \rangle, \tag{2.84}$$

$$\gamma(\mathbf{R}) \equiv \langle \phi(\mathbf{r}) | \delta V | \phi(\mathbf{r} - \mathbf{R}) \rangle. \tag{2.85}$$

Here we've kept the $\mathbf{r}$ label to remind you that the $\phi(\mathbf{r})$ are atomic orbitals centered on a particular location.

This is essentially a variation on the Hückel approach. The idea is that these parameters may be calculated for pairs of atoms within a particular model. The main result in the case of a crystalline solid ends up giving us the relationship between energy and wavevector. For the simple case above,

$$E(\mathbf{k}) = E_\phi - \frac{\beta + \sum_{nn} \gamma(\mathbf{R}) e^{i\mathbf{k}\cdot\mathbf{R}}}{1 + \sum_{nn} \alpha(\mathbf{R}) e^{i\mathbf{k}\cdot\mathbf{R}}}. \tag{2.86}$$

The first term is the energy eigenvalue of the particular atomic orbital we're using. The sums are over nearest-neighbors, if we assume that the overlap is small enough that other terms are negligible. If we assume we are only using s orbitals, and remember that the α terms in the denominator are small, we can find:

$$E(\mathbf{k}) \approx E_\phi - \beta - \sum_{nn} \gamma(\mathbf{R}) \cos \mathbf{k} \cdot \mathbf{R}. \tag{2.87}$$

Note that what had been a single energy level E_ϕ is now broadened out into a band of allowed energies, with the band width being given by γ. Notice, too, how this expression labels states with a variable $\mathbf{k}$, like the wavevector in the free electron gas case we talked about before. However, it's in the form of corrections to the atomic energy levels used in the calculation.

General remarks:

- When allowing more that one kind of atomic orbital per atom in a tight-binding calculation, Eq. 2.86 becomes a matrix equation. Using three p-orbitals means a three-by-three matrix; five d-orbitals means a five-by-five matrix, etc.
- The band width is proportional to the overlap integrals γ. This means that higher orbitals, which tend to be larger, lead to broader bands. Once the overlap of orbitals is substantial over an atomic spacing, our first-order treatment of the tight binding parameters is of questionable validity.
- Complicated crystal structures (and hence nearest-neighbor arrangements) in solids can split a single atomic orbital into more than one band.

2.4   More modern electronic structure methods

So far we have considered finding the single-particle electronic states for bulk solids from two directions: starting from the free electron gas on the one hand, and from the constituent atoms on the other. In the next chapter we shall consider in detail the importance of how the resulting manifolds of states are filled, and the kinds of material property that result. First, however, it is worthwhile to summarize very briefly other approaches for calculating electronic structure. We frame these in the form of how they address shortcomings in the treatments given above. This is meant to give a flavor of the different approaches, not to substitute for a real course in electronic structure methods.

2.4.1   Better treatments of the lattice and wavefunctions

So far we've treated the lattice as a weak perturbation, and we've also considered starting from essentially the atomic single-particle orbitals. There are a number of methods that interpolate between these extremes, and there is a veritable alphabet soup of acronyms associated with them.

The original *cellular method* tries to find solutions to the single-particle problem within a single Wigner–Seitz cell of the lattice in real space, with appropriate boundary conditions on the cell (rather than the atomic boundary condition that the wavefunction vanishes at large distances). That way a whole lattice may be built out of such cells. Rather than using the exact ion potential within the cell, the *muffin-tin* approach defines an atomic region (in the neighborhood of a particular ion) and the interstitial region (the rest of the cell). Within the atomic region, the ion potential (which, like the regular Coulomb atomic potential, diverges to $-\infty$ at the atom's site) is used, while the potential is set to a fixed constant value in the interstitial region.

In the *plane wave* method, the single-particle states are written as an expansion of planewaves, $|\psi\rangle = \sum_{\mathbf{k}} a_{\mathbf{k}} e^{i\mathbf{k}\cdot\mathbf{r}}$, and the potential throughout the crystal is a superposition of free-atom potentials. As an approximate wavefunction for variational calculations, this ends up being extremely inefficient.

One computational improvement [20] is the *augmented plane wave* (APW) method that uses the cellular idea of atomic and interstitial regions. In interstitial region an APW function is assumed to be plane-wave-like, while it must satisfy an atomic-like Schrödinger equation in the atomic region, and the APW is continuous across the atomic/interstitial boundary (this picks out **k** values). The single-particle states are then built out of superpositions of APWs, and a variational approach is done to optimize the states.

A further improvement [21] is the *orthogonalized plane wave* (OPW) approach. Conyers Herring pointed out that when considering the single-particle states of the conduction (non-core) electrons, it's smart computationally to start out not with plane waves, but with superpositions of plane waves already chosen to be orthogonal to the core states of the ion potential.

One other major approach [22, 23] has been that of *pseudopotentials*. Conceptually, the point is that the core electrons act to screen the nuclear charge, and the Pauli principle also acts like an effective repulsion. The result is that the effective potential seen by the conduction electrons is very different from the bare ion potential. Indeed, a common pseudopotential to use is one that is essentially *zero* in the core region, and a screened Coulomb potential ($\sim e^{-r/\lambda}/r$, where λ is some screening length scale) outside the core.

The pseudopotential method leads naturally into a discussion of the means to tackle the major shortcomings of the noninteracting single-electron approaches that we've used. What is the best way to account for the Coulomb repulsion of the electrons, and how does the requirement of wavefunction antisymmetry affect electronic structure? As we'll see in the next section, these are deeply interconnected questions. The two approaches described there are very widely known and applied.

2.4.2 Interactions and antisymmetry: Hartree and Hartree–Fock

As we mentioned at the beginning of this chapter, the interactions between the electrons in solids are not *a priori* weak. While independent electron treatments can give many of the qualitative trends seen in the electronic structure of bulk materials, quantitative accuracy requires some better way of treating the interactions than neglect.

At the nanoscale, the situation is just as bad, if not worse. In many-electron systems it is common to talk about *screening*. When considering the response of a particular electron to a fixed charge, the *other* electrons can rearrange themselves slightly in a way that mitigates the potential seen by the particular electron. In reduced dimensionality with small numbers of electrons, this screening process is less effective than in electron-rich bulk systems.

It is clear that a good treatment of screening must be self-consistent. That is, our decision to identify one particular electron of interest was completely arbitrary. There must be self-consistency between the predicted response of the one particular electron, and the assumed collective response of all the other electrons. One of the first attempts to address this leads to the *Hartree* approximation.

The Hartree approximation is an example of a *mean field* approach. Rather than trying to solve the many-electron problem, the idea is to turn it into an effective one-electron problem. Starting from a single-particle Schrödinger equation:

$$-\frac{\hbar^2}{2m}\nabla^2\psi(\mathbf{r}) + V_{\text{ion}}(\mathbf{r})\psi(\mathbf{r}) + V_{\text{e}}(\mathbf{r})\psi(\mathbf{r}) = E\psi(\mathbf{r}).$$ (2.88)

Here V_{ion} is due to the atomic nuclei, and is the appropriate Coulomb potential:

$$V_{\text{ion}}(\mathbf{r}) = \frac{-Ze^2}{4\pi\epsilon_0}\sum_{\mathbf{R}}\frac{1}{|\mathbf{r} - \mathbf{R}|},$$ (2.89)

while V_{e} is the Coulomb potential due to the other electrons:

$$V_{\text{e}}(\mathbf{r}) = \frac{e^2}{4\pi\epsilon_0}\int d\mathbf{r}'|\psi(\mathbf{r}')|^2\frac{1}{|\mathbf{r} - \mathbf{r}'|}.$$ (2.90)

The result is a set of nonlinear coupled single-particle equations:

$$-\frac{\hbar^2}{2m}\nabla^2\psi_i(\mathbf{r}) + V_{\text{ion}}(\mathbf{r})\psi_i(\mathbf{r}) + \left[\frac{e^2}{4\pi\epsilon_0}\sum_j\int d\mathbf{r}'|\psi_j(\mathbf{r}')|^2\frac{1}{|\mathbf{r} - \mathbf{r}'|}\right]\psi_i(\mathbf{r}) = E_i\psi_i(\mathbf{r}).$$ (2.91)

The nonlinearity in Eq. (2.91) means that even this approach to solving the many-electron problem is complicated. Iterative solution is the simplest method to try.

The Hartree approximation is something of an improvement over ignoring interactions entirely, but because it lacks the Pauli principle at a fundamental level, it is quantitatively poor and sometimes qualitatively wrong as well. The *Hartree–Fock* (HF) approximation tries to do better. In HF, the true solution to the many-particle Schrödinger equation is approximated by a single Slater determinant of one-electron states.[5] Labeling the (orthonormal) orbitals (including spin) that make up the Slater determinant as $\psi_i(\mathbf{r}, s_i)$, and variationally minimizing the energy of the many-body state, one is left with the Hartree–Fock equations:

$$-\frac{\hbar^2}{2m}\nabla^2\psi_i(\mathbf{r}) + V_{\text{ion}}(\mathbf{r})\psi_i(\mathbf{r}) + V_{\text{e}}\psi_i(\mathbf{r})$$
$$-\frac{e^2}{4\pi\epsilon_0}\sum_j\int d\mathbf{r}'\frac{1}{|\mathbf{r} - \mathbf{r}'|}\psi_j^*(\mathbf{r}')\psi_i(\mathbf{r}')\psi_j(\mathbf{r})\delta_{s_i s_j} = E_i\psi_i(\mathbf{r}).$$ (2.92)

The last term on the left side is the *exchange* term, and results from the antisymmetry of the Slater determinant. Note that the exchange term is actually an integral operator, rather than just a "simple" nonlinear potential multiplying $\psi_i(\mathbf{r})$.

Actually solving the Hartree–Fock equations in the presence of a periodic potential is very challenging. The exchange interaction is not something to be ignored, however. It can strongly affect electronic structure, particularly in systems where the aufbau way of thinking implies that there are unpaired electrons. Qualitatively, this term is the source of the Coulomb energy benefit that leads to Hund's rules, as we mentioned earlier.

---

[5] One can always write the *true* many-body solution as a *linear combination* of Slater determinants, if the basis of one-particle states is sufficiently large.

2.4.3 Density functional theory

Another modern electronic structure technique that has gained widespread importance is broadly known as *density functional theory* (DFT). The basic idea of DFT was first put forward by Pierre Hohenberg and Walter Kohn in 1964 [24], and then formulated into a methodology in a paper by Kohn and L. J. Sham the following year [25]. It is difficult to overstate the impact of this work: this latter paper is the most cited physics journal article.

Hohenberg and Kohn proved mathematically that, for any system of interacting particles in an external potential $V_{ext}(\mathbf{r})$, the *ground state* particle density $n_0(\mathbf{r})$ uniquely determines the full, interacting Hamiltonian, to within a constant. That implies that *all* properties of the system (many-body wavefunctions for ground and excited states, matrix elements for any operators) are completely determined by $n_0(\mathbf{r})$. They further proved that there is a universal functional[6] $E[n(\mathbf{r})]$ for the energy in terms of the density for any external potential. The density profile that minimizes $E[n(\mathbf{r})]$ is exactly $n_0(\mathbf{r})$.

The trick, then, is to try to recast questions of interest in terms of the interacting many-body ground state density, $n_0(\mathbf{r})$. One problem is that there is no nice way to write the *exact* functional $E[n(\mathbf{r})]$. Another is that it's generally not obvious how to use the density distribution to answer even simple questions, like whether a system is a metal or an insulator. Even if one has the exact $n_0(\mathbf{r})$, as a practical matter it remains a topic of research to find ways to calculate useful properties like the spectrum of excited states.

The Kohn–Sham approach takes the general problem and assumes that the interacting ground state density is the same as that for a particular effective *noninteracting* system (with long-range interactions effectively accounted for by a Hartree term). Since this auxilary system is noninteracting, all of the complications of the real interacting system are wrapped up in a term called the exchange-correlation functional. Conceptually, one approximates this exchange-correlation functional, solves the noninteracting problem, and this gives a good approximate solution of the the ground state energy and density of the real interacting system.

Since the long-range interactions have been wrapped up in a Hartree term, often one approximates the exchange-correlation functional as depending only on the local density. That is, it doesn't involve terms like $\int d\mathbf{r} d\mathbf{r}' n(\mathbf{r})n(\mathbf{r}')$, and instead is something like $\int d\mathbf{r} n(\mathbf{r})\epsilon_{xc}(n(\mathbf{r}))$. This is called the *local density approximation* (LDA). More refined approximations edge toward non locality by allowing the exchange-correlation functional to depend on the gradient of the local density; these are *generalized gradient approximations* (GGA). Like the other electronic structure approaches described above, actually solving the Kohn–Sham equations requires one to choose a basis of functions. As a result, the alphabet soup of OPW, APW, LCAO, etc. re-enters the discussion.

DFT and related techniques remain very active topics of research. Major areas of effort include: time-dependent DFT; the treatment of surfaces and interfaces; accurate treatments of open-shell and strongly correlated systems; accurate treatments of excited states; and computationally efficient approaches to large systems.

---

[6] A *functional* is a function of a function. That is, a functional $F[g(x)]$ takes some function $g(x)$ and maps it into some number $F[g(x)]$. A common functional in physics is the action, $S[q(t)]$, which takes a trajectory $q(t)$ and maps it onto a number, $S[q(t)] = \int_{t1}^{t2} L(q(t), \dot{q}(t)) dt$, where L is the Lagrangian.

2.5 Lattice dynamics: phonons

After this discussion of electronic structure and before moving on to addressing how electronic structure leads to different types of bulk materials, it is reasonable to consider the vibrational properties of bulk materials. Mathematics very similar to that used in counting the electronic states (Eqs. (2.19–2.21)) is relevant for counting vibrational modes, as we shall see below. Quantized vibrational modes in bulk solids are called *phonons*. Phonon modes correspond to displacements of the atomic lattice from its equilibrium, periodic configuration. Because of this disruption of periodicity, Bloch states are no longer perfect single-particle eigenstates in the presence of phonons, and acquire a finite lifetime. In other words, phonons can scatter electrons. Phonons are discussed extensively in many statistical physics and solid state physics texts.

2.5.1 Vibrations in the nano limit

Chemists are very familiar with the fact that molecules have vibrational modes. Indeed, the Born–Oppenheimer approximation in combination with the quantum chemistry electronic structure methods in the previous section can be used to calculate those modes relatively well in small to moderately large molecules. One can get a reasonable qualitative understanding of these modes by picturing the atoms as "balls" and the chemical bonds as "springs", with the effective spring constants of the different bonds depending on the chemistry of the participating atoms.

When confronted with a relatively small number of coupled mechanical oscillators, the standard physics approach would be to calculate the resulting *normal modes*. For example, Fig. 2.22 shows the normal stretching modes for a linear triatomic molecule. From this example the analogy to hyridization of energy levels should be clear. Each spring has some mechanical resonance frequency by itself (in the diatomic molecule case). Coupling the two mechanical resonators leads to a composite system (the triatomic molecule) with two normal modes. The total number of modes has not been altered by the formation of the coupled system, though in general the frequencies of the normal modes are shifted from their uncoupled values. This is analogous to the formation of bonding and antibonding σ orbitals from the hybridization of $1s$ orbitals. This sort of result shows up repeatedly, as we shall see.

It is often not a bad approximation that the vibrational normal modes of molecules are harmonic. That is, the springs are linear, and effectively decoupled normal modes is a good description. For each normal mode of frequency ω, one can treat that mode as a quantum harmonic oscillator with energies given by $(n + 1/2)\hbar\omega$, where n is the occupation number of that mode. At finite temperature, basic statistical mechanics can be used to find the average occupation,

$$\langle n \rangle = \frac{1}{e^{(\hbar\omega)/(k_B T)} - 1}. \tag{2.93}$$

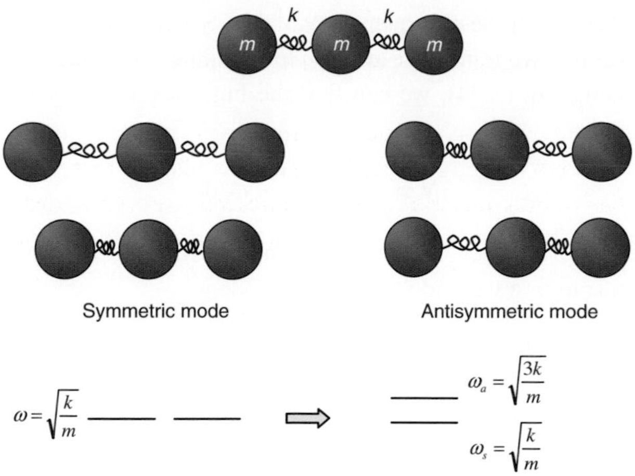

Top: toy model of a triatomic molecule. Middle left: symmetric vibrational mode of coupled system. Middle right: antisymmetric mode of coupled system. Bottom: uncoupled individual modes split when coupled, though the total number of modes is unchanged. This is analogous to hybridization of electronic orbitals and energy levels.

Fig. 2.22

In the high temperature limit, each vibrational mode ends up with $k_B T$ worth of energy, while in the low temperature limit the vibrational states "freeze out" as $k_B T$ falls below $\hbar\omega$.

2.5.2 Vibrations in the bulk: Debye

What happens in the limit of larger systems? For a generic molecule with N atoms, one would generally expect there to be $3N$ normal modes, since one can consider the x, y, and z degrees of freedom of each atom separately in a ball-and-spring picture. For systems constrained to move truly in 2d or 1d, the total number of modes would be $2N$ or N, respectively. In the high temperature limit, the 3d result above leads to a molar specific heat of $3R$, where R is the molar gas constant. In the low temperature limit, one again runs into the lack of occupation of the high frequency modes.

To keep track of this in some reasonable way, one counts acoustic modes. For a bulk, linear elastic medium, the dispersion relation is

$$\omega = c_s k, \tag{2.94}$$

where c_s is the speed of sound (here assumed to be isotropic) and k is the wavevector of the sound. By counting modes (acoustic ones this time) in k-space, we can use Eq. (2.94) to count up frequencies, and the occupation number from Eq. (2.93) to keep track of the energy content of each mode. Note, though, that the real disperson relation for phonons is more complicated than this, since sound with a wavelength comparable to the size of the unit cell is sensitive to the details of the atomic stacking. This is just the mechanical analog of the discussion in Section 2.2, where we saw that the dispersion relation for nearly free electrons in a lattice is differs at the Brillouin zone boundaries from that of free electrons.

When counting acoustic states there are two major differences from the electronic state counting done previously. First, the total number of normal modes is limited to $3N$, leading to a cutoff in the highest frequency available. Second, there are three independent polarizations for sound, two transverse and one longitudinal. Starting from counting modes in k-space, and using Eq. (2.94), we can find the highest physically significant acoustic mode frequency:

$$3N = \int_0^{\omega_D} \frac{3V}{2\pi^2 c_s^3} \omega^2 d\omega. \tag{2.95}$$

Here ω_D is the *Debye frequency*, which is material-specific, and V is the material volume. Integrating and rearranging we find

$$\omega_D = \left(6\pi^2 c_s^3 \frac{N}{V} \right)^{1/3}. \tag{2.96}$$

It is common to recast the Debye frequency as a temperature,

$$T_D \equiv \frac{\hbar \omega_D}{k_B}. \tag{2.97}$$

Roughly speaking, for a single-atom basis, the Debye frequency corresponds to a sound wave with a wavelength comparable to the lattice spacing. Since ordinary acoustic vibrations are the displacement of atoms from their equilibrium positions, it doesn't make sense to talk about wavelengths smaller than the atomic spacing.

Within the Debye model for phonons, the energy content in the vibrational degrees of freedom is

$$E = \frac{3V}{2\pi^2 c_s^3} \hbar \int_0^{\omega_D} \frac{\omega^3 d\omega}{e^{\hbar\omega/k_B T} - 1}. \tag{2.98}$$

This leads to the familiar Debye heat capacity at low temperatures that varies as T^3 as $T \to 0$. For temperatures less than T_D, the integrand in Eq. (2.98) is a good indicator of how the energy is distributed as a function of frequency in the phonons. By extremizing the integrand, one can find the phonon frequency that is the maximum in the phonon "black body" spectrum. This is exactly analogous to Wien's Law from statistical physics and black body radiation, and we find

$$\hbar \omega_{max} \approx 2.822 k_B T. \tag{2.99}$$

Combining this and Eq. (2.94), we can find the dominant thermal phonon wavelength, λ_T, when $T \ll T_D$:

$$\lambda_T \approx \frac{hc_s}{2.882 k_B T}. \tag{2.100}$$

2.5.3 Dimensionality, scattering, and optical phonons

Consider plugging in some numbers for Si. The speed of sound in Si is 8.4 km/s, and $T_D = 645$ K. At 77 K (liquid nitrogen temperature), that implies that the dominant thermal phonon wavelength is about 2 nm. As the temperature is decreased, the thermal phonon wavelength grows like $1/T$. Even when the constituent atoms are free to move in three dimensions, one can define a *phonon dimensionality* set by the ratio of λ_T and the system size. This is analogous to the electronic case described in Section 2.1.5.

The Si example raises other questions. Si actually has two atoms per lattice point, so the correct Debye frequency must account for this. Materials with complicated bases have additional branches of phonons. The simple acoustic vibrations in the basic Debye picture correspond to periodic displacements of the lattice points. For the complete case, one must consider relative displacements of the basis atoms *within* the unit cell. Such vibrational modes are called *optical phonons*, and have finite frequency even as $k \rightarrow 0$. For these modes, k describes the spatial variation of the intra-basis displacement. Physically a $k = 0$ optical phonon would correspond to an intra-basis vibration in every unit cell of the crystal.

Phonons with well defined **k** values can scatter off of defects in the lattice, including boundaries. Phonons are responsible for thermal conduction in insulators, and can be considered as carrying crystal momentum, like electronic quasiparticles.

2.6 Summary and perspective

In order to appreciate why nanoscale systems are different from bulk systems, it is necessary to have a sense of the origins of the properties of bulk materials. This chapter has introduced a lot of the concepts relevant to the electronic properties of matter. Starting from particle-in-a-box, we looked at the free, noninteracting electron gas, and introduced several ideas that persist in more realistic treatments.

- Boundary conditions on the wavefunction lead to a discrete spectrum of allowed single-particle states.
- The smaller the system, the larger the energy spacing between the allowed levels.
- It is useful to think in terms of k-space to count states described well as waves.
- The density of states describes how those allowed states are arranged in energy, and depends critically on dimensionality.
- The ground state is found by filling states "bottom-up" because of the Pauli principle. This leads to some wavelength, λ_F, for the highest occupied state. The effective electronic dimensionality of a system is set by the relationship of λ_F and the system size.

Extending our treatment to include the atomic lattice in crystalline solids, we introduced other ideas.

- The spatial lattice picks out special points in k-space called the reciprocal lattice.
- The reciprocal lattice leads to strong scattering (diffraction) of waves.
- In an infinite periodic potential electronic states must satisfy the Bloch condition. Bloch waves do so, and are a reasonable way of thinking about extended single-particle states.
- The interactions of Bloch waves with the periodic lattice leads to bands of allowed energies separated by gaps with no allowed single-particle states.

We also considered approaches to materials based more on atomistic pictures, and found:

- starting from atomic orbitals, one can construct molecular orbitals for collections of atoms;

- nearest-neighbor overlaps and on-site energies figure prominently;
- the chemical nature of the atoms modifies the weighting of the atomic orbitals in the molecular orbitals that minimize the total energy of the system;
- Hückel theory gives qualitatively correct trends of level spacings, etc., but is of limited use.

Finally, we discussed more sophisticated treatments of electronic structure, pointing out that this remains an area of active research.

We also looked briefly at mechanical modes of solids. In the nanoscale limit, one can think of the atoms and bonds like balls and springs, and consider particular normal modes of the coupled oscillators. In bulk systems, it is relevant to think about phonons, the quantized vibrations of the lattice.

Now we have a sense of the available electronic states in generic bulk systems. In the next chapter we are going to look at the filling of those states and the types of material that result, including some specific common materials particularly relevant for nanoscale science. Following that we will consider defects, interfaces, and the effects of electron–electron interactions.

2.7 Suggested reading

There are a number of excellent solid-state physics books available that treat this material at a greater depth. These include the following.

- *Solid State Physics*, by N. Ashcroft and N. D. Mermin (New York, Brooks Cole, 1976).
- *Introduction to Solid State Physics*, Eighth Edition, by C. Kittel (New York, John Wiley and Sons, 2004).
- *Solid-State Physics*, Third Edition, by H. Ibach and H. Lüth (New York, Springer-Verlag, 2006).
- *The Oxford Solid State Basics*, by Steven H. Simon (Oxford, Oxford University Press, 2013).
- *Condensed Matter Physics*, by M. P. Marder (New York, John Wiley and Sons, 2000).
- *Solid State Physics*, Second Edition, by J. S. Blakemore (Cambridge, Cambridge University, 1985).
- *Solid State Theory*, by W. A. Harrison (New York, Dover, 1980).
- *Electronic Structure and the Properties of Solids*, by W. A. Harrison (New York, Dover, 1989).
- *Bonds and Bands in Semiconductors*, by J. C. Phillips (New York, Academic Press, 1973).

The last two in particular are recommended for those with more of a chemistry background, since they work very hard to explain the atomic and molecular origins of electronic structure. For a great discussion of modern electronic structure methods, especially density functional theory, I recommend *Electronic Structure: Basic Theory and Practical Methods*, by R. M. Martin (Cambridge, Cambridge University Press, 2004).

Exercises

2.1   *The 2d density of states* Assuming a large 2d box of side length L with infinitely high walls and noninteracting particles, derive and confirm the 2d density of states as a function of energy, $\nu_{2d}(E) = \left(\frac{m}{\pi\hbar^2}\right)$. This is independent of energy all the way down to $E = 0$. Do you think this is physically reasonable? Suggest an approximate energy scale below which it makes little sense to use this expression, with a sentence or two to explain your reasoning.

2.2   *Level spacings and Fermi wavelengths*

(a) Let's consider a (too) simple approximation of a 3d metal. A reasonable packing density for atoms is about one atom per every $(0.5 \text{ nm})^3$, and we can assume that the electrons are an ideal Fermi gas of that 3d density. Assuming the free electron mass, what is the Fermi energy of this system?

(b) How small a piece of this metal do you need to have single particle level spacing at the Fermi level comparable to 1 K (i.e. what is the length L of the side of a cubic blob that gives this level spacing)? Recently, a team from Georgia Tech has shown (*Phys. Rev. Lett.* **93**, 147402 (2004)) that gold clusters of 18–22 atoms have level spacings large enough to luminesce at wavelengths of ~ 700 nm. How does this spacing compare with the simple 3d DOS result for a cube 1.5 nm on a side?

(c) Consider doping a 2d semiconductor with electrons. We'll cover this in more detail soon. For now, just assume that each dopant atom in the semiconductor contributes a single "free" electron of effective mass 0.067 m_0 (appropriate for GaAs) to a 2d electron gas. How many dopants per square centimeter are necessary to lead to a Fermi temperature of 20 K? Assuming a dielectric constant k for the semiconductor of around 13, how does the Fermi energy compare with the typical Coulomb energy (electrostatic potential energy, $e^2/(4\pi\epsilon_0\kappa r)$) between carriers? Do you expect electron–electron interactions to be important in a system like this?

(d) Consider doping a 3d semiconductor, Si, with electrons. We'll cover this in more detail soon. For now, just assume that each dopant atom in the semiconductor contributes a single "free" electron of effective mass 0.33 m_0 (appropriate for silicon) to a 3d electron gas. How many dopants per cubic centimeter are necessary to lead to a Fermi temperature of 400 K? Such a material is said to be "degenerately doped". How does the Fermi wavelength compare to the average interdopant separation? Intuitively, would you expect this material and more highly doped materials to act like a metal or an insulator?

2.3   *Confinement in 1d* Consider a "quantum wire", a region bounded in the y and z directions by hard barriers. Suppose those boundaries are absolute, so that in the y and z directions any particles are confined to a size a. In the x direction the particles are free.

(a) For free electrons, approximately how small must a be for the energy spacing between the ground state and the first y or z (they're degenerate) excited state to be 30 meV?

(b) Assuming $T = 0$ and an ability to tune the 1d density n_{1d} continuously, how many (noninteracting, spin-degenerate) electrons per cm can be placed into the well before the first transverse state is populated?

2.4   *Confinement in 2d* Consider a "quantum well", a region bounded above and below by two parallel planes. Suppose those boundaries are absolute, so that in the z direction any particles are confined to a size a. In the $x-y$ plane the particles are free.

(a) For free electrons, approximately how small must a be for the energy spacing between the z ground state and the z first excited state to be 30 meV?

(b) Assuming $T = 0$ and an ability to tune the 2d density n_{2d} continuously, how many (noninteracting) electrons per square cm can be placed into the well before the first excited z state is populated?

It's possible to prepare systems like this, though approximating the well potential as a 1d particle-in-an-infinite-box is not applicable. A system of electrons confined like this in the lowest z state is called a two-dimensional electron gas (2deg). Devices using 2degs are common.

2.5   *More Fermi gas fun* (Not really nano, but a fun problem.) Suppose I'm a deranged astrophysicist, and I want to store cold (that is, degenerate) neutrons in a coffee mug, and that the cup is 8 cm in diameter and 10 cm tall. Roughly how many neutrons can I drop into my cup before it overflows? One-word hint: gravity. Don't worry about the pesky fact that neutrons aren't stable outside of nuclei.

2.6   *From bonds to bands* This is a little *gedanken* experiment that should give you a bit more insight about the development of bands from atomic states. Consider a quantum system that has some set of energy eigenstates. If that system is weakly perturbed in some way (by connection to another quantum system, for example), there are two effects. (1) The single particle eigenstates of the original system will no longer be true eigenstates of the new system. Start the new system out in one of the unperturbed eigenstates, and on a timescale t one will find that the system has evolved such that the new system is no longer in that unperturbed eigenstate. (2) The true eigenstates will be shifted in energy from the original eigenstates by an amount proportional to the perturbation.

(a) The lifetime is simply related to the width in energy (zero before the perturbation) of what were the original eigenstates. If you had to guess, what do you think that width in energy would be? (This is a really simple expression, but the idea here will reappear later in the course.)

(b) Think about joining atoms together in a chain, one at a time. Suppose each atom, when isolated, has two bound states separated by a large amount of energy, Δ.

Crudely sketch the energies of allowed single-particle states for a chain of weakly coupled (compared to Δ) atoms for two through four atoms. Give a couple of sentences explanation of your sketch, including what you think will happen as the number of atoms is increased toward infinity.

(c) For the large number of atoms case, what do you think will happen to the band width and the density of states within each band as the coupling between the atoms is increased?

(d) From the above, how do you think the bandwidth and density of states in the bands will compare between a covalently bonded crystal like silicon, and a van der Waals bonded molecular crystal like C_{60}?

Bulk materials

This chapter continues our overview of bulk materials, and introduces the major materials systems relevant for the remainder of the book.

One sensible way to classify materials is by the arrangement of their constituent atoms – their structure. Macroscopic materials consisting of large numbers of atoms are readily grouped into two categories: *ordered* and *disordered*. More precisely, one can examine the density–density correlation function,

$$S(\mathbf{r}) \equiv \langle \rho(\mathbf{r}_0)\rho(\mathbf{r}_0 + \mathbf{r}) \rangle. \tag{3.1}$$

The intensity of elastic diffraction[1] at some wavevector transfer $\mathbf{q}$ from a bulk material is proportional to the Fourier transform of $S(\mathbf{r})$.

In a completely disordered system, the position of one atom is uncorrelated with the position of any of the other atoms. Thus $S(\mathbf{r}) \propto \delta(\mathbf{r})$. This is the case for a dilute, classical gas. For a liquid or supercritical fluid, particles are squeezed together closely enough that the finite size of the constituent atoms becomes relevant. In that case there is some typical nearest-neighbor distance, though there is no long range pattern to the arrangement of atoms. Mathematically $S(\mathbf{r})$ has a broad peak where $|\mathbf{r}|$ equals the nearest neighbor separation as well as at $\mathbf{r} = 0$, but little other structure. An essentially identical pattern results for a completely amorphous solid such as a glass. With increasing positional order, additional peaks develop in $S(\mathbf{r})$. The limit of this would be a perfect single crystal, in which $S(\mathbf{r})$ would have delta function peaks at each lattice vector, $\mathbf{r} = \mathbf{R}$.

An alternative classification scheme can be built around a material's response to mechanical stresses.[2] Fluids are defined by their inability to support shear stresses. That is, for a given patch of area at the boundary of a fluid, if a force is applied in the plane of that surface, the fluid will begin to deform, and will continue to deform at some rate as long as the shearing force is applied. In contrast, solids are said to resist shear. When a shear force is applied to the surface of a solid (that is anchored down, so the whole chunk of material doesn't begin accelerating), the solid deforms (undergoes strain) until internal forces within the solid balance the externally applied load. While the ability to resist shear is definitely related to the nanoscale structure of a material, that correlation is not always simple. Amorphous silica and single crystal quartz are both solids, though one is disordered and the other has long range positional order.

---

[1] Neutron diffraction is sensitive to nuclear positions; x-ray diffraction is sensitive to the distribution of electron density.

[2] Stresses are forces normalized by the areas over which the forces act. We discuss this extensively in Chapter 9.

A third approach, and the one that we will focus on in this chapter, is to group materials by their electronic properties. In Chapter 2 we established some basic concepts about electronic states in bulk solids. In particular, we concerned ourselves with general properties of the spectrum of allowed single-particle states. Now it is time to consider what happens as those states are filled. We shall categorize the resulting material types and give examples of each. At the same time we shall introduce concepts that will guide our thinking when considering the properties of nanostructured materials.

3.1 Electronic types of solids

Macroscopic solids can be fitted into a small number of categories based on their electronic ground states and low energy excitations. Even with a fixed spectrum of electronic states, the Fermi level (electrochemical potential) of the electrons within that spectrum can have dramatic effects on the resulting electronic properties. Three overarching groupings are as follows.

- **Band solids** In these materials the single-particle picture is a good approximation. The electronic states relevant for most processes are delocalized, and energetically reside in well-defined bands, determined by the spatial stacking and bonding of the constituent atoms. These are many of the most common materials of technological use, and we will discuss them extensively below.
- **Disordered solids** These are highly disordered materials in which most (if not all) of the single-particle states are *localized*, particularly the most energetic occupied single-particle states. Localization in this case results from disorder; the potential felt by the electrons is far from a perfectly periodic one, and thus extended Bloch-like waves are a poor approximation. One can have disorder without localization of all the states. Metallic alloys and liquid metals are examples of this. Interaction effects between particles may be weak or strong, depending on circumstances. In the weak interaction regime, single-particle excitations are still a useful description of the system. When the density of (localized) states near the chemical potential is high, electronic conduction in such systems can be possible and often proceeds via carrier hopping between localized sites.
- **Strongly correlated materials** These are systems in which trying to understand the electronic many-body ground state in terms of noninteracting single-particle states is a bad idea. Interactions between charge carriers are so strong that the ground state may disagree qualitatively with what would be the ground state in the single-particle case. One type of such a material is a *Mott insulator* [26]. Consider a lattice of sites, each occupied by a single electron. In the single-particle (noninteracting) picture, this system would be a metal (see below). In the limit of strong Coulomb repulsion, however, this system is an insulator: electronic conduction would require doubly-occupying sites, which would be very expensive energetically. Moreover, the excitations of a strongly correlated system do not have to resemble those of weakly interacting materials. In the strongly interacting limit the lowest energy excitations

above the many-body ground state may involve the complex, dynamic rearrangement of many electrons, with profound results: the excitations may have quantum numbers very different from those of noninteracting electrons. For example, the effective charge of the excitations in the fractional quantum Hall regime can be *fractional* [27, 28], rather than $-e$. Other examples of strongly correlated systems include the normal states of the high temperature superconductors, the colossal magnetoresistance compounds, heavy fermion compounds, and many transition metal oxides such as Fe_3O_4. These systems remain among the most challenging to understand even in bulk form.

Below, we largely restrict the discussion to band solids, in part because they are the most technologically relevant materials, and in part because they are also the best explored at the nanoscale. Recall the results from the previous chapter. The single-particle states that are good solutions for infinite periodic crystals are Bloch waves, with energies restricted to certain bands set by the boundary conditions and lattice arrangement. As we shall see, the electronic properties of these bulk materials depend on both the spectrum of states and how those states are filled.

3.2 Metals

The traditional condensed matter physics definition of a metal is a material with an electrical resistivity, ρ, that *decreases* with decreasing temperature. Traditional pure metals have, in the low–T limit, $\rho \sim T^2$, and an electronic specific heat $\sim T$. The electronic "fluid" formed by the conduction electrons in a normal metal has properties very similar to those of the free electron gas described in Chapter 2, as described below.

3.2.1 Ground state and excitations

The electronic properties of metals result from their band structure, which generally resembles that shown in Fig. 3.1. Above filled valence bands, there is a partially filled conduction band. At $T = 0$, there is a well defined Fermi energy separating filled single-particle states from empty ones. When $T > 0$, some of the electrons are promoted from filled states to empty ones above E_F; these are often called "particles", and the empty states left behind are *"holes"*. In the limit of a bulk material, the single-particle level spacing due to confinement is negligible (that is, the band is effectively continuous). That means that the energetic cost of creating a particle–hole excitation can be arbitrarily small. Such excitations are said to be *gapless*. We shall see that it is this characteristic that leads to metallic electrical conductivity.

Strictly speaking, the excitations (both electron- and hole-like) in real metals are *quasiparticles*, a term used to refer to low energy (compared to E_F) excitations with well defined quantum numbers (in this case, spin, charge, band index, and **k**, to name a few). Quasiparticles are not exact eigenstates of the full many-body problem. Each quasiparticle state has a finite lifetime, τ, and therefore a nonzero uncertainty in energy,

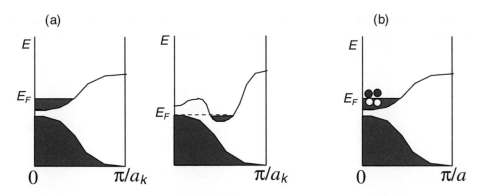

(a) Schematic band structure for a metal. Energies with filled states are shaded blue. The Fermi energy divides filled from empty states at $T = 0$. Conventional metals can result from partially filled bands (left) or overlapping bands (right). (b) Low energy excitations of a metal are (quasi)particles and holes. In a macroscopic metal, the number of states available near the Fermi energy is large, and electronic excitations may be made with little energetic cost, leading eventually to metallic conduction.

Fig. 3.1

$\Gamma \sim \hbar/\tau$. For the quasiparticle description to be a sensible way of characterizing the low energy excitations, the energetic cost of a typical quasiparticle must exceed Γ; otherwise it wouldn't make sense to talk about particular quasiparticles. In thermal equilibrium, this means that $\Gamma \ll k_\mathrm{B} T$ is necessary for meaningful quasiparticles.

Given the Coulomb interaction between the electrons, it is not at all obvious that this condition should be valid in general. Fortunately, careful calculations (first shown by Lev Landau [29]) demonstrated that, within some restrictions, the ground state of the interacting many-electron system is very similar to that of the noninteracting case – that is, there is a filled Fermi sea separated from empty states by a Fermi surface. Similarly, the low energy excitations of the interacting system are shown to have properties closely resembling those of noninteracting quasiparticles, and the lifetime relationship above is found to be obeyed as $T \to 0$ if τ^{-1} is dominated by electron–electron scattering. Such a system is called a *Fermi liquid*, to distinguish it from the noninteracting Fermi gas. The effects of those same interactions can be wrapped up in a small number of Fermi liquid parameters. A quasiparticle is then not just a bare electron excited above the Fermi surface, but also includes a "dressing" of collective response from all the other electrons that slightly modifies its properties.

Fermi liquid theory is the "standard model" of metals [30, 31], though there are some known exceptions. For example, in low-temperature superconductors the Fermi liquid ground state is unstable below the transition temperature, and the superconducting ground state takes over. Fermi liquid theory appears to be violated in the normal state of the high-T_c superconductors, as well as in some rare-earth "heavy fermion" compounds. Fermi liquid theory also has problems in systems that are truly one-dimensional, as we discuss later.

In the normal Bloch wave picture, the energy in the conduction (or valence) band depends on $\mathbf{k}$. Often it makes sense to expand that dependence in powers of $\mathbf{k}$ (this is particularly relevant for semiconductors, as we shall see in Section 3.3. When the primary dependence is quadratic, we use the coefficient of k^2 to define the *effective mass*:

$$E(\mathbf{k}) = Ak^2 \equiv \frac{\hbar}{2m^*}k^2. \tag{3.2}$$

The effective mass in general is tensorial, and in the noninteracting picture, m^* is determined by the details of the atomic lattice structure. In general, there can be Fermi liquid corrections to m^*. The effective mass can differ strongly from that of a free electron.

3.2.2 Types of metals

There are a few kinds of metals. Their unifying characteristic is that the Fermi level sits in the middle of a partially filled band.

- The *alkali metals* come from the first column of the periodic table. Like their neighbors in the second column, the alkali metals are generally very reactive. Their conduction bands are built out of s-orbitals, and exhibit weak spin–orbit effects.[3]
- The *noble metals*, such as Au, Ag, and Pt, are so named because they are much less reactive, and therefore are resistant to corrosion. Strictly noble metals have filled d-shells. Pd, Pt, and Ru have partially filled d-shells, and are often used as catalysts. The noble metals are generally heavier atoms and exhibit stronger spin–orbit interactions. Noble metals generally do not exhibit superconductivity by themselves.
- The *transition metals* include Ni, Fe, Co, and Cu. Transition metals have partially filled d-orbitals. These d electrons can be relatively localized to the ion core in the metal, or can have some itinerant character. Because of unpaired d electrons, the transition metals and their oxides often exhibit magnetic ordering (long range correlations in the spin states of the d electrons). Magnetism, its applications, and magnetic phenomena at the nanoscale will be discussed at length in Chapter 7.
- The *rare earth metals* are the actinides and lanthanides. These materials generally contain partially filled f-orbitals, and like transition metals can exhibit magnetic effects. They have very strong spin-orbit interactions, meaning that relativistic corrections to their band structure are important. Even elemental metals from this group (e.g., Pu) can have extremely complex properties.
- A particularly rare variety of metal is the *half-metal*, in which all of the electrons at the Fermi level are fully spin-polarized. The chemically simplest half-metal is CrO_2.

3.2.3 Superconductors

When discussing Fermi liquids, quasiparticles, and other properties of bulk metals, it seems appropriate to mention superconductivity. Discovered first by Onnes in 1911, superconductivity is a property of what is now called the BCS groundstate, after its

---

[3] Recall that spin–orbit coupling contributes to the fine structure of the hydrogen atom spectrum. From the point of view of the moving electron, the apparent motion of the nucleus is a current that produces a magnetic field which couples to the spin degree of freedom of the electron.

discoverers, Bardeen, Cooper, and Schrieffer [32, 33]. There are many good references that cover ordinary superconductivity in depth, including the well known books by Schrieffer [34], Tinkham [35], and Poole, Farach, and Creswick [36]. Rather than examining superconductivity in depth, let's look briefly at the main features of the superconducting state. This discussion, which assumes a bit more physics knowledge, is only relevant for certain specific later sections of the book, and may be skipped at a first reading.

Conventional superconductivity results from an instability in the Fermi liquid ground state arising from an effective attractive interaction between the electrons. This attraction comes about due to phonons, quantized vibrations of the ion lattice. These interactions lead to correlations between pairs of electrons of opposite spin (forming a singlet) and oppositely directed wavevectors, $\mathbf{k}$ and $-\mathbf{k}$ near the Fermi surface. These "Cooper pairs" act like composite bosons and below the critical temperature, T_c, are described by a single effective groundstate wavefunction. That wavefunction may be treated as the order parameter of the superconducting state and is complex, consisting of an amplitude and a phase.

The formation and condensation of the Cooper pairs dramatically affects the density of states for single electron quasiparticles. Since creating a quasiparticle requires breaking a Cooper pair, the density of states for these excitations is zero near E_F, as shown in Fig. 3.2. The result is an energy gap of size $2\Delta_0$ at the Fermi energy. Since excitations of single electrons are now gapped, many of the features of the normal metallic state are changed. The heat capacity, for example, is exponentially suppressed as $T/T_c \rightarrow 0$, rather than varying linearly with T.

Three main experimental traits associated with the superconducting state are:

- Zero electrical resistance. As long as the current density is kept below a material-dependent critical value, the dc resistance of a superconductor is zero.
- Exclusion of magnetic flux. This is known as the Meissner effect [37]. At sufficiently low material-dependent values of magnetic field, a superconductor develops screening currents at its surface when T is lowered through T_c. These screening currents lead to perfect diamagnetism.

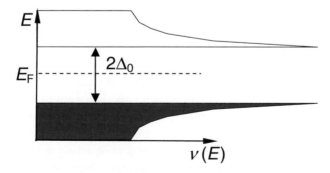

Density of states at $T = 0$ near the Fermi level of a Bardeen–Cooper–Schrieffer superconductor. The energy gap around the Fermi level appears below T_c, and is responsible for many of the remarkable properties of the superconducting state.

Fig. 3.2

- The quantization of magnetic flux. The magnetic flux threading a superconducting ring must be quantized in units of $h/2e$, often called the *flux quantum*. This originates from the vector potential and the requirement that the superconducting ground state wavefunction be single-valued. We will discuss this in more detail in the context of normal metals at the nanoscale in Chapter 5.

The superconducting state has received much attention in nanoscale systems. A number of fundamental questions have been of particular interest, including quantum size effects, the role of disorder, the interfaces between superconductors and nanoscale normal and magnetic materials, and superconductivity in reduced dimensionality. Furthermore, certain properties of the superconducting state have enabled remarkable experiments that have revealed otherwise inaccessible detailed information about nanoscale physics.

3.3 Inorganic semiconductors

Next to metals, the most technologically relevant bulk materials are probably inorganic semiconductors. Intrinsic semiconductors are materials with (at $T = 0$) some number of exactly filled bands of single-particle states. The highest filled band is called the *valence band*, and the most energetic single-particle state in the valence band ("at the top of the valence band") has energy E_V. The lowest empty band is called the *conduction band*, and its least energetic state ("bottom of the conduction band") has energy E_C. The *band gap*, E_g, is defined as $E_C - E_V$. The dividing line between semiconductors and band insulators is rather arbitrary. Typically when $0 < E_g \leq \sim 4$ eV, the material is considered to be a semiconductor.

Two schematic band structures are shown in Fig. 3.3, within the reduced zone scheme. In Fig. 3.3(a) (left), the maximum in the valence band happens at the same $\mathbf{k}$ point as the minimum in the conduction band. Such a system is said to be a *direct gap* semiconductor. Common direct gap materials are the III–V compound semiconductors, such as GaAs and InP. In contrast, Fig. 3.3(a) (right) shows the energetics of a semiconductor for the case where the global minimum in $E_C(\mathbf{k})$ occurs at a different $\mathbf{k}$ from the maximum in $E_V(\mathbf{k})$. Materials with this property are *indirect gap* semiconductors, and include Si and Ge.

The crucial difference between semiconductors and metals is the spectrum of low energy excitations. At $T = 0$, the minimum energy cost to produce an electronic excitation is E_g. Unlike the metallic case, the excited electrons and holes are in different bands, and therefore can have significantly different properties. One result of this is that in these materials the intrinsic electrical conductivity is exponentially suppressed as $T \to 0$ with an activation energy of E_g. That is, as the material cools, the number of "free" charge carriers available to carry current decreases exponentially because carriers only exist due to thermal activation over the band gap.

The direct/indirect distinction has significant implications for the properties of the material. As we shall see in Chapter 8, photons with $\hbar\omega > E_g$ have enough energy

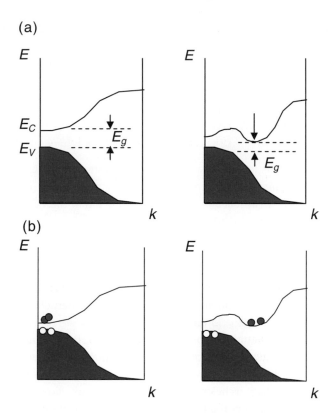

(a) Band structure for semiconductors, either direct gap (left) or indirect gap (right). In both cases there is an energy gap between the highest occupied state at the top of the valence band and the lowest unoccupied state at the bottom of the conduction band. (b) Low energy excitations of a semiconductor are also (quasi)particles and holes, but the band gap sets the energy scale for the lowest energy electronic excitations possible in this single-particle picture.

Fig. 3.3

to excite an electron–hole pair. However, total momentum must be conserved in such a process, and this restriction strongly suppresses that kind of optical activity in indirect gap semiconductors. An additional excitation is required to make up for the mismatch in **k** between the electronic state at the conduction band minimum and the hole state at the conduction band maximum. This is why direct gap semiconductors (InGaAs, InP) are commonly used in photodetectors and light-emitting diodes (LEDs), while indirect gap systems such as Si and Ge are generally not considered optically active. Interestingly, nanostructuring of such materials can bend these selection rules, since **k** is only perfectly defined as crystal momentum for an infinite periodic lattice. We discuss this further in Chapter 5.

As above, since $E_C(\mathbf{k})$ and $E_V(\mathbf{k})$ are no longer simple functions of $\mathbf{k}$, the effective masses of carriers in the conduction and valence bands are not equal to the free electron mass. Near the minimum in the conduction band, we can write

$$E_C(\mathbf{k}) \approx E_C(\mathbf{k}_{min}) + \frac{\hbar^2}{2m^*}(\mathbf{k} - \mathbf{k}_{min}) + \cdots . \tag{3.3}$$

where we define

$$m_{\mathrm{e}}^* \equiv \frac{1}{\hbar^2} \left(\frac{\partial^2 E_{\mathrm{C}}(\mathbf{k})}{\partial \mathbf{k}^2} \right)^{-1}, \tag{3.4}$$

the effective mass for electrons.[4]

Similarly, near the maximum in the valence band,

$$E_{\mathrm{V}}(\mathbf{k}) \approx E_{\mathrm{V}}(\mathbf{k}_{\mathrm{max}}) - \frac{\hbar^2}{2m^*}(\mathbf{k} - \mathbf{k}_{\mathrm{max}}) + \cdots . \tag{3.5}$$

For clarity, suppose $\mathbf{k}_{\mathrm{max}}$ is 0, which is often the case. Note that an electron sitting at the top of the valence band *decreases* its energy by increasing k. That is, it acts as if it has a negative effective mass! An *absence* of an electron at the top of the valence band acts like a positively charged hole, with an effective mass

$$m_{\mathrm{h}}^* \equiv -\frac{1}{\hbar^2} \left(\frac{\partial^2 E_{\mathrm{V}}(\mathbf{k})}{\partial \mathbf{k}^2} \right)^{-1}. \tag{3.6}$$

Accelerating a hole away from the top of the valence band costs energy, which makes perfect sense: removing a hole from that most energetic single-particle state means adding an electron to that state from some lower energy state.

Since the precise band curvature depends on the spatial arrangement of the atoms, the effective mass is generally tensorial, with symmetries analogous to those of the lattice. The effective masses of electrons and holes can be very different. Both of these points will become more clear in discussions below of particular semiconductors. From Eqs. (3.4) and (3.6), note that the more steeply the band energy curves with k, the lower the effective mass.

3.3.1 Some common semiconductors

Before discussing some more general semiconductor physics, it is illustrative to introduce some common materials. Two of the most well studied semiconductors are in Group IV of the periodic table, silicon and germanium. The common crystal structure of both of these materials is the "diamond" structure, consisting of two interpenetrating face-centered cubic lattices offset by one quarter of a primitive lattice vector. This can be described as an FCC lattice with a two-atom basis. Table 3.1 lists the values of some of the important parameters for Si and Ge.

Figure 3.4 shows the first Brillouin zone for these materials, with the high symmetry points labeled. These points are used as guideposts for tracing out the band structure of bulk materials. Because band diagrams can often be very complicated, they are sometimes nicknamed "spaghetti diagrams".

Figure 3.5(a) shows a small piece of a band diagram for Si. Notice that, as promised, the minimum in the conduction band is not at the Γ point in k-space; instead it lies along the symmetry direction between Γ and the X point (the center of the square face of the

---

[4] Here and in the following I will use "electrons" and "holes" rather than the more precise terms "electron quasiparticles" and "hole quasiparticles".

Table 3.1 Important parameters for Group IV semiconductors. All are indirect gap systems. Band gaps are given at 300 K. This information is compiled primarily from the electronic archive of New Semiconductor Materials at the Ioffe Physico-Technical Institute, St. Petersburg, Russia.

	Si	Ge	C (diamond)
Valley degeneracy	6	4	6
Valley direction	X	L	X
E_{g} [eV]	1.12	0.661	5.46
$m_{\mathrm{e,l}}^{*}$ [m_0]	0.98	1.59	1.4
$m_{\mathrm{e,t}}^{*}$ [m_0]	0.19	0.0815	0.36
m_{hh}^{*} [m_0]	0.49	0.33	2.12
m_{lh}^{*} [m_0]	0.16	0.043	0.7

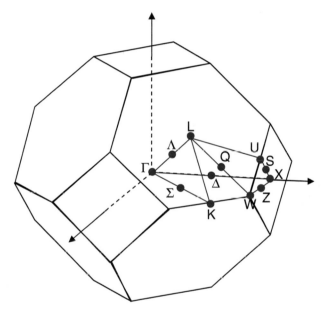

First Brillouin zone for the FCC lattice, with high symmetry points labeled.

Fig. 3.4

Brillouin zone). There are six equivalent minima, leading to a six-fold *valley degeneracy*. Figure 3.5(b) shows surfaces of constant energy taken slightly above the minimum in the conduction band. Figure 3.6 is an analogous figure for Ge. In this case the slight difference in the chemical properties of Ge and Si leads to a change in band structure. For Ge the minimum in the conduction band happens at the L point, at the edge of the first Brillouin zone. As a result, the valley degeneracy for Ge is four rather than six.

The k-space anisotropy in the conduction band minima leads directly to anisotropy in m_{e}^{*}. Calling the direction between the foci of the ellipsoids "longitudinal", we can define longitudinal and transverse effective masses, $m_{\mathrm{e,l}}^{*}$ and $m_{\mathrm{e,t}}^{*}$, respectively.

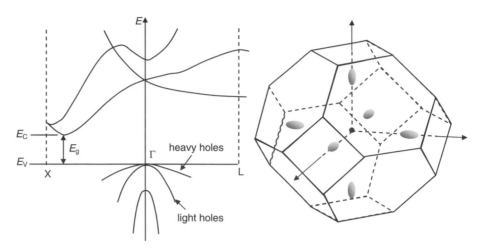

Fig. 3.5 (a) Part of the band diagram for Si, showing the dispersion of the bands from the Γ point toward the X and L directions in the first Brillouin zone. (b) A surface of constant energy slightly above the bottom of the conduction band, showing the six degenerate valleys.

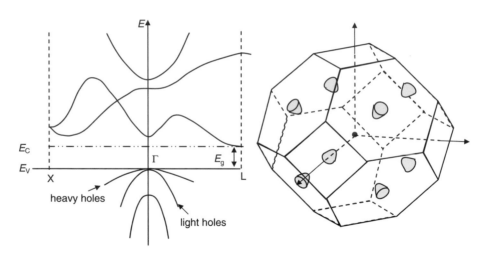

Fig. 3.6 (a) Part of the band diagram for Ge, showing the dispersion of the bands from the Γ point toward the X and L directions in the first Brillouin zone. (b) A surface of constant energy slightly above the bottom of the conduction band, showing the valley degeneracy at the L points. By symmetry there are four equivalent valleys.

Adapting our expressions for the density of states, Eq. (2.24), we can find the densities of states for electrons and holes in a straightforward way. In the conduction band in 3d,

$$\nu_e(E) = \frac{M_C}{2\pi^2} \left(\frac{2m_e^*}{\hbar^2}\right)^{3/2} (E - E_C(\mathbf{k}_{\min}))^{1/2}. \qquad (3.7)$$

Here M_C is the valley degeneracy. The correct effective mass to use in Eq. (3.7) is the direction-averaged effective mass in one valley, $m_e^* \equiv (m_{e,l}^*(m_{e,t}^*)^2)^{1/3}$. In the valence band, we can approximate

$$v_h(E) = \frac{1}{2\pi^2} \left(\frac{2m_h^*}{\hbar^2} \right)^{3/2} (E_V(\mathbf{k}_{max}) - E)^{1/2}. \tag{3.8}$$

For holes there is no valley degeneracy, since the top of the valence band is at the Γ point. However, in this crystal structure there are two valence bands with different energy dependences that meet at the Γ point. Their different curvatures as a function of energy correspond to differing effective masses; one band is the "light" holes (m_{lh}^*) and the other is the "heavy" holes (m_{hh}^*). That means that for some particular energy near (but below) the top of the valence band there are two contributions to the density of states, one from each of the hole bands. The approximation that the hole bands are parabolic and isotropic near $k = 0$ is often not terribly good.

Finally, there is an additional split-off band, seen in Figs. 3.5, and 3.6. This band is separated from the light and heavy holes due to spin–orbit interactions; these are relativistic effects that are relevant here in part because in the chemistry language these hole bands originate from p-orbitals with nonzero orbital angular momentum. We shall discuss spin–orbit interactions and the comparatively complicated nature of holes when we discuss quantum wells.

Another very common semiconductor for nanoscale devices is GaAs, an example of a III–V compound semiconductor. Recall that the diamond structure can be thought of as two interpenetrating FCC lattices. In GaAs, one sublattice is built from Ga and the other from As; this is commonly called a *zinc blend* structure. Figure 3.7 shows the band structure. Since GaAs is a direct gap system, a constant energy surface near the bottom of the conduction band is a sphere centered around $k = 0$.

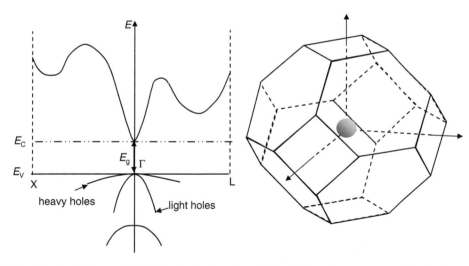

(a) Part of the band diagram for GaAs, showing the dispersion of the bands from the Γ point toward the X and L directions in the first Brillouin zone. (b) A surface of constant energy slightly above the bottom of the conduction band, showing the Γ minimum in the conduction band.

Fig. 3.7

Table 3.2 Important parameters for some III–V compound semiconductors. Band gaps are given at 300 K. This information is compiled primarily from the electronic archive of New Semiconductor Materials at the Ioffe Physico-Technical Institute, St. Petersburg, Russia.

	GaAs	AlAs	InAs	InP	GaP	InSb
Valley degeneracy	1	3	1	1	6	1
Valley direction	Γ	X	Γ	X	Γ	
E_{g} [eV]	1.42	2.15	0.35	1.34	2.27	0.18
$m_{\mathrm{e,l}}^{*}$ [m_0]	0.067	1.1	0.22	0.077	1.12	0.014
$m_{\mathrm{e,t}}^{*}$ [m_0]	0.067	0.19	0.22	0.077	0.22	0.014
m_{hh}^{*} [m_0]	0.5	0.5	0.41	0.79	0.4	
m_{lh}^{*} [m_0]	0.082	0.15	0.026	0.14	0.02	

3.3.2 Carriers in semiconductors

Semiconductors came by their name from the historical observation that while they have some electronic conductivity at room temperature, they do not exhibit metallic conduction at lower temperatures in their pure form. Instead their conductivity tends toward zero with an Arrhenius dependence as $T \to 0$. The dominant effect here is the "freezing out" of the free carriers. The mobile carriers in pure semiconductors arise from thermal activation of electrons from the top of the valence band into the conduction band.

Knowing the densities of states of electrons and holes in the valence and conduction bands, it is possible to find the equilibrium populations of electrons and holes as a function of temperature. Generically, the density of electrons n is given by

$$n = \int_{E_C}^{E_{\mathrm{top}}} \nu_{\mathrm{e}}(E) f(E, E_{\mathrm{F}}, T) \mathrm{d}E, \qquad (3.9)$$

where f is the Fermi–Dirac distribution function, ν_{e} is the density of states in the conduction band, and E_{top} is the energy of the top of the conduction band. In the nondegenerate limit, usually valid at 300 K, we can expand f to recover the usual Boltzmann factor. We will return to the value of E_{F} shortly. For the 3d electronic density of states from Eq. (3.7), we find

$$n = \left[2 \left(\frac{2\pi m_{\mathrm{e}}^{*} k_{\mathrm{B}} T}{\hbar^2} \right)^{3/2} M_{\mathrm{C}} \right] \cdot \exp \left(-\frac{E_C - E_{\mathrm{F}}}{k_{\mathrm{B}} T} \right)$$

$$= N_{\mathrm{C}}(T) \cdot \exp \left(-\frac{E_C - E_{\mathrm{F}}}{k_{\mathrm{B}} T} \right). \qquad (3.10)$$

Here we have defined $N_{\mathrm{C}}(T)$ as the effective density of states in the conduction band. Note that $N_{\mathrm{C}}(T)$ includes both the valley degeneracy and a factor of two for spin.

We can do the same analysis for the hole density, p, and find

$$p = \left[2 \left(\frac{2\pi m_e^* k_B T}{\hbar^2} \right)^{3/2} \right] \cdot \exp\left(-\frac{E_F - E_V}{k_B T} \right)$$

$$= N_V(T) \cdot \exp\left(-\frac{E_F - E_V}{k_B T} \right), \tag{3.11}$$

where we define the analogous effective density of states in the valence band, $N_V(T)$.

In thermal equilibrium for pure material, the number of electrons elevated into the conduction band must equal the number of holes left behind in the valence band. Equating n and p establishes E_F, which is usually very close to the middle of the band gap. The intrinsic carrier density, n_i, is defined as

$$n_i \equiv \sqrt{np} = \sqrt{N_C N_V} \cdot \exp\left(-\frac{E_g}{2k_B T} \right). \tag{3.12}$$

Plugging in numbers, this gives a carrier density (number per cm^3) of

$$n_i = 4.9 \times 10^{15} \left(\frac{m_e^* m_h^*}{m_0^2} \right) M_C^{1/2} T^{3/2} \exp\left(-\frac{E_g}{2k_B T} \right). \tag{3.13}$$

Notice that the number of carriers is strongly temperature dependent; when $2k_B T \ll E_g$ all carriers freeze out exponentially, and the system is a band insulator. Plugging in some numbers for common materials, we see that the bulk carrier density in intrinsic Si at 300 K is about 10^{10} cm^{-3}, while germanium, with its smaller gap, has around 2×10^{13} cm^{-3} at the same temperature. Intrinsic GaAs, by contrast, with a larger gap and no valley degeneracy, has approximately 2×10^6 cm^{-3} at room temperature, and is often called "semi-insulating".

3.3.3 Doping

Fortunately it is possible to modify semiconductor materials so that carrier densities other than the intrinsic level are possible. By **chemical doping**, substituting atoms of another type for those of the semiconductor, it is possible to change the number of available carriers. For example, As is a Group V element with five valence electrons. A single As impurity substitutionally placed in an Si lattice uses four valence electrons to satisfy the covalent bonds with the neighboring Si atoms. At $T = 0$, the fifth valence electron remains loosely bound to the As ion core. Energetically, this bound electronic state is about 54 meV below the bottom of the conduction band. At moderate temperatures, sufficient thermal energy exists to excite this bound electron into the extended single-particle states of the conduction band. Such an electron is "donated", and the arsenic atom is said to act as a donor.

Similarly, a boron atom (Group III) substituted into a Si crystal only has three valence electrons. To satisfy the covalent bonds with the neighboring Si atoms, the B atom "accepts" an electron from the silicon valence band. The result at $T = 0$ is a localized state with a hole bound to a negative B ion acting as an acceptor. That localized hole state exists about 45 meV above the top of the valence band; at moderate temperatures, the localized hole is excited into the extended states of the valence band.

It is not at all obvious that the energies of the bound electron (hole) states of donors (acceptors) are going to be at energies "shallow" enough to be useful at room temperature. The situation is particularly complex in compound semiconductors. For example, one can imagine a Group IV impurity atom in GaAs fulfilling either role depending upon the type of substitutional site. Indeed, Si is commonly used as an n-type dopant in interfacial GaAs nanostructures on (100) planes, while Si can be a p-type dopant on (311)A GaAs substrates. Similarly, C can be used as a p-type dopant in the same (100) GaAs material system.

A common approximation used to describe the bound states, particularly for electrons near donor ions, is to treat the donated electron bound to the donor ion in analogy to the hydrogen atom problem. In this simple picture the effect of the background lattice is lumped into an effective relative dielectric function, κ. This treats the surrounding semiconductor material as a homogeneous medium. This approximation is only self-consistent if the resulting effective Bohr radius of the donor's electron is large compared to the atomic spacing of the semiconductor lattice. Within this hydrogenic picture, the effective Bohr radius is

$$a_{\text{eff}} = \frac{4\pi\kappa\epsilon_0\hbar^2}{m_e^* e^2},$$

(3.14)

while the effective ionization energy for the donor's electron is

$$E_{\text{I,eff}} = \frac{e^4 m_e^*}{2(4\pi\kappa\epsilon_0\hbar)^2}.$$

(3.15)

Plugging in some typical numbers is instructive. For electrons in Si, $m_e^* \sim 0.26\, m_0$, $\kappa = 11.7$, giving $E_{\text{I,eff}} = 0.026$ eV, and $a_{\text{eff}} = 2.4$ nm. Similarly, for electrons doped into GaAs, $m_e^* \sim 0.067\, m_0$, $\kappa = 13.1$, giving $E_{\text{I,eff}} = 0.005$ eV, and $a_{\text{eff}} = 10.4$ nm. We see that in these cases, most donors are ionized at room temperature. This is not surprising, since if this doping scheme didn't work we would not be discussing it. We also see that in both of these examples, the effective Bohr radius is significantly larger than the interatomic spacing. This suggests that lumping the effects of the lattice into a single κ parameter is reasonable.

The large extent of the donor bound state wavefunctions raises a couple of natural questions. First, what happens when the donor states interact with each other? When considering increasing the doping density it is natural to consider dopant dimers. Using the hydrogenic treatment as a guide, it is not surprising that molecule-like states are possible, with the equivalent of bonding and antibonding levels pushed down and up in energy, respectively, relative to the isolated donor states. The more deeply bound states are more difficult to ionize. As a result, the number of free carriers no longer scales linearly with dopant density. This is a form of dopant *compensation*. Carrying this hydrogenic analog forward, you can see that at very high dopant densities the result will be an *impurity band*, and doping all the way into the metallic regime ("degenerate doping") is possible.

A second question is what happens when the bound dopant carrier wavefunctions run into material boundaries? We will discuss this further in the next chapter.

Doping of inorganic semiconductors is typically done by *diffusion* or *implantation*. Recall that diffusion is driven by a gradient in the concentration of the diffusing species,

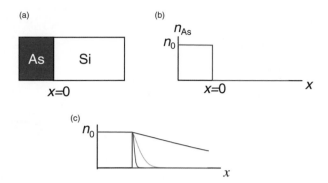

Doping via diffusion. (a) Idealized initial state. (b) Initial concentration profile. (c) Concentration profile at various times, showing the diffusion of dopant concentration into the Si in accordance with Eq. (3.18).

Fig. 3.8

as in Fick's Law:

$$\mathbf{J}_n = -D\nabla n. \tag{3.16}$$

Here $\mathbf{J}_n$ is the current density of diffusing atoms (number of atoms per area per second), D is the diffusion constant, and n is the concentration of the diffusing species. Conservation of particles leads to the diffusion equation,

$$\frac{\partial n}{\partial t} = D\nabla^2 n, \tag{3.17}$$

that determines the space and time dependence of the concentration.

A common situation is shown in Fig. 3.8(a). Here one considers (in the 1d limit) the spatial dependence of the diffused concentration as a function of time when a reservoir (concentration n_0) of dopants is in contact with an initially undoped material. The solution to Eq. (3.17) in this case is

$$\frac{n(x,t) - n(x, t=0)}{n_0 - n(x, t=0)} = 1 - \mathrm{erf}\left(\frac{x}{2\sqrt{Dt}}\right). \tag{3.18}$$

Roughly, the length scale of dopant diffusion is $\sim \sqrt{Dt}$.

The diffusion constant for atoms in solids usually depends *very* strongly on T. This isn't surprising, since moving atoms through solids typically involves either the creation and propagation of defects such as vacancies or interstitials. Often $D = D_0 \exp(-\Delta_n/k_B T)$, where D_0 is a material-dependent prefactor (often 0.1–10 cm^2/s) and Δ_n is an activation energy typically on the order of 3–4 eV. Controlling dopant concentrations on the nanometer scale is extremely difficult with these dependences, and achieving desired diffused densities for readily established concentration gradients and practical timescales can require high temperatures on the order of 1500 K.

Figure 3.9(a) shows ion implantation, the other major doping technlogy. The basic idea is very simple: the desired impurity atoms are ionized and accelerated up to energies that can be as high as hundreds of keV, focused via ion optics, and fired directly into the wafer. In principle this approach can implant nearly anything that is ionizable, and the depth of implantation is set by the acceleration energy. Figure 3.9(b) shows that the

(a)

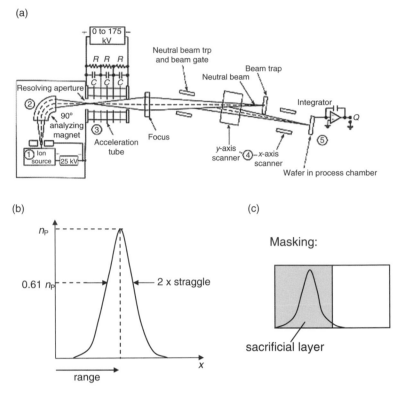

(b)   (c)

Fig. 3.9 (a) Schematic of an ion implantation system, from Jaeger [38]. (b) Depth profile of implanted dopants. (c) Sacrificial layer used to control doping depth in shallow implantations.

typical implanted dopant profile is approximated as gaussian, with the center position of the distribution called the "range", and the width of the gaussian termed the "straggle". To control doping at very shallow depths, it is possible to use sacrificial layers, as shown in Fig. 3.9(c).

Ion implantation is not without its own problems. There are practical limits to the ion currents that are achievable, with the consequence that reaching high dopant concentrations is difficult and slow. The crystallographic details of the host lattice can lead to strong anisotropies in the implantation range at fixed beam energy. Furthermore, the beam energies required to achieve implementation necessarily result in damage to the background lattice (broken bonds, vacancies, and interstitials) that must be annealed away. Similarly, since desired electronic properties of dopants can depend critically on where the dopant resides in the lattice, annealing can also be necessary to "activate" the implanted atoms.

3.3.4 Wide-gap semiconductors

Wide-gap semiconductors ($E_g \sim$ 2-4 eV) are of increasing technological interest. Such materials are well suited for applications that take advantage of the large interband energy scale. An example application is semiconductor power electronics, using semiconductor devices to manipulate large current loads. In high-power applications Joule heating due to the large current densities can be significant, with device operating temperatures readily exceeding 250° C and in some cases approaching 400° C. In the conventional semiconductors described above, these elevated temperatures would significantly alter the operating parameters of devices due to the thermal production of free carriers, as in Eqs. (3.10)–(3.12). SiC ($E_g \sim 2.4$ eV) in its 4H hexagonal crystal structure is extremely useful in power electronics devices.

Optoelectronic properties of wide-gap systems are also of much interest. Interband transitions in these materials occur for wavelengths ranging from red into the near ultraviolet. As a result, these materials can be of considerable use for display applications (e.g. green and blue light-emitting diodes). In the near UV emission may be used to excite phosphors for broadband re-emission ("white" LEDs). Compound semiconductors are often used in emissions applications, include GaN ($E_g \sim 3.2-3.4$ eV, depending on crystal structure) for blue and UV LEDs, InGaN for green LEDs, and ZnO ($E_G \sim 3.4$ eV) for UV LEDs.

Complementary to emission characteristics, optical absorption tuned to these wavelengths can be very useful for a variety of applications. Semiconductor photodetectors are common. The strong absorption of wide-gap materials in the ultraviolet has led to their use, particularly in nanoparticle form, in sunscreens and cosmetics. Titania (TiO_2) has been the subject of much interest as a wide-gap material for use in various solar cell approaches.

3.3.5 Magnetic semiconductors

Over the past two decades there has been an explosion of interest in somehow using the spin of charge carriers for information processing as well as their charge. While we will discuss magnetic materials and spin phenomena more generally in Chapter 7, we briefly summarize magnetic semiconductors here.

By doping of some semiconductors with transition metal atoms containing local magnetic moments (e.g., unpaired d-electrons), it has been possible to create materials that are simultaneously doped semiconductors and ferromagnetic [39, 40, 41]. The most extensively examined magnetic semiconductors (MSs) are based on II–VI and III–V materials. A common dopant is Mn (spin-5/2). Doped Group IV magnetic semiconductors are also possible [42].

In II–VI compounds such as ZnTe and CdTe, Mn tends to sit at the II sites and contributes its two outer s-electrons to satisfy the demands of covalent bonding within the crystal. As a result, while Mn acts as a local magnetic moment it does not act as a donor or acceptor, and therefore does not affect the density of free carriers. By contrast, Mn tends to occupy the III sites when introduced into III–V compound semiconductors (e.g., $Ga_{1-x}Mn_xAs$, $In_{1-x}Mn_xAs$). The Mn states exist energetically near the top of the valence band, and thus Mn acts as an acceptor.

The Mn fractions required to achieve ferromagnetic order in these systems are often as high as several percent. For example, $Ga_{1-x}Mn_xAs$ has been reported [43] with a Curie temperature, T_C, well over 100 K for $x \sim 0.05$, and $In_{1-x}Mn_xAs$ with $T_C \sim 90$ K for $x \sim 0.12$ has also been observed [44]. Differences in atomic diameter between Mn and the atoms of the host material can make homogeneous growth of such alloys extremely challenging. Often the alloyed forms are only metastable thermodynamically, and careful control of growth conditions is necessary to maintain good crystallinity.

The physics of magnetic ordering in MSs remains a topic of strong interest and debate. It is clear that mobile charge carriers play an important role in magnetic ordering. Manipulation of carrier density by electrostatic gating [45] and photoconduction [46] have been demonstrated to produce significant modulation of the magnetic ordering temperature in $Ga_{1-x}Mn_xAs$ and $In_{1-x}Mn_xAs$, for example. The promise of electronic and optical control over magnetic systems is explored in more detail in Chapter 7.

Much recent research has focused on the quest to achieve a MS with a magnetic ordering temperature greater than 300 K. Based on some theories of hole-mediated magnetism in MSs, there is an expectation that wide-gap semiconductor host materials should have higher ordering temperatures. A strong effort has therefore been expended in attempts to dope materials like GaN, ZnO, and TiO_2 with magnetic atoms like Mn and Co.

3.4 Band insulators

Band insulators are the extreme limit of wide-gap semiconductors, and have an exactly filled valence band separated from an empty conduction band by a large ($> \sim 4$ eV) gap. This filled band requirement typically implies an even number of valence electrons per unit cell of the lattice. These materials are routinely used as insulating layers in electronics applications.

3.4.1 SiO_2 and Si_3N_4

The best studied band insulator is probably amorphous SiO_2, which may be grown with extremely high quality by the thermal oxidation of Si. It has a number of remarkable properties [47], including an extremely high band gap (9 eV), very high resistivity at room temperature ($> 10^{14}$ Ω-cm), and excellent dielectric breakdown strength (the ability to withstand dc electric fields) exceeding 10^9 V/m. Its low frequency relative dielectric constant is $\kappa \approx 3.9$, and the large band gap leads to low optical absorption throughout the visible wavelength range.

Its amorphous structure possesses only short-range order and is built out of Si atoms tetrahedrally coordinated by oxygen atoms [48]. These tetrahedra form a corner-sharing network that is disordered because the Si-O-Si bond angle can be distorted between 100 and 170 degrees with relatively little cost in energy. The relatively open structure of the amorphous material is reflected by its low density (2.2 g/cm^3) compared to that of crystalline quartz (2.6 g/cm^3). Recent investigations [49] have shown (see Fig. 3.11) that the electronic structure of the amorphous oxide persists even down to distance scales of only a few tetrahedra.

A similar structure is shared by silicon nitride (Si_3N_4), which is often grown by plasma-enhanced chemical vapor deposition. Silicon nitride is a tough, strong material routinely

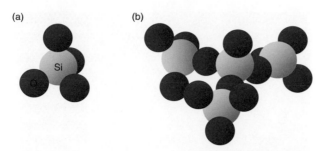

(a) Silicon—oxygen tetrahedron that is the building block of SiO_2. (b) Four such tetrahedra joined in a corner-sharing manner as in amorphous SiO_2. Note that the Si-O-Si bond angles and bond rotations vary.

Fig. 3.10

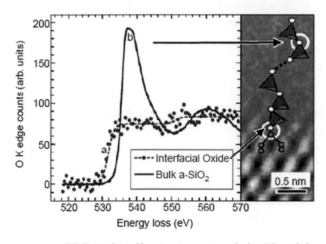

Electron energy loss spectroscopy (EELS) at indicated locations in an extremely thin SiO_2 oxide layer on a Si wafer. The TEM image at right shows the lattice structure of the underlying Si, and has the SiO_4 tetrahedra shown schematically. The inferred electronic structure is already bulk-like only four tetrahedra away from the interface. From [49].

Fig. 3.11

used in various nanoscience applications. For example, suspended Si_3N_4 membranes tens of nm thick supported by etched Si wafers are are sold commercially for use as TEM sample grids. Si_3N_4 is also used as a material for micro- and nanomechanical systems, and as an insulating dielectric in flexible electronics applications. While mechanically more robust than SiO_2, the nature of charge traps and defects in Si_3N_4 is less well explored.

3.4.2 Alumina

Aluminum oxide (Al_2O_3), sometimes known as "alumina", "corundum" or "sapphire", is another robust band insulator that has become extremely useful in nanoscience research and applications. Its structure can range from approximately amorphous to crystalline, with a preferred crystal structure also named "corundum", consisting roughly of an HCP array of oxygen atoms with Al atoms occupying 2/3 of the octahedrally coordinated interstitial sites. Its band gap is approximately 8.8 eV, making it attractive as an insulating barrier [50].

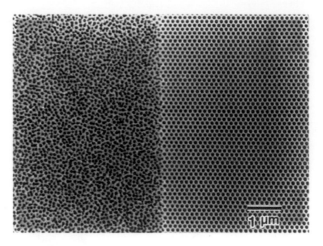

Fig. 3.12 Top view of a nanoporous anodic alumina film with (right) and without (left) prepatterning of the surface to trigger pore nucleation in predefined locations [52].

At room temperature the diffusion of oxygen through the alumina structure is very poor. As a result, under ambient conditions the oxidation of aluminum is self-limiting at a thickness of a few nanometers. The ease and reproducibility of the growth of Al_2O_3 has resulted in alumina becoming the standard tunneling barrier material used in low-temperature superconducting nanoelectronics.

It is possible to drive the oxidation of aluminum electrochemically, a process well known as "anodization". Thicker oxides are possible via anodization because it is possible to produce intrinsically porous alumina through this process. Porous alumina membranes can be readily produced with nanoscale pore sizes [51], and prepatterning of the surface can lead to controlled fabrication of ordered nanopore arrays [52] for a variety of purposes.

3.4.3 Other band insulator oxides

In recent years increasing attention has been paid to other oxide systems, often with an eye toward supplanting or replacing SiO_2 in microelectronics applications. In addition to the wide gap semiconductor TiO_2 mentioned above, other relatively simple transition metal oxides that are receiving increased scrutiny are HfO_2, ZrO_2, and Ta_2O_5.

Layered oxide compounds known as perovskites are also becoming better studied material systems. For example, $SrTiO_3$ and $BaTiO_3$ are both relatively wide gap systems. The latter is a *ferroelectric*, meaning that it develops a spontaneous dielectric polarization below its ordering temperature of 120° C. The former is an incipient ferroelectric. This means that $SrTiO_3$ is very close (thermodynamically) to a ferroelectric transition, and therefore has a very large relative dielectric constant; however, no ferroelectric ordering occurs even at very low temperatures.

Large and often anisotropic dielectric properties can also be useful for optoelectronic applications, as we shall see in Chapter 8. Another band insulator oxide material with such useful properties is lithium niobate, $LiNbO_3$. With a band gap of around 4 eV, $LiNbO_3$ is an essential part of many electro-optical and acousto-optical components.

3.4.4 Topological insulators

Until around 2005, it was assumed that all band insulators are comparatively boring. However, something interesting can happen in materials with very strong spin–orbit coupling [53, 54, 55]. Under the right circumstances, such materials can be *topological insulators*, with an insulating bulk possessing a band gap, and metallic (ungapped) states that live at the surface (we discuss surface states later in Section 5.2.2). There can be two-dimensional topological insulators, where the analog of surface states are one-dimensional "edge states" at the sample perimeter. There can also be 3d topological insulators [55, 56, 57, 58] such as Bi_2Se_3, where the surface states are two-dimensional. In both cases, the extremely strong spin–orbit coupling leads to orbital motion (**k**-direction) being effectively locked to the charge carrier's spin orientation, and vice versa. This feature and the (comparably) robust metallic nature of the surface states in the presence of disorder make these materials of much interest for eventual information technology applications.

Topological insulators are complex and rich, and a serious treatment of them is far beyond the scope of a broad book such as this one. Outstanding review articles exist ranging from the more popular [59, 60] to the in-depth [61, 62].

3.5 Correlated oxides

With increasing chemical and structural complexity, much more exotic materials physics is possible. We will mention these systems very briefly, as they have not been used extensively in nanoscale science at the time of this writing, and the literature on these materials in their bulk form is vast. However, nanoscale characterization tools have been applied extensively to some of these systems.

Strong electronic correlations are a hallmark of many oxide materials that incorporate rare earths and transition metals. As mentioned in Section 3.1, one possible state that results from strong electron–electron repulsion on localized sites is the Mott insulator. Mott insulators can arise when a lattice of singly occupied sites exists with such a strong on-site repulsion that electronic transport (which would require transient double occupation of each site) is suppressed. Unambiguously identifying Mott insulators in real materials can be challenging. The best known Mott insulators are the copper oxides that are the parent compounds of the high temperature cuprate superconductors, such as La_2CuO_4 and $YBa_2Cu_3O_7$. Both of those compounds are antiferromagnetic insulators in their ground state.[5] Another family of Mott insulating oxides are the rare earth manganites, e.g., $LaMnO_3$.

---

[5] The antiferromagnetism is actually a complicating factor in identifying Mott insulators. If the antiferromagnetic order is somehow "in the driver's seat", this leads to a doubling of the unit cell, and therefore the opening of a band gap, meaning that the material is a band insulator. If, however, on-site electron–electron interactions are

Chemical substitution makes it possible to dope such materials away from their insulating state. Depending on parameters such as doping, pressure, and magnetic field, these strongly correlated oxides can exhibit very rich phenomenology. In the copper oxides this includes a "normal" state often termed a "bad metal" because of the failure of Fermi liquid theory to describe its properties, as well as the superconducting state. The manganites show a number of magnetically ordered phases, and transitions between these phases can be accompanied by dramatic changes in electrical conductivity. Driving these conductance changes by varying an external magnetic field has led to the manganites also being known as "colossal magnetoresistance" (CMR) materials.

Even relatively simple compounds such as VO_2 and Fe_3O_4 can show surprising properties not readily inferred from a simple band picture. For example, VO_2 undergoes a transition between a high temperature metallic state and a low temperature insulating state at around 340 K. Optical measurements [63] of the electronic properties of VO_2 have shown that the quasiparticle picture of metals familiar from Fermi liquid theory does not apply to the metallic state of this material.

Similarly, charge transport in Fe_3O_4 (magnetite) is not at all a simple matter, and has been the subject of controversy for more than seven decades [64, 65]. At high temperatures, conduction is thought to occur via fluctuating valence (2+ or 3+) of the iron atoms, with some short-range correlations between nearby irons [66]. The result is a resistivity that increases weakly as T is decreased, though not like that in a semiconductor. Below the Verwey temperature ($T_V \sim$ 120 K in bulk), magnetite becomes significantly more insulating, and a charge ordered ground state has been suggested, though this remains controversial [64, 65].

As nanoscale materials synthesis and processing methods are extended and improved, these correlated materials are increasingly being examined in restricted geometries. Recent examples include studies of VO_2 nanowires [67] and magnetite nanoparticles [68]. Controlling stoichiometry in these correlated materials is very challenging at the nanoscale, and a problem of critical importance for future studies of these systems.

3.6 Molecular structures

Preceding sections have discussed strongly bound (covalently, ionically, or metallically), extended crystals. By some popular definitions one can consider a single crystal to be a single, enormous molecule. In contrast to these structures, there are materials of increasing relevance to nanoscale science and technology that are composed of smaller covalently bound entities held together in macroscopic forms by noncovalent interactions. Here we give a brief overview of some of these systems.

the dominant physics, then the appearance of antiferromagnetism is a consequence rather than a cause of the Mott gap. These distinctions remain an active area of research.

3.6.1 Graphene and carbon nanotubes

Graphene

One of the simplest examples of such a molecular material is graphite. Graphite is composed of graphene sheets of sp^2-bonded carbon atoms, with adjacent sheets held together by the van der Waals interaction, as shown in Fig. 3.13. Because of the simple geometry of each sheet, graphene is very amenable to tight-binding calculations of its electronic structure [69, 70]. The results are shown in Fig. 3.14.

Graphene is described with a basis of two carbon atoms per lattice site (and per unit cell), and may be thought of as consisting of a combination of two sublattices. The nearest neighbor distance is $a' = 0.142$ nm, while the lattice constant is $a = \sqrt{3}a' = 0.246$ nm. The symmetry of the lattice and its basis leads to a unique band structure. The valence and conduction bands touch at *Dirac points* in k-space. While Fig. 3.14 looks at first glance like there are six such points, the points shown occur at the zone boundary and are shared with neighboring zones. In the end there are exactly two inequivalent Dirac points, K and K', in the Brillouin zone. This two-fold degeneracy is analogous to the valley degeneracy discussed previously. With the zero of energy chosen as the energy of the Dirac points, the band dispersion [71] in the tight-binding approximation is

$$E(\mathbf{k}) = \pm \gamma \left[1 + 4\cos\left(\frac{\sqrt{3}k_x a}{2} \right) \cos\left(\frac{k_y a}{2} \right) + 4\cos^2\left(\frac{k_y a}{2} \right) \right]^{1/2}. \qquad (3.19)$$

Here γ is the overlap integral defined back in Eq. (2.85). One can readily see that the K (and K') points indicated in Fig. 3.14 satisfy $E(\mathbf{k}) = 0$.

Thus graphene is often termed a semimetal or a zero-gap semiconductor. Furthermore, in the single-sheet limit the dispersion $E(\mathbf{k})$ around the Dirac points is *linear* in $|\mathbf{k}|$. The implications of this are profound and are readily seen when considering the definition of effective mass, Eqs. (3.4) and (3.6). The effective mass is undefined at the Dirac points, and *zero* at other values of k_x, k_y nearby. This means that the electronic excitations of this material don't act like conventional massive quasiparticles; rather, they have properties

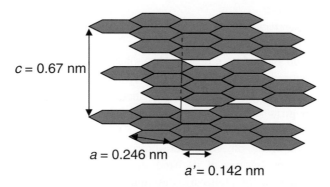

Structure of graphite, showing stacked graphene sheets held together by van der Waals interactions. The lattice constant is $a = a' \times \sqrt{3}$.

Fig. 3.13

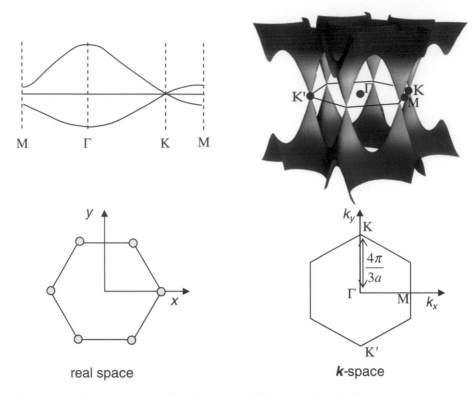

real space

k-space

Fig. 3.14 | Left: band diagram for single graphene sheet, based on tight-binding calculations. Right: band structure showing the high symmetry points in the Brillouin zone, and the two inequivalent Dirac points, K and K'. Adapted from [72].

analogous to approximately massless Dirac fermions, with a dispersion

$$E(\mathbf{k}) \approx \hbar k v_{\mathrm{F}}. \tag{3.20}$$

This resembles the relativistic expression for massless particles, $E = pc$, with the Fermi velocity playing the role of the speed of light. In graphene v_{F} is approximately 10^6 m/s. This unusual dispersion leads to profound consequences in the electronic conductivity, which we will discuss later.

There has been a *tremendous* burst of activity in the experimental study of graphene since 2004 [73, 72] and the Nobel Prize in Physics in 2010. Small patches of few-layer graphite [74, 75] and individual graphene sheets have been dispersed on surfaces and studied via their electronic propertics [76, 77, 78]. Patterning techniques have been applied to fabricate various device geometries [77, 79]. Chemical vapor deposition methods have been developed to grow large area graphene on polycrystalline metal foils [80], for mechanical transfer to other substrates. It is clear that graphene acts as a remarkable laboratory for examining the physics of two-dimensional electronic systems, with added richness due to the peculiarities of its band structure. Because of the rapid pace of advancement in the field, a comprehensive description of graphene physics could by itself fill a book which would immediately be out of date upon publication. The electronic properties of graphene are reviewed comprehensively by Das Sarma *et al.* [81].

It is worth noting that there are other materials that are similar in some ways to graphene, possessing covalently bound 2d layers held together by comparatively weak van der Waals forces. One large class of these materials is the *transition metal dichalcogenides* (TMDs), of the form TM_2, where T is a transition metal (e.g., Ti, Ta, Mo, W, Nb) and M is a chalcogen (e.g., S, Se, Te). These materials have been known and studied extensively for many years [82, 83], and have been the subject of enormously renewed enthusiasm with the discovery of exfoliation [84] and chemical vapor deposition methods to prepare samples down to the single-layer limit, similar to achievements in graphene. TMDs exhibit an impressively rich variety of phases depending on composition and polytype, including semiconductors (MoS_2), multiple competing charge-density wave phases (1T-TaS_2, $TiSe_2$, $NbSe_2$), Mott insulators (1T-TaS_2), and superconductors ($NbSe_2$). Intercalation of metal atoms and small organic molecules into the layered structure can dope the materials and tip the balance between these competing phases [85].

Nanotubes

Since their discovery[6] in 1991 [87], carbon nanotubes have been extensively studied as another realization of the remarkable electronic properties of graphitic carbon. Nanotubes may be either multiwalled or single-walled. The latter, first clearly isolated in 1993 [88, 89], have been studied more extensively and lend themselves better to theoretical analysis by virtue of their simpler structure.

Single-walled carbon nanotubes (SWNTs) can be described as individual graphene sheets "rolled up" and joined into cylindrical tubes. As shown in Fig. 3.15, there are many possible ways to join the edges of a graphene sheet and preserve the sp^2 graphene structure. Defining the two lattice translation vectors in the graphene plane as $\mathbf{a}_1$ and $\mathbf{a}_2$, the vector that shows which carbon atoms map onto each other in the rolling is written as $\mathbf{c}_h = (m+n)\mathbf{a}_1 + n\mathbf{a}_2$, and may be abbreviated simply as the ordered pair of integers (n, m). The diameter, d, of the resulting nanotube is

$$d = a\sqrt{n^2 + nm + m^2}/\pi, \tag{3.21}$$

while the chiral angle, defined as the angle between $\mathbf{c}_h$ and $\mathbf{a}_1$, is

$$\theta = \tan^{-1}\left(-\frac{\sqrt{3}m}{2n+m}\right). \tag{3.22}$$

The rolling imposes a new period boundary condition that must be satisfied by the single-particle electronic states of the carbon. The electronic wavefunction must be single-valued when going all the way around the circumference and returning to the starting atom. That is, for a single-particle state with a particular $\mathbf{k}$,

$$\mathbf{c}_h \cdot \mathbf{k} = 2\pi j, \tag{3.23}$$

---

[6] The unambiguous identification of carbon nanotubes in 1991 was preceded by an extensive history of research into the growth of carbon filaments dating back more than a century, including multiple prior incidents of possible nanotube observation [86].

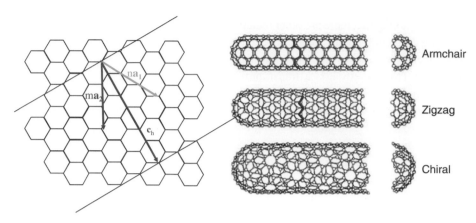

Fig. 3.15 Left: diagram showing the graphene lattice and the vector required to roll up a (3,3) nanotube. Right: particular tube types. "Armchair" tubes (such as the (3,3) tube) derive their name from the highlighted pattern of bonds as a tube circumference is traced, and are metallic. "Zigzag" tubes (such as (5,0)) are similarly named, and can be semiconducting. Most possible tubes are "chiral", with an inherent twist in their structure. While these tubes are depicted with closed ends, the electronic properties discussed in this section neglect end effects. Adapted from [90].

where j is an integer. The result is that the original 2d band structure of graphene from Fig. 3.14 is cut into a series of slices in k-space, with the k-space interval between the slices inversely proportional to the diameter of the nanotube. This is shown schematically in Fig. 3.16. If the allowed slices in k-space intersect the Dirac points, the nanotube has a nonzero density of states at all relevant energies and is therefore a metal. If the slices in k-space miss the Dirac points, then the nanotube has a band gap between the filled valence band states (originating from the filled π bands of the graphene sp^2 lattice) and the empty conduction band states (coming from the empty π^* bands), and the nanotube is a semiconductor. As expected, as tube diameters are increased the allowed slices in k-space become closer and closer together until the original graphene band structure is recovered; band gaps of semiconducting tubes decrease roughly inversely with diameter.

The densities of states shown in Fig. 3.16 have the characteristic $1/\sqrt{E}$ singularities expected for truly 1d electronic subbands (see Eq. (2.24) and Fig. 2.8). (The peaks really only diverge in the limit of infinite tube length.) Each additional subband corresponds to additional electronic momentum about the axis of the nanotube.

It can be shown (see Exercises) that from the periodic boundary conditions a (n,m) tube should be *metallic* when

$$n - m = 3j, \tag{3.24}$$

where j is an integer. In fact, when higher order effects due to curvature are included, only (n, n) armchair tubes are truly metallic, while a tiny energy gap (negligible at room temperature) forms in the other nominally metallic tubes [91] (nonzero j). The tube chiralities that do not satisfy Eq. (3.24) are all semiconducting. As a holdover from the K/K' valley degeneracy mentioned in graphene, at the charge neutrality level (the energy corresponding to the states at the Dirac points) conduction in metallic nanotubes is via

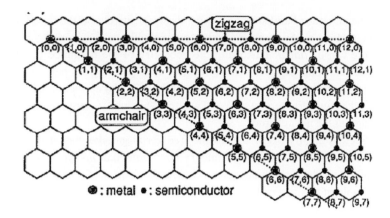

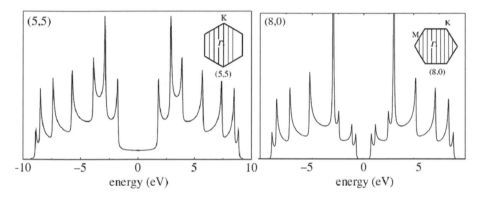

Fig. 3.16
Top: a large number of possible tube types, delineated as metals and semiconductors. Bottom: density of states as a function of energy for two different nanotubes, an armchair (5,5) at left that is metallic, and a zigzag (8,0) at right that is semiconducting. Insets: the periodic boundary conditions imposed on the graphene Brillouin zone by rolling into the tube geometry. Metallic nanotubes result when the allowed cuts through the Brillouin zone intersect the Dirac points. Adapted from [71, 91].

two single-particle spatial states, each with two-fold spin degeneracy. These states are effectively one-dimensional, with no energetically nearby states available as final states for small-angle scattering. One consequence of this very small density of states is a reduced scattering rate for charge carriers, as was already shown in Fig. 2.10.

Semiconducting nanotubes have band gaps that can range from a few hundred meV up to ~ 1 eV. Unlike the bulk inorganic semiconductors of Section 3.3, substitutional doping of SWNTs is difficult without destroying the sp^2 structure locally [92]. Doping by physisorption and charge transfer are more readily achievable [93, 94]. Semiconducting SWNTs tend to be p-type unless specific efforts are made to add donors or remove physisorbed oxygen believed to be responsible for acting as an acceptor. Charge transfer when nanotubes are contacted by metal electrodes also affects the position of the Fermi level (and hence the charge carrier type and population). These effects are discussed in Chapter 5.

Because the nanotube band structure depends so sensitively on the detailed periodicity of the single-particle states when traversing $\mathbf{c}_h$, it should not come as a surprise that the band structure can be *tuned*. Distorting the geometry of the nanotube cross-section via strain can alter the band gap of a semiconducting tube [95, 96] or open a band gap at the charge neutrality point of a previously metallic tube [97]. An applied axial magnetic field can also tune the band structure of a SWNT via the Aharonov–Bohm phase [98, 99], which we will discuss in the next chapter.

Back in Section 2.1.4 we saw that in the free Fermi gas the choice between periodic boundary conditions and hard wall boundary conditions had *no effect* on the *statistical* properties of the single-particle energy levels like the density of states. We can see a real material example of this insensitivity to the boundary details in a comparison of the band structures of SWNTs and graphene ribbons. Graphene ribbons may be classified by the angle of their edges using the same (n, m) labeling approach as in SWNTs.

Multiwalled nanotubes (MWNTs) are also readily produced, with structures formed by the coaxial growth of SWNTs. MWNTs can have diameters ranging from ~ 1 nm for the smallest double-walled tubes up to tens of nm for tubes with larger numbers of walls. The (n, m) types of the various individual walls within a MWNT are not necessarily correlated, and can be a mixture of metallic and semiconducting chiralities.

There is good evidence that the intertube coupling within a MWNT is comparatively weak. For example, when an individual MWNT is contacted by metal leads it would appear from magnetoresistive effects [100] that current is dominantly carried by the outermost tube. This is further supported by the demonstrated ability to vaporize the MWNT walls beginning with the outermost layer under high current conditions [101]. The interlayer mechanical coupling is also generally found to be very weak, especially when adjacent layers are of differing chiralities, such that their relative atomic stacking is incommensurate [102]. Neither of these observations is particularly surprising given the strong anisotropies of the electrical and mechanical properties of multilayer graphite.

We will discuss growth and characterization of nanotubes in Chapter 4. The remarkable mechanical properties of nanotubes, including the mechanism of how they eventually fail under tensile loading, will be addressed in Chapter 9. Similarly, optical properties of nanotubes will be considered in Chapter 8.

3.6.2 Molecular crystals

Graphene, with its covalently bonded 2d layers held together by van der Waals interactions, is in a sense the extreme 1d limit of a *molecular crystal*. A molecular crystal consists of a spatially periodic, extended array of individual molecules, with covalent intramolecular interactions and van der Waals intermolecular interactions. Just as one can consider the single-particle band structure as arising out of the hybridization of the periodically arranged atomic orbitals in a covalently bound crystal, a band structure originates from the coupling of the molecular orbitals of the molecules that make up the molecular crystal. Here we briefly survey the types of molecular crystals, in part for completeness and in part because some of the ideas from these systems are relevant to nanoscale organic and molecular electronics.

Because of the comparatively weak nature of the intermolecular coupling, the resulting single-particle bands are generally much narrower than those in covalently or metallically bonded crystals. The same weak intermolecular forces also lead to a lattice in molecular crystals that is comparatively "floppy" (low sound speeds; more prone to distortion) than in bulk covalent solids.

This relative lack of stiffness leads to important consequences for electronic conduction in these systems. In the description of electronic quasiparticles in Section 3.2.1, the background lattice was assumed to be essentially immobile; that is, the ion cores were assumed not to move or undergo dielectric polarization in response to the quasiparticles. This is often a poor assumption in molecular systems, where the lattice is less stiff and the constituting molecules are often highly polarizable. Low energy electronic excitations of bulk molecular systems are often described as **polarons**: an electronic excitation strongly coupled to an elastic deformation and/or dielectric polarization. Polarons were first considered in the context of slow charge carriers in insulating crystals [103, 104, 105] and were later extended to molecular crystals [106, 107].

Because polaron motion involves the carrying along of an associated lattice distortion (which may be pictured as a cloud of phonons), polarons are traditionally considered to be heavy. The polaron effective mass can be significantly enhanced relative to the bare band mass that one would obtain from a static lattice. Depending on the strength of the electron–vibrational coupling, in principle polarons can self-trap – the associated lattice deformation can act as a potential well from which the charge carrier itself cannot escape. In the context of individual molecules, chemists often use an alternative term (self-trapped radical cation or radical anion) to describe a charge excitation stabilized due to molecular distortion.

The spatial extent of the lattice distortion delineates between "large" and "small" polarons. A large polaron has an elastic or polarization field that extends over many lattice sites, while a small polaron's distortion is local compared to a lattice spacing. That is, optical phonons are the ones involved in small polaron formation. It is also possible for two charges to share a single distortion, leading to the formation the aptly named *bipolaron*. The question of motion of polarons in the presence of an applied electric field is complicated, particularly as applied to molecular solids [108].

3.6.3 Molecular (organic) semiconductors

Just as in more traditional systems, it is possible to have molecular systems that range from band insulators to semiconductors to metals. These materials can be molecular crystals with high structural order in 3d, ordered polymers with van der Waals bonding of nicely stacked essentially 1d chains, or strongly disordered (amorphous small molecule structures or polymer glasses).

Molecular crystals of small, conjugated molecules have been investigated as semiconductors since the 1950s [109, 110]. Photogenerated carriers in highly pure crystals of this sort have shown evidence of band-like transport [111]. Recently developed techniques have allowed the examination of single crystals of a number of similar materials, including rubrene, tetracene, pentacene, C_{60}, and phthalocyanines [112] via the fabrication of field-effect transistors. These materials tend to have relatively large band gaps (2-3 eV). Many basic properties of charge carriers in these systems (e.g., the effective masses) have yet to be determined experimentally, in part because of material purity. While single-crystal Si may be obtained with a purity of one part in 10^9, materials produced by chemical synthesis techniques can be difficult to purify beyond the part per thousand level.

There are a number of conjugated polymers that are also widely used as organic semiconductors. Examples include polythiophene and polyphenylvinylene. Forms of these materials have been developed that may be dissolved in a variety of solvents. The fact that these materials are solution-processable has led to much excitement about the possibility of inexpensive, printable electronic devices [113].

The electronic states in polymer materials can be complicated. In the limit of perfect crystalline stacking of the polymer chains, the physics relevant for molecular crystals should apply transverse to the chains. The π-conjugated bonding along the polymer chains leads to enhanced conduction along the chain direction. In the strongly disordered (glassy) limit, the single-particle electronic states at the top of the valence band and bottom of the conduction band are not extended Bloch-like states. Instead these systems have a density of *localized* states that depends very strongly on energy near the band edges [114, 115]. The electronic and optical properties of these materials remain a rich area of research.

3.6.4 Molecular metals

There are three main groups of molecular metals, all of which are essentially highly doped (or self-doped) semiconductors. One class of molecular metals is based on charge transfer salts [116, 117, 118]. These materials are based largely on 1d stacks of small, planar, π-conjugated molecules. The most well known is probably the combination of tetrathiofulvalene (TTF) and tetracyanoquinodimethane (TCNQ). TTF acts as an electron donor, while TCNQ is an electron acceptor. When crystallized, charge transfer takes place

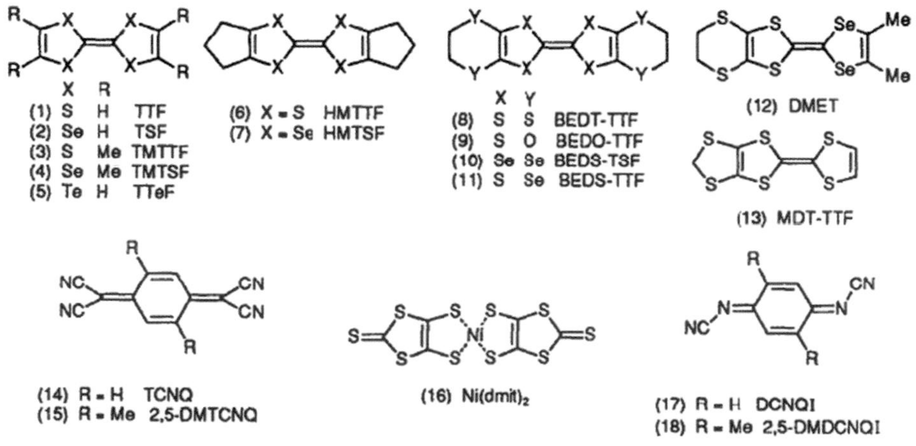

Fig. 3.17 A variety of charge transfer salt components. Combining small conjugated electron donors from the first row with electron acceptors from the second row in an ordered crystal can lead to metallic conduction. From [119].

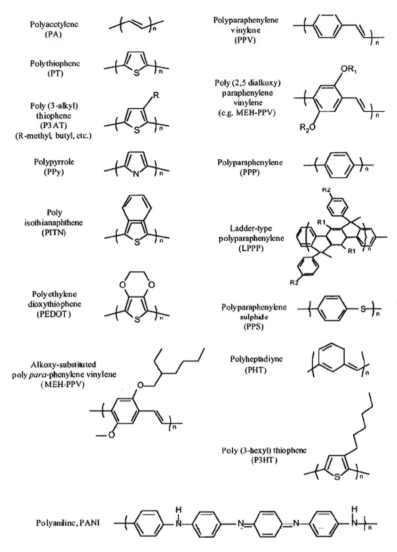

A variety of conjugated polymers. Undoped, these materials act as semiconductors. Some (polyacetylene, PEDOT, polyaniline) may be doped to conductivities approaching those of conventional metals. From [125].

Fig. 3.18

between the TTF and TCNQ, leading to the formation of segregated 1d stacks of TTF cations and TCNQ anions. Because of the face-to-face π stacking (not unlike adjacent graphene sheets in graphite), the unpaired charges are able to move through the stacks. Examples of charge transfer salt components are shown in Fig. 3.17.

Crystals of these materials can have conductivities (though highly anisotropic) that approach those of traditional metals. They have been seen to exhibit rich phenomena, including superconductivity [120, 118], the quantum Hall effect [121], and ferromagnetism [122]. A number of these materials (including TTF-TCNQ) have metallic conductivities near room temperature, but upon cooling undergo phase transitions to insulating ground states.

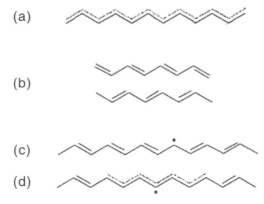

Fig. 3.19 (a) Ideal polyacetylene. (b) The two degenerate bond configurations for dimerized polyacetylene. (c) An idealized neutral soliton, with an unpaired electron. (d) A realistic soliton, showing the spatial extent of bond distortions. Adapted from [125].

The other major category of molecular metals consists of the conducting polymers, such as doped polyacetylene [123]. In its undoped form, *trans*-polyacetylene is a linear molecule, and as shown in Fig. 3.19, there is a degeneracy between two bond configurations (A and B), and by electron-counting arguments it would be expected to have a half-filled band and thus be a 1d metal. Peierls showed [124] that such systems tend to be unstable to lattice distortions. The real system can lower its total energy by picking one of the bond configurations and shortening the interatomic distance of the double bonds relative to the single bonds. This dimerization doubles the size of the unit cell so that there is no longer a half-filled band, opening a band gap at the Fermi level.[7] Undoped polyacetylene is a highly anisotropic semiconductor with a band gap on the order of 1.5 eV.

Adjacent regions of A and B bonding are separated by a *domain wall*, a boundary region that involves a lattice and electronic distortion that is spread over approximately 14 carbon lattice sites [126, 127]. This overall electrically neutral distortion is called a *soliton*, and is accompanied by an unpaired spin-1/2 moment localized to the boundary. Such solitons are detectable via their magnetic properties and are highly mobile along the polymer chains. It is also possible to have charged solitons.

Polyacetylene and other semiconducting polymers can be doped via charge transfer. In the case of polyacetylene, conductivities exceeding 10^5 S/cm (close to that of metals like platinum) are possible. To achieve *p*-type conduction electron-accepting dopants such as halides (e.g., iodine, bromine) are used, while *n*-type doping often involves exposure to alkali metals. Polyaniline is another semiconducting polymer that can be extremely highly doped through multiple possible chemical processes (e.g. oxidation, protonation). One increasingly common highly doped polymer is poly(3,4-ethylenedioxythiophene) (PEDOT), which has been doped *p*-type via charge transfer to the polymer poly(styrene sulfonate) (PSS). Together, these are called PEDOT-PSS.

---

[7] Metallic carbon nanotubes appear to be spared from this fate by their very high elastic stiffness.

Doping molecular crystals with halide or alkali metal dopants is also possible. For example, C_{60} crystals may be doped to metallic levels [128] by incorporation of alkali metals into the lattice, and the resulting material can even exhibit superconductivity [129]. The doped fullerenes are probably the most extensively studied doped molecular crystal [130].

3.7 Summary and perspective

Building on our look at electronic structure in the previous chapter, we classified materials by the filling of their electronic states, still assuming (infinite) bulk materials. The main categories were:

- Metals. Traditional metals have partially filled band(s), such that the density of states near the Fermi energy is nonzero. Excitations of normal metals are quasiparticles that strongly resemble the particle and hole excitations of the free Fermi gas. In some metals the Fermi liquid is unstable, leading to a superconducting ground state.
- Semiconductors. Semiconductors have a band gap less than a few eV separating the valence (highest filled) band from the conduction (lowest empty) band. The lowest energy electronic excitations are (quasi)particles and (quasi)holes that require an energy of at least E_g. These excitations generally have energies that vary quadratically with k, and can therefore be described by effective masses. Indirect gap systems have minima in the conduction band at different **k** points from the maximum in the valence band, while in direct gaps the **k** values coincide. Semiconductors may be doped via chemical substitution, and may even be made magnetic. Individual dopants may resemble hydrogenic atoms, though larger in size because of the effective dielectric background from the host material.
- Band insulators. The extremely large E_g values of band insulators means that they are very useful as insulating dielectric layers.
- Strongly correlated oxides. The simple single-particle band theory picture is inadequate for a number of oxide materials that can show dramatic phenomena (high temperature superconductivity, metal–insulator transitions).
- Graphene and nanotubes. These are particular examples of molecular materials. Because of their precise symmetries they have unusual electronic properties, including an effective mass of zero in graphene.
- Molecular materials. We also considered semiconductors and metals composed of van der Waals bonded molecules. These span from highly ordered single crystals of small molecules to strongly disordered systems made from polymer chains. Distortions of nuclear positions figure prominently in understanding the electronic excitations of these materials.

In the next chapter we begin to look at what can happen as we move away from the approximation of infinite, bulk materials. In particular we examine the effects of defects, interfaces, and electron–electron interactions.

3.8 Suggested reading

- *Normal metals* Many books and articles have been written about metals and their elementary excitations. Apart from the standard solid state physics texts, two more advanced physics treatments are *The Theory of Quantum Liquids*, Vol. I by Pines and Noziéres (New York, Addison Wesley, 1989) and *Landau Fermi-Liquid Theory: Concepts and Applications*, by Baym and Pethick (New York, Wiley, 1992).

- *Superconductors* A classic physics overview of superconductivity is *Introduction to Superconductivity* by M. Tinkham (New York, Dover, 1996).

- *Inorganic semiconductors* Because of the engineering applications of semiconductor devices, there are a huge number of books available that discuss the physical properties of bulk, inorganic semiconductors, though most do so from the device perspective. One encyclopedic work is *Physics of Semiconductor Devices* by S. M. Sze and K. K. Ng (New York, Wiley, 2006), and more concise introductions are *Semiconductor Fundamentals*, Vol. I and *Advanced Semiconductor Fundamentals*, Vol. VI by R. F. Pierret (New York, Prentice Hall, 1992).

- *Graphene and carbon nanotubes* A classic, though physicsy, reference for graphene and nanotubes is *Physical Properties of Carbon Nanotubes*, by R. Saito, G. Dresselhaus, and M. S. Dresselhaus (New York, World Scientific, 1998). Somewhat more accessible and up-to-date works are *Carbon Nanotubes and Related Structures* by P. F. Harris (Cambridge, Cambridge University Press, 2002) and *Carbon Nanotubes: Properties and Applications*, edited by M. J. O'Connell (New York, CRC, 2006). A great review of recent developments is "Electronic and transport properties of nanotubes", by J.-C. Charlier, X. Blase, and S. Roche, *Rev. Mod. Phys.* **79**, 677–732 (2007).

- *Molecular materials* An extremely complete and expensive reference work in this area is *Electronic Processes in Organic Crystals and Polymers* (2nd edn.), by M. Pope and C. E. Swenberg (New York, Oxford, 1999). Classic review articles on conducting polymers include: "Solitons in conducting polymers" by A. J. Heeger, S. Kivelson, J. R. Schrieffer, and W.-P. Su, *Rev. Mod. Phys.* **60**, 781–850 (1988); Alan Heeger's Nobel lecture, "Semiconducting and metallic polymers: the fourth generation of polymeric materials", *Rev. Mod. Phys.* **73**, 681–700 (2001); and Alan MacDiarmid's Nobel lecture, " "Synthetic metals": A novel role for organic polymers", *Rev. Mod. Phys.* **73**, 701–712 (2001). A review of recent work on single-crystal organic semiconductors is "Colloquium: Electronic transport in single-crystal organic transistors", by M. E. Gershenson, V. Podzorov, and A. F. Morpurgo, *Rev. Mod. Phys.* **78**, 973–989 (2006).

Exercises

3.1 *Metals and conduction* Even though we talk about electronic conduction in Ch. 6, we can get a qualitative feel for why metals are expected to be good conductors.

(a) Do you think filled bands contribute to conduction? Why or why not? Hint: electronic conduction is fundamentally based on the response of the electrons in a material to an applied electric field. Assuming that the applied electric field is weak compared to, e.g., the internal electric fields within the constituent atoms of the material, can the electrons in completely filled bands have a net response? What about the electrons in partially filled bands?

(b) An electron in a single-particle Bloch state labeled by $\mathbf{k}$ is subjected to an applied electric field along the $\hat{\mathbf{k}}$ direction. Does the electron remain in this state, assuming that states at higher k are unoccupied?

(c) You can think of an electron that encounters a deviation from the perfect infinite periodicity of the lattice (the lattice periodicity that picks out the Bloch states as good single-particle states) as scattering with some probability into another (empty) single-particle state. You can think of a quasiparticle with wavevector $\mathbf{k}$ propagating along some distance ℓ before it scatters. Here ℓ called a *mean free path*. If the deviations from perfect periodicity are due to phonons, then one might imagine ℓ would decrease with increasing temperature. What do you think would happen if ℓ became comparable to the Fermi wavelength, $\lambda_F = 2\pi/k_F$? For an interesting article about this issue, see Gunnarsson *et al.*, *Rev. Mod. Phys.* **75**, 1085–1099 (2003) [131].

3.2   *Metals and orthogonality*. Here's a mind-bender related to the idea of writing down the many-body wavefunction of the electrons in a metal as (a linear combination of totally antisymmetric Slater determinants) built out of single-particle states. Consider a lattice of around $N \sim 10^{22}$ atoms. In principle one can find the single-particle electronic states that satisfy the Schrödinger equation for that lattice potential, and these (in Slater determinant form) comprise a complete basis set. That is, it should be possible to write down any many-electron wavefunction on that lattice as some linear combination of the basis states we just found.

(a) Now consider suddenly removing one atom from the middle of the lattice somewhere (or changing its effective potential by, e.g, kicking out a core electron with an x-ray). Do you think the single-particle electronic states of the new lattice are, on the whole, very different from the single-particle electronic states of the old lattice? Why or why not? If a single-particle state in the old lattice is $|\psi_i\rangle$, and the corresponding single particle state in the new lattice is $|\psi_i'\rangle$, what is the approximate magnitude of $\langle \psi_i'|\psi_i\rangle$?

(b) The many-body wavefunction (even in the independent electron approximation) is built from Slater determinants of states like $|\psi_i\rangle$. That means that if we consider the overlap between the many-body wavefunction for the new lattice and that for the old lattice, we end up with a huge number of terms that look roughly like $|\langle \psi_i'|\psi_i\rangle|^N$. For large N, what does this approach? This is called the *orthogonality catastrophe* [132].

(c) In the sudden approximation, so that the lattice is changed instantaneously by the creation of a defect, the many-body wavefunction has no time to adjust to the new lattice. You can still always try to describe the old many-body wavefunction in terms of the new single-particle states (which form a complete basis set). What does the result of (b) imply about trying to do this? How many electron–hole excitations (down to vanishingly small energies) does this imply are created by such a sudden change? The signature of this is seen in "edge state singularities" in phenomena like x-ray absorption.

3.3 *Effective masses* The effective mass m^* is a measure of the curvature of the bands, in analogy with the free electron relationship $E(\mathbf{k}) = \frac{\hbar^2 k^2}{2m}$. One way to infer the effective mass experimentally in metals is through specific heat measurements. Practically, the specific heat of a solid may be found by adding a known amount of energy (such as through an electrical pulse of known size through a heater of known properties) δE to a thermally isolated system and measuring the resulting temperature rise, δT. If one works at temperatures far below the Debye temperature, the phonon contribution to the specific heat may be neglected, leaving the electrons as the only game in town.

(a) The total energy of the electrons for a free electron gas is, to within an additive constant, $E_{tot} = \int_0^\infty [(E - E_F)\nu_d(E) f(E, T, \mu)] \, dE$ (see Eq. (2.31)). (We can consider $(E - E_F)$ inside that integral rather than E since we are free to shift the energy by an overall constant, and this will make the math that follows simpler.) Let's work in 3d, and use the density of states expression given in Eq. (2.24). Assume that $k_B T \ll E_F$ so that $\mu \approx E_F$, find an integral expression for $\partial E_{tot}/\partial T$.

(b) Rewriting in terms of $x \equiv (E - E_F)/k_B T$, and using our low temperature approximation, show that the electronic specific heat $c_e \approx k_B^2 T \nu_{3d}(E_F) \int_{-\infty}^\infty (x^2 e^x/(e^x + 1)^2) \, dx = (1/3)\pi^2 k_B^2 T \nu_{3d}(E_F)$ for the free electron gas.

(c) The argument for a real metal is very similar. At low temperatures, filled bands below E_F don't contribute to the specific heat; why not? Assuming that the carriers in the conduction band have an effective mass m^*, argue that the linear in T part of the low temperature specific heat of the real metal is proportional to m^*. Often people compare c_e(measured) with c_e expected from a free electron gas, and find m^*/m_0.

(d) In the unusual "heavy fermion" compounds [133], people find effective mass ratios $m^*/m_0 \sim 1000$. In terms of the band at the Fermi level (that is, $E(k)$), what does this imply about the curvature of the band as a function of k? These systems ordinarily involve a band at the Fermi level that has a significant contribution from (highly localized) f electrons. Thinking about how bands form (see the discussion in Section 2.3.3), explain why comparatively localized orbitals would lead to a relatively flat band (and therefore a comparatively large m^*).

3.4 *Nanotube band gaps and metallicity* One way to find the band structure of nanotubes is to start with the band structure for graphene and consider different chiral wrapping vectors $\mathbf{c}_h = (m + n)\mathbf{a}_1 + n\mathbf{a}_2$. We can define $\mathbf{a}_1 = \sqrt{3}a'(1, 0)$ and $\mathbf{a}_2 = \sqrt{3}a'((-1/2), (\sqrt{3}/2))$, where $(1, 0)$, $(0, 1)$ are the unit vectors in the x and y directions in real space, and a' is the nearest neighbor interatomic distance, as defined in Fig. 3.14.

(a) Show that $\mathbf{b}_1 = \frac{2\pi}{a}(\frac{1}{\sqrt{3}}, \frac{1}{3})$ and $\mathbf{b}_2 = \frac{2\pi}{a}(0, \frac{2}{3})$.

(b) Argue that in these coordinates one can assign the independent K and K' points as $(\frac{-2\pi}{a'3\sqrt{3}}, \frac{2\pi}{3a'})$ and $(\frac{2\pi}{a'3\sqrt{3}}, \frac{2\pi}{3a'})$ in reciprocal space.

(c) We know from Eq. (3.19) that these corners of the Brillouin zone are those where the filled (valence) and empty (conduction) states come together to touch. To have a metallic tube would require these points in **k**-space to be compatible with the chiral periodic condition, $\mathbf{c}_h \cdot \mathbf{k} = 2\pi j$. With the K point as defined in (b), and the chiral vector defined above, show that the condition that must be met is $(2n + m) = 3j$, where j is some integer.

(d) Show that this is equivalent to $n - m = j'$, where $j' = 3(n - j)$ is also an integer.

3.5 *"Artificial graphene"* In the nearly-free electron picture, it was implied that the rough scope of band structure (bands with gaps at Brillouin zone boundaries, for example) was only weakly dependent on the details of the atomic potentials, and was much more a function of the arrangement of the atoms and the electronic density. This is an oversimplification, but it is interesting to see how far you can take this idea. First, take a look at Gomes *et al.*, *Nature* **483**, 306–310 (2012). In this work the authors create an "artificial graphene" system.

(a) What is the 2d electronic system employed by the authors? How do they spatially modulate the potential "seen" by those 2d electrons in a hexagonal lattice?

(b) What measurement do the authors perform that demonstrates the existence of Dirac cone-like features in the band structure of the modulated system?

(c) How can the authors effectively control the carrier density in their artificial graphene even though the actual carrier density in the unmodulated 2d electronic system is fixed?

(d) Now consider the system described in Gibertini *et al.*, *Phys. Rev. B* **79**. 241406 (2009) [134]. What is the 2d electronic system employed by the authors? What evidence do the authors present that their hexagonal patterning of the surface is creating a modulated band structure? What advantages do the authors think this approach will have in enabling the study of (effective) massless Dirac fermions?

3.6 *Molecular crystals* Molecular crystals are held together by van der Waals interactions between the molecules, rather than a global network of covalent bonds. Depending on molecular shape, the molecules can stack in interesting and complex ways.

(a) It is not a bad idea to think of electronic transport from molecule to molecule in terms of a tight binding picture, where the overlap integral (γ in Eq. (2.85)) is defined between adjacent molecules. How do you think the molecule shape, arrangement, and relative orientation will affect γ? What consequence do you think this may have for electronic transport in different directions within a molecular crystal?

(b) Given that van der Waals interactions are (to within an order of magnitude) ten times weaker than covalent interactions, what do you think would be a typical bandwidth (in eV) in a molecular crystal?

(c) Let's consider non-metallic molecular crystals. The relative dielectric constant κ for these materials tends to be around 3, in contrast to covalent semiconductors where $\kappa \sim 10$. Estimate the Coulomb potential between an electron and a hole on adjacent molecules 1 nm apart under these circumstances. Compare the resulting energy with the result of (b). In the absence of other physics (such as polaronic distortions), is it credible that electron–electron interactions could play an important role in the electronic properties of molecular solids?

Fabrication and characterization at the nanoscale

This chapter gives a necessarily brief overview of the techniques employed to create structures at the nanometer scale, and the tools used to examine matter down to that level. Entire books have been written about many of these topics. Here the emphasis will be broad exposure, and where possible explanations of the strengths and limitations of the various approaches.

4.1 Characterization

Characterizing materials at the nanoscale can be very challenging. It's easy to say what we want, as practitioners in this field. When presented with a sample or nanoscale structure, it would be wonderful if there were some tool or set of tools that could tell us, nondestructively, the position and composition of every atom in the sample, with atomic resolution. Unfortunately, this remains a fantasy. Over the years, however, a wide variety of experimental techniques have been brought to bear on the problem of materials characterization, applying different physical principles to acquire information about various aspects of the systems of interest. Some of these same techniques have been adapted or spun off into related approaches to nanoscale patterning and fabrication.

One can broadly divide the characterization techniques into two categories, bulk or global, and local. Bulk methods generally interrogate a volume or area of sample that is often much larger than the nanoscale, but nonetheless can reveal incredibly precise nanoscale information. Diffraction is the classic example. Local methods, in contrast, directly interrogate a nanoscale volume of sample, and somehow transduce those local interactions into macroscale signals.

4.1.1 X-ray techniques

Since their discovery in the late nineteenth century, x-rays have been incredibly useful for looking at the structure of materials. X-rays are electromagnetic radiation with photon energies in the keV to tens of keV range, and corresponding wavelengths down to subatomic scales. X-rays interact primarily with the inner electrons of atoms, and therefore their scattering is relatively greater for elements of higher atomic number. Because of their high energies, x-rays can penetrate through macroscopic amounts of material (hence their use in medical diagnostics). Their short wavelengths mean that they are ideally suited for examining materials on the nanometer scale.

External radiation interacts with samples in several ways. The material under investigation may scatter the incident radiation coherently, so that interference patterns can form; this is the basis for diffraction and reflection. Alternatively, the radiation may interact inelastically with the material. One may measure the energy loss of the radiation (absorption), or one may detect the effect of that energy loss on the material via the emission of more radiation (fluorescence) or electrons (photoemission).

Diffraction

As mentioned in Section 2.2.2, when coherent waves are scattered by a periodic lattice as in crystalline solids, the result is a diffraction pattern. The interference of the scattered waves results in greatly enhanced scattering in particular directions determined by the lattice spacing of the atoms. By looking at diffraction peaks from various crystal planes, it is possible to determine the *average* lattice spacing in three dimensions to better than 10^{-4} nm, an impressive level of precision. Because it is a coherent phenomenon involving superposition of scattered waves, if all other parameters are held fixed, the intensity of a diffraction signal scales quadratically with the number of atoms interrogated.

Assume that the sample in question is made by a periodic repetition of some unit cell, distributed in space according to a particular spatial (*Bravais*) lattice. The actual diffraction pattern is more complicated than that given simply by the Laue/Bragg condition (Eq. (2.40)) that the scattering wavevector ($\mathbf{k}' - \mathbf{k}$, where $\mathbf{k}$ and $\mathbf{k}'$ are the initial and final wavevectors of the x-rays, respectively) be a reciprocal lattice vector, $\mathbf{G}$. Assuming that the Laue condition is satisfied, the diffraction spot amplitude is proportional to

$$S_{\mathbf{G}} = \int n(\mathbf{r}) \exp(-i\mathbf{G} \cdot \mathbf{r}) d^3\mathbf{r}. \tag{4.1}$$

The quantity $S_{\mathbf{G}}$ is called the *structure factor*. Here $n(\mathbf{r})$ is the electronic density as a function of position within the unit cell, and the integral is over the unit cell volume. The actual diffracted intensity is proportional to $|S_{\mathbf{G}}|^2$. This can be a bit problematic, since the magnitude throws away phase information contained within the complex $S_{\mathbf{G}}$. The challenge of reconstructing the actual charge distribution within the unit cell without this information is called the "phase problem".

The structure factor is essential when interpreting the diffraction from materials with a complicated unit cell. By collecting diffraction information for large $\mathbf{G}$ peaks, the structure factor can be a powerful probe of atomic-scale information within the unit cell. This is the basis for using x-ray diffraction (XRD) to determine molecular structures. The most dramatic example of this is *protein crystallography*. Proteins are complicated organic molecules responsible for an enormous number of biologically important processes, and as we shall see in Chapter 11, they can contain tens of thousands of atoms. In a limited number of cases, it is possible to grow crystals of proteins, such that the unit cell contains a single protein molecule. By a combination of computational power, modeling, and exquisite XRD, it is possible to determine the detailed spatial structure of the protein. This demonstrates the tremendous power of XRD, since under the right (unfortunately rare) circumstances it can give 3d structural information about molecule-scale biological machines.

When looking at ensembles of micro- and nanocrystals, it is possible to use *powder diffraction*. Each small crystallite in a powder sample sits at some angle relative to the incident x-ray beam, and produces a particular array of diffraction spots on a detector. If there are so many crystallites present that the sample is a good ensemble average of possible orientations, then the superposition of the diffraction spots results in circles on the detector, at particular angles away from the axis defined by the incident radiation.

Fluorescence

X-ray fluorescence (XRF) is one common method of elemental analysis. A sample is interrogated with energetic x-rays, knocking out the inner electrons of some of the atoms in the sample. Higher electrons may then fall into the core holes, and the emitted x-ray radiation provides a fingerprint of the type of atoms in the sample. A similar technique, more sensitive to surfaces and readily capable of nanoscale lateral resolution, is *energy dispersive spectroscopy* (EDS) or *electron microprobe* (EMP), in which the sample is interrogated with a nanometer-scale, focused beam of energetic (tens of keV) electrons. The incident electrons again liberate core electrons in the sample, leading to fluorescing of x-rays, which may then be examined. The precise energy of the emitted x-rays may be determined by measuring their diffraction by known standard samples, a technique called *wavelength dispersive spectroscopy* (WDS).

Reflection

Just as one may perform optical reflectometry measurements to learn about the dielectric function and composition of some material films, *x-ray reflectometry* (XRR) may be used to determine composition and nanoscale flatness of thin films. While visible reflectometry of a supported thin film measures index of refraction contrast between the film and substrate, the contrast in XRR originates from differing electronic densities in different layers. As the angle of the incident radiation relative to the plane of the substrate, θ, is increased from zero, one passes from a regime of total reflection ($\theta < \theta_c$) to a regime where the radiation penetrates the surface ($\theta > \theta_c$). In the latter limit, the total reflected intensity is the result of interference between waves reflected from different interfaces within the material, and oscillates as a function of θ. These oscillations may be modeled to infer the densities of surface layers or multilayers, as well as the surface roughness. References [135, 136] provide references to a large number of works on this topic.

A further technique related to XRR is *x-ray standing wave* spectroscopy [138]. When x-rays are Bragg reflected from a crystalline material, interference between the incident and reflected waves leads to a pattern of nodal (and antinodal) planes within the material separated by the lattice parameter, as shown in Fig. 4.1(b). By varying the incident angle,

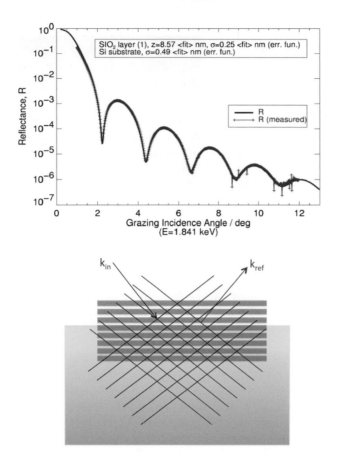

X-ray reflectometry. (a) Oscillations in the reflected x-ray intensity as a function of θ in a multilayer material, with a fit to a theoretical model used to infer the layer thickness and roughness. Adapted from [137]. (b) X-ray standing waves. Antinodal plane position (horizontal lines) may be shifted with great precision through small changes in θ.

Fig. 4.1

the standing antinodal planes may be shifted with subatomic precision through layers of the material, including the surface. Simultaneous measurement of XRF or photoemission can then give precise information about the surface and near-surface material. See, for example, [139].

Photoemission

X-ray photoelectron spectroscopy (XPS) (sometimes called electron spectroscopy for chemical analysis (ESCA)) has become a critical tool in characterizing material surfaces [140]. Incident x-rays have more than enough energy to photoemit electrons from sample surfaces. Electrons emitted from within 1–10 nm of the sample surface are unlikely to scatter inelastically and lose much of their kinetic energy before leaving the material. If the photoemission process is performed in UHV conditions, then a remote detector centimeters away from the sample may be used to measure the energies of the emitted electrons.

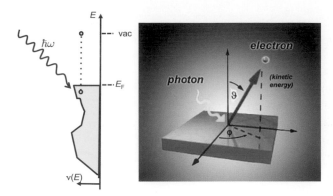

Fig. 4.2 Photoemission. An incident photon "kicks out" a bound electron, which is then detected. At right, angle-resolved photoemission allows determination of energy and momentum-dependence of the filled electronic states of a material. Image from [141].

For incident x-rays of energy $\hbar\omega$, and a detector[1] of work function ϕ, the binding energy of the photoelectron is $E_{\text{bind}} = (\hbar\omega + \phi) - E_{\text{det}}$. Here E_{det} is the photoelectron energy found at the detector.

Electron binding energies have been measured for all of the common elements. Energy resolution in XPS is set by both the energetic width of the source x-rays and the capabilities of the detector, and is typically ~ 0.25 eV. This energy resolution is sufficient to detect the changes in electron binding energy that result from changes in chemical bonding and oxidation state. This has made XPS an incredibly powerful tool for surface analysis, including the characterization of self-assembled monolayers and other surface functionalizations. A variation at lower energies, *ultraviolet photoelectron spectroscopy* (UPS), has been a powerful tool for examining the energetic alignment at metal/organic interfaces.

It is also possible to measure the direction of the outgoing photoelectron as well as its energy. In the plane of the sample, momentum is conserved in the photoemission process. Knowing the incident photon energy and direction, it is possible to use photoemission to characterize the electronic structure of the surface. This technique is called *angle-resolved photoemission spectroscopy* (ARPES) and has proven very useful for understanding the electronic structure of a number of materials, including high temperature superconductors, topological insulators, and graphene.

4.1.2 Electron diffraction

Just as in XRD, when a coherent source of electrons is available, it is possible to perform electron diffraction. Unlike XRD, UHV conditions are generally required for electron diffraction, to avoid unintentional scattering of the electrons by gas molecules. The major

---

[1] Remember, another way to think of work function is as the energy gained by an electron as it is brought from infinitely far away into a material. Therefore, the total energy of the electron at the detector includes the detector work function.

advantage of electron diffraction over XRD is the effective wavelength of the electrons. A 10 keV x-ray has a wavelength of 0.12 nm. In contrast, the deBroglie wavelength of an electron is h/p. At 10 keV we can neglect relativistic effects, and approximate $p \approx m_0 v$, while $m_0 v^2 /2 \approx 10$ keV gives $v \approx 5.7 \times 10^7$ m/s, $\rightarrow \lambda \approx 0.012$ nm, more than an order of magnitude smaller. At relativistic speeds achievable in common electron optics (e.g., 200 keV), picometer wavelengths are possible.

Two particular flavors of electron diffraction are favored in electron microscopy: diffraction via transmission through a thin ($<$ 100 nm thick) sample, and back-scattered diffraction, in which a detector on the same side of the sample as the incident beam is used to find diffraction spots.

In crystal growth, two other diffraction techniques are particularly popular. In *reflection high energy electron diffraction* (RHEED), a glancing beam of energetic electrons is grazed across the surface of a crystal, so that it interacts only with the interfacial atoms. A detector is then used to detect the scattered electrons and display the resulting diffraction spots. The diffraction pattern must be analyzed while accounting for extraneous signals due to inelastically scattered electrons and electrons scattered by the bulk of the crystal.

Alternately, one can achieve nanometer surface sensitivity by using *low energy electron diffraction* (LEED) . In LEED the electron energies are less than 1 keV, so that the electrons interact most strongly with only the surface few atomic layers. Back-scattered electrons are detected, with a biased grid of wires used as an energy analyzer, so that only elastically scattered electrons (those that preserve the initial beam energy) are able to be detected.

All of these electron diffraction techniques, even those that are extremely surface-sensitive, are ensemble surveys of the material. They give only statistical information about the material, averaged over many unit cells, rather than acting as a direct local probe. Moreover, because electrons are charged, it is important to work with comparatively conducting samples to avoid charging effects.

4.1.3 Electron microscopy

The ability to create bright electron beams with small (sub-nm) spot sizes has given us the most widely used nanoscale microscopy, electron microscopy. There are two main varieties, *scanning electron microscopy* (SEM) and *transmission electron microscopy* (TEM). Both depend on the creation of a collimated, focused beam of electrons, typically created via thermionic emission from a filament or field emission from a sharp tip (either a cold cathode or a heated emitter). After extraction, the electrons are accelerated to a desired beam energy, ranging from as low as 1 keV in some SEM applications to as high as 300 keV in some high resolution TEMs. The beam is passed through a series of magnetic condensor lenses and apertures in high vacuum or UHV conditions, and is focused by a magnetic "objective lens" to impinge on a sample surface. Spot size can be sub-nm.

In SEM, the objective lens placed in close proximity (a "working distance" ranging often in the mm range) to the sample surface. The incident electrons strike the sample, producing a shower of *secondary electrons*, which are then detected by a nearby detector such as a scintillator/photomultiplier. While the beam is rastered over the sample, the secondary electron signal is recorded, building up an image of the surface. Heavier elements are more

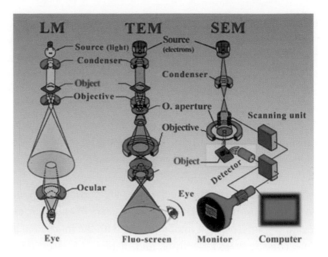

Comparison of optical (left), transmission (center), and secondary electron microscopy. Image from [142], where it was adapted from drawings by JEOL.

effective at producing secondary electrons, so that SEM images show good contrast when the samples consist of high-Z material such as Au on a low-Z substrate such as Si or C. Because of the kinetics of the process that produces secondary electrons, and the ease with which those electrons escape the sample surface, SEM also highlights topography: elevated features and edges are comparatively brighter. High accelerating voltages often lead to improved resolution, since the transit time between the final lens and the sample is reduced. However, more energetic electrons penetrate the sample further (tens of nm), implying that lower acceleration voltages are better for examining surface features. Lateral resolution in standard SEM can be as small as 1 nm.

With modified detectors and specially designed pumping schemes, it is possible to run some SEMs in "high pressure"/"low vac"/"environmental"[2] modes, so that imaging may be performed with the sample under a pressure as high as several millibar. This can permit the imaging of wet samples (biological materials or curing concrete, for example). Residual gas pressure can disperse surface charge, also allowing the imaging of uncoated, insulating samples.

Several other imaging modalities are possible within SEM. Using electrodes as energetic filters and carefully controlling detector geometry, *back-scattered electron imaging* (BEI) records a signal due to incident electrons that are strongly scattered elastically backward from the sample. The rate of this back-scattering varies strongly with Z, so that BEI shows brighter features from heavier elements. With the right sample/detector arrangement, it may be possible to perform *electron backscatter diffraction* (EBD), such that the back-scattered electron signal is dominated by those electrons which Bragg diffract from the sample. As mentioned previously, the incident electron beam has sufficient energy to knock core electrons out of the sample. The sample then fluoresces in the x-ray, and the resulting radiation may be studied via EDS, enabling elemental analysis. Finally, for some sample

---

[2] The name choice depends on the SEM vendor.

materials, the incident electrons can cause the sample to emit visible light (e.g., incident electrons can promote carriers from the valence to the conduction band of a semiconductor, and eventual relaxation gives emission at the band gap energy), which may be detected separately for *cathodoluminescence* imaging.

In TEM, very thin ($<\sim$ 100 nm) samples are prepared via various means (mechanical cutting and polishing; ion milling; focused ion beam cutting) and mounted on porous "grids" (often lacey carbon films) or suspended silicon oxide or nitride membranes. The sample is imaged by having a highly energetic (100–300 keV) electron beam transmit through the sample into a detector on the far side. In addition to the condensor and objective lenses of SEM, a further "projector" lens behind the sample expands the transmitted beam onto a detector (typically a CCD array). In conventional TEM the beam and detector are fixed, and imaging is performed by translating the sample back and forth through the beam. In scanning TEM (STEM) instruments, the beam is rastered in a manner similar to SEM.

There are several imaging modalities in TEM. Bright field imaging is the analog of traditional transmissive optical microscopy, with bright regions indicating high transmission, and dark regions indicating occlusion (scattering or absorption) of the incident beam. Diffraction of the incident beam by the sample opens up several other possibilities, including dark field imaging (via clever geometry simply transmitted electrons are not detected), phase-contrast imaging (using interference of electrons transmitted by different portions of the sample to get added contrast), and electron holography. With careful correction of the electron optics for aberration of the lenses, phase contrast can give ultrahigh resolution, down to sub-Ångstrom levels. It is important to bear in mind that contrast mechanisms in high resolution TEM can be complex because of phase effects. Columns of atoms or columns of interstitial sites may appear as bright spots, depending on

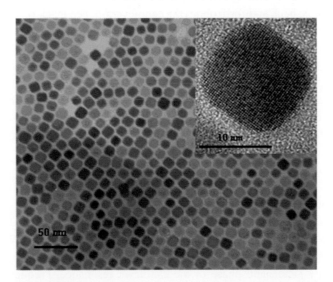

PbS nanocrystals as imaged via TEM. The apparent grayscale contrast between different nanocrystals is set by their crystallographic orientation relative to the electron beam. Inset shows a high-resolution image of an individual nanocrystal, with individual lattice planes visible. From [143].   Fig. 4.4

the details of the measurement. An example of this is shown in Fig. 4.4, where some of the PbS nanocrystals appear bright and others appear dark, depending on their crystallographic orientations relative to the electron beam.

Electron energy loss spectroscopy (EELS) is an additional characterization technique enabled by TEM. By precisely measuring the energies of transmitted electrons, it is possible to identify excitations in the sample that can be excited via inelastic interactions with the beam. The energy resolution of EELS may be as fine as 0.1 eV.

4.1.4 Atomic force microscopy

Atomic force microscopy (AFM) is the first example of a scanning probe microscopy. AFM has its roots in profilometry, in which a sharp stylus is dragged over a surface of interest, with the stylus' deflection giving a map of the surface topography. Developed originally by Binnig, Quate, and Gerber [144], AFM uses a microfabricated tip rather than a macroscopic stylus. The tip is typically at the end of a cantilever that acts as a spring. Forces between the tip and the surface can lead to deflections of the cantilever, which are measured either optically (looking at the deflection of a laser beam bounced off the cantilever, for example) or electrically (such as measuring strain-induced changes in the electrical resistance of the cantilever). The AFM and its variants rely on the fact that in addition to any long-range interactions between the tip and the sample surface, there is a short-range repulsive force that develops when the tip sufficiently close (atomic-scale distances) to the surface. Typical cantilever spring constants can range from $10^{-2} - 10^2$ N/m, while tip radii of curvature range from a few nm to tens of nm. An alternative approach is to attach a sharp tip to a "tuning fork" rather than a cantilever.

"Contact" mode is the analog of traditional profilometry. A piezoelectric stack is used to sweep the sample back and forth under the cantilever tip. Feedback of the sample height piezo is used to maintain a constant cantilever deflection. By tracking this z feedback signal as a function of sample $x - y$ position, a map of the sample topography may be found. Because it is possible to make quite floppy cantilevers and to measure very small angular deflections, the z resolution of contact mode AFM can be 0.1 nm or better, under controlled conditions. Lateral resolution, however, is typically limited to a distance scale comparable to the radius of curvature of the AFM tip. As shown in Fig. 4.5, in general even atomically sharp surface relief will be apparently broadened due to these tip convolution effects. One variant of contact-mode AFM is lateral force (LF) microscopy. LF-AFM takes advantage of the *lateral* forces that result from friction as the tip is scanned across the sample surface. The cantilever's resulting twist provides a map of tip–sample frictional interactions, and therefore sample elastic response.

In "noncontact" (occasionally called "tapping") mode, an additional piezo is used to oscillate the cantilever at or near its natural resonance frequency. When the tip is sufficiently close to the sample surface, the forces between tip and sample shift the frequency, quality factor, and phase of the cantilever resonance. Frequency-mode AFM (FM-AFM) works by getting the cantilever to self-resonate and monitoring the resonance frequency, which can be done with fantastic precision. Feedback on the z piezo is performed to try to keep the frequency constant. FM-AFM has been demonstrated to give

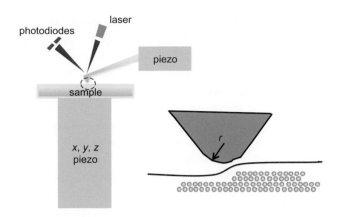

Typical atomic force microscope arrangement, with piezo-scanned sample and piezo-oscillated cantilever tip. Tip deflection here is detected optically. At right, a detail view of the tip–sample region, showing that the tip radius of curvature generally places an effective limit on lateral resolution.

Fig. 4.5

atomic resolution images when samples are prepared carefully and imaged in vacuum. Typical cantilever natural frequencies are on the order of 100 kHz.

Amplitude mode AFM (AM-AFM) instead drives the cantilever at a fixed frequency, f_{dr}, slightly deviating from the cantilever's natural resonance, f_0. If the natural frequency shifts due to interactions between the tip and the sample, the amplitude of the driven oscillations is correspondingly affected, as is the phase relative to that of the drive. Feedback may be done to hold the tip oscillation amplitude constant, and the resulting z height and phase shift may be plotted as a function of $x - y$ position. While amplitude- and frequency-feedback-based maps are conventionally interpreted as showing the sample topography, maps of the phase are more difficult to interpret quantitatively. Nonetheless, phase maps can highlight some details that are difficult to resolve in the topographs, in part because phase can be a very steep function of oscillator parameters near resonance.

Because of its comparatively modest requirements (e.g., samples may be insulators or conductors; measurements may be performed under ambient conditions if needed, or over a broad range of temperatures and vacuum quality if desired; fully functional AFM systems are readily available commercially for moderate cost), AFM has been spectacularly successful as a nanoscale characterization technique. Moreover, since the operating principle of AFM does not depend in detail on the nature of the tip–sample forces, many variants of AFM have been created to examine specific pieces of physics.

One set of AFM approaches leverages the fact that some forces of interest (magnetic, electrostatic) are long-ranged, compared to the short-ranged van der Waals and repulsive interactions that contribute to the topograpy signal. A scan is made (say in the $+x$ direction) across the sample with a tip–sample distance small enough to guarantee short-ranged interactions and acquire a good topography map. When the scan is complete, the tip–sample distance is increased by a preset "lift" height (say 10 nm), and the tip is rescanned over the same line (this time in the $-x$ direction), with the computer control using the previously acquired topography data to maintain the nominally fixed lift height.

Any residual forces, detected through their effects on frequency, amplitude, or phase, are then assumed to result from long-ranged interactions. For example, in *electrostatic force microscopy* (EFM) [145, 146], a conducting AFM tip (often as simple as a conventional Si tip coated by a thin metal film) is employed. By applying a well-defined voltage to the tip relative to the sample surface, electrostatic tip–sample forces develop. Mapping these forces gives information about the local electrostatic potential beneath the tip, which may be modified by surface charges, polar adsorbates or monolayers, flowing currents, etc. The lateral resolution of these force microscopies is a convolution of the tip radius of curvature, lift height, and distance dependence of the long-ranged force. *Magnetic force microscopy* [147] is a conceptually similar technique, in which an AFM tip coated with a ferromagnetic material interacts with a magnetic sample.

A further variant of EFM is *scanning Kelvin probe microscopy* (SKPM) [148]. In SKPM, the electrostatic potential of the tip is adjusted via feedback to null out any electrostatic force between the tip and sample. For an unbiased sample, changes in the feedback tip voltage during a scan are directly related to the local sample work function.

Alternatively, one may simply use the conducting AFM tip as a relatively local electrical probe. For example, an ac voltage may be applied to the tip and, with appropriate equipment, the mutual tip–sample capacitance may be measured. This is the basis of *scanning capacitance microscopy* (SCM) [149], a technique that has proven very useful in the semiconductor industry for measuring local carrier concentrations with ~ 10 nm resolution.

Chemical force microscopy (CFM) [150, 151] may be performed as a variant of LFM, or as an example of a "pulling" techique related to AFM. The cantilever tip may be functionalized with a desired chemical group, and the interactions between that group and the sample surface may then be mapped. In the pulling mode, using the carefully calibrated cantilever spring constant, the cantilever force as a function of tip–sample distance may be measured with high precision. These pulling experiments, also known as *force spectroscopy*, have also been used to look at single molecule processes [152], including unfolding of complicated biomolecules such as proteins [153].

One common issue in scanned probe microscopies such as these AFM variants is the challenge of understanding contrast mechanisms. In the cases mentioned above, it is possible, with appropriate calibrations (and, in the case of MFM, micromagnetic modeling), to acquire quantitative information from the contrast in these microscopies. Great care must be exercised, however. When AFM and its variants are performed under ambient conditions, another challenging factor is a ubiquitous layer of adsorbed moisture and hydrocarbon contamination. Such surface contamination can modify tip–sample interactions, including frictional and chemical interactions and surface work functions.

One approach to high resolution AFM imaging has been the development of special sharp tips with high aspect ratios. A particularly good example of this is the use of an individual single-walled carbon nanotube as an AFM tip. The resulting lateral resolution has been sufficient to examine internal structures of folded proteins [154]. Another resolution enhancement method is discussed in the Exercises at the end of the chapter.

4.1.5 Scanning tunneling microscopy

Scanning tunneling microscopy (STM) is, in a sense, the ultimate nanoscale microscopy, allowing atomic-resolution imaging of conducting surfaces, spectroscopic capabilities, and manipulation of individual atoms and molecules. The STM technique was demonstrated for the first time in 1982 [155], enabled by the development of stable piezoelectric translation stages for sample positioning and amplifiers for feedback control. The instrument consists of a metal tip held very close to a conducting sample surface. When the tip–sample distance is sufficiently small (1–2 nm), a tip–sample voltage can drive a measurable tunneling current. Suppose that the tip is negatively biased relative to the sample, so that electrons flow out of the tip. Classically, the electrons of the tip are energetically "below" the vacuum level, because they are bound by the work function of the tip. Therefore, at low bias voltages, they can only make it to the sample via quantum mechanical tunneling through the classically forbidden vacuum region between tip and sample. As discussed in Section 6.4.3, the probability for an electron to tunnel decreases exponentially with increasing tunneling distance.

This extreme sensitivity of tunneling current to tip–sample separation is the key to the fantastic resolution of STM. A more careful analysis [156] confirms that the vertical resolution of STM can be smaller than 0.04 nm, while the lateral resolution is roughly $[(0.2 \text{ nm})(R + d)]^{0.5}$, where R is the radius of curvature of the tip, and d is the vacuum gap. A rough approximation (assuming typical tip and sample work functions of around 5 eV) gives a tunneling conductance that decreases by a factor of 7.2 ($= e^2$) for every 0.1 nm increase in d. This steep distance dependence means that often it is not necessary to struggle to produce atomically sharp tips; there is very often some nanoscale protrusion sufficiently sharp to give atomic resolution.

A number of imaging modalities exist in STM. In constant current mode, the z position of the tip (or sample) is controlled via feedback to maintain constant tunneling current as the tip is rastered over the sample. The image is then created by recording the z position as a function of x and y. In addition to the exponential distance dependence, the actual tunneling conductance (dI/dV) of STM is proportional to the local densities of states of both the tip and the sample. This means that constant-current STM images at a particular bias condition are a convolution of the true sample topography and the local electronic properties of the sample, assuming that the tip density of states is "boring" (that is, smooth) near the Fermi level.

This electronic sensitivity leads to another approach, particularly for highly stable STMs in a vacuum environment, called *scanning tunneling spectroscopy* (STS). At each $x - y$ position, the z feedback loop is temporarily turned off, and the tip–sample bias voltage is swept, allowing a measurement of the tunneling $I-V$ characteristic. Using lock-in amplifier techniques, it is possible to acquire dI/dV and d^2I/dV^2 data simultaneously with I as a function of V. As shown in Fig. 4.7, peaks in the sample's local density of states as a function of energy manifest themselves as peaks in dI/dV at particular tip–sample biases. By imaging at different bias voltages, one maps the sample density of states at those energies. This can be directly useful, for example, in identifying particular molecular orbitals of molecules adsorbed on the sample surface [157], or mapping other

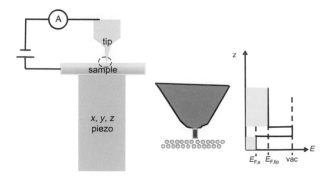

Fig. 4.6 A schematic of scanning tunneling microscopy. A voltage bias is applied between the tip and a conducting sample. At right, a close-up view of the tip–surface interaction. Beause of the exponential dependence of tunneling probability on distance, the tunneling current is localized to an atomic-scale volume between the tip and surface. Tip atoms are not shown. The energy diagram shows that, when biased as shown, electrons tunnel from the Fermi level of the tip, through the vacuum region, into the sample. The diagram does not show corrections due to screening effects in the tip and sample, which would tend to round the vacuum "barrier".

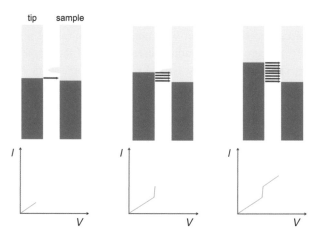

Fig. 4.7 Tunneling conduction in STM. The usual assumption is that the tip's density of states is comparably smooth, while the sample's may have features at particular energies. The tunneling current at a particular energy is proportional to the product of the two densities of states at that energy. As bias is swept, peaks in the sample density of states show up as steps in the tunneling $I-V$ characteristics, and therefore as peaks in dI/dV.

spectral features such as variations in a sample's superconducting energy gap [158]. By performing spectral measurements in a very fine, regular grid of points in real space, it is also sometimes possible to Fourier transform this spatial information about the energy spectra to get information about reciprocal space quantities such as the Fermi surface [159].

The d^2I/dV^2 information that may be recorded in STS measurements can highlight inelastic tunneling processes. As the energy of tunneling electrons exceeds the threshold necessary for the excitation of some local mode of the sample, there is typically a kink in the $I-V$ that shows up as a peak in d^2I/dV^2. Such inelastic electron tunneling spectroscopy

(IETS; see Section 6.7.3) has been used to resolve molecular vibrational modes [160] as well as spin-flip excitations of single magnetic atoms [161].

Spin-polarized STM (SP-STM) is one important example of leveraging the density of states of the tip as well as that of the sample. As discussed in Sections 7.2 and 7.4, a magnetically ordered metal often has significantly different densities of states for electrons of different spin orientations (relative to the magnetic ordering direction). This means that the tunneling conductance between a magnetically ordered tip and a magnetically ordered sample will depend on the relative orientations of the tip and sample magnetizations. The result is atomic-scale magnetic contrast [162].

On sufficiently clean surfaces STM is also useful for the manipulation of individual atoms [163]. One mechanism for manipulation is field-assisted diffusion, in which the electric field between the tip and the sample surface induces a dipole moment in the object to be manipulated, and the electric field gradient exerts a net force on the induced dipole, biasing surface diffusion toward regions of strong field (that is, toward the closest tip–sample separation). Another mechanism makes use of chemical interactions between the tip and the atom or molecule to be manipulated. The tip is lowered close enough to the object of interest that the object is picked up by the tip. Through a combination of voltage and tip position manipulations, the object may be placed back on the surface in a desired location.

Beyond simply placing atoms or adsorbed molecules in particular locations, STM has also been demonstrated as a tool for performing chemistry at the atomic scale [164]. Chemical bonds may be formed spontaneously, or with assistance in the form of a voltage or current pulse. The detailed mechanisms whereby voltage and current can precipitate chemical processes are complex. Electric fields and polar forces can be relevant, as well as vibrations driven by the inelastic tunneling processes mentioned above.

One limitation of STM is speed. Acquisition of atomic-resolution surface maps in constant-current imaging over moderate areas (tens of nm^2) usually requires minutes or longer. More complex imaging modalities such as STS require considerably longer acquisition times, with consequently more stringent requirements on stability and drift. Similarly, these timescales also limit the kinds of structures that may be built, one atom at a time, via STM manipulation.

4.1.6 Near-field scanning optical microscopy (NSOM)

A final scanned probe microscopy that must be included in any discussion of nanoscale characterization is *near-field scanning optical microscopy* (NSOM) [165]. The relevant physics will be discussed in detail in Chapter 8, but the situation is summarized here. Conventional optical microscopy is limited in resolution to features on the order of λ, the wavelength of light being used. However, nanoscale objects can affect the electromagnetic field on scales much smaller than the wavelength of visible light. For example, consider light shining on a nanoscale aperture (much smaller in diameter than λ) in an otherwise opaque material. In the "far field", at distances $\gg \lambda$, very little light gets through the aperture. However, the electromagnetic field very close to the aperture is nonzero. Such a field, which decays on a distance scale generally smaller than λ and does not survive

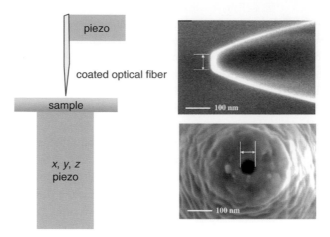

Fig. 4.8 Near-field scanning optical microscopy. Left: a schematic, in which a coated optical fiber is used as the NSOM tip. Right: images of a coated, drawn optical fiber tip, adapted from [166].

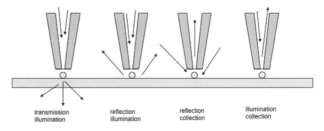

Fig. 4.9 Various modes of NSOM operation, distinguished by illumination and collection strategies.

into the far field, is called "evanescent". Evanescent fields can encode short distance scale ($\ll \lambda$) information about the electromagnetic field, and therefore the optical properties of underlying materials.

NSOM is a means of probing such evanescent fields, to gain access to that latent information. In traditional NSOM, a coated, tapered optical fiber (Fig. 4.8) is prepared for use as a tip. The metal coating on the sidewalls of the fiber is intended to be somewhat opaque, to minimize leakage of light through the sides, and to provide a sub-wavelength aperture at the fiber tip. The fiber is then mounted on a tuning fork, as in tuning fork-style AFM. The force interactions between the fiber tip and the substrate affect the tuning fork resonance, so that AFM-style feedback is used to maintain tip–sample separation. "Crashing" the tip in NSOM is particularly annoying, since fabrication of the coated fibers is delicate and time consuming.

There are multiple operating modes for NSOM, as shown in Fig. 4.9. Initial forays into NSOM [167, 168, 169, 170] were in collection mode, as in standard optical microscopy. Signal strength is often an issue, as coupling of the near field of the sample into propagating modes within the fiber is not always easy. An alternative approach is to dispense with the aperture altogether. "Apertureless" NSOM [171] relies on the excitation of local modes in the metal covering the tip as a means of coupling optical energy either into the fiber, or from the fiber/sample interface into the far field.

In general, NSOM images show sources of near-field scattering. Contrast often comes from differences in the local (optical frequency) dielectric polarizability of the sample. This can be particularly useful in biological settings, when many biological structures are transparent in the visible, yet have differing indices of refraction.

NSOM is also ideally suited for examining photonic devices [166]. Many photonic systems such as photonic crystals, optical waveguides, and plasmonic structures have near-field radiation patterns that provide valuable diagnostics of the desired photonic functionality. NSOM characterization is one way of examining these patterns *in situ*. As with other scanned probe methods, interpretation of the contrast in NSOM images is tricky. Unless great care is taken in calibration and modeling, it is generally difficult to extract absolute quantitative information from NSOM images.

4.2 Materials growth

In general, there are two ways to prepare nanostructured materials: "bottom-up" chemical means, relying on thermodynamics, self-organization, and self-assembly; and "top-down" physical approaches, beginning from bulk or thin-film materials, and creating nanostructures through patterning and material deposition or etching.

There are many ways to grow materials relevant for nanoscale science. Bulk materials growth (e.g., silicon ingots) is important, of course, but much of the critical work in nanoscience and nanotechnology depends on thin film growth and the direct growth of 2d, 1d, and 0d nanostructures. We begin with a brief overview of the deposition and growth of thin films, the subject of a number of books [172, 173, 174].

4.2.1 Physical vapor deposition

One of the earliest forms of nanoscale materials engineering to be developed was the deposition of thin metal films, such as those used to silver mirrors. *Evaporation* is the technique of choice, and Fig. 4.10 shows the usual configuration. A substrate is positioned above a source of material to be deposited, in a chamber typically pumped to high vacuum (10^{-5} mBar or better). The source material is then heated to a sufficiently high temperature that the vapor pressure of the source material is non-negligible. A flux of metal atoms then impinges on the substrate. This is *physical vapor deposition* (PVD).

In *thermal evaporation*, the source material is placed in a "boat" or basket made of some refractory metal (tungsten, molybdenum) that is resistively heated. In *electron beam evaporation*, a high power (several kW) beam of energetic (several keV) electrons is formed by accelerating electrons produced by thermionic emission from a heated filament. The beam is steered through 270 degrees and guided onto source material held in a crucible in a water-cooled copper hearth, depositing its energy.

The optimal evaporation technique depends on the source material of interest. Many elemental metals have relatively high vapor pressures at the moderate temperatures accessible in thermal evaporation, as do a handful of dielectrics (SiO_x, MgF_2). Thermal

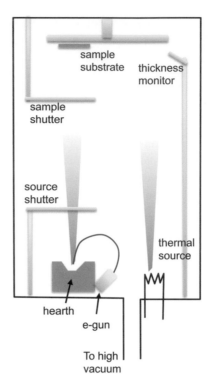

Fig. 4.10 An evaporator for physical vapor deposition. Vacuum is typically provided by a turbopump, diffusion pump, or cryopump. Deposited material thickness is monitored by a quartz crystal thickness monitor. As shown, there is a resistively heated thermal evaporation source and an electron beam source.

sublimation of small molecule organic systems (pentacene) is also possible, though care must be taken to avoid temperatures that would result in decomposition. Other materials, particularly highly refractory metals (Pt, W, Nb) and dielectrics (Al_2O_3, HfO_2), are better suited to e-beam deposition. Note that e-beam evaporated dielectrics often have final compositions that differ in stoichiometry from that of the starting material. For example, evaporated aluminum oxide is usually oxygen-poor relative to the Al_2O_3 source stoichiometry. Evaporated alloys can also differ in composition from PVD source material, depending on the thermodynamic properties of the constituents. The alloy phase diagram determines the composition of vapor produced by either sublimation of a solid source or evaporation of a liquid source.

Materials compatibility is also of crucial importance, since molten source material can often be highly reactive. For example, molten aluminum can react with heated tungsten, forming a brittle combination such that Al/W thermal sources are typically only used once. Similarly, for e-beam evaporation some source materials should be placed in a crucible liner rather than the bare hearth. Common liner materials are graphite, glassy vitreous carbon, alumina, tungsten, and molybdenum. Improper selection of liner material can lead to much excitement. For example, molten nickel reacts exothermically with carbon to form nickel carbide; thus, a graphite crucible liner would be an unwise choice. Fortunately, evaporation guides are freely available from several sources (including vendors of PVD materials and supplies), providing compatibility information and recommendations.

High vacuum in the chamber is necessary for several reasons. Evaporation source temperatures can exceed 3000 K for refractory metals, and chemical reactions between residual gas and sources/boats/crucible liners/filaments can degrade performance and film quality. A low chamber pressure can ensure that the mean free path for the evaporated source material greatly exceeds the source–sample separation, so that the source vapor does not scatter significantly from residual gas molecules. Furthermore, depending on the composition of the substrate and the deposited material, residual gas could be incorporated into the growing film. As will be seen in one of the Exercises, a residual chamber pressure of 10^{-6} mBar implies a flux of gas molecules sufficient to produce approximately one monolayer per second if those molecules adhered to the substrate upon impact. This suggests that to obtain a pure film one should deposit source material at a rate that significantly outstrips the arrival of residual gas atoms.

The deposition process is commonly monitored *in situ* via a thickness monitor, in some systems allowing the feedback control of the deposition rate. The standard thickness monitor is based on a quartz crystal resonator. Quartz crystals are piezoelectric, so that application of an ac voltage can induce a strain response. In a thickness monitor, such a crystal is driven near a mechanical resonance frequency that depends on the crystal geometry. As source material is deposited, the increased mass loading on the surface of the crystal shifts the resonant frequency as $\delta f / f \sim \delta m / m$. The resulting change in oscillation amplitude, combined with knowledge of the source material density and acoustic properties (parametrized by the "z-ratio"), allows the calculation of deposited film thickness. Impressively, sub-Ångstrom precision is readily possible. Mechanical shutters coupled with a thickness monitoring system allow highly reproducible film production.

The morphology of the deposited film depends in detail on the surface interactions between the source material and the substrate, the deposition rate, and the subtrate temperature. Except in very special circumstances, deposited films are either polycrystalline or amorphous. In the former, grain size depends on the number of initial nucleation sites and the ease with which surface atoms can diffuse. If every impinging source atom sticks where it hits the substrate, films with multilayer surface roughness form simply due to the statistical distribution in arrival times of the atoms. In systems with comparatively weak substrate/source interactions and stronger source/source affinity (e.g., Au on SiO_2), the deposited material may adhere poorly, tending to ball up at low thicknesses into a discontinuous group of nanometric islands. Frequently additional layers of material are used to tailor film adhesion and morphology. For example, a few nanometers of Cr or Ti as an underlying adhesion layer dramatically increases the uniformity and adhesion of an overdeposited Au film.

4.2.2 Molecular beam epitaxy

The ultimate limit of evaporation-based PVD is molecular beam epitaxy (MBE), designed to grow (ideally) single crystal thin films and multilayers with atomic precision. An MBE

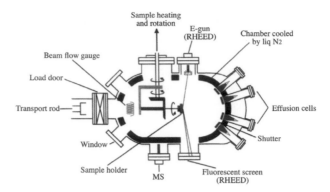

Fig. 4.11   Schematic of a molecular beam epitaxy system for III–V semiconductor growth, from [175].

system is usually a stainless steel growth chamber maintained at UHV conditions by a combination of pumping techniques. Extremely stringent vacuum requirements (better than 10^{-11} mBar) are necessary to achieve the desired degree of purity and structural perfection. Rigorously cleaned substrates for growth are introduced to the chamber via a load lock system, with users avoiding the venting of the main chamber. The most common materials grown by MBE are compound semiconductors, though group IV and metal MBE are also increasingly performed.

Exceedingly pure source material is placed in special deposition cells attached to a source flange. These so-called Knudsen cells are designed to produce a highly uniform beam of source material vapor, from either a thermal or an e-beam source, with an extremely stable and calibrated flux. Cells are controlled precisely, with temperature monitoring and individual shuttering. During the growth process a computer can actuate a sequence of shutter operations, allowing the growth of complicated multilayer structures. In some cases gas sources of growth material are also provided, either in pure form or as precursors that decompose on the growth surface. For example, the nitrogen in III-nitride materials may be supplied from a N_2 plasma source or decomposition of ammonia.

To facilitate the growth of highly crystalline films, substrates for MBE are cleaned in UHV and maintained at a significantly elevated temperature during deposition. Enhanced surface diffusion at high temperatures can allow incident atoms to migrate to their equilibrium lattice positions. Proper growth conditions depend in detail on the materials and their thermodynamic surface properties. Various crystallographic surfaces can have different surface energies (equivalent to surface tensions for liquids), such that diffusion can be enhanced along some directions and suppressed along others. These dependences can result in growth instabilities that lead to roughness and anisotropic morphologies. An example of this is the "orange peel" texture often seen on the (001) surface of MBE-grown GaAs/AlGaAs structures. Growth modalities include "island growth" (the formation of islands which coalesce into each succeeding layer) and layer-by-layer growth at step edges (which can sometimes be encouraged by a vicinal growth surface).

It is critical to monitor the rate and quality of film growth during the process. This is most commonly done via *in situ* electron diffraction techniques such as RHEED and LEED. In systems that exhibit layer-by-layer growth, the diffraction pattern can oscillate in sharpness and intensity with the addition of each subsequent layer. This allows the determination of the growth rate with atomic precision.

The surface energy issues relevant to ordinary evaporation are of critical importance to MBE. To avoid crystallographic defects, the lattice constant of the substrate must be as close as possible to commensurate with that of the growing film. In homoepitaxy, the growth of one material on itself, this is not an issue. In heteroepitaxy, however, this can be quite challenging, due to the differing sizes of various atoms. For example, while Al covalent bonding is electronically satisfied when occupying the Ga site in GaAs, the lattice constant of AlAs is 0.5662 nm while that of GaAs is 0.5653 nm. These differences in lattice constant can be difficult to accomodate, because for some systems there are no convenient substrates.

Lattice mismatches can, however, be useful things, as shown in Fig. 5.17(a). The mismatch between an InAs overlayer and the GaAs substrate is accomodated by the the spontaneous formation of InAs islands rather than a continuous InAs film. The strain (lattice distortion) from each island in turn affects the migration of the surrounding material. The result is a long-ranged ordered array of islands.

4.2.3 Sputtering

Sputtering is another physical vapor deposition technique. Rather than relying on quasi-equilibrium thermodynamic properties (vapor pressure) of source material, however, sputtering is a fundamentally nonequilibrium approach. Near a target of source material, a carrier gas such as argon is struck into a plasma, either by rf or dc electrical discharge. The energetic ions of the plasma then strike the surface of a target of source material. Their kinetic energy is transferred to atoms or groups of atoms of the source, which are liberated from the target. Some fraction of the source flux impinges on the substrate of interest, losing kinetic energy via inelastic interactions with the substrate. Conventional dc sputtering is used for metallic source material, while rf sputtering is more commonly employed in deposition of dielectrics and insulators.

As in evaporative techniques, morphology and adhesion of the deposited material may be influenced by the surface physics at the substrate interface. However, sputtering is a process that is always far from thermodynamic equilibrium, and the carrier gas (pressure, temperature) and plasma parameters (rf intensity and/or discharge voltage; dc bias between source and sample) provide additional degrees of freedom for tuning. The impact of gas molecules from the plasma can aid in the production of flat, dense films, with momentum transfer from the gas to the deposited material acting to assist surface diffusion. As a case in point, the metal multilayers (discussed in Section 7.4) used in magnetoresistive sensors can be produced with atomic thickness resolution and nearly atomic-scale flatness.

Sputtering generally tends toward conformal coating, with deposition taking place relatively isotropically on the substrate. A consequence of this is that sputtering is usually not well suited for lift-off processing, since the deposited material tends to coat the sidewalls of developed resist patterns. Subsequent lift-off is then impeded, with tearing or roughness at the edges of the desired deposited pattern. Adhesion of sputtered films may be improved by a pre-deposition sputter cleaning process, in which the substrate is briefly exposed to the plasma in the absence of the deposition source material. This cleaning process can remove the usual monolayer of physisorbed water and organic contaminants present on ambient surfaces.

Deposition rates in sputtering are comparable to those in evaporation, fractions of a nanometer per second. Because of the nonequilibrium, kinetic nature of the process, deposition rates are comparatively unrelated to the equilibrium thermodynamic properties of the source material. This allows sputtering to deposit alloys without significant changes in their composition. Similarly, sputtered oxides suffer fewer issues with loss of source stoichiometry than in the evaporation case. It is also possible to *cosputter* using multiple sources simultaneously. This is a standard means of preparing some mixed-composition alloy and dielectric films. Indeed, sometimes cosputtering is performed in a geometry designed to produce gradients in final composition across the substrate. This is an approach taken in combinatorial materials science, allowing the parallel production of patches of film with wide-ranging composition.

4.2.4   Pulsed laser deposition

Pulsed laser deposition (PLD) shares some features of sputtering and others of MBE. Like sputtering, PLD is very much a nonequilibrium process. In an ultrahigh vacuum environment, a powerful pulsed laser is fired at a piece of source material, depositing enough energy locally to produce a plume of vaporized material. The plume then impinges on the targer substrate. Because the vaporization process is so far from equilibrium, the plume composition tends to reflect the source composition rather than the volatility of the constituents.

As in MBE, with careful choice of lattice-matched substrate material, and with control of substrate temperature, it is possible to grow single crystal thin films of some materials via PLD. For example, PLD is a preferred method of producing oxide heterojunctions [176] with atomically sharp interfaces. Film growth is slow (sub-nm/sec) and is monitored *in situ* via the same electron diffraction techniques applied to MBE.

4.2.5   Chemical vapor deposition

The deposition techniques described so far rely on the transport of compositionally correct source material to the substrate, where it subsequently adheres. An alternative approach is to introduce precursor compounds and produce the desired final material via *in situ* chemical reactions at the subtrate surface. This is called *chemical vapor deposition* (CVD), and has proven to be extremely versatile and useful. Materials commonly deposited by CVD include compound semiconductors, silicon dioxide, polycrystalline silicon, and carbon. One important variant of CVD is metalorganic vapor phase epitaxy (MOVPE), in which reactions proceed preferentially on step edges, allowing epitaxial layer-by-layer growth.

CVD is commonly performed at significantly elevated substrate temperatures (200–1000 °C, depending on the materials). This is necessary so that the precursor compounds (for example, methane for carbon and silane for silicon) spontaneously decompose upon contact with the substrate. Clearly not every material may be grown by CVD. The process requires the availability of precursor gases with appropriate chemical stabilities and reaction rates. Furthermore, in addition to the desired material to be grown, the decomposition products of the precursors must be gas-phase, so that they may be carried away by a pumping system. Depending on the particular reactants and the desired product, CVD is often performed with additional gases present, to control the reaction rates and minimize unintended side reactions. For example, growth of silicon via decomposition of silane must be performed in a reducing environment (such as excess H_2) to avoid incidental formation of silicon oxides (and, in the extreme limit, explosion hazards due to the pyrophoric nature of SiH_4).

Growth rates in CVD can be high, in the range of tens to hundreds of nm per minute. These high rates have made CVD a preferred technique for the growth of III–V compounds for use in semiconductor lasers. While MBE produces higher purity material with better electronic transport properties, CVD growth has proven better suited to mass production of consumer electronics components. Note that the precursors used in this growth (e.g., trimethylgallium, arsene) can be rather nasty.

Two variations of CVD have become of particularly wide use. In *atomic layer deposition* (ALD), the self-limiting growth of thin film materials takes place via two half-reactions at the substrate surface. The growing surface is exposed to a first reactant that coats via monolayer formation. After a purge by an inert gas such as argon, a second precursor reactant is introduced which completes the formation of a single complete layer of the final product on the substrate. Temporally separating the arrival of the two reactants at the substrate ensures monolayer growth. ALD can put down metals, but its primary use so far has been in the deposition of high quality dielectric oxides for use as critical insulating layers.

A second major variation is *plasma-enhanced CVD* (PECVD). In PECVD reactants at comparatively low pressures along with carrier gas are fed into a reaction chamber and subjected to a discharge to form a partially ionized plasma. Thanks to radio frequency power pumped into the chamber, the electrons are accelerated out of thermal equilibrium with the ions and the neutral gas molecules. These energetic electrons can then interact with precursors/reactants, enhancing the rates of chemical reactions that would otherwise be comparatively slow. PECVD is a common method for growing SiO_2 and Si_3N_4 as well as oxynitride blends. During the growth process, it is possible to apply a dc bias between the plasma source and the platen that holds the deposition substrate. This leads to some control of the anisotropy of the growth process. As in sputtering, momentum transfer between the gas species and the growing surface can help control morphology.

Graphene and nanotubes

Carbon nanomaterials are now commonly grown via CVD. In the case of carbon nanotubes, carbon-containing feedstock, often methane or acetylene in a reducing carrier gas, is flowed at low pressures over transition metal nanoparticles that act as catalysts for growth in either a plasma environment [177, 178] or in ordinary high-temperature CVD [179, 180]. Much research effort has gone into investigating the nucleation and growth processes and their kinetics. In brief, the transition metal nanoparticles catalytically decompose the hydrocarbon feed gas, and at the high growth temperatures (600–900 C, typically) the carbon atoms are dissolved into the nanoparticle metal. When the solubility limit of carbon in the particular catalyst particle is exceeded, carbon begins to precipitate out onto the surface of the particle. The precise growth mechanism remains a topic of debate, though recent evidence [181] supports a picture of growth in which carbon dimers are added sequentially around the base of a growing chiral nanotube, causing the whole tube to rotate during growth. Recall that nanotube type (semiconductor vs. metal) depends critically on tube diameter and chirality, and it would be extremely desirable to be able to grow only a single, specific type of nanotube on demand. Much effort has been expended in trying to engineer catalyst nanoparticles for specific growth, but success has so far been limited.

It is challenging for CVD to achieve kg-scale quantities of nanotubes (which would be necessary for large scale use of nanotubes in structural composites, for example). However, CVD is very well suited to growing long tubes at comparatively high growth rates for surface-based applications such as electronics and sensing. A key to producing long nanotubes is the control of conditions at the catalyst surface, so that the catalytic particles do not "go dead" and terminate the growth process. CVD-grown nanotubes tens of centimeters long have been reported [182].

When catalyst particles are sufficiently close together that the growing nanotubes may interact, one sees "forest growth", as shown in Fig. 4.12. In that case, growth exceeding 100 μm/min was reported for single-walled tubes. Tuning growth parameters and catalyst size and composition can favor double-walled [183] or multiwalled tube growth. Sparse catalytic sites can yield individual, isolated nanotubes that tend to end up lying down on the substrate surface. In some cases interactions between the growing nanotubes and the substrate surface can lead to greatly enhanced alignment of the nanotubes [184, 185].

Alternative (non-CVD) methods of nanotube growth include carbon arc electrical discharge, high power laser ablation of carbon targets, and catalytic growth in a reactor using high pressure carbon monoxide (HiPCO) feed gas. In all cases, transition metal catalyst particles have been found to enhance the yield of nanotubes significantly. In the

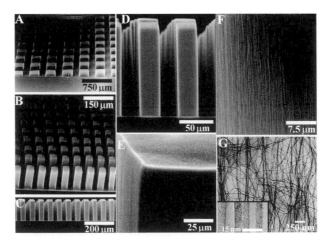

Fig. 4.12   Carbon nanotube "forest" grown by CVD using patterned catalyst nanoparticles. Adapted from [179].

HiPCO process [186], catalyst nanoparticles are formed *in situ* in the growth reactor by the decomposition of $Fe(CO)_5$ or similar compounds.

The same physical chemistry used for nanotube growth, based on catalysis at transition metal surfaces and the limited solubility of carbon in transition metal matrices, has been applied to the growth of graphene. Nickel films [187] and copper foils [188] have both been used as templates for CVD growth of single-layer graphene, which grows in registry with the crystallographic orientation of the underlying metal. Graphene grown this way is therefore multidomain in terms of lattice orientation, with domain size set by the crystallinity of the annealed metal film. Analogous to the nanotube case, growth conditions (temperature, gas pressure) may be tuned to optimize for single- or multi-layer graphene growth. The resulting films may then be transferred from the metal growth substrates onto a working substrate.

An alternative means for growing graphene on a large scale is through the thermal decomposition at very high temperatures (1200–1400 C) of the carbon-rich (0001) face of 6H-SiC [189]. This results in epitaxial growth of graphene, with the number of layers controlled by careful tuning of process details. A reducing or UHV environment is essential.

Semiconductor nanowires

Like nanotubes, it is also possible to grow single-crystal semiconductor nanowires via CVD. The standard growth mechanism often used is *vapor–liquid–solid* growth [190, 191]. A metal catalyst particle (say a gold cluster or droplet produced via laser ablation [191]) is heated in the presence of vapor-phase precursors (e.g., silane). The catalyst particle is chosen so that it solvates the decomposition product of the precursor (in our example, Si). When the concentration exceeds the solubility limit in the catalyst particle, a crystal of the material of interest is nucleated, with a lateral extent comparable to the catalyst size. Growth, ideally of single crystal filaments, continues as reactants are supplied. This approach has also been implemented for compound semiconductors [192], though in that case the reactants may be supplied via laser ablation of solid targets rather than vapor phase precursors.

The catalyst particles do not have to be molten. Temperature-dependent studies of compound semiconductor nanowire growth [193, 194] have shown that growth may proceed from solid catalyst particles. As in nanotube growth, it is also possible to pattern arrays of catalyst particles and grow nanowires in predetermined locations [195, 196].

By varying the flow of reactants as a function of time, researchers have demonstrated the growth of superlattice structures [197, 199] with atomically sharp heterointerfaces. Similarly, by performing sequential growth processes, it has been possible to grow core-shell structures with different semiconductor compositions [198]. The particular attraction of the core-shell geometry is the possibility of using the different band structures of the shell and core to perform [200] the one-dimensional version of modulation doping (Section 5.5.4).

These nanostructures are of tremendous fundamental and applied interest. However, they are not without challenges. Because of the growth method, the purity of the final nanowire

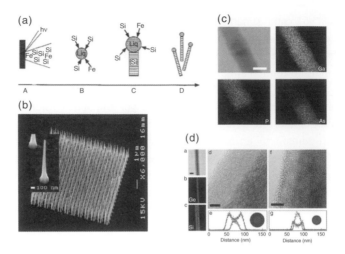

Fig. 4.13 Semiconductor nanowires. (a) VLS growth of Si nanowires, in this case assisted by laser ablation of growth material, adapted from [191]. (b) Arrays of CVD-grown InP nanowires using patterned Au catalyst particles, from [196]. (c) A nanowire containing a GaAs/GaP heterojunction, as shown by EDS obtained in a TEM examination, adapted from [197]. (d) Germanium core, silicon shell nanowire, adapted from [198].

material can be of some concern. For example, gold is known to act as a deep trap in Si, so incorporation of catalyst material into the final nanowire must be controlled carefully. Likewise, care must be taken with growth conditions, growth chamber cleanliness, and source material purity to minimize unintentional doping due to impurities.

4.2.6 Nanocrystals

Many nanocrystals are most commonly grown via wet chemistry techniques [201, 202, 203]. Chapter 3 of [204] is an excellent reference on the physical chemistry at work in the growth process. The basic principle at work is control of the relative rates of nucleation and growth. For semiconductor nanocrystals, a heated, stirring solvent and surfactant bath is prepared, as shown in Fig. 4.14. Organometallic precursors of the desired final material are injected into the bath. The now-heated precursors decompose via pyrolysis, rapidly increasing the concentration of growth material in the solvent up to supersaturated conditions. The rate of homogeneous nucleation of "seeds" for nanoparticle growth depends strongly on the temperature, material parameters, and most importantly, concentration of the solute material (see Exercises). Under proper conditions, many seeds nucleate extremely rapidly (essentially simultaneously), which then lowers the local concentration of solute sufficiently that the nucleation rate drops precipitously. The existing seeds then grow until the reaction conditions are changed or until the solute concentration reaches the equilibrium solubility. With good mixing, so that conditions are homogeneous throughout the reaction chamber, this colloidal approach can give extremely monodisperse size distributions of particles (variations of a few %).

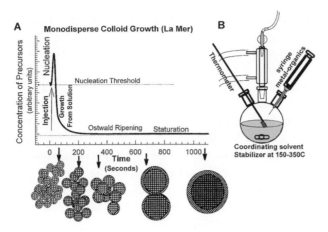

Solution-phase "molecular beaker epitaxy" approach to monodisperse chemical growth of nanoparticles, adapted from [203].

Fig. 4.14

Further narrowing of the size distribution may be achieved via *Ostwald ripening*. The same surface vs. bulk free energy considerations come into play as in the nucleation process. Over time, depending on conditions, smaller nanoparticles tend to dissolve at the expense of furthered growth of the larger particles. This tends to narrow the size distribution.

Metal nanocrystals may also be grown in solution. Rather than pyrolysis, the solid material comes from reduction of dissolved metal salts. As in the semiconductor nanocrystal case, however, the surfactant present in the growth medium plays multiple critical roles. The surfactant keeps the nanoparticles solvated and separated so that they do not flocculate and fall out of solution. With appropriate selection of hydrophilic or hydrophobic surfactant, it is possible to solubilize nanoparticles for many diverse environments. The surfactant also stabilizes the nanoparticle surfaces. Finally, the coordinating surfactant coating can be tuned and altered to promote or hinder growth of particular crystallographic surfaces [205].

4.3 Material removal

Beyond growing or depositing substances, the controlled removal of material with nanometer precision is an essential capability for nanofabrication. Material removal approaches are either selective or nonselective in their abilities.

A standard nonselective method is polishing, the use of abrasive means to rub away undesired material. In standard polishing, a sample is gently, uniformly pressed against a rotating polishing wheel, and often a slurry is used to act as the polishing medium. For coarse polishing, the slurry would contain abrasive particles with a size comparable to the desired final roughness. If true nanometer or Ångstrom-scale flatness is required, the slurry contains chemical agents designed to weaken the surface bonds of the sample. This approach is called *chemical mechanical polishing* or *planarization* (CMP), and is widely used in the semiconductor industry. The slurry encourages the polishing process and carries away debris.

For small-area samples, *ion milling* is another (relatively) nonselective material removal method. Ion milling uses a collimated beam of inert (typically Ar) ions to sputter away material. Ion milling etch rates depend on the energy of the ions (often a few hundred eV), the density of the ion beam (typically formed from a plasma struck at a gas pressure of $\sim 10^{-4}$ mBar), and the angle made by the ion beam relative to the sample surface. Etch rates can be on the order of nanometers per second. Ion milling rates do have some dependence on the sample material. In the case of single-crystal substrates, "channeling" can take place, in which ions are favored to penetrate the material further along certain crystallographic orientations. Sputtered material may also be redeposited elsewhere on the sample surface, depending on geometry. Moreover, the ion milling process can produce damage (vacancies, interstitials, broken bonds) in the nominally unetched sample surface left behind. To avoid charge buildup of insulating samples, ion mills often have an additional filament to act as a "neutralizing current" source of electrons.

A much more precise and spatially localized relative of ion milling is etching via *focused ion beam* (FIB). FIB systems typically use field ionization of molten gallium wetting a tungsten emission tip to produce a flux of Ga^+ ions. These ions are accelerated to tens of keV energies and focused electrostatically into a nm-scale spot that may be rastered over a surface via computer control. FIB is commonly used for sectioning of materials for TEM examination. FIB is often integrated with an electron beam column and detector, for combined SEM imaging and FIB processing. To avoid surface contamination with Ga ions, sometimes helium ions are used instead, particularly for imaging in a method analogous to SEM called ion microscopy.

4.3.1 Wet etching

Wet etches are also common in the fabrication of nanoscale structures, and they offer chemical selectivity based on the reactants chosen and the sample material. This chemical selectivity is often used to limit the etching process, with some materials acting as "etch stops". Like any solution-based chemistry, the reaction rates depend strongly on both etchant concentration and reaction conditions such as temperature. Reproducible results require care and control of both.

Unsurprisingly, some of the most well studied etching recipes come from the semiconductor industry and involve materials like Si, SiO_2, and GaAs and related compound semiconductors. Common etches include hydrofluoric acid or buffered HF (often called buffered oxide etch, or BOE) for SiO_2; concentrated KOH for the anisotropic etching of Si; and $H_2O_2/NH_4OH/H_3PO_4$ etching for GaAs.

The KOH etch of Si is a great example of how chemical reactivity depends on crystallographic surface. The Si $\langle 100 \rangle$ plane is etched approximately 400 times more rapidly than the $\langle 111 \rangle$ plane. The result is anisotropic etching that can self-terminate (see Fig. 4.15(a)), allowing the ready fabrication of grooves or pyramidal pits. In the absence of these reactive anisotropies, wet etching tends to be comparatively isotropic.

(a) (b)

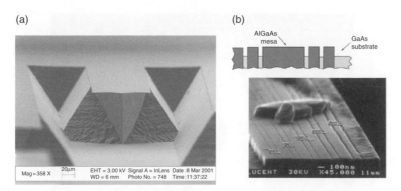

Anisotropic etching. (a) Si after etching with KOH, showing exposed ⟨111⟩ faces, from [206]. (b) AlGaAs/GaAs
multilayer, cleaved, with GaAs selectively etched, from [207].

Fig. 4.15

Figure 4.15(b) shows the result of selective etching of cleaved GaAs/Al$_{0.3}$Ga$_{0.7}$As
multilayers with the phosphoric acid solution mentioned above. The aluminum content
in the AlGaAs layers is sufficient to act as an etch stop via the formation of aluminum
oxide.

The final part of any wet etching process is the removal of the etchant and drying of the
surface. Often this is accomplished by simple rinsing in deionized water and blowing dry
with N$_2$ gas. However, for cases when the etching is meant to produce micro- or nanoscale
freestanding structures, this can be problematic. Surface tension forces due to the liquid–
vapor interface can be strong enough to cause structural problems. As will be discussed
further in Chapter 9, the solution is supercritical drying of the sample surface.

4.3.2 Dry etching

Just as CVD and PECVD use surface chemical reactions to decompose precursors and
deposit material, one can instead take advantage of surface reactions to decompose sample
material into gas-phase products and thus perform etching. Plasma-based techniques
include *reactive ion etching* (RIE) and a variant, inductively coupled plasma (ICP) RIE. In
standard RIE the sample sits on a platen that is the bottom member of a pair of electrodes
used to strike the plasma via a few hundred Watts of radio frequency power. Applying a dc
bias across the two plates can provide additional directionality for the incident ions, leading
to a comparatively anisotropic etch process. As shown in Fig. 4.16, etch anisotropies (which
determine the slope of trench sidewalls) of more than 40:1 are readily achievable.

Gas-phase etchants are chosen depending on the material system to be attacked, and are
often combined with inert gases (Ar) to tune relative concentrations. Examples include
CF$_4$ as an etchant for SiO$_2$, SF$_6$ as an etchant for Si, and Cl$_2$ and BCl$_3$ as etchants for
GaAs and related compound semiconductors. Chlorine-based chemistries are also used
to etch metals. Etching is often done with the sample at elevated temperatures, and
one must exercise care when mixing etch chemistries, as not all of the etch products
are chemically compatible with one another. For example, hydrogen gas is a great

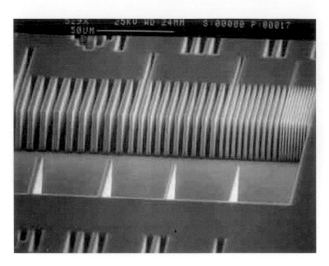

Fig. 4.16   Deep reactive ion etching of silicon. Image from [208].

reducing agent when present as a component of the plasma, but in the presence of fluorine-based compounds it can lead to the formation of fluoropolymers that can coat chamber walls. Likewise, aluminum fluoride can be an annoyance if fluorine chemistries are mixed with III–V semiconductors. An extremely common and rather isotropic etchant is the plasma formed from O_2, which is outstanding for stripping organic compounds, including resist residues, fluoropolymers, and carbon-based nanomaterials like nanotubes or graphene.

The typical etch rate for standard RIE and ICP processes is on the order of ten nanometers per minute (though oxygen plasma will consume resists faster than that). This relatively slow speed and the geometric challenges of getting reactants in and products out efficiently tend to limit the accessible aspect ratios available for etched structures. It is extremely challenging to use standard RIE to etch trenches with vertical sidewalls and aspect ratios exceeding 20:1.

"Deep RIE", also known as the Bosch Process, was developed to overcome these limitations. The Bosch Process, so named because it was originally developed by Bosch, is a multistep, cyclic etch recipe developed especially for Si. An initial (rather isotropic) Si etch such as SF_6 removes some Si material, followed by the use of a hydrocarbon to coat the sidewalls and floor of the etched region with a fluoropolymer that acts as a protective layer. The fluoropolymer is then etched away anisotropically from the bottom of the etched region (leaving the sidewalls coated), and the process is repeated. Etch speeds of several μm/min are possible, aspect ratios exceeding 40:1 have been demonstrated, and it is possible to use the Bosch process to etch vias[3] entirely through 500 μm-thick silicon wafers. Deep RIE has become a particular staple of micromachining and microfluidic device fabrication.

---

[3] A "via" is a vertical pathway filled with metal used to link different layers of devices in the commercial semiconductor fabrication process.

4.4 Patterning

For many applications it is desirable to pattern features with high resolution and fidelity across some two-dimensional surface. The actual patterning is a prelude to either material deposition or material removal. Traditionally, since most of the well developed nanoscale patterning techniques have evolved from the microelectronics industry, the substrate of interest is usually flat and smooth. Common patterning processes, whereby the finest user-designed lateral features are defined, are mostly known as *lithographic* procedures. Lithography takes its name from the Greek words for "stone" and "writing".

4.4.1 Photolithography

The vast majority of microelectronics device fabrication is accomplished through *photolithography*. The basic process is shown in Fig. 4.17. A substrate of interest is coated with a photoactive polymer *photoresist* via spin-coating. "Positive-tone" photoresists, when controllably exposed to ultraviolet light, typically produce photoacids which then attack the polymer backbone. The exposed substrate is then washed in a developer that etches or rinses away the exposed regions. The result is a polymer stencil on the substrate, where the holes in the stencil are defined by the regions exposed to the UV light. Conversely, "negative-tone" photoresists use different photochemistry so that light exposure cross-links

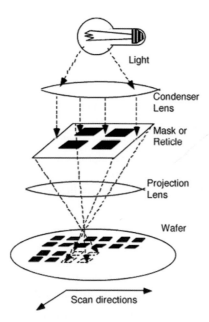

Schematic of a projection photolithography system (courtesy Rice Connexions project). Light is passed through a reticle or mask of patterned opacity, and focused onto a target substrate coated with a layer of photochemically active polymer resist.

Fig. 4.17

or otherwise modifies the local polymerization, making the exposed material more resistant to a developer than unexposed resist. In either case, a crucial step is the creation and use of a *photomask*, to pattern UV exposure.

Photolithography is a multibillion dollar industry. In large scale industrial fabrication, *projection lithography* is the standard, in which reducing optical elements (lenses) are used to project a demagnified (usually by a factor of ~ 5) image of a larger-scale mask (a "reticle") onto a region of the substrate (a "die" or a "field"). A "stepper" shifts the projection relative to the substrate, so that the exposure process may be repeated many times across a substrate. Clearly, for multistep processes, registration and alignment of subsequent exposure operations relative to previous steps is of critical importance.

The ultimate resolution limit for projection photolithography is often set by the optical system and the wavelength of UV light used for illumination, and the properties of the photoresist itself. In general, for standard far-field projection lithography, it is extremely difficult to fabricate features smaller than $\lambda/4$, where λ is the UV wavelength, often 193 nm. As discussed further in Section 6.6.1, a number of clever advances and tricks have enabled optical lithography to function down to a few tens of nanometers resolution.

Contact lithography is much more common at the university level. In this approach a mask with features patterned at 1:1 scale relative to the desired finished exposure is aligned and placed into contact with the substrate via a "mask aligner" tool. Standard masks are quartz coated with Cr films patterned appropriately to be transparent and opaque. These masks are either produced in-house using a mask fabrication tool, or are ordered from commercial vendors. Contact lithography has a common resolution limit of around 1 μm because of the difficulties of getting resist films to be sufficiently flat for truly intimate contact with the mask over large areas.

A major advantage of projection and contact lithography is that they are *parallel* techniques, so that comparatively large areas (square centimeters) of substrates may be exposed at a time. In contrast, *direct-write photolithography* is a common approach used to produce custom contact lithography masks. A resist-coated substrate is placed on a calibrated translation stage, and a controlled UV source (such as a semiconductor laser) is activated in conjunction with computer-controlled stage movement to draw an exposed pattern directly on the resist. Unlike the other photolithography methods described here, direct-write is a *serial* approach, such that individual critical features are drawn one at a time.

4.4.2 Electron-beam lithography

There are other alternatives to UV-based photochemistry when it comes to patterning polymer resist layers. A common approach uses an electron beam to damage a polymer resist (positive tone) or cross-link a polymer resist (negative tone). As in photolithography, after electron beam exposure the resist-coated substrate is rinsed for a prescribed time in a developer that dissolves away the exposed (unexposed, in the negative tone case) resist. Poly(methyl methacrylate) (PMMA) is a common positive-tone e-beam resist, with methylisobutylketone (MIBK) as a common developer.

A schematic of an e-beam lithography (EBL) system is shown in Fig. 4.18. An electron beam is produced using either thermionic emission from a filament, cold cathode field

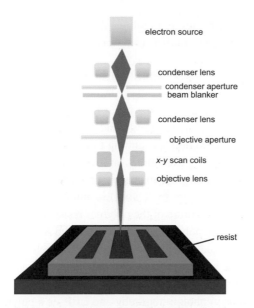

electron source

condenser lens

condenser aperture
beam blanker

condenser lens

objective aperture

x-y scan coils

objective lens

resist

Fig. 4.18

Schematic of an electron beam lithography system, which may be a dedicated machine or a converted electron microscope. A computer controls the $x-y$ scan coils to raster the electron beam over the sample surface in a prescribed pattern. A computer-controlled electrostatic beam blanker is used to deflect the beam into a dump to avoid sample exposure during repositioning of the beam.

emission, or thermally assisted field emission. The electrons are collimated and accelerated up to a desired beam energy, typically in the tens of keV range or higher in commercial instruments. The beam spot size in such a system, after a series of magnetic lenses, can be on the order of one nanometer in diameter. In standard e-beam systems, the beam is then rastered over the substrate surface, drawing the desired features.

The typical resolution possible with EBL is ∼20 nm, with a few nm being possible under rather heroic efforts. One limiting factor is the spatial spread of secondary electrons, the same secondaries that make electron microscopy possible. This leads to the "proximity effect", in which features written close to one another (on the tens to hundreds of nanometers scale) affect the optimum local exposure dosage (in units of charge per area), because of stray secondaries. Other limitations are the development process and the interaction of the developer with the exposed resist material. In academic environments, EBL is often done via converted scanning electron microscopes (SEMs) with writing fields on the order of square millimeters. Commercial EBL systems generally use sophisticated sample translation stages to automate lithography over larger (wafer-scale) areas, allowing individual small writing fields to be "stitched" together with nanometer registry.

As with direct-write photolithography, conventional EBL is a serial technique, and thus *slow*, and therefore impractical for full wafer-scale commerial patterning. There has been significant work on developing projection EBL tools [209], but advances in photolithography have made it very difficult for these systems to be economically competitive.

A further limitation of EBL is the need for the substrate to be somewhat conductive, so that charging does not affect patterning. While this is not really an issue for many electronics applications, photonics and biological studies increasingly require fabrication of nanoscale structures on glass or other insulators.

4.4.3 Imprint lithography

Imprint lithography is an alternative to the usual lithographic patterning strategy of local chemical damage. A hard template master is made with surface relief of the fine features desired, perhaps through a more time consuming process like direct-write EBL. Ridges on the master will correspond with the locations for resist removal on the substrate. In imprint lithography's original incarnation, the master is brought into contact with the polymer-coated substrate. Moderate pressure ($\sim$ 1 atm.) is applied while the substrate is warmed above the glass transition temperature of the polymer resist. The resist then flows, allowing conformal contact between the polymer and the master, and squeezing out most of the resist from the substrate regions contacted by the master surface relief. After the resist hardens, the master is removed. The imprint process does not entirely remove the polymer from the desired areas; this is accomplished in a subsequent anisotropic etching step via reactive ion etching. The end result, as in photolithography and EBL, is a polymer stencil suitable for material deposition or patterned etching of the underlying substrate. There are a number of variations of imprint lithography, including the use of a photocurable polymer as the imprint layer.

Demonstrated resolution of imprint lithography is very good [210, 211], down to the 10 nm scale [212]. One advantage of imprint, pointed out in the original papers, is that it dispenses with the critical exposure and development steps of conventional photo or EBL. Thus there are no concerns about diffraction limits, exposure proximity effects, or chemical diffusion.

There are other challenges. Adhesion of the polymers to the hard master must be avoided, often involving chemical modification of the master surface [213] with some nonstick functionalization. In multistep fabrication that is standard for microelectronics, alignment and registry of the imprinting with previously patterned features are critical. Achieving the appropriate positioning precision, and then maintaining the correct geometry throughout the imprint process, is not a simple task. At the nanometer level, even comparatively minor effects (differential thermal expansion of the master and the substrate; flexure under loading) must be considered [214].

4.4.4 "Soft" lithography

There has been tremendous progress in recent years using nontradiational patterning methods, particularly for the creation of flexible structures, patterned chemical functionalization, and working on non-planar surfaces. The most widely practiced methods involve the use of silicone rubber as a molding or stamping material; hence the name *soft lithography* [215, 216]. The most common material of interest is polydimethylsiloxane (PDMS), a silicone polymer related to that used in soft contact lenses. PDMS is available commercially at low cost as a two component system (monomer and polymerization agent), cures at modest temperatures, and is optically transparent in the visible. A flat PDMS

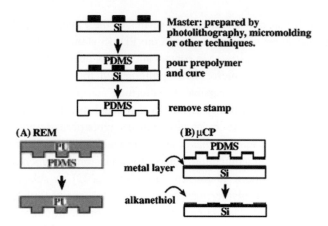

The basics of soft lithography, showing replica molding (A) and microcontact printing (B), adapted from [218]. Fig. 4.19

surface can form a good mechanical seal against another smooth, flat surface such as glass or silicon.

The initial step of soft lithography is the creation of some master surface relief using some combination of lithography, etching, and/or deposition. Unpolymerized PDMS is then poured onto the master and allowed to cure, forming a silicon rubber stamp with surface relief inverted relative to the master. The cured PDMS is removed from the master, and several avenues are then available to pursue.

One approach is to use the PDMS as an actual rubber stamp, to transfer material from raised areas of the PDMS onto a target surface. For example, the PDMS surface can be functionalized with a self-assembled monolayer of some molecule of interest, such as an alkane chain that is terminated to bind strongly to a desired metal surface. The stamp is then applied to a wafer coated with the metal film in question and peeled away, leaving behind patterned patches of monolayer. The unprotected metal may then be etched via wet etching, leaving behind the protected pattern. This process is called *microcontact printing* [215]. Rather than printing a molecular monolayer, it is also possible to print colloidal metal nanoparticles [217], which may then be used as seeds for electroless deposition of metal films in the printed pattern.

Similarly, one may use relative surface energies cleverly to enable to wholesale *transfer printing* of metal films directly [219, 220]. A PDMS stamp may be coated with a thin film of the metal of interest via physical vapor deposition. Because no covalent bonds are formed between the metal and the PDMS, the physical interaction between the two materials is comparatively weak. If such a stamp is brought into contact with a surface for which the metal has a greater affinity (that is, a lower surface energy), then when the stamp is peeled off, the metal will be left behind.

An even fancier, more general approach [221] of transfer printing takes advantage of the kinetics of the adhesion process. On a "donor substrate", desired nanoscale objects are created or patterned via some desired technique, preferably without extremely strong adhesion between the objects and the substrate. Then a PDMS stamp us placed in contact with the donor substrate and peeled away rapidly. The adhesion of the PDMS to the objects is rate-dependent because of the elastic properties of the PDMS. Rapid peeling can pull the objects off the donor substrate and onto the PDMS stamp, so that they may then be transferred to a receiving substrate via contact and a slow peeling removal of PDMS.

A major selling point of microcontact printing and transfer printing is that, unlike traditional lithography, both are capable of patterning non-planar target substrates. However, the very mechanical flexibility of PDMS that enables these printing techniques can be problematic. Placing PDMS in precise registry with prepatterned features without distortion is not easy. PDMS also has a large thermal expansion coefficient (3×10^{-4} per °C), necessitating good temperature control for precision placement.

PDMS itself may also be used for replica molding [222]. A substrate coated with a liquid-phase photocurable polymer may be brought into contact with a premade PDMS stamp having surface relief. The stamp acts as a mold for the liquid, with raised areas of the PDMS coming into contact with the hard substrate surface. The photocuring process may then be performed and the PDMS removed, leaving behind a replica of the original template for the PDMS stamp [223].

4.4.5 Scanned probe lithography

The scanned probe characterization methods in Sections 4.1.4, 4.1.5, and 4.1.6 are all potentially adaptable to patterning techniques. One major limitation of scanned probe approaches to lithography is the generally slow scan speed inherent in the need to perform careful feedback of tip positions. With the exception of massively parallel cantilever-based approaches [224], scanned probe methods are also serial, such that patterning large areas with nanoscale structures is usually impractical. However, for research purposes, scanned probe methods can be extremely versatile and useful.

Dip-pen nanolithography (DPN) takes advantage of the ever-present meniscus of physically adsorbed water that exists at the tip–sample junction in AFM when performed in a humid environment [225]. When the AFM is in contact mode, the layer of liquid water allows the transfer of molecular materials, particularly those with a chemical affinity for the target surface, to be deposited in scan-directed patterns with a resolution on the order of the radius of curvature of the AFM tip ($\sim$ 10 nm). By having multiple reservoirs of various molecules available for pick-up by the tip, DPN can be very effective for patterning chemical functionality on the nanoscale.

Lithography via AFM-based "plowing" of resist has also been demonstrated [226]. A sharp tip may be used to scratch lines in polymeric resist, allowing access to the underlying substrate for either etching or deposition. Liftoff processing with this approach is difficult, however, since the resist profile of the scribed region edges tends to favor deposition of material on sidewalls. Multilayer resists [227] can mitigate these issues.

Rather than using the probe tip as a passive conveyance, as in DPN, or as a mechanical scribe, with conducting tips the AFM may be used as a controlled, local environment for electrochemistry. Local oxidation of metal films via AFM [228] has been used to create nanoscale devices including tunnel junctions and single-electron transistors. Local oxidation has also been demonstrated as a means of producing semiconductor nanostructures from MBE-grown substrates [229]. Single-walled carbon nanotube AFM

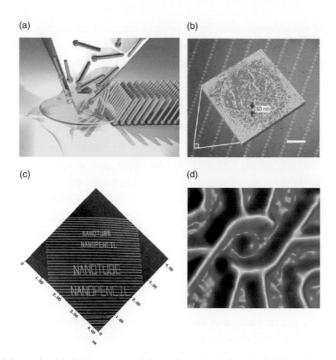

Examples of AFM lithography. (a), (b) Dip-pen nanolithography, adapted from [231]. (c) Local oxidation using a carbon nanotube AFM tip, adapted from [230]. (d) Local oxidation used to define a double quantum dot, gates, and point contacts starting from a 2d electron gas GaAs/AlGaAs structure, adapted from [232].   **Fig. 4.20**

tips [230] can perform controlled oxidation with sub-10 nm lateral definition on Si surfaces.

Conducting AFM tips have also been used as localized sources of (not very) energetic electrons for lithography using e-beam resist (PMMA) on conducting substrates [233]. With a voltage bias $\sim -10 - 20$ V applied to the tip relative to the underlying substrate, a field-emission-like process took place, with fields enhanced by the sharp tip. At tip–sample contact forces low enough (~ 10 nN) to avoid mechanical damage, an electron dose comparable to that used in e-beam lithography (~ 100 μC/cm^2) succeeded in chemically altering the PMMA sufficiently to allow for development.

NSOM-based photolithography [234] has been demonstrated, using an optical fiber optimized for UV transmission. More recently, plasmon interactions between the coated NSOM fiber and the underlying conducting substrate have enabled NSOM-based photolithography with resolution in the tens of nanometer range, comparable to e-beam lithography [235].

STM-based lithography is also possible, since STM provides a means of using electrons to perform chemical modifications of the (conducting) sample surface at the atomic scale. For example, via controlled current and voltage applications, STM may be used to strip hydrogen from the surface of hydrogen-terminated Si (in vacuum) in controlled patterns [236, 237]. This exposes the underlying Si for material deposition or chemical modification.

Self-assembled monolayers have been popular choices as resists for scanned probe lithography [238, 239], including STM methods. Voltage pulses can displace particular SAM molecules, allowing patterned replacement of selected molecules via chemisorbtion [240].

4.4.6 Templated methods

Beyond lithography, there are a number of means of fabricating nanostructures using templates for either deposition or etching. In some cases, templating's advantage is that it enables the large-scale fabrication of many nanostructures without the need for sophisticated lithographic infrastructure or effort. In others, templating of individual nanostructures is performed to achieve nanoscale dimensions in configurations that would otherwise be extremely challenging with standard e-beam or photolithography.

Nanosphere lithography [241] is an example of large-area templating. Silica or latex colloidal spheres are allowed to form a close-packed layer on a substrate of interest, which is then placed in a physical vapor deposition system. After deposition of a metal film, the spheres are stripped away chemically, leaving behind a pattern of cuspy corner-sharing triangles defined by the projection of the interstitial spaces between the spheres. This has proven a useful approach for creating periodic metal structures over large areas for use in optical applications.

Novel nanomaterial templates may be used as substrates for deposition as well. Suspended carbon nanotubes [243, 244] and DNA molecules [245] have been employed to fabricate metal nanowires by acting as scaffolds for metal deposition.

Rather than using a template to mask deposition, one may also use nanostructured templates to mask etching processes. For example, both nanotubes [246] and oxide nanorods [247] have been dispersed on top of thin metal films, and the substrates were subsequently exposed to ion milling. Unprotected regions of the film are sputtered away, leaving behind metal nanowires defined by the elongated nanotube or nanorod masks.

Substrates grown by molecular beam epitaxy have particular interest as templates, since layer-by-layer atomic precision is possible in their growth. By working on the cleaved edges of selectively etched MBE-grown materials, metal nanowires with sub-10 nm widths and very large aspect ratios have been made [248, 249]. Such templates have also been used to transfer arrays of metal nanowires onto separate target substrates [250].

Pores in membranes have also proven extremely useful as templates for the growth of nanowires and nanotubes of various kinds. The nanomaterials may then be used *in situ* or released from the template via wet etching. Etched particle tracks [251] may be filled via electrochemical deposition to produce long, very narrow nanowires [252]. Anodic aluminum oxide [253] may be produced with a hexagonal close-packed array of nanoscale diameter pores. This porous alumina film may then be used as a template for the growth of many different materials, including metals, polymers, carbon nanotubes, and other oxides. In the case of electrochemical growth, single-crystal metal nanowires can be achieved, and cyclic changes in the electrochemical solution may be used to grow multilayer metal wires.

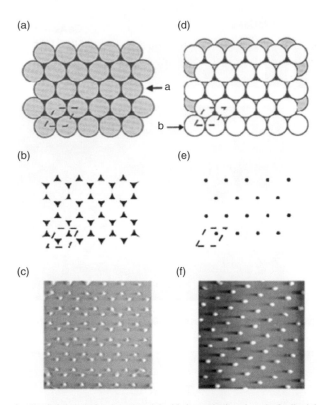

Fig. 4.21

Nanosphere lithography. Single layers (left column) and double layers (right column) of colloidal spheres may be used as templates for metal deposition and then stripped, yielding nanostructured films (bottom row). Adapted from [242].

Templating may also be used for the creation of three-dimensional structures. One particular example of this is the fabrication of 3d period (often dielectric) structures ("photonic crystals" or photonic bandgap materials; see Chapter 8) using colloidal crystal templates [254]. A colloidal crystal is formed from polystyrene or silica spheres. Electroless deposition or some other chemical technique is used to infiltrate the desired final material into the interstitial space between the spheres. The spheres are then removed via etching or calcination of the sphere material, leaving behind a system as shown in Fig. 4.23. Because the iridescent properties of the mineral opal originate from a microstructure that is an array of silica spheres, the templated interstitial lattices are often called "inverse opals". These structures highlight a challenge of 3d templating, the need to have a sufficiently connected or accessible structure that the template material may be removed.

4.4.7 Self-assembly

A "Holy Grail" of nanofabrication is the idea of *self-assembly*, in which desired patterns are formed spontaneously by designed constituents, generally as a consequence of thermodynamically favored processes over multiple energy and lengthscales. "Natural" examples of self-assembly include the growth of crystals and the organization of biological

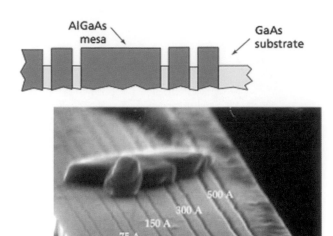

Fig. 4.22   Selectively etched, cleaved, MBE-grown substrates as nanoscale templates, adapted from [207].

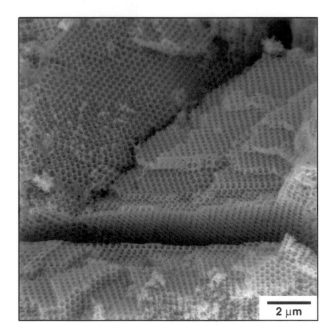

Fig. 4.23   Inverse opal formed by infiltration of a colloidal crystal template, adapted from [254].

molecules such as amino acids into larger structures such as proteins and protein fibers. Self-assembly can be driven by (comparatively) static interactions between the components themselves and between the components and their environment. For example, suspended colloidal particles in water can take on net electrical charges and spontaneously crystallize.

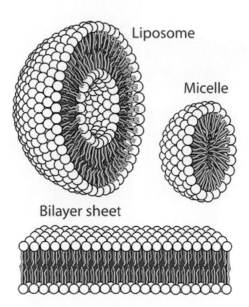

Fig. 4.24

Self-assembly of lipids. When placed in water, lipids spontaneously arrange themselves into structures based on inter-lipid couplings and the interaction between hydrophilic head groups (white spheres), hydrophobic tails, and the surrounding water. Image from adapted from Wikimedia Commons, public domain (Mariana Ruiz Villareal).

Alternately, self-organization may take place due to dynamic effects, as in the arrangement of sand dunes into ordered patterns due to wind and vortex shedding. The static case is of the most common relevance to nanoscale science. A number of reviews of self-assembly have been written [255, 256, 257, 258, 259], and here we touch only briefly on a couple of examples and key ideas.

Lipid bilayers and self-assembled monolayers (SAMs) are the most famous examples of self-assembly. In the former example, directly relevant to many biological processes, the components are small lipid molecules that are *amphiphilic*, meaning that each molecule possesses a hydrophilic (polar) head group and a hydrophobic tail group. When placed in water at an appropriate concentration, configurations that minimize the overall free energy are those that arrange the lipids with their hydrophilic groups exposed to the water, and the hydrophobic groups away from water and in proximity to one another. While there is some entropic benefit in having the lipids dispersed throughout the water, there is an energy gain in "satisfying" the hydrophobic and hydrophilic interactions. The result, as shown in Fig. 4.24, is the spontaneous formation of sheets of lipid bilayers. The overall energy may be lowered further by eliminating the edges of such bilayers, since the molecules at the edges have unsatisfied interactions. The edges may be eliminated by the formation of vesicles, as shown.

In the case of SAMs, small molecules with chemical termination (on at least one end) that has a chemical affinity for a target surface can spontaneously assemble on surfaces [258]. Depending on the relative strength of the intramolecule interactions and the molecule/surface interactions, molecules can diffuse around on the surface to form ordered regions, in which various techniques (scanned probe; small angle x-ray scattering; etc.) can

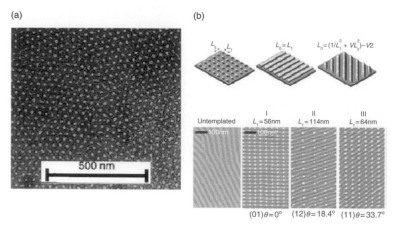

Fig. 4.25 Diblock copolymer self-assembly. (a) Ordered domains of polybutadiene inclusions in a polystyrene matrix, adapted from [260]. (b) Top-down patterning of surface features can template long-ranged order in self-assembled diblock copolymer films, adapted from [261].

demonstrate crystalline molecular domains. As in the lipid bilayer case, the basic process is driven by thermodynamics. The total free energy contains energy terms proportional to the molecule/surface and molecule/molecule interactions. At the same time, there is an entropic term that tends to favor positional and orientational disorder (whether, where, and how the molecules are bound on the surface). The degree of order is set by a competition between these tendencies.

It is possible to engineer intramolecular interactions by design, to manipulate self-assembly. Diblock and triblock copolymers, for example, are formed by the polymerization of "blocks" of chemically distinct monomers (such as styrene and butadiene). As in the amphiphilic lipid case, the different blocks have distinct interactions with their surroundings, leading to the spontaneous formation of complex arrangements. The relative lengths of the blocks may be controlled in the synthesis process, allowing some control over the kinds of patterns that may self-assemble [262, 263]. Figure 4.25(a) shows an example of such pattern formation in a thin film of diblock copolymer, with the spontaneous formation of an array of cylindrical polybutadiene inclusions in a polystyrene matrix [260]. Macroscale ordering of such arrays is difficult, for the same entropic reasons as discussed above. These self-assembled copolymer systems can be used as templates for etching or material deposition.

Much effort has gone into trying to find ways to guide self-assembly, particularly to overcome or circumvent the entropic tendency toward disorder. Some amount of top-down patterning, to act as a constraint or tempate for the self-assembly process, can be remarkably effective [261] (Fig. 4.25(b)). On larger length scales, a combination of (top-down patterned) shape constraint (lock-and-key geometric patterning) and capillary forces has been used to self-assemble entire integrated circuits from discrete components [264]. It seems abundantly clear that self-assembly will continue to become an increasingly powerful tool of the nanofabrication trade in coming years.

4.5 Summary

This chapter surveyed materials characterization methods, particularly those that revealed information about the atomic-scale structure of materials, as well as local nanoscale approaches. The discussion then turned to fabrication methods for the production of nano-structures, including materials growth, removal, and lithographic patterning, concluding with a brief look at self-assembly. Some key points follow.

- There are many characterization and fabrication techniques, each with their own domain of utility, determined by the underlying physics, chemistry, and materials science of the systems involved. As a result, all choices of materials and methods are tradeoffs.
- X-ray and electron diffraction are standard methods relying on scattering of interfering waves to infer microscopic structural detail about bulk (and nanoscale) materials. X-ray fluorescence, absorption, and photoemission techniques can give elemental speci-ficity. Photoemission is surface-sensitive and can reveal band energetics of occupied states.
- Transmission electron microscopy can reveal atomic-scale information about structure, with the caveat that samples must be in vacuum, of nanoscale thickness, and able to withstand the electron beam. Combined with electron energy loss spectroscopy, TEM can map out excitations at the nanoscale. TEMs enable electron diffraction studies.
- Scanning electron microscopy is a standard workhorse for nanoscale imaging, com-monly making use of secondary electrons and back-scattered electrons. Ordinarily samples must be in vacuum, and imaging resolution can be on the single nanometer scale. SEM may be combined with x-ray detection for elemental mapping.
- Scanned probe microscopies such as STM, AFM, and NSOM rely on the precise, feedback-controlled motion of a tip over a planar sample surface with atomic-scale precision. STM requires conductive samples, images a convolution of topography and local electronic density of states, and can deliver spectroscopic information. AFM relies on the mechanical interaction between the tip and the sample, may be performed on insulating materials, and can be broadened to other modalities via electrostatic, magnetic, and frictional tip-sample interactions.
- Nanomaterials may be grown in a bottom-up mode (chemical vapor deposition for nanotubes; VLS growth for nanowires; solution methods for nanocrystals) or through top-down patterning of large-area deposition (evaporation, sputtering, CVD, PECVD, MBE). Many factors affect material morphology, and growth of single-crystal materials requires great care, often including excellent vacuum, pure starting material, and lattice-matched substrates.
- Materials are commonly removed through wet chemical etching or gas-phase reactive ion etching. The latter can be highly directional, though the former tends to be much more rapid.
- Lateral patterning down to the nanoscale usually takes place through forms of lithogra-phy, in which a (usually polymer) stencil is created by optical or electron-beam exposure of a "resist" and subsequent chemical development. Ultimate resolution limits depend

on resist composition and material structure. Other forms of lithography include those based on scanned probe microscopies, physical imprint of a rigid master, and "soft" lithography based on rubber-stamp-like chemical patterning.

- Self-assembly takes advantage of the energetic minimization trend toward self-organization in some material systems. While self-assembly may be guided and constrained, there is a constant tension with the entropic tendency toward some amount of disorder.

4.6 Suggested reading

- *Scattering methods* There are a large number of books available on both x-ray and electron diffraction. Two popular x-ray diffraction books are A. Guinier's *X-Ray Diffraction in Crystals, Imperfect Crystals, and Amorphous Bodies* (Dover, 1994) [265] and Als-Nielsen and McMorrow's *Elements of Modern X-Ray Physics* [266]. Another inexpensive book introducing diffraction methods is Sands' *Introduction to Crystallography* (Dover, 1994).
- *Electron microscopy* Many books cover the basics of electron microscopy, though some introductory books focus on biological applications and sample preparation techniques. For a more physical view, Goldstein *et al.*'s *Scanning Electron Microscopy and X-Ray Microanalysis* [267] is strong. For TEM, Williams and Carter's *Transmission Electron Microscopy: a Textbook for Materials Science* [268] is well-regarded.
- *Scanned probes*. Scanned probe microscopy also has a number of books dedicated to its various forms. These include C. Julian Chen's *Introduction to Scanning Tunneling Microscopy*, 2nd edn. [269] and Roland Wiesendanger's *Scanning Probe Microscopy and Spectroscopy: Methods and Applications* [270]. In addition to books, some good review articles exist in the literature on various forms of scanned probe microscopy. These include *Rev. Mod. Phys.* articles, "Advances in atomic force microscopy" by Giessibl [271], and "Theories of scanning probe microscopes at the atomic scale" by Hofer, Foster, and Schluger [272].
- *Materials growth and removal* Materials deposition, particularly thin films, is treated exhaustively in several books [172, 273, 173, 174]. Particularly encyclopedic are Smith's *Thin Film Deposition: Principles & Practice* (McGraw Hill, 1995) and Seshan's *Handbook of Thin Film Deposition Processes and Techniques* (William Andrew, 2002).

  Guozhong Cao's *Nanostructures & Nanomaterials: Synthesis, Properties, and Applications* [204] is a good overview of nanomaterials growth. Geoffrey Ozin and André Arsenault's *Nanochemistry: A Chemical Approach to Nanomaterials* [274] is a great treatment of chemical growth techniques. Review articles of particular note are that by Murray, Kagan, and Bawendi [203] on nanocrystal growth, "Synthesis and characterization of monodisperse nanocrystals and close-packed nanocrystal assemblies", and that by Law, Goldberger, and Yang [275] on 1d nanostructure growth, "Semiconductor nanowires and nanotubes".
- *Nanoscale patterning* Two books that present modern photolithography in some detail are Suzuki and Smith's *Microlithography: Science and Technology*, 2nd edn.

[276] and Mack's *Fundamental Principles of Optical Lithography: The Science of Microfabrication* [277]. More nanoscale fabrication techniques are emphasized in Brodie and Muray's *The Physics of Micro/Nano-Fabrication* [278] and Zheng Cui's *Nanofabrication: Principles, Capabilities, and Limits* [279].

The *Nanomanufacturing Handbook* [280] is a strong collection of review articles that covers more nano-oriented techniques, including soft lithography, microcontact printing, and nanoimprint methods. Excellent review articles covering soft lithography include those by Xia and Whitesides [215] and by Quake and Scherrer [216].

- *Self-assembly* In addition to the review articles cited previously, two recent books that give overviews of the physics and chemistry of self-assembly are Yoon S. Lee's *Self-Assembly and Nanotechnology: a Force Balance Approach* [281] and John Pelesko's *Self Assembly: the Science of Things That Put Themselves Together* [282].

Exercises

4.1 **Near-field effects and the stethoscope** You have undoubtedly encountered near-field resolution enhancement effects analogous to those in some forms of NSOM, even if you did not realize it at the time.

    (a) What is the wavelength range for typical acoustic sound (say 1 kHz) in your body? Approximate your body as mostly water. To within a factor of 2.3, this wavelength tells you the spatial resolution of far-field acoustic methods of locating your heart.

    (b) Estimate the spatial precision with which a stethoscope can locate your heart. Compare this with the answer to (a). Explain in words why you think that the stethoscope can do so much better than the far-field acoustic resolution. This acoustic analogy for aperture-based NSOM was pointed out explicitly in the title of [168], "Optical stethoscopy: image recording with resolution $\lambda/20$".

4.2 **Nanocrystal nucleation** The free energy difference per unit volume between having a dissolved material with concentration C and a solid nucleus of that material precipitated out is given by $\Delta G_V = \frac{k_B T}{V_0} \ln(C/C_0)$, where V_0 is the atomic or molecular volume of the material, and C_0 is the *equilibrium* solubility of the material. Thus, when $C/C_0 > 1$, the volume free energy of the system is lowered by nucleating solid. If the *surface* free energy cost of having an interface between the solid and the solvent is γ Joules/m^2:

    (a) Show that there is a critical radius, $r_* = -2\gamma/\Delta G_V$, for nucleation of a spherical blob of solid. For blobs with $r < r_*$, the free energy of the system is lowered by shrinking the blob, while for $r > r_*$, the free energy is lowered by growing the blob.

    (b) Show that the free energy "barrier" to nucleation is $\Delta G_* = 16\pi\gamma/(3\Delta G_V)^2$ in this case. The nucleation rate will be proportional to $\exp(-\Delta G_*/(k_B T))$ (with a prefactor that involves temperature and viscosity of the solvent).

    (c) If subsequent growth is through diffusion of solute to the particle surface, argue that $\delta r(t) \propto \sqrt{D(C - C_s)t}$, where C_s is the solute concentration at the particle surface and D is the diffusion constant for the solute in the solvent. Assume that this part of the growth process is steady-state, so that C in the bulk of the solution is not changing with time.

4.3   Photoemission This exercise is meant to give a little intuition to the numbers relevant for photoemission (inspired by a similar exercise devised by Prof. James Kolodzey of the University of Delaware).

(a) For potassium, the work function for electrons is around 2.3 eV. What is the maximum wavelength of incident light that will lead to photoemission?

(b) Suppose the bottom of the conduction band lies a further 2.1 eV below the Fermi level of the metal. For incident light of wavelength 193 nm, what are the minimum and maximum kinetic energies of the photoelectrons that would be ejected by the metal? Assume no additional scattering of the photoelectrons.

(c) All other things being equal, the rate of photoemission is proportional to the density of (filled) states at the initial electron energy times the density of (empty) states at the final electron energy, based on a Fermi's Golden Rule. Consider sending in 193 nm light, as in (b). Show that the ratio of the photocurrents for photoelectrons with an emitted energy of 4.1 eV to those with an emitted energy of 3.1 eV is about 1.6. Assume parabolic bands with an effective mass the same as the free electron mass, m_0, and assume that the photoemitted electrons also have a parabolic dispersion.

4.4   Driving an AFM In non-contact mode, AFM cantilevers oscillate vertically over the sample. The standard way to do this is to use the cantilever's own inertia. The base of the cantilever is driven (using piezo-actuators) sinusoidally at some frequency, so that its vertical position is $A \cos(\omega t)$.

(a) In the (accelerating) frame of the cantilever base, what is the effective driving force experienced by the tip, approximating the tip as a mass m (and assuming the cantilever itself is massless)? Write this force as $F_0(t)$.

(b) The tip's vertical motion relative to the cantilever base is $z(t) = z_0(\omega) \cos(\omega t + \phi)$. Treating the cantilever as a spring of force constant k for vertical motion, and assuming that damping of the tip motion is provided by a force proportional to the velocity, $b\dot{z}$, what is the amplitude of the tip's motion relative to the cantilever $z_0(\omega)$, and what is the phase angle $\phi(\omega)$? You may define $\omega_0 \equiv \sqrt{(k/m)}$, and assume $b/m \ll \omega_0$. Worry only about the steady state, after any initial transients have died away. Hint: the problem may be simpler if you use complex notation.

(c) What happens to the phase angle $\phi(\omega)$ as the cantilever driving frequency is swept through ω_0?

(d) One means of transduction in AFM is to monitor the phase of the tip's motion as the tip is scanned over the surface. If the interaction between the tip and the surface acts like a slight, position-dependent change in spring constant, and the drive is at a fixed frequency $\omega \approx \omega_0$, how big a change in phase angle is expected for a sample interaction that changes the spring constant by $\delta k \ll k$?

(e) You should have found in (c), (d) that the weaker the damping (the smaller b), the sharper the change in $\phi(\omega)$ around ω_0. Therefore, for this transduction mechanism, you might think that smaller damping (a higher Q, quality factor) would be better. However, really light damping can lead to a limitation on scanning speeds. What is the characteristic timescale for transient changes in the cantilever response to damp away, in terms of b and m?

4.5 **AFM and force sensing** One common extension of atomic force microscopy is to use the tip as a force sensor. Here we shall see that there are advantages to resonant sensing as opposed to static (dc) force detection.

(a) Similar to the preceding exercise, consider the tip as a mass m attached to a cantilever that acts like a spring of force constant k. Suppose a static dc force of magnitude F_0 is applied to the tip. What is the static tip displacement expected? (Yes, this is very simple.) Now consider applying a driving force of that same magnitude, but oscillating at the resonant frequency $\omega \approx \omega_0$. As found in part (b) of the previous exercise, what is the amplitude of the driven displacement? The ratio between the resonant and static displacements, which you should have found to be $m\omega_0/b$, is the Q or quality factor of the resonator.

(b) Real experiments make use of this effect – the fact that a force acting at the resonant frequency can produce a much larger displacement than one acting off-resonance. Magnetic resonance force microscopy (MRFM) is a technique developed by scientists at IBM (and described in Chapter 7) to try to look at the magnetic properties of materials approaching the atomic scale. A cantilever tip (effectively a lightly damped oscillator) with a natural frequency of $\omega_0 \approx 2\pi \times 2$ kHz is made from single-crystal silicon. A tiny piece of magnetic material is at the very tip of this resonator. Magnetic forces between the tip and some nearby sample of interest are around 2.0 aN (attoNewtons, 10^{-18} N) when those forces exist. In MRFM, those forces are applied and oscillate at the resonance frequency ω_0. If the Q of the resonator is 20,000, and the mass of the resonator is 2×10^{-13} kg, find the amplitude of the cantilever displacement at resonance. Given that this displacement is what is detected in the actual experiment, do you think a higher Q would be desirable? Explain.

(c) At comparatively high temperatures, the Equipartition Theorem tells us that thermal fluctuations should lead to mechanical displacement with a mean-square amplitude given by $(1/2)k\langle z^2_{\text{thermal}}\rangle \approx (1/2)k_B T$. One can convert into effective fluctuating forces via the spring constant k. At 300 K, how do the thermal fluctuating forces compare to the magnetic forces of (b)?

4.6 **AFM, STM, and enhanced resolution** Enterprising researchers continue to find ways to improve the resolution of scanned probe microscopies. One approach that has proven effective in both AFM and STM has been the attachment of individual small molecules to the apex of the relevant tip. The detailed mechanism for the resulting resolution enhancement is not always obvious, but it is instructive to look at the remarkable results.

(a) First consider the paper by L. Gross *et al.*, *Science* **325**, 1110–1114 (2009) [283]. These investigators terminate their tip with a single CO molecule, and compare side-by-side the images acquired in constant-height STM microscopy and non–contact AFM. To what tip–sample interaction mechanism do the authors ascribe the remarkable resolution that they achieve? What do they say about the effects of electrostatic and van der Waals interactions? What theoretical calculation do they use to justify these statements?

(b) Similar in spirit is the paper by R. Temirov *et al.*, *New J. Phys.* **10**, 053012 (2008) [284]. The authors are able to resolve intramolecular details of sample molecules via STM when the microscopy is performed in the presence of molecular hydrogen adsorbed on the tip and sample. How do the authors demonstrate experimentally that hydrogen is in the tunneling junctions? What arguments do the authors make regarding their images showing sharp discontinuities (related to underlying molecular structure) on the submolecular scale as the tip is moved across the sample?

(c) Finally, inelastic electron tunneling spectroscopy can be used to enhance the effective resolution of STM, as in the work by C.-L. Chiang *et al.*, *Science* **344**, 885–888 (2014) [285]. As in (a) the authors attach a CO molecule to their STM tip. What inelastic modes then appear in tunneling spectroscopy measurements made between the functionalized tip and a plain metal surface? Explain in a couple of sentences how the authors then construct images of sample molecules on that surface, making use of the inelastic modes to enhance the resolution beyond the local electronic density of states seen by the background tunneling current.

Real solids: defects, interactions, confinement   5

Now that we have discussed the physics of infinite, perfectly ordered, bulk materials with essentially noninteracting electrons, it is time to consider what happens in systems that are less "ideal".

5.1 Defects

No real material is ever perfect. The basic argument for the existence of some nonzero density of defects is an entropic one, and makes perfect sense from the perspective of free energy. Even if there is a moderate energetic cost for the creation of a defect, if there are a very large number of possible configurations for the defect (e.g., sites where the defect can reside), the system's total free energy can be lowered through defect formation. Common defects can be zero-dimensional (*point defects*), one-dimensional (*line defects*), or two-dimensional (*grain boundaries* or interfaces).

What are the effects of these defects, particularly on the electronic properties of the materials? The Bloch condition and our picture of Bloch waves as single-particle eigenstates of the lattice are predicated on the assumption of the perfect, infinite spatial periodicity of the lattice. That is, the single-particle Hamiltonian in crystalline solids is assumed to be invariant under discrete translational symmetry. Strictly speaking, once that symmetry is broken by defects, the Bloch states are no longer exact solutions to the single-particle problem.

If the defect density is *low*, then we generally don't care. It's hard to imagine that a handful of defects in a crystal could have a large impact on the electronic structure of a crystal containing 10^{22} atoms (though see Exercises). The vast majority of single-particle states in the crystal are still approximated very well as Bloch waves in this case. However, the defects do alter the spectrum of allowed states, and these changes can be significant if the number of atoms near defects becomes a substantial fraction of the total number of atoms. We discuss this further below.

One can get a crude estimate of the effects of the defects on the electronic wavefunctions. Define ℓ as a typical distance between defects. If $k_F\ell \gg 1$, then a Bloch wave would still be a reasonable description of the extended electronic states near the Fermi level. If $k_F\ell \ll 1$, then defects are spaced more closely than the effective wavelength of a putative Bloch wave. One would expect the Bloch wave description of the extended states to be poor in this limit.

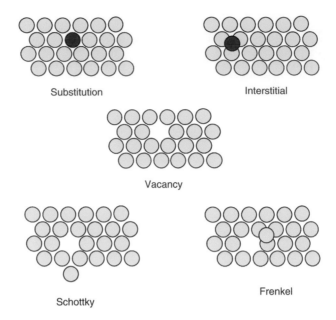

Fig. 5.1   Various point defects in crystals.

5.1.1 Point defects

Common point defects are shown in Fig. 5.1. Chemical impurities are an obvious deviation from the infinite, bulk crystal. Impurities residing on an atomic site are *substitutional*, while impurities squeezed into what would ordinarily be empty volume within the lattice are *interstitial*. Unsurprisingly, interstitial impurities are more common when the impurity is small compared to the size of an atom in the bulk, such as the case of hydrogen in the FCC lattice of palladium.

Other point defects are possible without introducing impurities. These include vacancies. In crystals with strongly polar bonding (the extreme being ionic salts), vacancies are produced in pairs to maintain overall charge neutrality of the lattice. Each vacancy can then diffuse around the lattice independently. These are *Schottky defects*. A vacancy formed by the creation of an interstitial is called a *Frenkel defect*.

Producing a point defect takes energy. Moving an existing point defect generally requires a different amount of energy than forming a new defect. This is easily visualized from the cartoons in Fig. 5.1. Defect motion is often thermally activated with a diffusion constant having an Arrhenius form, $D \sim \exp(-\Delta_n/k_\mathrm{B}T)$, as discussed in Eqs. (3.16)–(3.18). The activation barrier Δ_n can be significantly lower along grain boundaries and free surfaces. This is intuitively clear, since moving a defect within the bulk crystal requires breaking or distorting many bonds, while the undercoordinated atoms at the surface are much less constrained.

Point defects may act as scatterers for charge carriers, since they are a perturbation away from the perfect lattice periodicity that guarantees conservation of crystal momentum $\mathbf{k}$ for a Bloch state. This same physics can allow defects to act as centers for other processes that would otherwise be forbidden by **k** conservation, such as nonradiative recombination of electrons and holes. Point defects may also act as traps for charge carriers. Because of the disruption of the regular chemical bonding of the crystal, individual point defects can have a net charge, leading to an effective potential well for charge carriers of the appropriate sign.

5.1.2 Line defects

There can be an energetic benefit for point defects to group together, since this can lower the number of distorted or broken bonds per defect. More extreme versions of this are line defects, 1d extended disruptions of normal crystalline stacking. Two common types of line defects are shown in Fig. 5.2. These are examples of *dislocations*. A dislocation can be characterized by a *Burgers vector*, **b**. One can find **b** by starting out at some lattice site, and making single-site steps that would (in the absence of the defect) return to the starting site while enclosing the defect. The Burgers vector of the defect is the vector in real space that connects the initial site with the final site.

An *edge dislocation* may be pictured as an extra row of atoms that terminates in the crystal. Near the dislocation, there is a local elastic strain, with the lattice in compression on one side of the dislocation, and in tension on the other side. When external forces are applied to the material as shown, an edge dislocation propagates in the direction parallel to the mechanical load. Note that propagating an edge dislocation is one means of plastically (irreversibly) deforming a material.

Suppose you wanted to displace the half of the material above the dislocation horizontally by one lattice spacing relative to the half of the material below the dislocation. It takes far less energy to do this by propagating the dislocation through the material than it would to break all of the bonds at that interface and re-form them. This is analogous to moving a rug along the floor by pushing a bump in the material, rather than attempting to slide the whole rug uniformly.

A *screw dislocation* is a topological defect such that a complete circuit around the dislocation results in a translation along the direction of the dislocation by one lattice spacing. Under applied loads as shown, screw dislocations propagate perpendicular to the

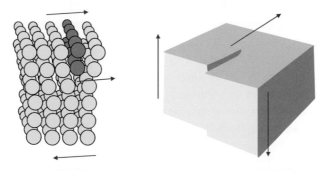

Line defects. Left: edge dislocation, with the extra row of atoms indicated in dark gray. Right: screw dislocation. In both cases, blue arrows indicate the direction of defect propagation when stresses are applied directed along the red arrows. **Fig. 5.2**

direction of the applied load. This is analogous to the direction of motion of the edge of the torn edge when a stack of paper is ripped. Near the screw dislocation there is residual shear strain in the material.

Like point defects, dislocations can be charged and/or act as scattering sites for mobile charge carriers. A discussion of the effect of dislocations on the mechanical strength of bulk and nanoscale materials is left for Chapter 9.

5.1.3 Two-dimensional defects

There are also defects that are extended in two dimensions. A *stacking fault* is a two-dimensional defect caused by a one or two layer mis-ordering of a lattice stacking pattern. For example, as shown in Fig. 2.17, the difference between HCP and FCC lattices is the location of subsequent close-packed layers of atoms. A stacking fault along the (111) direction of an FCC crystal would happen when the layer structure is ABC/AB/ABC/ABC....

An energetically more costly planar defect would be an *antiphase boundary*, usually seen in ordered alloys. Suppose that the ordinary compositional stacking of layers of A and B atoms is ABABAB. An antiphase boundary would occur when stacking instead is ABABBABA. That is, the crystallographic direction is unchanged across the boundary, but there is a "phase shift" of one layer moving across the defect.

Maintaining composition but changing crystallographic direction across a boundary is an example of a *twin defect*. For example, FCC stacking could run ABCAB-CABCBACBACBA. In this case the (111) plane is a mirror or twin plane, with the structure having reflection symmetry about that plane. Twin defects can be formed during the crystal growth process.

A *grain boundary* occurs when two crystals with the same structure but arbitrary crystallographic orientation run into one another. Generically the chemical bonding of the atoms at the grain boundary is disrupted relative to the interior of the two crystals. Atoms are identified as belonging to one crystal or the other, though like all atoms at surfaces, the coordination of those atoms by neighbors differs from those in the bulk.

5.2   Interfaces and surfaces

In some sense the next step beyond a grain boundary is an *interface* (the boundary between one type of material and another) or a *surface*.[1] We consider surfaces first, since they are simpler and should be particularly relevant for nanoscale structures, since *for smaller systems a higher fraction of the constituent atoms are at the surface* (and therefore undercoordinated) than are in the bulk (and fully coordinated) – see the Exercises at the end

---

[1] There is an entire discipline called *surface science* that studies precisely the interesting physics and chemistry that is relevant at material surfaces. Much of traditional surface science has been relabeled "nanoscience" in recent years.

5.2 Interfaces and surfaces   155

of this chapter. This fact, coupled with the fact that surface atoms have different binding energies to the solid from bulk atoms, can lead to big changes in thermodynamic stability of nanoparticles.

5.2.1 Work function

Once we have decided to make our crystal finite rather than infinite, one question that naturally arises is, how tightly bound are the electrons in that crystal? Clearly the Coulomb interaction between the electrons and the ion cores does a decent job of holding the electrons inside the material. Physical chemistry experience gives us some guidance on how to quantify this binding. In a single atom or molecule, we would speak of an *ionization energy* as the amount of energy required to remove one electron from the highest occupied orbital and carry it infinitely far away. Similarly, we could talk of an *electron affinity*, the amount of energy released when a neutral atom or molecule acquires an additional electron (from infinitely far away) in the lowest unoccupied orbital.

These terms have been adapted for use in semiconductors by analogy. The ionization potential is the energy needed to take an electron from the top of the valence band away to infinity (the *vacuum level*). The electron affinity is then the energy released if an electron is brought from the vacuum level and placed in the bottom of the conduction band.

In bulk metals the single-particle electronic spectrum is essentially continuous near the Fermi level, and therefore the electron affinity and ionization potential converge to a single quantity, the *work function*, the amount of energy required to take an electron at the Fermi level and remove it to outside the material.[2] If we define the vacuum level to be the zero of our energy scale, then the work function is positive and E_F must be *negative*.

The electronic density at the surface of a crystal is not identical to that within the crystal. We can define a surface layer that extends from some distance within the crystal out some distance away from the last layer of atoms. If the energy required to pull an electron through this surface layer is W_s, then the work function Φ is given by $-E_F + W_s$. Figure 5.3 shows why, even at an ideal, clean surface, there is an interfacial electric dipole layer: there is some leakage of the electronic density out into what would otherwise be empty space.

This contribution W_s shows why precisely measuring the intrinsic work function of some material is extremely challenging and requires extremely clean (ultrahigh vacuum) conditions. Single layers of atomic or molecular adsorbates can strongly affect Φ by altering this surface charge distribution. This fact has been used to technological advantage. Alkali metals have been used to reduce the work function of electrodes in electron emitting applications [286], while self-assembled monolayers of molecules with electric dipole moments have been used to raise and lower metal electrode work functions in organic electronics applications [287].

In practice work functions are measured empirically, most commonly using photoemission. The clean metal surface in question is illuminated with incident light of a

---

[2] Strictly speaking, the work function involves moving an electron from within the material to "just outside" the material, but not a macroscopic distance away. In general, there is extra work due to screening and image potential effects required to move an electron from just outside the surface all the way to infinity.

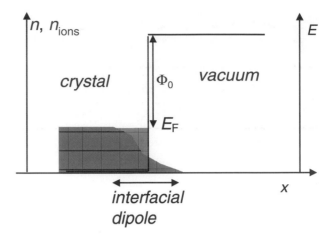

Fig. 5.3 Illustration of the work function, showing the intrinsic interfacial dipole at a metal surface.

precisely known photon energy while being held at a constant, reference electrostatic potential. When that photon energy is sufficiently high (typically in the ultraviolet or x-ray range), electrons are photoemitted from the surface. These photoelectrons are collected and their kinetic energies measured precisely, allowing their original binding energies to be determined. The photoelectrons with the highest kinetic energies were emitted from the Fermi level and allow the determination of the work function. Work functions may also be inferred from other emission processes (e.g. thermionic emission, field emission, both described later) and precise measurements of the *contact potentials*.

Consider two metals with differing work functions, $\Phi_1 > \Phi_2$, joined by a wire such that electrons can flow between them. In thermal equilibrium, the total electrochemical potential, μ, for the electrons must be uniform between the two metals, by definition. However, the work functions should be intrinsic properties of the metals in question. Suppose we remove an electron from material 2 and drop it back into material 1. This should not change the total energy of the combined system if μ is uniform. The only way for this to work is if an electrostatic potential difference, V_{cp} exists between the two metals so that $eV_{cp} = \Phi_1 - \Phi_2$. That contact potential is developed by slight rearrangements of the electron positions when the two materials are connected. One can measure V_{cp} by seeing what external voltage supply is required for there to be no charge flow between two metals when they are connected. This is the Kelvin probe method, named for its inventor, Lord Kelvin. A modern variant of this is Kelvin probe force microscopy (KPFM), which uses the modern ability to measure extremely small forces at the nanoscale in concert with the ideas of the Kelvin probe to enable nanometer-resolution measurements of contact potential differences [288]. This was described in more detail in Section 4.1.4.

5.2.2 Surface states

Since surfaces violate the discrete translational symmetry assumed in the derivation of the Bloch condition, what *does* happen to the spectrum of single-particle states? Defining the

A "quantum corral" on the surface of Cu(111) formed by iron atoms positioned using an STM. The interference pattern results from scattering of delocalized 2d copper surface states by the iron atoms. Image via Don Eigler, IBM Almaden Research Center.

Fig. 5.4

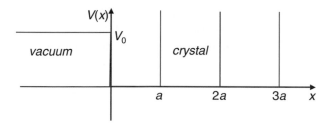

Tamm's model, used to show the existence of surface states. The model is essentially a terminated Krönig–Penney model with δ-function barriers.

Fig. 5.5

z direction as normal to the surface, new *surface states* arise that are *localized* in z, and pinned to the surface. These particular states are generally delocalized in x and y, and can have significant consequences in experiments. Figure 5.4 shows one of the most famous images of such delocalized 2d surface states, in this case imaged by a scanning tunneling microscope (STM) on the (111) surface of copper.

This problem was first treated by Igor Tamm [289, 290], who considered the 1d model shown in Fig. 5.5. This is the δ-function version of the Krönig–Penney model, but truncated with a constant potential for $z < 0$. The solutions at $z > 0$ must look like those of the untruncated problem, where the allowed Bloch wavevectors are given by Eq. (2.61),

$$\psi(z > 0) = Au_k(z)\exp(ikz) + Bu_k(z)\exp(-ikz). \tag{5.1}$$

However, for $0 < E < V_0$, the solutions must also look like

$$\psi(z < 0) = C\exp\sqrt{(2m(V_0 - E)/\hbar^2)}(z). \tag{5.2}$$

One must then match boundary conditions at $z = 0$, with continuity of $\psi(z)$ and $\psi'(z)$.

For the real k given by Eq. (2.61), this results in two equations for three unknowns, meaning that it is always possible to find a solution. So, all single-particle states allowed in the infinite 1d crystal are also permitted in the truncated one.

Complex k values would correspond to states that lie in the energy gaps of the infinite crystal model. If k has some imaginary part, then either A or B must be zero – otherwise the wavefunction would diverge exponentially as $z \to \infty$. Matching boundary conditions at $z = 0$ then leaves two equations with two unknowns, implying that there is only a single solution (one unique value of E) for which Eqs. (5.1) and (5.2) are satisfied. For these states, the wavefunction decays exponentially away from the interface into both empty space and the crystal. *There is one Tamm state per band gap, and that state is localized to the crystal surface.*

The density of surface states can be significant. In the 1d Tamm model above there is one surface state per band of states extended in z. Extending the analysis to higher dimensionality and more realistic lattices is rather complicated, but the basic intuitive result remains correct: one should expect a band of surface states to develop within each energy gap, with roughly one surface state for each atom at the surface, around 10^{14} states per cm^2.

Other kinds of surface states are also possible, including ones that are localized in all three dimensions (e.g., bound electronic states associated with point defects or adsorbed impurities on the surface). Like any other single-particle state, surface states may be occupied or empty depending on their energy and the chemical potential of the electrons. Understanding and controlling surface states has already had a strong impact on the development of macroscale semiconductor technologies, as we shall see in more detail in Chapter 6. In nanoscale systems these states are even more relevant, since (see Exercises) it is not difficult to have the number of surface states be comparable to the number of bulk single-particle states!

5.3 Screening

We need an additional physics ingredient before we can properly discuss interfaces between different materials: *screening*, the response of the electrons in a solid to the presence of a local excess or deficit of charge and the resulting electric field.

5.3.1 Dielectrics

We start with the simple case of a fixed impurity of charge q. In bulk systems without mobile charges, the bound electrons shift their positions slightly, leading to a dielectric polarization. This polarization, parametrized by the relative dielectric constant κ, leads to a reduction in the Coulomb potential experienced by a test charge some distance r from the impurity relative to the free space case:

$$\phi(\mathbf{r}) = \frac{q}{4\pi\epsilon_0 r} \to \frac{q}{4\pi\kappa\epsilon_0 r}. \tag{5.3}$$

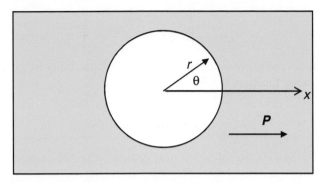

Fig. 5.6

Polarized dielectric. Considering the field at the center of a spherical void in a uniformly polarized dielectric is the way to derive the Clausius–Mossotti relation Eqs. (5.9) and (5.10).

We can relate κ to the microscopic polarizability, α, of the atoms of the material. Recall that the induced dipole moment, $\mathbf{p}$, of an atom in the presence of a local electric field $\mathbf{E}_{\text{loc}}$ is approximated $\mathbf{p} = \alpha \mathbf{E}_{\text{loc}}$. If the 3d density of the ith atomic species is n_i, then the induced polarization, $\mathbf{P}$, defined as total dipole moment per unit volume, is given by

$$\mathbf{P} = \sum_i n_i \alpha_i \mathbf{E}_{\text{loc}}. \tag{5.4}$$

The divergence of $\mathbf{P}$ is related to the density of bound charge, so that

$$\mathbf{D} \equiv \kappa \epsilon_0 \mathbf{E} \equiv \epsilon_0 \mathbf{E} + \mathbf{P}, \tag{5.5}$$

where $\mathbf{E}$ is the *average* electric field. Reshuffling,

$$\mathbf{P} = \epsilon_0 (\kappa - 1) \mathbf{E}. \tag{5.6}$$

Figure 5.6 shows a slab of dielectric material with a uniform polarization, with a spherical void of radius r centered on the origin. Because of the polarization the inside surface of the void has an effective surface charge density, $\sigma(\theta)$ given by $-P\cos(\theta)$, where $P = |\mathbf{P}|$ and θ is the angle shown. Adding up the field contributions from that surface charge, we find that the x-component of the field due to the polarization (other components vanish by symmetry) is:

$$E_P = \frac{-1}{4\pi\epsilon_0 r^2} \int_0^\pi (-P\cos\theta) \cdot \cos\theta \cdot [2\pi r^2 \sin\theta\, d\theta] = \frac{P}{3\epsilon_0}. \tag{5.7}$$

Since $\mathbf{E}_{\text{loc}} = \mathbf{E} + \mathbf{E}_P$, we can combine Eqs. (5.4, 5.6 and 5.7) and get

$$\mathbf{P} = \left[\frac{\sum_i n_i \alpha_i}{1 - \frac{1}{3\epsilon_0} \sum_i n_i \alpha_i} \right] \mathbf{E}. \tag{5.8}$$

That leads to the *Clausius–Mossotti relation*, which can be written two ways:

$$\kappa = \left[1 + \frac{\frac{1}{\epsilon_0} \sum_i n_i \alpha_i}{1 - \frac{1}{3\epsilon_0} \sum_i n_i \alpha_i} \right] \mathbf{E}, \tag{5.9}$$

$$\frac{\kappa^2 - 1}{\kappa^2 + 2} = \frac{1}{3\epsilon_0} \sum_i n_i \alpha_i. \tag{5.10}$$

Now we have a relationship between the microscopic polarizability of the atoms in a solid and the bulk dielectric function.

5.3.2 Conductors

Systems with free charges can do more than polarize; the free carriers can move long distances in response to local charged impurities. As a result, the long-range bare Coulomb potential of a point charge is modified to a form that drops off more rapidly with distance than $1/r$. Often one can define a characteristic *screening length* associated with the range of the screened potential. The screening length is also of critical importance when considering interfaces between materials. The general idea of screening is also relevant in systems with mobile ions, such as electrolytes, and this is discussed further in Chapter 10.

To see where this distance scale comes from, we can still think in terms of a dielectric function, κ, and start by working in three dimensions. Let's further assume that the dielectric behavior in the previous subsection contributes a factor κ_0 to the total κ. What we care about is the electrostatic potential experienced by an electron somewhere near a charged impurity, compared to what that potential would be in the absence of a material response. Schematically, here's what happens if we drop a charged impurity into a material with free carriers.

- The ion cores polarize, as in the previous section. This already reduces the Coulomb potential from the impurity by a factor of κ_0 relative to the bare impurity.
- The mobile charges rearrange themselves, but must do so *self-consistently*. For a positive impurity and mobile electrons, the potential energy for an electron decreases near the impurity. This would tend to increase the local electronic concentration, $n(\mathbf{r})$, near the impurity, but there is a cost to doing this because of the density of states of the electrons near the Fermi level. Recall that $\nu(E) = \partial n / \partial E$.
- Self-consistency means that any change in the local carrier density near the impurity affects, via Poisson's equation (see below), the electrostatic potential farther away from the impurity.

Screening in 3d

We know Gauss' law:

$$\nabla \cdot \epsilon_0 \mathbf{E} = \rho, \tag{5.11}$$

where ρ is the *total* charge density (including any impurities plus the response of the free carriers), and we know

$$\nabla \cdot \epsilon_0 \kappa \mathbf{E} = \rho_{\text{ext}}, \tag{5.12}$$

where ρ_{ext} is the charge density of the impurities only. Since $\mathbf{E} = -\nabla\phi$, From the definition of the electrostatic potential, we can write Poisson's equation:

$$\nabla^2\phi = -\frac{1}{\epsilon_0}\rho. \tag{5.13}$$

Of course, $\phi = \phi_{ext} + \phi_{ind}$, where ϕ_{ind} is the potential due to the induced charge density and ϕ_{ext} is the potential due to the impurities.

If we want to consider arbitrary configurations of impurities it is useful to work in Fourier space. Ignoring dynamics for now, we assume $\phi = \phi(\mathbf{K})$, where $\mathbf{K}$ is some wavevector describing the spatial periodicity of one harmonic of the electrostatic potential, $\phi(\mathbf{K}) \sim \exp(i\mathbf{K} \cdot \mathbf{r})$. Poisson's equation then becomes

$$K^2\phi(\mathbf{K}) = \frac{1}{\epsilon_0}\rho(\mathbf{K}). \tag{5.14}$$

In Fourier language, then,

$$\kappa(\mathbf{K}) = \frac{\phi_{ext}(\mathbf{K})}{\phi(\mathbf{K})} = 1 - \frac{\phi_{ind}(\mathbf{K})}{\phi(\mathbf{K})}. \tag{5.15}$$

The solution to Eq. (5.14) is simple for the case of a point charge $-e$. Since $\rho(\mathbf{r}) = -e\delta(\mathbf{r})$, the Fourier transform $\rho(\mathbf{K}) = -e$. That means that

$$\phi(\mathbf{K}) = \frac{-e}{\epsilon_0\kappa_0 K^2}, \tag{5.16}$$

where we've already accounted for the background polarization contribution.

Poisson's equation must hold for the induced potential as well:

$$\phi_{ind} = \frac{\rho_{ind}}{\epsilon_0\kappa_0 K^2}. \tag{5.17}$$

If $\rho_{ind} = -e\delta n$, where δn is the induced change in electronic density, and $\delta E = -e\phi$ is the change in energy of an electron due to the (total) electrostatic potential, from Eq. (5.15) we can write

$$\kappa(\mathbf{K}) = 1 - \frac{e^2}{\epsilon_0\kappa_0 K^2}\frac{\delta n}{\delta E}. \tag{5.18}$$

All we have to do now is figure out what $\delta n/\delta E$ is, do an inverse Fourier transform, and we can find $\phi(\mathbf{r})$.

The *Thomas–Fermi* model of screening assumes that the chemical potential for electrons is constant throughout a material with mobile carriers, as in Figure 5.7, even when the local band energies are perturbed by things like charged impurities. That means that, for a degenerate system (that is, assuming that the Fermi distribution function is pretty sharp), we can plug in our good old density of states, $\nu_{3d}(E)$ from Eq. (2.24). Notice that pulling the local band energy *down* ($\delta E < 0$) ends up *increasing* the local electronic density. Accounting for that sign, we then find

$$\kappa(\mathbf{K}) = 1 + \frac{1}{r_{TF}^2 K^2}, \tag{5.19}$$

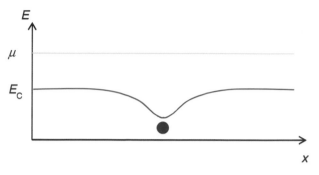

Fig. 5.7 Chemical potential and Thomas–Fermi screening. A positively charged impurity shifts the local electrostatic potential. When the chemical potential for the electrons equilibrates self-consistently, the result is a Yukawa-type screened potential from the impurity and an excess in electronic density proportional to the density of states at the Fermi level.

where we have defined the Thomas–Fermi screening length:

$$r_{\text{TF}} \equiv \left(\frac{e^2}{\epsilon_0 \kappa_0} \nu_{3d}(E_{\text{F}}) \right)^{-1/2}. \tag{5.20}$$

Putting E_{F} into $\nu_{3d}(E)$ tacitly assumes that the local change in band energies is relatively small.

Let's see what the actual potential from a point charge of size Q. From Eq. (5.16), we know

$$\phi_{\text{ext}}(\mathbf{K}) = \frac{Q}{\epsilon_0 \kappa_0 K^2}. \tag{5.21}$$

So,

$$\phi(\mathbf{K}) = \frac{\phi_{\text{ext}}}{\kappa(\mathbf{K})} \rightarrow$$

$$\phi(\mathbf{K}) = \frac{Q}{\epsilon_0 \kappa_0 (K^2 + r_{\text{TF}}^{-2})}. \tag{5.22}$$

Inverse Fourier transforming, this gives back

$$\phi(\mathbf{r}) = \frac{Q}{4\pi \epsilon_0 \kappa_0} \frac{\exp(-r/r_{\text{TF}})}{r}. \tag{5.23}$$

This is the *Yukawa* potential, originally developed in modeling nuclear physics. Instead of generating the unscreened long-range $1/r$ Coulomb potential, the impurity's influence is exponentially suppressed on a distance scale comparable to r_{TF}. Also, the higher the density of states at the Fermi level, the shorter the screening length.

Let's plug in some numbers. Assume $\kappa_0 = 2$ for now. For a typical noble metal like Au, $\nu_{3d}(E_{\text{F}}) \sim 10^{47}$ m^{-3}J^{-1}, giving $r_{\text{TF}} \sim 0.1$ nm. That's impressively short! In other words, there are so many free electrons (with accessible states) that only a very slight shift in the positions of the electrons right near the impurity are enough to completely shield the rest of the electrons.

The *non*degenerate limit is also important to consider, and is treated in one of the exercises. The end result is that $\delta n/\delta E$ should be replaced by $n/k_B T$, where n is the average 3d density of carriers. This leads to a nondegenerate expression for the screening length:

$$r_{\rm DH} \equiv \left(\frac{e^2}{\epsilon_0 \kappa_0} \frac{n}{k_B T} \right)^{-1/2}. \tag{5.24}$$

Point charges are still screened, but the relevant length scale is the *Debye–Hückel* screening length, $r_{\rm DH}$.

Note that these screening lengths are derived assuming that there are plenty of carriers around to do the screening. If all of the mobile carriers with a region are moved elsewhere as a consequence of trying to screen, that region is said to be *depleted*.

Reduced dimensionality

Screening is significantly *worse* in reduced dimensionality, for multiple reasons. First, as we have discussed previously, the effective density of states ($\delta n/\delta E$ from above, where now n is the d-dimensional density of carriers) is generally smaller than in 3d. Second, and more importantly, the exponential functional form for screening from Eq. (5.23) is not valid for systems with carriers constrained to move in reduced dimensionality.

The physical reason for this is apparent in Fig. 5.8(a). The unscreened Coulomb potential from a point charge still extends out into all three dimensions. If the carriers that rearrange in response to this potential are constrained to move in only 2d (or 1d), then it is not possible for those carriers to effectively attenuate the unscreened electric field that points out of the plane (or line).

One consequence of poor screening can be enhanced *noise* in the electrical conductivity. Figure 5.8(b) shows an example of this using the channel of a Si MOSFET at low temperatures. As a localized trap state near the 2d conducting layer is alternately filled and emptied, the mobile carriers in the channel are affected, with the result being "telegraph noise" in the electrical resistance of the device. Such noise problems can be particularly extreme in 1d conductors such as carbon nanotubes. With a truly 1d system, every carrier that traverses the length of the device must interact with every impurity.

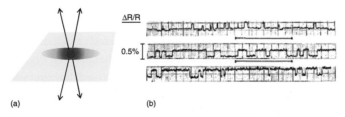

(a)   (b)

Screening in 2d. (a) Screening is less effective in 2d because screening charges cannot compensate for field lines that pass out of the 2d plane. (b) "Telegraph" noise in the resistance of a 2d charge layer caused by stochastic charging and discharging of a nearby trap [291].   Fig. 5.8

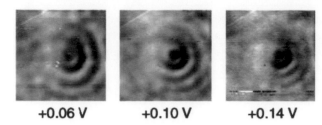

+0.06 V   +0.10 V   +0.14 V

Fig. 5.9   Friedel oscillations seen via STM near a charged defect on an InAs surface at three different tip voltages. Increasing positive voltage leads to accumulation of a higher electron density and correspondingly a shorter spatial period of the oscillations [292].

Friedel oscillations

There is one further correction to our screening treatment that we must consider. The electrons in a conductor are not free to respond to perturbations on arbitrarily short length scales. That is, in our Fourier treatment of screening of Eq. (5.14), it doesn't make sense to think about electrons completely screening at wavevectors $|\mathbf{K}| \gg k_F$. The result of terminating the Fourier series for $\rho(\mathbf{K})$ at $K = k_F$ is spatially oscillatory "ringing" in the screening charge density, oscillating with a wavevector $\sim k_F/2$, superposed on top of the decay of the screened response at longer distance scales.

Figure 5.9 shows an example of these *Friedel oscillations*, detected by scanning tunneling microscopy (STM) near a charged defect on an InAs structure. In this case the conducting system is a two-dimensional electron gas at the InAs surface. Here the voltage bias between the metal STM tip and the InAs allows the experimenters to change the average 2d carrier density. As expected, at higher accumulated carrier densities (increasing k_F for the 2d electrons), the wavelength of the oscillations decreases.

5.3.3 The importance of electron–electron interactions

In our discussion of screening, we have seen that mobile charges redistribute themselves in response to the presence of a charged object, consistent with the Poisson equation. The result is a configuration of charges that minimizes electrostatic potential energy. When carrier densities (and dimensionality) are high, only a small spatial rearrangement is required for full screening of charges. This implies that *the gas of electrons must have screening correlations within itself.* On average, each electron within a metal should, for example, have a "correlation hole" as a result of this dynamical screening.

How important are these electron–electron interactions? Back in Section 3.2.1 we introduced the concept of a Fermi liquid, and mentioned that many of the essential features of a noninteracting treatment (the single-particle state approach that gives us band theory) work remarkably well even in the presence of interactions a lot of the time. Here we will examine briefly why this is so, and when electron–electron interactions must be considered.

To quantify the importance of interactions to the many-particle ground state, it is natural to compute the ratio of the average electron–electron potential energy to the average electron kinetic energy at $T = 0$. When that ratio is large, one would expect interaction effects to dominate. For the *bare* Coulomb interaction, the typical interelectron Coulomb interaction scales like $1/r \sim (n_d)^{1/d}$, the d-dimensional electron density to the $1/d$ power. The typical electron kinetic energy scales like E_F. The ratio of potential to kinetic energy therefore scales like $n_{3d}^{-1/3}$ in 3d, $n_{2d}^{-1/2}$ in 2d, and n_{1d}^{-1} in 1d. In other words, in all dimensionalities, *at sufficiently low carrier densities, interaction effects dominate*, and as density is decreased, *systems of lower dimensionality are expected to start showing interaction effects first*. This counterintuitive result, that interaction effects are *less* important when interparticle separations are small, is a direct consequence of the Pauli principle via the Fermi energy.

A similar quantity to the aforementioned ratio is the dimensionless interaction parameter r_s, which is the radius of a d-dimensional circle enclosing one particle, in units of the Bohr radius:

$$r_{s,3d} \equiv \left(\frac{4\pi}{3}n_{3d}\right)^{-1/3} \bigg/ \frac{4\pi\epsilon_0\hbar^2}{me^2},$$

$$r_{s,2d} \equiv (\pi n_{2d})^{-1/2} \bigg/ \frac{4\pi\epsilon_0\hbar^2}{me^2},$$

$$r_{s,1d} \equiv n_{1d}^{-1} \bigg/ \frac{4\pi\epsilon_0\hbar^2}{me^2}. \tag{5.25}$$

As you can see from the density dependence, r_s is proportional to the ratio of the previous paragraph, and a simple interpretation would suggest that interactions may be relevant once $r_s >\sim 1$. In simple, bulk elemental metals, $r_s \sim 2-4$.

In the limit of large r_s, electrons are predicted to order spatially, forming a *Wigner crystal*. In 2d, calculations suggest [293] that this transition should occur at $r_s \sim 37$. In 3d, the electrons have more ways of getting out of each other's way, and stronger interactions are required to produce ordering. Predictions [294, 295] range from $r_s \sim 65$ to 100.

It's reasonable to think that electron–electron interactions should have significant consequences even at higher densities (lower values of r_s) than those required for dramatic effects like Wigner crystallization. However, these effects tend to be relatively subtle under most conditions. One example to consider is the lifetime of a particular Bloch state. Once electron–electron interactions are in the picture, the single-particle Bloch states are no longer energy eigenstates, and therefore acquire some finite lifetime due to electron–electron scattering. The fermionic nature of electrons restricts this scattering rate, however.

To see this, start with a full Fermi sea at $T = 0$ and imagine dropping in an extra electron with some energy $E_1 > E_F$. That electron interacts via the Coulomb interaction with another electron with energy E_2, and we know that $E_2 < E_F$ because all the other Bloch states are empty. After this scattering process, the final state energies of the two electrons, E_3 and E_4, must be greater than E_F for those states to be available. Conservation of energy further requires $E_1 + E_2 = E_3 + E_4$. It turns out that as $E_1 \rightarrow E_F$, the number of available states that satisfy these restrictions drops to *zero*, and therefore so does the scattering rate due to electron–electron interactions. The electron–electron scattering time,

τ_{e-e}, at $T = 0$ for an electron in a single-particle state with energy E_F is infinite. This is called *Pauli blocking*.

At nonzero T there are some empty states available near the Fermi level to allow electron–electron scattering. The Pauli principle still leads to a reduction of the scattering rate by a factor of $(k_B T/E_F)^2$. Screening (also an interaction effect, of course) also reduces the cross-section, σ, for Coulomb scattering to an effective value something like r_{TF}^2 in 3d. That means that the mean free path due to electron–electron scattering in 3d is something like

$$\ell_{e-e} \approx \frac{1}{n_{3d}\sigma} \approx \frac{1}{n_{3d}r_{TF}^2}\left(\frac{E_F}{k_B T}\right)^2. \tag{5.26}$$

At room temperature in a metal like copper, this works out to something like 10^{-6} m, which is much longer than the mean free path for scattering by other interactions. *Other* scattering processes (e.g., electron–phonon) limit electron motion at room temperature, not electron–electron scattering. That is why these effects are not readily noticeable in day-to-day existence.

At low temperatures, however, when other inelastic processes like phonon scattering freeze out, it is possible to see the effects of both electron–electron scattering and the fermionic statistics of electrons. We shall see some examples of this when we discuss the importance of quantum corrections to classical electronic properties. As mentioned in Chapter 3, Fermi liquid theory results from a more detailed treatment of electron–electron interactions in 3d. The interactions lead to certain corrections to the noninteracting picture, but the basic qualitative features (a Fermi surface at $T = 0$, quasiparticles as a useful description of the low energy excitations) survive. In reduced dimensionality, electron–electron interactions and reduced screening can lead to other electronic states, such as the Luttinger liquid state in 1d, and the Laughlin liquid state in 2d and high magnetic fields.

5.4 Excitons

In non-metals it is often necessary to worry about Coulomb interaction effects between carriers in *different bands*. The classic example of such an effect is the **exciton**, an electron–hole pair bound together by their mutual Coulomb attraction. An exciton is also a quasiparticle, in the sense that it is an excitation of the whole system with well defined quantum numbers.

In materials with comparatively large electronic polarizabilities like most inorganic semiconductors, exciton states may be described as hydrogenic, just as electrons bound to donor ions were treated in Section 3.3.3. These are *Mott–Wannier* excitons, with effective Bohr radii that span many interatomic spacings. As in the hydrogenic impurity case we will ignore the lattice momentarily (that is, assume a direct gap and approximate the effect of the lattice electrons via a relative dielectric permittivity, κ), and consider the interacting two-body problem of the electron and the hole, working in 3d for now. This is analogous to the problem of positronium treated in standard undergraduate quantum mechanics.

The total effective mass relevant for the center-of-mass motion of the exciton is $M \equiv m_e^* + m_h^*$. Similarly the reduced mass relevant for the relative coordinate motion (that is, $\mathbf{r} \equiv \mathbf{r}_e - \mathbf{r}_h$) is

$$m_{rm}^* = \frac{m_e^* m_h^*}{m_e^* + m_h^*}. \tag{5.27}$$

Similarly, the center-of-mass coordinate is

$$\mathbf{R} = \frac{m_e^* \mathbf{r}_e + m_h^* \mathbf{r}_h}{m_e^* + m_h^*}. \tag{5.28}$$

In these variables, a reasonable effective time-independent Schröedinger equation for the system is (see Exercises)

$$\left[E_C - E_V - \frac{\hbar^2}{2M} \nabla_{\mathbf{R}}^2 - \frac{\hbar^2}{2m_{rm}^*} \nabla_{\mathbf{r}}^2 - \frac{e^2}{4\pi \kappa \epsilon_0 |\mathbf{r}|} \right] \Phi(\mathbf{R}, \mathbf{r}) = E \Phi(\mathbf{R}, \mathbf{r}). \tag{5.29}$$

Writing $\Phi(\mathbf{R}, \mathbf{r})$ as a product of a function depending on $\mathbf{R}$ and one depending just on $\mathbf{r}$, we find that the center-of-mass motion is plane-wave-like, while the relative coordinate wavefunction $\phi(\mathbf{r})$ is hydrogenic:

$$\Phi(\mathbf{K}, \mathbf{r}) \propto \int d\mathbf{R} \exp(i\mathbf{K} \cdot \mathbf{R}) \phi(\mathbf{r}). \tag{5.30}$$

The allowed energies for excitons in bulk are therefore

$$E_{n,K} = E_g + \frac{\hbar^2 K^2}{2M} - \frac{1}{n^2} Ry^*, \tag{5.31}$$

where n is the principal quantum number for the exciton, and Ry^* is the effective Rydberg energy for the exciton in the host material:

$$Ry^* \equiv \frac{1}{(4\pi \kappa \epsilon_0)^2} \frac{m_{rm}^* e^4}{2\hbar^2} = Ry \times \frac{m_{rm}^*}{\kappa^2 m_0}. \tag{5.32}$$

Recall that the hydrogen atom Rydberg, Ry, is 13.6 eV. The characteristic size of the ground state of an exciton is therefore the equivalent of the Bohr radius:

$$a_{ex} = \frac{4\pi \kappa \epsilon_0 \hbar^2}{m_{rm}^* e^2}. \tag{5.33}$$

Recall from our discussion of donors and acceptors that κ for common inorganic semiconductors can be substantial, ~ 10. Combined with the low effective mass compared to free electrons, this results in large values of a_{ex}, approaching 10 nm in GaAs. Thus the assumption of a bulk average κ is self-consistent in these materials. In other systems (e.g., molecular crystals) κ can be lower, $\kappa \sim 2-3$, and the lattice spacing can be larger. An exciton that is smaller than the typical lattice spacing is called a *Frenkel* exciton.

5.5　Junctions between materials

Before looking at the effects of nanoscale confinement on the electronic properties of systems, let's start with a problem that is inherently "nano" even if it is usually not presented as such. What happens when a junction is formed between two different materials? This question is most interesting when the materials are permitted to exchange electrons, and for simplicity for now we consider the case of uniform temperature.

As we know from statistical mechanics, two systems that can exchange particles are only in thermodynamic equilibrium if their chemical potentials are equal. Therefore, if states are available one would expect electrons to diffuse from the material with (initially) a higher chemical potential to the material with (initially) the lower chemical potential, until the chemical potentials are equalized and equilibrium is reached. However, it makes sense that each of the two materials should maintain its own defining characteristics far from the interface. If material A has a workfunction Φ_A, then far from the junction it should still take Φ_A Joules to remove an electron from the Fermi level and carry it away to infinity. Furthermore, one must worry about the electrostatics of the situation, and treat that appropriately. If mobile electrons flow out of one material they leave behind an ionic background charge, leading to a nonuniform electrostatic potential, all of which must be solved self-consistently via the Poisson equation. If any surface states are relevant at the interface, their electrostatic contribution must be considered as well.

The upshot of all of this is that the general question, "How do the electronic levels of material A and material B align at and near the interface?" is an extremely difficult one to answer in general. Fortunately, there are some simple rules for drawing the resulting band diagrams that work pretty well a lot of the time. Figure 5.10 gives an example of *Anderson's Rule* [296]: find band alignment at an interface by assuming that the vacuum levels for the two materials align. This gives the band offsets between valence and conduction bands of A and B. Charge transfer then takes place to equilibrate the chemical potentials, and this is shown schematically by bending the bands over some length scale (the screening length). When drawn with the chemical potential as uniform across the combined system,

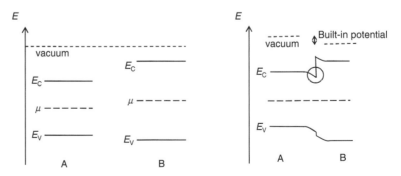

Fig. 5.10　Band diagram. At left, two materials with differing work functions and chemical potentials. At right, the result of bringing these materials together, after interfacial charge transfer (either forming an interfacial dipole or producing a mobile charge layer in the circled region) equilibrates the chemical potential.

the A and B vacuum levels far from the junction are now offset vertically by an amount corresponding to the "built-in" potential between the two materials.

This approach assumes that surface states at the interface play a negligible role compared to other populations of charge carriers. We should also remember what Poisson's equation tells us: in the absence of charge density, the electric field ($\nabla\phi$, where ϕ is the electrostatic potential) must be uniform.

5.5.1 Screening and charge transfer

Let's look at this in more detail for a specific example, the interface between two identical semiconductors *doped* such that they have differing equilibrium Fermi levels when unjoined. This is the classic *p-n* junction, and the above procedure tells us that there are no band offsets across the junction. This case is shown in Fig. 5.11. Intuitively the result is clear. The electrons from the *n*-doped side have a higher chemical potential before junction formation than those on the *p*-doped side. Therefore the electrons diffuse across the interface, combining with some of the holes. The result is an interfacial region with a reduced number of free carriers, but containing uncompensated dopants (ionized acceptors on the *p* side, donors on the *n* side). This is called the *depletion layer*, and we can figure out how big it is by using our knowledge of electrostatics and the thermodynamics of the charge carriers.

Recall our expressions for the thermal equilibrium electron and hole populations in doped semiconductors, Eqs. (3.10) and (3.11). Once the spatial distribution of carriers has equilibrated, there is a *uniform* chemical potential, μ, everywhere. Far from the junction region, μ needs to sit near the conduction band on the *n* side, and near the valence band on the *p* side, with the spatial band bending described by a *local* electrostatic potential, $\phi(x)$. That is,

$$n(x) = N_C(T) \cdot \exp\left[-\frac{[E_C - e\phi(x) - \mu]}{k_B T}\right],$$

$$p(x) = N_V(T) \cdot \exp\left[-\frac{[\mu + e\phi(x) - E_V]}{k_B T}\right]. \tag{5.34}$$

Far from the junction position ($x = 0$), we assume that all the donors and acceptors are ionized, and this yields an expression for the built-in potential difference across

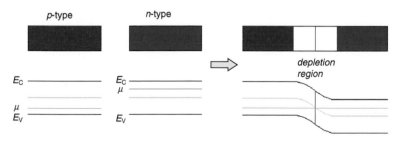

The *p-n* junction. Bringing together *p* and *n*-doped semiconductors results in charge transfer and the formation of a depletion region with no mobile carriers and dopant ions are uncompensated.

Fig. 5.11

the junction:

$$e\Delta\phi = e\phi(x = \infty) - e\phi(-\infty) = E_C - E_V + k_B T \ln\left(\frac{N_d N_a}{N_C N_V}\right). \tag{5.35}$$

One can plug into the Poisson equation and find the characteristic size of the depletion region. At $T = 0$,

$$d_{n,p} = \left[\left(\frac{N_d + N_a}{N_a N_d}\right)\frac{2\kappa\epsilon_0\Delta\phi}{e}\right]^{1/2}. \tag{5.36}$$

The physics at work here is that equilibrium is a balance between *drift* – the motion of the charge carriers in response to the built-in electric field – and *diffusion* – the tendency of carriers to diffuse away from regions of high density to regions of low density.

The general trend is that higher carrier densities and higher densities of states lead to shorter depletion lengths. Plugging in typical silicon numbers above, for light doping d can be as large as a micron. As we saw previously, for metallic carrier densities, the equivalent length in a metal is comparable to r_{TF}, which is usually fractions of a nanometer.

Direct interfaces between metals and semiconductors may be considered as a limiting case of the $p-n$ junction scenario. For a junction with an n-type semiconductor, one can arrive at the right expression for barrier width by taking the limit of Eq. (5.36) with $N_a \to \infty$. The corresponding depletion width in the metal is very small, as above, $\sim r_{TF}$. In the case described, an electron in the metal attempting to enter the semiconductor would encounter a *Schottky barrier* of energetic "height" $\Delta\phi$, with an accompanying depletion region extending into the semiconductor.

5.5.2   Heterojunctions and quantum wells

Let's look at *heterojunctions* without free carriers for a moment, systems where different semiconductor materials are joined, as in Fig. 5.12. (These kinds of system can be prepared via crystal growth techniques like MBE or MOCVD). Without any lateral patterning, the band structure is modulated along the growth direction, which we will take to be the z axis.

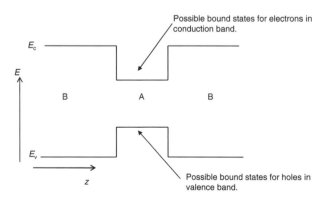

Fig. 5.12   Quantum well. By varying chemical composition during crystal growth, band alignment results in spatially varying conduction and valence band edge energies, acting as confining potentials for charge carriers.

In the GaAs/AlGaAs material system, for example, these structures may be grown with atomic precision. For the example shown, if we simply take the band offsets by using Anderson's Rule, we find for this structure an apparent potential well for conduction band states. This kind of structure is called a *quantum well*. Likewise, from the band structure shown it looks like the quantum well also defines a potential well for holes.

How do we describe the electronic states of inhomogeneous materials like this? Within the single-particle noninteracting electron picture, we can take Bloch wavefunctions to be reasonable trial wavefunctions for electrons in these materials. The x, y, and z components of the problem separate nicely, as long as we make the assumptions that (a) the conduction (and valence) band states in the two materials are similar (that is, have similar dispersions in k-space), and (b) the interface potential at the junctions between materials doesn't mix the valence and conduction band states.

The single-particle trial states for electrons look like

$$\psi(\mathbf{r}) = \sum_{A,B} \exp(i\mathbf{k}_\perp \cdot \mathbf{r}) u_{c\mathbf{k}}^{A,B} \chi_n(z). \tag{5.37}$$

Here A and B label the different materials, $\mathbf{k}_\perp$ is the wavevector in the x–y plane, and $u_{c\mathbf{k}}^{A,B}$ is a Bloch function for material A or B (depending on where the particle is in z). The function $\chi_n(z)$ is called an *envelope function*.

This form of trial wavefunction is useful because, since it's constructed from Bloch states appropriate for the various materials, it takes care of the periodic lattice potential effects already. When it's plugged into the full Hamiltonian, what ends up falling out are an equation for χ_n:

$$\left(-\frac{\hbar^2}{2m_e^*(z)}\frac{\partial^2}{\partial z^2} + E_C\right)\chi_n(z) = \epsilon_n \chi_n(z), \tag{5.38}$$

and boundary conditions that $\chi_n(z)$ and $(1/m_e^*)\partial\chi_n/\partial z$ must be continuous. This is basically a single-particle time-independent 1d Schrödinger equation for an electron experiencing a spatial potential given by the local conduction band energy. (The spatial variation in the effective mass is a comparatively small detail for our considerations at the moment.) This kind of description is extremely useful, so much so that people often don't talk about "envelope functions" at all and use the words "wavefunction" to mean the same thing. We will do the same, but remember that when we talk about calculating matrix elements for transition rates we need to use the actual total wavefunctions. The eigenvalues ϵ_n correspond to the contribution to the total energy due to the z degree of freedom. The *total* energy of the single-particle solutions will have an additional contribution due to the kinetic energy in the x–y plane, proportional to $k_\perp^2$.

The system shown in Fig. 5.12 is then a finite square well for electrons, at least from the envelope function perspective. We can readily guess some qualitative features of the solutions. The energies[3] of bound states deep in the well should look a bit like those of the 1d box with infinitely high walls, though lower because in the finite well case the bound state wavefunction is able to leak into the classically forbidden region. Since the

---

[3] Here we're talking about the eigenvalues of Eq. (5.38) rather than the total energy.

wavefunction is more spread out than in the hard wall case, its curvature is reduced and the kinetic energy is lower. For higher excited z states, the bound state energy levels should become more closely spaced in energy near the top of the well.

Set the zero of energy at the bottom of the well and have a well of thickness L and height V_0 centered on $z = 0$. For energies $\epsilon_n < V_0$, inside the well χ_n should look like free particle states (sines and cosines of $k_z z$, where k_z is some wavevector describing the z motion), while outside the well the envelope function should decay exponentially. Applying the boundary conditions at the edges of the well tells us what values of k_z are allowed. Assuming that the effective masses in A and B aren't very different (see Exercises), we find that k_z must satisfy

$$\tan \frac{k_z L}{2} = \sqrt{\frac{2m_e^*}{\hbar^2 k^2} - 1},$$

$$-\cot \frac{k_z L}{2} = \sqrt{\frac{2m_e^*}{\hbar^2 k^2} - 1}, \tag{5.39}$$

where the upper condition is for solutions of even symmetry about the center of the well, and the lower is for solutions of odd symmetry about the center of the well. The energies are given by

$$\epsilon_n = \frac{\hbar^2 k_z^2}{2m_e^*}, \tag{5.40}$$

with the restriction that $\epsilon_n < V_0$. This means that there are only a finite number of z bound states, given by

$$\text{no. of bound states} = 1 + \text{int} \left[\left(\frac{2m_e^* V_0 L^2}{\phi^2 \hbar^2} \right)^{1/2} \right], \tag{5.41}$$

where "int" means the largest integer less than the argument in brackets.

5.5.3 Holes and confinement

The example in the previous section dealt with electrons in the conduction band as the carriers of interest. One might imagine applying the same sort of treatment to holes in the valence band, but in many materials (including GaAs, for example) the situation is considerably more complicated. In *bulk* GaAs, there is no valley degeneracy of the conduction band near $\mathbf{k} = 0$, while there is degeneracy at the top of the valence band at that wavevector.

Furthermore, those valence states have reasonably strong spin–orbit coupling. This means that spin is no longer a "good" quantum number,[4] though $\mathbf{J}$, the total angular momentum, still is. In tight-binding treatments it's clear that the valence states can be considered as being built from p-orbitals, which have orbital angular momentum $L = 1$. Adding that to $S = 1/2$ for the spin, it turns out that the relevant states near $\mathbf{k} = 0$ are $J = 3/2$.

---

[4] That is, spin operators don't commute with the Hamiltonian, so that spin eigenstates are not energy eigenstates.

At a level more advanced than this book (see, e.g., [297]), one can do degenerate perturbation theory for the valence states near $\mathbf{k} = 0$. The result looks rather messy:

$$H = \frac{\hbar^2}{2m_0}\left[\left(\gamma_1 + \frac{5}{2}\gamma_2\right)k^2 - 2\gamma_2(k_x^2 J_x^2 + k_y^2 J_y^2 + k_z^2 J_z^2) - 4\gamma_3((k_x k_y)(J_x J_y + \cdots))\right].$$

(5.42)

The γ coefficients here are the Luttinger parameters. If motion is along z, the [100] crystallographic direction, things simplify a bit, with eigenstates having energies:

$$E = \frac{\hbar^2 k_z^2}{2m_0}(\gamma_1 - 2\gamma_2), \quad J_z = \pm\frac{3}{2},$$

$$E = \frac{\hbar^2 k_z^2}{2m_0}(\gamma_1 + 2\gamma_2), \quad J_z = \pm\frac{1}{2}.$$

(5.43)

The upshot of all this is that the spin–orbit couplings (the γs) are related to the effective masses of the light and heavy holes in *bulk* GaAs.

In confined systems such as the quantum wells above, things get even more complicated. To meet the boundary conditions on the envelope function at the well interfaces, it is necesary to consider mixing between the $J_z = 3/2$ and $J_z = 1/2$ states. The result is that *the effective masses and angular momentum properties of confined holes can be very different from those in bulk.* This has far-reaching implications, since this means that exciton properties are therefore tunable through confinement, as are selection rules for various transitions, including optical processes.

5.5.4 Triangular wells

The term "quantum well" is more generally applied than just to square wells like Fig. 5.12. Triangular potential wells as shown in Fig. 5.13 are also important to consider, since they occur in multiple situations. The classic examples are the channel region of Si MOSFETs and at doped GaAs/AlGaAs heterojunctions. The sloping side of the well can be modeled as an electric field along the z direction. In the MOSFET case this field is due to a voltage

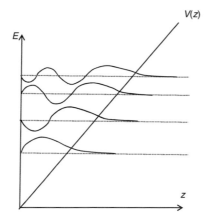

Infinite triangular well, with first four energy eigenstates sketched.

Fig. 5.13

applied to the gate electrode, as we shall see in Chapter 6. In the GaAs/AlGaAs case the field results self-consistently from charge transfer from donors in the AlGaAs into the GaAs conduction band followed by band bending.

The envelope function for states in the well must be a solution of

$$\left(-\frac{\hbar^2}{2m_e^*}\frac{\partial^2}{\partial z^2} + eFz\right)\chi_n(z) = \epsilon_n\chi_n(z), \tag{5.44}$$

where F is the electric field. For a hard well boundary (not a bad approximation for the MOSFET case, for example), the appropriate conditions are

$$\chi_n(z = 0) = 0,$$

$$\chi_n(z \to \infty) = 0, \tag{5.45}$$

$$\frac{\partial\chi_n}{\partial z} \text{ continuous.} \tag{5.46}$$

The solutions to this are

$$\chi_n(z) = \text{Ai}\left(\left(\frac{2m_e^*}{\hbar^2 e^2 F^2}\right)^{1/3}(eFz - \epsilon_n)\right), \tag{5.47}$$

where $\text{Ai}(x)$ is the Airy function.[5] Once again, the boundary condition (in this case, that $\chi_n(0) = 0$) sets the allowed eigenvalues. Forcing the argument of Ai to be a_n, the nth zero of the Airy function, gives

$$\epsilon_n = -\left(\frac{e^2 F^2 \hbar^2}{2m_e^*}\right)^{1/3} a_n \approx \left(\frac{\hbar^2}{2m_e^*}\right)^{1/3}\left(\frac{3\pi eF}{2}\left(n + \frac{3}{4}\right)\right)^{2/3}, \quad n = 0, 1, \dots \tag{5.48}$$

Real triangular wells and 2d electron gas

One of the most important realizations of a triangular quantum well is that which occurs at the single interface between doped AlGaAs and otherwise undoped GaAs. The situation is shown schematically in Fig. 5.14. Donated electrons from the $\text{Al}_{0.3}\text{Ga}_{0.7}\text{As}$ move across the interface into the GaAs layer, where band bending and electrostatics result in the formation of a (finite) triangular well, trapping the electrons up against the interface. Because the conduction band offset between GaAs and $\text{Al}_{0.3}\text{Ga}_{0.7}\text{As}$ is only about 0.2 eV [298], there is some penetration of the donated electron envelope function into the AlGaAs.

The confined electrons are unconstrained in the x–y plane, and can be described well by free 2d Bloch waves, forming a *two-dimensional electron gas*. One exciting aspect of this approach is that the mobile 2d electrons in the GaAs conduction band are physically displaced from their donor ions. Scattering of the carriers by those donor ion potentials is then much reduced compared to conventional doped semiconductors, where doped carriers and dopant ions are co-located. This strategy to produce extremely clean 2d electron

---

[5] The Airy function is another special function that shows up with some regularity in boundary value problems. Like the Bessel functions and the amusingly named confluent hypergeometric function, it's really no more unusual than sine or cosine, except that it's less common and lacks a simple geometric interpretation.

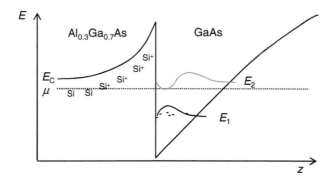

Modulation doping and 2d electron gas. A real (finite) triangular well is formed by the conduction band edge at the AlGaAs/GaAs interface, with doped carriers in the GaAs separated from their donor ions in the AlGaAs. The real potential $V(z)$ must be solved self-consistently, accomodating the charge transfer.

Fig. 5.14

gases is called *modulation doping* and was pioneered by Horst Stormer, John Dingle, and collaborators at Bell Labs in the late 1970s.

To get a first-order understanding of this system, assume that the electric field is due to the n_{2d} electrons per unit area that have been transferred across the boundary. Via Gauss' Law, the resulting electric field is constant in magnitude:

$$F = \frac{n_{2d}e}{\epsilon_0 \kappa}. \tag{5.49}$$

The result for the infinite triangular well is that the energy of the lowest z state (measured relative to the bottom of the well) is then

$$E_1(n_{2d}) \approx \left(\frac{\hbar^2}{2m_e^*}\right)^{1/3} \left(\frac{9\pi e^2 n_{2d}}{8\epsilon_0 \kappa}\right)^{2/3}. \tag{5.50}$$

So, the shape of the potential well and the energy of the bound states both depend on n_{2d}. We must solve this self-consistently, and remember that each z bound state carries with it a 2d density of states (see Eq. 2.24), ν_{2d}. If the electrochemical potential (Fermi level) of the donated electrons lies in the first z subband, and we measure energies with zero defined as the bottom of the GaAs conduction band, then

$$E_{F,\text{GaAs}} = E_1(n_{2d}) + \frac{n_{2d}}{\nu_{2d}} = E_1(n_{2d}) + \frac{\pi \hbar^2}{m_e^*} n_{2d}. \tag{5.51}$$

This must equal $E_{F, \text{AlGaAs}}$ for equilibrium.

Assume that there is a depletion width, W, on the AlGaAs side of the interface – that is, that all the donor ions ($N_d \equiv$ the 3d density of donors) within a distance W of the interface have donated their electrons into the 2d layer that is now on the GaAs side of the interface. This implies $n_{2d} = N_d W$. The electrostatic potential of the unscreened donor ions within the depletion width leads to a voltage built up across W within the AlGaAs:

$$V = -\int_0^{-W} F(z)\,dz = \int_0^{-W} \frac{eN_d z}{\epsilon_0 \kappa}\,dz = \frac{eN_d W^2}{2\epsilon_0 \kappa}, \tag{5.52}$$

where we've assumed that κ is the same in both GaAs and AlGaAs. If the conduction band offset is ΔE_C and the donor levels lie ϵ_d below the AlGaAs conduction band edge, we can equilibrate the Fermi levels in the two materials:

$$\Delta E_C - \frac{en_{2d}^2}{2\epsilon_0 \kappa N_d} - \epsilon_d = E_1(n_{2d}) + \frac{\pi \hbar^2}{m_e^*} n_{2d}. \tag{5.53}$$

This is an implicit equation for n_{2d} at $T = 0$.

Modulation doping in the GaAs/AlGaAs system has been continuously improved over the last three decades. There are several complications associated with the technique, some related to the details of material growth (e.g., GaAs and $Al_x Ga_{1-x} As$ have different solubilities for impurities; there can be monolayer fluctuations in interfacial roughness) and others to the detailed nature of the dopants (e.g., ϵ_d depends on x, and is around 10 meV for $x \approx 0.25$; not all Si donors are necessarily electrically active). Typical densities that may be achieved in modulation-doped GaAs/AlGaAs 2d electron gas range from $n_{2d} \approx 10^9 - 5 \times 10^{11}$ cm^2. Similar densities have recently become possible with modulation-doped 2d hole systems at the same interface, using C as a dopant in the AlGaAs rather than Si. We will return to this system later, when we discuss electronic transport. It turns out that GaAs/AlGaAs 2d electron gas has been an electronic system of critical importance in our developing understanding of quantum effects in electronic devices, physics which is particularly relevant at the nanoscale.

5.5.5 Multiple quantum wells

There is nothing special about the finite square well structure of Fig. 5.12. With the crystal growth techniques now available (particularly in the III–V compound semiconductors such as GaAs), it is possible to grow multiple wells while controlling well thickness and spacing at the atomic scale. Wells may be placed sufficiently close together that the "tails" of the z-bound states of adjacent wells can overlap. Combined with the ability to tune conduction band offsets via the chemical composition, it is then possible to engineer the hybridization of the z-components of the well bound states.

Let's consider the specific case shown in Fig. 5.15, and worry only about electrons in the conduction band, as before. The effective Schrödinger equation for the z envelope function, which we will now re-label $|\Psi(z)\rangle$,[6] is

$$\left(\frac{\hbar^2}{2m_e^*} \frac{\partial^2}{\partial z^2} + V_1(z) + V_2(z) \right) \Psi(z) = \epsilon \Psi(z). \tag{5.54}$$

Here $V_1(z)$ is the finite square well potential of just the left well in isolation; similarly for $V_2(z)$.

Assume a trial wavefunction for the double-well problem that is a linear combination of the states of the *isolated* finite square wells:

$$\Psi(z) = a_1 \Psi_1(z) + a_2 \Psi_2(z), \tag{5.55}$$

---

[6] When talking about hybridization of quantum well bound states, it is convenient to act as if the envelope function really is a 1d quantum state in the position representation.

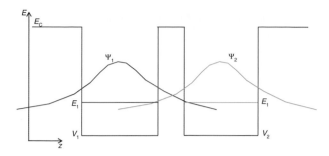

Fig. 5.15

Double quantum well, conduction band shown. The wells are close enough that lowest states of individual wells have nonzero overlap with the adjacent wells.

where the well states satisfy

$$\left(\frac{\hbar^2}{2m_e^*} \frac{\partial^2}{\partial z^2} + V_1(z) \right) \Psi_1(z) = E_1 \Psi_1(z),$$

$$\left(\frac{\hbar^2}{2m_e^*} \frac{\partial^2}{\partial z^2} + V_2(z) \right) \Psi_2(z) = E_2 \Psi_2(z). \tag{5.56}$$

If we further assume that each well in isolation is identical, then we can define an overlap matrix element:

$$V_{12} \equiv \langle \Psi_1(z)|V_2(z)|\Psi_2(z)\rangle = \langle \Psi_2(z)|V_1(z)|\Psi_1(z)\rangle, \tag{5.57}$$

and an on-site matrix element,

$$V_0 \equiv \langle \Psi_1(z)|V_2(z)|\Psi_1(z)\rangle = \langle \Psi_2(z)|V_1(z)|\Psi_2(z)\rangle. \tag{5.58}$$

We can then rewrite Eq. (5.54) as

$$\begin{pmatrix} E_1 + V_0 - \epsilon & V_{12} \\ V_{12}^* & E_1 + V_0 - \epsilon \end{pmatrix} \begin{pmatrix} a_1 \\ a_2 \end{pmatrix} = \begin{pmatrix} 0 \\ 0 \end{pmatrix}. \tag{5.59}$$

The resulting energy eigenvalues are

$$\epsilon = E_1 + V_0 \pm |V_{12}|. \tag{5.60}$$

This is precisely analogous to the simple diatomic molecule problem, with solutions for the z envelope function that look like bonding and antibonding orbitals. In the x and y degrees of freedom the solutions to the full single-particle problem still look like free Bloch waves.

The ability to design and control crystal structures like this implies the remarkable conclusion that *we can engineer band structures for electronic states that otherwise would not exist naturally*. For example, with precise layer-by-layer growth one can implement a real-life version of the Krönig–Penney model via a series, or *superlattice*, of quantum wells.

A simple and natural way to treat a superlattice of identical quantum wells is to use the same tight-binding approach that we applied in Section 2.3.5 to a 1d chain of atoms. Suppose the wells are uniformly spaced by a distance d. If we define $\psi_0(z)$ as the ground bound state envelope function localized to a single, isolated well centered on $z = 0$, then we can find an overlap matrix element:

$$\gamma \equiv \langle \psi_0(z)|V(z)|\phi_0(z-d)\rangle, \qquad (5.61)$$

and an on-site term,

$$\beta \equiv \langle \psi_0(z-d)|V(z)|\phi_0(z-d)\rangle. \qquad (5.62)$$

Following the same method that gave us Eq. (2.86), we find (apart from the energy related to transverse motion in the $x-y$ plane) for a superlattice of N wells a *miniband* of z states, with energies labeled by an effective wavevector q in the z direction:

$$E(q) = E_1 + \beta + 2\gamma \cos(qd). \qquad (5.63)$$

Here E_1 is the ground state energy of a single isolated well, and q runs from $1/Nd$ to $(N-1)/Nd$. The respective envelope functions look like

$$\Phi_q(z) = \frac{1}{\sqrt{(N)}} \sum_j e^{iqjd} \psi_0(z-jd). \qquad (5.64)$$

This solution has assumed periodic boundary conditions and an effective mass that doesn't vary with z. Remember, each of these z envelope functions carries with it an associated manifold of 2d $x-y$ Bloch states. The resulting total density of states for the miniband of Eq. (5.63) is

$$\nu_{\text{miniband}}(E) = N \frac{m_e^*}{\pi \hbar^2} \cos^{-1} \left(\frac{E - E_1 - \beta}{2\gamma} \right). \qquad (5.65)$$

The factor $m_e^*/\pi \hbar^2$ is the 2d density of states associated with each z state.

One could just as well carry out the procedure of Eqs. (5.61)–(5.64) using single-well states higher than ϕ_0, leading to multiple minibands. The resulting miniband structure has gaps between the minibands controlled by the well dimensions and growth. This has enabled unprecedented engineering of optoelectronic devices. While Nature may not have provided any homogeneous semiconductors with a band gap in the far infrared energy scale, crystal growers have been able to use this kind of "bandgap engineering" to develop diode lasers that operate at that wavelength via optical transitions between minibands. These devices are called *quantum cascade lasers* [299, 300] and are briefly discussed in Chapter 8.

5.6 Quantum wires

The quantum well states and minibands are examples of what happens when electronic structure is manipulated to confine electrons in one dimension, while leaving them free to move in the other two. One can carry this confinement business further, and ask what happens when charge carriers are confined on a scale comparable to their wavelength in *two* dimensions; the result is a *quantum wire* that has electronic states with energetics resembling those of the idealized 1d case. Often one is interested in the particular electronic states of these systems and their ability to serve as electronic or optoelectronic devices. Here we briefly discuss different implementations of quantum wires.

5.6.1 Metal quantum wires

Quantum wires fitting this definition may be fabricated in several ways. First, from ordinary metals there are times when it is possible to fabricate structures extended in one dimension composed out of essentially a single chain of atoms. This may be done in a free-standing geometry using a mechanical break junction, as discussed in Chapter 1. Some metals in particular (e.g., gold) are able to form extended atomic chains in this manner that are sufficiently long (several atoms) compared to their diameter (one atom) to merit the word "wire" [301, 302].

Longer atomic-width metallic wires are possible in systems supported by a substrate. For example, some single-crystal substrates can be cut and polished slightly (a few degrees) off-axis from a high symmetry direction. Under appropriate high temperature annealing conditions, this leads to the formation of atomic-level terraces, as shown in Fig. 5.16(a). Such a substrate is called a *vicinal surface*. When surface energetics and kinetics are favorable, coating a vicinal surface with a sub-monolayer coverage of metal leads to the spontaneous formation of metal wires at step edges.

Alternatively, a mismatch in equilibrium lattice spacing between the substrate and the deposited (and possible reacted) material can lead to spontaneous nanowire formation in the presence of surface steps. Figure 5.16(b) shows a particularly pretty example, with the formation of dysprosium disilicide nanowires on annealed, miscut Si(001).

One challenge in experimentally assessing the instrinsic electronic properties of these supported metal nanowires is understanding the influence of the substrate. This is particularly an issue when the nanowires are characterized via STM, since this requires at least moderate electronic coupling between the nanowires and a conducting substrate. An additional requirement is usually ultrahigh vacuum conditions, since atomic-scale structures can be affected by even a physisorbed monolayer of atmospheric contaminants.

The electronic structure of these supported nanowires has been probed through local measurements, including scanning tunneling spectroscopy [11], and large-area techniques

(a) (b)

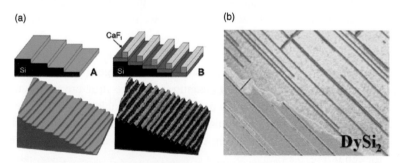

Two types of supported metal nanowire. (a) Vicinal wires grown on step edges of a miscut substrate [303]. (b) Strain-assembled rare earth silicide nanowires [304].

Fig. 5.16

such as photoemission. Of central importance is the evolution of electronic structure as a function of nanowire geometry and composition.

5.6.2 Semiconductor quantum wires

Using semiconductors there are (at least) three ways to produce extended structures that function as quantum wires. The first is via supported crystal growth of different chemical compositions of semiconductor. This is in close analogy with the quantum well structures described in Section 5.5.2. As in the quantum well case, many of the essential features of the single-particle electronic states confined to the quantum wires are described via an envelope function approach that leads to an effective Schrödinger equation.

Strain-based self-assembly of quantum wires is possible [305], as are other growth methods based on textured or grooved substrates [306]. Some of the cleanest 1d heterostructures defined by materials growth are those made with the cleaved-edge overgrowth method (CEO) [307]. In CEO, an initial heterostructure (say a quantum well) is grown on the (001) surface of a GaAs wafer. The wafer is then cleaved *in situ* in the MBE system, rotated 90 degrees, and another heterostructure is grown on the freshly exposed surface ((010), for example). The intersection of the two heterostructures can produce a 1d quantum wire.

A second approach to semiconductor quantum wires is to produce elongated semiconductor structures of very small transverse dimension (relative to the Fermi wavelength) with no encasing material. This would be the 1d analog of producing a free-standing quantum well. This can be done via lithographic patterning of a quantum well structure followed by anisotropic etching, as is commonly done to produce mesa-structures for measurements of the Hall effect. However, when the lateral dimensions of the unetched semiconductor material approach those needed for 1d, such surface processing can result in a relatively disordered system. An alternative that works well is to define the nanowires by local surface oxidation [308] rather than etching.

Alternatively, there are bottom-up techniques for growing large numbers of free-standing semiconductor nanowires by chemical vapor deposition, using group IV semiconductors as well as II/VI and III/V compound semiconductors. Extensive reviews have been written about the growth process and the variety of materials and structures that are possible [275, 309, 310]. In addition to homogeneous nanowires, wires may be grown with their chemical composition modulated along the length of the wire [197, 311]; the resulting structures are like the heterojunctions and quantum wells discussed previously, but with lateral confinement set by the wire diameter.

Core-shell nanowire structures are also possible with these vapor-phase growth methods. With appropriately chosen materials, the shell may have a larger band gap than the core, resulting in an effective *radially* confined quantum well wire structure for electrons. This has two advantages: electrons within the core are less vulnerable to direct scattering from surface imperfections; and a form of modulation doping may be performed, such that doped charge carriers are separated radially from their donor ions via the same charge transfer energetics as seen in Fig. 5.14.

Top gating is the third main method of producing effective quantum wire structures. Beginning with a 2d electron gas in a quantum well or heterointerface structure, metal gate electrodes may be patterned lithographically onto the semiconductor surface. By applying a negative voltage to those top gates relative to the potential of the 2d electron gas, the mobile carriers are repelled, depleting the 2d gas in the vicinity of the gates. With a starting 2d gas relatively shallow (close to the surface), top gates may be used to confine the 2d electron gas. If two top gates are relatively long, smooth, parallel, and spaced such that their respective depletion regions are separated by an amount on the order of the Fermi wavelength of the free carriers, a gate-defined quantum wire may be formed. A particularly short example of this is the quantum point contact introduced previously in Fig. 2.9. The advantage of top gating is that the resulting confined structures are *tunable*, while the other structures discussed above have static confining potentials.

5.7 Quantum dots

The logical extreme of confinement would be a "zero-dimensional" structure, with charge carriers constrained in all three dimensions to a region comparable in size to the effective λ_F. Achieving this in metals can be very challenging, and the resulting materials are usually clusters containing a relatively small number of atoms. Metal clusters, often fabricated via condensation in expanding jets of metal vapor, have rich physics and chemistry. Because of their extremely small size they can exhibit properties that differ greatly from those of bulk metals. Chemical bonding and electron–electron interactions lead to certain preferred cluster sizes containing "magic numbers" of constituent atoms. These magic number clusters are rather analogous to noble gas atoms with filled electronic shells. The electronic, magnetic, and optical properties of metal clusters have been studied extensively and were an inroad into nanoscale science long before the word "nanotechnology" became trendy.

Such structures in semiconductors are called *quantum dots*, and as with nanowires they may be made through various approaches. Figure 5.17 shows examples of these. The crystal growth of lattice-mismatched materials can lead to strain-mediated self-assembly of dots, where electronic confinement is again provided by the conduction and valence band offsets between the dot material and the surrounding semiconductor.

The confining potential can also result from the physical boundaries of the semiconductor system. This may be achieved by patterning and etching away of the surrounding material, as shown in Fig. 5.17(b), where GaAs/AlGaAs mesas have been left behind. The quantum dot here consists of the remaining 2deg within each mesa. These dots may be accessed electronically via top (and bottom) contacts to doped layers above (and below) the 2deg. Those contacting layers must be isolated from the 2deg via tunneling barriers (high band gap layers grown with higher aluminum concentration), or confinement in the *z* direction would be lost.

Alternatively it is possible to use chemical means to produce free-standing semiconductor nanocrystals, as shown in Fig. 5.17(c). Techniques for growing these nanocrystals have become highly refined in the last decade, and these are the materials referred to by the term "quantum dots" as used by chemists, chemical engineers, materials scientists, biophysicists, and bioengineers. These nanocrystals are typically grown by wet chemistry

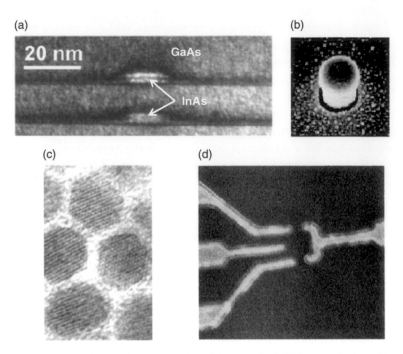

Fig. 5.17 Quantum dots. (a) Self-assembled InAs dots grown within a GaAs matrix [312]. (b) Quantum dot formed by wet etching of a GaAs/AlGaAs heterostructure [313]. (c) CdSe nanocrystal quantum dots formed by wet chemical methods [314]. (d) Quantum dot formed using gates to deplete GaAs 2deg [315].

techniques [201, 202], with surfactant molecules used to control growth kinetics and to prevent the nanocrystals from floculating or coalescing. Depending on the nanocrystal material, these chemical coatings can passivate electronic surface states and drastically improve the electronic and optical properties of these materials.

Nanocrystal-type quantum dots have become very useful as fluorophores. The effects of quantum confinement on charge carriers, particularly on the excitons discussed in Section 5.4, allow tuning of the optical absorption and emission of nanocrystals via dot diameter [316]. Detailed discussions of optical processes in semiconductor nanocrystals are beyond our scope here, but are discussed in books such as those by Gaponenko [317] and Klimov [318].

Nanocrystal morphology (both semiconductor and metal) can also be controlled, taking advantage of the fact that the binding affinity of surfactant ligand molecules depends on the crystallographic face of the growing nanomaterial. As a result, a menagerie of nanocrystal shapes is possible, including rods, tetrapods, and hyperbranched structures.

Study of the electronic transport in semiconductor nanocrystals is rather challenging. The surfactant ligands on the nanocrystal surface are commonly saturated hydrocarbon chains that act as rather effective insulating barriers. In thin films of nanocrystals longer chains (such as those based on hexadecane) can result in intercrystal distances of several nanometers, sufficiently large that charge carriers cannot readily hop. Methods do exist (e.g., ligand replacement [319] or ligand decomposition [320, 68]) that allow relatively efficient crystal-to-crystal conduction. Measuring the electronic properties of *individual* nanocrystals is even more challenging, since it requires either a scanning probe approach or the fabrication of conducting leads positioned with nanometer precision relative to the nanocrystal. This latter requirement has been accomplished via electrostatic trapping [321] and electromigration [322].

Finally, quantum dots may be defined from 2d electron gas via suitably arranged electrostatic top gates, as in Fig. 5.17(d). A key advantage of this approach is that the dot geometry is broadly tunable by adjusting the relative gate voltages. The amount of contact between the dot and the surrounding 2deg may be adjusted with high precision by having adjacent gates form quantum point contacts. With comparatively small (sub-micron) dots and shallow, low density 2deg, it is possible to make quantum dots that can be completely emptied of free carriers.

5.7.1 Single-particle states in quantum dots

Just as z-confinement from band structure leads to an effective Schrödinger equation for the Bloch envelope function in the z direction, lateral confinement has the same effect in the x and y directions. If one assumes some form for the confining potential and solves to find the allowed energies and single-particle states, the result is a series of discrete bound single-particle (envelope) wavefunctions and an accompanying energy spectrum. For this reason, quantum dots are sometimes called *artificial atoms*, and the allowed single-particle states are orbitals.

For confining potentials with a high degree of symmetry, the allowed spectrum can be easy to interpret, as we shall see below for a circular dot. The allowed states can be identified with classical periodic orbits possessing quantum numbers with simple interpretations. For dots of reduced or no symmetry, the spectrum can be very complicated, corresponding to classical chaotic motion.

Electronic transport measurements may be used as a form of spectroscopy to study the allowed single-particle states within quantum dots. The specific technique most commonly used involves measuring the conductance in the so-called Coulomb blockade regime, explained in detail in Section 6.7.2. This makes it possible to track the evolution of the single-particle states as a function of external parameters like a magnetic field.

For example, consider a circular quantum dot formed from a confined 2deg, as in Fig. 5.17(b). Modeling the detailed confining potential is challenging, but we shall follow the analysis of Kouwenhoven *et al.* and assume a harmonic potential. The time-independent effective Schrödinger equation for the electron (envelope) wavefunction looks like:

$$-\frac{\hbar^2}{2m_e^*}\nabla^2\Psi(\mathbf{r}) + \frac{1}{2}m_e^*\omega^2 r^2\Psi(\mathbf{r}) = E\Psi(\mathbf{r}), \qquad (5.66)$$

where ω describes the strength of the confining potential, and $\mathbf{r}$ is restricted to lie in the $x-y$ plane. The resulting spectrum of solutions is described (ignoring spin for a moment) by two quantum numbers, n and l, the principal quantum number, and the orbital angular

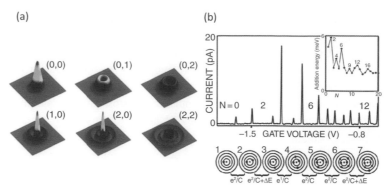

Circular quantum dot states. (a) Single-particle wavefunctions for the states described by Eq. (5.67). (b) Electronic transport data in the Coulomb blockade regime (Section 6.7.2) showing a shell structure expected from such a system. Adapted from [313].

momentum due to motion in the plane, respectively:

$$E_{n,l} = (2n + |l_z| + 1)\hbar\omega. \tag{5.67}$$

This spectrum implies a shell structure similar to that seen in atomic spectra. The spatial ground state of the system has $n = l = 0$, and in our *aufbau* way of thinking about single-particle states it can hold two electrons, one of each spin. The next possibility is to have $n = 0$ and $|l| = 1$, for which l can take on two values, $l_z = \pm 1$. Again, each such orbital state can hold two electrons, leading to a "shell" that can hold a total of four electrons. Things get more complicated in the third shell, where the $n = 1, l = 0$ spatial states are degenerate with $n = 0, l_z = \pm 2$ spatial states. This shell can hold six more electrons.

Note that we have been ignoring electron–electron interactions in our discussion of shell-filling so far. This is a crude approximation, and it is precisely these interactions that allow us to use electronic conduction as a kind of spectroscopy to check for this shell structure in real devices. Figure 5.18 shows the ideal (envelope) wavefunctions that describe the solutions to Eq. (5.66) as well as conduction data with evidence for the existence of such a shell structure. In the next chapter we introduce the concepts of electronic transport, and will then return to the discussion of this quantum dot.

5.8 Summary and perspectives

This chapter centered on three main deviations from the "noninteracting electrons in an infinite periodic potential" picture of solids that we used to understand much electronic structure. All of these are of critical importance at the nanoscale and allowed us to introduce key concepts.

- Structural effects. We began by discussing point, line, and other defects, emphasizing that deviations from an infinite periodic potential have specific consequences for electronic structure. We introduced the work function and pointed out that atomic-scale changes to surfaces can have large electronic consequences (e.g., changing the energy required to add or remove an electron from a system by an eV). We also saw that the breaking of the lattice symmetry introduces electronic single-particle states that can look very different from the Bloch waves. Surface states are localized perpendicular to the interface at which they exist and may be occupied or empty depending on electronic population. In nanoscale systems a large fraction of the total single-particle states can effectively be surface-related.

- Electron–electron interactions. We discussed the response of the electrons within a material to the presence of charge, showing that charge rearranges in a way that reduces (or screens) the long-range electric field. In dielectrics we introduced the idea of polarizability and self-consistently found the Clausius–Mossotti relation, an expression for bulk dielectric effects based on the microscopic response of the constituents of a material. We then moved to the case of conductors, when charges are free to move, and introduced the concept of a screening length, an exponential decay of the Coulomb potential far from a given point charge. We saw the general trends that screening becomes worse when the density of mobile charge carriers is low, when the dimensionality available for carrier motion is reduced, and (in the nondegenerate limit often appropriate for doped semiconductors) when the temperature is increased. We also introduced excitons, bound electron–hole pairs in semiconductors.

- Confinement. Our last major concept was the idea of confinement, proceeding from single interfaces and quantum wells to quantum wires all the way down to quantum dots. The envelope function was introduced as an effective wave function useful for describing single-particle states in confined systems. As a result we were able to talk about coupled quantum wells and the formation of minibands. When confinement takes place on length scales comparable to the Fermi wavelength, a dimensional crossover takes place. The resulting single-particle states (envelope functions) can be strongly affected. The next chapter will show some of the consequences of these confinement effects for electronic properties such as conductance.

The next chapter presents the physics of electronic conduction, starting with classical physics, gradually adding in quantum corrections, and ending up in the full quantum limit. After discussing quantum transport and related nanoscale phenomena, we will examine the operating principles of field-effect transistors, the technological underpinning of a huge swath of modern electronics. We will consider the state of the art and study in some detail the nanoscale physics that is relevant in ultrascaled devices as well as possible nanoelectronic options to augment or replace transistors.

5.9   Suggested reading

- *2d electron systems and quantum wells* The classic early review article about 2d electronic systems is that by Ando, Fowler, and Stern, "Electronic properties of two-dimensional systems", *Rev. Mod. Phys.* **54**, 437–672 (1982).

Good descriptions of screening as well as issues specific to semiconductor quantum wells and superlattices are found in John H. Davies's book, *The Physics of Low-Dimensional Semiconductors* (Cambridge University Press, 1997), as well as Weisbuch and Vinter's *Quantum Semiconductor Structures: Fundamentals and Applications* (Academic Press, 1991).

Other useful references for these systems include A. Y. Shik, *Quantum Wells: Physics & Electronics of Two-dimensional Systems* (World Scientific, 1997) and, for a more computational approach, Paul Harrison, *Quantum Wells, Wires, and Dots: Theoretical and Computational Physics of Semiconductor Nanostructures* (Wiley, 2005).

- *Quantum wires* Several reviews have been published about semiconductor quantum wires, including Law *et al.*, "Semiconductor nanowires and nanotubes", *Ann. Rev. Mat. Res.* **34**, 83–122 (2004) and Lu and Lieber, "Semiconductor nanowires", *J. Phys. D: Appl. Phys.* **39**, R387–R406 (2006).

- *Quantum dots* A good review of semiconductor dots is the one by Yoffe, "Semiconductor quantum dots and related systems: electronic, optical, luminescence, and related properties of low dimensional systems", *Adv. Phys.* **50**, 1–208 (2001), and an emphasis on electronic effects in such structures is given by Efros and Rosen, "The electronic structure of semiconductor nanocrystals", *Ann. Rev. Mat. Sci.* **30**, 475–521 (2000). A more chemistry-oriented approach is taken by Trindade *et al.*, "Nanocrystalline semiconductors: synthesis, properties, and perspectives", *Chem. Mater.* **13**, 3843–3858 (2001).

An extensive look at quantum confinement effects in both semiconductor and metal nanocrystals is Moriarty, "Nanostructured materials", *Rep. Prog. Phys.* **64**, 297–381 (2001). A detailed look at the electronic states in metal nanocrystals may be found in von Delft and Ralph, "Spectroscopy of discrete energy levels in ultrasmall metallic grains", *Phys. Rep.* **345**, 61–173 (2001).

Semiconductor quantum dots defined from 2deg are described well in Kouwenhoven *et al.*, "Few-electron quantum dots", *Rep. Prog. Phys.* **64**, 701–736 (2001).

Exercises

5.1 *Defect densities* Suppose the energetic cost of creating a point defect in a crystal lattice is $U = 3$ eV.

    (a) For a simple cubic crystal with $N = 10^{23}$ lattice sites, with atomic diameter $a = 0.3$ nm, based *purely on U*, how many such point defects would you expect to exist in thermal equilibrium at 300 K?

    (b) Now consider the *free* energy per defect, $U - k_B T \ln(N)$, and again determine the expected equilibrium population of defects at 300 K.

5.2 *Crystal structure and small particles* Consider building up a face-centered cubic crystal using atoms of diameter a.

    (a) What is the length of one side of a unit cell?

(b) What is the packing fraction of such a crystal? That is, what fraction of a crystal volume is made up of the atoms themselves, the remainder being the interstitial spaces?

(c) What is the 2d packing density (number of atoms per unit area) of the (100) (and equivalent) crystal plane?

(d) Consider building up cubes of this material from complete unit cells. The smallest cube would be a single FCC unit cell (with complete atoms at the corners and faces, of course). The next possible cube would have eight unit cells, and the one after that 27. Make a table showing the number of surface atoms and the number of interior atoms for these first three cubes (you have to either count or deduce the algebraic rules). What trend do you see? Roughly how large a cube of material would you need for the interior atoms to outnumber the surface atoms, if a is 0.3 nm? The energetic differences between surface atoms (unfavored – fewer satisfied bonds) and interior ("bulk") atoms are the driving reasons behind phenomena like melting point suppression and size-dependent structural transitions.

5.3   *Crystal structure and small particles II* Redo the previous exercise, but for a BCC crystal structure.

5.4   *Screening in nondegenerate systems* When computing the Thomas–Fermi screening length, we tacitly assumed that all states below the chemical potential were full, and those above were empty – essentially we assumed degeneracy of the electron gas. To generalize to the nondegenerate limit, we can look at the analysis in our discussion of screening. In general,

$$n = \int \nu(E) f(E, \mu, T) dE, \qquad (5.68)$$

where f is the Fermi–Dirac distribution. Starting from this equation and remembering that what we really care about in Eq. (5.18) is $\delta n/\delta E = -\partial n/\partial \mu$,

(a) Show that at $T = 0$ when f is step-like, $\partial n/\partial \mu$ reduces to $\nu(E_F)$.

(b) Show that in the nondegenerate limit when f looks like the Maxwell–Boltzmann distribution, $\partial n/\partial \mu \approx n/k_B T$.

5.5   *Finite square well states* Consider the finite square well of Fig. 5.19. Assume that the effective mass for electrons, m_e^*, is the same in the well as in the wall material.

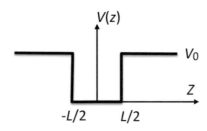

Finite square well.

Fig. 5.19

(a) For a trial wavefunction for the ground state, assume $\chi(z) = C \cos k_z z$ inside the well, where C is a normalization constant, and $\chi(z) = D \exp(\pm \kappa z)$, where

$$\frac{\hbar^2 \kappa^2}{2m_e^*} = V_0 - \epsilon$$

for the trial wavefunction in the walls, with the sign appropriate for $\chi(z) \to 0$ as $z \to \pm\infty$. By applying the appropriate boundary conditions at the edges of the well, derive the upper equation of Eq. (5.39).

(b) Assuming $\chi(z) = C \sin k_z z$ for the first excited bound state of the well, follow the same procedure as in (a) and derive the rest of Eq. (5.39).

PART II

Charge transport and nanoelectronics   6

The importance of the infinitely little is incalculable.

Dr. Joseph Bell

In this chapter we examine nanoscale science as applied to nanoelectronics. We begin with a review of electronic transport in general, starting from a classical treatment and then introducing the effects of quantum mechanics on our concepts of conduction. After examining quantum corrections to (semi)classical conduction and the ideas of quantum coherence, we discuss in detail the fully quantum mechanical limit, starting with quantum tunneling. This allows us to introduce the idea of conduction as a quantum scattering problem, and bring in the elegant Landauer–Buttiker formulation of quantum transport that is applicable in the extreme nanoscale limit.

Having laid this groundwork, we delve into an overview of the metal-oxide-semiconductor field-effect transistor (MOSFET), the heart of modern computing technologies. After examining the device physics relevant for MOSFETs we discuss the current state-of-the-art and explore how the physics that arises at the nanoscale will strongly influence the form of future devices. We discuss some of the physical limits that apply to ultrascaled electronic devices; we then examine some of the alternative device designs enabled by nanoscale structures, including nanotube and nanowire-based transistors, single-electron transistors, and molecular electronic systems.

6.1 Transport terminology

One natural technique used to study the consequences of nanoscale confinement on the electronic properties of materials is electronic transport. For example, an electrostatic potential difference, eV, is applied across the material, and the resulting motion of charge carriers is measured as a function of one or more control parameters (V, T, B). In this section we briefly review the basic terminology of electronic transport measurements.

The parameters most commonly measured in electronic transport at low frequencies (e.g., dc) are the conductance, G,

$$G \equiv \frac{\partial I}{\partial V}, \tag{6.1}$$

and the resistance, R,

$$R \equiv \frac{1}{G} = \frac{\partial V}{\partial I}, \tag{6.2}$$

where I is the current through the system and V is the voltage drop across the system of interest. In the so-called linear response regime, $G \to I/V, R \to V/I$.

When interpreting experimental results it can be crucial at times to understand what was actually measured, the conductance or the resistance. A simple true conductance measurement applies a well-defined voltage across the sample and measures the resulting current response. A device with a voltage deliberately applied across it is said to be *biased*, and that voltage is a *bias voltage*. Conversely, a simple true resistance measurement pushes a known current (sometimes called a *bias current*) through the sample and measures the resulting electrostatic potential difference across the sample.

One challenge in probing electronic transport is discerning between the intrinsic properties of the material under investigation and the extrinsic details associated with the measurement itself. Measurements on nanoscale systems can be particularly vulnerable to this issue. For example, with a particularly tiny sample such as a single semiconductor or metal nanocrystal, it may be necessary to perform a *two-terminal measurement*, using the same electrodes to carry current as well as measure the potential drop across the sample. The problem is that there can be potential drops at the contacts that are a function of the nature of the contact, rather than being simply related to the intrinsic electronic conduction of the nanoparticle.

The standard approach to mitigating these challenges is the *four-terminal measurement*, as shown in Fig. 6.1. In the ideal four-terminal measurement, two contacts are used to deliver and sink the current, while two other contacts are used as voltage probes.

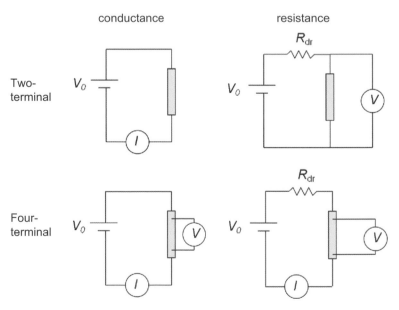

Fig. 6.1   Two- and four-terminal measurements. Left column: conductance measurements (source voltage, measure current). Right column: resistance measurements (source current, in this case using a large "ballast" resistor and a voltage source, measure voltage). The use of separate voltage and current leads can allow the measurement of electrical resistance without contributions from the contacts, though there are some subtleties (see text).

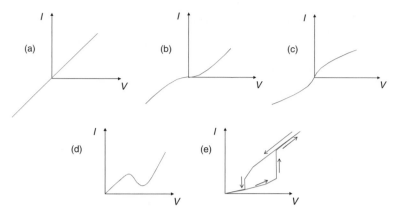

Various $I-V$ characteristics that are often seen when measuring the electronic properties of nanostructures. (a) Ohmic conduction, where dI/dV, the differential conductance, is constant as a function of V. (b) Zero-bias suppression of the conductance. (c) Zero-bias enhancement of the conductance. (d) An example of negative differential resistance (or negative differential conductance), where the dI/dV is negative over some range of V. (e) Hysteretic conduction, or "switching", such that the measured conductance depends on the past history of V.

Fig. 6.2

Implementing this approach faces practical challenges, not the least of which is the geometric constraint of attaching multiple conducting wires to what may be a nanoscale sample. For an ideal four-terminal measurement it is also necessary that the voltage probes be noninvasive – that they function purely as detectors of the local potential without disturbing the current flow. In principle this can be accomplished using conducting scanned probe microscopy tips and careful measurements of electrostatic force. In practice, one must be conscious of these concerns, as we shall see in Section 6.4.

There are various types of $I-V$ characteristics that commonly arise in studies of nanostructures, as shown in Fig. 6.2. An $I-V$ characteristic is said to be "Ohmic" when the current depends linearly on the voltage. One common non-Ohmic $I-V$ curve has a weaker slope near $V = 0$ called a "zero-bias suppression" of the conductance. Alternatively, one can have an enhanced slope near zero bias. These "zero-bias anomalies" and their dependence on temperature are important clues to the underlying conduction mechanisms at work.

Two other more exotic $I-V$ features are often discussed in nanoscale systems: negative differential resistance (NDR) and hysteretic switching. Both are of interest because of their potential technological utility. NDR, when the $I-V$ curve actually has a negative slope over some region of voltage bias, can be used for amplification. Hysteretic conduction may be used in switching or memory applications if appropriately stable and characterized.

The basic transport coefficients relevant to conduction are the conductivity and resistivity tensors, as defined here for three dimensions:

$$\mathbf{J} = \tilde{\sigma} \cdot \mathbf{E},$$

$$\begin{pmatrix} J_x \\ J_y \\ J_z \end{pmatrix} = \begin{pmatrix} \sigma_{xx} & \sigma_{xy} & \sigma_{xz} \\ \sigma_{yx} & \sigma_{yy} & \sigma_{yz} \\ \sigma_{zx} & \sigma_{zy} & \sigma_{zz} \end{pmatrix} \begin{pmatrix} E_x \\ E_y \\ E_z \end{pmatrix}, \tag{6.3}$$

$$\mathbf{E} = \tilde{\rho} \cdot \mathbf{J},$$

$$\begin{pmatrix} E_x \\ E_y \\ E_z \end{pmatrix} = \begin{pmatrix} \rho_{xx} & \rho_{xy} & \rho_{xz} \\ \rho_{yx} & \rho_{yy} & \rho_{yz} \\ \rho_{zx} & \rho_{zy} & \rho_{zz} \end{pmatrix} \begin{pmatrix} J_x \\ J_y \\ J_z \end{pmatrix}. \tag{6.4}$$

The diagonal elements of $\tilde{\sigma}$ and $\tilde{\rho}$ are often called the *longitudinal* conductivity and resistivity, respectively, since they connect electric fields and current densities pointing along the same direction. The off-diagonal elements are then *transverse* components.

The resistivity tensor is the inverse of the conductivity tensor. Things get a bit subtle in reduced dimensionality. Considering only two dimensions, one finds that the resistivity tensor is

$$\tilde{\rho} = \frac{1}{(\sigma_{xx}\sigma_{yy} - \sigma_{xy}\sigma_{yx})} \begin{pmatrix} \sigma_{yy} & -\sigma_{xy} \\ -\sigma_{yx} & \sigma_{xx} \end{pmatrix}. \tag{6.5}$$

This implies a counterintuitive situation: in two dimensions a system with zero longitudinal conductivity also has zero longitudinal resistivity.

The various components of the $\tilde{\sigma}$ and $\tilde{\rho}$ tensors are independent. In the presence of a magnetic field, B, underlying symmetries of nature and the restriction that entropy production must be greater than or equal to zero lead to restrictions on the components, as pointed out by Lars Onsager [323]. These restrictions give what are now called the Onsager relations, such as

$$\sigma_{xy}(\mathbf{B}) = \sigma_{yx}(-\mathbf{B}). \tag{6.6}$$

The intensive quantities $\tilde{\sigma}$ and $\tilde{\rho}$ are often not what is measured directly; rather, extensive (that is, geometry dependent) parameters like the conductance and resistance are typically the direct output of experiments. In a multiterminal measurement scheme, it is customary to define a four-terminal resistance. $R_{ab,cd} \equiv$ the resistance calculated by dividing the voltage difference measured between leads c and d by the current sourced at lead a and sunk at lead b. In terms of these resistances, the Onsager relations become reciprocity relations:

$$R_{ab,cd}(\mathbf{B}) = R_{cd,ab}(-\mathbf{B}). \tag{6.7}$$

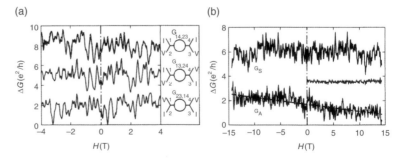

Fig. 6.3   Onsager relations and reciprocity in conduction. (a) Conductance as a function of magnetic field for different lead configurations in a mesoscale metal ring. The fluctuations are a quantum mechanical effect, not noise.
(b) Combinations of the conductance traces from (a) that are predicted by the Onsager relation to be symmetric (top) and antisymmetric (bottom) in B. Adapted from [324].

This is a rather remarkable result. For many nanoscale systems, the magnetoresistance can be quite complicated. Equation (6.7) implies that the magnetoresistance for some lead configuration and some orientation of **B** should be the same as the magnetoresistance for the opposite direction of **B** with the current and voltage leads swapped. This kind of reciprocity is demonstrated experimentally in Fig. 6.3, where linear combinations of four-terminal resistances have been computed that are predicted by reciprocity to be symmetric or antisymmetric in B.

6.2 Kinetic concepts

The classical way to think about electronic transport is in terms familiar from the kinetic theory of gases [325]. Treat electrons like a classical gas of noninteracting particles of mass m that happen to have charge q. One can get detailed and assume some distribution function that describes the probability of finding a particle with a particular velocity; proceeding down that route leads to the Boltzmann equation formalism to describe the evolution of that distribution as a function of time when driven out of equilibrium. Instead, one can get a sense of many of the important results by assuming that the particles travel with some typical speed v that will turn out to be something like v_F, the Fermi velocity. Further assume that the directions of motion are randomly distributed and therefore the average velocity is zero.

Since in this picture the particles are noninteracting, the only scattering processes that we need to worry about are the particles bouncing off disorder – static defects of some kind. Assume that the characteristic time between these scattering events is τ, leading to a mean free path between scattering events that is $\ell = v_F \tau$. Since the disorder is static, this scattering has to be elastic, so that its effect is to randomize the direction of the particle motion rather than affect the speed of the particles.

The mean free path concept immediately divides motion into two regimes. On distance scales short compared to ℓ, the particles do not scatter and motion is *ballistic*. On distance scales long compared to ℓ, the individual particle trajectories look like a random walk and motion is *diffusive*.

Consider what happens when such a gas is put into an electric field. Because they're charged, the particles accelerate in response to **E**, but experience scattering events in a typical time τ that randomize the direction of their momentum. The result is that the ensemble of particles acquires an average *drift velocity*:

$$\mathbf{v}_d = \frac{q\mathbf{E}}{m}\tau,$$

(6.8)

and there is a resulting current density,

$$\mathbf{J} = n_{3d}q\mathbf{v}_d = \frac{n_{3d}q^2\tau}{m}\mathbf{E}.$$

(6.9)

This allows us to define what is commonly known as the Drude conductivity,

$$\sigma = \frac{n_{3d}q^2\tau}{m},$$ (6.10)

and the mobility,

$$\mu \equiv \frac{v_d}{E} = \frac{q\tau}{m}.$$ (6.11)

Because it contains the scattering time, mobility is a particularly useful figure of merit for evaluating the relative cleanliness of semiconductors. The temperature dependence of mobility can also reveal the dominant scattering mechanisms for free charge carriers, since τ, the momentum relaxation time, can have contributions from inelastic scattering processes (e.g., electron–phonon scattering) at higher temperatures. The colloquial units of mobility are cm^2/Vs. The typical mobility for high purity single-crystal Si at 300 K is approximately 1000 cm^2/Vs, and increases as temperature is decreased, consistent with phonon scattering playing a role in setting τ. In contrast, the mobility in amorphous Si at 300 K is more like 10 cm^2/Vs. In GaAs 2deg at low temperatures, mobilities exceeding 10^7 cm^2/Vs have been reported, implying a scattering time exceeding 4×10^{-10} s; for typical carrier densities in this system, $v_F \sim 10^5$ m/s, implying that $\ell \rightarrow 40$ μm, a remarkably long mean free path.

In the diffusive regime motion of individual carriers is through a random walk of average step size ℓ. The typical distance diffused by a carrier in a time t is given by $\sqrt{Dt}$, where D is the diffusion constant, $D = (1/d)v_F\ell = (1/d)v_F^2\tau$. Here d is the dimensionality of the diffusion process, determined by the size of the system compared to the mean free path. If the length, width, and thickness of a system all exceed ℓ, the system is 3d for diffusion; by contast, if only the length exceeds ℓ, diffusion is essentially a one-dimensional process. The utility of D here is in translating time scales into length scales in moderately disordered conducting systems. For example, if an inelastic process has some rate τ_i^{-1}, then the typical length scale associated with that process in a diffusive system would be $\sqrt{D\tau_i}$.

Fortunately one can infer the diffusion constant experimentally. As derived in the Appendix, Section A.3, for metals the Einstein relation connects the measured electrical conductivity with the electronic diffusion constant: $\sigma = q^2 D v_{3d}(E_F)$. Here $v_{3d}(E_F)$ is the density of states at the Fermi level, inferred experimentally from measurements of the electronic heat capacity. For a typical metal like Au at room temperature, $\rho \approx 2$ μΩ-cm and $v_{3d} \approx 10^{47}$ $J^{-1}m^{-3}$, implying $D \approx 0.02$ m^2/s. For Au $v_F = 1.5 \times 10^6$ m/s, leading us to infer that $\ell \approx 40$ nm and therefore $\tau \approx 2.7 \times 10^{-14}$ s.

From these numbers it is clear that it is possible to make thin metal films that are effectively one or two-dimensional as far as diffusion is concerned. Boundary scattering is an example of a purely classical finite size effect. One limit is specular scattering, where the particles bounce off flat, smooth surfaces such that their momentum is not randomized (rather, the component of momentum normal to the surface is reversed in sign). This kind of scattering does not affect ℓ. The opposite limit is diffuse scattering, where the momentum after scattering depends so sensitively on the incident momentum prior to scattering that the boundary scattering effectively randomizes the particle direction. Diffuse scattering is characteristic of rough surfaces and does reduce ℓ from what would be its bulk value. If this scattering is happening at the boundaries of a 2d film of thickness w, then

$$\ell_{\text{eff}}^{-1} \approx \ell^{-1} + w^{-1},$$

$$\sigma_{\text{eff}} \approx \sigma \frac{w}{\ell}. \tag{6.12}$$

6.3 Hall effect

Introducing a magnetic field into the problem leads to a classical correction to the conductivity and resistivity tensors via the Hall effect. As shown in Fig. 6.4, when current flows perpendicular to the magnetic induction, a transverse voltage develops in steady state. Normalizing this Hall voltage by the current yields the Hall resistance, R_H.

The microscopic origin of the Hall resistance is straightforward. Moving charge carriers are acted upon by both the electric field and the Lorentz force. The Lorentz force pushes carriers toward the sample edges where they pile up until a suffcent transverse electric field develops to balance the transverse Lorentz force and establish a steady state. Setting equal the contributions of momentum change due to scattering and momentum change due to the forces,

$$\frac{m\mathbf{v}_d}{\tau} = q[\mathbf{E} + \mathbf{v}_d \times \mathbf{B}]. \tag{6.13}$$

Writing out the components and recalling the definition of mobility, in two dimensions the usefulness of the Hall effect is clear:

$$\begin{pmatrix} E_x \\ E_y \end{pmatrix} = \begin{pmatrix} m/q^2 n_{2d}\tau & -m\mu B/q^2 n_{2d}\tau \\ m\mu B/q^2 n_{2d}\tau & m/q^2 n_{2d}\tau \end{pmatrix} \begin{pmatrix} J_x \\ J_y \end{pmatrix}. \tag{6.14}$$

We can identify $\rho_{xx} = \rho_{yy}$ as the inverse of the Drude conductivity, and find that $\rho_{yx} = -\rho_{xy} = B/|q|n_{2d}$. For current flowing in the x direction in steady state, $J_x = I/x$, $J_y = 0$, and $V_x = E_x L$, $V_H = E_y w$, where L and w are the length and width of the sample, respectively.

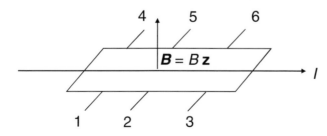

Geometry for the Hall effect. When a current flows as shown in the presence of a transverse magnetic field, a Hall voltage will develop (e.g., between leads 2 and 5) with a sign that depends on the type of charge carriers. In the geometry shown, for holes, $V_2 - V_5 > 0$, while for electrons $V_2 - V_5 < 0$. This transverse potential difference (produced by a slight change in the spatial distribution of charge) develops because in steady state the force from the transverse electric field must balance the transverse Lorentz force acting on the carriers.

Fig. 6.4

Combining these makes it clear why Hall measurements are so useful. First, the sign of the Hall voltage can be used to identify the sign of the charge carriers. Second,

$$n_{2d} = \frac{I/|q|}{dV_H/dB},$$
$$\mu = \frac{I/|q|}{n_{2d}V_x w/L}. \tag{6.15}$$

So, from measurements of the Hall resistance as a function of magnetic field, it is possible to infer the carrier density. Knowing sample dimensions it is then possible to infer the mobility (known in this context as the Hall mobility) from measurements of the longitudinal resistance.

Notice that the Hall voltage is inversely proportional to the carrier density and directly proportional to the mobility. With the advent of high mobility 2deg systems, this has led to the development of powerful magnetic field sensors.

What happens in the presence of a magnetic field in the "clean" limit, when the charged particles only scatter rarely? In the absence of an electric field, classical charged particles with a speed v_F would have their trajectories curved into circular orbits; the radius of those paths is known as the cyclotron radius:

$$r_c = \frac{mv_F}{qB}. \tag{6.16}$$

This is a new length scale that can have interesting effects. It is clear that a requirement to see unambiguous effects of this motion is that $r_c >\sim \ell$, providing a guideline for our definition of the clean limit. In the presence of an electric field as well, the instantaneous center of the cyclotron orbit drifts in the direction of $\mathbf{E} \times \mathbf{B}$.

There is a corresponding frequency scale associated with this cyclotron motion:

$$\omega_c = \frac{qB}{m}. \tag{6.17}$$

This is the angular frequency of the motion of the charged particle around its orbit.

As B is varied, r_c can cross other length scales. Consider the effect on a relatively clean sample with diffuse boundary scattering. As B is turned on, carrier trajectories begin to curve, and more particles hit the boundaries than when $B = 0$. As a result ℓ_{eff} is reduced, and the electrical resistance increases. In the strong field limit $r_c \ll w$, the width of the sample. This means that particles move in such tight spirals that boundary scattering is reduced. In this case the classical conductivity is not monotonic in B.

The magnetic field introduces another length scale and two more energy scales that become particularly relevant when discussing quantum effects in transport. These are discussed below and in Section 6.4.

6.3.1 Semiclassicality

Starting from the relatively (and in some sense unreasonably) successful description of classical transport, the next task is to incorporate what we know about realistic electronic structure. Remember that electrons are actually quantum mechanical objects, and in the limits when ordinary band theory is appropriate they occupy single-particle states that look like Bloch waves. In fact the Bloch wave nature of the states and the Pauli principle are two of the primary reasons that the classical Drude picture works as well as it does. Classically it was not clear why charge carriers do not scatter off the charged ions of the lattice. Bloch waves are constructed to accommodate the periodicity of the lattice and scatter from breaks in that periodicity. The Pauli principle further reduces the scattering rate by preventing carriers from ending up in states that are already occupied. As a result, the classical picture of carriers propagating relatively long distances between scattering events is often not bad.

Further incorporating quantum effects into a classical picture of charge transport results in the *semiclassical* description of electron motion. The residual classicality is largely two-fold: neglecting quantum interference (treating probabilities classically) and neglecting complementarity (allowing particles to have a specific $\mathbf{r}$ as well as a specific wavevector $\mathbf{k}$). In addition to $\mathbf{k}$, the electronic states are associated with a band index, n, and a spin, s. The energy of the electronic states is assumed to be given by the band structure and satisfies $E_n(\mathbf{k}) = E_n(\mathbf{k} + \mathbf{G})$, where $\mathbf{G}$ is a reciprocal lattice vector. In analogy to the velocity of a wave, a semiclassical velocity is defined by

$$\frac{\partial \mathbf{r}}{\partial t} = \mathbf{v}_n(\mathbf{k}) = \frac{1}{\hbar} \frac{\partial E_n(\mathbf{k})}{\partial \mathbf{k}}. \tag{6.18}$$

This gives the evolution of $\mathbf{r}$ as a function of $\mathbf{k}$. The semiclassical equation of motion for $\mathbf{k}$ is patterned on classical physics:

$$\hbar \frac{\partial \mathbf{k}}{\partial t} = q[\mathbf{E}(\mathbf{r}, t) + \mathbf{v}_n \times \mathbf{B}(\mathbf{r}, t)]. \tag{6.19}$$

These equations assume that the band index is a constant of the motion. Electrons in states not near extrema of $E_n(\mathbf{k})$ are always in motion in this picture. Also note that there are no scattering or forces from the ion cores here; all of that is taken into account by our earlier assumption of Bloch waves. The rate of change of the crystal momentum, $\hbar\mathbf{k}$, is given by the total external forces acting on the electron. The semiclassical regime is discussed extensively in several solid state physics texts, with particularly thorough treatments in Ashcroft and Mermin [6] and Marder [15].

6.3.2 Scattering

In the semiclassical approach scattering from defects may be understood in a method not too different from the classical treatment of carriers. To simplify a complex process, one can define a momentum *relaxation time*, τ, as before. To define things more formally than in Section 6.2, consider a distribution function $f(\mathbf{r}, \mathbf{k}, t)$ that is the probability of finding a particle with a position within a volume $d^3\mathbf{r}$ of $\mathbf{r}$ and with a wavevector within $d^3\mathbf{k}$ of $\mathbf{k}$. If this distribution function differs from the equilibrium distribution $f_0(\mathbf{r}, \mathbf{k})$ by an amount δf, then the relaxation time approximation says that the collisional contribution to the return to equilibrium looks like

$$\frac{df}{dt}\bigg|_{\text{coll.}} = -\frac{\delta f}{\tau}. \tag{6.20}$$

For the electrons described by the semiclassical picture, the relevant equilibrium distribution function is the Fermi–Dirac distribution,

$$f_0(\mathbf{r}, \mathbf{k}) = \frac{1}{1 + e^{(E(\mathbf{k}) - \mu(\mathbf{r}))/k_B T}}. \tag{6.21}$$

Here $\mu(\mathbf{r})$ is an effective local chemical potential (which can vary in space, as we have seen in our discussion of screening). Since it's an equilibrium quantity, f_0 has no explicit time dependence.

To study the general time dependence of the distribution function, we must keep in mind the restriction that no particles are actually created or destroyed. Any probability lost in some region of phase space, $(\mathbf{r}, \mathbf{k})$, must be recovered by an increase in the distribution function elsewhere in phase space. The equation that results is the *Boltzmann equation* for the total rate of change of the distribution function:

$$\frac{df}{dt} = \frac{\partial f}{\partial t} + \frac{\partial \mathbf{r}}{\partial t} \cdot \frac{\partial f}{\partial \mathbf{r}} + \frac{\partial \mathbf{k}}{\partial t} \cdot \frac{\partial f}{\partial \mathbf{k}}. \tag{6.22}$$

We will derive a very similar looking equation when we discuss fluid mechanics in Chapter 10. Inserting the collisional term from (6.20) for the left side of (6.22), we arrive at a set of partial differential equations that can be used to find the evolution of the distribution function under various conditions. The equations of motion, Eqs. (6.18) and (6.19), describe the response of particles to fields, for example, and feed into the evolution of f.

Actually solving the Boltzmann equation in general can be very challenging, even in the purely classical limit. We will return to the discussion of equilibration of electronic distributions later when describing electronic transport in nanoscale systems. For now, we present the *steady state* solution to a common question: what happens to the Fermi sea and Fermi surface of a conductor in the presence of an applied electric field? As in the classical approach, the electrons acquire an average drift velocity, $\mathbf{v}_d$, that, in the absence of weird anisotropies, points along the direction of the applied electric field. In the relaxation time approximation the magnitude of $\mathbf{v}_d$ is still given by the mobility as in Eq. (6.11). The remarkable result of the Boltzmann equation is that the overall shape of the Fermi sea and Fermi surface are unchanged in the steady state. Rather, they are translated uniformly in $\mathbf{k}$-space so that they are centered on $m^* \mathbf{v}_d / \hbar$ rather than $\mathbf{k} = 0$. This is shown in Fig. 6.5.

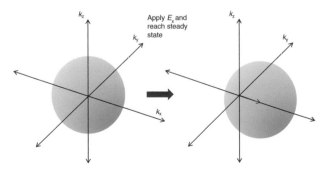

Fig. 6.5   Fermi sphere translation. In steady state after the application of an electric field in the x direction, the Fermi sphere is shifted by $m^* v_d / \hbar$ in $\mathbf{k}$-space. The equilibration process that reaches this steady state is beyond the scope of this book.

6.3.3 Fermi surface effects

There is one further quantum mechanical condition that is the reason behind the key triumph of the semiclassical picture, its ability to explain magnetic field effects related to the shape and size of the Fermi surface in metals and doped semiconductors. From Eqs. (6.18) and (6.19), in the presence of **B** but an absence of **E**, it is possible to have particles that trace orbits in **k**-space; that is, $\mathbf{k}(t)$ can evolve in a periodic way with time within the Brillouin zone ("closed orbits") or periodic to within a reciprocal lattice vector ("open orbits"). In the "old quantum theory" of Bohr and Sommerfeld, the quantization condition for valid states was that the total phase gained following a periodic orbit is an integer multiple of 2π. For the semiclassical orbits that implies

$$\int_{\text{orbit}} \mathbf{k} \cdot d\mathbf{r} = j2\pi, \tag{6.23}$$

where j is an integer.

In the presence of a magnetic field, the vector potential **A** leads to an additional quantum mechanical phase shift[1]. The quantum mechanical phase accrued along some path by a charged particle moving in a vector potential $\mathbf{A}(\mathbf{r})$ is

$$\delta\phi = \frac{q}{\hbar} \int_{path} \mathbf{A} \cdot d\mathbf{r}. \tag{6.24}$$

Recall that **A** is defined such that $\nabla \times \mathbf{A} = \mathbf{B}$. This implies that the phase acquired by such a particle when traversing a closed loop is proportional to the magnetic flux through the loop:

$$\begin{aligned}
\phi_{\text{loop}} &= \frac{q}{\hbar} \int_{\text{loop}} \mathbf{A} \cdot d\mathbf{r} \\
&= \frac{q}{\hbar} \int_{\text{surf}} \nabla \times \mathbf{A} \cdot d\mathbf{S} \\
&= \frac{q}{\hbar} \int_{\text{surf}} \mathbf{B} \cdot d\mathbf{S}.
\end{aligned} \tag{6.25}$$

Combining this with Eq. (6.23), we see that the new quantization condition is that the total phase, orbital plus vector potential contribution, must add up to an integer multiple of 2π.

This restriction is only satisfied by certain of the single particle states defined in the absence of **B**. As **B** is changed, new sets of **k**-states are brought into and out of consistency with this condition. As places on the Fermi surface with large densities of states (and therefore large numbers of states with similar orbits in real space) are shifted into and out of this condition, various measurable parameters oscillate. The **k**-space states compatible with this condition are spaced such that the oscillations are periodic in $1/B$.

Such oscillations of the electronic heat capacity are called *de Haas–van Alphen* oscillations, while similar effects in the resistivity are *Shubnikov–de Haas* oscillations. Careful analysis of these oscillations in clean materials as a function of magnetic field orientation and crystallographic direction has been a boon in deciphering the Fermi surface morphology in metals. The requirement of "cleanliness" is discussed in the next section,

---

[1] See the Appendix for a brief discussion of the vector potential and its interpretation.

but it should be intuitively clear that these effects should only be measurable if carriers are able to complete the orbits in question without much scattering.

Another way to think about these oscillations is in terms of *Landau levels*. In the absence of a lattice potential, spin effects, or confinement effects, charged particles in the presence of a magnetic field have energy levels spaced by quanta of the cyclotron energy, $E(\nu) = (\nu + \frac{1}{2})\hbar\omega_c$. Here ν is the Landau level index. These levels are still important even in realistic metal and semiconductor systems.

Each such level can be hugely degenerate – many, many single-particle states can have the same energy. The easiest way to see this is to consider the limit of very high fields, when $\hbar\omega_c$ far exceeds even E_F. In this case every electron contributes essentially the same amount, $\frac{1}{2}\hbar\omega_c$, to the total energy of the system – every electron "sits in the lowest Landau level".

In a 2d electron gas that is very clean (so that cyclotron motion is not disrupted by disorder scattering), at high magnetic fields, the density of states becomes a series of peaks at the cyclotron energies $E(\nu)$, and zero elsewhere. The width of those peaks is set by the disorder. The transport consequences are profound. If the Fermi level sits between the Landau levels, there are no states into which electrons can scatter, implying $\rho_{xx} = 0$ (and $\sigma_{xx} = 0$), and $\rho_{xy} = $ constant (in particular, $R_{xy}^{-1} = (e^2/h)\nu$, ignoring any Zeeman splitting of the Landau levels). This is the integer *quantum Hall effect* [326]. Further discussion is beyond the scope of this book (as it is not particularly "nano"), but there are several outstanding pedagogical references, including lecture notes by S. M. Girvin [327] (also available as arxiv.org/abs/cond-mat/9907002) and the book by Weisbuch and Vinter [328].[2]

At lower values of B in real systems, the Landau levels manifest themselves weakly, as an enhanced density of states for energies corresponding to the Landau levels. In **k**-space, the single-particle states associated with a particular Landau level correspond to a tube with its axis along the direction of **B**. As B is increased, the radii of the tubes increases. As each tube becomes tangent to the Fermi surface, the relatively enhanced density of states leads to a fluctuation in the electronic heat capacity (de Haas–van Alphen oscillations).

6.3.4 Validity

The semiclassical approach is valid when the main underlying assumptions (Bloch-like states, band structure giving the dominant energy scales) hold true. For example, we are tacitly assuming that electric fields are too small to disrupt the single-particle band structure $(e|\mathbf{E}|a \ll (E_C(\mathbf{k}) - E_V(\mathbf{k}))^2/E_F$, where a is the lattice spacing). Similarly, magnetic energy scales need to be too small to destroy the essential band structure $(\hbar\omega_c \ll (E_C(\mathbf{k}) - E_V(\mathbf{k}))^2/E_F$. Likewise, the semiclassical equations of motion are not equipped to handle inherently quantum mechanical effects like interband transitions. Such effects at high frequencies $(\hbar\omega \sim E_C - E_V)$ must be added in "by hand" rather than following from Eqs. (6.18) and (6.19).

---

[2] In exceedingly clean 2d systems, electron-electron interactions that we have hitherto ignored can lead to the more exotic fractional quantum Hall effect [329, 330, 331], which is fascinating but even farther beyond the scope of this book.

One more subtle criterion for validity is the limit of weak disorder. Since ℓ is one way to parametrize the effects of defects, grain boundaries, and other sources of scattering, the disorder condition is sometimes written as $k_F \ell \gg 1$. Recall that k_F^{-1} is proportional to the wavelength of the plane wave part of the Bloch wavefunction at E_F. Physically the disorder condition means that the semiclassical picture of electronic motion is valid as long as the electronic wave propagates through many wavelengths between scattering events. If $k_F \ell \sim 1$ or smaller, then describing the single-particle states as Bloch-like makes little physical sense.

6.4 Quantum transport

Electrons are quantum mechanical objects, and yet we know that classical and semiclassical treatments of conduction are often adequate for most purposes. For example, no-one worries about quantum interference effects when they turn on the light switch. On the other hand, we do know that quantum mechanics is essential to understand matter on the very small scale, such as the bonding and energy levels of diatomic nitrogen. A full treatment of quantum transport must be able to handle both limits (the very small, when quantum mechanical effects dominate; and the macroscale, when classical physics generally rules the day), as well as the intermediate (often called *mesoscopic*) regime, when quantum corrections to the (semi)classical conductivity begin to appear.

6.4.1 Coherence and probabilities

A central feature of quantum mechanics is the nonclassical behavior of probabilities. To find the classical probability for an event, say a particle traversing from an initial location to a final location, one adds the probabilities of mutually exclusive ways for the event to transpire. Quantum mechanically, one adds the complex *amplitudes* for the different trajectories, with the probability given as the magnitude squared of the sum of amplitudes. These are very different procedures, and understanding to what degree each procedure is appropriate and why has been a long-standing issue in the interpretation of quantum mechanics.

Consider a concrete example, the two-slit experiment shown in Fig. 6.6. Particles are emitted from a source and propagate through the two-slit aperture, eventually being detected at the screen. If the amplitudes for the respective paths to a point on the screen are ϕ_A and ϕ_B, then the total probability of measuring the particle at that point is $|\phi_A + \phi_B|^2$. If we consider $|\phi_A|^2$ and $|\phi_B|^2$ as the classical probabilities for the two paths, respectively, we find that the quantum result differs by the cross-term $2\text{Re}\phi_A^*\phi_B$. This cross-term is exactly what leads to the interference pattern seen when this experiment is performed using electrons or photons traveling through vacuum – effectively isolated from any other degrees of freedom.

Think of the wavefunctions as plane waves with the same k that start off in-phase, propagate, diffract (elastically, so that $|\mathbf{k}|$ is unchanged) through the slits, and propagate

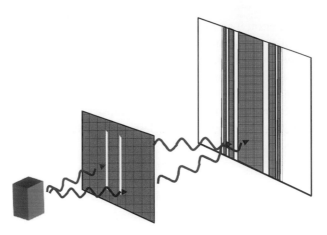

Fig. 6.6 Two-slit experiment. Coherent quantum particles propagate from a source, through a two-slit aperture, with their positions recorded on a screen. The interference term (resulting from the summation of amplitudes rather than classical probabilities) leads to an oscillatory pattern on the screen.

the rest of the way to the screen. The phase accumulated along trajectory A is then kL_A, where L_A is the total path length of trajectory A. Likewise, the phase accumulated along trajectory B is kL_B, and the cross-term that gives the interference pattern is proportional to the cosine of the phase difference, $kL_A - kL_B$. Notice that the interference pattern results because the phases of the two trajectories have a well defined relationship.

Now consider the case when there is a detector at one of the slits that indicates when a particle has passed through that slit. The correct way to write the amplitudes now is to include the state of the detector. Suppose the detector state can take on two values, $|\chi_A\rangle$ when the particle has followed trajectory A, and $|\chi_B\rangle$ in the other case. This can only happen if there is some sort of interaction that couples the particle and the detector. While the exact coupling and a mathematical description of the detector matter when considering a particular implementation, the main result is insensitive to the details. The total probability for the particle to end up at a particular position on the screen is then

$$P = |\phi_A|^2 \langle\chi_A|\chi_A\rangle + |\phi_B|^2 \langle\chi_B|\chi_B\rangle + 2\mathrm{Re}[\phi_A^*\phi_B\langle\chi_A|\chi_B\rangle]$$
$$= |\phi_A|^2 + |\phi_B|^2 + 2\mathrm{Re}[\phi_A^*\phi_B\langle\chi_A|\chi_B\rangle]. \tag{6.26}$$

Now the interference term is multiplied by $\langle\chi_A|\chi_B\rangle$. If the detector states are orthogonal (that is, they have no overlap, such as a needle that can point in two distinct directions), then this factor is zero. The interference term is then suppressed, and the probability is given by the classical result. To put this in more rigorous language, *we see classical probabilities here because interactions entangle the particles with the detector, and we trace over detector final states which, in this case, are orthogonal.*

This reduction to classical probabilities is commonly called *decoherence*. Rather than considering the process from the point of view of what happens to the detector state, one can take a complementary approach. In the plane wave picture mentioned above, the interaction between the detector and the particles introduces an *uncertainty in the relative phases* ("dephasing") of the two trajectories. It is far from obvious, but it turns out that the entangling interaction is sufficient to dephase the trajectories and wash out the interference pattern (except in very unusual circumstances). Imry [332] considers specific realizations of detectors and discusses this in detail.

Notice that this destruction of the interference pattern does not involve an obvious, instantaneous "collapse of the wavefunction", an expression that has its roots in what is known as the Copenhagen interpretation of quantum mechanics. While the "collapse" way of viewing decoherence has certain virtues and, if handled consistently, can be used to describe reality,[3] it is not always the most useful way to solve problems, particularly subtle ones relating to quantum measurement. Don't worry – there are still plenty of interpretational and philosophical issues remaining to bother us about quantum mechanics.

The interaction between the particle and the detector is a particular case of the general problem of a particle interacting with some environment. Pure elastic scattering interactions (like diffraction off the slits) do not lead to decoherence; the phase relationship between incoming and outgoing waves is well defined and can, in principle, be determined from measurements of interference patterns. Instead it is the interactions that entangle the particle with *environmental degrees of freedom* that are relevant for decoherence. The essential issue is whether the particle *changes* the state of the environment, leaving an "imprint" of its passage. This could be through the creation of an excitation of some kind (emitting a photon or phonon) or more subtle physics (undergoing magnetic exchange, flipping an electronic or nuclear spin). A weak interaction with the environment, one that perturbs the environment into states with some remaining overlap, does not completely decohere the particle in one go.

When considering quantum effects in conduction, it then makes sense to speak of a *coherence time*, τ_ϕ. If one considers dropping a particular electron into a solid, τ_ϕ is the typical timescale over which the phase of the electron becomes uncorrelated with its initial phase due to interactions with environmental degrees of freedom.

Clearly τ_ϕ will depend on the environment. In a metal or semiconductor the natural decoherence mechanisms include electron–electron scattering, electron–phonon scattering, and possible spin-flip scattering off of paramagnetic impurities. The rates of these processes depend on temperature. For example, electron–electron scattering via the Coulomb interaction may only proceed if there are empty electronic states available into which the carriers may scatter, as we discussed in Section 5.3.3. The electron–electron scattering rate approaches zero as $T \to 0$. Similarly, the electron–phonon scattering rate also approaches zero as $T \to 0$ due to freeze-out of the phonon populations. Measurements of $\tau_\phi(T)$ therefore provide valuable information about the dominant inelastic processes at work. At absolute zero, the expectation is that the properly defined coherence time should approach infinity. That is, at $T = 0$ a quantum mechanical system will remain forever in its ground state.

Unfortunately, direct measurements of τ_ϕ or its corresponding length scale, L_ϕ, are not possible. The situation is rather analogous to that of entropy, S, in thermodynamics.

---

[3] That is why the different interpretations of quantum mechanics are "interpretations". In calculation they all predict the same things.

There is no direct means of measuring the entropy; rather, one must infer S based on its theoretically defined relationship to an experimental observable, the heat capacity. For τ_ϕ, the best approach is to quantum interference corrections to the (semi)classical electron transport. With a proper theoretical foundation the magnitude of those corrections provides a quantitative means to infer coherence times for the charge carriers.

6.4.2   Quantum corrections to conduction

Quantum interference effects present themselves in several forms in the conduction through metal and semiconductor nanostructures. A detailed theoretical analysis of these effects often involves many-body physics techniques such as Feynman diagrams and is beyond the scope of this book. However, it is possible to acquire reasonable insight into these phenomena and physical intuition with a more qualitative treatment. A similar discussion that is more in-depth may be found in [332].

The magnitude of those corrections often depends on the ratio of the system size, L, to the coherence length, L_ϕ. Recall that $L_\phi \equiv \sqrt{D\tau_\phi}$ in diffusive conductors, where D is the diffusion constant for the charge carriers. To give you a sense of the numbers, L_ϕ is typically on the order of 1 nm at room temperature in ordinary metals, while at liquid helium temperatures (4.2 K), L_ϕ can be on the micron scale. For a metal with a resistivity in the few-mΩ-cm range, this corresponds to $\tau_\phi(300\text{ K}) \sim 10^{-16}$ s, $\tau_\phi(4.2\text{ K}) \sim 10^{-10}$ s.

The coherence length introduces another form of "dimensionality": dimensionality with respect to quantum coherence. Samples with $L, w, t \gg L_\phi$ are said to be quasi-three-dimensional with respect to coherence phenomena, while $L, w > L_\phi > t$ and $L > L_\phi > w, t$ are quasi-2d and quasi-1d, respectively. The use of the prefix "quasi" is used colloquially as a reminder that a long wire can be 1d as far as coherence phenomena are concerned, while still having a three-dimensional electronic structure (that is, $L, w, t \gg \lambda_F$.).

If a system is much larger than L_ϕ, one effect that may be relevant is *ensemble averaging*. Consider a 1d example, and suppose that within a single coherent region there is a characteristic correction to the conductance δG. Suppose that correction can be either positive or negative and depends critically on the microscopic details of the coherent region (e.g., the exact arrangement of grain boundaries and elastic scattering sites). In this situation it's better to characterize the correction as δG_{rms}, where "rms" means "root mean square" of the possible corrections. Imagine dividing the sample into $N = L/L_\phi$ coherent regions, with the details of each region *uncorrelated* with those of the others. In this situation the end-to-end conductance change that would be seen would be reduced from δG_{rms} by a factor of $N^{-1/2}$ (in the limit of large N, of course). This is basic statistical averaging, and goes a long way toward explaining why quantum interference effects have no impact on our typical daily experience with electrical conduction.

A second kind of ensemble averaging that can be important is thermal averaging. In this case the relevant distance scale that must be taken into account is the *thermal length*, $L_T \equiv \sqrt{\hbar D/k_B T}$. Thermal averaging has a simple origin if we think about the semiclassical Bloch wave description of carriers and the idea that propagating carriers accrue quantum mechanical phase via integrating $\mathbf{k} \cdot d\mathbf{r}$ along trajectories. Only at $T = 0$ do all the carriers at the Fermi surface have perfectly defined kinetic energies given by E_F and perfectly crisp values of $|\mathbf{k}|$. At any nonzero temperature, the carriers relevant to conduction have their energy distributed within a window of size $\sim 2k_{\mathrm{B}}T$; therefore two carriers with nominally identical $\mathbf{k}$ vectors can differ in wavevector magnitude slightly. If those two carriers started off in some direction initially in-phase, their phases would eventually become uncorrelated because of the slight difference in wavevector magnitude due to finite T. One interpretation of L_T is as the length scale over which this happens.

A related kind of ensemble averaging is more general *energy averaging*. For a given coherence time one can find an effective energy scale, $E_\phi = \hbar/\tau_\phi$, or in a diffusive system, $\hbar D/L_\phi^2$. One way to think about this energy scale is that single-particle Bloch states with energies that differ by less than E_ϕ have quantum mechanical phases that are correlated, and can have nontrivial interference effects. Conversely, single-particle states that differ in energy by much more than E_ϕ have phases that are effectively uncorrelated. If carriers are distributed over a range of energies ΔE, then one can think of this as $N = \Delta E/E_\phi$ independent statistical ensembles, and the $N^{-1/2}$ reduction factor for interference-related conductance changes comes into play. Note that doing this analysis for thermal energies, $k_{\mathrm{B}}T$, leads back to the thermal length as defined above. Another way to get carriers with a large distribution of energies is in short devices under a significant voltage bias, as we shall discuss in Section 6.4.2.

Aharanov–Bohm oscillations

The Aharonov–Bohm (AB) effect provides dramatic evidence for quantum interference. The basic idea is shown in Figure 6.7(a). Consider a metal wire in the shape of a ring, with current and voltage leads as shown, and initially think about the $B = 0$ case. Suppose further that the ring circumference is much smaller than L_ϕ. An incoming electron can traverse the ring via the right arm, scattering off some impurities and the wire boundaries, eventually ending up at point X having acquired some phase ϕ_{R}. Alternatively, that electron could have travelled around the left arm and similarly acquired a phase ϕ_{L}. The resulting probability of finding the electron at point X would be the classical contribution plus an interference term proportional to $\cos(\phi_{\mathrm{R}} - \phi_{\mathrm{L}})$. The conductance would differ from the classical result by this interference term, and could be enhanced or reduced, depending on the particular phases.

Now consider the case of nonzero magnetic field B through the loop, limiting our discussion to the regime where B is sufficiently weak that it does not significantly alter the classical trajectories. As we discussed in Eqs. (6.24) and (6.25), there is an additional term to the phase difference proportional to the magnetic flux, Φ, through the loop: $\cos(\phi_{\mathrm{R}} - \phi_{\mathrm{L}} - \frac{e}{\hbar}\Phi)$. If the magnetic field is swept, there will be oscillations in the probability of ending up at point X as a function of B. The period of those oscillations will be inversely proportional to the area of the loop. In fact the conductance will be periodic in the flux in units of $\phi_0 \equiv h/e$, often called the *flux quantum*. Figure 6.7(c) shows both the raw resistance data as a function of B and a Fourier transform, showing pronounced periodicity as expected.

Thermal ensemble averaging acts to suppress the amplitude of the AB oscillations. The idea here is that electrons with energies that differ by $\sim k_{\mathrm{B}}T$ pick up slightly different phases (say ϕ_{R} and ϕ'_{R}) when elastically scattering while going around the ring. When

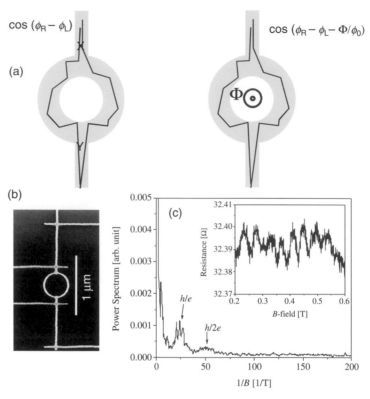

$\cos(\phi_R - \phi_L)$

$\cos(\phi_R - \phi_L - \Phi/\phi_0)$

(a)

Φ

(b)

(c)

Fig. 6.7 Aharonov–Bohm effect. (a) The quantum correction to the conductance through a ring is determined by the phase difference between left- and right-branch trajectories. In the presence of magnetic flux, Φ, through the ring, that phase difference is changed. (b) An electron micrograph of a silver ring configured for four-terminal measurements [333]. (c) Raw resistance vs. B (inset) taken at $T = 0.05$ K and its Fourier transform as a function of $1/B$, showing the periodicity of the conductance in units of $\phi_0 = h/e$.

averaging over all allowed electron energies this smears out the $\phi_R - \phi_L$ part of the phase difference, and hence averages away some of the conductance variation. Similarly, the AB oscillations are rapidly quashed if L_ϕ becomes comparable to or shorter than the circumference of the loop. The dependence of the conductance oscillation amplitude on L_ϕ is discussed in multiple places [334, 335].

Additional ensemble averaging can take place if multiple loops are considered, either in series or in parallel. Recall that the phase shift $\phi_R - \phi_L$ depends on the details of the elastic scattering in each loop, and therefore so does the phase of the AB oscillations as a function of B. In a large ensemble of loops the ϕ_0-periodic oscillations of $G(B)$ therefore average away to zero. The same sort of averaging suppresses the AB oscillations in thin conducting cylinders when B is directed along the axis of the cylinder; one can think of the cylinder as an ensemble of rings. To see clear ϕ_0-periodic $G(B)$ oscillations requires measurements on a single loop.

Now look at the trajectories in Fig. 6.8 and consider the probability of ending up back at point Y, initially in the absence of a magnetic field. The left- and right-directed

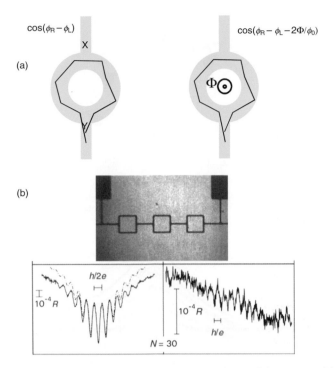

$\cos(\phi_R - \phi_L)$

X

(a)

$\cos(\phi_R - \phi_L - 2\Phi/\phi_0)$

Φ

(b)

h/2e

$10^{-4} R$

$10^{-4} R$

h/e

N = 30

Aharonov–Aronov–Spivak oscillations. (a) Now we consider trajectories that go all the way around the ring, starting in **Fig. 6.8** either the left or right arm. In the presence of magnetic flux, Φ, through the ring, the phase difference is changed by twice as much as in the Aharonov–Bohm case. (b) An electron micrograph of a three rings in series, and resistance vs. B traces for an ensemble of 30 rings, showing that the $h/2e$-periodic AAS oscillations survive ensemble averaging while the AB oscillations are suppressed. Adapted from Umbach *et al.* [336].

paths are a pair of time-reversed trajectories. The phase shifts picked up during elastic scattering from structural defects are independent of the order in which the scattering events take place. This implies that the phase shifts ϕ_L and ϕ_R are *equal*. The result is constructive interference for *all* such pairs of time-reversed trajectories. This implies a quantum interference enhanced probability for the particle to be at point Y, in any such loop, provided that L_ϕ is much larger than the circumference of the loop. With this enhanced probability for back-scattering, the conductance due to these trajectories is at a *minimum* in the absence of an external magnetic field.

A magnetic field threading the loop again leads to a phase difference between the left and right paths when considering these trajectories. Now, however, the phase difference is proportional to *twice* the flux through the loop, since each trajectory encloses the entire loop. Therefore, the conductance $G(B)$ due to these trajectories is periodic in B (flux) with a period of $\phi_0/2 = h/2e$, and always starts with a minimum in G at $B = 0$. Because of this zero-field condition (a consequence of the time-reversed nature of the paths under consideration here), these $h/2e$-periodic magnetoconductance oscillations are *not* suppressed by ensemble averaging in systems with multiple loops or in cylinders. These are *Altshuler–Aronov–Spivak* (AAS) oscillations [337], and their contribution is apparent in the power spectrum shown in Fig. 6.7, as well as in Fig. 6.8.

(a)

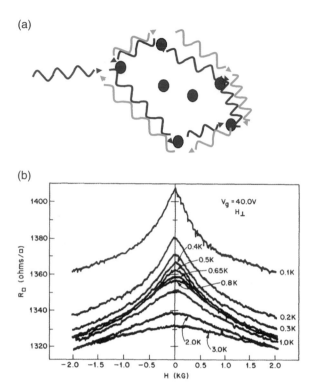

(b)

Fig. 6.9 Weak localization magnetoresistance. (a) An example of a pair of time-reversed loop trajectories for electrons scattering from disorder (red circles). With no spin–orbit effects, these trajectories interfere constructively, leading to enhanced back-scattering and a resistance higher than the classical result. (b) Magnetoresistance data in an Si MOSFET at various temperatures, showing an enhanced $B = 0$ resistance that increases as $T \rightarrow 0$. This is the weak localization magnetoresistance. Adapted from Bishop *et al.* [338].

Weak localization magnetoresistance

The AAS oscillations are not the only quantum correction to electronic conduction that result from the special properties of time-reversed trajectories. Remember, the key fact is that (in the absence of additional physics[4]) a closed loop trajectory involving only elastic scattering and its time-reversed conjugate pick up the *same* quantum mechanical phase. Figure 6.9 shows two such classical paths and their time-reversed partners in a disordered metal. As in the AAS case, in the absence of a magnetic field, each pair of time-reversed trajectories should interfere constructively to produce enhanced back-scattering. For a finite L_ϕ, only closed paths shorter than L_ϕ should contribute to the overall reduction in conduction. This specific process – the reduction in conduction due to the constructive interference of time-reversed back-scattered trajectories – is called *weak localization*. This name is historical, and is a contrast to *strong localization*, when disorder in a conductor is so strong that the appropriate single-particle quantum states are localized, with exponentially decaying tails.

---

[4] We will discuss the role of spin and its effect on this discussion shortly.

Like AAS oscillations (and unlike AB oscillations), weak localization is not suppressed by ensemble averaging. It is, however, suppressed in the presence of magnetic flux threading the loop-like trajectories. Aharonov–Bohm phase due to such flux introduces a phase difference between clockwise and counterclockwise motion around a particular loop trajectory, proportional to the flux through the loop. Each loop trajectory pair acquires its own particular phase difference since each possible loop differs in detail. As a result the addition of a magnetic field eliminates the consistent contributions of the ensemble of time-reversed trajectories, eliminating the weak localization as B is increased through some field scale that we discuss shortly. The resulting change in resistance is often called the *weak localization magnetoresistance* (WLMR).

The WLMR may be used to infer the size of L_ϕ. To see how this works, first realize that increasing B affects the WL contributions of larger loop trajectories first, since it is the total magnetic flux through the loop that is relevant to the quantum mechanical phase shift. Bear in mind that there are more large loops than small loops, simply because there are more possible large loop trajectories. While detailed calculations are required to determine the precise functional form of $R(B, L_\phi)$, a rough estimate of L_ϕ may be found by considering only the largest loops that are relevant, those with a typical linear dimension of L_ϕ. Defining B^* as the typical field (normal to the loops in question) required to suppress the WL, this implies:

$$L_\phi^2 B^* \sim \frac{h}{e}, \qquad \text{quasi-2d, 3d,}$$

$$L_\phi w B^* \sim \frac{h}{e}, \qquad \text{quasi-1d.} \tag{6.27}$$

To infer $L_\phi(T)$ one then performs a series of measurements of $R(B)$ at various temperatures and fits each $R(B)$ curve to the appropriate functional form where L_ϕ is a fit parameter. As Eq. (6.27) shows, a shorter L_ϕ implies that a larger B is necessary to suppress the WL. This trend is readily apparent from the data shown in Fig. 6.9; as the temperature decreases, L_ϕ increases and the WL feature in the magnetoresistance becomes increasingly prominent.

Including the effects of spin–orbit scattering complicates matters. Just as there is a term in the Hamiltonian that couples the spin of a bound electron to its orbital angular momentum in an atom, a similar term can appear when a Bloch electron (in a band not arising out of s-orbitals) scatters elastically off of a lattice defect. That term introduces an additional phase shift to the scattering event. The spin of the particle is also rotated with each scattering event. Unlike the pure potential scattering case, however, the spin–orbit term swaps sign if one considers reversing the direction of the particle's trajectory.

One can define a timescale, τ_{SO}, and an associated distance, L_{SO}, related to the magnitude of the spin–orbit coupling. If an electron propagates a distance much larger than L_{SO}, it's spin direction is uncorrelated with its initial spin (though in a deterministic way set by the particular trajectory under consideration). As mentioned in Chapter 3, spin–orbit effects are generally stronger in high-Z metals such as Au. For reference, a typical τ_{SO} for Au is on the order of picoseconds [339], while it may be 50 times longer in a low-Z metal such as magnesium [340]. The magnitude of τ_{SO} is set by both the band structure of the material and the microstructure (through τ, the elastic scattering time).

We now see another kind of dimensional crossover. When $L_\phi \ll L_{SO}$, spin–orbit effects are weak on the timescales associated with decoherence, and the weak localization discussion above applies well. However, when $L_\phi \gg L_{SO}$, spin–orbit effects are strong and the result is a *change in sign* of the weak localization correction, or *weak antilocalization*. For this reason in Au films at low temperatures the WLMR is *positive*, with the resistance having a *minimum* at $B = 0$. Metals with intermediate values of the coupling, such that $L_\phi(T)$ crosses L_{SO} as T is decreased, can show a crossover from weak localization at higher temperatures to weak antilocalization at lower temperatures. This is exactly what is observed in Cu and Ag.

Universal conductance fluctuations

The discussion of the loop trajectories in the context of weak localization illustrates a particular example of a general principle: electronic conduction through a disordered material in the presence of quantum coherence is an interference experiment analogous to diffraction from an array of random scatterers. The basic idea is shown in Fig. 6.10, where different trajectories are considered for electron propagation from one side of a disordered material to the other. Assuming that an electron at the left begins with a phase of zero, each trajectory shown corresponds to an amplitude with a different quantum mechanical phase at the destination point on the right.

The final probability of transiting from left to right will be determined by the interference of many cross-terms, most of which will cancel. Without knowing the microscopic details of the trajectories, it is not possible to say whether the resulting conductance will exceed or be reduced from the classical expectation. An analogous effect is seen in the "speckle" interference pattern that results when a laser beam is diffracted from a random array of scatterers. The resulting diffraction pattern, or speckle, is complicated and determined by the precise positioning of the scatterers.[5]

There are four different ways to examine the complicated interference that underlies the conduction process here. All four involve means of changing the interference pattern as a function of some parameter, and thus producing conductance fluctuations. Because those fluctuations, when corrected for ensemble averaging effects, are all of the same magnitude as $T \to 0$, $\langle \delta G \rangle_{rms} \sim 2e^2/h$, independent of the microscopic details of the system, these are called *universal conductance fluctuations* (UCF). The universality of the fluctuation magnitude is examined in an exercise at the end of the chapter. The quantity $G_0 \equiv 2e^2/h$ is often called the *quantum of conductance*, and appears in many places in nanoscale physics.

The most common UCF discussed in the literature are UCF as a function of external magnetic field (MFUCF). Here the Aharonov–Bohm phase is used to shift the relative phases of the many trajectories that contribute to the conductance. This produces changes in conductance as a function of B that look to the eye somewhat like random noise, since the detailed positions of all the scattering sites determine the trajectories; indeed, two nominally identical (same resistivity, same overall dimensions) metal nanostructures

---

[5] The remainder of this section is a detailed discussion of this, and may be of greatest interest to specialized readers.

Universal conductance fluctuations. The two different trajectories shown for electrons scattering from disorder both end at the same position but have different quantum mechanical phases. Their interference as their relative phases are varied results in UCF.

Fig. 6.10

will show very different $G(B)$ traces because their scattering sites differ. However, $G(B)$ is completely reproducible in a given nanostructure, provided that there are no structural changes such as thermal annealing. Furthermore, as complicated as the MFUCF appear, the Onsager relations (Eq. (6.7)) still hold (see Fig. 6.3).

There is a characteristic field scale associated with the MFUCF that may be found by looking at the correlation function,

$$F(\Delta B) = \langle g(B)g(B + \Delta B)\rangle - \langle g\rangle^2. \tag{6.28}$$

Here the angle brackets represent an average taken over a large range of magnetic field, and $g \equiv G/G_0$. This correlation function peaks at $B = 0$ and falls off over a field scale that is related to L_ϕ, in a manner very similar to the WLMR. By analyzing $G(B)$ traces at different temperatures it is possible to extract $L_\phi(T)$. Precise theoretical expressions exist for $F(\Delta B)$ and its implicit dependence on L_ϕ and T for various dimensionalities and strengths of spin-orbit scattering.

In semiconductor nanostructures it is also possible to use electrostatic gating to shift the local chemical potential of the electrons. By tuning the carrier concentration and hence the Fermi wavelength, the interference terms related to scattering from disorder again fluctuate in a sample-specific, complicated, yet deterministic manner. This is an example of universal conductance fluctuations as a function of Fermi energy (EFUCF). These fluctuations can be analyzed with a simple generalization of Eq. (6.28):

$$F(\Delta E_F, \Delta B) = \langle g(E_F, B)g(E_F + \Delta E_F, B + \Delta B)\rangle - \langle g\rangle^2. \tag{6.29}$$

If the scattering sites within the nanostructure change their positions or scattering cross-sections as a function of time, this can also result in fluctuations of the conductance. This is analogous to rearranging the optical disorder that leads to laser speckle. This is easy to understand even from the two-slit diffraction thought experiment; altering the spacing between the slits such that the relative path length of the two dominant trajectories varies by half a wavelength can lead to a shift at a particular location from constructive to destructive interference. In ordinary metals the relevant wavelength is λ_F, implying that motion of a single (strong scatterer) impurity atom by a distance comparable to an atomic diameter can lead to a detectable change in conductance.

Even at cryogenic temperatures it is possible to have motion of scattering sites on experimentally relevant timescales. It is often reasonable to approximate these mobile

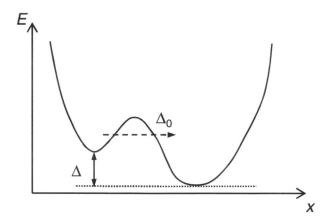

Fig. 6.11 Tunneling two-level system. Potential energy of some atom or group of atoms as a function of some configurational coordinate for a TLS in a disordered solid. Assuming the existence of an ensemble of these with a simple distribution of asymmetries (Δ) and tunneling matrix elements (Δ_0) explains many of the universal properties of disordered solids.

defects by a model incorporating *two-level systems* (TLSs). Each TLS is represented schematically in Fig. 6.11 as described by a double-welled potential as a function of some configurational coordinate. The asymmetry between the wells is Δ, and the tunneling matrix element connecting the wells is Δ_0. At low temperatures the TLS can either be kicked from well to well via thermal excitation, or it can tunnel through the barrier. The microscopic nature of TLS is generally not well known in most materials, though they are ubiquitous in disordered insulators and metals. TLSs were originally proposed [341] to explain the universal thermodynamic properties observed in amorphous insulators at low temperatures [342].

The standard TLS model assumes that a wide variety of TLSs exist in disordered materials, with a statistical distribution of Δ that is flat and ranges from 0 out to some cutoff energy scale (e.g., $k_B \times$ 100 K, above which other processes are assumed to be important). Likewise a distribution of Δ_0 is assumed that is flat in $\ln \Delta_0$ with low and high end cutoffs (corresponding to the limits of isolated wells and one big well, respectively). With these assumptions the TLS are expected to have a broad range of relaxation times. The *time-dependent UCF* (TDUCF) that result from the effect of these TLS on the quantum interference of conduction electrons then has a characteristic frequency (f) dependence. If a constant voltage is applied to such a nanostructure, the conductance fluctuations, $\delta G(t)$, will lead to a current noise power that varies like $1/f$. TDUCF were first observed in bismuth thin films [343], and have since been measured in a variety of metal nanostructures.

It is well known that $1/f$ noise is also ubiquitous and can arise from many mechanisms [344, 345]. The TDUCF have two very particular features that allow them to be distinguished from non-interference-related $1/f$ noise. First, TDUCF *increase* with decreasing temperature because of ensemble averaging effects (growing L_ϕ and reduced $k_B T$ as $T \to 0$). This is shown dramatically in Fig. 6.12(a). Second, the magnitude of TDUCF

(a)   (b)

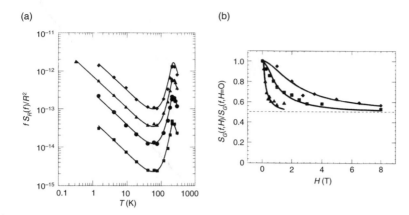

Time-dependent universal conductance fluctuations. (a) Magnitude of $1/f$ noise in a series of Bi nanowires as a function of temperature. The dramatic increase in noise as $T \to 0$ is indicative of TDUCF due to motion of TLS. (b) TDUCF as a function of magnetic field, showing the predicted factor of two drop over a field scale related to the coherence length. Adapted from Birge *et al.* [346].

Fig. 6.12

vary with B in a universal way. In analogy with WLMR and MFUCF, the precise functional form of the conductance noise power $S_G(B)$ depends on L_ϕ and L_{SO}.

The detailed $S_G(B)$ also depends on the overall magnitude of the TDUCF. If the motion of a typical single TLS gives a conductance change $\delta G_{1,\text{rms}} \ll G_0$, then the TDUCF are said to be in the "unsaturated" regime. Conversely, if $\delta G_{1,\text{rms}} \sim G_0$ then the TDUCF are "saturated". The effect of single-TLS motion is determined by the microscopic nature of the TLS, which can vary between materials and is not known *a priori*.

$S_G(B)$ drops by exactly a factor of two over a field scale that is expected to be comparable to $B*$ of the weak localization case. Qualitatively, this is because there are two equal contributions to the TDUCF in diffusive systems: one from paths reminiscent of the time-reversed loops in WL ("cooperon" contribution), and another from paths that have no special phase relationship at $B = 0$ ("diffuson" contribution). The former contribution is suppressed at $B > B*$. Figure 6.12(b) shows this field dependence of the noise power. In the weak spin–orbit case, an additional factor of two drop in $S_G(B)$ takes place at high fields when the Zeeman splitting due to electronic spin, $g\mu_B B$ exceeds $k_B T$.

The UCF considered so far, MFUCF, EFUCF, and TDUCF, are all essentially quasi-equilibrium phenomena. That is, all may be characterized by using a very tiny bias voltage to excite a very tiny current barely big enough to measure the conductance, but not large enough to drive the Fermi distribution functions of the electrons way out of thermal equilibrium. An additional type of UCF can take place when a phase coherent conducting system is driven far from equilibrium due to applied voltage bias. These nonlinear UCF (NLUCF) are sample-specific, reproducible fluctuations of the *differential conductance*, dI/dV, as a function of V [347]. An example of this [348] in a diffusive system is shown in Fig. 6.13(a), while the same phenomenon in a metal point contact [349] is shown in Fig. 6.13(b).

NLUCF originate from the same energy-dependent interference patterns that are responsible for the EFUCF. In EFUCF the relevant energy is E_F as tuned by a gate voltage;

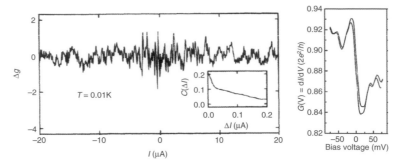

Fig. 6.13   Nonlinear universal conductance fluctuations. (a) Differential conductance as a function of bias current of a 600 nm long diffusive Sb nanowire. Inset: autocorrelation function that may be analyzed to estimate L_ϕ. Adapted from Webb *et al.* [348] (b) Differential conductance as a function of bias voltage in an atomic-scale Au point contact, also showing NLUCF. Adapted from Ludoph *et al.* [349]

in this case the relevant energy scale is set by the chemical potential difference across a phase coherent volume of the sample, $\propto eV$. Energy averaging is a clear concern in such measurements, as is heating, since by definition the conducting system is driven out of equilibrium by the voltage bias.

Persistent currents

An additional quantum conduction phenomenon that can appear in phase coherent nanostructures is *persistent current*. The basic idea of persistent current is familiar from superconductivity and superfluidity. Consider a system described by a quantum mechanical wavefunction $\psi(\mathbf{r})$. The requirement that the wavefunction be single-valued if one goes around the ring leads to the conclusion that any quantum mechanical phase accumulated traveling around the ring must be an integer multiple of 2π.

The question is, what happens when a magnetic field is applied that threads a small amount of flux through the ring? We know from our earlier discussion that the phase due to the externally imposed magnetic flux (via the Aharonov–Bohm effect, Eqs. (6.24), (6.25)) is given by $2\pi\,\Phi/\phi_0$, where ϕ_0 is the flux quantum. Since B is continuously tunable, this AB phase does not have to be quantized, while the *total* phase does. To enforce the quantization condition, spontaneous currents develop in the ring, such that the total magnetic flux through the ring is indeed quantized.

These currents do *not* decay in time, and are just as "persistent" as currents established in superconductors. The existence of these currents can also be argued from general thermodynamics considerations [332]. If L_ϕ is smaller than the circumference of the ring, the persistent currents are suppressed, since couplings to other degrees of freedom make the phase of the wavefunction on the ring not well defined.

These persistent currents have actually been detected [350, 351] via their resulting magnetic moments using extremely sensitive magnetometry (based on superconducting quantum interference devices (SQUIDs) that happen to take advantage of very similar physics). The precise magnitude and sign of the currents, and the joint roles of disorder, ensemble averaging, and decoherence processes complicate the interpretation of persistent current measurements. This striking demonstration of quantum coherence remains an active topic of research.

6.4.3 Tunneling

Now let us consider a fully quantum mechanical phenomenon, the "tunneling" of a particle through a classically forbidden region. Figure 6.14 shows an example, where we consider tunneling through a rectangular barrier in vacuum. Our treatment of this problem is patterned on that by Ferry and Goodnick [352], though they consider the problem of an electron in the conduction band moving through a 1d semiconductor barrier structure, of the kind discussed in the previous chapter. The essential mathematics are identical to the vacuum tunneling case; however, in crystalline solids one must recall that our wavefunction, $\phi(x)$, is really an envelope function for Bloch waves, and one must keep track of effective masses.

The appropriate Schrödinger equation for our situation is

$$\left(-\frac{\hbar^2}{2m}\frac{\partial^2}{\partial x^2} + V_{\text{eff}}(x)\right)\phi(x) = E\phi(x), \tag{6.30}$$

and the usual boundary conditions ($\phi(x)$ and $\partial\phi/\partial x$ must be continuous everywhere) apply. We choose our zero of energy such that $V_{\text{eff}} = V_0$ for $-a < x < a$ and zero elsewhere.

Let's contend first with the case where $0 < E < V_0$. It is then easy to guess the form of the solution:

$$\phi(x) = \begin{cases} Ae^{ikx} + Be^{-ikx}, & x < -a, \\ Ce^{\gamma x} + De^{-\gamma x}, & -a < x < a, \\ Fe^{ikx} + Ge^{-ikx}, & x > a. \end{cases} \tag{6.31}$$

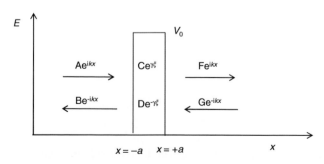

Schematic of a simple 1d tunneling problem. The trial solution components shown assume $E < V_0$. For $E \geq V_0$, the wavevector in the barrier region would be real rather than imaginary.

Fig. 6.14

As we learned to do in undergraduate quantum mechanics, here

$$k \equiv \sqrt{\frac{2mE}{\hbar^2}},$$

$$\gamma \equiv \sqrt{\frac{2m(V_0 - E)}{\hbar^2}}. \tag{6.32}$$

We then apply our boundary conditions at $x = -a$ and $x = a$. This will give us a total of four equations which we will then rewrite in matrix form. A warning: this will be algebraically rather messy, but this will introduce a useful way of thinking about both quantum conductors and eventually optical systems. The four equations are:

$$Ae^{-ika} + Be^{ika} = Ce^{-\gamma} + De^{\gamma a}, \tag{6.33}$$

$$ik[Ae^{-ika} - Be^{ika}] = \gamma[Ce^{-\gamma a} - De^{\gamma a}], \tag{6.34}$$

$$Ce^{\gamma a} + De^{-\gamma a} = Fe^{ika} + Ge^{-ika}, \tag{6.35}$$

$$\gamma[Ce^{\gamma a} - De^{-\gamma a}] = ik[Fe^{ika} - Ge^{-ika}]. \tag{6.36}$$

Rewriting these as two matrix equations,

$$\begin{pmatrix} A \\ B \end{pmatrix} = \begin{bmatrix} \frac{ik+\gamma}{2ik} e^{(ik-\gamma)a} & \frac{ik-\gamma}{2ik} e^{(ik+\gamma)a} \\ \frac{ik-\gamma}{2ik} e^{-(ik+\gamma)a} & \frac{ik+\gamma}{2ik} e^{-(ik-\gamma)a} \end{bmatrix} \begin{pmatrix} C \\ D \end{pmatrix}, \tag{6.37}$$

and

$$\begin{pmatrix} C \\ D \end{pmatrix} = \begin{bmatrix} \frac{ik+\gamma}{2\gamma} e^{(ik-\gamma)a} & -\frac{ik-\gamma}{2\gamma} e^{(ik+\gamma)a} \\ -\frac{ik-\gamma}{2\gamma} e^{-(ik+\gamma)a} & \frac{ik+\gamma}{2\gamma} e^{-(ik-\gamma)a} \end{bmatrix} \begin{pmatrix} F \\ G \end{pmatrix}. \tag{6.38}$$

We can plug Eq. (6.38) into (6.37), and the result is

$$\begin{pmatrix} A \\ B \end{pmatrix} = \begin{bmatrix} M_{11} & M_{12} \\ M_{21} & M_{22} \end{bmatrix} \begin{pmatrix} F \\ G \end{pmatrix}, \tag{6.39}$$

where the matrix elements result from the products of the matrices in Eqs. (6.37) and (6.38).

The **M** matrix here relates the wavefunction coefficients on the left side of the barrier to those on the right side of the barrier. Often it is useful to think in terms of a scattering process, where we consider a particle flux incident on the left side of the barrier and a resulting transmitted particle flux leaving on the right side of the barrier. Remembering our definition of particle current density from quantum mechanics,

$$\mathbf{J}_n = \frac{-i\hbar}{2m}[\psi^* \nabla \psi - \psi \nabla \psi^*], \tag{6.40}$$

and applying it to our plane-wave-like states $\phi(x)$ on either side of the barrier, we can see that the incident particle flux on the left side of the barrier is

$$f_{inc} = |A|^2 \frac{\hbar k}{m}, \tag{6.41}$$

and the outgoing (transmitted) particle flux on the right side of the barrier is

$$f_{tran} = |F|^2 \frac{\hbar k}{m}. \tag{6.42}$$

We can then define a *transmission coefficient*:

$$T(E) = \frac{|F|^2}{|A|^2} = \frac{1}{|M_{11}|^2}$$

$$= \frac{1}{1 + \left(\frac{k^2 + \gamma^2}{2k\gamma}\right)} \sinh^2(2\gamma a). \tag{6.43}$$

In the limit of a wide barrier ($\gamma a \gg 1$),

$$T(E) \rightarrow \left(\frac{4k\gamma}{k^2 + \gamma^2}\right)^2 e^{-4\gamma a} \propto e^{-4a\sqrt{2m(V_0 - E)}/\hbar}. \tag{6.44}$$

The transmission coefficient for particles incident from the left side of the barrier varies exponentially with the barrier thickness. As we shall see shortly, in an experiment involving electrons (e.g., a scanning tunneling microscope, quantum wells placed close together, or two metals separated by a thin insulating layer) this will translate into an electrical conductance that is exponentially sensitive to the geometry.

Tunneling is fundamentally a quantum mechanical phenomenon that flies in the face of everyday (classical) physical intuition. Classically, if you run directly at a solid brick wall, your probability of getting through the wall is zero. Quantum mechanics, however, implies that you have a nonzero (though exponentially small) chance of transmitting through the classically forbidden region to the other side of the wall. Since the mass of the tunneling particle, the barrier height, and the barrier width all appear in the exponential part of $T(E)$, we see that the probability of transmission is greatly enhanced for narrow barriers and very light particles, exactly the regime relevant at the nanoscale.

We can repeat the derivation of Eq. (6.43) for the $E > V_0$ case as well, arriving at

$$T(E) = \frac{1}{1 + \left(\frac{k^2 - k'^2}{2kk'}\right)} \sin^2(2k'a), \tag{6.45}$$

where $k' \equiv \sqrt{2m(E - V_0)}$. This is also a remarkable result considering classical physics intuition. Classically if a particle has enough energy to surmount a potential barrier (e.g., a bullet that has been fired from a gun has enough kinetic energy to punch through the barrier presented by a sheet of paper), the probability of transmission is 100% ($T(E > V_0) = 1$). In the quantum mechanical case, however, reflections from such a barrier are still possible even in the classically allowed regime. One can interpret this as resulting from interference between waves scattered off the front of the barrier and those scattered off the back of the barrier.

For tunneling through a generic barrier $V(x)$, one can imagine dividing up the classically forbidden portion of the barrier into a series of thin, rectangular barriers of width dx. Defining a local $\gamma(x)$, the transmission coefficient would then be the product of all of the individual transmission coefficients:

$$T(E) = \exp\left[-\frac{2}{\hbar} \int_{x_1}^{x_2} (2m(V(x) - E))^{1/2} dx\right], \tag{6.46}$$

where x_1 and x_2 are the edges of the classically forbidden region ($V(x) > E$ if $x_1 < x < x_2$). This is an example of the *WKB approximation*, named for Wentzel, Kramers, and Brillouin.

This is a reasonable approximation to make as long as x_1 and x_2 are far apart compared to the effective wavelength of the tunneling particle, and as long as $V(x)$ varies slowly on the scale of a wavelength.

One particular tunneling calculation often done via WKB is the problem of a triangular barrier. This is physically relevant when one considers field emission, for example, in which the triangular potential results from a large electric field tilting an otherwise flat potential. If the initial barrier height $V(x = 0) - E \equiv e\phi_B$, then in an electric field F we find $V(x) - E = e\phi_B - eFx$ for an electron of charge $-e$. Plugging in to the WKB expression,

$$T(E) = \exp\left[-\frac{2}{\hbar}\int_0^{\phi_B/eF} (2m(e\phi_B - eFx))^{1/2}dx\right],$$

$$= \exp\left[-\frac{4}{3}\frac{(2me)^{1/2}\phi_B^{3/2}}{\hbar F}\right]. \tag{6.47}$$

This predicts exponentially enhanced transmission at large electric fields, as the effective barrier becomes thinner.

Note that a relatively general treatment of tunneling through barriers as a function of bias was developed by Simmons [353]. Moreover, when considering tunneling of electrons between metal electrodes, the true potential felt by the tunneling electron is modified from the ideal "rectangular barrier" situation by image potential effects. The screening response of the metal electrodes lowers the effective barrier near the electrode surface, rounding over the corners of the original potential. This, too, is discussed in the Simmons model.

One question that has prompted much discussion over the years is how long does it take a tunneling particle to traverse the classically forbidden region? This turns out to be a very subtle issue [354, 355, 356], but one particular feature is worth noting. For tunneling through a 1d barrier of length d and height V_0, one can define a tunneling time based on the magnitude of the (imaginary) velocity of the particle in the barrier region [357]:

$$\tau_{\text{tun}} = \sqrt{\frac{m}{2(V_0 - E)}}d. \tag{6.48}$$

While this approach is not valid in the limit that $E = V_0$, it is true that as the particle becomes closer to the resonant tunneling limit ($E \to V_0$), the tunneling time *increases*. In more complicated barriers with internal degrees of freedom, such as molecules, this means that tunneling particles have more time to interact with those degrees of freedom when closer to resonant tunneling.

Multiple barriers

We can easily consider the case of two barriers in a row, as shown in Fig. 6.15. The region between two barriers is sometimes considered a potential well. The mathematics is straightforward. Since we have reduced the single-barrier problem to a matrix relating left- and right-side amplitudes, we just need to work out the matrix that connects the complex amplitudes F and G (just to the right of the left-most barrier) with the amplitudes A' and B' (just to the left of the right-most barrier). Here we work with the $E < V_0$ case, for

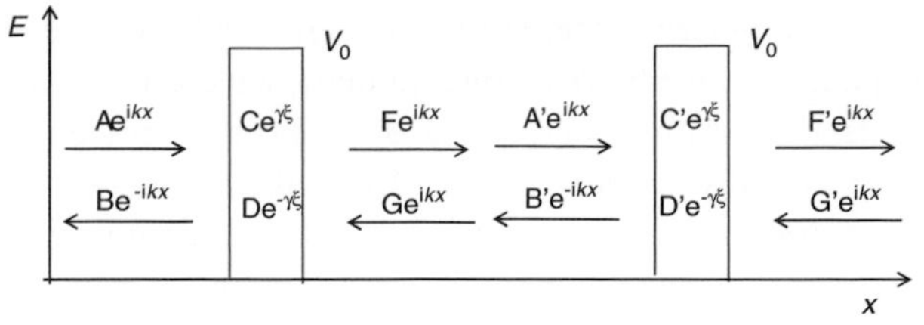

A double-barrier tunneling problem (for $E < V_0$). In quantum transport interference between different possible trajectories affects the total transmitted amplitude and hence the transmission probability.

Fig. 6.15

concreteness, but one can do the same approach for the classical over-barrier limit, as above.

Since the potential is flat and zero in the space between the barriers, the complex amplitudes are related by simple phase factors associated with propagation of plane waves across the interbarrier distance b:

$$\begin{pmatrix} F \\ G \end{pmatrix} = \begin{bmatrix} e^{-ikb} & 0 \\ 0 & e^{ikb} \end{bmatrix} \begin{pmatrix} A' \\ B' \end{pmatrix} = \mathbf{M_W} \begin{pmatrix} A' \\ B' \end{pmatrix}. \tag{6.49}$$

Here $\mathbf{M_W}$ is the matrix that accounts for the propagation in the well. The combined double-barrier problem may then be written

$$\begin{pmatrix} A \\ B \end{pmatrix} == \mathbf{M_R M_W M_R} \begin{pmatrix} F' \\ G' \end{pmatrix}, \tag{6.50}$$

where $\mathbf{M_L}$ and $\mathbf{M_R}$ are, respectively, the matrices for the left and right barriers taken individually. One can write the combined matrix as $\mathbf{M}_{tot} = \mathbf{M_L M_W M_R}$. As in the single-barrier case, we can consider fluxes incident on the left and outgoing on the right of the double barrier, and find (after some algebra) a transmission coefficient. Assuming that the barriers are identical and individually would each have a transmission coefficient $T_1(E)$, one finds

$$T(E) = \frac{|F'|^2}{|A|^2} = \frac{1}{|M_{tot11}|^2} = \frac{T_1}{T_1^2 + 4(1 - T_1) \cos^2(kb - \theta)}. \tag{6.51}$$

Here

$$\theta = -\arctan\left[\left(\frac{k^2 - \gamma^2}{2k\gamma}\right) \tanh(2\gamma a)\right] + 2ka. \tag{6.52}$$

Here k and γ are defined as in the single-barrier problem.

The most important thing to note here is that transmission through this structure has resonances. When $\cos(k(E)b - \theta(E)) = 0$ the transmission is *perfect*, regardless of the transmission of the individual barriers taken in isolation. Similarly the minimum transmission through this identical double-barrier system is $\approx T_1^2/4$ when T_1 is small, considerably worse than the classical expectation of T_1^2. These results are a consequence of quantum interference, as we shall see shortly.

There is another way to interpret these resonances, particularly when T_1 is small (thick and/or tall barriers). Suppose the barriers were infinitely wide yet remained separated by a well of width b of depth V_0 with a floor at $E = 0$. That (finite) potential well would have a spectrum of bound states with energies $E < V_0$. In the double-barrier structure it can be shown that the transmission resonances for $E < V_0$ take place when the energy of the incident particle is approximately equal to the energy of one of those bound states. In a semiclassical handwave, picture a particle incident from the left tunneling through the first barrier, bouncing around for a while in the bound state of the well, and then eventually tunneling through the second barrier to the right. This phenomenon, a transmission $T(E)$ that has peaks when E lines up with the energies of pseudo-bound states, is called *resonant tunneling*. More clearly now than in our discussion of tunneling time, in a semiclassical picture, while they are ultimately transmitted with high probability, particles that resonantly tunnel spend a comparatively long time in the well structure.

One can do a more general treatment of an asymmetric double-barrier structure (e.g., Ferry and Goodnick [352], pp. 108 ff.). While the calculation is algebraically more complex, the basic result remains: transmission has pronounced resonances, described by generalized forms of Eqs. (6.51) and (6.52). One can consider shifting the baseline of the potential between the left and right sides of the double-barrier structure by an amount ΔE. This is one way of modeling an applied bias voltage between the two sides. Depending on the magnitude of ΔE, the bound states of the well can be brought into and out of resonance with the energy of the incident particle. The result is a dramatic peak in the transmission as a function of ΔE.

Such 1d barrier structures can be implemented readily in semiconductor multilayers of the type discussed in Section 5.5.2. In these solid state incarnations, the formulae derived above for wavefunctions often may be translated into equivalent expressions for the envelope functions of Bloch states of electrons or holes in the semiconductor materials. The barriers for electrons (or holes) in $V(x)$ then originate from the spatial variation of the conduction (or valence) bands. One must be careful to apply the correct boundary conditions (continuity of the envelope function and $1/m^*$ times its derivative) at the interfaces. The peak in transmission as a function of ΔE in the asymmetric double-barrier case is the basis for a semiconductor device known as the resonant tunneling diode [358, 359].

Scattering matrix formalism

The formalism that we developed in the preceding section may be rewritten in the language of *scattering matrices*. In the 1d case, rather than relating components of the wavefunction on the left and right sides of the scattering region using $\mathbf{M}$, as we have done so far, we instead consider a matrix, $\mathbf{S}$, that relates incident and scattered components of the wavefunction. That is,

$$\begin{pmatrix} A \\ B \end{pmatrix} = \begin{bmatrix} M_{11} & M_{12} \\ M_{21} & M_{22} \end{bmatrix} \begin{pmatrix} F \\ G \end{pmatrix} \rightarrow \begin{pmatrix} B \\ F \end{pmatrix} = \begin{bmatrix} S_{11} & S_{12} \\ S_{21} & S_{22} \end{bmatrix} \begin{pmatrix} A \\ G \end{pmatrix}. \tag{6.53}$$

From the matrix elements we can define transmission and reflection coefficients for particles incident from the left and right sides. For example, the transmission coefficient from left to right is

$$T_{\mathrm{LR}}(E) = |S_{21}|^2. \tag{6.54}$$

Likewise, $R_{\mathrm{LR}}(E) = |S_{11}|^2$, $T_{\mathrm{RL}}(E) = |S_{12}|^2$, and $R_{\mathrm{RL}}(E) = |S_{22}|^2$.

These definitions assume that the effective velocities ($\hbar k(E)/m$) of the particles far to the left and right sides of the scatterer are the same. Suppose the scattering region is confined to $-a < x < a$. For the general asymmetric case, such that the potential $V(x \ll -a)$ differs from $V(x \gg a)$, one must keep track of the resulting velocity difference. Define $v_1 \equiv \hbar k_1(E)/m$ and $v_2 \equiv \hbar k_2(E)/m$, where $k_1 = \sqrt{2m(E - V(x \ll a))/\hbar^2}$ and $k_2 = \sqrt{2m(E - V(x \gg a))/\hbar^2}$. Then the matrix elements in Eq. (6.53) should be modified so that $S_{ij} \rightarrow S_{ij}\sqrt{v_i/v_j}$.

When properly defined, the **S** matrix is unitary, which has the important consequence that probability is conserved; all probability flowing into the scattering region is balanced by probability flowing out of the scattering region. We shall see echoes of the **S**-matrix approach in our treatment of quantum coherent conduction below. The analog of the transfer matrix, **M**, will appear in our discussion of photonics in Chapter 8.

6.4.4 Landauer–Büttiker formalism

The scattering treatment presented so far may be expanded into a general approach for understanding the conduction properties of small systems with quantum coherence. Like the discussion of tunneling, this approach will be presented in terms of single-particle physics, though there are ways of accounting for electron–electron interactions. This scheme, known as the *Landauer–Büttiker formalism*, considers quantum coherent scattering regions connected via contacts to classical (decohering) reservoirs that can serve as current sources or sinks, or as probes of the voltage (local chemical potential of the charge carriers). The main insight is to think about conduction through a quantum system as transmission in a scattering problem. The brief treatment here is modeled on that by Datta [360] and Ferry & Goodnick [352]. It is meant to give some intuition in how to think about quantum transport rather than to be a rigorous exposition.

Ahead of his time, in the late 1950s [361] Rolf Landauer began examining the physics of a *ballistic* conductor. Consider the 1d model of two classical contacts linked by a piece of conductor much shorter than the mean free path of the carriers. Is the conductance of such a system infinite? Experimentally the answer is "no"; the measured conductance remains finite (the two-terminal resistance is nonzero) even when $L \ll \ell$, as in a point contact between two pieces of metal. Given that no scattering takes place within the ballistic region, the nonzero resistance must somehow originate from the contacts, as we shall see below.

For ease of calculation, let's assume that the ballistic region is a narrow 1d conductor. We can describe the single-particle states of this region with specific, discrete 1d subbands, each corresponding to a particular transverse mode, as in Fig. 2.7. Figure 6.16 shows the geometry under consideration as well as a cartoon of the parabolic $E_j(k)$ dispersion relation

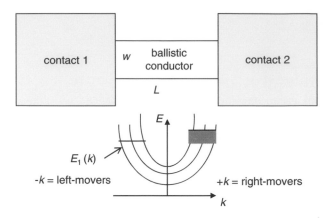

Schematic for Landauer formula discussion. Classical left and right contacts are connected by a segment of ballistic conductor. As shown in the lower diagram, there are three 1d subbands occupied below the chemical potentials of the two contacts, and the left contact (which supplies the right-moving carriers) has a chemical potential, μ_L, slightly higher than the right contact. This chemical potential difference is what drives electrons from left to right.

for each subband. For any particular energy E we can define a function $M(E) \equiv \sum_j \Theta(E - E_j)$ that uses the theta function to count the number of modes with energies $E_j < E$. As shown in the figure, for the indicated energy $M(E) = 3$.

Saying that our contacts are classical objects that do not permit reflections has a major consequence. All of the right-moving carriers in the wire (those with $k > 0$) must have originated at the left contact. Similarly, all of the carriers in the wire with $k < 0$ must have started out in the right contact. If we assume that a battery is used to establish a voltage difference (a bias, and therefore a chemical potential difference) between the two contacts, then we find that the right- and left-moving carriers have different population distributions as a function of energy. The right-moving carriers have an occupation probability given by the Fermi function appropriate for the left contact, $f_+(E, T, \mu_L)$, where μ_L is the chemical potential of the carriers in the left contact. Working with electrons, the total current carried from left to right is then given by

$$I_+ = \frac{-e}{L} \sum_k v(E(k)) f_+(E, T, \mu_L) M(E(k)), \qquad (6.55)$$

where $v \equiv (1/\hbar)\partial E/\partial k$ is the semiclassical velocity of the particles. We can convert this sum over k values into an integral, knowing that traveling wave states are spaced by $2\pi/L$ in 1d k-space (see Eq. (2.13) and following). Realizing that we can write $(\partial E/\partial k) \cdot dk \rightarrow dE$, we find

$$I_+ = \frac{-2e}{h} \int_0^\infty f_+(E, T, \mu_L) M(E) dE, \qquad (6.56)$$

where the 2 originates from the spin-1/2 degeneracy of the electrons. This expression implies that the current carried per mode per unit energy is $2e/h$ for spin-degenerate electrons. One can do the same analysis for *left*-moving electrons (those that originate in the right contact), and find the left-moving current:

$$I_- = \frac{-2e}{h} \int_0^\infty f_-(E, T, \mu_R) M(E) dE,$$ (6.57)

where labels have been changed appropriately.

The net current is the difference, $I = I_+ - I_-$. At $T = 0$, the resulting integral becomes very simple since f_+ and f_- become step functions at μ_L and μ_R, respectively. The result, for $M(\mu_L) = M(\mu_R) = M$, is

$$I = \frac{-2e}{h} M \cdot (\mu_L - \mu_R) = \frac{2e^2}{h} M \left(\frac{\mu_L - \mu_R}{-e} \right) = \frac{2e^2}{h} MV,$$ (6.58)

where V is the voltage difference between the left and right contacts. Net positive current exists because the difference in chemical potentials between the contacts means that there are more occupied modes going left-to-right than right-to-left, as shown in Fig. 6.16.

From this we see that the Landauer forumla for the (two-terminal) conductance of an M-channel ballistic conductor at $T = 0$ is

$$G = \frac{2e^2}{h} M.$$ (6.59)

Since the conductor here was assumed to be ballistic, the resistance $1/G$ must somehow be a contact resistance, associated with the interfaces between our idealized contacts and the ballistic conductor. Indeed, the modern interpretation is exactly this. The contacts contain a very large number of modes (propagating electronic states), yet only some small number of these are able to couple well into the modes that propagate through the ballistic conductor. Wavepackets built from states that couple poorly throught the conductor are reflected from it. This is analogous to an electromagnetic waveguide rcflccting most incident radiation except that matched to the waveguide's propagating modes.

The implication of the Landauer formula is clear: if we could somehow deform a ballistic conductor to vary the number of subbands or modes, M, that lie below the Fermi level of the leads, then we should see steps in the conductance. The conductance should be quantized in amounts of size $2e^2/h \equiv G_0$, approximately ($1/12.9$ kΩ).

Conductance quantization was first observed experimentally in the late 1980s [362, 363] in two-dimensional electron gas systems at GaAs/AlGaAs interfaces. By tuning the voltage on a pair of gate electrodes, the investigators were able to deplete the 2deg under the gates, leaving a narrow *quantum point contact* (QPC) through which electrons could flow. Figure 6.17(a) shows the conductance of such a QPC as the width of the electronic channel is varied. Clear steps are seen in the two-terminal conductance. These steps are smeared out with increasing T due to the broadening of the Fermi distribution functions.

Similar steps are observed in the conductance through an atomic-scale contact between two pieces of metal, as in Fig. 6.17(b). While the chemical nature of the metal atoms affects the precise quantization [365] (the number of channels available and the transmission of each channel), the basic principle still applies. We will discuss some details further below. Note that conductance quantization signatures can be seen in metal junctions even at room temperature, since $k_B T \ll E_F$ for most metals at 300 K.

Where is the voltage actually dropped in such ballistic systems? This is a rather subtle question. Right-moving carriers in our model remain at μ_L until they reach the right

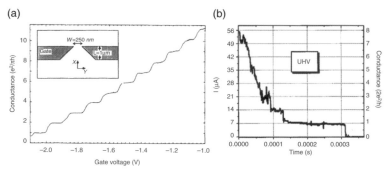

Fig. 6.17 Conductance quantization. (a) Steps in the conductance of a 2deg in a GaAs/AlGaAs heterostructure at 0.6 K as a function of voltage applied to gates acting to pinch off the constriction. Adapted from van Wees *et al.* [362]. (b) Similar steps in the conductance during breaking of an atomic-scale contact between gold electrodes. Adapted from Costa-Krämer *et al.* [364].

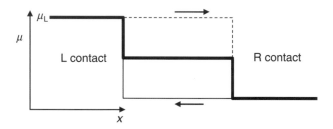

Fig. 6.18 Chemical potential as a function of position for the ballistic conductor situation in Fig. 6.16. The dashed line shows the chemical potential for right-moving carriers ($+k$), while the thin line shows the chemical potential for left-movers ($-k$). The thick solid line is the local averaged chemical potential. The fact that the chemical potential drops all happen at the interface between the ballistic and contact regions emphasizes that these drops are contact effects.

contact. Left-moving carriers remain at μ_R until they reach the left contact. Clearly this energy distribution of electrons within the ballistic region is *not* what one would find in thermal equilibrium. The *average* chemical potential lies in between, and as shown in Fig. 6.18. This emphasizes that the effective two-terminal resistance somehow originates at the contacts, while the average chemical potential is uniform within the ballistic conductor. The discontinuity of μ_{avg} at the contacts is an artifact of our idealized model; in any real structure the electrostatic potential contribution is smeared out due to screening from our so-far neglected electron–electron interactions.

It is also important to recognize that the *dissipation*, the actual macroscopic irreversibility that we associate with resistive heating, takes place far from the ballistic constriction. We have been assuming that inelastic processes are not relevant in the ballistic region, and those are precisely the processes that re-establish thermal equilibrium distribution functions for the electrons. These inelastic interactions typically take place a distance $\sqrt{D\tau_i}$ away, assuming diffusive contacts and an inelastic scattering rate τ_i^{-1}. The result appears macroscopically irreversible because energy that had been effectively concentrated in a small number of propagating modes ends up being redistributed in a large number of

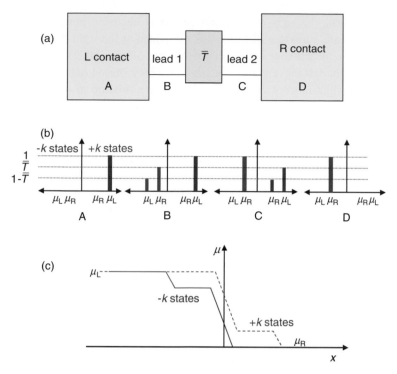

Fig. 6.19

A more realistic treatment of a quantum conductor. (a) Classical contacts attached by mesoscopic, ballistic leads to a tunneling barrier of transmittance $\overline{T}$. (b) Assuming carriers are incident from the left contact, this shows, at each point (A–D), the fraction of $+k$ and $-k$ states that have a particular chemical potential. For example, at B all of the right-moving carriers originated at the left contact, and therefore have $\mu = \mu_L$. (c) Spatial distribution of the chemical potentials for the $-k$ (solid line) and $+k$ (dashed line) states. Discontinuities at boundaries have been smeared spatially. Adapted from Fig. 2.3.1 of [360].

modes in the contacts. One practical consequence of the remote character of the dissipation is that atomic-scale metal contacts (and other ballistic conductors) can pass surprisingly enormous current densities, much higher than bulk, diffusive systems.

Generalizing our discussion to imperfect conductors can better illuminate the origins of the Landauer–Büttiker contact resistance. First, one can re-do the analysis of Eqs. (6.55)–(6.58) and assume that the region between the contacts has a transmission coefficient $\overline{T}(E)$ that is roughly constant over the energy range from μ_L to μ_R for all the subbands. The result is a two-terminal conductance

$$G = \frac{2e^2}{h} M\overline{T}, \qquad (6.60)$$

where $\overline{T}$ is the average transmission coefficient.

Figure 6.19(a) shows a slightly more realistic model. We still have our classical, reflectionless contacts, but this time they are attached to ballistic "leads" which are coupled to a central scattering region with some transmission coefficient $\overline{T}$. (This notation avoids later confusion with the temperature, T.) Looking at the chemical potential distribution now, in the left lead, the right-moving carriers are at chemical potential μ_L. The left-movers, however, are a mixture of those that originated at the right contact (at μ_R) and those that started in the left contact (at μ_L) but reflected off the scattering region. Similarly, in the *right* lead, the left-movers are all at μ_R since they started out in the reflectionless right contact; the right-movers, on the other hand, are a mixture of those that originated at the left contact (μ_L) and were transmitted through the scattering region, and those that started at the right contact (μ_R) and were reflected from the scattering region.

Again we see that the energetic distributions of the carriers are far from thermally equilibrated. One can average the chemical potentials appropriately, as shown in Fig. 6.19(c). Ignoring screening for a moment, we now see three discontinuities in the average chemical potential for the electrons that translate into voltage drops. In addition to the expected drop across the partially transmitting region, there are chemical potential shifts at each contact/lead interface.[6] This idealization makes clear that the two-terminal conductance of Eq. (6.60) can be written as the inverse of a two-terminal resistance that has clear contributions from the *contacts* and the *scatterer*:

$$G^{-1} = \frac{h}{2e^2 M}\frac{1}{\overline{T}} = \frac{h}{2e^2 M} + \frac{h}{2e^2 M}\frac{1-\overline{T}}{\overline{T}}. \tag{6.61}$$

The first term on the far right represents the contact resistance piece; that results from the same bulk-contact-to-few-mode-lead coupling issue that we discussed when considering a ballistic conductor. The second term is the scatterer contribution to the series resistance, which vanishes when $\overline{T} \to 1$ and diverges as $\overline{T} \to 0$. In other words, the scatterer resistance contribution acts as one would naively expect.

Pondering how one could actually measure just the scatterer contribution to the overall resistance is instructive. The first thing that comes to mind is some kind of four-terminal measurement approach, with voltage probes somehow "picking up" the chemical potential difference between left ballistic lead and the right ballistic lead in Fig. 6.19(a). In practice, this is very challenging. One wants noninvasive voltage probes that do not perturb the underlying current distribution in the structure, while at the same time the probes must be able to "talk" to the local charge carriers to equilibrate at the appropriate chemical potentials. For example, in a real 1d quantum device it is difficult to ensure that physical voltage probes couple equally well to both left- and right-moving carriers. Looking at the energy distributions in Fig. 6.19(c), it is clear that one could end up with problems. If the left probe only coupled to left-moving states, and the right voltage probe only coupled to right-moving states, one would compute an incorrect four-terminal resistance.

Thanks to advances in materials growth, this kind of comparison of two- and four-terminal resistances has actually been performed in a real physical system [366]. Figure 6.20(a) shows a semiconductor quantum wire structure made by cleaved-edge overgrowth (see Section 5.6.2) using the GaAs/AlGaAs system. The source and drain regions of 2deg are the contacts, while the narrow "probe A" and "probe B" regions of undepleted 2deg are the voltage probes. By tuning the voltages on the tungsten gates, the authors are able to adjust the invasiveness of probes A and B. When the system is properly

[6] This analysis is performed in detail by Datta [360].

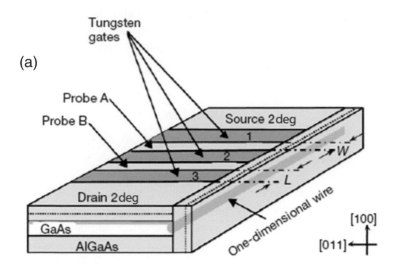

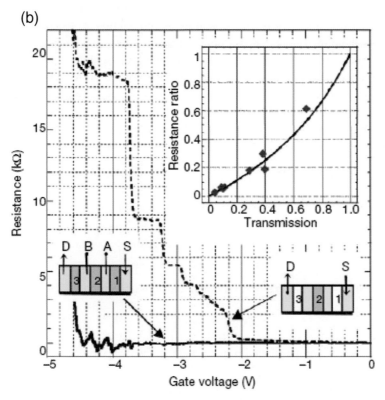

Fig. 6.20

True four-terminal measurement of quantum conductance. (a) Schematic of a 1d quantum wire formed by
cleaved-edge overgrowth using GaAs/AlGaAs heterostructures. (b) Two-terminal measurements (upper trace) show
quantized conductance as the wire is depleted, while four-terminal measurements (lower trace) show *perfect
conduction* along the wire. This clearly demonstrates that the finite resistance seen in two-terminal measurements of
ballistic quantum conductors is a contact effect. Adapted from [366].

adjusted, the situation approaches that of the theoretical idea. Two-terminal measurements of the conductance between source and drain show the usual conductance quantization, while four-terminal measurements using voltage probes A and B show *zero* resistance for the intervening ballistic portion of 1d conductor.

Ohm's Law

In quantum transport, what has become of Ohm's Law? As we've shown in Section 6.4.3, placing two tunnel barriers in series can lead to resonances in the transmission rather than a simple multiplication of the individual barrier transmission probabilities. These resonances result from the phase-coherent interference of transmitted and reflected waves.

To recover Ohm's Law, first we suppose that scatterers (barriers) are farther apart than the typical coherence length. Then we need only consider adding the *probabilities* (not amplitudes) of various processes. For example, consider the 1d case and two scatterers with transmission probabilities $\overline{T}_1$ and $\overline{T}_2$ respectively. What is $\overline{T}_{12}$, the total probability for a particle to make it past both scatterers? One process that is relevant is direct transmission past both scatterers; this should have a probability $\overline{T}_1 \overline{T}_2$. A second process that will contribute to conduction is a double bounce, where the particle is transmitted through the first barrier, reflected off the second, reflected again off the first, and then finally transmitted through the second. The classical probability for this would then be $\overline{T}_1(1 - \overline{T}_2)(1 - \overline{T}_1)\overline{T}_2 = \overline{T}_1 R_2 R_1 \overline{T}_2$, writing in terms of the transmission and reflection probabilities. The total probability of transmission is found by summing the probabilities of all the multiple bounces. The interference of competing paths, which gives resonances in the fully coherent quantum transport limit, is absent here. We find

$$\overline{T}_{12} = \overline{T}_1 \overline{T}_2 + \overline{T}_1 \overline{T}_2 R_1 R_2 + \overline{T}_1 \overline{T}_2 R_1^2 R_2^2 + \cdots = \overline{T}_1 \overline{T}_2 \frac{1}{1 - R_1 R_2}, \qquad (6.62)$$

where we've summed the infinite series of multiple bounce possibilities. Rewriting in terms of transmission probabilities,

$$\frac{1 - \overline{T}_{12}}{\overline{T}_{12}} = \frac{1 - \overline{T}_1}{\overline{T}_1} + \frac{1 - \overline{T}_2}{\overline{T}_2}. \qquad (6.63)$$

For N identical scatterers in a row, each with transmission probability T, this implies

$$\overline{T}_N = \frac{\overline{T}}{N(1 - \overline{T}) + \overline{T}}. \qquad (6.64)$$

If those N scatterers are distributed over a length L, then we write the total transmission probability in the form

$$T_N(L) = \frac{L_0}{L + L_0}, \qquad (6.65)$$

where $L_0 \equiv \overline{T}L/(N(1 - \overline{T}))$ is essentially a mean free path for scattering. When written this way, it's clear that in the large L limit, $G \propto \overline{T}_N(L) \propto 1/L$, as one would expect for Ohm's Law. Note: to get Ohm's Law in its traditional sense, explicitly ignoring large quantum interference contributions to the conductance is required.

Interaction effects

As discussed in Fig. 6.19, the discussion of the chemical potential profile of the electrons in a biased quantum junction is subtle. In our simple idealization we showed abrupt changes in the chemical potential at the interfaces between the contacts and the leads, as well as across a scatterer of transmittance T. We know, however, that the physical electrostatic potential *cannot* change discontinuously in space, since that would imply an infinite local electric field. In any real system, this situation is resolved through the screening properties of the charge carriers – in other words, the electron–electron interactions that we have so far neglected play an essential role in establishing the actual spatial distribution of charge carriers.

A quick, crude way of accounting for averaged electron–electron interactions is to solve Poisson's equation self-consistently for the junction region, including screening length effects (see Section 5.3) in the leads. The result is that a biased junction containing a scatterer exhibits an electric dipole moment, a *resistivity dipole*, with "excess" electronic density piling up on the high chemical potential (for the electrons) side of the scatterer. The larger the current, the larger the dipole moment. A more detailed, realistic discussion of this effect in a particular junction geometry is beyond the scope of this book. An interesting example of this effect in candidate molecular junctions is discussed by Lang and Avouris [367].

There are more general strategies for incorporating electron–electron interactions into the Landauer–Büttiker approach to quantum transport. However, these tactics [368], typically describing the scattering of electrons from noninteracting leads, through an interacting region, and back into noninteracting leads, are beyond the scope of this book.

Thermal and bias effects

So far we have been working at zero temperature, assuming that all of the electrons incident from the left have energies exactly at μ_L. We've also been assuming small biases, so that we haven't had to worry about a the number of channels varying with bias, or the possibility that left- and right-moving carriers may see significantly different transmission coefficients. Note that we're still going to assume fully coherent transport for now.

To address this, we need to start from the expressions in Eqs. (6.56) and (6.57) and arrive at a finite temperature version of Eq. (6.58) for a particular energy E. That gives us

$$I(E) = \frac{2e}{h} \left(M(E)\overline{T}(E)f_+(E, T, \mu_L) - M'(E)\overline{T}'(E)f_-(E, T, \mu_R) \right), \tag{6.66}$$

where we have explicitly considered that the transmission and number of accessible channels to be functions of energy.

In the absence of inelastic scattering in the region of interest (which would redistribute the electron energies), it turns out that this can be simplified in the two-terminal case because $M(E)\overline{T}(E) = M'(E)\overline{T}'(E)$. For a proof, see [360], section 2.6. We can then integrate over energy to arrive at an expression for the total current:

$$I = \int I(E)dE = \frac{2e}{h} \int M(E)\overline{T}(E)[f_+(E, T, \mu_L) - f_-(E, T, \mu_R)]dE. \tag{6.67}$$

Clearly when $\mu_L = \mu_R$ and temperature is uniform, then the two distribution functions are identical and no net current flows, as expected.

Consider when a small voltage (that is, $|e\,\delta V|$ is small compared to the energy scale over which $M(E)\overline{T}(E)$ varies) is applied to the two-terminal device. Then we can Taylor expand the difference between the two Fermi functions:

$$\delta[f_+ - f_-] \approx \left(-\frac{\partial f_0(E,T,\mu)}{\partial E}\right)_{\mu=\mu_L,T} [\mu_L - \mu_R]. \tag{6.68}$$

The conductance is given by the change in current divided by the applied voltage:

$$G = \frac{\delta I}{(\mu_L - \mu_R)/e} = \frac{2e}{h}\int M(E)\overline{T}(E)\left(-\frac{\partial f_0(E,T,\mu)}{\partial E}\right)dE. \tag{6.69}$$

The thermal broadening function (Eq. (2.32)) has reappeared.

What does this mean? Imagine that $M(E)\overline{T}(E)$ has lots of variation as a function of energy. By assuming that $|e\,\delta V| = \mu_L - \mu_R$ is small compared to that variation scale ($\sim E_\phi$), we are in the *linear response* regime. In this case Eq. (6.69) shows that finite temperature smears out the electron energies a bit, letting the electrons sample different transmission coefficients through the system. Slightly changing the applied voltage will not change the conductance (δI will remain linearly proportional to δV, hence the name "linear response"), since the transmission properties are set by the thermal smearing. Conversely, at large voltages one can be in a regime where the applied bias dominates the sampling of the energy landscape.

Note that throughout this whole discussion we have been assuming that the contacts themselves have "nice" flat densities of states and can supply or take up electrons readily. This is a reasonable assumption when, e.g., the contacts are good, bulk metals. However, in real nanoscale systems one may have to be careful.

The multichannel case

The Landauer approach can be generalized to multiterminal conductors, placing all terminals on equal footing. Each terminal sits at some chemical potential μ_i, and can have a net current flowing out of itself, I_i. Dropping explicit references to $M(E)$, and working again at zero temperature for simplicity, the appropriate notation would then be

$$I_i = \frac{2e}{h}\sum_j (\overline{T}_{j\leftarrow i}\mu_i - \overline{T}_{i\leftarrow j}\mu_j). \tag{6.70}$$

The net current out of terminal i is set by summing over the contributions of all the other terminals, each of which can accept carriers scattered from terminal i (the first term on the right side) or scatter carriers into terminal i (the second term).

We can rewrite the chemical potentials in terms of voltages and define $G_{ij} \equiv (2e/h)\overline{T}_{i\leftarrow j}$, leading to

$$I_i = \sum_j (G_{ji}V_i - G_{ij}V_j). \tag{6.71}$$

This immediately leads to a sum rule. If all of the voltages applied to all of the terminals of some multiterminal system are the same, then no net current should flow. This implies that

$$\sum_j G_{ji} = \sum_j G_{ij}. \tag{6.72}$$

Therefore,

$$I_i = \sum_j G_{ij}(V_i - V_j). \tag{6.73}$$

It's also worth noting that the Onsager relations (Eq. (6.7)) as a function of magnetic field, B, manifest themselves here as $G_{ij}(+B) = G_{ji}(-B)$, though proving that in general is difficult [360].

Consider setting various $j \neq i$ to various voltages. If the current out of lead i is zero, then Eq. (6.73) implies that

$$V_i = \frac{\sum_{j \neq i} G_{ij} V_j}{\sum_{j \neq i} G_{ij}}. \tag{6.74}$$

So, when a terminal is left "floating", its potential is some average of the potentials at the other terminals, weighted by the relevant transmission coefficients that connect the terminals.

For an n-terminal device, one can rewrite Eq. (6.73) as a matrix equation, connecting an n-component vector of currents to an n-component vector of voltages by an $n \times n$ conductance matrix, $\tilde{G}$. We can simplify this a little further, since we know current is conserved ($\sum_i I_i = 0$) and we're always free to measure all voltages relative to one terminal which we can designate as "ground". Therefore, we really only need $n - 1$ currents and voltages, and an $(n - 1) \times (n - 1)$ matrix, as in this example for the four-terminal case:

$$\mathbf{I} = \tilde{G} \cdot \mathbf{V},$$

$$\begin{pmatrix} I_1 \\ I_2 \\ I_3 \end{pmatrix} = \begin{pmatrix} G_{12} + G_{13} + G_{14} & -G_{12} & -G_{13} \\ -G_{21} & G_{21} + G_{23} + G_{24} & -G_{23} \\ -G_{31} & -G_{32} & G_{31} + G_{32} + G_{34} \end{pmatrix} \begin{pmatrix} V_1 \\ V_2 \\ V_3 \end{pmatrix}. \tag{6.75}$$

Here we have chosen $V_4 = 0$ to be our ground. Equivalently, it is possible to invert $\tilde{G}$ into a resistance matrix, $\tilde{R} \equiv \tilde{G}^{-1}$, such that $\mathbf{V} = \tilde{R} \cdot \mathbf{I}$.

The point here is that it is possible to determine the elements of $\tilde{G}$ (or $\tilde{R}$) *experimentally* through a series of measurements. These matrices (at a given magnetic field value) give a complete description of coherent transport through such a system. Once the particular elements of $\tilde{G}$ are known, it is then possible to predict the currents and voltages that result from some given configuration of conditions applied at the contacts.

Exactly this procedure was carried out in a multiterminal 2deg structure by Shephard *et al.* [370, 369]. Those investigators cyclically permuted simple current measurements in a four-terminal structure while applying a known voltage to one contact. They performed these measurements over a range of magnetic fields. Then, using the inferred coefficients, they were able to predict the result of a Hall measurement, and subsequently found very

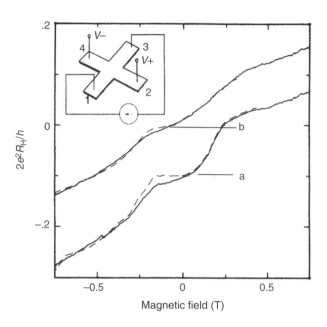

Fig. 6.21 The power of the Landauer–Büttiker approach. After performing a series of measurements to determine $\tilde{G}(B)$ for two different gate voltages in a four-terminal 2deg structure, Shepard *et al.* were able to predict (solid lines) the expected Hall resistance as a function of B. Measurements (dashed lines) show experimental confirmation of these predictions. Adapted from [369].

good agreement with experiment. This is shown in Fig. 6.21. The remarkable result is that the actual mesoscopic Hall trace, with its bumps and wiggles that depend on the details of the sample (such as the particular arrangement of defects and impurities), can be predicted using this formalism based on the results of other measurements, without direct knowledge of those details.

The Landauer–Büttiker approach and its conceptual framework can also provide a means of accessing those microscopic details under some circumstances. Scheer *et al.* [365] examined electronic transport through atomic-scale metal contacts of the type discussed in the Introduction of this book. In these junctions there are only two terminals (corresponding to the two macroscopic electrodes), but it is not obvious how many quantum channels determine the junction conductance. Can one tell whether a junction with conductance $2e^2/h$ involves a single perfectly transmitting channel, or multiple channels with transmittances that sum to one? By working with superconducting junctions and using a phenomenon known as "subgap structure" [371], Scheer *et al.* were able to determine precisely the number of channels and the transmission coefficient of each channel relevant in a series of atomic-scale metal junctions. The results are not just an impressive demonstration of the power of this formalism. They demonstrate an essential feature of the underlying nanoscale physics: the microscopic chemical structure of the junction atoms determines the number of channels participating in transport. For example, conduction through single-atom aluminum junctions involves three channels, originating from the $s - p_z, p_x$, and p_y orbitals of the aluminum valence electrons.

Scattering matrices

Given the similarity between the Landauer–Büttiker approach to coherent transport and our discussion of tunneling, it should come as no surprise that one can re-cast the coherent transport problem into the language of scattering matrices. As in Section 6.4.3, instead of talking about transmission matrix elements, $\overline{T}_{ij}$, connecting amplitude *out* of channel i due to amplitude *into* channel j, we can instead choose to write scattering matrices.

For an n-terminal coherent scattering region, label the incident amplitudes as $a_1,...,a_n$, and the outgoing amplitudes as $b_1,...,b_n$. Then we can think of a scattering matrix $\tilde{S}$ that connects these:

$$\mathbf{b} = \tilde{S} \cdot \mathbf{a}. \tag{6.76}$$

For the simplest case where there the scattering region has just two terminals, then

$$\begin{pmatrix} b_1 \\ b_2 \end{pmatrix} = \begin{pmatrix} r & t' \\ t & r' \end{pmatrix} \begin{pmatrix} a_1 \\ a_2 \end{pmatrix}. \tag{6.77}$$

Reading this directly, the amplitude for an outgoing particle in terminal 1 has two contributions, one ($a_1 r$) due to reflection of an incoming particle in terminal 1, and the other ($a_2 t'$) due to transmission of an incoming particle from terminal 2.

Suppose there were two such scattering regions, each described in isolation by scattering matrices $\tilde{S}^{(1)}$ and $\tilde{S}^{(2)}$, respectively. We can imagine connecting them together coherently in some way by a coupling matrix, $\tilde{S}^{(1-2)}$, that links particular incident amplitudes of one region with outgoing amplitudes of another with appropriate phase shifts for propagation. This is completely analogous to our discussion of transfer matrices (Eq. (6.39)). However, because of the directionality of propagation (some of region 1's outgoing channels become region 2's incoming channels), we cannot simply multiply scattering matrices. Abbreviating the correct operation using the $\otimes$ symbol, the combined scattering matrix is then given by $\tilde{S}^{\text{tot}} = \tilde{S}^{(1)} \otimes \tilde{S}^{(1-2)} \otimes \tilde{S}^{(2)}$.

To see how this works, consider two of our two-terminal regions from Eq. (6.77). Assuming that the two scattering regions are directly connected ($\tilde{S}^{(1-2)} = \tilde{I}$, the identity matrix), the correct total scattering matrix has elements given by [372, 360]:

$$\begin{aligned}
t^{\text{tot}} &= t^{(2)}(1 - r'^{(1)}r^{(2)})^{-1}t^{(1)}, \\
t'^{\text{tot}} &= t'^{(1)}(1 - r^{(2)}r'^{(1)})^{-1}t'^{(2)}, \\
r^{\text{tot}} &= r^{(1)} + t'^{(1)}r^{(2)}(1 - r'^{(1)}r^{(2)})^{-1}t^{(1)}, \\
r'^{\text{tot}} &= r'^{(2)} + t^{(2)}(1 - r'^{(1)}r^{(2)})^{-1}r'^{(1)}t'^{(2)}.
\end{aligned} \tag{6.78}$$

While the rules for combining scattering matrices coherently look complicated,[7] interpretation is relatively straightforward. Looking at the expression for t^{tot}, we can expand this to get

$$t^{\text{tot}} = t^{(2)}t^{(1)} + t^{(2)}(r'^{(1)})r^{(2)})t^{(1)} + t^{(2)}(r'^{(1)})r^{(2)})^2 t^{(1)} + \cdots. \tag{6.79}$$

---

[7] Generalizing Eq. (6.78) to the multiterminal case is straightforward, if tedious. Each t, t', r, amd r' would then be a matrix itself involving some number of terminals, and expressions like $(1 - r'^{(1)}r^{(2)})^{-1}$ would be matrix inversions, with $\tilde{I}$ of the appropriate rank taking the place of the 1.

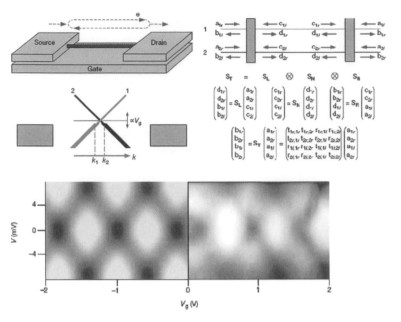

Fig. 6.22 Scattering matrices in real life, adapted from [373]. A single-walled carbon nanotube contains two propagating electronic modes with |k| values that depend on carrier density. Interference between those modes as they reflect from and transmit through the two contacts at the tube ends lead to an oscillatory conductance as a function of gate voltage (which tunes carrier density) and source–drain voltage. Treating this in the Landauer–Büttiker approach with scattering matrices, the authors are able to reproduce (calculation results, bottom left) the measured conductance properties (bottom right) quantitatively.

The first term corresponds to the amplitude for direct transmission through both scattering regions; the second includes one bounce at each region prior to transmission; the third, two bounces, and so on. This is analogous to our expression in Eq. (6.62), but there we were summing probabilities; here we are summing *amplitudes* since we are considering a fully coherent situation. One would need to consider $|t^{\text{tot}}|^2$, etc. to calculate probabilities or the coherently coupled system. As we saw in the discussion of tunneling, it's the phase factors in the cross-terms in $|t^{\text{tot}}|^2$ that give the resonances that we commonly think of as interference effects.

An outstanding example of using this exact approach in practice is shown in Fig. 6.22. The experimental system [373] is a single-walled carbon nanotube contacted at two ends. Because of the band structure of nanotubes (see Section 3.6.1), there are expected to be two propagating electronic modes that run the length of the tube. The authors formulate the scattering problem so that the left and right electrodes are considered to be coherent scattering regions (analogous to $\tilde{S}^{(1)}$ and $\tilde{S}^{(2)}$ above), and the nanotube itself provides the coherent propagating connection between the two regions ($\tilde{S}^{(1-2)}$). The authors are able to use an underlying gate voltage to tune the carrier density in the nanotube, thus altering the |k| values for the propagating electrons (which show up in phase factors in $\tilde{S}^{(1-2)}$). The end result is an oscillatory interference pattern in $|t^{\text{tot}}|^2 \propto G$ as a function of gate voltage, as shown in the figure. The scattering matrix formulation of the Landauer–Büttiker treatment of conduction explains the observed conductance oscillations quantitatively.

It is also possible to modify this formalism to include the effects of decoherence. The limit of complete decoherence between scattering regions is simple. Instead of considering a scattering matrix, one considers a *probability matrix*:

$$\begin{pmatrix} r & t' \\ t & r' \end{pmatrix} \rightarrow \tilde{P} \equiv \begin{pmatrix} |r|^2 & |t'|^2 \\ |t|^2 & |r'|^2 \end{pmatrix} \equiv \begin{pmatrix} R & T \\ T & R \end{pmatrix}, \tag{6.80}$$

where T and $R = 1 - T$ are transmission and reflection probabilities (not amplitudes!), respectively.[8] The rules for incoherently combining probability matrices of two scattering regions are similar to the rules in the amplitude case, Eq. (6.78):

$$T^{\text{tot}} = \frac{T^{(1)} T^{(2)}}{1 - R^{(1)} R^{(2)}}. \tag{6.81}$$

Since $T^{(1)}$, $T^{(2)}$, $R^{(1)}$, and $R^{(2)}$ are all real numbers, there are no resonances or other interference effects here.

The case of *partial* decoherence (e.g., when two scattering regions are spatially separated by a distance comparable to L_ϕ) is more subtle. Büttiker's initial approach [374] is to introduce fictituous leads that act as sources of decoherence. Since each lead is fictitious, it can have no *net* current flowing in or out of itself. An electron has some probability of entering the fictitious lead, with a phase-uncorrelated electron (of identical energy and, ideally, momentum as the original) taking its place. Of course, one can also perform a detailed physics-based calculation involving a specific decoherence mechanism, but that moves beyond the spirit of the Landauer–Büttiker treatment.

The bottom line

The Landauer–Büttiker approach is an impressive intellectual achievement that accounts for many of the features observed in quantum transport. When one is concerned primarily with elastic processes in fully quantum coherent systems, it provides an elegant abstraction, wrapping up the essential physics in transmission coefficients (elements of scattering matrices). Other formalisms (e.g., non-equilibrium Green's functions [360, 375, 376]) are better suited for approaching the full nonequilibrium quantum statistical mechanics problem, and for explicitly worrying about electron–electron interactions. Still, at the single-particle level, the Landauer–Büttiker picture of quantum-conductance-as-scattering serves as an excellent source of physical intuition in nanoscale electronic systems.

It is possible to determine transmission coefficients and scattering matrices in real physical systems through experiments. One can then use this information to predict other phenomena and test whether other physics is relevant in these devices. The quantitative origins of those coefficients are often rooted in the microscopic details of the devices. As we shall see when we discuss topics like molecular electronics, a first-principles understanding of the origins of those parameters remains a major goal for part of the theory community.

---

[8] It should be clear from context that T here does not mean temperature, and R does not mean resistance.

6.5 The classical MOSFET

The metal-oxide-semiconductor field-effect transistor (MOSFET) is the basis for a huge fraction of modern electronics technology. A transistor is a three-terminal device that is the electronic analog of a valve. There are many kinds of transistor (junction transistors; bipolar transistors), but the type most relevant to nanoscale physics is the *field-effect transistor*. In an FET, the flow of charge from a source electrode through a semiconductor "channel" to a drain electrode is controlled by the voltage applied to a gate electrode that is capacitively coupled to the channel. The term "field-effect" refers to the fact that the device operates because the gate imposes an electric field at the channel that is normal to the semiconductor surface.

The ability to modulate conduction allows a properly designed transistor to serve as an amplifier, so that a small voltage signal applied to the gate may be transduced into a large voltage signal measured across an external resistor in series with the source and drain. It was in this context that transistors were first developed by Bell Laboratories in the 1940s, as a potential solid-state replacement for vacuum tubes such as the vacuum triode. A figure of merit in this application is *power gain*, the ratio of the output signal power over the input signal power.

One can also think of the transistor as a voltage-controlled switch, and therefore an ideal building block for electronic logic elements. This is the context in which MOSFETs have drastically changed the world. A typical microprocessor now contains more than 10^9 silicon MOSFETs. The trend toward miniaturization of the MOSFET, packing in ever more computing power per unit area, has been the industrial driver of nanoelectronics research for decades.

The original concept of the FET was proposed by Lilienfeld in the 1930s, but practical realization required the development of the high purity semiconductor materials discussed in Chapter 3. For excellent reviews of the history of the transistor, see *Crystal Fire* by Riordan and Hoddeson [377] and *The Story of Semiconductors* by Orton [378].

Here we review the basic physics of the standard MOSFET, discussing the various operating regimes and parameters. We then point out the device physics challenges associated with aggressive scaling of the MOSFET to very small sizes, and we consider variations of the MOSFET meant to mitigate these issues. We will briefly look at manufacturing concerns; while the particular details are likely to be out of date by the time you read this, the general areas of concern are likely to remain relevant.

6.5.1 CMOS, accumulation, depletion, inversion

Modern silicon electronics are based on *complementary metal-oxide-semiconductor (CMOS)* technology. The term "complementary" means that both *n*- and *p*-type devices are used on the same substrate. After briefly describing types of FET, we will look at the physics of the standard MOSFET.

There are several types of FET. In the *junction FET* (JFET) the source, drain, and the bulk of the channel are all the same type of doped semiconductor (say *n*-type Si). The gate electrode makes contact to a local region that is doped with carriers of the opposite sign (in this case, p-type). The $p-n$ junction formed at the interface of this doped region is depleted of charge carriers (see Section 5.5.1). This largely isolates the mobile carriers in the two regions, constricting the space available for the direct drift of carriers between source and drain. As described, this device is "normally on", allowing conduction between source and drain with no bias applied to the gate. For an n-channel (p-channel) device, applying a negative (positive) voltage to the gate (relative to the source) acts to "pinch-off" the channel and turn the device off.

The same basic idea is at work in the metal-semiconductor FET (MESFET), another device that is normally on. Instead of doping the gate region to form a local $p-n$ junction, a gate metal is chosen that forms a significant Schottky barrier with the channel material. As in the JFET, current flow through the channel is restricted by the resulting depletion zone.

A final normally-on device design is the *high electron mobility transistor (HEMT)*. In an HEMT, the channel is a two-dimensional electron gas formed in a semiconductor heterostructure typically made from III–V compound semiconductors, as in Section 5.5.2. HEMTs are commonly used in similar applications as JFETs and MESFETs, high frequency amplification as in radar and microwave systems. All three of these normally-on devices are difficult to integrate into common circuit architectures used for logic.

While modern MOSFETs are an example of nanotechnology by any reasonable definition of the term, it is remarkable how well they can be described by simple, classical models. To begin, consider a metal-oxide-semiconductor (MOS) capacitor, as shown in Fig. 6.23(a). The "oxide" here is assumed to be a wide band gap insulator (e.g., SiO_2 in traditional CMOS). The corresponding band structure, showing the single-particle electronic states as a function of position, is shown in Fig. 6.23(b) in the idealized case of *flat bands*. The flat band limit assumes negligible charge transfer at the interfaces, and no significant contribution of surface states. As a result, there is no band bending, and the density of charge carriers in the semiconductor should be its full, doped value right up to the interface between the semiconductor and the dielectric.

Now we consider what happens when a voltage is applied to the metal gate, relative to the potential of the semiconductor (imagined to be set by a "back contact" somewhere far away from the junction). The mobile charge carriers within the semiconductor respond to screen the electric field that would otherwise result from the voltage difference between the gate and back contact. Local changes in the charge density within the semiconductor self-consistently shift the local electrostatic potential, bending the bands.

When the gate bias bends the bands such that the free charge density in the plane at the semiconductor-oxide interface is increased, the device is said to be operated in *accumulation* mode. Conversely, gate bias of the opposite polarity bends the bands in the other direction, reducing the density of free charge carriers at the oxide interface below the nominal level from doping; this is *depletion* mode. If one operates in depletion and continues to increase the magnitude of the gate bias, eventually the chemical potential in the semiconductor at the interface crosses the middle of the gap. This implies *inversion*, with the buildup of free carriers at the interface having the opposite sense as the bulk doping. Similar to the situation in a $p-n$ junction, the inversion layer is separated from the free charges in the bulk of the semiconductor by a depletion zone.

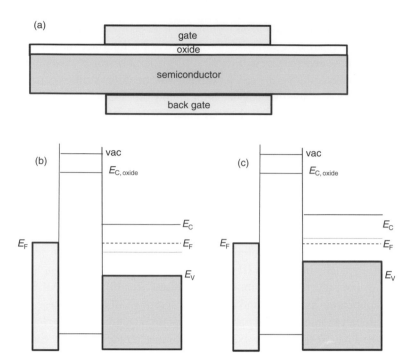

Fig. 6.23   (a) Metal-oxide-semiconductor capacitor, in cross-section. (b) Energy diagram for such a device, assuming an *n*-type semiconductor and the flat band condition. (c) Equivalent diagram for the *p*-type semiconductor case.

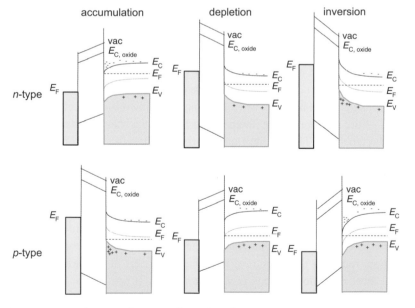

Fig. 6.24   MOS capacitor energetics under various biasing conditions. Note that a *negative* voltage on the gate relative to the semiconductor moves the Fermi energy of the metal *upward* relative to that of the semiconductor.

MOSFETs are generally operated in inversion mode, with the inversion layer carrying the current. The depletion at the edge of the inversion region naturally isolates the device from its surroundings. The *threshold voltage*, V_T, is defined as the gate voltage required to produce inversion: V_T in a real device depends on many things, including the semiconductor properties and the device geometry (oxide thickness, κ for the dielectric and the semiconductor, etc.). In test structures that may contain nonidealities difficult to model, V_T is often determined empirically. For an intrinsic MOS stack with flat bands and no surface states, $V_T = 0$. In the following, we will take V_T to be a device parameter.

From our definitions of V_T and inversion, it is easy to see that the accumulated 2d charge density in the MOS capacitor is just

$$n_{2d} = \frac{1}{e}C_x|(V_G - V_T)|, \tag{6.82}$$

where C_x is the capacitance per unit area of the oxide, and V_G is the gate voltage.

6.5.2 Shallow channel regime

Now consider the diagram of a MOSFET shown in Fig. 6.25. As drawn, the bulk of the semiconductor is p-type, while the source and drain contacts are heavily doped n-type, contacted by metal electrodes. The heavy doping of the contacts is intended to minimize contact resistance as well as to reduce the thickness of any metal/semiconductor Schottky barrier. Unless explicitly noted otherwise, standard formulae for transistor electrical response generally assume negligible contact resistances. Notice that the gate electrode and gate dielectric extend such that they overlap with the source and drain contacts. This is essential for good device performance in inversion.

Consider grounding the source and referencing all voltages to that electrode, and suppose that the device is gated into inversion ($V_G > V_T$). With no voltage applied to the drain, we are back to the MOS capacitor limit, and V_G tunes the density of electrons in the inversion layer that connects the n-doped source and drain regions. In the *shallow channel*, *gradual channel*, or *linear* regime, $V_D \ll (V_G - V_T)$, and we can assume that the charge density in the channel is essentially unperturbed from the MOS result. The *channel length*, L, is defined as the distance between the inner edges of the doped source and drain regions, and the channel width, w, is the extent of the device in the plane transverse to

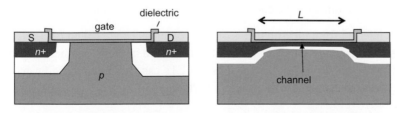

Cross-section of a MOSFET. Left: the $n+$ highly doped source and drain contacts are isolated from the p-type body by depletion regions. Right: when the gate is biased at sufficient positive voltage relative to the body ($V_G > V_T$), an inversion layer is formed at the semiconductor–dielectric interface that acts as a conducting channel connecting the source and drain.

Fig. 6.25

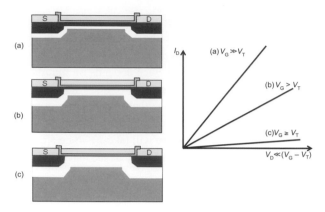

Shallow channel regime for a MOSFET, in which the device acts as a gate-modulated variable resistor.

the source-drain direction. Often w/L is called the *aspect ratio* of the device. Using these parameters and our definition of mobility, we find

$$I_D = -wen_{2d}\mu\frac{V_D}{L} \approx -\frac{w}{L}\mu C_x(V_G - V_T)V_D. \qquad (6.83)$$

Here the drain current, I_D, is defined to be positive if current is flowing *into* the drain. From this equation we see that a FET in the shallow channel regime operates as a gate-controlled variable resistor. The source-drain $I_D - V_D$ characteristics are linear, and biasing V_G to increase the carrier density in the inversion layer lowers the source–drain resistance.

We can introduce a figure of merit for transistors called the *transconductance*,

$$g_m \equiv \frac{\partial I_D}{\partial V_G}. \qquad (6.84)$$

The transconductance is a measure of how sensitively the transistor responds to the gate, and larger is almost always better. For the shallow channel limit,

$$g_m = -\frac{w}{L}\mu C_x V_D. \qquad (6.85)$$

Since this depends on V_D it's not as useful as it might be, but this is instructive. It tells us:

- the larger the aspect ratio, the higher the transconductance;
- the higher the capacitance per unit area, the higher the transconductance;
- knowing device dimensions, one can measure I_D vs. V_G and infer μ.

Mobilities inferred from measurements of transconductance are called field-effect mobilities. These are generally lower than mobilities measured in bulk via other techniques, and the intuitive reason for this is that carriers in FETs can experience significant boundary scattering at the dielectric–semiconductor interface.

6.5.3 Saturation regime

A dramatic and extremely useful change in device characteristics takes place when higher source–drain voltages are applied ($V_D >\sim (V_G - V_T)$). The physical picture is shown

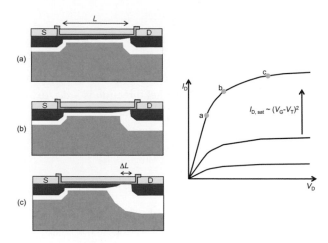

Fig. 6.27

Saturation regime for a MOSFET. At higher source–drain biases, the channel pinches off-near the drain electrode in this device. Point (b) is approximately at $V_{D,sat} = V_G - V_T$. At higher drain voltages such as (c), the drain current has saturated.

in Fig. 6.27. Think about the local conditions in the channel near the drain as $V_D \rightarrow (V_G - V_T)$, particularly in the limit that V_T is close to zero. In this case, since the drain and gate are at approximately the same potential, locally there is very little gate field in the channel. As a result, the charge density in the inversion layer there becomes very small, vanishing at a particular value of $V_D \equiv V_{D,sat}$ called the "pinch-off" or saturation voltage.

What happens beyond pinch-off? As V_D is increased further, the inversion layer is depleted over some distance ΔL from the drain electrode, and I_D becomes approximately constant. The device is said to be in the *saturation regime*. To see intuitively why the current saturates, at least in a long device, think about what is happening to the local voltage at the channel surface as a function of position between the source and drain. The depleted region near the drain must have a very high resistance, since it contains no free charge carriers. This means that nearly all of the applied source–drain bias, V_D, is dropped across that region. When $\Delta L/L$ is small, this suggests that a small change in V_D in this regime doesn't really change the geometry of the conducting part of the channel very much, nor the voltage dropped across that conducting part. Therefore, the current through the conducting part should be approximately constant beyond pinch-off in long channel FETs. That is, $I_D \rightarrow I_{D,sat}$.

Within simple models we can perform an analytical calculation to obtain a closed-form expression for the saturation current. The simplest approach is to assume that the dominant charge in the channel is that due to the inversion layer, and see what happens in the run-up to pinch-off. If we define the coordinate running along the channel from source to drain as x, and label the local electrostatic potential in the channel (at the semiconductor/dielectric interface) as $\phi(x)$, then the local 2d charge density is $C_x(V_G - V_T - \phi(x))$ before we reach pinch-off. We are explicitly allowing the local electrostatic potential to vary, since $\phi(x)$ depends implicitly on V_D, and $\phi(x = L) = V_D$ by definition. Based on our definition of current, the local current

$$I(x) = w\mu C_x(V_G - V_T - \phi(x))\frac{d\phi}{dx}. \tag{6.86}$$

Here we have accounted for the fact that in our example the charge carriers have charge $-e$, and the local electric field along the channel direction is $-d\phi/dx$. In steady state this must hold everywhere in the channel, and in particular

$$\int_0^L I(x)dx = I_D L = w\mu C_x \int_0^{V_D} (V_G - V_T - \phi)d\phi. \tag{6.87}$$

This gives

$$I_D = \frac{w\mu C_x}{L}\left[(V_G - V_T)V_D - \frac{V_D^2}{2}\right], \tag{6.88}$$

which should be valid when $0 \leq V_D \leq V_{D,\text{sat}}$ (that is, until pinch-off takes place) and $V_G \geq V_T$.

Since I_D becomes approximately constant when $V_D \geq V_{D,\text{sat}} = V_G - V_T$, we can use Eq. (6.88) to find the saturation current:

$$I_{D,\text{sat}} = \frac{w\mu C_x}{2L}(V_G - V_T)^2. \tag{6.89}$$

This is the so-called *square law*: neglecting contact resistances, assuming constant μ, and ignoring changes in depletion width down the length of the channel, the saturation current scales quadratically with $V_G - V_T$. This provides another way of inferring mobility from experimental data.

One can perform more realistic analyses of the approach to saturation. For example, in addition to the charge in the inversion layer (included in Eq. (6.86)), one can explicitly keep track of the local variation in the thickness of the depletion region below the inversion layer. This is called the "bulk charge" approach, and results in an expression that has additional correction terms inside the large brackets of Eq. (6.88) (and therefore predicts deviations from the simple square law). Neither the square law nor the bulk charge approaches actually predict saturation; that has to be inserted by hand into the model. A complete numerical treatment of the whole system (self-consistently solving the 2d electrostatics of the channel including the mobile charges, the fixed dopants, and the local band structure) does give nice results, including saturation.

It should be clear that transistors are tremendously versatile devices. Depending on how they are placed in a circuit, they can be used for either analog signal amplification or digital logic. The saturation regime of the MOSFET is particularly useful for digital applications, since the output current (readily converted into an output voltage by a resistive load) is insensitive to small deviations in V_D.

6.5.4 Short channel effects and other scaling issues

Now we consider what can happen when devices are made with progressively shorter channels. There are a number of "short channel effects", and the desire to mitigate the

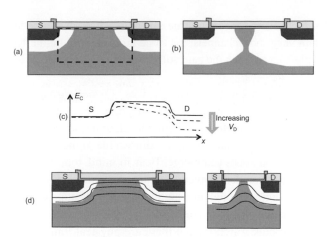

Fig. 6.28

Short channel concerns. (a) Dashed box shows the region in which the charge distribution is controlled by the gate, in an electrostatically "good" FET. (b) Punch-through due to contact between depletion regions is one hazard of short channel devices. (c) The potential barrier seen by an electron leaving the source is lowered at high V_D values. (d) Solid lines indicate rough electrostatic equipotentials, showing that, for identical oxide thicknesses and doping levels, gate influence is weaker in shorter channel devices.

resulting problems has driven the semiconductor industry's approach to scaling. Here we give a very brief overview of the physics at work.

Consider the dashed box shown in Fig. 6.28. The *gate charge approximation* assumes that the charge distribution in this box is predominantly set by the gate electrode. That is, the electric field normal to the gate is assumed to be much larger than the electric field along the channel direction. "Well-designed" transistors (such as those in the limit where the square law is a good description of their operating characteristics) operate in this limit. Anything other than the gate electrode that affects the charge distribution in our around this region will produce changes away from the simple, idealized square law model.

One obvious issue is *channel shortening* in the saturation regime, when the size of the pinched-off region is not small compared to the channel length ($\Delta L/L$ no longer $\ll 1$). Remember, because the inversion layer is depleted in the pinched-off region, most of V_D is dropped across ΔL, leading to large local electric fields. This large electric field enhances drift of carriers that get injected into this region from the inversion layer, leading to a boost in I_D as V_D is increased beyond $V_{D,\text{sat}}$. A semi-empirical treatment of this leads to modified versions of the square law expressions:

$$I_D = \frac{w\mu C_x}{L}\left[(V_G - V_T)V_D - \frac{V_D^2}{2}\right](1 + \lambda V_D) \tag{6.90}$$

for $V_D < V_{D,\text{sat}}$, where λ is a parameter that depends on the device geometry and semiconductor properties. Likewise, for current nominally in the saturation regime,

$$I_{D,\text{ sat}} \approx \frac{w\mu C_x}{L}\left[\frac{(V_G - V_T)^2}{2}\right](1 + \lambda V_D). \tag{6.91}$$

The result is that there is no longer any true saturation of I_D as a function of V_D. Indeed, in real devices with very short channels, the same large electric field effects at sufficiently high V_D can take place even when the gate voltage is below threshold, leading to significant I_D in nominally "off" devices.

Another challenge (Fig. 6.28(b)) in small devices is *punch-through*. Recall that, below threshold, the doped source and drain are surrounded by depletion regions (with sizes set by the screening length, and therefore the doping level, of the semiconductor in the channel and the source/drain). If the channel length is comparable to the depletion region size, the source and drain depletion zones can merge. Bear in mind, too, that the size of the drain (assuming the source is grounded) depletion region depends on V_D. Current can still flow through this depletion region, though in a highly nonlinear fashion characteristic of space-charge limited currents ($\sim V_D^2$). The simplest way to avoid punch-through is to increase the doping level in the bulk wafer, though there are physical limits to doping levels (see Section 3.3.3).

Figure 6.28(c) is a cartoon of an *n*-MOSFET below threshold, showing the spatial variation of the energy of the bottom of the conduction band as a function of position along the channel. At low V_D, from the point of view of a conduction electron in the source, there is an effective potential barrier preventing the electron from transiting the channel. However, increasing V_D tilts that potential due to the electric field along the channel, V_D/L. This distortion of the potential is increased for larger V_D and smaller L. At sufficiently large source–drain fields, the height of the effective potential barrier is reduced, a phenomenon called *drain-induced barrier lowering*. Thermionic emission over the barrier is relevant, and in sufficiently small devices tunneling can also play a role. The net effect is that even in nominally "off" devices there can be substantial currents and significant power dissipation. As we shall see, heating is a major problem in ultralarge scale integrated circuits, exacerbated by effects like this.

Figure 6.28(d) shows a basic issue of electrostatics relevant to shorter channel lengths. For a fixed gate dielectric thickness, a shorter channel implies that the electrostatic potential in the channel region is influenced relatively more strongly by the source and drain. To avoid this when reducing the channel length, it is necessary to reduce the (effective) dielectric thickness by the same geometric factor as L.

At fixed operating voltages, a further consequence of short channels is an increase in the electric field along the channel, E_x. At sufficiently high fields, the simple linear relationship between drift velocity and electric field fails. This can be described as a field-dependent mobility, so that $v_d = \mu(E_x) \cdot E_x$. In the high electric field limit, carrier drift velocity saturates to some value v_{sat}. The carriers lose energy by exciting optical phonons, and inelastic process with a rate that depends on material band structure as well as electric field. When energy gain due to the electric field is balanced by energy loss due to phonon scattering, velocity saturates. These effects are seen clearly in Si [379, 380] and GaAs [380]. Recently it has become clear that the same physics is relevant in nanoscale systems such as carbon nanotubes [381] and graphene transistors [382].

In the saturated limit (that is, high fields and $L \gg$ the relevant inelastic scattering length), one finds a saturation in transistor $I_D - V_D$ characteristics that is not driven by pinch-off:

$$I_D(V_D) \to wC_x(V_G - V_T)v_{sat}. \tag{6.92}$$

This can be quite a bit lower than the nominal square law value of $I_{D,sat}$. Note further that the saturated current now depends *linearly* on $V_G - V_T$ rather than quadratically. For a sense of numbers, in Si at room temperature, $v_{sat} \sim 10^7$ cm/s, when $E_x > 3 \times 10^6$ V/m for electrons or 10^7 V/m for holes [47].

It is possible for drift velocities in very short devices to exceed v_{sat}. Scattering of carriers off optical phonons has some characteristic (field- and temperature-dependent) rate, and if carriers can cross the channel before scattering, then *velocity overshoot* can take place [383, 384]. This can lead to funny looking features (e.g., nonmonotomic behavior) in I_D–V_D traces under some circumstances. However, it can be beneficial to device performance, allowing carriers to cross the channel faster than one would expect from velocity saturation concerns.

A further potential problem due to inelastic effects in short channel devices is *avalanche breakdown*. In places where carriers move in very high electric fields (pinch-off regions; very short channel devices at high source–drain biases), highly energetic carriers can, at some rate, inelastically scatter to produce electron–hole pairs. If the local field is sufficiently large, the pair is ripped apart (instead of binding to form an exciton), and the resulting carriers can also accelerate in the local field. If this leads to even more pair formation, the result is a runaway I_D at high V_D that is not controlled by V_G.

As transistors have been scaled to smaller sizes, the need to preserve good gate control over the charge in the channel region has led to thinner and thinner gate oxide (dielectric) layers, well below 10 nm. This presents several challenges. First, at a given V_G, decreasing the oxide thickness, t_{ox}, increases the electric field across the oxide. If V_G/t_{ox} exceeds the oxide's dielectric breakdown strength, in general the oxide will fail. This failure results from the build-up of defects in the oxide due to chemical damage by electrons from the gate electrode. If an electron from the gate has sufficient energy, it can break chemical bonds within the oxide layer. The resulting defect can be associated with a localized "trap" state that typically has an energy somewhere in the band gap of the oxide. Such a defect can act as a site for tunneling or thermally driven hopping of additional carriers, leading to further damage. A sufficient density of these defects can lead to "hard" breakdown of the oxide, with Ohmic conduction via an effective impurity band. At lower densities, "soft" breakdown, with weaker, nonOhmic conduction is possible.

A related concern is that the dielectric's properties will deviate strongly from their bulk values at nanoscale thicknesses. For example, the traditional dielectric is SiO_2, with a breakdown field exceeding 10^9 V/m. SiO_2 layers thinner than 2 nm have been used as gate dielectrics. At this thickness, the layer is only a dozen or so atoms thick; it is not at all obvious that the bulk band gap and breakdown properties of SiO_2 should hold true at this thickness.

Finally, at the few-nm layer level, direct tunneling from the gate through the dielectric becomes a serious issue. Leakage currents on the order of 10–100 A/cm^2 are possible near single-nm oxide thicknesses. This leakage is responsible for significant power dissipation, a major problem in integrated nanoelectronics.

These issues are sufficiently grave that for a number of years there was great concern that the gate oxide limitations would prevent further scaling [385]. The main path to mitigation has been the development of "high-κ" dielectrics. Replacing SiO_2 ($\kappa = 3.9$) with an alternative material such as hafnium oxide (HfO_2, $\kappa = 25$) means that t_{ox} can be increased by the ratio of the κ values without losing capacitance. Increasing the physical oxide thickness exponentially reduces tunneling leakage, while maintaining a very small *effective* oxide thickness (referenced to SiO_2). Integrating high-κ materials into CMOS processing has proven to be quite challenging [386], but in 2007 Intel Corp. began selling chips using a hafnium-based high-κ gate oxide.

6.5.5 Transistor speed

Often people are interested in the inherent response time of a transistor. In analog applications this sets the bandwidth available for signal modulation, while in digital applications it contributes to limiting the rate at which logic operations may be performed.

Physically, there are two intrinsic timescales that are relevant. The first is the time it takes for a charge carrier to cross the channel. Clearly it would not make a lot of sense to try to modulate the current faster than the time it takes a carrier to go from source to drain. This timescale is roughly $\tau_{xing} = L/v_d = L^2/(V_{DS}\mu)$, and in a modern Si device can be extremely short. For $\mu = 1000$ cm^2/Vs, $L = 200$ nm, $V_{DS} \sim 1$ V, this is around 4×10^{-13} s. Note, too, that the mean free path for electrons at moderate doping in Si at room temperature is tens of nanometers. Transistors with $L < \ell$ are in the ballistic limit, and therefore the relevant crossing time is given by L/v_T, where v_T is the thermal velocity of (nondegenerate) charge carriers:

$$v_T \equiv \sqrt{\frac{3k_B T}{m_e^*}}. \tag{6.93}$$

For $m_e^* = 0.26\, m_0$ and $T = 300$ K, $v_T \approx 2.3 \times 10^5$ m/s. For a 45 nm channel length, the crossing time would then be 2×10^{-13} s. Ballistic transistors are discussed in Section 6.6.4.

While the crossing time may be an ultimate speed limit, a more practical timescale is that associated with the formation of the inversion layer. This "delay time" is basically the RC charging time of a capacitor formed by the channel and the gate electrode. The capacitance intrinsic to the transistor is dominated by that of the gate dielectric, and is given by $C_x Lw = \kappa\epsilon_0 Lw/t_{ox}$. Particularly in smaller devices, other contributions to the capacitance can be important as well. In any realistic device, the capacitance and resistance of the gate line (the metal wire that applies bias to the gate electrode) are likely to dominate the intrinsic capacitance and resistance of the transistor itself.

One approach to mitigating the stray capacitance of the device wiring is the use of "low-κ" dielectric material as an insulating layer under and around the copper interconnects. The most common low-κ materials are based in some way on SiO_2 ($\kappa = 3.9$). Doping SiO_2 with carbon and making the resulting material somewhat porous (that is, with some void fraction) can reduce κ as low as 2.7. Intel introduced low-κ interlayers with the debut of their 65 nm manufacturing process in 2004.

6.6 State-of-the-art

We now turn to a brief discussion of the state-of-the-art in MOSFETs. These devices are the basis of modern digital logic and are closely related to devices used for memory and data storage. The purpose is not to give a detailed description of these devices and their specific operational parameters, since those will certainly be out of date by the time you read this. Instead, we will emphasize the current trends, the important drivers for further development, and concepts put forward for future nanoelectronics approaches.

Figure 6.29(a) shows a dominant trend that drives the semiconductor electronics industry. Coined by Gordon Moore, founder of Intel, Moore's (First) Law states that the number of components in an integrated circuit doubles roughly once every 18 months. This remarkable trend is not an accident; for the last three decades it has actually been a target set by the semiconductor industry itself. Maintaining this aggressive pace has only been possible through geometric scaling of components, especially MOSFETs. Figure 6.29(b) shows a typical minimum lateral feature size (e.g., MOSFET channel length) over the same time period. Notice that the 100 nm "barrier" was crossed in the early years of this century. By this and any other reasonable definition, the semiconductor industry has been developing and selling nanotechnology for some time now.

Of course, this scaling process cannot continue indefinitely. As discussed, the physics of conduction at the nanometer scale is quite different from the classical effects relevant for standard transistors. The atomic character of matter itself sets an ultimate limit on

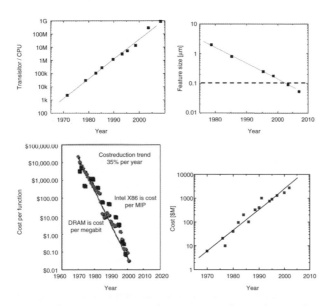

Moore's Laws. (a) The number of components on a typical microprocessor chip as a function of year. (b) The typical minimum lateral feature size over the same time period. (c) Cost per "function" (millions of instructions per second for logic; megabit for memory) over the same period. (d) The cost of a typical semiconductor fabrication facility, also over the same period. Information for (a), (b) for www.intel.com; for (c), (d) from ICKnowledge.com.

Fig. 6.29

the size of electronic devices. Indeed, as of 2001 the International Technology Roadmap for Semiconductors (ITRS – put together by a global consortium of semiconductor manufacturers to outline targets and approaches to common engineering challenges) proclaimed that we had entered the era of "materials limited device scaling", meaning that further scaling would require changes in the basic constituent materials used in chip manufacturing. The Roadmap (available at www.itrs.net) is updated annually, and provides detailed information on the challenges facing the nanoelectronics industry.

Furthermore, Fig. 6.29(c) shows what some would call Moore's Second Law. While the cost per unit of computing power or memory capacity has been falling exponentially, the cost of a semiconductor fabrication plant has been *rising* exponentially. The technical demands of industrial-scale nanofabrication are extreme and therefore costly. The economics at work here are beyond the scope of this book. However, when considering nanoelectronics, it is important to realize that (a) the per-transistor cost of Si MOSFETs is extraordinarily low while their reliability is incredibly high; and (b) decisions about commercialization of technologies are ultimately economic.

Historically, the ITRS has defined a "technology node" by a label connected to the transistors used in dynamical random access memory (DRAM). Thus, when a company talks about "the 45 nm technology node" or "45 nm manufacturing process", they are referring to a process that can produce DRAM with a half-pitch (spacing between leads connected to the driving transistors) of 45 nm. Physical gate lengths in transistors correlate with this designation, but are generally smaller. For example, Intel's 65 nm manufacturing process actually involved MOSFETs with 30 nm physical channel lengths.

The ITRS declares that a given node or milestone has been reached when two or more manufacturers are shipping more than 10,000 chips (roughly 50 wafers) per month with that particular process. Interestingly, the first conference papers on a given process historically appear approximately two years prior to the production threshold, and an additional two years is often required for adoption of a process throughout the industry. This is *very* aggressive compared with slower industries such as automotive manufacturing. The take-home message here is an important one, given the prevalence of hype associated with nanotechnology: if a given nanoelectronics technology cannot be demonstrated in a manufacturable setting today, it is very unlikely that that technology will be widely available in commercial products within five years.

6.6.1 Fabrication

A typical high end processor chip (Intel quadcore I7) contains on the order of 10^9 transistors on a total chip area of around two hundred mm^2. The active area of all the transistors is only 15% of the total chip area, the rest being consumed by associated wiring and other components. Figure 6.30 is a cross-sectional image of a state-of-the-art MOSFET circa 2009. In this device the geometric channel length (between the near edges of the diffusion-doped source and drain) is around 32 nm (though the center-to-center distance between the source and drain is around 100 nm). The gate oxide thickness is approximately 1.8 nm. The gate itself is a metallic silicide. For this device to work properly, the gate must be positioned relative to the source and drain with an accuracy of only a few nanometers.

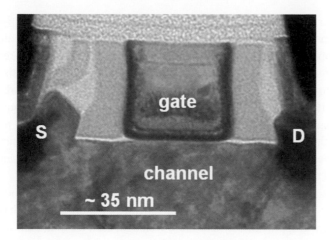

State-of-the-art MOSFET, from www.intel.com, 2009 (IEDM abstract by Packan *et al.*) [387].   Fig. 6.30

Fabrication of a wafer full of such processor chips requires hundreds of distinct process steps, tens of different photolithographic masks, and exquisite process control. At least eight layers of interconnects are defined, with copper being electrodeposited into vials lined with a barrier material that prevents the copper atoms from diffusing into the Si wafer. Many steps are followed by planarization, involving the use of chemical-mechanical polishing to achieve remarkable surface flatness prior to the next step.

The initial wafer material is sliced from a large single crystal called a boule. The boule is grown by the Czochralski method, in which a seed crystal mounted on a rod is introduced into a bath of molten silicon (held in a quartz crucible). With appropriate control of temperature gradients, rod rotation, and the rate at which the rod is withdrawn from the melt, it is possible to extract a large, cylindrical single crystal from the melt. Clearly this whole process is performed under an inert gas atmosphere, to avoid oxidation of the Si or introduction of unwanted impurities. The crystal is ground to a standard diameter, with flats ground onto the cylindrical surface to indicate particular crystallographic directions. The crystal is then sliced into wafers, and each wafer is meticulously polished to an incredibly high degree of flatness. Often additional silicon may be grown on the working surface via epitaxial chemical vapor deposition. This is particularly important for novel materials processes.

The requirements on starting material are outlined in the "Front End Processes" section of the Roadmap. Current wafer processes are based on 300 mm-diameter material, though there are plans to migrate to 450 mm-diameter wafers in the next several years. The requirements are daunting. Over a 26 mm × 8 mm site, wafers must be flat to within 40 nm. Moreover, there must be less than 0.15 contaminant particles (of diameter exceeding 65 nm) per cm^2 on the working surface. On epitaxial wafers, there must be less than 0.008 structural defects per cm^2. These requirements are difficult to achieve, but more importantly, they are difficult even to *evaluate* at the necessary precision, particularly in an efficient way. This is an example of why *metrology* – particularly the ability to discern nanoscale details over macroscopic areas – is an enormous challenge.

Lateral patterning is done via advanced forms of photolithography. In standard photolithography, a photochemically active polymer (*photoresist*) is spin-coated to uniform thickness on top of a wafer. Ultraviolet light is then projected through a carefully designed mask called a reticle onto the surface, exposing areas of the underlying resist. In a positive tone resist the polymer interacts with the light generating photoacids which attack the polymer network, making it soluble in a developer. Subsequent development washes away the exposed regions of resist. Conversely, in negative tone resists, the photochemistry crosslinks or otherwise stabilizes the polymer network, so that exposed resist is left behind after development. The underlying surface is then patterned, e.g., by a selective etching process, before the resist is stripped.

Traditionally optical elements are used to reduce the projected shadows from the reticle onto the resist surface. Conventional diffraction limits the resolution of simple photolithography to approximately half the wavelength used for illumination. The standard illumination source during early 2000s was the ArF laser line at $\lambda = 193$ nm. Moving to shorter wavelength sources is very challenging for a number of reasons. A major difficulty is that the optical elements used for projection tend to have rapidly increasing absorption as λ is decreased deeper into the ultraviolet. Deep or extreme ultraviolet lithography will likely require a transition from transmittive to all-reflective optics. Exposure tools based on reflective optics are currently being tested.

To achieve higher resolution a number of clever tricks have been employed. For example, two-photon photochemistry can be employed so that the effective resist exposure scales like the square of the optical intensity rather than linearly with intensity. This has the effect of sharpening features. *Phase-shift masks* use diffraction and interference themselves to generate sharper features, by engineering the interference pattern that results from the passage of light through the mask. Recent advances have been made by immersing the last stage of the optics in high purity water, using the index of refraction of the liquid to shorten the effective wavelength. There is also an increasing emphasis on multistep techniques such as double exposure patterning (exposing resist twice with different masks before development, so that critical feature sizes are set by the differences between the masks), double patterning (a similar idea), and self-aligned spacers. Finally, there is much discussion of non-optical methods such as nanoimprint lithography, particularly for certain niche applications.

Beyond lithography, advanced scaling of CMOS electronics has relied on the development and integration of novel materials in the fabrication process. Doping of the semiconductors to the desired levels requires annealing at temperatures in excess of 800° C, as do other processing steps. This temperature range and the need for materials compatibility with Si greatly restrict the choice of available materials.

One of the most well known transitions in processing involved the change from aluminum-based interconnects to copper-based wiring on chips. Copper has a number of appealing properties, including excellent electrical and thermal conductivity, the ease with which it may be deposited either by physical vapor deposition or electrochemistry, and its ability to anneal away grain boundaries even at room temperature. However, isolated copper atoms can act as very deep traps (that is, acceptor-type dopants with bound states well below the bottom of the silicon conduction band), "poisoning" silicon. There adoption of copper as an interconnect material required great care in process development, including the integration of an additional new material to act as a diffusion barrier, lining the inside of interconnects and vias to prevent copper diffusion into the bulk. A low-κ diffusion barrier is desirable, to limit stray capacitance of the leads. Copper interconnects were introduced in large scale in 1998 by IBM.

The transition from SiO_2 to high-κ gate dielectrics was mentioned previously. In this case some requirements were a high dielectric breakdown strength, temperature stability, and appropriate band alignment between the oxide and silicon. Temperature stability was particularly important, since if any of the high-κ oxide decomposed at the Si interface to form a layer of mixed SiO_x, that greatly reduce the benefits of having a high-κ layer in the first place. The transition in earnest to high-κ dielectrics began in earnest in 2007 with Intel's adoption of hafnium-based gate oxide for the 45 nm manufacturing node.

An additional materials challenge regarding silicon dioxide is the increasing desire for silicon-on-insulator (SOI) wafers. For a number of applications, including integrated optoelectronics, it can be very useful to have high performance CMOS devices and their accompanying crystalline Si sitting on top of a thick insulating layer of SiO_2 (which then sits on a substrate wafer). The two main approaches to preparing SOI material are oxygen implantation and wafer bonding. In the former, oxygen ions are fired into a wafer surface, with their final depth determined by their incident energy. Annealing then forms a subsurface oxide layer and heals (ideally all) the damage to the overlying Si lattice caused by the ions. In the latter, two wafers with oxide surfaces are bonded together, typically through a combination of high temperatures and pressure. Most of the upper wafer is then removed, leaving behind a thin single-crystal Si layer above the oxide. A process called Smart Cut has been developed by the French firm Soitec to do this cleanly, using ion implantation of hydrogen to force crack formation at a particular depth in the bonded wafer.

Deliberate engineering of Si to enhance its materials properties has also become prevalent. Straining Si layers can increase the local mobility of charge carriers relative to that in unstrained bulk Si. This can be done by growing strained Si epitaxially on Si material that has been alloyed at the few percent level with Ge. Strain-engineered silicon was first introduced at the 90 nm manufacturing node by Intel in 2002.

6.6.2 Scaling

Historically there have been two approaches to scaling of CMOS electronics: *constant field* scaling and *constant voltage* scaling. In the former, as device geometry is modified so are operating voltages, with the intent of maintaining constant electric fields from one scaling generation to the next. In the latter, one maintains constant voltages as device sizes are reduced, while modifying other device properties (e.g., doping) to mitigate effects of the resulting higher electric fields.

Both scaling regimens have some appeal. Constant field scaling makes a great deal of physical sense. It avoids high field problems such as oxide breakdown and other short channel effects. However, there are realistic limits on the operating voltages in CMOS circuits. For example, threshold voltages are strongly affected by intrinsic material properties such as work functions and interfacial band alignments. Threshold voltages can

Table 6.1 Scaling of MOSFET parameters under different device scaling approaches.

Parameter	Symbol	Const. field	Const. V	Const. V with v_{sat}
Channel length	L	$1/\alpha$	$1/\alpha$	$1/\alpha$
Channel width	w	$1/\alpha$	$1/\alpha$	$1/\alpha$
Field	$|\mathbf{E}|$	1	α	α
Oxide thickness	t_{ox}	$1/\alpha$	$1/\alpha$	$1/\alpha$
Si doping	$N_{A/D}$	α^2	α^2	α^2
Oxide cap.	C_{ox}	α	α	α
Crossing time	τ_{xing}	$1/\alpha$	$1/\alpha^2$	$1/\alpha$
Voltage	V	$1/\alpha$	1	1
Current	I	$1/\alpha$	α	1
Power		$1/\alpha^2$	α	1

also have extrinsic variations from device to device due to materials processing. Therefore one cannot lower operating voltages arbitrarily, limiting the range of field scaling. Constant voltage scaling, on other hand, generally runs into all of the high field issues discussed in Section 6.5.4.

If one scales the channel size by a factor α, so that $L \to L/\alpha$, then the two different scaling procedures mandate other changes in device geometry, properties, and operating parameters, assuming standard MOSFET physics remains valid. Table 6.1 shows these relationships [388]. A couple of the exercises at the end of the chapter are meant to make you think further about the physics behind these.

In current practice, scaling requires tradeoffs, ever more so as device sizes become much smaller than 100 nm. An established industrial approach is to study device properties empirically, looking at trends in ensembles of "good" devices to infer empirical scaling laws. For example, dopant density needs to be adjusted upward as channel length decreases, to avoid punch-through. In the early 1990s this variation was observed to follow the trend $N_A \approx 4 \times 10^{16} L^{-1.6}$, where the acceptor density, N_A, is in units of atoms per cm^3 and L is in units of Ångstroms [389]. These kinds of scaling rules are updated continuously as device sizes progress toward ever smaller scales. Furthermore, different scaling approaches may be optimal depending on the device application (e.g., memory vs. logic) [390]. As nanoscale physics becomes more important to the operation of MOSFETs and related devices, scaling procedures are certain to evolve.

6.6.3 Advanced CMOS MOSFETs

A number of alternatives to standard MOSFET designs have been proposed and examined. These designs are motivated by the desires to continue scaling and mitigate potential fabrication challenges. Figure 6.31 shows examples of particular advanced MOSFETs being considered.

Given the development of high quality SOI technology, it should come as little surprise that people have considered silicon-on-insulator MOSFETs. Placing the channel material

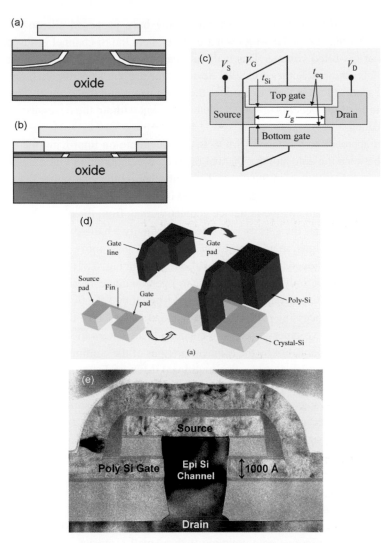

Advanced FET structures. (a) Partially-depleted and (b) fully depleted SOI devices. (c) Double-gate FET, adapted from [392]. (d) FinFET, adapted from [392]. (e) Vertical replacement gate FET, adapted from [393].

Fig. 6.31

on an insulating substrate has the benefit that the "body" of the transistor is reduced in size. This cuts down on the overall capacitance of the device, potentially leading to higher speeds. At the same time, the confining effect of the insulating layer influences the size of the depletion region around each contact, helping to avoid punch-through. SOI devices can be "partially" depleted, so that under no gate bias there remains some volume of Si between the source and drain that is not in a depletion region of either contact. Alternatively, for very thin upper Si layers, SOI devices can be "fully" depleted. In the latter, parasitic punch-through is not possible, since there is essentially no body region in parallel with the channel – there is only the channel.

Extrapolating SOI to its logical extreme, some have proposed silicon-on-"nothing" devices. Rather than an underlying insulating layer, here the Si channel is suspended over a vacuum or gas-filled volume. This is the ultimate in low-κ environments. Furthermore, there are fabrication schemes for SON devices that use the selective etching of a well-controlled sacrificial SiGe layer to produce the void underneath the channel. This approach may enable all of the benefits of fully depleted SOI, but without the necessity of full-wafer SOI fabrication.

Another approach to advanced CMOS is to concentrate on improving the electrostatic coupling between the gate and the channel region. As mentioned in Fig. 6.28(d), ensuring that the charge distribution in the channel is controlled by the gate rather than the source and drain is a recurring challenge in scaled MOSFET design. A major direction being explored is a significant change in device geometry, away from the traditional asymmetric planar structure that has the gate on one side of essentially a half-space of semiconductor.

In *double-gate* designs, in addition to the usual "top" gate above the channel, a second "bottom" gate is positioned below the channel, as shown in Fig. 6.31(c). The channel region here is intended to be only 5–10 nm thick. In one candidate fabrication scheme [392], this channel is grown epitaxially through a void formed by selective etching (a recurring theme in proposed processing paths). At these thickness levels, quantization effects become important. Threshold voltages for electron conduction are determined not by the bottom of the bulk conduction band, but by the (higher) energies of the confined subbands.

One can also consider turning the traditional MOSFET on its side, as shown in Fig. 6.31(d). In this "FinFET" design, the channel material is a "fin" of Si between two large source/drain pads. The gate is then wrapped around the fin, providing gate coupling from both of the large planar fin surfaces as well as one edge. In addition to strong gate coupling, this design has comparatively large contacts to minimize contact resistance. FinFETs are already being deployed in high end processors.

The logical endpoint of this progression would be a channel completely surrounded by a gate electrode. Exactly this idea has been suggested, and an example is shown in Fig. 6.31(e). This is the vertical replacement gate (VRG) MOSFET [393], in which the channel is a vertical pillar of Si grown epitaxially into a predefined hole. The VRG design provides extremely effective gate coupling with a small footprint, and all of the critical dimensions and alignments are determined by nonlithographic processes. The vertical FET geometry has received renewed interest in recent years, with the development of techniques for growing semiconductor nanowires [394, 395].

While none of these variations in CMOS design have entirely replaced the orthodox MOSFET, it is clear that there is vigorous interest in finding ways to continue the CMOS MOSFET paradigm that has been so successful.

6.6.4 Ballistic MOSFETs

Suppose, either through extensions of current manufacturing techniques or alternative designs such as those mentioned in the previous section, that Si devices are scaled down toward their ultimate size limits. What kind of physics and device performance should one expect to see in Si MOSFETs as channel lengths fall toward the molecular scale, below 10 nm?

Here we follow one treatment of this problem, put forward by Prof. Mark Lundstrom of Purdue University [396], that contains much of the essential physics. A few assumptions underlie this approach. First, we consider channel lengths no shorter than $\sim$ 10 nm. For shorter channels we would need to take into account direct source–drain tunneling as well as possible significant deviations from Si band structure. Second, we neglect disorder or possible traps at the interfaces and within the channel.

Strong confinement in the directions transverse to the channel are also very likely in such ultraminiaturized devices, as one would find in very skinny surround-gate structures. As in the discussion of double-gate FETs, under these conditions the allowed 1d subbands in the channel can be significantly higher in energy than the floor of the bulk Si conduction band.

The profile of a 1d subband as a function of position x along the channel is shaped like a barrier, as shown in Fig. 6.32. As gate bias is swept, the effective barrier height is modulated. Likewise, as V_{DS} is changed, the barrier is affected, with significant barrier lowering possible at large source–drain bias.

The basic physics of ballistic transport that was outlined in Fig. 6.19 remains correct. With no scattering taking place in the channel region, the distribution functions that describe the energy and population of the charge carriers are *not* those of thermal equilibrium. First we consider the high temperature (nondegenerate carrier) limit that is typically relevant for CMOS. In this case we should think about classical carrier velocities

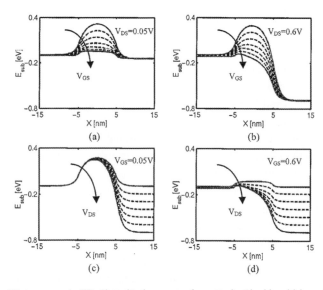

The effect of V_{DS} and V_G in nanoscale FETs. Plotted is the energy of a particular 1d subband (above the bottom of the bulk conduction band due to quantum confinement effects) as a function of position in the channel. Top row: the effect of sweeping V_G at low (left) and high (right) V_{DS}. Bottom row: the effect of sweeping V_{DS} at low (left) and high (right) V_G. Adapted from [396].

Fig. 6.32

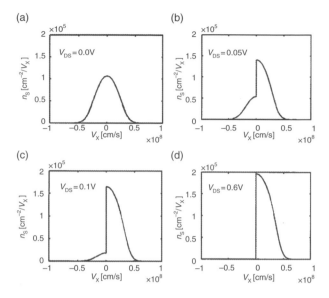

Fig. 6.33 Computed distribution of the x-component of the carrier velocity in ballistic nanoFETs for various V_{DS} at fixed $V_G = 0.6\,V$ as a function of position in the channel, assuming room temperature. The extremely nonthermal velocity distribution is analogous to the situation seen in Fig. 6.19, and leads to velocity saturation at high V_{DS}, even in the absence of scattering. Adapted from [396].

and the equilibrium Maxwell–Boltzmann distribution (as opposed to the degenerate Fermi–Dirac case relevant at $T \approx 0$ in the Landauer–Büttiker formalism).

Figure 6.33 shows snapshots of the x-component of the velocity distribution at different points along the channel at moderate V_G and large V_{DS}. Within the source itself, the distribution function is symmetric about $v_x = 0$ – there are as many right- as left-moving carriers. Moving along the channel toward the top of the barrier, the distribution function is crowded into the $+v_x$ side. The electric field in the channel has been accelerating carriers into the $+x$ direction, and because of the tilting of the potential, no left-moving carriers from the drain have enough energy to make it over the barrier toward the source. Farther toward the drain, the carriers are intensely concentrated into the $+v_x$ part of phase space. Finally, just inside the drain itself, the usual symmetric velocity distribution is augmented by "hot" electrons from the channel at large $+v_x$.

Similarly, one can consider looking just at the top of the barrier at fixed gate bias and seeing what happens to the carrier distribution as V_{DS} is ramped increased. Initially, for $V_{DS} = 0$ and the barrier not too high, we have the symmetric situation with equal numbers of left- and right-moving carriers. With increasing V_{DS} and the corresponding tilting of the potential, fewer and fewer left-moving carriers from the drain are able to reach the top of the barrier. At a sufficiently large V_{DS}, no carriers from the drain reach the top of the barrier, and further increases in V_{DS} do not significantly change the velocity distribution of carriers at the top of the barrier. This leads to the somewhat surprising result of *velocity saturation* in the channel, this time at the barrier peak near the source electrode, even *without scattering*.

A full treatment of the ballistic FET in the nondegenerate semiconductor limit is possible, though not simple [396]. We define a Fermi–Dirac integral of order s as

$$F_s(\eta) \equiv \int_0^\infty \frac{x^s \, dx}{\exp(x - \eta) - 1}.$$

$$(6.94)$$

Assuming n-type conduction and a grounded source, we also define $U_D \equiv eV_{DS}/k_B T$ and a renormalized Fermi energy, $\eta_F \equiv (E_F - E_1)/k_B T$, where E_1 is the energy of the first 1d subband in the conduction band. In this case, we find the operating characteristics:

$$I_D = ewC_x(V_G - V_T)\tilde{v}_T \left[\frac{1 - F_{1/2}(\eta_F - U_D)/F_{1/2}(\eta_F)}{1 + F_0(\eta_F - U_D)/F_0(\eta_F)} \right].$$

$$(6.95)$$

Here there is an effective velocity,

$$\tilde{v}_T = \sqrt{\frac{2k_B T}{\pi m_*^e}} \frac{F_{1/2}(\eta_F)}{F_0(\eta_F)}.$$

$$(6.96)$$

In the saturation regime, this reduces to

$$I_{D, \text{sat}} = ewC_x(V_G - V_T)\tilde{v}_T.$$

$$(6.97)$$

It is quite remarkable that there is, in fact, a saturation regime in the ballistic case. This is a direct result of effective velocity saturation, even in the absence of scattering. As in the ordinary velocity saturation case, $I_{D,\text{sat}}$ is linear in $V_G - V_T$.

At low V_{DS}, one recovers a linear regime:

$$I_D = \left[wC_x(V_G - V_T) \frac{\tilde{v}_T}{2(k_B T/e)} \left(\frac{F_{-1/2}(\eta_F)}{F_0(\eta_F)} \right) \right] V_{DS}.$$

$$(6.98)$$

If one considers lower temperatures such that the doped semiconductor is fully degenerate, Eq. (6.98) reduces exactly to the equivalent of Eq. (6.60), and the Landauer–Büttiker result is recovered [397].

6.7 Beyond CMOS

Silicon CMOS-based electronics has been a truly revolutionary set of technologies, responsible for a large part of what we call the Information Age. As scaling approaches its physical limits, it is possible that we are also nearing the end of CMOS, *provided* that some alternative can be found that makes economic sense. Here we take a brief look at potential components of post-CMOS technologies, with an emphasis on those that leverage nanoscale physics. Approaches based on magnetic materials are discussed in the next chapter, in the context of magnetoelectronics.

6.7.1 Nanowire, nanotube, and graphene devices

With the advent of semiconductor nanowires, carbon nanotubes, and graphene, there has been a great deal of interest in employing these nanomaterials for transistors. Because of their reduced dimensionality (and in the case of carbon nanotubes and graphene, their particular band structures), these materials can have transport properties that are attractive, such as enhanced mobilities. All three material types face major challenges. Rather than review these extensively, let's briefly consider the physics relevant to each in turn.

Semiconductor nanowires are typically grown through versions of vapor–liquid–solid processes (though it is possible to start from an SOI material and arrive at nanowire dimensions through lithography and etching). FETs based on chemically grown nanowires have been demonstrated extensively in III–V and group IV materials, with variations including in-wire superlattices and core-shell structures. Core-shell nanowires are of particular interest since they raise the possibility of modulation doping in a 1d geometry, with dopants residing in the shell transferring carriers to the single-crystal nanowire core.

The main challenges in adopting these materials in high density nanoelectronics are practical in nature. Residual doping is not uncommon, and passivation of surface states is also important. Fortunately, this latter issue has already been faced in structures based on the same materials in macroscopic form. Likewise, recipes for making contact to these materials often may be adapted from macroscale studies.

Integration (*"the wiring problem"*) is a huge issue in any attempt to create systems employing large numbers of individualized components (nanowires, nanotubes, nanocrystals, single molecules, etc.). In conventional integration, the positioning and wiring of billions of components are accomplished with (relative) ease, since all of the devices are fabricated from a single underlying semiconductor wafer. In the nanocomponent limit, instead one is faced with the prospect of somehow arranging billions of discrete components with high precision, making electrical contact to them reliably, and ending up with high yields of functioning, reproducible devices. There are no general solutions to this issue. Self-assembly techniques, letting thermodynamics and surface interactions arrange things, are one approach. An alternative that has been explored for both nanowires and nanotubes is *in situ* growth, where the novel materials are grown in preplanned positions via patterned catalyst or seed particles.

Carbon nanotubes have additional challenges of their own. As discussed in Section 3.6.1, nanotube band structure depends critically on tube chirality. As of this writing, all nanotube growth techniques produce a distribution of tube types and diameters, often with a significant fraction of metallic tubes mixed in with semiconductors of various band gaps. Selective "burn out" or chemical functionalization offer paths toward eliminating metallic tubes in devices. *Tour de force* efforts have demonstrated logic gates [93] and ring oscillators [398] painstakingly fabricated on individual nanotubes. However, there remains a long path to realizing wafer-scale integration of nanotube devices that have channel lengths comparable to the most advanced Si CMOS FETs and realizing the full benefits of the exciting transconductance-per-unit-width performance possible in $\sim$1 nm diameter nanotubes.

Very recently, graphene has arisen as a candidate for post-Si-CMOS nanoelectronics. In principle, because of its band structure, graphene can have many of the same benefits of nanotubes (reduced back-scattering, low effective mass, high mobilities). Graphene also appears to be easier to contact. There are multiple growth approaches that may permit either wafer-scale transfer of graphene to Si substrates, or selective CVD growth of graphene in targeted areas. However, the basic device physics (role of edge defects; optimal doping strategies; dominant scattering mechanisms) is at present an active area of investigation.

6.7.2 Single-electron devices

Rather than looking at more alternative materials, consider a change away from FETs to a device type better suited for the extreme nanoscale, where Si CMOS faces its most formidable physics challenges. Imagine a transistor in which the source–drain current is controlled by the presence or absence of a single electron on a controlling gate. Using nanoscale structures it is possible to do this, though significant tradeoffs must be made in terms of performance and operating temperature, as we shall see. Still, this sensitivity to charge can be incredibly useful for a number of applications, even if it is very unlikely to be a direct successor to CMOS.

The physics of *single-electron devices* is the physics of *Coulomb blockade*. Extensive discussions of Coulomb blockade physics may be found in [399, 352]. Now is the time when the electron–electron interactions that we have been so effective at ignoring come back into play. First we will treat the situation classically. Recall that classical conductors that can influence each other electrostatically may be described by mutual capacitances. For example, two parallel plates attached to the opposite terminals of a battery with voltage V build up net charges, $Q = CV$. One plate has an excess of positive charge, the other an excess of negative charge, and the space between and around the plates contains an electric field. The work done in charging the capacitor is $Q^2/2C$. In this situation, the net charge on the conducting plates does not have to be precisely quantized in units of e, since it is possible to shift the positions of the electrons continuously in the conductors.

This gets interesting when considering the simple geometric capacitances of very small objects. For the case of a sphere of radius a embedded in a dielectric κ a distance d from a ground plane, the capacitance is

$$C = 4\pi\kappa\epsilon_0 a \left(1 + \alpha + \frac{\alpha^2}{1 - \alpha^2} + \cdots \right),$$

(6.99)

where $\alpha \equiv a/2d$. Similarly, for a disk of radius R at the same distance,

$$C = \frac{\kappa\epsilon_0\pi R^2}{d}, \qquad d \ll R,$$
$$= 8\kappa\epsilon_0 R, \qquad d \gg R.$$

(6.100)

Looking at these expressions, it's clear that smaller objects yield smaller capacitances. This means that we can make systems where the single-electron charging energy, $E_c \equiv e^2/2C$, exceeds the thermal scale, $k_B T$. At room temperature, this threshold is crossed at $C \sim 3 \times 10^{-18}$ F. For a conducting sphere in vacuum, that corresponds to $a \sim 28$ nm. This charging energy is simply a macroscopically averaged way of accounting for the electron–electron repulsion energy that must be overcome to add one more electron's worth of charge to a capacitor. Single-electron devices use this charging energy to control the flow of current.

Coulomb blockade (CB) takes place when the energy E_c required for moving a single electron through some system exceeds the available thermal (or bias) energy. Ideally, *no current* flows at $T = 0$ in a fully blockaded circuit. The blockade may be "lifted" if a sufficiently large bias is applied, so that eV exceeds the charging energy, or if an additional capacitively coupled electrode is used to offset the charging energy, as we describe in more detail shortly.

Initially we discuss "classical" Coulomb blockade, in the sense that we neglect high order tunneling processes and quantum interference effects. We will also neglect the actual single-particle level spacing, Δ, of the island, assuming that it is small compared to both E_c and $k_B T$. This goes along with assuming that classical capacitances are reasonable parameters, and that any leads or electrodes are "good metals".

Single junctions

Consider a single tunnel junction [400] at $T = 0$, modeled as a capacitor, C. Suppose the capacitor initially has no charge Q on the plates, and is attached to a current source of size I. What happens? Polarization charge, Q, begins to accumulate on the capacitor at I Coulombs/s while simultaneously the voltage across the capacitor, V, increases at a rate I/C. The energy stored in the capacitor increases quadratically like $Q^2/2C$.

Now let's allow tunneling processes that can move individual electrons from one plate to the other. What happens to the energy stored in the capacitor if we instantly move one electron from the negative plate onto the positive plate?

$$\Delta E = \frac{(Q - e)^2}{2C} - \frac{Q^2}{2C}$$
$$= \frac{e^2}{2C}\left(1 - \frac{Q}{e/2}\right). \tag{6.101}$$

So, if the accumulated polarization charge $Q = CV < e/2$, then moving an electron across the capacitor actually increases the stored energy. However, once $V > e/2C$, the stored energy in the capacitor is *lowered* by tunneling an electron in the right direction. In this picture, the voltage across the junction would oscillate in a sawtooth fashion, as shown in Fig. 6.34. Such "single-electron tunneling oscillations" have been detected [400], but this is generally challenging because stray capacitances make it very difficult to set up a current-biased junction like this.

To model tunneling, we can consider some resistor, R, in parallel with the capacitor. The characteristic timescale for relaxing the charge distribution between the capacitor plates is then $\tau = RC$. For these Coulomb charging effects to be clear at $T = 0$, that relaxation time must be sufficiently slow that quantum smearing of the energy, $\delta E \sim h/\tau$ is negligible compared to E_c. That implies $R \gg G_0^{-1}$, a requirement that is typical for single-electron device structures in general.

In a *voltage*-biased junction, the same sort of physics *can* be relevant. To see CB, we need to assume that the voltage source cannot instantly maintain constant V on the capacitor. If $V < e/2C$, then tunneling a single electron across the junction would lead to a net instantaneous *increase* in the energy stored in the system (as well as an instantaneous deviation between the voltage across the capacitor and the supply voltage, V). Thus no current would flow. Once $V > V_c \equiv e/2C$, current can flow since transferring an electron

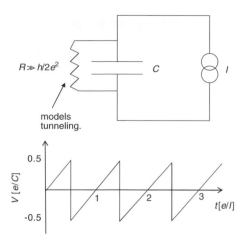

A current-biased single junction, modeled as a resistor in parallel with a capacitor. Below: the expected single-electron **Fig. 6.34**
tunneling oscillations in this ideal situation at $T = 0$. The voltage across the junction ramps up until it is energetically
favorable to tunnel across a single electron.

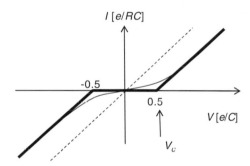

The current–voltage characteristic for the idealized system shown in Fig. 6.34. The dashed line is the result without **Fig. 6.35**
Coulomb blockade effects. The thick line is the $T = 0$ result, and the thin line shows what one would expect as
$T > 0$ thermal smearing begins to take place.

across the junction lowers the system energy. The result in this case is an $I-V$ curve with a
slope $= 1/R$, but offset from $V = 0$ by V_c (sometimes called the "Coulomb gap"), as shown
in Fig. 6.35. At elevated temperatures, thermal energy is available to the system to offset
the Coulomb energy cost, so these effects are smeared out. In this case, when $k_B T >\sim E_c$,
the $I-V$ characteristics look Ohmic with slope $1/R$, with no zero-bias suppression of the
conductance.

Note that observing this single-junction blockade is not easy; if the electrodes and wiring
have fast dynamics and are able to maintain the voltage across the junction at V very
efficiently, the CB effects are washed out. The details of how a particular electromagnetic
environment affects the CB physics in a voltage-biased junction are fascinating [401] and
beyond the scope of this book.

Two-junction devices

This discussion is patterned on that found in Ferry [352]. Consider an "island" connected to the outside world via two tunnel junctions, with capacitances and (large) tunneling resistances C_1, R_1 and C_2, R_2, respectively. We will think about attaching this system to a voltage source and ask under what conditions charge will flow through the circuit. We approach this problem similarly to how we did the single-junction case. We will find an expression for the total energy of the system and consider what happens to that energy when an electron tunnels across one of the junctions.

When attached to a voltage source of size V, at $T = 0$ the system will settle into its ground state. The charge on the capacitors will be given by $Q_1 = C_1 V_1$ and $Q_2 = C_2 V_2$, and the voltage drops across the two capacitors must add to the total applied voltage, $V_1 + V_2 = V$. While the charges on the two capacitors may be tuned continuously, since they include polarization effects from the leads that go to the voltage source, the *net* charge on the island must correspond to an integer number of electrons. That is,

$$Q \equiv Q_2 - Q_1 = -ne, \tag{6.102}$$

where $n = n_1 - n_2$ is an integer. Here n_1 is the number of electrons tunneled through the first junction onto the island, and n_2 is the number of electrons tunneled through the second junction off the island.

Defining the total capacitance $C_{\text{tot}} = C_1 + C_2$, we can rearrange terms to find

$$V_1 = \frac{C_2 V + ne}{C_{\text{tot}}},$$
$$V_2 = \frac{C_1 V - ne}{C_{\text{tot}}}. \tag{6.103}$$

The total electrostatic energy stored in the capacitors is then

$$E_{\text{es}} = \frac{Q_1^2}{2C_1} + \frac{Q_2^2}{2C_2} = \frac{C_1 C_2 V^2 + Q^2}{2C_{\text{tot}}}. \tag{6.104}$$

We need to keep track of the work done by the voltage source in setting the polarization charge on the capacitor electrodes, too. Look at the first junction, and consider what happens if an electron tunnels off the island via the second junction (so that $n \to n - 1$). Since no electron flows across the first junction in this process, any change $\Delta Q_1 = -eC_1/C_{\text{tot}}$ must be purely due to changes in polarization charge. The work done *by* the voltage source here is then $\Delta Q_1 V$. Therefore, the change in energy of the voltage source (due to work that it does) is eVC_1/C_{tot} for every charge that tunnels through junction 2. One can do exactly the same argument for an electron hopping onto the island through junction 1, and find for that case that the change in energy of the voltage source per charge through junction 1 is eVC_2/C_{tot}.

Therefore, we can write down the total energy cost (including voltage source contributions) of bringing the entire circuit into a configuration with net island charge

$Q = -ne = -(n_1 - n_2)e$, where n_1 electrons have tunneled onto the island through junction 1 and n_2 electrons have tunneled off the island through junction 2:

$$E_{tot} = \frac{C_1 C_2 V^2 + ((n_1 - n_2)e)^2}{2C_{tot}} + \frac{eV}{C_{tot}}(C_1 n_2 + C_2 n_1). \qquad (6.105)$$

With this expression, now we can worry about what happens to the energy of the system if we consider changing n_1 or n_2 by one. If the system is at $T = 0$, then the only allowed tunneling processes are the ones that *lower* the total energy of the system. Consider the particular case where the island is initially neutral ($n_1 = n_2$), and we want to tunnel an extra electron onto the island ($n_1 \rightarrow n_1 + 1$). From Eq. (6.105), the change in energy would be

$$\Delta E_{tot} = E_{tot}(n_1 + 1, n_2) - E_{tot}(n_1, n_2) = \frac{e^2}{2C_{tot}} + \frac{eVC_2}{C_{tot}}. \qquad (6.106)$$

For this to be negative, $V < -e^2/2C_2$. Similarly, if we want the system to lower its total energy by tunneling an electron off the island through junction 2, we find $V < -e^2/2C_1$. For current flow the other direction ($n_1 \rightarrow n_1 - 1$ or $n_2 \rightarrow n_2 - 1$), we find $V > e^2/2C_2$ and $V > e^2/2C_1$, respectively. *Current only flows in a double-junction system at $T = 0$ if the bias voltage exceeds a threshold determined by the capacitances!* If $C_1 = C_2$, then the threshold voltage for charge flow through the island is $|V_c| = e/C_{tot}$. This is the double-junction version of Coulomb blockade.

Figure 6.36 shows the idea schematically, and provides a way to think about such systems that is more intuitive than energetic book-keeping. The initially neutral island has no accessible electronic states near the Fermi level of the source and drain when the applied bias, V, is zero. Rather, the nearest states available (consisting of having an excess or deficit of one electron) are energetically distant. Applying V shifts the levels relative to one another, and when $V = V_c$, the next electronic level of the island is brought into alignment, allowing current to flow. The balance of how V is split between the two junctions is set by the relative capacitances, as in Eq. (6.103). Stronger tunnel couplings lead to lifetime broadening of the island levels. If that broadening is large compared to $E_c = e^2/C_{tot}$, then blockade is destroyed. Likewise, if thermal broadening of the source and drain occupations, $\sim k_B T$, is comparable to E_c, again blockade is destroyed.

Recall that there was nothing particularly special about our choice of the island being initially neutral; it just made the arithmetic simpler. At a bias just greater than V_c, as shown the island electron population increases by one. To increase the island population further would require an additional e/C_{tot} of voltage drop between the source electrode and the island. Instead, an electron must tunnel off the island into the drain before another electron can tunnel on from the source. This is the *sequential tunneling* mechanism of current flow. A major experimental consequence of this is the *Coulomb staircase*, a step-like $I-V$ as shown in Fig. 6.36(d) and discussed in one of the Exercises. Without Coulomb blockade effects, the $I-V$ curve would simply be linear, with a slope of $R_1 + R_2/R_1 R_2$.

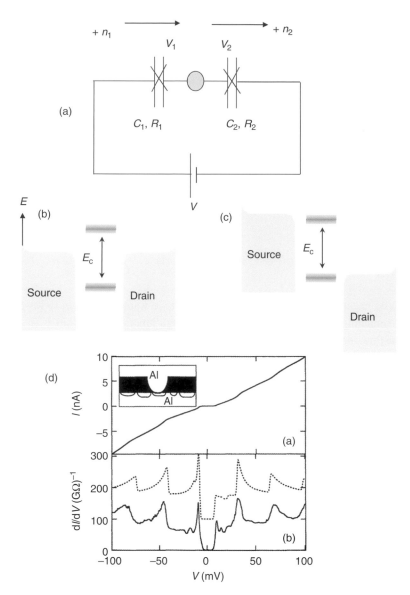

Fig. 6.36 The double-junction situation. (a) Schematic of a voltage-biased island connected via tunnel junctions. (b) The energetic situation at zero bias. A slight finite-temperature roll-off in occupation is shown for the essentially continuous states of the source and drain. The finite width of the discrete island states is due to lifetime broadening. As shown, no current flows. This is Coulomb blockade. (c) When $V = e/C_{tot}$, blockade is lifted as the island states become energetically aligned with the source and drain Fermi levels. The situation is shown for a symmetric two-junction system. (d) An example of a Coulomb staircase, adapted from [10].

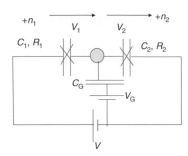

A single-electron transistor. A positive n_1 (n_2) corresponds to tunneling electrons onto (off) the island via junction 1 (2).

Fig. 6.37

The single-electron transistor

Now consider adding a gate electrode, capacitively coupled by a capacitor C_G to the island, as shown in Fig. 6.37. Before delving into the energetic analysis, consider how Fig. 6.36(b) would be affected by a gate. The additional capacitance, C_G, would result in a slightly lower charging energy, e^2/C_{tot}. Applying a bias to the gate would induce a polarization charge on the gate while shifting the electrostatic potential of the island. That is, the gate would shift the discrete island states vertically in Fig. 6.36(b) relative to the source and drain chemical potentials. The precise amount of shifting will again depend on the relative capacitances. To see how this works physically, think about adding an electron to the island, with or without a positive voltage in the gate relative to the source and drain. Without a positive V_G, not much has changed relative to the two-electrode case. With $V_G > 0$, however, the energy of that electron is lowered due to the coupling to the gate (that is, the electron is attracted to the positive gate electrode).

We can define a gate charge, $Q_G = C_G(V_G - V_2)$. With this sign convention, we then have the net island charge

$$Q = Q_2 - Q_1 - Q_G. \tag{6.107}$$

As before, the actual physical excess charge on the island must be quantized, so $Q = -ne$. Ferry adds in an explicit term for "polarization charge" due to work function mismatches and charged defects near the island. From our perspective, we wrap all of that into an effective gate, so that V_G in the following equations means the actual gate voltage that we apply plus these other capacitive effects. As expected, $C_{tot} = C_1 + C_2 + C_G$. Now that we have a three-terminal device, let's refer to the source–drain voltage as V_{DS}, for clarity and consistency with our earlier work on conventional transistors.

The analog of Eq. (6.103) is now

$$V_1 = \frac{1}{C_{tot}} \left((C_G + C_2)V_{DS} - C_G V_G + ne \right),$$

$$V_2 = \frac{1}{C_{tot}} \left(C_1 V_{DS} + C_G V_G - ne \right). \tag{6.108}$$

We can again compute the electrostatic energy of the system from all the capacitors:

$$E_{\text{es}} = \frac{1}{2C_{\text{tot}}} \left(C_G C_1 (V - V_G)^2 + C_G C_2 V_G^2 + C_1 C_2 V^2 + Q^2 \right). \tag{6.109}$$

Now when we consider an electron tunneling, we have to worry about the work done pushing polarization charge around by both the voltage supply that gives us V and the one that gives us V_G. The work done by the voltage sources per electron for tunneling onto the island through junctions 1 is

$$W = - \left(eV_{\text{DS}} \frac{C_2}{C_{\text{tot}}} + e(V - V_G) \frac{C_G}{C_{\text{tot}}} \right), \tag{6.110}$$

and the work done by the voltage sources per electron tunneling off the island through junction 2 is

$$W = - \left(eV_{\text{DS}} \frac{C_1}{C_{\text{tot}}} + eV_G \frac{C_G}{C_{\text{tot}}} \right). \tag{6.111}$$

Keeping track of this as well as E_{es}, the total energy change for the whole system involved in tunneling an electron onto the island via junction 1 is, in analogy to Eq. (6.106),

$$\Delta E_{\text{tot}} = \frac{e}{C_{\text{tot}}} \left(\frac{e}{2} + (en + (C_2 + C_G)V_{\text{DS}} - C_G V_G) \right). \tag{6.112}$$

Recall that the only processes allowed at $T = 0$ are those for which $\Delta E_{\text{tot}} < 0$. Note that making V_G more positive makes ΔE_{tot} for the electron addition process more negative, and therefore more favored, exactly as one would expect. One can do the same energy balance for the other three tunneling processes (removing an electron from the island via junction 1, and adding/removing an electron via junction 2). In compact notation, the tunneling of electrons requires

$$\mp[en + (C_G + C_2)V_{\text{DS}} - C_G V_G] > \frac{e}{2} \quad \text{or}$$
$$\pm[en - C_1 V_{\text{DS}} - C_G V_G] > \frac{e}{2}. \tag{6.113}$$

The edges of these inequalities define regions in the V–V_G plane, as shown in Fig. 6.38(b). This kind of plot is a *stability diagram*. When none of the inequalities are satisfied, at $T = 0$ no current can flow within the black diamond-shaped regions shown. These are blockaded, "stable" parts of parameter space, within which the number of electrons on the island is fixed. The points where adjacent diamonds touch are *charge degeneracy points*. At those values of V_G, the total energies of having the island electron population at n and $n+1$ (for example) are equal. From the energy diagram point of view, a degeneracy point is a V_G value that brings one island state into energetic alignment with the source and drain Fermi levels. When moving from one diamond to another along the V_G axis, the island occupancy changes by one electron for every degeneracy point traversed.

In a real gated double-junction device it is comparatively simple to map this out experimentally (e.g., compiling a series of I–V_{DS} characteristics at a variety of V_G values). Plotting a color map of the differential conductance, $\partial I/\partial V_{\text{DS}}$, as a function of V_{DS} and V_G, produces an empirical stability diagram. The slopes of the edges of the blockaded region are determined by the ratios of the capacitances. The positively

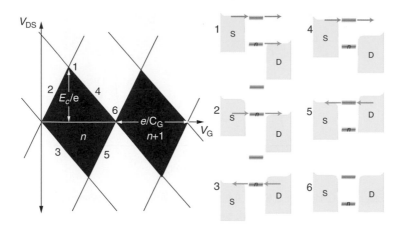

Stability diagram for a SET. At right are the energy level diagrams showing the situation at the various numbered points on the stability diagram.

Fig. 6.38

sloping lines have slopes $dV_{DS}/dV_G = C_G/(C_G + C_2)$, while the negatively sloping lines have $dV_{DS}/dV_G = -C_G/C_1$. The width in V_G of each Coulomb diamond, from degeneracy point to degeneracy point, establishes C_G, since that corresponds to changing the island charge by one electron. The intensities and widths of the lines on the differential conductance map are, at $T = 0$, set by the junction resistances, which determine the lifetime broadening of the island levels. A complete and generalizable way of modeling such maps, including finite temperature effects, is through rate equations [402].

Gated double-junction structures like this are known as *single-electron transistors* (SETs). They are transistors because they are three-terminal devices where the conductance between the source and drain is modulated by the voltage applied to the gate electrode. They are single-electron devices because shifting V_G to change the island electronic population by a single charge carrier is enough to modulate the device conductance. Clearly SETs have very different $I-V_{DS}-V_G$ response than conventional FETs. There is no saturation of the current as V_{DS} is increased. In conventional FETs (with the exception of ambipolar devices), I at fixed V_{DS} is a monotonic function of V_G. In contrast, near zero $V_{DS} = 0$ the current is a *periodic* function of V_G. Plotting $G = dI/dV_{DS}$ at $V_{DS} = 0$ as a function of V_G, one finds a series of "Coulomb blockade peaks", one at each charge degeneracy point with a maximum $G_{max} = 1/(R_1 + R_2)$, with blockade in between.[9] This periodicity is the key to using SETs as charge sensors, and also makes possible interesting circuit architectures that differ from those of conventional CMOS.

The effect of finite temperature is relatively straightforward. Thermal energy broadens the electronic Fermi distribution in both the source and drain. In turn, this broadens the widths of the stability diagram edges. As in the single-junction and double-junction cases, when k_BT becomes comparable to e^2/C_{tot}, Coulomb blockade effects get washed out

---

[9] The peak conductance here does *not* go all the way to the perfect transmission limit that one sees in resonant tunneling as in Section 6.4.3, because this discussion is in the classical limit. The quantum interference effects that give perfect resonant transmission are being neglected here.

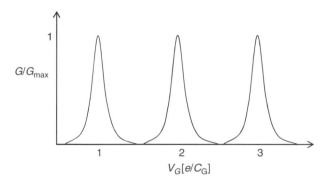

Fig. 6.39 Coulomb blockade peaks. Conductance at $V_{DS} = 0$ as a function of V_G in a SET, for a temperature where $k_B T$ is considerably less than E_c. The shape of each peak is given by Eq. (6.114).

entirely. The Coulomb blockade peaks in G vs. V_G acquire a width, as shown in Fig. 6.39. When this thermal width dominates, the shape of an individual peak at a degeneracy point V_G^0 is:

$$\frac{G(V_G, T)}{G_{max}} \approx \cosh^{-2}\left[\frac{e(C_G/C_{tot})(V_G^0 - V_G)}{2.5 k_B T}\right]. \tag{6.114}$$

To use a SET as a charge detector, imagine deliberately biasing the gate to sit on the "side" of such a blockade peak. In that situation, any change in the charge environment of the island acts like an effective change in V_G (see Eq. (6.107)), shifting the conductance up or down as the peak position V_G^0 moves. This is analogous to the way superconducting quantum interference devices (SQUIDs) are used to sense tiny changes in magnetic fields [403]. The sensitivity of SETs as charge sensors [404] is quite remarkable, better than $10^{-5} \, e/\sqrt{\text{Hz}}$. As a result, SETs are commonly used in many advanced nanostructure-based measurement schemes, including the sensing of tiny mechanical displacements [405] and the readout of candidate quantum computation circuits [406].

In the limit of a very small island, the effects of finite single-particle level spacing, Δ, become apparent, as we alluded in Section 5.7.1. Rather than the island states being spaced evenly in energy by an amount e^2/C_{tot}, instead they are then typically spaced by $e^2/C_{tot} + \Delta$, though one must pay attention to spin. Ignoring magnetic exchange processes, if the next island electron would go into a new single-particle level, then one must overcome both E_c and Δ. However, if the next electron would go into a singly-occupied island spatial state, then one only has to overcome E_c. This makes Coulomb blockade a powerful form of spectroscopy, sensitive to the detailed electronic structure of the island.

Moreover, in general we should also be concerned with tunneling via excited states of the island, as shown in Fig. 6.40(a), if there is enough source–drain bias to overcome E_c and relevant Δs. The signature of such tunneling is the appearance of lines in the differential conductance stability diagram that parallel the edges of the Coulomb diamonds yet are offset (by Δ/e) in V_{DS} from the ground-state tunneling features. In SETs with metal or semiconductor islands, the relevant excited states are typically electronically excited, involving electron–hole excitations of the island. Other excited states are allowed

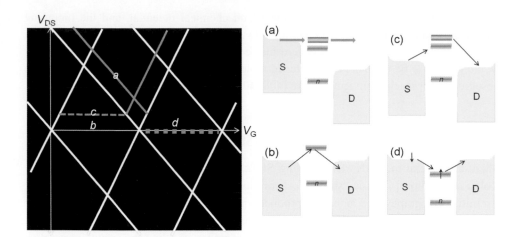

Fig. 6.40

Other tunneling processes. (a) Sequential tunneling through an excited state of the next electronic level. (b) Elastic cotunneling in the blockaded regime. (c) Inelastic cotunneling in the blockaded regime. (d) The Kondo process, only possible when there are an odd number of electrons on the island.

as well. For example, in nanotube-based SETs or SETs based on other molecular structures, tunneling can proceed through *vibrationally* excited states of particular electronic levels.

Moving further beyond the classical SET, we can consider higher order tunneling processes. A full treatment of these is well beyond the scope of this book, but it is not difficult to get a flavor for the possibilities. For example, rather than treating tunneling onto the island and tunneling off the island as incoherent events (such that probabilities would add), we can consider coherent second-order tunneling processes as in Fig. 6.40(b). Such a process is sometimes called *elastic cotunneling*, and as shown, it contributes to conduction even in the classically blockaded regime. The intermediate state (in which the island charge has changed by one electron) is a virtual state, and is allowed ephemerally via quantum mechanical uncertainty. If sufficient energy is available via source–drain bias, *inelastic cotunneling* processes may also take place, as shown in Fig. 6.40(c). In this case, an electron tunnels on to the ground state of the island, and tunnels off from an excited state of the island, all in one coherent process. (Clearly for this to take place, the island ground and excited states must not be completely orthogonal.)

One particular higher order process is important to mention because its consequences can be very dramatic. The *Kondo effect* was originally observed in bulk, nonmagnetic metals containing trace amounts of magnetic impurities. In the context of SETs [315, 407], a Kondo scattering process is shown in Fig. 6.40(d). It is energetically unfavorable due to Coulomb charging considerations for the unpaired electron on the island to leave for either the source or drain. However, an allowed higher order tunneling process is one in which the local electron leaves and is replaced by a source/drain conduction electron with oppositely directed spin. The superposition of such processes leads to a $T = 0$ ground state with the island electron forming a singlet with a "screening cloud" of source/drain conduction electrons. This many-body state builds up below a characteristic Kondo temperature,

T_K, that depends exponentially on E_c, the island chemical potential, and the tunneling couplings, and leads to *resonant* tunneling between the source and drain. In the limit that the two junctions are symmetric, the zero-bias conductance $G \to G_0$ as $T \ll T_K$ in what would otherwise be the blockaded state. In molecule-based Kondo systems, T_K has been observed to approach room temperature [408, 409].

Making SETs

There are several methods of fabricating SETs. The traditional approach, pioneered by Dolan and Fulton [410], is "shadow evaporation" of aluminum, as shown in Fig. 6.41. An initial lithography step is used to produce a resist stencil elevated above the substrate surface by some distance. A first evaporation is used to produce either the source or grain. Some oxygen is introduced into the evaporation chamber, enabling the growth of a native oxide on that electrode for use as a tunnel barrier. A second evaporation at a different angle deposits the island using the same stencil, with the evaporation geometry determining the overlap area of the tunnel junction. A further oxidation produces the second tunneling oxide, and a third evaporation at an additional angle deposits the other member of the source/drain pair. An additional evaporation may be used to deposit a proximal gate electrode, or an underlying trace or the substrate itself may be used as the gate.

This traditional approach, with all components made using lithography and metal as a working material, has the advantage of simplicity. However, the tunneling oxides can be very fragile and static sensitive, and the readily accessible geometries result in energy scales like $E_c \sim 10^{-4}$ eV, $\Delta \sim 10^{-7}$ eV. These small scales necessitate dilution refrigerator (sub-1 K) operating conditions to resolve clean Coulomb charging effects. Extremely small metal SETs can be fabricated from metal nanoparticles, either through physical vapor deposition of metal grains [10], or solution-based deposition of chemically synthesized metal nanoparticles onto nanoseparated electrodes [411, 412].

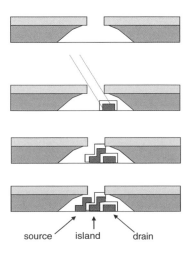

Fig. 6.41 The shadow evaporation method for fabricating single-electron transistors.

SETs have also been studied extensively using semiconductor nanostructures. While semiconductor nanocrystals may be trapped between metallic electrodes [321], it is much more common to fabricate SETs from a two-dimensional electron gas. This can be done using top-gates to deplete the underlying 2deg [413], or using physical etching [414] or oxidation [415] to define the island and tunnel barriers. One-dimensional semiconductor structures (nanotubes [416] and semiconductor nanowires [417, 199, 418]) have also proven to be excellent systems for examining SET physics. In semiconductor-based SETs, the energy scales tend to be a bit larger, with $E_c \sim 10^{-3}$ eV, and $\Delta \sim 10^{-4} - 10^{-3}$ eV. The larger single-particle level spacing when compared to metal SETs originates directly from the lower carrier density and correspondingly lower density of states.

Extreme SETs, with energy scales of tens or hundreds of meV, are possible, but only through very strong confinement. A simple estimate (see Exercises) suggests that fewer single-nanometer characteristic lengths are required to reach these energies, and this is consistent with experiment. SETs have been made that incorporate individual small molecules [419, 420, 421, 422, 423], and these structures can exhibit Coulomb blockade with very large energy scales, as shown in Fig. 6.52. Similarly, in very challenging experiments investigators have succeeded in examining Coulomb blockade through individual dopant atoms embedded in Si nanostructures [424, 425].

SETs for metrology

The SET and related devices provide previously unprecedented control of electronic populations. For example, consider a multijunction, multigate "turnstile" device like that shown in Fig. 6.42. In a conventional double-junction SET, operating at nonzero temperature can result in some small rate of tunneling events even when in the nominally blockaded regime of V_G and V_{DS}. Similarly, higher-order cotunneling processes can also allow charge to flow at some small rate. In a turnstile device, however, the net effect of such processes is strongly suppressed as the number of junctions is increased. This is most clear for cotunneling processes, where each island increases the order (and thus decreases the amplitude) of the relevant source-to-drain tunneling process. One use of such a turnstile is as a precision source of a known number of electrons. By cycling the gate electrodes sequentially in the presence of a small source–drain bias, it is possible to drive electrons through the device one (and only one!) at a time. This is the SET limit of the same "bucket brigade" process used in charge-coupled device (CCD) detectors [47].

This has two possible applications to *metrology*, the science of measurement. One major concern of metrology is the establishment of fundamental standards or definitions for physical quantities. For example, the second is defined as the amount of time it takes for 9192,631,770 periods of the radiation resonant with the transition between the two hyperfine levels of the ground state of ^{133}Cs. Likewise, the Ampére is the amount of current required to produce a force of 2×10^{-7} N/m between ideal parallel wires one meter apart. These are examples of "base units". Other quantities (e.g., the Coulomb, defined as the amount of charge transported by one Ampére of current in one second) are "derived units".

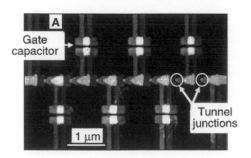

Fig. 6.42   A single-electron turnstile, consisting of a multi-island structure fabricated via shadow evaporation. Adapted from [426].

Using a SET electron turnstile, it is possible to create a laboratory standard for capacitance. Recall that capacitance is defined as the ratio between changes in the amount of charge on some object and changes in that object's electrostatic potential. Using a turnstile, one can pump a precisely known number of electrons onto a charge reservoir (capacitor plate), and measure (using a potentiometer calibrated back to base standards) the change in that reservoir's electrostatic potential. This is exactly what was done using the device [426] shown in Fig. 6.42.

A second use of turnstile capabilities is as possible fundamental current sources. If one could cycle the gates at a particular frequency, f, then the turnstile would produce a current of magnitude ef. If this could be done with a sufficiently low rate of errors and high stability of f, such a current source would act as an outstanding consistency check of our system of units. For example, the volt and the Ohm may be experimentally determined with high precision through measurements of the Josephson and quantum Hall effects, respectively [427, 428]. If any inconsistencies appear when comparing between the measured quantities, it could be the signature of new physics.

Figure 6.43 shows an actual implementation of a multiple-junction-based current source capable of driving currents in the pA range [429]. Notice that the detection portion of the circuit, the part that determines the current by counting the individual electrons as they pass by, is also based on an SET! Above we had briefly discussed SETs as sensitive electrometers, but we had been speaking in the dc limit. For this electron counting task, it is necessary to detect single electrons at comparatively high speed; a current of 1 pA corresponds to counting 6.2×10^6 electrons per second. From the dc circuits point of view, this is problematic. Good Coulomb blockade requires tunnel junction resistances on the order of a few times $h/e^2 \sim 26 \text{ k}\Omega$, and stray capacitances on the order of a few pF are difficult to avoid. This implies RC times comparable to the typical time interval between electron arrivals, suggesting that it would be very difficult indeed to use standard dc measurement techniques for this kind of electron counting.

Fortunately, the *rf-SET* (radio frequency SET) provides an alternative means of using SETs for high bandwidth charge detection [430]. Rather than a dc measuring circuit, the rf-SET incorporates the SET as an element in a high frequency ($\sim$ 1 GHz) LC resonant circuit that terminates a high frequency transmission line. As the charge capacitively coupled to the rf-SET island changes, the dc conductance effects expected from Eq. (6.114)

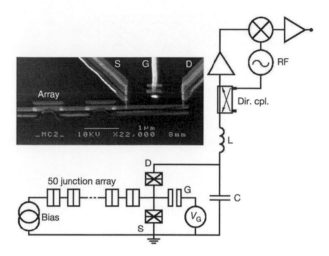

SET current source. An array of islands joined by tunnel junctions is used as a source of current. Detection of the charge arriving at the end of an array is done using a rf-SET, as discussed in the text. Adapted from [429].

Fig. 6.43

translate into changes in the rf impedance of the resonant circuit. If a high frequency carrier signal ($\sim$ 1 GHz) is fed into the rf-SET-terminated transmission line, the charge-induced impedance shifts manifest themselves as changes in the amount of power reflected back by the resonant circuit, which may be detected with high bandwidth ($\sim$ 10 MHz). These rf-SET techniques have been used to examine individual electrons tunneling on and off a semiconductor SET in real time [431] and in many other measurement applications [432].

It is worth noting that the *RC* problem of using SETs in dc circuits is only one example of a challenge that is general to *many* very small nanoelectronic devices, one of impedance matching. Many high performance nanoelectronic applications (CMOS-style logic, for example) require circuit elements that perform well at frequencies in the radio frequency range (MHZ through low GHz). While the nanoscale circuit elements themselves are much smaller than the wavelength corresponding to such high frequencies, it may not be possible to neglect propagation delays of signals down wires in the overall circuit. That is, wires must be treated as electromagnetic transmission lines, and one must be concerned with reflections and interference effects when propagating voltages and currents.[10] Many nanoelectronic devices (quantum point contacts and SETs, for example) have characteristic impedances that are quite large compared to typical transmission lines (often designed to have an impedance $Z_0 = 50\ \Omega$). The rf-SET approach is one example of integrating a high impedance nanodevice into a circuit with a relatively high operating bandwidth.

SETs for logic

We began discussing SETs because they have been suggested as a possible successor to CMOS Si-based MOSFETs for logic applications. There are some obvious technological

---

[10] For a good reference on rf design, consider Pozar [433].

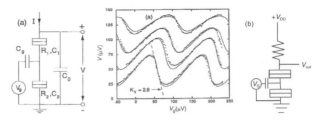

Fig. 6.44 (a) Demonstration of voltage gain using a single SET, adapted from [436]. (b) An SET inverter, where V_{DD} is the supply voltage, and V_G is the input.

hurdles that would have to be overcome for this to be realistic. Large-scale integration of SETs into electronics packages that function at or near ambient temperature would require reliable fabrication of sub-5 nm structures with little or no variation. Likewise, stray capacitances and extremely sensitive structures such as tunnel barriers would have to be produced with great fidelity and reliability at the full wafer scale. The control of single electronic offset charges would also likely be necessary, since such trapped charge would have an effect identical to random offset voltages applied to gate electrodes.

There are some benefits to consider as well. The nonmonotonic $I-V_G$ characteristic means that new architectures are now possible that were not available via CMOS devices. This can mean that extremely dense integration may be possible. Furthermore, elastic tunneling processes are nondissipative, and the motion of single electrons can be all that is necessary to tune conduction. Together, these suggest that ultralow power dissipation may be possible in some SET configurations.

Likharev [434, 435] has presented a detailed discussion of SETs for use in digital logic, which is summarized briefly here. *Voltage-based logic* is employed today in conventional CMOS digital electronics, where a one or zero is represented, respectively, by a high or low voltage. One necessary ingredient for a scalable voltage-based SET logic is the ability to have voltage gain, so that voltage signals are not attenuated across a chip. This has been demonstrated with using a single SET [436], as shown in Fig. 6.44(a).

In principle, sticking with voltage-based logic means that circuit designers can treat SET-based logic gates like conventional logic gates, leaving the details of implementation up to hardware engineers. Figure 6.44(b) shows an example of using a single SET and a resistive load (which could be a second SET) as an inverter. This would substitute for the conventional CMOS inverter that requires two transistors of complementary types.

Unfortunately, for device performance (in terms of error rates) comparable to conventional CMOS, the component SETs would need to be very small, such that $e^2/C_{\text{tot}} \sim 0.01 k_B T$. This is an extremely stringent requirement. It is possible to mitigate this to some degree by using arrays of tunnel junctions rather than single junctions, but that remains a fabrication challenge and sacrifices areal density.

An alternative approach would be to use *charge-based logic*. This is more reminiscent of how flash memory works, and uses the presence or absence of stored charge (ideally individual electrons) to represent ones and zeros. For logic purposes, switching action must depend on the charge states of particular islands. Figure 6.45 shows an example of a charge-based logic device called a quantum cellular automaton (QCA) [437]. The idea is that a

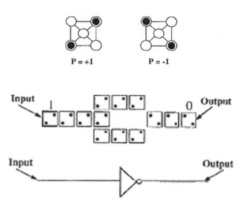

Using charge positions for logic in single-electron devices. Top: one can define a cell consisting of islands coupled capacitively, and further define two particular charge configurations to represent "0" and "1". Bottom: by coupling a series of such cells cleverly, the system can act like a logic gate as it finds its ground state. Adapted from [437].
Fig. 6.45

controlled gate application makes it energetically favorable for a particular input state to be set. Then electrostatic interactions between different islands/cells will favor a sequence of tunneling events so that the whole coupled system will find its new ground state. By careful design of the island configurations and interactions, the configuration of the final output islands/cell is related to the input configuration by a desired logical operation. SETs would then be used to sense the output charge configuration. Very simple versions of this approach have been implemented [438] using aluminum islands and SETs.

Some problems with charge-based logic are readily apparent. Charge defects, while annoying in voltage-based logic, can be fatal in charge-based systems, pinning a particular bit at a value regardless of inputs. In the case of QCAs, an unfortunately located charge defect will stop propagation of signals along a line of cells in a logic network. A further concern is that of speed of operations, for the same RC reasons mentioned above.

While SETs are unlikely to become a direct successor technology to CMOS logic, it is clear that they are extremely interesting. As tools for basic science they have enabled studies of quantum systems and charge dynamics at the single electron scale. Technologically they have uses in electrometry, metrology, and potential for applications where speed is not necessarily a major driver, such as nonvolatile memory.

6.7.3 Molecular electronics

For decades [439] people have considered using chemically synthesized molecules as building blocks for electronics. Aviram and Ratner [440] are widely cited as the modern progenitors of the idea that single-molecule electronic devices are a logical endpoint of micro/nanoelectronics scaling. Figure 6.46 shows the prototypical molecular device, as conceived by these chemists. An electron-donating chemical group is linked to an electron-withdrawing group via a nonconjugated spacer group. This is meant to be a direct analog of a pn junction, with the result being a single-molecule structure that can act as a diode or rectifier.

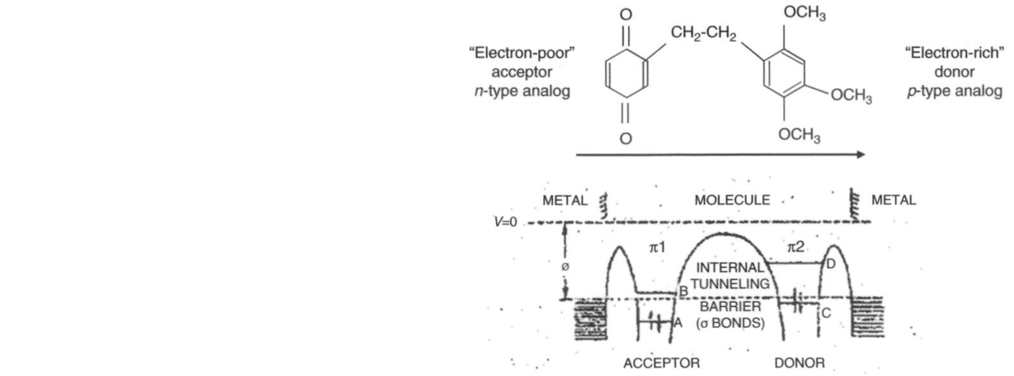

Fig. 6.46 A single-molecule rectifier concept, adapted from Aviram and Ratner [440].

In its most ambitious interpretation, true molecular electronics would use specially engineered individual molecules as nontrivial electronic components (more than just wires). Ideally, molecules could be designed to have intrinsic functionality (e.g., acting as a switch or a logic gate) such that a single molecule could replace many conventional components. There are several appealing aspects to this idea. Chemical synthesis techniques mean that one can imagine producing $\sim 10^{23}$ components in parallel. Moreover, for a given molecular component, every one that is correctly synthesized should be exactly alike. Synthetic chemists have at their disposal an incredible designer "toolkit", with more than a century of background to draw upon. Recent advances in electronic structure calculations (see Section 2.4.3) give considerable predictive power for energy levels and vibrational degrees of freedom. The tools of chemistry are also very well suited to working with systems that are intrinsically small (on the sub-10 nm scale), exactly the size range of interest.

There are several obstacles to achieving commercially relevant molecular electronics. First and foremost, while we do have tools that allow us to study electronic transport at the scale of single molecules, the "wiring problem" facing dense integration of molecular-scale devices remains an enormous challenge. Atomic-scale details of molecular attachment can greatly influence conduction through molecules, and achieving that level of control is exceptionally difficult. Quantum chemical electronic structure tools have steadily improved, but correctly treating the problem of molecules on surfaces driven out of equilibrium remains open, in general.

While these issues are daunting, molecular electronics remains an active, exciting area of study for (at least) two main reasons. In terms of basic science, electronic conduction through one or a few molecules is an excellent test problem for nonequilibrium quantum mechanics. The flow of energy, the origins of dissipation, the importance of electronic correlations, and the roles of local molecular degrees of freedom (magnetic, vibrational, electronic) are all of interest. Technologically, molecular devices are test cases for device physics likely to be relevant at the few-nm scale. Even Si CMOS will show some of the same "molecular" physics if scaled down to 2 nm channels. Several recent reviews have been written about molecular electronics [441, 442, 443, 444, 445].

Conduction vs. electron transfer

Chemists have long been concerned with the transfer of electrons within and between molecules. An exhaustively large literature exists concerning chemical electron transfer, much of which is not readily understood by physicists and electrical engineers due to differences in vocabulary. Chemical electron transfer and molecular conduction are related, and it is important to understand what parameters determine whether the chemical or physical point of view is more appropriate to a given problem.

Figure 6.47 shows a typical situation of interest to chemists. The initial state is a (neutral) donor/bridge/(neutral) acceptor (DBA) molecule, interacting with a fluctuating environment, and the final state is one in which an electron has been transfered, D^+BA^-. One example of such an environment would be surrounding water molecules, in the case of a solution-based experiment. In that case the interactions could be steric (short-ranged forces due to water molecules physically bumping into the DBA) or electrostatic (long-ranged forces due to the electric dipole moments of the tumbling water molecules). This environment is characterized schematically as a "reaction coordinate", representing, for example, the arrangement of the atoms. Figure 6.47(b) sketches the potential energy of both the initial state (DBA) and the final state (D^+BA^-) as a function of this reaction coordinate. When drawn this way, one sees that there is a "barrier" to be overcome for transfer to take place, and that fluctuations of the environment can play a crucial role in the transfer process. The difference in energy between the initial and final states at the fixed (final state equilibrium) reaction coordinate is the "reorganization energy", λ_r, and includes contributions due to mechanical relaxations of the molecule. Computing the actual transfer

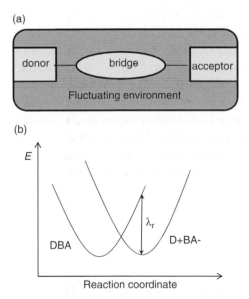

Electron transfer from the chemistry perspective. (a) A donor–bridge–acceptor molecule surrounded by a fluctuating environment. (b) (Free) energy as a function of environment reaction coordinate for the initial DBA molecule and the post-transfer molecule. As shown, there is a reorganization energy associated with relaxing the molecule and environment after charge transfer.

Fig. 6.47

rate involves both the environment and the detailed quantum description of the electronic states. In the electron transfer case, nuclear degrees of freedom (atomic positions) drive the initial state and final state energies into degeneracy; electron transfer occurs rapidly; and then the (comparatively slow) nuclear degrees of freedom relax to accomodate the new arrangement of electronic charge.

In contrast, when physicists and electrical engineers discuss conduction, they often are considering a *static* environment. (Measurements made in solution are, of course, an exception to this.) Moreover, in the electron transport case we are considering an *open* system. Because the electrons that are transmitted through the molecule escape "to infinity" via the leads, nuclear reorganizations tend to be less important in the transport case. In other words, typical electronic timescales are $<\sim 10^{-15}$ s, while motion of atoms is much slower, $\sim 10^{-12}$ s, comparable to phonon vibrational periods. If the transmitted electron does not reside in the system for a time comparable to vibrational timescales, vibrational effects on transport are generally small.

As in the other nanostructures we have considered, electronic transport can be coherent, in which case amplitudes of different processes can give rise to interference effects in the conductance. Transport can also cross over into an incoherent regime, in which inelastic interactions between the transporting electrons and the environment can dominate. Incoherent transport tends to be most relevant in longer molecules, where naively the electronic transit time can be sufficiently long that coupling to the environment has a chance to take place.

Ways to study molecular conduction

Several experimental techniques have been developed to examine electronic conduction through molecules, as shown schematically in Fig. 6.48. Some methods are best for examining many molecules in parallel. This has the advantage that it allows ensemble averaging over different molecular configurations. Such approaches do require caution, however, since single highly conducting defects can dominate the total measured conductance. Other approaches are suited to measuring individual molecules one at a time. A challenge in this situation is that device-to-device variation can be considerable, and ensemble averaging requires the study of many devices in a serial fashion. In all cases, great care must be taken to avoid contaminants. While a monolayer of adsorbed water and hydrocarbon contaminants, unavoidable in ambient conditions, doesn't matter when connecting macroscopic wires, such an interloping layer would be a disaster in an attempt to measure molecular conduction.

To look at many molecules at once, one can consider preparing a self-assembled monolayer (SAM) directly on one electrode, and depositing a second electrode on top to form a two-terminal device. The defect density in SAMs can be relatively high, even on single-crystal metal substrates, and therefore in large area junctions there is a risk that the top electrode can short directly to the bottom electrode. One approach to mitigating this risk is to work with very small patches of SAM, such as in nanopore devices (Fig. 6.48(b)). The method of top electrode deposition is also critical. While physical vapor deposition (metal evaporation) is simple, the resulting metal/molecule interface is very

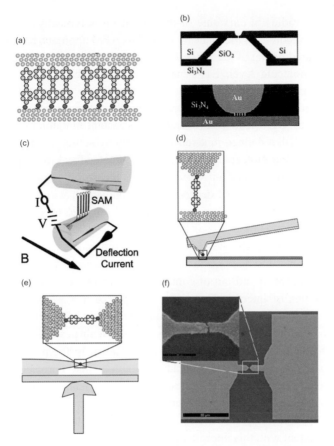

Ways to make electrical measurements on molecules. (a) An idealized layer of molecules between ideal metal electrodes. Achieving this in practice is extremely difficult. (b) A nanopore approach to making small patches of monolayer contacted by metal electrodes. Adapted from [446]. (c) Crossed wires as a means of contacting patches of molecules. Adapted from [447]. (d) Conducting AFM as a means of contacting one (shown) or, more commonly, multiple molecules. (e) A mechanical break junction as a means of contacting single molecules. (f) A planar, electromigrated break junction.

Fig. 6.48

difficult to characterize after the fact, and one must be concerned about possible molecular degradation upon exposure to hot metal atoms. Recent developments to mitigate these difficulties include soft contact or microcontact printing methods. For studies near room temperature, conducting polymers (PEDOT-PSS) have been used as "gentle" intermediary layers between the molecular layer and the top metal electrode [448].

Moving away from potential device configurations and focusing instead on scientific studies, simple mechanical junctions between crossed metal wires (Fig. 6.48(c)) have been used to great effect. One wire is coated with the molecules of interest, and a second wire is gently brought into mechanical contact. The contact area in crossed wire junctions typically contains on the order of 1000 molecules.

Likewise, the development of scanned probe microscopy has been a boon to molecular electronic studies. A conducting (metal-coated) AFM tip can be used as a top contact

(Fig. 6.48(d)). The radius of curvature of the coated tip is typically $\sim$ 10–20 nm, again giving a contact to hundreds of molecules at once. The AFM approach has the added benefit of allowing calibrated tuning of the applied force at the contact.

At the single molecule level, STM has proven invaluable. With STM techniques, particularly in ultrahigh vacuum, it is possible to prepare well-defined single-crystal substrates decorated with molecules of interest, image those surfaces, and then interrogate individual molecules. STM feedback (operating at constant tunneling current) allows the investigator to establish a defined junction asymmetry by setting the tunneling resistance between the tip and the molecule. By turning off the feedback and sweeping the tip–substrate voltage, two-terminal $I-V$ curves may be measured, as well as dI/dV and, if desired, d^2I/dV^2. The exponential dependence of conduction with distance restricts the volume probed by the current to a region comparable in size to a small molecule, for reasonably sharp tips.

STM-style junctions may also be used to examine molecular conduction even without imaging. Much progress has been made through conductance measurements while repeatedly moving a metal tip in and out of contact with a metal substrate in an environment containing molecules of interest. This is an example of a *mechanical break junction*, similar to that used to examine conductance quantization (Fig. 6.17(b)). A histogram of the measured conductance over many breaking cycles reveals peaks at conductance values corresponding to recurring, stable junction configurations. In the presence of molecules it is often possible to determine conductance features that correspond to individual molecules bound between the two electrodes. Peaks in conductance histograms that appear at conductances well below 1 G_0 have been inferred to arise from individual molecules. As discussed in Section 6.7.3, these molecular peaks have systematic trends with molecule type that are consistent with this interpretation.

A planar form of mechanical break junction is also possible (Fig. 6.48(e)). A notched wire or lithographically defined metal constriction is located on top of a bendable substrate. Because of the system geometry and the elastic properties of the bending member, a vertical shift of the pushing pin translates into a much smaller change in the lateral separation of the two sides of the constriction. Once the constriction is broken, incredibly precise (10^{-11} m) control of the resulting interelectrode gap has been demonstrated. As in the STM configuration, breaking and reforming metal junctions in the presence of molecules has enabled single molecule electrical transport measurements.

It is also possible to form three-terminal single-molecule devices. Electrostatic gating of the region between two nanometer-separated electrodes is a challenge simply from geometric considerations. The electric field from a proximal gate electrode must penetrate into the source/drain gap, despite efficient screening from the source and drain electrodes. One approach that has been relatively successful is based on electromigrated break junctions (Fig. 6.48(f)). A lithographically defined metal constriction is made on top of a gate electrode such as an oxidized aluminum trace. By ramping the voltage across the constriction, the resulting high current density transfers momentum to atoms in the constriction located at grain boundaries and defects in the polycrystalline film. This process, called electromigration, can be controlled, so that the constriction pinches off to form distinct source and drain electrodes separated by a nanometer-scale interelectrode gap. The result is a situation akin to two STM tips facing each other, with interelectrode conduction dominated by tunneling through a molecular-scale volume at the point of closest electrode separation [449]. If this migration process is performed in the presence of molecules, it is possible for the interelectrode gap to contain a molecule of interest for study [419, 450].

As should be clear from these descriptions of approaches to single-molecule electrical measurements, these techniques are a long way from practical technologies. Scaling fabrication to large numbers of devices appears difficult. Assessing the atomic-scale details of any such junction is not generally possible, and such details differ from junction to junction without particular control. Still, these studies have led to an emerging understanding of the physics relevant in conduction at these scales.

Molecules as tunnel barriers

The most simple, passive role for a molecule between two electrodes is as an effective tunnel barrier. Figure 6.49 shows a simplified picture of the energetic situation for a generic molecule bridging two electrodes. The discrete molecular orbitals are broadened by an amount Γ due to the coupling of the molecular orbitals and the conduction electron states of the source and drain. The HOMO and LUMO align in some way relative to the chemical potential (Fermi level) of the electrons in the electrodes. Note that it is extremely challenging to perform a realistic calculation of this energetic alignment from first principles, even for very simple molecules and idealized metal surfaces.

This picture should look familiar; it is analogous to that shown in Fig. 6.40(b) for off-resonant transport through a single-electron transistor. In the sense of a SET, the molecule is in the blockaded regime, though as described so far the energy scale dominating the blockade in this molecular case is Δ, the single-particle level spacing (here, the HOMO–LUMO gap), rather than a Coulomb charging energy. For small molecules, the HOMO–LUMO gap is typically in the range of eV. Transport in this regime for low V_{DS} is through second-order tunneling processes. In an SET this would be called cotunneling; in the language of chemical electron transfer theory, this process is called *superexchange*.

It can be shown [451] that one may model transport in this regime as tunneling through an effective barrier. One key prediction of this model is that, for fixed HOMO–LUMO

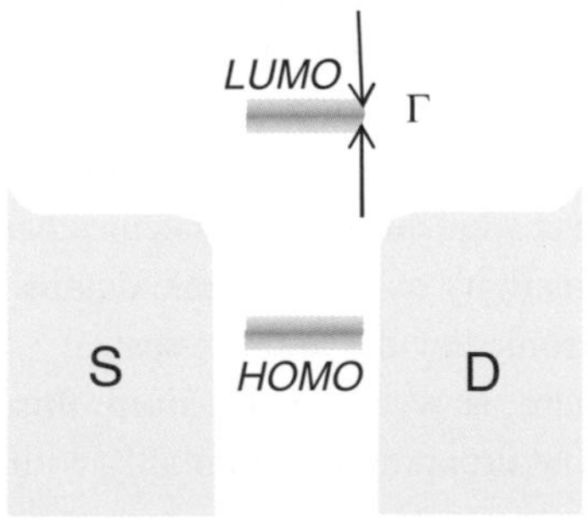

A molecule as an effective tunnel barrier.

Fig. 6.49

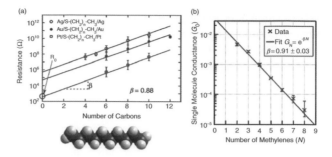

Fig. 6.50 Length dependence of conduction in alkane chains. The molecule shown is decane. (a) Alkanes with thiol (sulphur) endgroups on various metals, measured in ensembles of molecules via conducting AFM. Adapted from [454]. (b) Alkanes with amine (NH_2) endgroups, measured in single molecules via mechanical break junction. Adapted from [455]. In both cases, the conductance decay constant, β, is approximately 0.9 Å^{-1}.

gap (and fixed level alignment with the electrodes), the tunneling conductance through a molecule or molecular layer should decrease exponentially with increasing molecular length. That is,

$$G(d) = G_c \exp(-\beta d), \tag{6.115}$$

where β parametrizes this distance dependence, and d is the molecular length. Alkane chains, where carbon–carbon bonding is via saturated sp^3-hybridized orbitals, are expected to be a good example of a system with a HOMO–LUMO gap that is nearly independent of molecular length. As is shown in Fig. 6.50, measurements made using different techniques, both ensemble and single-molecule, show good agreement with Eq. (6.115), with an effective $\beta \sim 0.8$–0.9 Å^{-1}. For comparison, the typical β found in vacuum tunneling between Au electrodes is close to 2 Å^{-1}. Interpretations of tunneling, such as the Simmons model [353, 452], are sometimes used to extract an effective mass, m^*, from β values. This is questionable, however, since the standard concept of effective mass originates with the parabolic band dispersion of periodic solids (see Chapter 2) and is not directly relevant in a small molecule.

The prefactor, G_c, is the conductance in the limit of zero molecular length, and therefore may be interpreted as a signature of the metal–molecule contacts. This is clearly seen in Fig. 6.50(a), where a systematic variation of $1/G_c$ with the contacting metal work function is apparent. The metal/molecule contact involves the details of chemical bonding, charge transfer, and energetic alignment, and remains a topic of current research. Electronic structure calculations have shown that changing the molecular binding from one kind of metal atom site to another can change the metal/molecule charge transfer by a large fraction of an electron [453]. This sensitivity to atomic-scale details is one major challenge faced in trying to make useful electronic devices at these scales.

For longer molecules, there is a crossover from this coherent tunneling picture (G temperature independent but dropping exponentially with increasing molecular length) to incoherent hopping (G having some Arrhenius activated temperature dependence but only weakly dependent on molecular length). One way to think about this crossover is that the tunneling electron experiences decohering interactions with local molecular degrees of freedom (e.g., vibrational modes) when the tunneling process is slow compared to those timescales (picoseconds). Recalling our handwave picture of tunneling time (Eq. (6.48)), we see that this crossover is likely to happen sooner for longer molecules and those with smaller effective barrier heights (and therefore smaller HOMO–LUMO gaps). This crossover is seen in long, conjugated molecules [456].

In the coherent regime it is still possible to detect interactions between electronic and vibrational degrees of freedom in molecules. In analogy with the inelastic cotunneling process of Fig. 6.40(c), when there is sufficient source–drain bias, it is possible to have a second-order tunneling process in which an electron tunnels on to the vibrational ground state of the molecule, and tunnels off a vibrationally excited state of the molecule. This process leaves behind vibrational energy in the molecule, and can only take place in the presence of an appropriate electron–vibrational coupling. This process becomes an available channel for transport as V_{DS} exceeds the vibrational energy $\hbar\omega$, and there is a corresponding increase in the slope of the $I-V$ curve. This is usually measured as a peak in d^2I/dV_{DS}^2 as a function of V_{DS}. This is IETS, as discussed in Section 4.1.5 and is shown in Fig. 6.51.

IETS is closely related to another topic of much interest in ultrascaled nanoelectronics: the origins of dissipation and the flow of energy at the nanoscale. In IETS the electronic distribution at the molecule is nonthermal, having been set by the application of a voltage bias. Some of the electronic energy is redistributed into local vibrational modes, driving those vibrational populations out of equilibrium. The rest of the electronic energy derived

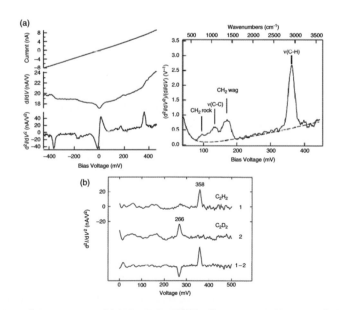

Inelastic electron tunneling spectroscopy. (a) An example of IETS in alkanes measured in a crossed wire junction, adapted from [457]. (b) IETS at the single molecule level, measured by STM, showing the shift in molecular vibrational frequencies due to isotopic substitution, adapted from [160].

Fig. 6.51

from eV_{DS} is eventually dissipated via inelastic scattering among the electrons and phonons in the electrodes, on distance scales ranging from a few to tens of nanometers. The molecular vibrational energy may be redistributed among the different molecular modes, and eventually finds its way into the bulk phonons of the substrate. Understanding this flow of energy and how irreversibility enters into this driven quantum problem is critical to the eventual design of electronics technologies that operate on these scales.

It is worth noting that molecular rectifiers of the type envisioned by Aviram and Ratner have actually been implemented [458, 459]. In the context of passive tunneling devices, these systems are based on molecular designs such that two halves of the molecule are relatively decoupled, with quite different HOMO–LUMO levels. As a result, one can consider these molecules as presenting very asymmetric barriers to transport, leading to rectification.

Molecules as ultrasmall SETs

Beyond passive tunneling barriers, single molecule devices have also been demonstrated acting as ultrasmall SETs. There are two main differences between molecular SETs and the metal or semiconductor SETs discussed in Section 6.7.2. First, small molecules can have extremely large energy scales compared to other SET implementations. Modeling a molecule as a classical conductor ~ 2 nm in diameter, separated from nearby metal surfaces by ~ 1 nm, one arrives at a Coulomb charging energy on the order of hundreds of meV. While a much more realistic treatment would look at the quantum capacitance of a real molecular system, this order of magnitude estimate is not too bad. Likewise, the single-particle level spacing, Δ, is essentially the HOMO–LUMO gap, and clearly can be quite large, on the order of eV. These large energy scales mean that in principle it should be possible to operate molecule-based SETs at or near room temperature.

These large values of E_c and Δ lead to the second major difference, the relative importance of internal degrees of freedom. In semiconductor and metal SETs, these energy scales for electronic excitations are much lower than the tens of meV relevant for local vibrational modes such as optical phonons. The consequence of this is that local vibrational degrees of freedom usually do not strongly affect such devices. In molecular SETs, however, this hierarchy is reversed, with local molecular vibrational modes often being "cheaper" than electronic excitations. As a result, vibrational effects are much more prominent in the conduction of molecular SETs.

Figure 6.52 shows examples of Coulomb blockade in a molecular SET based on a single C_{60} molecule [419]. One challenge in having such a large electron addition energy is that it is extremely difficult to access more than two charge states of this molecule. At quite low energies there are a series of excited state tunneling features (in analogy with Fig. 6.40(a)) approximately equally spaced in V_{DS} by 5 meV. This energy scale is far lower than the electronic energy scales, and therefore the channels in question are believed to be vibrationally excited states, in this case due to a "bouncing ball" vibrational motion of the entire molecule on the Au electrode. Also shown is data from another device, where a ~ 33–35 meV mode is clearly present. This energy scale corresponds to a known vibrational mode of C_{60}.

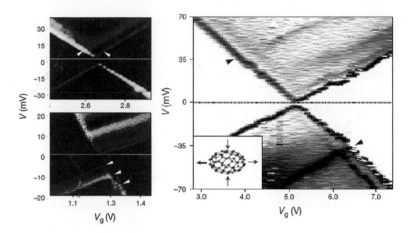

Stability diagrams from SETs based on individual C_{60} molecules. At left, diagrams show excited state tunneling with low-lying excitations believed to be center of mass modes of the whole molecule. At right, signatures of excited state tunneling at an energy consistent with the indicated molecular vibrational mode. Adapted from [419].

Fig. 6.52

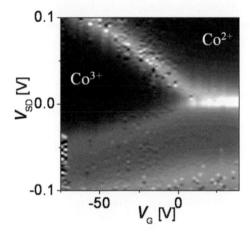

Coulomb blockade and a zero-bias Kondo resonance in a single transition metal complex molecule, adapted from [460].

Fig. 6.53

Figure 6.53 shows an example of a Kondo process (Fig. 6.40(d)) in a single-molecule SET based on a transition metal complex [460]. The large energy scales associated with the small molecule lead to a Kondo temperature exceeding 60 K in this device. The possibility of Kondo temperatures near 300 K [408, 409] suggests that actual devices based on Kondo physics are not out of the question. Note that in an ideal, symmetric Kondo device well below the Kondo temperature, gating through charge degeneracy would modulate the zero-bias conductance from its minimum (blockaded) value to perfect resonant transmission, $G \rightarrow G_0$.

Single-molecule SETs have, of course, the same challenges that all other SETs face in terms of the critical need to control the immediate charge environment of the molecular island. Achieving significant gate couplings at molecular scales is also very difficult. Finally, it is not obvious how particular chemical bonding motifs translate into effective tunneling resistances and capacitances in the SET geometry. As in the molecular tunnel barrier situation, the nature of the metal/molecule contact and level alignment remains a topic of current research interest. Still, single-molecule SETs have demonstrated that it is possible to create potentially useful devices with active regions comprising individual, chemically synthesized and assembled units.

Molecules as switches

Beyond SET-like properties, for years there has been a strong interest in molecules as active, switching devices. For example, there are many molecular motifs that are known to take on different geometric configurations depending on charge state or available optical energy. Likewise, chemists have long known of molecules that could take on different isomeric configurations. It is not a stretch to imagine a junction based on such a molecule, so that a particular voltage history or other electrical or optical stimulus could switch the molecule controllably between its available forms. Provided the possible molecular states had significantly different conductances, such a two-terminal switch could be the basis for either molecule-based logic or memory (see Section 6.7.4).

Figure 6.54 shows candidates for such molecules. The first, an oligophenylene ethynylene (OPE) derivative, was intended to show switching based on a "polaronic" mechanism. The polaronic idea is that at sufficiently high V_{DS} to align a molecular level with either the source or the drain, the average charge on the molecule changes, as in a SET biased outside the blockaded regime. In the presence of strong electron–phonon interactions, that excess charge can be stabilized by a distortion of the molecule, effectively lowering the energy of

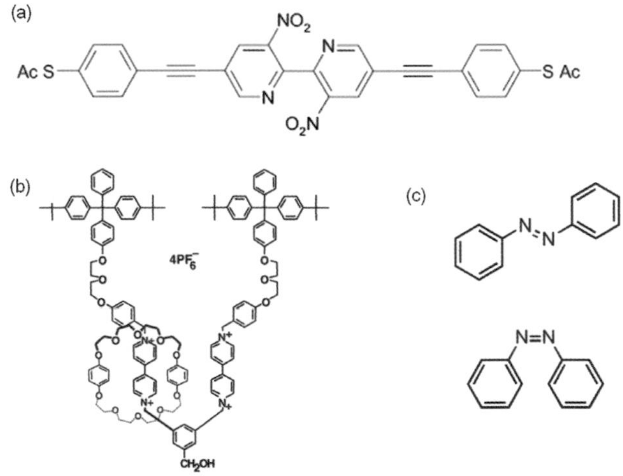

Fig. 6.54 Candidate molecular switches. (a) An oligophenylene derivative suggested as a polaron-based molecular switch. (b) A rotaxane used as a switch in [461]. (c) Azobenzene, which may be switched by light or heat between the *trans* (top) and *cis* (bottom) isomers.

the now-occupied molecular level, and changing the conductance. While there have been intriguing experimental hints, so far there is no unambiguous experimental evidence for this mechanism in practice.

Figure 6.54(b) shows another candidate switch called a rotaxane. The circular portion of this molecule can be moved back and forth along the linear part via alterations of the molecular charge distribution. This bistability has been shown clearly in solution-based electrochemical measurements [462]. Early experiments incorporating such molecules into metal/molecule layer/metal sandwiches were encouraging [461]. Later experiments [463] showed, however, that great care must be taken to avoid the voltage-driven growth of metal structures that could mimic apparent molecular switching. Recent work on rotaxane-based switching has continued to show some promise [464].

Another common switching motif is shown in Fig. 6.54(c), based on azobenzene. This molecule can exist in either the *cis* or *trans* forms, as shown, and can be switched reversibly between these two isomers via photoactivation. The idea is to use such molecules as photo-switches, since the different isomers are likely to have different conductances [465]. Figure 6.54(d) is a similar concept, based on diarylethene, which can be switched optically between conjugated and nonconjugated isomers [466].

These various switching approaches are intriguing. Important concerns common to all of them include: stability of the switched states; reversibility of the switching process; ultimate timescale of the switching; and the efficiency of the switching, particularly the effect of molecular environment. Given the wealth of possibilities available to synthetic chemists, it is no surprise that research continues on these systems.

Molecule-assisted nanoelectronics

Given the incredible staying power of silicon-based electronics, it is natural to consider whether molecule/semiconductor hybrid structures may have a role to play in future technologies. Integration of engineered molecules with Si nanoelectronics may benefit from the best of both worlds: the highly developed state of CMOS engineering and the array of possibilities open to synthetic chemistry.

Initial forays into hybrid Si/molecule electronics have largely been in trying to understand the Si/molecule interface's charge transfer properties [467] and in using molecules as tools to influence the local electronic properties of the Si. In the latter case, one can imagine molecules as surface-based dopants, providing (or accepting) electrons from the semiconductor [468]. One can also consider molecular layers as a means of engineering surface electric dipoles and surface band bending, acting rather like a local gate [469].

The major challenges in store for molecule-assisted semiconductor nanoelectronics are ones of surface chemistry and stability. Grafting molecular layers to semiconductor surfaces in an ordered way, while not enabling the formation of defects or charge traps, is not easy. Furthermore, standard CMOS device processing frequently involves excursions to high temperatures, while most organic molecules decompose at less than 300° C.

6.7.4   Novel architectures and nanoeconomics

Recall that the current paradigm for integrated micro/nanoelectronics is one of incredible reliability, with billions of components on a typical CMOS processor chip, essentially all functioning perfectly for years on end. We have learned that the electronic properties of truly nanoscale systems can be influenced by atomic-scale details, and we do not in general have the ability to control materials with atomic precision. It seems clear that we are likely to run into a clash between our fault-intolerant approach to electronics and the physical and practical limits of our fabrication abilities.

One possible approach has evolved out of thinking about this problem in the context of molecular electronics. The idea is to develop fault-tolerant computer architectures, meaning logic designs that can function with large (certainly by comparison with current CMOS standards) numbers of defective components. Novel self-assessing and self-programming software schemes are equally important, so that such a computer could adjust on the fly to the demise of components. The now-classic example of this is Teramac [470], a small-scale supercomputer built from field-programmable gate arrays, logic elements with a high defect rate. The reported Teramac hardware contained more than 220,000 defective components, approximately 3% of the total logic resources available to the machine. The designers' response to this was to develop a software approach for assessing the state of the components, configuring the resources, and then compiling the program that one desired to run. These ideas, reconfigurable and self-assessing devices designed to work even in the presence of large numbers of defects, are powerful and represent enormous shifts away from the traditional paradigms of microelectronics design.

A specific architecture that has been mentioned repeatedly in the context of nanoelectronics, especially two-terminal switches of the type discussed in Section 6.7.3, is the crossbar latch [471]. Using a very simple (and thus comparatively easy to pattern) crossbar design, where wire "rows" are overlayed on wire "columns", and the intersection of each wire pair is a switch in combination with a resistor and a diode, it is possible to perform all the logic operations necessary to build a computer. As we have seen, switching, diode, and resistor functionality can all be reduced to the molecular scale. While this exact approach may not be deployed in consumer electronics, it certainly represents an interesting limit to consider.

Concluding our look at nanoelectronics, it is important to remember that the future of this field will be determined as much by economics as by physics and chemistry. There is tremendous economic momentum behind silicon-based electronics, and no competing technology will take over unless it offers performance benefits at a competitive price. Individual Si MOSFETs cost less than 10^{-6} cents, operate at room temperature, and can be manufactured with ridiculous levels of precision and reliability. This is an incredible barrier to entry for any competing material or technology.

Exponential growth of performance, as in Moore's Law, is a target rather than a rule. It is instructive to note that air travel speeds followed a similar exponential growth from 1900 until approximately 1960. Economics, rather than a technological limit, is the primary reason for the plateau in airplane speed since then. Whether electronics performance will have to undergo a similar transition remains unclear. While none of the nano-based alternatives to CMOS may take flight in large scale technologies, they provide a fascinating look at the physics that is relevant to electronic transport as systems approach the atomic scale.

6.8 Summary and perspective

Before delving into nanoelectronics, we began with an overview of classical and semi-classical electronic conduction, defining a number of useful concepts and parameters. We considered quantum corrections to classical electronic conduction, introducing various physical effects often called "mesoscopic". At the nanoscale, one cannot ignore these effects. We then discussed tunneling as an example of a coherent process, and a general scattering approach to quantum coherent systems. Having set the stage, we moved on to look at conventional transistors and CMOS-based nanoelectronics, including recent innovations like the fin-FET. Looking beyond conventional FETs, we examined single-electron transistors and molecular electronics as examples of interesting nano-based gadgets that may play a role in future electronics technologies. We also finished up with a look at unconventional architectures and the economics of nanoelectronics.

- Charge transport. We introduced transport coefficients which must obey certain symmetries. It's often useful to think about classical concepts like the mean free path and diffusion. Electronic transport is a powerful experimental observable.
- Quantum corrections to conduction. Inelastic interactions of the electrons lead to loss of quantum coherence on some time (and therefore length) scale. Quantum interference effects that can modify classical conduction include the Aharonov–Bohm effect, universal conductance fluctuations, and weak localization.
- Tunneling and transport as transmission. Electrons can move through classically forbidden regions. In the coherent regime, it makes sense to think of transport in terms of transmission and reflection (or scattering) of quantum channels, as described by Landauer and Büttiker. Perfect transmission of a single channel leads to a conductance of $G_0 = 2e^2/h$, the quantum of conductance.
- Field-effect transistors. These three-terminal devices are the workhorses of modern electronics, functioning as switches and operating based largely on classical semiconductor physics. Modern electronics is already in the nano-world, with billions of sub-100 nm transistors per chip. Continued scaling is approaching fundamental limits, requiring changes in device design and approach. The economic barriers to dramatic change are considerable, however.
- Single-electron devices and molecular electronics. It is possible to create switches controlled by motion of a single electron, or with current flowing through a single small molecule. While direct application of these schemes faces many hurdles, the physics of these devices is potentially relevant to any ultrascaled nanoelectronic technology.

6.9 Suggested reading

- *Conventional semiconductor devices* For standard CMOS electronics, it is hard to do better than the gold standard of reference texts, *Physics of Semiconductor Devices*, 3rd Edn. by S. M. Sze and K. K. Ng (Wiley Interscience, 2006), and the somewhat more pedagogical *Physics of Semiconductor Devices*, M. Shur (Prentice Hall, 1990).
- *Meso/nanoscale conduction* In recent years a number of excellent books have appeared that specialize in quantum transport. These include *Introduction to Mesoscopic Physics* by Y. Imry (Oxford University Press, 2008), *Transport in Nanostructures* by D. K. Ferry, S. M. Goodnick, and J. Bird (Cambridge University Press, 2009), *Electrical Transport in Nanoscale Systems* by M. Di Ventra (Cambridge University Press, 2008), *Quantum Transport: Atom to Transistor* by S. Datta (Cambridge University Press, 2005), and *Molecular Electronics: an Introduction to Theory and Experiment* by J. C. Cuevas and E. Scheer (World Scientific, 2010). A more mathematical book is *Quantum Transport: Introduction to Nanoscience* by Y. V. Nazarov and Y. M. Blanter (Cambridge University Press, 2009).

  Classic review articles on mesoscopic electronics include "The single-electron transistor", by M. A. Kastner, *Rev. Mod. Phys.* **64**, 849–858 (1992) [472], "Random-matrix theory of quantum transport", by C. W. J. Beenakker, *Rev. Mod. Phys.* **69**, 731–808 (1997) [473], "Conductance viewed as transmission", by Y. Imry and R. Landauer, *Rev. Mod. Phys.* **71**, S306–S312 (1999) [474].
- *The semiconductor electronics industry* Two scientifically literate books about the electronics industry and how we got here are *Crystal Fire: the Birth of the Information Age* by Michael Riordan and Lillian Hoddeson (W. W. Norton, 1997) and *The Story of Semiconductors* by J. W. Orton (Oxford University Press, 2009).

Exercises

6.1 *Transistor scaling* Suppose we are considering scaling the typical linear dimension of a field-effect transistor by a factor α, so that the channel length L, for example, becomes L/α.
  (a) At fixed V, show that the source–drain and gate electric fields scale like α.
  (b) Show that the timescale associated with a carrier crossing the device scales like $1/\alpha^2$ if V is fixed.
  (c) Show that the power dissipated scales like $1/\alpha$ for constant V, and $1/\alpha^2$ for constant field scaling.

6.2 *Transistor scaling and oxide thickness* Transistor scaling needs in the early 2000s drove traditional gate dielectric thicknesses progressively lower. Assume that the threshold voltage in this problem is 0 V.
  (a) Assuming that $\kappa = 3.9$ for SiO_2 even at very low thicknesses, what is the capacitance per unit area of a gate dielectric 1.8 nm thick? Assuming a

breakdown field of 2×10^9 V/m, what is the maximum operating voltage for the gate? What is the gated charge density at that voltage?

(b) Treat the gate dielectric like a rectangular tunneling barrier, with barrier height of about 3 eV. From Eq. (6.44), assuming an effective mass equal to the free electron mass, what is the transmission probability of an electron to get through such a barrier?

(c) A not-crazy estimate of the conductance "per channel" for tunneling is $2e^2/h \times T$, where T is the transmittance found in (b). Supposing that there are about 10^{14} effective channels per cm^2, what is the estimated tunneling conductance per unit area of the gate oxide layer? At a gate bias of 100 mV (low enough that the barrier can still be considered approximately rectangular), what is the resulting tunneling current?

(d) Suppose there was a dielectric that had a comparatively high κ, at the cost of having a lower (bulk) breakdown field than SiO_2. (This is actually the case for HfO_2, for example.) If the bulk breakdown field was 2×10^7 V/m, and the dielectric was 10 nm thick, do you think an applied bias of 0.3 V would cause destructive damage? Why or why not? Hint: think about the kinetic energy available to the electrons in this situation.

6.3 *Threshold voltages in double-gate and vertical MOSFETs* One contribution to the threshold voltage for inversion is the requirement to deplete away the carriers present in the channel of the would-be transistor.

(a) Consider a slab of p-type silicon of thickness t_{Si} with density of acceptors N_A. Above and below this slab are oxide layers each of thickness t_x with relative dielectric constant κ_x. Above and below those oxide layers are top and bottom gate electrodes, respectively. For this symmetric case, show that for full depletion (when the entire charge in the Si slab is balanced by opposite charges on the gates) the voltage across the oxide is $eN_A t_{Si} t_x/2\kappa_x\epsilon_0$.

(b) Now consider a *cylindrical* vertical MOSFET channel of radius r_{Si}, again of p-type silicon with the same acceptor density, with the outside of the cylinder coated in oxide of thickness t_x. Using Gauss' Law to find the electric field in the oxide region at full depletion, show that the voltage dropped across the oxide layer is $\frac{eN_A r_{Si}^2}{2\kappa_x\epsilon_0} \ln(1 + t_x/r_{Si})$.

(c) Speculate (a couple of sentences) on which of these you think would be better, from the engineering perspective, and why. Notice that this has all used just classical electrostatics.

6.4 *Quantum confinement effects in nanoscale MOSFETs* When calculating the threshold voltage for MOSFET devices, we have been using an entirely classical picture. For extremely small channels, as may appear in sub-100 nm SOI, SON, or vertical FET structures, quantum confinement effects can become important. In addition to the classical energy scales (like the threshold voltage contributions above), one must be concerned with quantum confinement energies.

(a) Consider first the double-gate MOSFET from the previous problem, where now the Si channel slab is moderately long (in the y direction), but has transverse (x and z) dimensions of $t_{Si} \times t_{Si}$. The shift in threshold voltage from the above result due to quantum confinement will just be the energy of the first transverse subband. Assuming hard-wall boundary conditions and an effective mass m^*, write an expression for the change in threshold voltage from the classical result.

(b) Now consider a long cylindrical channel of radius r_{Si}, again with hard-wall boundary conditions. The solution to the radial Schrödinger equation for the cylindrically symmetric case is the zeroth order Bessel function, $J_0(x)$, where x is defined as kr (here k is the radial component of the wavevector for the electrons). For the lowest confined state, the first zero of $J_0(x)$ must occur at the boundary to meet the boundary condition. Find the quantum confinement change in threshold voltage from the classical value for this case. (Estimate the numerical value of x for the first Bessel function zero; the energy measured from the bottom of the conduction band will be given by the usual relationship between E and k in a semiconductor, and the minimum energy case is when the longitudinal component of the wavevector is essentially zero.)

(c) Consider comparing the cases of (a) and (b) with $2r_{Si} = t_{Si}$. Assuming identical effective masses for the two cases, which has a bigger quantum mechanical confinement effect? Can you handwave a physical reason why (beyond the non-answer that the formulae say so)? Suppose $t_{Si} = 5$ nm and $m^* = 0.31m_0$. How big a voltage shift are we talking about for these two cases? Do you think this is significant? How small a value of tSi is required before the quantum voltage shift is worth worrying about in room temperature devices?

6.5 *Persistent currents* Consider a wavefunction that describes a particle constrained to move on a ring, $\psi(\mathbf{r}) = Ae^{i\phi(\theta)}$. Here ϕ is a quantum mechanical phase, and θ is the coordinate describing the particle's azimuthal position on the ring. From the definition of probability current density,

$$\mathbf{J}_n = \frac{-i\hbar}{2m}[\psi^*\nabla\psi - \psi\nabla\psi^*],$$

show that $\phi(\theta) = \theta$ results in a nonzero circulating current density. These persistent currents have actually been measured, as in Lévy *et al.* [350].

6.6 *Tunneling & the WKB approximation* Consider a VW Beetle (curb weight $m = 1340$ kg) approaching a speed bump of length a, the left edge of which is located at $x = 0$. Treat the Beetle as a point particle (!).

(a) First, assume the speed bump potential is well modeled as a half-sine wave: $V(x) = mgh \sin(\pi x/a)$ where h is the height of the bump (30 cm) and $a = 60$ cm. Naturally, g is 9.807 m/s². Assume that the VW is traveling the in $+x$ direction at 10 kph. Between what values of x would the automobile be classically forbidden?

(b) Numerically integrating using your favorite software, in the WKB approximation what is the probability that the Beetle will be transmitted through the speed bump? This is what physicists call a rare event.

(c) Consider a wall instead of the speed bump. Suppose that the wall is rectangular in cross-section, and is 2 m high. How thin would it have to be for the Beetle to have a 0.1% chance of tunneling through the wall, again assuming the car starts out moving at 10 kph?

6.7 *Why are universal conductance fluctuations universal? (a bit of statistics for you)* (Caution: math ahead.) In the coherent regime conductance is basically a diffraction and interference experiment, and fluctuations due to varying the relative phases of different trajectories are universal in magnitude. That is, the rms size of the fluctuations in the conductance within a coherent volume is always around $G_0 = 2e^2/h$. This problem lets you get a sense of the argument that universal conductance fluctuations are, in fact, universal. Assuming M input and output channels, we define the total transmittance $\bar{T} = \sum_m \sum_n T(m \to n)$ and the total reflectance as $\bar{R} = \sum_m \sum_n R(m \to n)$, where m and n run from 1 to M.

(a) Show that, for some quantity X with average value $\langle X \rangle$, the mean square fluctuations $\langle (X - \langle X \rangle)^2 \rangle \equiv \langle \delta X^2 \rangle = \langle X^2 \rangle - \langle X \rangle^2$.

(b) Start from the (two-terminal) Landauer formula for M channels, and define the normalized conductance $g \equiv G/G_0$. Working with g in terms of the total reflectance, use part (a) and the definition of reflectance to show that $\langle \delta g^2 \rangle = \langle \delta \bar{R}^2 \rangle$.

(c) Assume that the M^2 terms of the sum $\bar{R} = \sum_m \sum_n R(m \to n)$ are uncorrelated. In that case we can write $\langle \delta \bar{R}^2 \rangle = M^2 \langle \delta R(m \to n)^2 \rangle$. That is, the total mean square fluctuations include M^2 contributions, one for each combination of modes connecting inlet channel m with outlet channel n. Suppose there are P paths that connect inlet mode m to outlet mode n. If each path corresponds to an amplitude A_i, we can write $R(m \to n) = |\sum_i A_i|^2$. If the phases of the paths are completely random, we can write $\langle R(m \to n) \rangle = \sum_i \sum_j \langle A_i A_j^* \rangle = \sum_i |A_i|^2$. That is, the cross terms vanish unless $i = j$. Show, then, that $\langle R(m \to n)^2 \rangle = \sum_{i,j,k,l} \langle A_i A_j A_k^* A_l^* \rangle = 2 \langle R(m \to n) \rangle^2$.

Thus, $\langle \delta R(m \to n)^2 \rangle = \langle R(m \to n) \rangle^2$. For a sample that is many mean free paths long, $\langle R(m \to n) \rangle \approx 1/M$. So, $\langle \delta R(m \to n)^2 \rangle \approx 1/M^2$. Remember, though, that there are M^2 terms like that one that sum to give the total fluctuations. Therefore, $\langle \delta g^2 \rangle \approx 1$, and we see that the typical conductance fluctuation size is $2e^2/h$, under very general conditions.

6.8 *Quantum effects at room temperature?* Atomic-scale junctions are a way to make nanodevices with conduction dominated by a truly nanoscale region. Conductance fluctuations are an example of a measurable phenomenon that can be used to assess the importance of quantum effects in conduction.

(a) Read the paper by Ludoph *et al.* [475]. These authors measured conductance fluctuations as a function of bias voltage in metal nanojunctions. Dont worry if the detailed math is over your head. What is the origin of the fluctuations? What is their main observation? This paper effectively makes an important point often neglected in the world of molecular electronics: you can't just worry about the molecule when trying to understand the conduction process; rather, you have to worry about the details of the metal electrodes as far as a coherence length away from the molecule!

(b) Now look at Untiedt *et al.* [476]. In particular, look at the lower panel of Fig. 1. Is there nonlinearity in the conduction? Assuming you could measure this kind of data in many junctions of different conductances, in a couple of sentences describe how could you test for the existence of conductance fluctuations in this system at room temperature. Quantum effects like this can be detected even at room temperature, provided conduction is controlled by a sufficiently small region. Do you think these effects would be detectable in ultrascaled (few-nm) Si transistors?

6.9 *Coulomb staircase* Consider the double-barrier structure shown in Fig. 6.36. Suppose that one tunnel barrier is considerably more transparent than the other (that is, $R_1 \gg R_2$), but assume that the capacitances are equal ($C_1 = C_2 = C$). Consider what happens at $T = 0$ when the applied bias is increased across $V_c = e/2C$. When $V < V_c$, no current flows. When V just exceeds V_c, the system enters the sequential tunneling regime discussed in the text, where the island population is increased by one. Considering Eq. (6.103), *find an expression* for the jump in the current as V moves across V_c. Note that since $R_1 \gg R_2$, junction 1 controls the flow of current. As V is increased further, each time there is a change in steady-state island population, n, there is a jump in the current of roughly the same magnitude. This leads to the "Coulomb staircase", a major experimental signature of Coulomb blockade physics. Note that something qualitatively identical takes place when $R_1 = R_2$ but there is a large mismatch in C_1 and C_2.

6.10 *Electrometry with SETs* We've seen that the source–drain conductance of an SET can be strongly modulated by the movement of a single electron on or off the gate electrode. This may be used for electrometry by capacitively coupling the gate electrode to the object whose charge we want to monitor. Now let's plug in the numbers to see just how good an electrometer one can make.

(a) Suppose that the SET in question is of the Al/Al$_2$O$_3$ variety, made by shadow evaporation onto an insulating substrate that we'll ignore for now. Suppose that the tunnel junctions connecting the island to the source and drain are identical, and that each has a tunneling resistance of 1 MΩ, a cross-sectional area of 20 nm $\times$ 10 nm, and an oxide ($\kappa = 9$) thickness of 2 nm. The nearby gate is coupled to the island with a capacitance of Cg $= 10^{-18}$ F. What is the charging energy of this SET? What is the maximum conductance of the SET?

(b) Suppose $T = 4.2$ K (liquid helium temperature). Assume that the conductance is at a Coulomb blockade maximum when $V_G = 0$. As the gate voltage is increased, the conductance drops for a while, obeying Eq. (6.114). Using that formula, at what gate voltage is $\partial G/\partial V_G$ maximized? What is $\partial G/\partial V_G$ there? A nice way to use an SET as an electrometer is to sit at this gate voltage, and monitor the conductance of the SET as the system whose charge is to be measured is varied in some way.

(c) The electrometer is capacitively coupled to a small test object with self-capacitance $C_{\text{self}} = 10^{-18}$ F and a coupling capacitance to the island of $C_{\text{couple}} = 10^{-19}$ F (small compared with C_G.). Now we can test our sensitivity. If we could place 0.01 electrons onto the test object, the test object potential changes by $0.01e/(C_{\text{self}} + C_{\text{couple}})$. That induces a polarization charge on the island of $Q_p = 0.01eC_{\text{couple}}/(C_{\text{self}} + C_{\text{couple}})$. That polarization charge, in turn, shifts the voltage position of the Coulomb blockade conductance peak by

Q_p/C_G. Using the result from part (b), how much does the conductance of the SET change as a result of changing the test object charge by 0.01 e? This is measurable!

6.11 *Molecular-scale SETs* Suppose one wanted to have an SET that operates (at least to some extent) at room temperature. One approach is to use a small molecule as the island.

(a) For E_c 0.02 $k_B T$ at room temperature, estimate the capacitance of the island. Assuming for the moment that the island can be modeled as a classical metal sphere of radius a located $d = 0.5$ nm away from three ground planes (the source, drain, and gate electrodes), what a is required to achieve the necessary capacitance, to first order in $a/2d$? Assume $\kappa = 3.9$, appropriate for SiO_2.

(b) How does a compare with a typical electronic wavelength in a metal?

(c) Using the mass of a free electron, what is an estimate for Δ in such a system?

7 Magnetism and magnetoelectronics

In our discussion of the electronic properties of bulk and nanoscale materials, we have so far concentrated on the charge carried by the electrons. However, the spin degree of freedom is also tremendously important, from the standpoints of fundamental science and technological impact. The collective response of electronic spins (in concert with their orbital motion) gives rise to the magnetic properties of materials. Magnetism's technological impact cannot be overstated, from the first primitive compasses thousands of years ago to advanced magnetic data storage and magnetic imaging techniques today.

In this chapter we introduce magnetism and magnetic materials, beginning with bulk systems. After discussing how to characterize magnetic response down to very small scales, we look at the effects of surfaces and interfaces, including confinement down to the nanoscale. After reviewing the history and evolution of magnetic data storage and magnetoelectronic devices, we consider nanoscale magnetism and its potential impact.

7.1 Definitions and units

One unfortunate and unavoidable challenge of discussing magnetism in materials is the unwieldy nature of magnetic units. This is largely the result of the historical development of our understanding of electricity, magnetism, and their relationship.

Let's begin by introducing the different types of vector field relevant to magnetism. The truly fundamental field is $\mathbf{B}$, the *magnetic induction* or *magnetic flux density* (though, truth be told, most physicists would call this the magnetic field, as was done in the previous chapter). This is the field that shows up in the Lorentz force law and determines the force on a moving charge. It is also the field that determines magnetic resonance frequencies, and is related to the vector potential by $\nabla \times \mathbf{A} = \mathbf{B}$. The SI unit of $\mathbf{B}$ is the Tesla. The Earth's magnetic field is about 3×10^{-5} T over most of the planet. The cgs unit of $\mathbf{B}$ is the Gauss, and is more conveniently sized: $1 \text{ G} = 10^{-4}$ T. An important condition to recall is one of Maxwell's equations:

$$\nabla \cdot \mathbf{B} = 0. \tag{7.1}$$

This is another way of saying that there are no free magnetic monopoles (the magnetic analog of free charges). The magnetic induction can include contributions from spins, free and bound currents, etc.

A second field that crops up all the time is $\mathbf{H}$, the *magnetic field*, which is in some sense the change in $\mathbf{B}$ due to materials. The unit of $\mathbf{H}$ is Ampères/m. One source of $\mathbf{H}$ is free currents, as shown by Ampère's Law:

$$\nabla \times \mathbf{H} = \mathbf{J}, \tag{7.2}$$

where $\mathbf{J}$ is the current density. In general, in SI,

$$\mathbf{H} = \frac{\mathbf{B}}{\mu_0} - \mathbf{M}, \tag{7.3}$$

where $\mathbf{M}$ is the magnetization, as we shall discuss below, and $\mu_0 = 4\pi \times 10^{-7}$ N/A^2 is the *permeability of free space*. In a material where $\mathbf{M}$ is simply proportional to $\mathbf{B}$, then one can recast Eq. (7.3) as

$$\mathbf{B} = \mu\mathbf{H} = \mu_r\mu_0\mathbf{H}. \tag{7.4}$$

Here μ is the permeability of the material, and μ_r is the *relative permeability*, in analogy with the relative dielectric constant. In cgs, the unit for $\mathbf{H}$ is the Oersted. Note that cgs units are designed so that the permeability of free space is one; therefore in vacuum one Oersted is formally the same as one Gauss.

The remaining vector field, $\mathbf{M}$, is the *magnetization* of some material. The units of $\mathbf{M}$ are also A/m, and it has a very physical interpretation: magnetization is the magnetic dipole moment per unit volume of a material. We will discuss magnetic dipoles below. As we shall see, some materials develop spontaneous, nonzero $\mathbf{M}$ even in the absence of an external magnetic field.

Other ("nonmagnetic") materials only develop $\mathbf{M}$ in response to an externally applied $\mathbf{H}$-field. The *magnetic susceptibility*, χ, is defined as the coefficient of the linear term in their relationship:

$$\chi \equiv \frac{\partial \mathbf{M}}{\partial \mathbf{H}}. \tag{7.5}$$

If this is the only important term, then the permeability introduced above is a sensible quantity:

$$\mu = \mu_0(1 + \chi). \tag{7.6}$$

In general, χ and μ are tensorial, though we won't worry about that for now. Nonmagnetic materials are often classified as *diamagnetic* ($\chi < 0$) or *paramagnetic* ($\chi > 0$). Note that susceptibility as defined in Eq. (7.5) is dimensionless in SI units. Often χ is reported as the dimensionless χ times the molar volume of a material (m^3/mole).

Magnetic dipole moment, $\mathbf{m}$, may be modeled as a little loop of current, with $\mathbf{m} = IA\mathbf{n}$, where $\mathbf{n}$ is a unit vector normal to the loop, I is the current, and A is the loop area. The units of $\mathbf{m}$ are Am^2 = J/T. Particles with intrinsic angular momentum (spin) have associated magnetic dipoles as well. A useful quantity to recall is the *Bohr magneton*, $\mu_B = e\hbar/2m = 9.27 \times 10^{-24}$ J/T. The magnetic moment of the electron is related to the Bohr magneton by the *g*-factor,

$$\mathbf{m}_e = -g\mu_B(\sigma/\hbar), \tag{7.7}$$

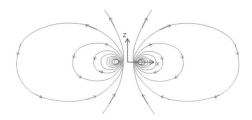

Fig. 7.1   B-field from a magnetic dipole pointed in the $+z$ direction. The dipole is shown as a current loop.

where σ is the electron spin angular momentum. For free electrons, $g = 2.002319....$

The torque on a magnetic dipole is

$$\tau = \mathbf{m} \times \mathbf{B}, \tag{7.8}$$

while the force on a magnetic dipole (in the current loop picture) is

$$\mathbf{F} = \nabla(\mathbf{m} \cdot \mathbf{B}). \tag{7.9}$$

This latter result is related to the potential energy of a dipole in an externally applied B-field,

$$U = -\mathrm{m} \cdot \mathrm{B}. \tag{7.10}$$

The magnetic induction due to a dipole at distances large compared to the dipole's size (in the case of a current loop) is given by

$$\mathbf{B(r)} = \mu_0 \frac{3(\mathrm{m} \cdot \mathbf{r})\mathbf{r} - r^2\mathbf{m}}{4\pi r^5}. \tag{7.11}$$

Note that the magnitude of this field drops off like $1/r^3$. This dipolar field is shown in Fig. 7.1.

Because of Eq. (7.8), a magnetic dipole placed in a non-aligned B-field will precess. For a particle of mass m, charge $|e|$, and g-factor g, that precession takes place at the *Larmor frequency*, $\omega_{\mathrm{L}} = (ge/2m)B$. In electron spin resonance or nuclear magnetic resonance experiments, the observable is often the magnetic field due to the precessing dipoles.

The current loop picture of magnetic dipoles is useful. A group of identical dipoles in a plane can be replaced by one big effective dipole with the same current circulating around the perimeter, as shown in Fig. 7.2. Likewise, a 3d stack of aligned dipoles (representing a block of uniformly magnetized material) can be replaced by thinking about a sheet current running around the perimeter, for the purposes of calculating what the far-field **B**-field looks like.

Alternately, one can think of a magnetic dipole as analogous to an electric dipole, comprising fictitious magnetic monopoles of opposite "magnetic charge" separated by a small distance. In this picture, a block of uniformly magnetized material can be represented by sheets of magnetic charge density at the top and bottom surfaces, again for the purpose of calculating the far-field **B**.

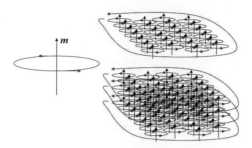

Fig. 7.2

A single magnetic dipole envisioned as a current loop. One can consider a 2d or 3d array of aligned loops as a larger loop with a circulating surface current.

What is B at the surface of a long, uniformly magnetized body with a magnetization **M** aligned along the system's length, in the absence of other fields? Just outside the top surface of the magnetic material, $\mathbf{B} = \mu_0 \mathbf{M}$. This follows from Eq. (7.1).

Some common materials (e.g., silicon) are diamagnetic. The classical explanation for diamagnetism lies in Lenz's Law, which says that when an external magnetic field is ramped up on a system with a circulating current, the response due to the induced circumferential electric field (from Faraday's Law) changes the current to oppose the external magnetic flux penetrating the system. Quantum mechanically, the right way to think about this is from the point of view of Eq. (7.10). For a diamagnetic system, increasing an applied **B** should increase the magnetostatic energy by generating additional magnetic dipole moment *opposed* to **B**. One can look at the electronic structure of a material and see what happens to the energies of the filled electronic states in this situation. In second order perturbation theory (see, e.g., Marder [15], chapter 25), there are two terms in the **B**-induced changes in the energy of opposite sign. If the positive term dominates (as is common when all orbitals are filled and there are no unpaired electrons), then the system exhibits *Larmor diamagnetism*. If the negative term dominates, then the system exhibits *van Vleck paramagnetism*.

In the case of materials with mobile charge carriers (metals or doped semiconductors), there is a further possibility called *Landau diamagnetism*. Landau diamagnetism results from quantization of electron orbits into Landau levels separated by the cyclotron energy scale, $\hbar\omega_c = \hbar eB/mc$. For a detailed explanation, see Marder, chapter 25, section 25.3.2 [15].

In addition to van Vleck paramagnetism, there are two other common origins of paramagnetic response. *Curie paramagnetism* results from the tendence of localized spins to align with an externally imposed magnetic field. Consider n noninteracting, localized moments per unit volume of spin J, g-factor g, at temperature T. In a B-field, the magnetic moments will tend to align with B to lower their energy, with the average alignment improving as T is decreased. As is derived in many texts on statistical physics, the magnetization of such a system is:

$$M = n g \mu_B J \mathcal{B}_J(x), \tag{7.12}$$

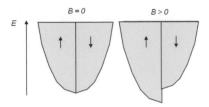

Fig. 7.3 Pauli paramagnetism. An applied magnetic field shifts the spin-resolved electronic density of states at the Fermi level via the Zeeman effect, leading to a net magnetization proportional to B.

where $x \equiv g\mu_B JB/k_B T$, and $\mathcal{B}_J(x)$ is the Brillouin function:

$$\mathcal{B}_J \equiv \frac{2J+1}{2J}\coth\left(\frac{2J+1}{2J}x\right) - \frac{1}{2J}\coth\left(\frac{x}{2J}\right). \tag{7.13}$$

In the limit $x \ll 1$, we can expand $\mathcal{B}_J$ in Eq. (7.12) and get

$$M \approx ng^2\mu_B^2\frac{\mu_0 H}{k_B T}\frac{J(J+1)}{3}, \tag{7.14}$$

where we've assumed that the contribution of M to the total B is small (consistent with our original assumption of noninteracting spins). From this we see that Curie paramagnets have susceptibilities like

$$\chi \approx ng^2\mu_B^2\frac{\mu_0}{k_B T}\frac{J(J+1)}{3} \sim \frac{1}{T}. \tag{7.15}$$

Curie susceptibilities grow inversely with temperature as T is reduced (as long as $k_B T$ remains large compared to $g\mu_B JB$).

For itinerant electrons (e.g., the Fermi gas discussed in Section 2.1.4), there is an additional effect called *Pauli paramagnetism*. As shown in Fig. 7.3, one can separately consider the densities of states for spin-up and spin-down electrons. Applying a **B**-field makes it energetically favorable for electrons to flips their spins and align with the field. However, only those near the Fermi surface are free to do this; electrons deep in the Fermi sea have no states available to them. Using the density of states, $\nu(E)$, the Fermi distribution function, $f(E)$, for the electrons (suppressing the notation for the chemical potential, for clarity), and approximating $g \approx 2$, we find

$$n_{\text{up}} = \int_0^\infty \frac{\nu(E)}{2}f(E + \mu_B B)\mathrm{d}E,$$
$$n_{\text{down}} = \int_0^\infty \frac{\nu(E)}{2}f(E - \mu_B B)\mathrm{d}E. \tag{7.16}$$

Assuming that the electrons are degenerate, Taylor expanding each (assuming that the magnetic energy shift is small compared to the Fermi energy) and subtracting gives

$$M \approx \mu_B^2\nu(E_F)\mu_0 H. \tag{7.17}$$

In Pauli paramagnetism, χ is proportional to the density of states at the Fermi level, and is approximately temperature independent.

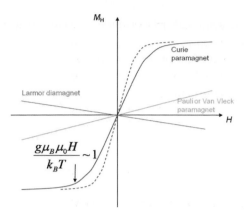

Magnetization vs. magnetic field for different magnetic responses. The dashed line shows the Curie paramagnet at a lower temperature than the relevant solid line.

Fig. 7.4

Figure 7.4 shows **M** as a function of **H** for the kinds of material that we have been discussing so far. As you can see, diamagnets are linear with negative slopes on such a plot, while Pauli/van Vleck paramagnets are linear with positive slopes. Curie paramagnets have a linear regime at comparatively low H values, with a temperature-dependent rollover to saturation at high fields. Lowering T steepens the slope at the origin and pushes the saturation to lower field values.

7.2 Magnetic order

Collective magnetic effects can emerge when spins are able to "talk" to each other. The simplest example of this is the exchange term in Eq. (2.92), which favors the alignment of electron spins. One way to think about this kind of exchange is that electrons with aligned spins cannot occupy the same spatial state (Pauli principle); thus aligning spins keeps electrons far from one another, lowering the overall total energy by reducing the contribution from the Coulomb repulsion. This kind of interaction is an example of *ferromagnetic exchange*, which would tend to align spins. It is also an example of *direct exchange*, because the term in the energy involves the direct spatial overlap of the wavefunctions of the two electrons whose spins are of interest.

There are other electronic processes that can lead to effective spin–spin interactions. For example, one can consider two electronic spins on atoms separated by an intervening nonmagnetic (no unpaired electrons) atom. There are "superexchange" processes (related to the higher order tunneling processes discussed in the previous chapter) that can couple the spins. Likewise, in materials with conduction electrons, there can be indirect exchange processes that couple localized spins via their mutual interactions with itinerant carriers. Depending on the precise system and processes involved, spins of interest can also experience interactions that favor *antialignment*. Such interactions are termed *antiferromagnetic*.

There are many, many models of magnetism, in which the complicated real-world interactions are often replaced by simple parametrized problems. Very often the spin–spin interactions are written in a form like

$$\mathcal{H}_{\text{spin}} = \sum_{i,j} J_{ij}\sigma_i \cdot \sigma_j, \tag{7.18}$$

where σ_i is the spin of the ith electron and J_{ij} is some constant that describes the effective exchange interaction between the ith and jth spins. Often one is concerned with a lattice of spins in space, and the sum runs only over neighboring sites. If J_{ij} is less (greater) than zero, then the interaction has ferromagnetic (antiferromagnetic) tendencies.

Rather than providing a comprehensive look at the theory of magnetism, I refer readers to any one of several books on the subject [477, 478, 479] as well as solid state physics texts. Instead, the important message to take away here is: *magnetic interactions between electronic spins (1) are fundamentally quantum mechanical in nature; (2) depend sensitively on the details of electronic wavefunctions and therefore atomic positions; and (3) typically involve electron–electron interactions and the Pauli principle at their root.*

One can get a sense of how effective spin–spin interactions can lead to some kind of long-ranged magnetic order (in the sense that orientations of spins within a material can be correlated with one another across many lattice spacings) from a simple toy model. Start from Curie paramagnetism, with some susceptibility χ_0, and add in a "molecular field", $\mathbf{H}_m$, in addition to any externally applied field. The molecular field is a manifestation of interactions between the spins, so that $\mathbf{H}_m = \eta\mathbf{M}$, where η is some proportionality constant that hides all the deep physics. Assuming that $\mathbf{M} = \chi_0(\mathbf{H} + \mathbf{H}_m)$, we get

$$M(1 - \eta\chi_0) = \chi_0 H. \tag{7.19}$$

This implies that there is an effective susceptibility,

$$\chi_{\text{eff}} = \frac{\chi_0}{1 - \eta\chi_0}. \tag{7.20}$$

Since $\chi_0 \sim 1/T$ at high temperatures, this implies that

$$\chi_{\text{eff}} \sim \frac{1}{T - T_c}, \tag{7.21}$$

where T_c is some critical *Curie temperature* (related to η). This expression is called the *Curie–Weiss Law*, and it provides a valuable hint to the effect of spin–spin interactions. The effective susceptibility diverges as $T \to T_c$. A truly divergent χ would imply that the system develops a spontaneous $\mathbf{M}$ even in the absence of $\mathbf{H}$. Positive values of T_c as inferred from Curie-Weiss fits to the susceptibility are associated with ferromagnetism (and closely related ordered states, as we'll discuss shortly). Negative values of T_c inferred from χ data imply an onset of antiferromagnetism near $|T_c|$.

In ferromagnetic materials that are conductive, the mobile charge carriers can possess a net spin polarization. This is shown in Fig. 7.5, for three common ferromagnetic metals, Fe, Ni, and Co. This shifting of the spin-up and spin-down (relative to the magnetization direction) densities of states, $\nu(E)$, relative to one another takes place

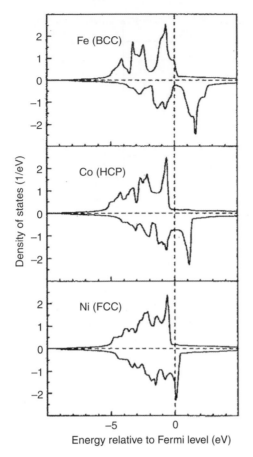

Density of states (1/eV)

Energy relative to Fermi level (eV)

Fe (BCC)

Co (HCP)

Ni (FCC)

Spin-resolved density of states for three common itinerant ferromagnets. From [480].

Fig. 7.5

when the ferromagnetic (FM) exchange interaction is sufficiently strong that it can offset changes in the electron kinetic energy. In these FM transition metals, the primary sources of the magnetism are the unpaired d electrons. Some of the bands crossing the Fermi level in these materials originate with those same d orbitals. The spin splitting of the conduction band is sometimes called Stoner splitting. One can integrate $\nu_\uparrow(E)$ and $\nu_\downarrow(E)$ up to the Fermi level, and determine the respective spin populations. It's not easy to see from the figure, but in Co, for example, the *majority spin* population (upper, as shown) actually has a smaller density of states at the Fermi level than the *minority spin* population.

An extreme version of this is a *half-metal*. As shown in Fig. 7.6, one spin band lies entirely below E_F, and is completely full, while the other spin band crosses E_F. This means that the mobile charge carriers are fully spin polarized (and, as shown, antialigned with the majority spin carriers that dominate the total magnetization). As we shall see when discussing magnetoresistive effects, half-metals [481] are of much technological interest.

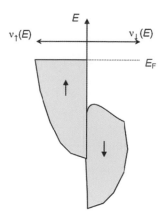

Schematic spin-resolved density of states for a half-metal.

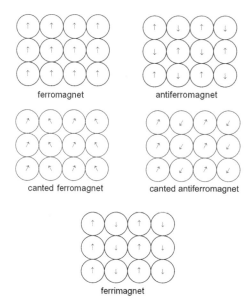

Simplified depictions of various types of magnetic order, shown via two-dimensional lattices of localized spins.

In addition to ferromagnetism and antiferromagnetism, Fig. 7.7 schematically shows several other possible forms of long-range magnetic order. The type of order and its relation to a material's crystallographic axes depend in detail on the electronic structure. Predicting such quantities from first principles is a major theoretical challenge.

For now, we focus on the phenomenology of ferromagnetism, as it has been of the greatest technological utility. Consider sweeping an externally applied magnetic field **H** along some axis, and measuring the magnetization response **M**. In a macroscopic ferromagnet, the result would be something like what is shown in Fig. 7.8. This is an example of a *hysteresis loop*. The material's response, **M**, depends not just on the

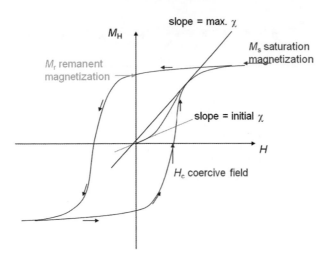

A typical hysteresis loop of a ferromagnetic material.

instantaneous value of $\mathbf{H}$, but on its past history. At sufficiently high fields $\mathbf{M} \to \mathbf{M}_{\text{sat}}$, a *saturation magnetization*. If the field is swept back to zero, the material retains a *remanent magnetization*, $\mathbf{M}_r$. The value of applied $\mathbf{H}$ required to flip the sign of $\mathbf{M}$ is called the *coercive field*, $\mathbf{H}_c$. In general, *hard* magnetic materials have comparatively large value of $\mathbf{H}_c$ and $\mathbf{M}_{\text{sat}}$, while the opposite is true for *soft* magnetic materials.

Because of the electronic origins of ferromagnetism, in bulk FM materials it can be energetically favorable to have $\mathbf{M}$ lie along particular crystallographic directions. This is called *magnetocrystalline anisotropy*, and the energetically favored direction is called the *easy axis*. There are two particularly common forms of this. Define u as the energy density specifically due to the misalignment of $\mathbf{M}$. In the case of cubic anisotropy, it is useful to define the unit vector $\hat{\mathbf{m}} = \mathbf{M}/|\mathbf{M}|$. If the axes of a cubic lattice are identified as 1, 2, and 3, then

$$u = K_{c1}(\hat{m}_1^2 \hat{m}_2^2 + \hat{m}_1^2 \hat{m}_3^2 + \hat{m}_2^2 \hat{m}_3^2) + K_{c2}(\hat{m}_1^2 \hat{m}_2^2 \hat{m}_3^2). \qquad (7.22)$$

Depending on the choices of K_{c1} and K_{c2}, the minimum energy direction can be made to lie along the principal axes, or along the diagonals. For example, for iron the anisotropy constants are $K_{c1} = 4.8 \times 10^4$ J/m^3 and $K_{c2} = -1.0 \times 10^4$ J/m^3, while in nickel they are $K_{c1} = -4.5 \times 10^3$ J/m^3 and $K_{c2} = -2.3 \times 10^3$ J/m^3.

Another common magnetocrystalline anisotropy is *uniaxial*:

$$u = K_{u1} \sin^2 \theta + K_{u2} \sin^4 \theta. \qquad (7.23)$$

For positive K_{u1}, K_{u2}, the system pays an energetic penalty if $\mathbf{M}$ is misdirected from the particular easy axis by an angle θ.

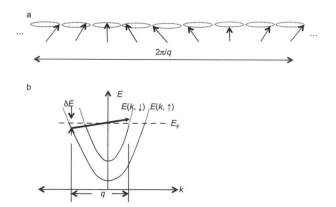

Fig. 7.9 Excitations of ferromagnets. (a) A 1d example of a spin wave or magnon, where the local spin is tilted away from the overall magnetization direction in a manner periodic in space with wavevector q. (b) A Stoner excitation, when an electron in one spin-split band makes a transition into an empty state of the other spin-split band, at some energy cost δE and some crystal momentum change q.

7.2.1 Excitations

One can ask, what are the low energy excitations of the ferromagnetic state? One type of excitation is a *spin wave* or *magnon*. Figure 7.9(a) shows a cartoon of such an excitation at some instant in time. The local spin direction is tilted away from the direction of **M** in a spatially periodic manner that is characterized by some wavevector, **q**. As a function of time, each local moment precesses about the overall **M** direction. In the long wavelength limit, neighboring spins have orientations that hardly differ, and hence the exchange energy cost of such a spin wave is comparatively low. A quantum treatment of spin waves [482] finds a dispersion relation between the spin wave energy, $\hbar\omega$, and the wavevector **q** given roughly by

$$\omega = Aq^2, \tag{7.24}$$

where A is called the spin wave stiffness. In the $q \to 0$ limit, the energy cost of a spin wave excitation goes to zero. As with phonons and electronic quasiparticles, spin wave dispersion is modified by the spatial periodicity of the material's crystal lattice (see Section 2.2.6). In the sense that they are well defined collective modes with distinct quantum numbers, spin waves are another kind of quasiparticle, hence their other name, magnons. Due to magnetocrystalline anisotropy, spin waves can also be gapped, meaning that Eq. (7.24) should be modified by adding an extra constant term to the right side. In the gapped case, there is a nonzero energy cost to $q = 0$ spin waves.

In the case of ferromagnets with itinerant electrons, for example the metals Ni, Fe, and Co, there is an additional set of excitations that is allowed. A *Stoner excitation* involves the scattering of an electron from the majority spin band into the minority spin band, as shown in Fig. 7.9(b). A Stoner excitation requires the relevant electron to flip its spin, and may also require a finite amount of crystal momentum, $\hbar \mathbf{q}$, to make up for the difference between the $\mathbf{k}$-vectors of the initial majority spin and final minority spin electronic states. There are a continuum of these excitations, and the $\mathbf{q} = 0$ excitations have an energy that corresponds to the exchange splitting of the bands.

7.3 Energy and magnetic configurations

Given ferromagnetic material with some geometry, one is often interested in figuring out the likely equilibrium configuration of the magnetization. Unsurprisingly, this is often a question of minimizing the total energy of the magnetic system. The total energy includes contributions intrinsic to the magnetic material, such as the magnetocrystalline anisotropy of Eqs. (7.22 and 7.23). Likewise, the total energy, in the presence of an external field, $\mathbf{H}_{\text{ext}}$, includes the interaction energy of the magnetic material with that external field:

$$U_{\text{ext}} = -\mu_0 \int \mathbf{H}_{\text{ext}} \cdot \mathbf{M} dV,\qquad(7.25)$$

where we integrate over the volume of the magnetic material. This follows naturally from Eq. (7.10) and our definition of magnetization.

We must also keep track of the energy from the magnetic induction produced by the magnetized material itself. For example, a uniformly magnetized bar magnet produces a roughly dipolar $\mathbf{B}$-field that extends throughout space, and that magnetic energy density must be taken into account. Generically, the energy due to a magnetic field in some volume is

$$U = \frac{1}{2} \int_{\text{all space}} \mathbf{H} \cdot \mathbf{B} dV = \frac{1}{2} \mu_0 \int_{\text{all space}} H^2 dV.\qquad(7.26)$$

As mentioned in Section 7.1, we can model a magnetized piece of material either by current sheets or through fictitious magnetic surface charge, as shown in Fig. 7.10. We can find the magnitude of that fictitious charge density by remembering Eq. (7.1). Defining the *demagnetizing field*, $\mathbf{H}_d$, as the magnetic field produced by the effective magnetic charge, Eq. (7.1) gives us

$$\nabla \cdot (\mathbf{H}_d + \mathbf{M}) = 0 \rightarrow \nabla \cdot \mathbf{H}_d = -\nabla \cdot \mathbf{M}.\qquad(7.27)$$

This resembles the Maxwell equation for Gauss' Law, for an effective magnetic (volume) charge density of $-\nabla \cdot \mathbf{M}$. At the surface of a magnetized material, the effective magnetic surface charge density reduces to $\mathbf{M} \cdot \mathbf{n}$, where $\mathbf{n}$ is the unit vector normal to the surface.

So, for a particular geometry of magnetized material, one can compute the effective magnetic charge density, and then solve Eq. (7.27) with appropriate boundary conditions to find $\mathbf{H}_d$ everywhere. Doing this for the case of a magnetized bar, as in Fig. 7.10, shows that $\mathbf{H}_d$ inside the magnetized material is in the opposite direction of $\mathbf{M}$; hence the name

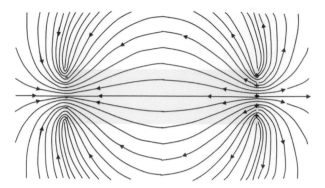

H-field from a bar uniformly magnetized to the right, as approximated by an effective magnetic surface charge. Note that the field inside the bar points to the left, the opposite direction as the magnetization; hence the term "demagnetizing field".

"demagnetizing field". The energy associated with the demagnetizing field is

$$U_{\mathrm{d}} = \frac{1}{2}\mu_0 \int_{\text{all space}} \mathbf{H}_{\mathrm{d}}^2 \mathrm{d}V, \tag{7.28}$$

which may not be easy to evaluate. Fortunately, this integral is exactly equal to

$$U_{\mathrm{d}} = -\frac{1}{2}\mu_0 \int_{\text{mag mater}} \mathbf{H}_{\mathrm{d}} \cdot \mathbf{M} \mathrm{d}V, \tag{7.29}$$

where this integral is only over the volume of magnetized material. (We can think of this latter integral as keeping track of the work required to "put together" all the little effective, interacting dipoles that make up the magnetic material. The factor of 1/2 ensures that we don't double-count interdipole interactions.)

In general, one can related $\mathbf{H}_{\mathrm{D}}$ to $\mathbf{M}$ by the *demagnetizing factor*, which is really a tensor, $\hat{\mathbf{N}}$:

$$\mathbf{H}_{\mathrm{d}} = -\hat{\mathbf{N}} \cdot \mathbf{M}. \tag{7.30}$$

One can calculate $\hat{\mathbf{N}}$ for given geometries of magnetic material. For certain common shapes, $\mathbf{M}$ is parallel to $\mathbf{H}_{\mathrm{d}}$, so $\hat{\mathbf{N}}$ reduces to a simple scalar quantity. For a uniformly magnetized sphere, $N = 1/3$. For long rods (length $L \gg$ radius r) with $\mathbf{M}$ parallel to the rod axis, $N \to 0$, while for $\mathbf{M}$ perpendicular to the rod axis, $N \to 1/2$. The parallel-to-axis result is intuitively clear. For a very long rod, the surface "charges" at the poles are separated by L, and so the resulting H_{d} inside the rod falls like $\sim 1/L$, clearly approaching zero as L becomes infinite. Similarly, for a large, thin plate, for $\mathbf{M}$ perpendicular to the plate, $N \approx 1$, while for $\mathbf{M}$ in the plane of the plate, $N \to 0$ as the plate's lateral extent approaches infinity.

Note that $N = 0$ implies, via Eq. (7.29), an energy cost due to stray fields that is *zero*. Conversely, a positive N implies (via Eq. (7.30)) $\mathbf{H}_{\mathrm{d}}$ directed opposite to $\mathbf{M}$, which in turn gives a *positive* energy cost in Eq. (7.29). This dependence of stray field energy cost on magnetization orientation is the origin of *shape anisotropy*. All other things being equal, it is energetically favorable (because of this stray field energy) for $\mathbf{M}$ to be oriented along the axis of long, thin rods. Likewise, shape anisotropy implies that thin magnetic films "prefer" to have $\mathbf{M}$ lie in the plane.

This is easy to see in the thin plate case. If the plate material has saturation magnetization M_{sat} oriented at an angle θ away from the normal of the plate, then the resulting $\mathbf{H_d} = -M_{\text{sat}} \cos \theta \, \hat{\mathbf{n}}$. Therefore, the resulting energy density is $u_d = (\mu_0/2) M_{\text{sat}}^2 \cos^2 \theta$.

The situation in real magnetic materials can be quite complicated, with both magnetocrystalline and shape anisotropies coming into play. Moreover, because of the role of crystal structure in influencing electronic structure and therefore exchange processes, there can often be a strong coupling between magnetization and elastic deformation (strain). One consequence of this magnetoelastic physics is *magnetostriction*, in which reorienting the magnetization of a ferromagnetic material results in an elastic deformation, a change in the lattice constant of the material and/or the associated accumulation of internal stresses. Magnetostriction may also be used intentionally. For example, built-in strain in magnetic thin films deposited on non-lattice-matched substrates can alter the magnetocrystalline anisotropy, reorienting the easy axis or "pinning" the magnetization in a particular direction. It is possible to describe magnetoelastic physics in terms of some effective $\mathbf{H}$-field that is due to the local stress and strain of the material.

Finally, in the general case of material with a *nonuniform* magnetization $\mathbf{M}(\mathbf{r})$, there is an energy cost for having a spatially varying $\mathbf{M}$, due to the exchange physics of Eq. (7.18). This is often written as

$$u_{\text{ex}} = \tilde{A}(\nabla \mathbf{m})^2, \tag{7.31}$$

where $\tilde{A}$ is the *exchange stiffness constant* and u_{ex} is the energy density from the magnetization nonuniformity.

7.3.1 Domains

Given that the Curie temperature for iron is 1073 K, and its saturation magnetization is $\mu_0 M_{\text{sat}} = 0.17$ T, why don't we see large-scale magnetic effects from any iron bar or nail all the time? The demagnetizing field analysis above doesn't address this because it was presented assuming uniformly magnetized material. The answer is that real magnetic materials often are not uniformly magnetized. Instead the material spontaneously breaks up its magnetization into *domains*. Inside each domain the magnetization is uniform. However, the relative orientation of $\mathbf{M}$ varies from domain to domain.

The formation of domains does have some energetic cost. The exchange interaction in a ferromagnet aligns spins because that reduces the total energy of the system. At a *domain wall*, the boundary between two domains, perfect ferromagnetic alignment clearly cannot hold, and therefore costs some exchange energy. However, domain formation can be favored if that exchange cost is balanced by a net savings in stray field energy. Figure 7.11 shows the effect of domain formation in a macroscopic ferromagnet.

There are different kinds of domain walls that are possible. Figure 7.12 shows two common examples of 180 degree domain walls. In a *Bloch wall*, the magnetization reverses direction by rotating out of plane, over some number of lattice sites. In a *Néel wall*, the

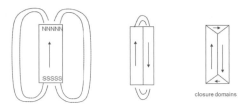

Fig. 7.11 H-field from a bar uniformly magnetized upwards, as approximated by an effective magnetic surface charge. Note that the field inside the bar points up, the opposite direction to the magnetization; hence the term "demagnetizing field".

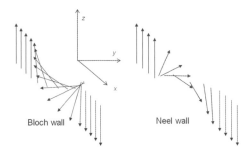

Fig. 7.12 Domain walls. With **M** initially pointing along the z-axis, in a Bloch wall, **M** locally rotates in the yz-plane to reverse direction, as one moves along x. In contrast, in a Néel wall, **M** instead rotates in the xz-plane as one moves along x.

magnetization rotates in its plane to reverse direction. The type of wall observed and the thickness of the wall depend on the detailed energetics of the material.

Because of the strength of the exchange interaction, which wants to align neighboring spins, abrupt (on the atomic scale) changes in magnetization direction cost a lot of energy. As a result, the local magnetization spreads the change in direction out over some distance, the domain wall thickness. Likewise, magnetocrystalline and geometric anisotropies also contribute to the energetic cost of altering the local direction of **M**. The domain wall thickness and the type of domain wall that result are the outcome of competition between these various contributions to the energy.

We can do a model calculation to get some feel for the numbers at work here. The exchange energy for a pair of spins is

$$U_{ex} = (2JS^2) - 2J\mathbf{S}_1 \cdot \mathbf{S}_2, \tag{7.32}$$

where we are working with spin S moments and we have chosen the zero of energy to correspond to aligned moments. A not-crazy estimate relating J to the Curie temperature is $J \approx 0.3k_B T_c$. For a small change in angle $\delta\theta$ between nearly-aligned neighboring spins, $U_{ex} \approx JS^2(\delta\theta)^2$. If a total rotation of angle π is spread out over a 1d array of $N + 1$ spins, then the total exchange cost for that array would approximately be

$$U_{ex,\ 1d} \approx NJS^2(\pi/N)^2. \tag{7.33}$$

Assuming a crystal built out of such 1d chains spaced in two dimensions by the lattice spacing a, Eq. (7.33) implies an energy density per domain wall area of $\pi^2 JS^2/(Na^2)$ Joules/m^2 for a domain wall N atoms thick.

Suppose we have a simple uniaxial anisotropy in our model, so that the initial and final (oppositely directed) magnetizations point along the easy axis, but our in-the-domain-wall spins do not. Working from the first term in Eq. (7.23), the anisotropy energy per unit volume would be given by the average of $K_{u1} \sin^2 \theta$ as θ passes from zero to π through the domain wall. Doing this average gives an anisotropy contribution to the energy per unit wall area of $K_{u1} Na$ for an N-atom thick wall, with lattice parameter a. The total areal energy density of the model domain wall is then

$$u_{\text{2d, tot}} = \frac{\pi^2 JS^2}{Na^2} + K_{u1} Na. \tag{7.34}$$

The ground state of the system would be the one with wall thickness N that minimizes Eq. (7.34). That gives

$$N_{\min} \approx \sqrt{\pi^2 JS^2/(K_{u1}a^3)}, \tag{7.35}$$

and the domain wall thickness would be $N_{\min}a$. If one plugs in reasonable numbers for iron, one finds a domain wall thickness of around 100 nm. Notice that this model predicts that materials with lower anisotropy would have thicker domain walls. This is, indeed, what is observed. Nickel/iron alloys designed to have nearly zero magnetocrystalline anisotropy can have domain walls as thick as 1 μm [478]. In addition to the magnetic energies used in this calculation, magnetoelastic contributions can further complicate the issue.

When an external **H**-field is swept slowly, domains will continuously rearrange themselves if possible so that the total energy of the system will be minimized. For example, see Fig. 7.13. The relative sizes of different domains can change via the propagation of domain walls. Alternately, individual domains can reorient their magnetizations.

Sometimes it can be useful to think of domain walls as discrete entities of interest, rather than thinking about the magnetization of the domains. It may be possible, for example, to recast the problem of magnetization reversal in a nanowire instead as the problem of the motion of a domain wall through the system. The domain wall experiences some effective potential. An externally applied **H**$_{\text{ext}}$ that encourages growth of one domain can be thought of as tilting that potential, exerting an effective force on the domain wall that bounds the domain, pushing that wall along. We will return to this picture later.

This domain wall-centric point of view does provide insight into one particular property of domains in ferromagnets. Domain walls do not always move smoothly under the action of external fields. In the potential language above, domain walls can interact with disorder and surfaces. Microscopically, this isn't surprising, since surface atoms have different coordination and therefore different exchange couplings from bulk atoms. Likewise, because of magnetoelastic couplings, surfaces and lattice defects, with their associated strain fields, can modify the effective potential that influences the domain wall. The net effect of this is that domain walls can be *pinned*, stuck in place even in the presence of an **H**-field that tries to push them.

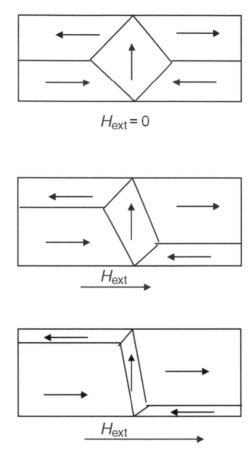

Fig. 7.13   Domain reconfiguration. Domain walls move as an external **H** is increased, to minimize the total energy.

One consequence of domain wall pinning is shown in Fig. 7.14. Zooming in closely on the $\mathbf{M}-\mathbf{H}$ curve of a typical bulk ferromagnet, it becomes clear that the hysteresis loop is not smooth. Rather, it consists of many small, discrete jumps of $\mathbf{M}$, as domain walls are pinned and then snap free. If a pickup coil is wound around a piece of the magnetic material as this takes place, currents are induced in the pickup coil as the domain walls unpin and move. Plugging such a coil into a loud speaker produces a clear, audible noise made up of clicks from the motion of individual domain walls. This is called the *Barkhausen effect*, and was discovered in 1919 [483].

7.3.2 Dynamics of magnetization

Thus far we have been discussing magnetostatics, looking at the stable magnetization configurations that are favored under various conditions of geometry and external fields. For later topics we need to have some sense of the dynamics of magnetization, the time-dependent response of the system when it is not in static equilibrium.

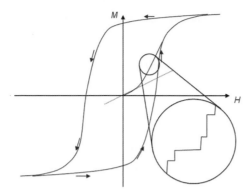

The Barkhausen effect. Magnetization changes in macroscopic, multidomain materials can show discrete jumps as   **Fig. 7.14**
domain walls unpin.

The basic dynamics come from Eq. (7.8), and the relationship between magnetic moment and angular momentum. If the magnetization is not colinear with the local effective magnetic field, then

$$\frac{\partial \mathbf{M}}{\partial t} = -\gamma \mathbf{M} \times \mathbf{H}_{\text{eff}}, \tag{7.36}$$

where γ is the *gyromagnetic ratio*,

$$\gamma \equiv \frac{\mu_0 g e}{2 m_{\text{e}}^*}. \tag{7.37}$$

Here g is the g-factor of the electrons that contribute to the magnetization, and can differ from the free electron value due to spin–orbit physics and crystal field effects. The local effective magnetic field that acts on a small region of magnetization is

$$\mathbf{H}_{\text{eff}} = \mathbf{H}_{\text{ext}} + \mathbf{H}_{\text{d}} + \mathbf{H}_{\text{ex}} + \mathbf{H}_{\text{aniso}} + \mathbf{H}_{\text{el}}. \tag{7.38}$$

The first two terms are the external and demagnetizing fields, respectively. The third term is an explicit way of accounting for local exchange effects due to nearby spins deviating in orientation from the bulk magnetization. The fourth term recasts the anisotropy energies of Eqs. (7.22 and 7.23) in terms of effective $\mathbf{H}$-fields. The final term is a contribution to the effective field due to elastic/magnetostrictive effects.

The prediction that comes from Eqs. (7.36) and (7.38) is that the dynamics of $\mathbf{M}(\mathbf{r})$ is predominantly a precession around the local $\mathbf{H}_{\text{eff}}(\mathbf{r})$. This kind of precession is analogous to the precession of electron magnetic moments that takes place in electron spin resonance, or the precession of nuclear magnetic moments in nuclear magnetic resonance. A piece of ferromagnet can exhibit *ferromagnetic resonance* (FMR), with a resonance frequency determined by $\mathbf{H}_{\text{eff}}$, and thus in part by the shape of the material via $\mathbf{H}_{\text{d}}$ [484].

Equation (7.36) cannot be the full story, however. It contains no dissipation term, no way to model how a real system would eventually approach some magnetostatic equilibrium. Landau and Lifschitz [485] and Gilbert [486, 487] proposed specific ways to incorporate dissipation. Defining the local unit vector $\hat{\mathbf{m}} = \mathbf{M}/|\mathbf{M}|$, the Landau–Lifshitz equation is

$$\frac{\partial \hat{\mathbf{m}}}{\partial t} = -\gamma_{LL}\hat{\mathbf{m}} \times \mathbf{H}_{eff} + \alpha_{LL}\hat{\mathbf{m}} \times (\hat{\mathbf{m}} \times \mathbf{H}_{eff}), \tag{7.39}$$

and the Gilbert equation is

$$\frac{\partial \hat{\mathbf{m}}}{\partial t} = -\gamma_G\hat{\mathbf{m}} \times \mathbf{H}_{eff} - \alpha_G\hat{\mathbf{m}} \times \frac{\partial \hat{\mathbf{m}}}{\partial t}. \tag{7.40}$$

Here γ_{LL} and γ_G are effective gyromagnetic ratios (both reduce to γ in the absence of any damping), and α_{LL} and α_G are effective damping constants. One can go back and forth between the Landau–Lifschitz and Gilbert representations. In general, the damping coefficients are empirical numbers determined from experiment for a given system. Damping mechanisms can include interactions with magnetic impurities, radiation of spin waves and/or Stoner excitations, and eddy currents due to changing magnetic flux in the conducting material. We will see the importance of these dynamics equations when we consider magnetization reversal and nanoscale magnetic systems.

In the cases of nuclear magnetic resonance or electron spin resonance, usually applied when there is *not* a magnetically ordered state, the equations of motion for the magnetization are usually called the *Bloch equations* [488]. Suppose that there is a magnetic induction $\mathbf{B}(t) = (B_x(t), B_y(t), B_z(t) + B_0)$. Here B_0 is a large, static field that picks out a particular direction; in magnetic resonance experiments this is usually applied via a solenoid. The static magnetization that would result from B_0 alone is M_0, directed along the z-axis. Component by component, the time evolution of the magnetization is given by

$$\frac{dM_x}{dt} = \gamma(\mathbf{M} \times \mathbf{B})_x - \frac{M_x}{T_2},$$
$$\frac{dM_y}{dt} = \gamma(\mathbf{M} \times \mathbf{B})_y - \frac{M_y}{T_2},$$
$$\frac{dM_z}{dt} = \gamma(\mathbf{M} \times \mathbf{B})_z - \frac{M_z - M_0}{T_1}. \tag{7.41}$$

Here, the precession of the magnetization is again clear, and there are two particular relaxation effects considered. The time T_1 is the *longitudinal relaxation time* and sets the timescale for the magnetization to relax back to its equilibrium value, M_0, if perturbed away (e.g., by the temporary application of a transverse (to z) $\mathbf{B}$ that causes precession away from the z-axis). Another name for T_1 is the *spin–lattice relaxation time*.

In contrast, the time T_2 characterizes the decay of transverse (to z) magnetization. Imagine reaching into the material system and tipping the static magnetization away from z and into the x direction (via a transverse magnetic induction, as in the previous paragraph). In the limit that T_2 and T_1 are infinite, that magnetization will precess coherently about z. In common magnetic resonance experiments, that precessing magnetization may be detected via various means. However, a finite T_2 will cause a net decay of the transverse magnetization. One mechanistic way to think of this [488] is as *dephasing*; while each individual spin may be precessing about z, if the spins get out of sync with one another, there is no *net* transverse magnetization. The T_2 time often called the *spin–spin relaxation time* or spin dephasing time, the former because it can contain a dominant contribution due to spin–spin interactions. In systems with mobile spins, there can be an additional contribution [489] to the dephasing rate due to diffusion of the spins and inhomogeneities in B_0. The resulting net dephasing time in that situation is called T_2^*.

7.4 Magnetism at small scales

The interplay of magnetic energy scales and the intrinsic size scales inherent in magnetic phenomena make for very interesting magnetic phenomenology at the micro- and nanoscale. A question that arises immediately is, how do we assess and characterize the magnetic response of materials on these scales? There are a number of techniques, each with their own advantages and drawbacks. Here is a representative sampling.

- **Bitter decoration** As we all learned in elementary school, one way to trace the spatial arrangement of magnetic field is to sprinkle a region of interest with iron filings, or, more generally, small magnetically polarizable particles. Those particles will feel forces as described in Eq. (7.9) and will concentrate in regions of strong fields. One can use the same approach to examine the field distribution caused by stray fields near domain boundaries on the surfaces of magnetic materials. This process, pioneered by Bitter [490] using colloid suspensions of magnetite, is not well suited for quantitative analysis or micro/nanoscale magnetism. However, it has been of major importance in understanding magnetic phenomena over the past century.

- **Hall sensors** We considered the Hall effect back in Section 6.3 as a means of extracting knowledge of a conducting system's parameters (2d carrier density and mobility) from longitudinal and Hall measurements in a known magnetic field. It is possible to come at this from the other direction, and use a well characterized semiconductor system's Hall response to infer the stray magnetic field at some location. For a fixed Hall measurement geometry, the Hall response provides a measure of the component of the magnetic field normal to the plane of the 2d electronic system. Operating temperatures can be constrained by the type of Hall bar material. For example, GaAs/AlGaAs heterojunction electron gases typically need to be cooled well below room temperature to optimize their transport characteristics.

  There are some challenges in trying to implement this as a general probe technique, particularly if the objective is to map magnetic fields as a function of position over the surface of some sample. The spatial resolution will be limited by two scales: the separation between the Hall measurement region and the sample surface, and the spatial extent of the Hall geometry itself.

  It is possible to prepare 2d electron system-based Hall bars at the corners of wafer pieces that may be mounted on AFM-style cantilevers or tuning fork probes. An example [491] is shown in Fig. 7.15(a), so that the final distance between the Hall probe and the surface can be as small as tens of nanometers. Minimizing the size of the Hall region is also not a trivial issue. In Hall bars from clean 2d electron systems it is not difficult to have electronic mean free paths exceed the desired size of the Hall

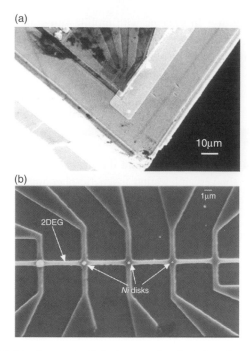

(a)

(b)

10μm

1μm

2DEG

Ni disks

Fig. 7.15   Hall probes. (a) GaAs-based Hall bar for scanning Hall microscopy, from [491]. (b) Ballistic Hall crosses for measuring the magnetization of prepositioned samples, in this case Ni nanodisks, from [492].

measurement region. In this limit one is in the ballistic regime, and the diffusive transport relationships (Eq. (6.15)) must be re-examined [493, 494].

Typical field sensitivities can be in the range of 10 nT/$\sqrt{\text{Hz}}$, for readily achievable carrier densities and measuring currents. Rather than strict $|\mathbf{B}|$ measurement, often people are interested in characterizing magnetic flux. For a measurement region $\sim$ 1 μm on a side, the example number implies a flux sensitivity of $\sim 10^{-5}\phi_0/\sqrt{\text{Hz}}$. Measurements are usually limited to fields where quantum effects (Landau level quantization) can be neglected in the Hall bar itself, say below a few Tesla.

Rather than using movable Hall probes to examine surfaces, a complementary approach is to place a nanoparticle or nanowire directly on the measurement region of a static Hall bar [493, 494]. An example of this is shown in Fig. 7.15(b), where the investigators were examining the magnetization properties of several nanoscale Ni disks [492].

• **SQUIDs** Superconducting quantum interference devices (SQUIDs) are based on quantum effects similar in spirit to the Aharanov–Bohm effect discussed in Section 6.4.2. Figure 7.16 shows the basic geometry for a typical DC SQUID. There is a superconducting loop with two "weak links" where the critical current is suppressed relative to the bulk. Such links can be tunnel junctions or constrictions. The superconducting order parameter includes a phase factor proportional to the vector potential, as in Eqs. (6.24 and 6.25), though the charge carriers in the superconducting state are pairs of electrons, and thus have charge $-2e$. For the order parameter to be single-valued as one goes

Fig. 7.16

Diagram of a typical DC superconducting quantum interference device (SQUID). The crosses are "weak links" (constrictions or tunnel barriers), where the critical current for superconductivity is reduced relative to the rest of the structure. The superconducting order parameter must be single-valued going around the loop, and magnetic flux (ϕ) threaded through the loop induces screening currents to meet this condition.

around the loop, the total phase shift going around the loop must be an integer multiple of 2π. This means that the total magnetic flux through the loop must be quantized in units of $\phi_0/2$ (where the 2 comes from the Cooper pair charge). If the *externally* applied magnetic flux (from a sample near the loop, for example) does not satisfy the quantization condition, then superconducting currents will be induced going around the loop to bring the total flux to the nearest integer multiple of $\phi_0/2$.

In one resistive DC configuration, a bias current exceeding the critical current of the weak links is applied to the loop as shown. The screening currents required by flux quantization, superposed on the bias current, then lead to a voltage that develops across the loop. That voltage is periodic in flux through the loop with a periodicity of $\phi_0/2$. Knowing the loop geometry, one can then infer the average **B** component normal to the loop. SQUIDs are often run in more sophisticated configurations. For example, an additional coil can be used for feedback, so that the flux through the SQUID is maintained at a value that maximizes voltage change with flux. The feedback current needed is then monitored while the SQUID operates under constant sensitivity conditions.

SQUIDs can be exquisitely sensitive magnetic field sensors. In terms of flux sensitivity, they can readily achieve $10^{-6}\phi_0/\sqrt{\text{Hz}}$ levels. An obvious complicating factor is that SQUIDs are based on superconductivity, and therefore require a low temperature environment. Furthermore, they cannot operate in magnetic fields that exceed the critical field required to kill superconductivity.

Calibrated SQUIDs are routinely used as magnetometers to measure the magnetic response of materials of interest. In such a configuration, the SQUID loop is often a counterwound pair of coils, one containing the material of interest and the other empty. In the presence of a known externally applied field, in the absence of any magnetic response from the sample, the net flux through the counterwound coils would be zero; any magnetic response of the material leads to net flux of one sign or the other. Commercial SQUID magnetometers for bulk materials are designed so that the SQUID is in a cryogenic environment while the sample is in a temperature-controlled region enclosed by the SQUID. Another metric of sensitivity of such a device is to ask how many net Bohr magnetons could be detected. Commercial magnetometers can typically measure a few billion Bohr magnetons. Specially designed SQUID assemblies have been used to detect the magnetic fields associated with brain function [495] and to detect nuclear magnetic resonance signals in very low magnetic fields comparable to the Earth's [496].

For characterizing magnetic response on the nanoscale, two main implementations have evolved. The scanning SQUID microscope is similar in spirit to scanning Hall magnetometry. A small SQUID coil is fabricated on some tip which is then scanned in close proximity to a surface of interest. As in the Hall case, spatial resolution is limited by a combination of the loop diameter and the loop–surface separation. One well known use of scanning SQUID microscopy has been in the examination of the pairing symmetry of high temperature superconductors [497].

For examining magnetic processes in nanoscale particles, the microSQUID has been quite successful. In this technique [498] a micro- or nanoscale SQUID is fabricated on a substrate, and the nanoparticle of interest is placed directly on top of the SQUID. The main challenge of microSQUIDs is the comparative difficult in threading net flux through the loop. Sensitivities have been such that reversal of $\sim 1000 \mu_B$ are detectable.

- **Magneto-optical response** The dielectric function of materials controls their optical response, the rules that the electromagnetic field must follow in both reflection and transmission of light. The dielectric function of most materials includes terms that couple to the magnetization. Without going into extensive detail here (more in Chapter 8), one major effect of the magneto-optical couplings is to rotate the polarization of incident light. In transmission, this is called *Faraday rotation*. Faraday rotation studies of magnetic materials are comparatively rare, since most materials of interest are not transparent. In reflection, the polarization rotation relative to the incident light is called *Kerr rotation*, or the magneto-optical Kerr effect (MOKE). With a polarization-sensitive optical microscope and digital image processing techniques (taking explicit differences of polarized and unpolarized images, for example), MOKE is an outstanding tool for examining magnetization and domain walls on scales accessible to optical microscopy.

  Quantitative sensitivity in MOKE is challenging to achieve, requiring detailed calibrations of the optical system and the sample optical properties. Other complications include the requirement for minimal surface roughness, the limitation of optical resolution ($\sim$ 0.3 μm), and the fact that MOKE is often only sensitive to the surface magnetization because of the limited penetration ($\sim$ 10 nm) of light into metal surfaces.

  In its favor, MOKE can produce impressive images of large scale magnetization structure, and video-rate movies of domain dynamics. Near-field optical techniques have also been developed [499] to improve the lateral resolution available.

- **Electron microscopy approaches** As discussed in Chapter 4, scanning electron microscopy uses incident energetic electrons to knock secondary electrons out of a sample surface. In *scanning electron microscopy with polarization analysis* (SEMPA) [500, 501], a device called a polarimeter [502] is used to measure the magnetic polarization of the secondary electrons. The secondary electron polarization is set by the magnetization of the sample surface that is the source of the secondary electrons [503]. With the right kind of polarimeter, it is possible to obtain all three vector components of **M**.

SEMPA has a spatial resolution similar to that of ordinary SEM. However, because the difference spin populations of the emitted electrons is typically not very large, comparatively long integration times are required to achieve clear magnetization determination. Like conventional SEM, SEMPA only works on electrically conducting substrates that can dissipate the charges left by the incident electron beam and the emitted electrons. SEMPA is sensitive to the surface magnetization only, since secondary electrons come only from the volume struck by the incident beam. One must also be concerned about scattering processes that can take place at surfaces that can change the polarization of the outgoing electrons. For this reason, surface preparation is stringent, and ultrahigh vacuum conditions are required. Trace amounts of surface oxide can be very deleterious here. For example, while Ni is a good ferromagnet, NiO is an antiferromagnetic insulator, and a submonolayer coating of NiO that interacts with the outgoing secondary electrons can degrade the polarization signal.

Transmission electron microscopy may also be used as a tool to examine nanoscale magnetic structure. In *Lorentz microscopy*, the electron beam is deflected when passing through a magnetic sample due simply to the Lorentz force law [504, 505]. The presence of **M** in the sample leads to **B** in the sample, and electrons are defected by an amount proportional the component of **B** perpendicular to the beam direction. Taking the resulting electron image, it is possible to back-calculate the magnetic configuration. This is analogous to looking at the pattern of sunlight on the bottom of a swimming pool, and deducing a map of the water surface height.

A considerably more sophisticated approach is *electron holography*, which is based on interference of coherent electron beams [506, 507]. The ability to do interference measurements in a TEM should come as no surprise, since the electron beams in a TEM typically have enough coherence to perform electron diffraction. In electron holography, the electron beam is split so that different portions of the beam interact with regions of the sample, and the resulting interference pattern on the detector is recorded. In addition to phase differences caused simply by differences in the electron path length, there are further phase shifts due to the Aharanov–Bohm phase. If two electron paths enclose an amount of flux Φ, then they pick up a relative phase shift of $2\pi \Phi/\phi_0$. While TEM techniques, particularly electron holography, can be quite sensitive and have nanoscale resolution, the requirement that samples be thin enough for electron transmission is quite restrictive.

- **Magnetic force microscopies** The AFM, with its ability to detect sub-nanoNewton forces, can be an invaluable tool for magnetization studies at material surfaces. In magnetic force microscopy [147, 508], the AFM tip is coated with or fabricated from a ferromagnetic material, and magnetized. Typical imaging procedures are as follows. An initial scan over the sample uses contact or noncontact mode AFM to map out the sample's surface topography via the short-ranged atomic forces between the tip and sample. The tip is then raised to a desired tip–sample separation and moved back across the sample surface under computer control, using the just-acquired topography information to hold the tip–sample separation nominally constant. The tip still interacts with the sample via longer ranged magnetic forces, however, leading to changes in the amplitude, phase, and/or frequency of the tip's vertical oscillations (depending on the detailed imaging technique).

Magnetic force microscopy (MFM) provides detailed sub-100 nm images of stray field distributions at sample surfaces. Quantitatively interpreting such images, however, is challenging, requiring a detailed knowledge of the magnetization distribution of the tip itself, as well as a calibration of the tip's response to external forces. One can model the tip itself as some spatial distribution of $\mathbf{M}(\mathbf{r})$, and find the total force in a given tip position by integrating the local dipole-like forces (Eq. (7.9)) on each volume element due to the local magnetic field gradient [508].

There are natural tradeoffs in MFM. A smaller tip would imply higher spatial resolution; however, the magnetic forces scale roughly like the volume of magnetic material on the tip, so that smaller tips are inherently less sensitive for a given tip material. Another way to improve spatial resolution is to perform MFM with the tip as close as possible to the sample surface. However, just as the tip is immersed in the stray fields from the surface, the sample is immersed in the stray fields from the tip itself. If the local field at the sample due to the tip exceeds the coercive field of the sample, then the MFM measurements themselves can strongly perturb the magnetization.

One approach to boosting the sensitivity of MFM is to use the resonance properties of the cantilever to greater advantage, combined with the precession of magnetic moments in a noncolinear magnetic field. The result is *magnetic resonance force microscopy* (MRFM), conceived with the goal of producing three-dimensional images of the distribution of (nuclear) magnetic moments in a sample near a magnetic tip [509]. The local magnetic field in the sample is provided by a dc applied field plus the stray field from the magnetized nanoscale tip. A radio-frequency microwave electromagnetic field is applied, so that magnetic moments within a slice of material near the tip experience a resonant condition (the microwave frequency coinciding with the Larmor precession frequency for those moments). As the tip oscillates mechanically, this magnetic interaction leads to a shift in the tip mechanical resonance frequency [510] that may be detected by modulating the amplitude of the microwave radiation at a low frequency. The ability of MRFM to detect single electronic spins has been demonstrated [510]. Recently it has been possible to perform three-dimensional magnetic resonance imaging of the hydrogen distribution within individual viruses at ~ 10 nm resolution with this approach [511].

The main challenge of MRFM as a characterization technique lies in its sheer technological complexity. The magnetic forces due to the desired resonant interaction are generally dwarfed by fluctuating forces on the probe tip in thermal equilibrium. Low temperatures, rf electronics expertise, special floppy cantilevers, and a good vacuum are all required for MRFM to achieve its highest sensitivity and resolution.

- **Spin-polarized STM** Through a mechanism that we will explore in more depth in Section 7.4, there is a variant of STM that is capable of nanoscale discrimination of magnetization information at the surface of conducting samples. Spin-polarized STM takes advantage of the idea shown in Fig. 7.5. In a STM tip with a well-defined

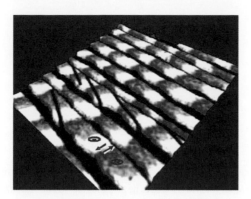

A spin-polarized STM image of iron nanostripes on a tungsten (11) substrate. Topography is rendered as apparent height, while brightness shows the spatially varying magnetic contribution to the tunneling density of states. A domain wall is indicated. Adapted from [162, 512].

Fig. 7.17

magnetization, there is a spin splitting of the bands, so that the densities of states of electrons with spins parallel or antiparallel to the tip magnetization differ. Recall that the tunneling conductance is proportional to the product of the local density of states of the tip and the sample. This gives a mechanism for imaging [162], as shown in Fig. 7.17. Extracting quantitative information from such images requires detailed knowledge of the magnetic properties of the tip. The relevant physics will be the tip. The relevant physics will be discussed further below.

- **Magnetoresistive effects** The tunneling physics above is a specific example of a magnetoresistive effect. When the electrical resistance of a system reflects the orientation of the magnetic order of the system's constituents, the application of an external magnetic field can strongly alter the resistance. In a well defined system, it is possible to use such magnetoresistive signatures to infer much information about the system's magnetic states.

The *anisotropic magnetoresistance* (AMR) is a simple example of this idea. In many ferromagnetic materials the presence of spin–orbit coupling leads to differences in the scattering properties of conduction electrons that depend on the orientation of the magnetization. The result is an intrinsic resistivity of the material such that, for a current density $\mathbf{J}$ and a magnetization $\mathbf{M}$, $\rho(\mathbf{J}||\mathbf{M}) \neq \rho(\mathbf{J} \perp \mathbf{M})$. In the case of Ni, $\rho(\mathbf{J}||\mathbf{M}) > \rho(\mathbf{J} \perp \mathbf{M})$ by about 1–2%, depending on the relative importance of other scattering mechanisms. AMR is particularly apparent in materials when the geometric anisotropy dominates. For example, Fig. 7.18 shows the magnetoresistance of a nickel nanowire. When no external field is applied, $\mathbf{M}$ lies along the wire axis. In the presence of a transverse $\mathbf{H}$, the magnetization may be coerced away from the axis, leading to a decrease in the wire resistance.

The *giant magnetoresistance* (GMR) was discovered in 1988 in work [514, 515] that was recognized by the 2007 Nobel Prize in Physics. GMR is not a property of pure materials, but requires a nanostructured composite of FM and nonmagnetic metals. Consider the multilayer of FM and nonmagnetic metals shown in Fig. 7.19. The original

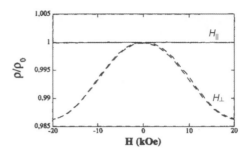

Fig. 7.18   Anisotropic magnetoresistance (AMR) for a Ni nanowire. When the external field is directed along the wire axis, there is no resistance change, since **M**||**J**. When **M** is forced to be perpendicular to **J** due to a transverse magnetic field, the resistance decreases by ∼1.5%. Adapted from [513].

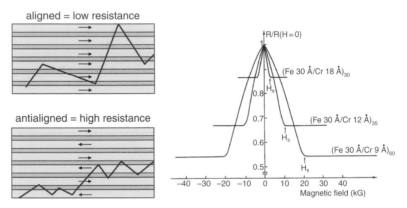

Fig. 7.19   Giant magnetoresistance. An electron moving through a multilayer of ferromagnetic materials separated by nonmagnetic material undergoes increased scattering when the magnetic layers are antialigned. At right, in Fe/Cr multilayers, an external magnetic field forces magnetic alignment of the Fe layers at high fields, leading to a large (tens of percent) drop in resistance for the multilayer. From [514].

GMR configuration considered the current to be in the plane of the multilayer, but the same physics is relevant regardless of current direction. physics behind GMR is latent in Fig. 7.5. In the FM layers layers there is some net spin polarization of charge carriers, and that polarization (and therefore the density of states as a function of spin orientation) is affectcd by **M**. If subsequent FM layers have identical properties and parallel orientations of **M**, then majority spin carriers from one FM layer can scatter into the same majority spin DOS in the next FM layer, leading to some relatively low electrical resistance. However, if subsequent FM layers are oppositely directed, as spin-polarized carriers from one FM layer try to enter the next FM layer, the mismatch in spin-dependent densities of states at the Fermi level leads to enhanced scattering and a correspondingly higher electrical resistance. The manitude of GMR depends on the difference in the spin densities of states at E_F and can be tens of % or more; hence, relative to AMR, the label "giant".

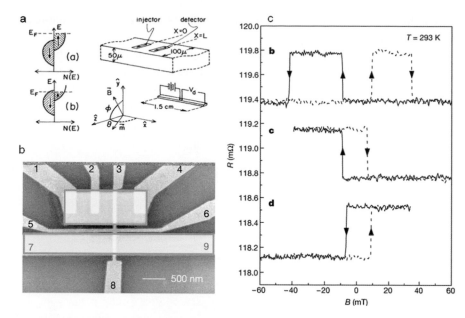

Spin valves. (a) The first spin valve experiment, with two ferromagnetic electrodes separated by a nonmagnetic material (Al in this case). The two-terminal resistance is low if the magnetizations are aligned, and high if antialigned. From [517]. (b),(c) A lateral spin valve, with two permalloy ($Ni_{0.8}Fe_{0.2}$) electrodes (shaded) and copper wiring. In (c), crisp switching in the resistance is clear as the two permalloy magnetizations reorient at different external field values. Adapted from [518].

Fig. 7.20

Nontrivial physics leads to a remarkable fact: the exchange coupling between subsequent FM layers may be tuned in an oscillatory way [516] by the thickness of the nonmagnetic spacer layers, with a spatial period on the nm scale. This allows the growth of GMR multilayers that tend to be antialigned in the absence of any external **H**. Further, good GMR requires spin-preserving (and therefore high quality and clean) interfaces between FM and nonmagnetic metals, and nonmagnetic layer thicknesses small compared to the *spin diffusion length*, the characteristic distance over which spin orientation is preserved in a nonmagnetic metal.

An example of a GMR device is a *spin valve*, schematically shown in Fig. 7.20 along with two examples. By controlling the coercive field of a switchable FM layer relative to a second FM layer, the electrical resistance of the spin valve has a clear, sharp dependence on external **H**, making such a spin valve an excellent sensor of magnetic field. In the second example, a lateral spin valve, by measuring the size of the GMR response as a function of the spacing between the FM electrodes, the spin diffusion length may be inferred.

The *tunneling magnetoresistance* (TMR) [519] is similar in spirit to GMR, and in some sense is conceptually simpler. The magnitude of GMR may be influenced by a number of complicating factors (interfacial roughness, the precise band structure of the FM and nonmagnetic metals) that affect the interfacial scattering of carriers. In contrast,

TMR is based, in its simplest limit, on just the difference between the spin-resolved electronic densities of states.

Choosing our reference direction in the FM along the magnetization direction, we can define $\nu_\uparrow^S$ as the density of states of spin-up electrons at the Fermi level for the source electrode, and $\nu_\downarrow^D$ as the equivalent spin-down DOS. We follow the model proposed by Julliere [520]. For parallel $(\uparrow, \uparrow)$ and antiparallel $(\uparrow, \downarrow)$ source and drain magnetizations, respectively:

$$
\begin{aligned}
I_{\uparrow,\uparrow} &\propto \nu_\uparrow^S \nu_\uparrow^D + \nu_\downarrow^S \nu_\downarrow^D, \\
I_{\uparrow,\downarrow} &\propto \nu_\uparrow^S \nu_\downarrow^D + \nu_\downarrow^S \nu_\uparrow^D.
\end{aligned}
\tag{7.42}
$$

The tunneling magnetoresistance is defined:

$$
\mathrm{TMR} = \frac{I_{\uparrow,\uparrow} - I_{\uparrow,\downarrow}}{I_{\uparrow,\downarrow}} = \frac{2 P_S P_D}{1 - P_S P_D},
\tag{7.43}
$$

where we have defined the spin polarization of the source electrode:

$$
P_S = \frac{\nu_\uparrow^S - \nu_\downarrow^S}{\nu_\uparrow^S + \nu_\downarrow^S},
\tag{7.44}
$$

and equivalently for the drain. This approach assumes that spin is conserved at interfaces and during the tunneling process. TMR measurements in well-defined structures are one means of inferring spin polarizations experimentally. Some measured values of P for some common ferromagnetic materials are [521], $P(\mathrm{Ni}) = 23\%$, $P(\mathrm{Fe}) = 40\%$, and $P(\mathrm{Co}) = 35\%$. With nanoscale tunnel junctions, as in spin-polarized STM, small scale information about local magnetic properties is accessible. Note that Eqs. (7.42–7.44) assume that the local density of states is the only relevant factor in determining the tunneling current. In real structures, the detailed electronic states determine the actual tunneling matrix elements that lead to the tunneling rates [522]. By clever engineering of the tunnel barrier interfaces, these matrix element effects can lead to greatly enhanced TMR values relative to those obtained purely from Eq. (7.44).

For completeness, even though they are not strictly useful for determining nanoscale magnetic properties, it is worthwhile enumerating three remaining forms of magneto-resistance. *Colossal magnetoresistance* (CMR) is the name given to an extremely large magnetoresistive effect in certain strongly correlated oxides. CMR materials [523, 524], typically manganites such as $\mathrm{La}_{1-x}\mathrm{Ca}_x\mathrm{MnO}_3$, are on the cusp of a phase transition between a ferromagnetic metal and a paramagnetic insulator. An applied $\mathbf{H}$ can push the material out of the insulating state and into the metallic state, leading to an extremely large ($\gg 100\%$) change in the resistivity.

A proposed relative of GMR and TMR is the *ballistic magnetoresistance* [525] (BMR). Rather than multilayers or tunnel junctions, BMR was reported in atomic-scale junctions between ferromagnets made through break junction methods. The idea was that a domain wall thinner than the elastic mean free path for carriers could be trapped at the atomic-scale constriction, leading to very large differences in resistance depending on the relative orientations of $\mathbf{M}$ in the two FM electrodes. While some controversy remains, measurements [526, 527, 528] suggest that BMR effects are consistent with

TMR measurements of spin polarization, and that great care must be taken to avoid magnetostrictive issues in such atomic-scale junctions.

Finally, a very large (thousands of %) magnetoresistive effect, *extraordinary magnetoresistance* (EMR), has been reported [529] in hybrid semiconductor/metal structures. EMR is based on the fact that in the limit of large Hall angle, current flows perpendicular to lines of electric field. Therefore, at high magnetic fields, current will flow *around*, rather than through, comparatively highly conducting metal portions of the EMR device. EMR structures may be useful in nanoscale magnetic sensing [530], and are comparatively unusual in being magnetoresistive devices that do not, themselves, contain ferromagnetic material.

7.4.1 Physics of magnetization in reduced dimensionality

Now that we can probe the local magnetization, $\mathbf{M}$, at small scales under a variety of conditions, many interesting questions arise concerning the fate of magnetic order at surfaces and in response to various perturbations. These include:

- how are the origins and low-lying excitations of the magnetically ordered state affected by geometric constraints?
- how do surfaces affect magnetization and domain dynamics?
- how do domains move and reorient?
- how do *single domain* particles behave?
- how does electronic transport affect magnetic domains?

These last two questions are especially important from the point of view of magneto-electronics, while the earlier questions have special relevance to magnetic data storage technologies.

To give a qualitative answer to the first of these questions, the presence of surfaces affects the magnetic exchange processes responsible for the ordered state. First, as we have discussed previously, even at ordered, crystalline interfaces, atoms at the surface are undercoordinated compared to atoms in the bulk. As a result, they experience different net exchange interactions. A classic example of this is the phenomenon of *exchange bias*, as shown in Fig. 7.21. Because of the presence of an interface with an antiferromagnetically ordered material, the hysteresis loop of the adjacent ferromagnetic material is shifted significantly away from $\mathbf{H} = 0$. In general the situation may be further complicated by the presence of interface or surface electronic states – that is, by the alteration of the underlying material's electronic structure.

In the presence of disorder at surfaces, the effect of local coordination on exchange interactions may be more profound. Magnetic moments at such surfaces may not fully participate in the magnetic order of the bulk material. Instead, such moments, experiencing exchange interactions that depend on the atomic-scale details of the surrounding material, may form a *glassy* state. A spin glass is characterized by a lack of long-range order, but a comparatively static spin configuration below a spin glass transition temperature. Interaction between moments are sufficiently strong that the kinetics of reorientation of the moments below the spin glass transition is strongly hindered. From the perspective of a

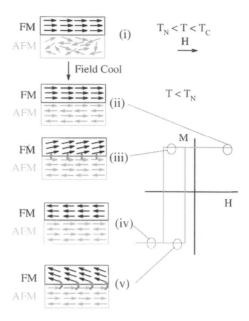

Fig. 7.21 Exchange bias. The presence of an intimately coupled antiferromagnetic layer can shift the hysteresis loop of a ferromagnetic layer. From [531].

potential energy "landscape" as a function of moment configuration, there are many, many local minima that are nearly degenerate. As a result, relaxation of the moments is extremely slow and takes place across many timescales. Such disordered spin configurations could clearly have a negative impact on surface-sensitive effects such as GMR and TMR.

As with every other wave phenomenon that we have encountered, there are also geometric confinement effects on the spectrum of the spin waves that are the low-lying excitations of magnetically ordered states. A particularly striking example of this takes place in the antiferromagnetic metal chromium. Experiments have examined the magnetic properties of Cr as a function of metal thickness down to the atomic level, and clearly reveal a discreteness in the spin wave spectrum that is the expected signature of geometric confinement.

The discussion of demagnetization and geometric anisotropy in Section 7.3.1 contains the essential ingredients for the effects of geometric constraint on magnetization. The more severe the geometric constraint, the greater the importance of demagnetization fields in determining the configuration of $\mathbf{M}(\mathbf{r})$. Competition between geometric anisotropy and magnetocrystalline anisotropy, combined with disorder-induced perturbations on the exchange process, can lead to a more macroscopic cousin of the spin glass situation. Consider the 2d film shown in Fig. 7.22. Many domain configurations, such as those shown, have nearly identical total energies. The motions of the domains, however, are influenced and restricted by disorder and magnetoelastic interactions between domain walls. The result is a system with many nearly-degenerate almost-ground states, kinetically trapped in some configuration at low temperatures. The FM magnetization may be coerced and aligned out of the plane by a strong external $\mathbf{H}$, but when the external field is removed, the

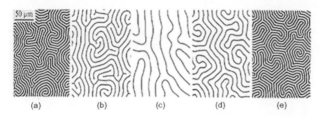

"Configurational hysteresis" in 2d domains. The domain configuration in this thin doped yttrium garnet film has many possible configurations with very similar energies. Domains coalesce as an out-of-plane **H** field is applied (a)–(c), and when the field is removed (c)–(e), the final configuration is similar but distinct from the starting arrangement. From [532].

Fig. 7.22

system (slowly) relaxes back into a complex configuration, likely different from that prior to the procedure.

In one effective dimension (that is, when the magnetic material is confined in the transverse direction to a size smaller than or comparable to the thickness of a domain wall), the situation becomes somewhat simpler, particularly if magnetocrystalline anisotropy is neglected. It is energetically favorable for **M** to lie along the wire's long axis. Within the wire there can exist multiple domains separated by domain walls, and those domain walls may move as the bulk magnetization adjusts in response to external fields. In such wire structures, the motion of domain walls may be detected by their impact on electronic transport, for example spin-dependent scattering contribution of the domain walls [533, 534].

Consider such a wire, uniformly magnetized along its long axis, and an external magnetic field **H** antialigned to **M** is gradually ramped up. At some point it will be energetically favorable for the magnetization of the wire to reverse its direction, most likely through the nucleation and propagation of one or more domain walls. The physics of this reversal process may be quite rich, in part due to the strong interaction between the domain wall and disorder (grain boundaries, surface defects). Domain wall motion may be thermally assisted [533], as thermal energy is used to overcome the energetic "barrier" presented by disorder, magnetocrystalline anisotropy, and geometric anisotropy. More interesting and exotic is the possibility of the quantum tunneling of the domain wall (or, equivalently, the quantum tunneling of the magnetization) [535, 536]. We will discuss this further below.

Even in the absence of strong pinning effects, the detailed motion of domain walls involves rich physics. Nanowires patterned from GMR multilayers [537] are one tool used to examine domain wall propagation. In these structures, with a fixed magnetization direction for the bottom layer and a switchable magnetization for the top, the change in resistance of the whole nanowire is directly related to the fraction of the upper magnetization that has been reversed, as shown in Fig. 7.23. Investigations such as these have resolved the velocity of domain wall propagation, v_{DW}, as a function of applied magnetic field, and led to the phenomenological definition of a domain wall mobility, via $v_{DW} = \mu_{DW} \times (H - H_0)$, where H_0 is some threshold field for domain wall motion. The details of magnetization reversal in nanowires can be quite complex, as shown in Fig. 7.24,

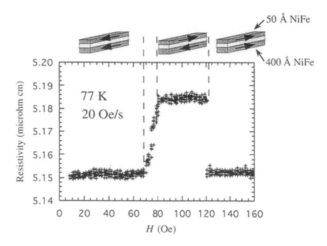

Fig. 7.23   Inferring domain wall position in a magnetic nanowire, via GMR. As the wall moves, the fraction of antialigned magnetization changes and may be detected via the resistance. From [537].

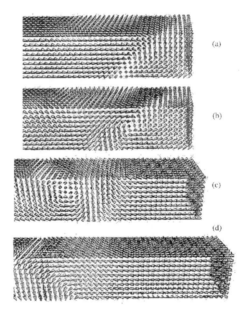

Fig. 7.24   Micromagnetic simulation of magnetization reversal in a nanowire. Note that the process in this case is much more complicated than simple domain wall motion. From [538].

meaning that μ_{DW}'s dependence on geometry and the magnetic material properties is not simple [538]. Of critical importance are the dynamic effects, including the damping, described in Eqs. (7.39 and 7.40). Unraveling the micromagnetic details is complicated by the lack of direct nanoscale probes of the time-varying magnetization.

We should also consider "zero-dimensional" systems, where the overall magnetization of an entire particle tends to move as a unit. As with our other discussions of dimensionality, it is important to compare a characteristic linear dimension of the system in question with a physically relevant length scale. The energetic cost of distorting $\mathbf{M}$ within a small particle can be quite large, due to the exchange stiffness term of Eq. (7.31). The stiffness coefficient, $\tilde{A}$, has units of J/m, while the anisotropy coefficients K have units J/m^3. Combining these, we can define an *exchange length*,

$$L_{\text{ex}} \equiv \sqrt{\tilde{A}/K}. \tag{7.45}$$

To within a numerical factor of order 1, this is the same as the domain wall thickness estimated in Eq. (7.35). Ferromagnetic particles with $L < L_{\text{ex}}$ are effectively zero-dimensional in terms of their magnetization.

Since magnetically zero-dimensional structures are typically a single domain, their physics may often be described with relatively simple analytical expressions. For a general magnetic particle in an external magnetic field $\mathbf{H}$, with $\theta = 0$ defined as the direction of the particle's easy axis, the energy cost per unit volume for having the magnetization in some direction θ is

$$u_{\text{tot}} = g(\theta) - \mu_0 H_{\parallel} M_{\text{sat}} \cos\theta - \mu_0 H_{\perp} M_{\text{sat}} \sin\theta. \tag{7.46}$$

Here $H_{\parallel}$ and $H_{\perp}$ are the components of $\mathbf{H}$ that are parallel and perpendicular, respectively, to the easy axis, and $g(\theta)$ describes the crystallographic anisotropy. For the simple uniaxial case, $g(\theta) = K \sin^2\theta$. Because the particle is a single domain, $|\mathbf{M}| = M_{\text{sat}}$. The equilibrium direction of $\mathbf{M}$ is the value of θ that would minimize Eq. (7.46), such that $du_{\text{tot}}/d\theta = 0$. This energy analysis tells when it is energetically favorable for a simple, uniaxial anisotropy particle to switch its magnetization, but does not describe the actual mechanism of magnetization reversal.

When we worry about magnetization reversal for such a single-domain nanoparticle, we care about the stability of such an energy minimum. The edge of stability, the crossover between stable and unstable equilibrium, happens with the *second* derivative changes sign from positive (stable) to negative (unstable). For a given θ, finding $d^2 u_{\text{tot}}/d\theta^2$, setting it equal to zero, and solving for the components of H gives the critical field necessary for switching the magnetization. For uniaxial anisotropy, the result is called the *Stoner–Wohlfarth astroid*, and is shown in Fig. 7.25. The critical fields are:

$$H_{\parallel}^* = -\frac{2K}{\mu_0 M_{\text{sat}}} \cos^3\theta,$$
$$H_{\perp}^* = \frac{2K}{\mu_0 M_{\text{sat}}} \sin^3\theta. \tag{7.47}$$

Note that when $H > H^*$, the magnetization no longer exhibits hysteresis; it is energetically favorable for $\mathbf{M}$ to point along $\mathbf{H}$ all the time.

Suppose the external field is below the critical fields of Eq. (7.47), such that it is energetically favorable for the magnetization to reorient itself; that is, system is trapped in a *local* minimum of $u_{\text{tot}}(\theta)$, but not the *global* minimum. There are three different ways for the microscopic spin degrees of freedom within the nanoparticle to realign themselves. In *incoherent reversal*, analogous to the formation and propagation of a domain wall, magnetization in different regions of the particle reorient comparatively independently. The

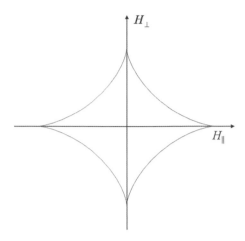

The Stoner–Wohlfarth astroid, showing the critical field for magnetization reversal of a single-domain particle with uniaxial anisotropy, as described by Eq. (7.47).

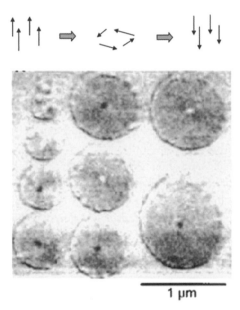

Curling mode of magnetization reversal. Top: spins rotate to form a vortex-like pattern midway through the reversal process. Bottom: in permalloy disks, energy is minimized by a similar curled pattern, and spins at the vortex core point out of the plane, leading to bright or dark dots in the center of this MFM image. Adapted from [539].

opposite extreme would be *coherent rotation*, with **M** throughout the particle rotating all at once, as if the magnetization were due to a single effective spin degree of freedom. Finally, there can be a topologically nontrivial transition through a *curling* mode of reversal. In the curling case, at the intermediate state midway through the reversal of the direction of **M**, the magnetization has a vortex-like distribution, with a small core, as shown in Fig. 7.26.

Consider a single-domain particle when external fields are small, and assume uniaxial anisotropy. In that case, $u_{\text{tot}}(\theta)$ has two minima, at $\theta = 0, \pi$, with a barrier of magnitude K, the anisotropy constant. While it is energetically neutral for the particle to flip its magnetization along the easy axis, to do so requires overcoming an energy barrier of KV, where V is the volume of the particle. We can define a *blocking temperature*, $T_{\text{b}} \equiv KV/k_{\text{B}}$. When $T \gg T_{\text{b}}$, there is enough thermal energy available that the magnetization reorients freely. For an ensemble of many (independent) particles, this means that even though each particle may be far below its Curie temperature (and thus is fully magnetized, with $|\mathbf{M}| = M_{\text{sat}}$), the whole ensemble has no net magnetization. If an external magnetic field is applied, the total magnetization of the ensemble will respond, with each particle playing the role of a single spin as in Eq. (7.12). The ensemble will exhibit no hysteresis even though $T < T_{\text{c}}$ for the constituent nanoparticle material. This lack of hysteresis in ensembles of (individually) ferromagnetically ordered nanoparticles at high temperatures is *superparamagnetism* [540], and as we shall see, poses limitations in magnetic data storage.

Conversely, if $T \ll T_{\text{b}}$, the magnetization of each particle is blocked energetically from reorienting in the presence of an external field. Hysteresis would then be observable in overall $\mathbf{M}$ vs. $\mathbf{H}$ for an ensemble of such blocked nanoparticles, with coercive fields set by Eq. (7.47). The blocking temperature may be inferred experimentally from the comparison of $M(T)$ data taken in with $|\mathbf{H}| = 0$ ("zero-field-cooled") and that with $|\mathbf{H}| \neq 0$ ("field-cooled"). The temperature at which these data sets diverge is T_{b}, as shown in Fig. 7.27.

In the blocked regime, one can often characterize the rate of magnetization reversal [542, 543] or relaxation with the approximate form

$$\tau^{-1} = \Omega \exp(-KV/k_{\text{B}}T), \qquad (7.48)$$

where Ω is some attempt frequency. The attempt frequency, typically $\sim 10^9 - 10^{10}$ Hz, is related [542] to the ferromagnetic resonance timescale associated with precession of such a domain.

The energy barrier language and the discussion of thermal activation suggests that one may need to consider quantum processes as well, such as the "tunneling" of the magnetization "through" the barrier. There is theoretical [544] and experimental [545]

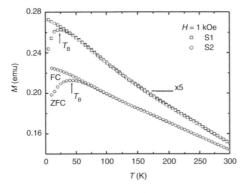

Superparamagnetism in silica-coated Ni nanoparticles. The blocking temperatures for two different batches of particles are set by particle size. From [541].

Fig. 7.27

support for this idea. More recently, there has been considerable interest in the quantum tunneling of magnetization [546, 547] in single-molecule magnets [548], chemical compounds containing transition metal ions so that each molecule acts like a single large spin (e.g., $S = 10$). This is, indeed, a tunneling process, in the sense of a quantum system traversing from one allowed state to another via a range of parameters classically forbidden by conservation of energy. However, one essential piece of physics to recall is that the magnetization, $\mathbf{M}$, is related to real angular momentum, dominantly in the form of the spin of the electrons in this case. For magnetization to tunnel, total angular momentum must still be conserved (through coupling to external magnetic field, the motion of the ions, and/or the nuclear degrees of freedom).

7.4.2 Spin currents

We are well aware that it is possible to push charge, in the form of mobile charge carriers, from one material to another, and those charge carriers possess spin degrees of freedom. In a material with itinerant charge carriers having a nonzero spin polarization, an electrical current will result in a net flow of spin polarization. Forcing that current into a nonmagnetic material will produce magnetization inside the nonmagnet near the interface. This is *spin injection*. However, since the nonmagnetic material lacks the exchange interaction responsible for ferromagnetic order, the spins are eventually randomized through relaxational processes, such as the T_1 spin–lattice process mentioned below Eq. (7.41).

Figure 7.28 shows two specific spin relaxation mechanisms relavent to moving carriers in solids. The *Elliott–Yafet* mechanism [549, 550] is a direct consequence of spin–orbit interactions, and is commonly relevant in metals or small gap semiconductors. Because of the spin–orbit coupling, the z-component of spin is no longer a "good" quantum number, meaning that the correct single-particle electronic Bloch eigenstates are linear combinations of spin-up and spin-down. Every scattering event that changes $\mathbf{k}$ a transition from one such state to another, and alters the relative likelihood of the particular spin being up or down. The more scattering, the more mixing of the spins, so that one would expect $T_1 \sim \tau$, the mean free time between scattering events, with some coefficient of proportionality that would depend on the material type.

The *D'yakonov–Perel* mechanism [551, 552] is also spin–orbit related, but is relevant in semiconductor materials that lack an inversion symmetry, such as GaAs or ZnSe. In the presence of the spin–orbit interaction, internal electric fields within the material act, from the point of view of the carrier, like an effective internal *magnetic* field that depends upon $\mathbf{k}$. As carriers propagate, their spin effectively precesses about that internal field. Each elastic scattering event that randomizes the direction of $\mathbf{k}$ also therefore randomizes the spin precession, leading again to a $T_1 \sim \tau$ proportionality.

For completeness, an additional mechanism, the Bir–Aronov–Pikus mechanism [553] (not shown) is appropriate to consider in semiconductors when both electrons and holes are present. Recall that holes have considerably different spin–orbit interactions from electrons. If there is an effective exchange coupling between electrons and holes, and the holes experience strong spin–orbit coupling, this shortens the lifetime of spin states of the electrons.

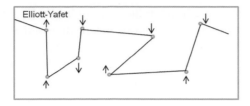

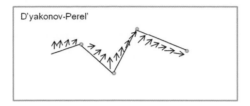

Spin relaxation mechanisms. Top: Elliott–Yafet, where the z-component of spin has some probability to flip in elastic   Fig. 7.28
scattering events. Bottom: D'yakonov–Perel, in which the spin of a carrier precesses about an effective k-dependent
magnetic field. Both effects originate through spin–orbit coupling.

When considering the injection of a spin current into a material with diffusive motion of charge carriers (that is, sample size $\gg \ell$), it is often more useful to talk about a *spin diffusion length*, $L_s \sim \sqrt{T_1}$, rather than a lifetime or relaxation time. One can visualize L_s as the typical distance random-walked by a carrier before its spin is randomized. Another way to think about this lengthscale is to imagine a ferromagnetic metal with some net spin polarization joined ideally (with no energy barrier at the interface) with a nonmagnetic metal. In diffusive equilibrium, one would expect a net spin polarization to "leak" into the nonmagnetic metal, decaying spatially away from the interface on a distance scale $\sim L_s$. In a material like gold, with $T_1 \approx 4.5 \times 10^{-11}$ s [554], and a typical electronic diffusion constant in a polycrystalline metal film of $D \approx 10^{-2}$ m^2/s, then $L_s = \sqrt{DT_1} \approx 670$ nm, quite a long distance.

Suppose we drove a current of spin polarized electrons from a ferromagnet into a nonmagnetic metal. How much excess magnetization would actually result in the nonmagnetic metal? If the polarization of the FM is P, then a current I would pump PI/e net excess spins into the normal metal. Those spins would remain polarized for a time $\sim T_1$ for the normal metal. For the Au case considered above, with iron ($P = 0.4$) as the ferromagnet, and a current of one microamp, the total number of excess spins driven by the current would be $\sim P(I/e)T_1 \approx 100$. That's a small number of spins, very difficult to detect directly.

It is possible to detect spin polarized currents, however. One approach [517], shown in Fig. 7.20(a), employs the same essential physics as GMR (though it's proposal predates GMR's discovery). One FM contact is used as an "injector" of spin-polarized electrons into a nonmagnetic (paramagnetic, aluminum in this case) material, while a nearby second FM contact (with magnetization that may be aligned or antialigned relative to the first) acts as an "analyzer". The buildup of nonequilibrium magnetization in the paramagnet due to spin injection should lead to a chemical potential difference (manifested as an open-circuit voltage difference) between the *detector* and the paramagnet, proportional to the projection of the spin's direction onto the polarization of the detector. If the two FM electrodes are much farther apart than L_s, no voltage signal will be observed. A further refinement is to use an analog of a technique known as the Hanle effect. For a single-domain injector, all the injected spins are aligned in the plane of the paramagnet. A small magnetic induction B perpendicular to that plane will cause the injected spins to precess as they diffuse around. The statistical average of this precession and diffusion means that there there will be a nontrivial dependence of the voltage signal on the perpendicular B-field, the spatial separation and relative magnetization directions between the FM contacts, and T_2 processes, which will naturally cause spins to become out-of-sync with one another [517].

A similar approach is shown in Fig. 7.20(b), in a device design known as a lateral *spin valve*, also a GMR device. The two single-domain permalloy[1] electrodes are designed (with different aspect ratios) so that they have different geometric anisotropies and therefore different **H**-fields required for reversal of **M** along the electrode long axis. Parallel (antiparalel) magnetization leads to easier (more difficult) transmission of electrons through the nonmagnetic (copper) material. For current in lead 1 and out lead 5, more difficult transmission due to spin accumulation would correspond to a larger voltage difference between lead 6 and lead 9. By performing measurements on a series of devices, the authors are able to infer L_s from the decay of the voltage signal as the distance between the FM electrodes is increased.

These examples highlight some of the challenges inherent in spin transport. For example, to understand the data in both papers, it is necessary to consider the efficiency of spin injection across the FM/nonmagnetic metal interface. Surface defects with magnetic character (e.g., paramagnetic copper oxide on the copper surface) at the interface can lead to spin-flip scattering and loss of injected polarization. The situation is further complicated in attempts to inject spin into semiconductors via ferromagnetic metals. In the typical semiconductor situation, the relatively low carrier densities can lead to non-negligible Hall voltages due to fringing **B**-fields from the ferromagnets, confusing the situation.

7.4.3   Spin transfer torque

Recall that spin is real angular momentum. The flow of a net spin current into a volume of material exerts a literal *torque* on that material. This was described decades ago by Feynman [9] in a thought experiment, in which a beam of spin-polarized electrons hitting a torsion pendulum would produce a measurable deflection, as the torque from the influx of spin angular momentum is balanced by the mechanical torque from the torsion fiber. The exact analog of this thought experiment has actually been performed [555], with spin injection from a cobalt wire into a gold wire (larger than L_s) attached to a nanomechanical torsional resonator. The mechanical torque results from the polarized spin angular momentum being redistributed to the lattice as a whole.

---

[1] Permalloy is a name for a nickel–iron alloy with very low magnetostriction and crystallographic anisotropy.

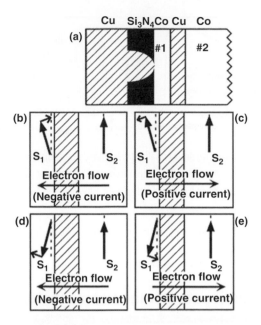

Cu   Si$_3$N$_4$Co Cu   Co

(a)

#1   #2

(b)

S$_1$ S$_2$

Electron flow

(Negative current)

(c)

S$_1$ S$_2$

Electron flow

(Positive current)

(d)

S$_1$ S$_2$

Electron flow

(Negative current)

(e)

S$_1$ S$_2$

Electron flow

(Positive current)

Spin transfer torque. Depending on the direction of current flow, the motion of spin polarized electrons exerts a torque on the magnetization of the thinner (and more readily reoriented) cobalt layer. From [556].   **Fig. 7.29**

A spin-polarized current into a magnetically ordered material also exerts a torque, and thanks to the exchange interaction, the angular momentum of the incoming spins can be delivered to the moments in the magnetic material. Likewise, a spin-polarized current leaving a magnetically ordered material (with charge conserved via an influx of unpolarized carriers) exerts a torque on magnetization of the magnetic material. This is called *spin transfer torque*, and it means the magnetization of ferromagnets may be manipulated *by electrical means* rather than requiring externally applied **H**-fields. Figure 7.29 shows the basic concept, with a thick cobalt layer having a magnetization comparatively fixed in direction, and a thin cobalt layer having lower anisotropy and thus a magnetization easier to reorient. Because of this asymmetry, there is an inherent directionality in the magnetic response as a function of current. Thanks to the GMR effect, the electrical resistance of the device gives comparatively direct information about the relative orientation of **M** for the two layers [557, 556, 558]. Note that nanostructures are required to observe these effects. As in the cases of magnetization reversal and spin injection, surface quality and disorder can pin magnetization and affect the efficiency of spin injection, strongly influencing current-driven magnetization effects.

In a system with lower anisotropy, so that **M** of the reorientable layer is particularly free to turn, continuous application of an appropriate spin transfer torque under chosen **H** conditions leads to precession of **M**, just as in ferromagnetic resonance. Figure 7.30 shows such a structure [559], using permalloy layers with one pinned by adjacent IrMn, and the resulting precession, the frequency of which depends on both the external **H** and the applied current [560]. This precession raises the possibility of coherent electrical manipulation

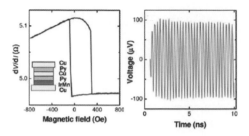

Fig. 7.30 Left: a particular spin-valve structure, with one permalloy layer pinned by an adjacent IrMn, and the other "free" to reorient its magnetization, shown with its GMR/spin valve response. Right: current-driven precession of the free layer, detected via the GMR response, and driven by injection of spin-polarized current. Adapted from [559].

of $\mathbf{M}$ on sub-nanosecond timescales, with great potential technological applications (see Section 7.6).

7.5 Magnetic data storage and magnetoelectronics

As we did in the case of electronics, we will briefly consider the history and current state of the art in one particular technological niche relevant to magnetism: magnetic data storage and magnetoelectronics. It is not an exaggeration to say that the growth of data storage has profoundly reshaped modern society, particularly when coupled with the advances in computing power that have come over the last two decades. While nonmagnetic technologies are increasingly useful for certain applications, the vast majority of rewritable high capacity data storage has at its root the physics of magnetic materials, as discussed earlier in this chapter. While we concentrate on the role of magnetism, it is worth noting that other nanoscale challenges are latent in magnetic storage technologies, including issues of materials flatness, friction, and nanoscale-accuracy positioning.

7.5.1 Historical trends and state-of-the-art

Magnetic data storage has roots [561] tracing back to the late nineteenth century. Once there was an appreciation for the nature of magnetization and domains, it did not take long for people to consider using $\mathbf{M}(\mathbf{r})$ as a means of recording information. Oberlin Smith was one of the first, with the idea of using an electrical microphone to pattern magnetic domains in a magnetic filament or wire (c. 1878). Valdemar Poulsen actually demonstrated the wire recorder in 1898, and went on to invent the reel-to-reel metal tape recorder (and the telephone answering machine!). Sony introduced the reel-to-reel tape recorder using polymer tape coated with magnetic particles in 1948. Further advances in sound recording came with the putting multiple tracks on a single tape (RCA, 1958) and the eventual introduction of cassettes (1962).

All of these technologies were analog. A microphone would convert acoustic signals eventually into currents that would be passed through a coil to generate a time-varying coercive magnetic field acting on a piece of recording medium passing nearby. Those fields would reorient the magnetization of the magnetic medium, leaving a spatially distributed pattern of magnetization (along the axis of the wire or in the plane of the tape). Readout would involve moving the medium past a close-by pickup coil, with fringing fields from the magnetization producing a time-varying magnetic flux that induced voltage signals in the coil. After amplification, those voltage signals could be used to drive speakers, reproducing sound.

Contemporaneous with this analog technology, magnetic materials were making their way into digital information, with magnetization direction standing in as a single bit (one or zero) in binary storage. In addition to magnetic tape, small ferromagnetic rings were used as "core" memory. Current pulses through wires threaded through the rings could orient the magnetic domains within the FM core. Similarly, changes in the magnetization direction would be detected via the wires through the voltage pulses induced by the reorientation. Reading core memory is "destructive", in the sense that bits are flipped to (or remain at) zero during the read process. Multiplexed operations are readily possible by threading two wires (a "word" line and "bit" line) through each FM ring. Current pulses in word lines and bit lines are, by themselves, too feeble to reorient $\mathbf{M}$, but together accomplish the task. Core memory, like other magnetic data storage, is *nonvolatile*; thanks to hysteresis, well-designed magnetic storage retains its state once the currents used to set that state cease. Core memory was eventually supplanted by charge-based storage, but there may yet be a resurgence of nonvolatile memory based on magnetic materials, as we shall see.

The first hard disk drive (HDD) was a component of the IBM RAMAC computer system (1957). The drive, the IBM 350 disk storage unit, consists of 50 platters each 60 cm in diameter. In a HDD, rigid platters coated with a magnetic thin film are employed as the storage medium.[2] By rotating the platters and having a read-head on an arm that can shift radially from the axis of rotation, it is possible to give rapid access to particular bits of information, stored as circumferential "tracks" spaced radially on the disk. This first HDD stored a total of 5 MB of data, at an areal density of 2 kb/in^2, and the data transfer rate was 8800 char/s.

Clearly we have come a *long* way. Figure 7.31(a) shows the trends [562] in HDD storage density from the inception of the technology until 2003, when IBM's storage division was acquired by Hitachi, and the exponential growth has continued since then. Areal density of storage has increased by nearly 10^9 over fifty years. The result, as shown in Fig. 7.31(b), is that the cost per bit of storing data magnetically has fallen far below that of traditional paper media.

Increasing the density of magnetic data storage has been achieved (obviously) by decreasing the physical size that each bit occupies on the disk surface. As we have seen, the physics relevant to magnetic properties evolves as material lengthscales are decreased into the nanoscale regime. As in the case of transistor-based nanoelectronics and scaling, in the end there are physical limitations to how far the scaling process may be carried. Those

---

[2] We will focus on hard disks; the now nearly extinct technology of "floppy disks" is similar, though with thin, flexible, polymer disks coated by magnetic material. One key reason for the (much) lower capacity of floppy disks was the challenge of positioning the read head relative to the flexible medium.

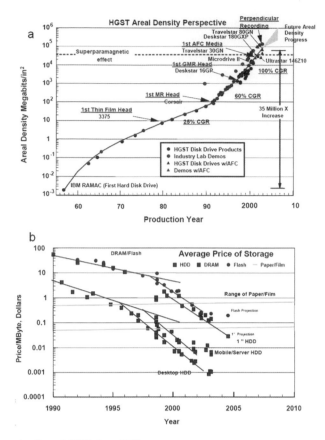

Fig. 7.31   Trends in data storage (up through 2003). From [562].

limitations are primarily associated with the disk media itself and the read and write head technology.

The interplay between physics, material properties, and scaling is best seen in the historical evolution of the HDD and its components. Originally, HDD platters were large (tens of cm) in diameter, and the read/write head consisted of a C-shaped piece of magnetically soft material (e.g., ferrite) wrapped by a wire coil. The gap in the head would be positioned over the disk medium. The core metal would enhance the magnetic field generated by the coil (for writing), which would reorient the in-plane magnetization of a patch of disk surface. For reading, the core metal would enhance the inductive pickup of the coil as stray magnetic induction from the disk surface passed through the gap. The physical size of the gap in the core set the spatial size of the bits. As demand for higher density storage grew, more advanced read and write heads were manufactured via photolithographic processes, shrinking the footprint of the heads and the size of bits.

At the same time, the need to increase the efficiency of head/medium coupling also drove down the physical separation ("flying height") between the heads and the rapidly rotating (thousands of revolutions per minute) disk surface. The flying height is now in the few-nm range (!), and relies critically on the fluid dynamics of the air inside the platter enclosure. Unsurprisingly, there are stringent requirements on disk flatness and cleanliness, and special areas of the disk surface reserved as "landing pads" for the head, in case of power outage or shutdown. The term "crash" to describe the sudden, unanticipated failure of a computer system has its origins in the physical crashing of drive heads into disk media.

Read heads based on inductive pickup were supplanted by those based on AMR by the mid-1990s. Writing information to the disk surface requires a separate head, since magnetoresistive heads are not intended to produce the required magnetic fields. Indeed, layers of soft magnetic materials are required to divert stray magnetic fields produced by the write head away from the highly sensitive read head. Around the same time, GMR-based read heads were demonstrated and introduced in commercial products over the next several years. A GMR read head consists of a magnetically fixed layer pinned by proximity to an antiferromagnetic film, and a magnetically "free" layer with a magnetization that may be reoriented by the stray fields from the disk medium. GMR read heads have largely been based on a "current in plane" design.

The disk medium itself has evolved. Platters were originally coated with magnetic oxide powder, and that technology was supplanted by thin film magnetic materials, such as CoPtCr alloys, deposited by sputtering. As bit size has decreased, the properties of the magnetic materials have assumed paramount importance. A central issue is superparamagnetism. If the areal density of data storage is 100 Gb/in^2 (a favorite unit), that implies an individual bit ~ 80 nm on a side. An individual crystal grain in the bit may be around 10 nm on a side, and a typical crystallographic anisotropy constant for disk material is around $K = 1.5 \times 10^5$ J/m^3. This implies an anisotropy barrier to spontaneous flipping of $\mathbf{M}$ of $KV \approx 1.5 \times 10^{-19}$ J, almost an electron volt. That may sound large, but if the typical attempt frequency Ω in Eq. (7.48) is 10^9 Hz, then at $T = 300$ K, there would be an effective relaxation rate of

$$\tau^{-1} \sim \Omega \exp(-KV/k_B T) \sim 1.8 \times 10^{-7} \text{ s}. \tag{7.49}$$

The typical timescale for thermal reversal of magnetization in particular grains would then be only two months (!), far shorter than a desirable period for archival data storage.

One obvious way to circumvent this problem is to use materials with larger values of K. However, larger values of K generally translate into larger coercive fields, H_c, and therefore bits that are more difficult to set or reset. One approach that has been successful has been the development of disk media incorporating other layers that change the energetics of reversal without drastically altering coercive fields. For example, Fig. 7.32 shows the use of a 0.6 nm thick (!) ruthenium layer sandwiched between FM media layers, a strategy developed by IBM, who referred to the ruthenium as "pixie dust". The ruthenium leads to antiferromagnetic coupling between the two FM layers. This coupling increases the effective volume of the bit (in the vertical direction) without drastically hurting the write head's ability to reorient its magnetization.

In 2006, Seagate and other storage companies introduced another innovation in the disk medium that also required a redesign of the write head: *perpendicular recording* [563]. Rather than having bits with magnetization in the plane of the disk, perpendicular media have $\mathbf{M}$ oriented normal to the disk surface. One clear advantage is that this geometry should produce more fringing fields for interaction with the read head. This allows

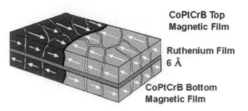

CoPtCrB Top
Magnetic Film

Ruthenium Film
6 Å

CoPtCrB Bottom
Magnetic Film

Fig. 7.32 Magnetic recording media using an ultrathin ruthenium layer to couple two layers antiferromagnetically. This approach helps suppress superparamagnetism. From IBM press release, 2001.

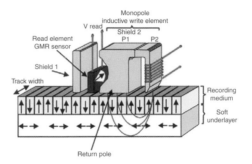

Fig. 7.33 Disk head design for perpendicular magnetic recording. Adapted from [564].

an increase in recording density, though it carries with it additional requirements: a crystallographic anisotropy large enough to overcome geometric anisotropy's tendency to favor $\mathbf{M}$ lying in the plane; textured deposition of the disk medium so that the appropriate crystallographic orientation is favored; and redesign of the write head shape and disk medium for efficient writing of bits. As shown in Fig. 7.33, the write head has a sharp, highly coercive tip to channel magnetic flux, and the disk medium has a soft magnetic underlayer for flux return. The trailing edge "return pole" of the write head provides a spread out path for the magnetic flux, so that the local $\mathbf{H}$ field there is too low to affect the bits not directly under the write pole. Even with this approach, the highest flux densities possible at the sharp write tip are set by the magnetization of available materials, and are limited to $\sim \mu_0 M_{\text{sat}} \approx 2.5$ T.

Present state-of-the-art HDDs have proprietary disk media (see Fig. 7.34), complex multilayer structures tailored to give precisely the right magnetic properties, and operate via perpendicular recording. The fly height of the heads over the disk surface is only a few nanometers. Storage densities are now approaching 500 Gbit/in^2, implying individual bits $\sim$30 nm on a side.

Read heads have also evolved to meet these challenges, switching from GMR to TMR for two main reasons. First, by engineering the oxide tunnel barrier, it is possible to achieve even larger magnetoresistances than GMR [522]. Second, TMR structures are naturally appropriate for current-perpendicular-to-plane (CPP) operation, and this geometry allows a smaller form factor for the read head [565].

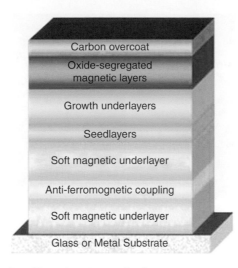

Disk media, showing the complicated layered structure used to fine-tune magnetic properties. Adapted from [564].   Fig. 7.34

7.5.2 Technological demands and next directions for data storage

While solid-state storage based on flash memory continues to increase in capacity and fall in price, truly high capacity, long-term storage will likely be dominated by magnetic technologies for a long time to come. The pressures to scale toward ever higher storage densities continue, and the technical challenges remain similar to those that have reshaped HDD technologies over the last decade.

Superparamagnetism remains a paramount concern as bit sizes are reduced. Several major HDD manufacturers are considering making the move to storage media with considerably higher anisotropy constants, even though such a transition would seem to make the writing task more difficult. To circumvent this problem, the manufacturers are considering *heat-assisted magnetic recording* (HAMR) [566]. In HAMR, a laser would be focused onto the bit to be written as the bit passes beneath the write head. This focusing would ideally take place via near-field optics or plasmonic means (see Chapter 8), to confine the optical interactions with the disk medium to a region comparable to the bit size and far smaller than an optical wavelength. The light would locally heat the disk medium. Based on a carefully chosen material, the medium would have an anisotropy constant that decreases markedly with increasing temperature, allowing the write process to take place without the application of unreasonably large magnetic fields.

HAMR requires the solution of a number of significant engineering challenges, including materials development, integration of near-field optics with magnetic storage technology, and careful control over heat transfer. The heat transfer problem is clearly a concern in the disk medium, where it is essential to avoid inadvertent heating of surrounding bits. It is also a major issue in the write head itself. The near-field and plasmonic optics techniques required to concentrate light into a deep subwavelength spot are quite inefficient, meaning that a significant fraction of the optical energy fed into the

Patterned disk media. A topograph (left) and an MFM image (right) of patterned disk media made through nanoimprint lithography. The full width of each image is four microns. Adapted from [567].

write head is likely to be absorbed there rather than in the disk medium. At issue is the thermal expansion of the write head, and the fact that the flight height of the head is only a few nanometers.

Another direction being examined is *patterned media* [567]. Having bits defined out of a continuous polycrystalline film by the location of the write head has worked so far, but size scales are approaching the limit where the number of grains per bit is no longer statistically large. That can lead to statistical fluctuations in the properties of individual bits. One way to avoid this is to have every bit be an individual single-domain magnetic grain, isolated by spacing from neighboring bits. The fabrication challenges inherent in this approach are formidable, requiring alignment and registration at the few-nm level, on a par with the needs of high density patterning in microprocessors. To achieve this accuracy and reproducibility over whole disk platters inexpensively is very difficult. Nanoimprint lithography is a front-running approach.

Magnetic data storage may have a future beyond disk media as well. There has been increased interest in *magnetic random access memory* (MRAM), and there are several different approaches that would marry modern nanophysics (GMR, TMR, spin torque) with some of the old virtues of magnetic core memory [568]. The appeal of very robust, nonvolatile data storage is great.

Some examples of MRAM designs are shown in Fig. 7.36. The simplest version of MRAM uses current pulses though bit and word lines to generate the local magnetic fields necessary to reorient the "storage" FM layer of a GMR multilayer. The state of the bit is defined by the resistance of the GMR stack, which is low (high) when the free and fixed magnetization layers are aligned (antialigned). A related approach uses GMR or TMR multilayers in the shape of rings, so that the two states are defined with **M** of the storage layer oriented circumferentially, either clockwise or counterclockwise.

Spin transfer torque has also been proposed as an alternative method of writing to MRAM bits, with spin-polarized current pulses driving precession of the magnetization in the storage layer. Measurements of the TMR or GMR stack resistance at lower currents would then be used for readout. This approach has the advantage of requiring fewer wires,

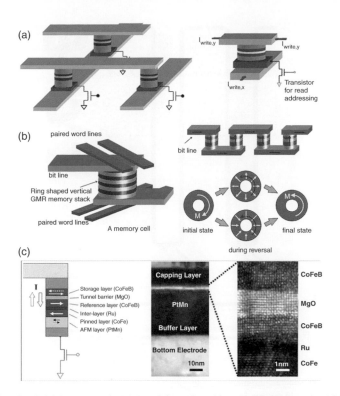

MRAM designs, from [568]. (a) Conventional MRAM, with bit and word lines used to generate local fields that affect GMR multilayers. (b) Toroidal MRAM, designed to mitigate effects of stray fields and writing errors. (c) A version of spin transfer torque MRAM involving a TMR multilayer. The cross-sectional TEM image clearly demonstrates that this is without doubt an example of nanotechnology.

Fig. 7.36

and thus being better suited to compact layout. However, the materials engineering required to get the optimal level of damping in the precession of the storage layer is not simple.

An alternative architecture called *racetrack memory* has been proposed by Stuart Parkin and coworkers at IBM, and is shown in Fig. 7.37. A racetrack memory cell is not random access memory; rather, it is a shift register that can store ~ 100 bits. The bits are represented by the location of domain walls in a magnetic nanowire. These walls may be introduced by a write line, either directly via coercive current pulses or through the fringing fields of domain walls in a ferromagnetic write wire. The walls may pushed forward or backward via current pulses, taking advantage of spin transfer torque. Local magnetization may be sensed via GMR or TMR detection. Racetrack memory is somewhere between MRAM, where one GMR/TMR stack is required per bit, and HDD storage, where a single GMR/TMR read head is used for 10^{13} bits. Dense integration may be possible since the magnetic nanowires may be oriented vertically, and none of the moving parts or mechanical tolerances associated with HDD storage are necessary.

As in our discussion of nanoelectronics, the wide-spread adoption of any of these advanced magnetic storage technologies depends on economics at least as much as it does on engineering sophistication and technical performance. The commercial adoption

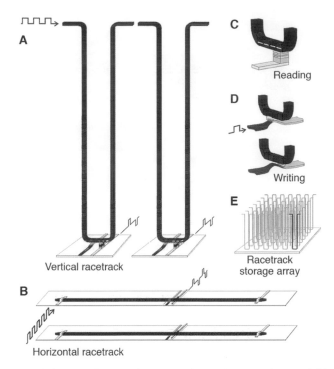

Fig. 7.37   Racetrack memory, in which domain walls are used to represent bits, and current pulses can shift bits along each nanowire, which acts as a shift register. From [569].

of MRAM has been long in coming because of materials, fabrication, and reliability challenges. In the meantime, charge-based flash memory has seen plummeting prices and soaring capacities. While magnetic memory has an edge in both speed and data retention, it remains to be seen whether it will be able to displace flash as a widely deployed form of nonvolatile storage.

7.6   The future: spintronics

Beyond data storage, there is tremendous interest in somehow taking advantage of the electron's spin degree of freedom for information processing. Every electron carries spin, and one can imagine storing and manipulating information by using the spin to represent a bit. Moreover, if one can transduce back and forth between spin states and more readily measurable charge-related quantities (currents and voltages), it may be possible to use spin manipulation to enable novel switching and logic devices. One potential advantage of *spintronic* approaches [570, 571] is that spin currents do not have to be dissipative. Since one can have net flow of spin without net flow of charge, the prospect exists of performing information processing with significantly reduced power consumption.

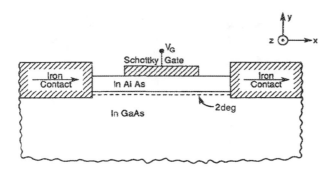

Fig. 7.38

The Datta–Das concept for a spin transistor [572]. In the presence of spin–orbit coupling, the transverse electric field from the gate electrode will cause the injected polarized spins in the channel to precess, modulating the source–drain conductance.

Major challenges remain, however. As discussed in Section 7.4.2, unlike charge, spin is not conserved, meaning that spin currents persist only over finite distances. Coherent manipulation of spin is limited by T_1 and T_2 processes. For interfacing with conventional technologies, electrically generating spin currents of significant polarization in conventional semiconductor materials would be very helpful. It turns out, however, that this is quite challenging.

The Datta–Das spin transistor [572], shown in Fig. 7.38, is an example of the promise of spintronics. The idea is straightforward. A spin polarized current is injected from a ferromagnetic source and meant to be collected at a ferromagnetic drain. The channel material is chosen so that the carriers exhibit spin–orbit couplings. Current flow between source and drain is modulated by the electric field from the gate. Because of spin–orbit coupling, from the point of view of the carriers the gate electric field acts as an effective magnetic field, causing voltage-tunable precession of the spins of the carriers. For a perfectly polarized current and flawless interfaces, one could then use precession to tune from full transmission to complete suppression of current.

7.6.1 Generating and detecting spin currents

Implementing spintronic ideas requires both the injection and detection of spin currents in nonmagnetic materials. In metals, we have seen that this has been accomplished in conjunction with electrical currents simply by having the carriers flow from an itinerant ferromagnet (with nonzero polarization at the Fermi level) into a nonmagnetic metal. For maximal spin injection from a polarized material, one would ideally want to work with a half-metal, such as CrO_2.

The situation is more complex in semiconductors. The experimental situation can be very challenging to unravel purely from magnetoresistive effects, since Hall voltages due to fringing fields from ferromagnetic materials are not necessarily small. More seriously, while semiconductors have higher charge mobilities than metals, their actual conductivities are generally much lower because of their much smaller carrier density. This

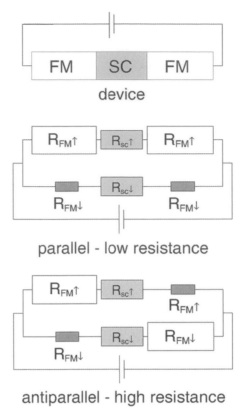

Fig. 7.39   The challenge of spin injection from ferromagnetic metals into semiconductors [574]. The mismatch between metal and semiconductor conductivity greatly reduces the effectiveness of spin injection compared to the metal/metal case.

mismatch has serious consequences for attempts to inject or collect spin directly across the ferromagnetic/semiconductor interface [573, 574]. The situation is shown in Fig. 7.39.

Let's work in terms of the particular spin channels, and assume that the majority spin is pointed down. The electrical resistance of the spin-up channel in the FM metal is $R_{FM}^{\uparrow} = 2R_{FM}/(1 - P)$, where R_{FM} is the total resistance of the metal, and P is the polarization of the ferromagnet. Similarly, $R_{FM}^{\downarrow} = 2R_{FM}/(1 + P)$. This makes sense, so that the parallel resistance of the two spin channels is R_{FM}, and the channel with the larger density of states has the lower effective resistance. In the semiconductor, the two spin channels ordinarily have identical densities of states and scattering properties, so $R_{SC}^{\uparrow} = R_{SC}^{\downarrow} = 2R_{SC}$.

With these resistances, it is now possible to compute the total polarization of the resulting current in the semiconductor:

$$P_{SC} = P \cdot \left(\frac{R_{FM}}{R_{SC}}\right) \left(\frac{2}{2(R_{FM}/R_{SC}) + (1 - P^2)}\right). \qquad (7.50)$$

While the spin polarization of the current in the semiconductor is proportional to that in the ferromagnet, it is reduced by a factor of R_{FM}/R_{SC}, which can be $\sim 10^{-4}$. This conductance mismatch between the materials leads to a strong suppression of GMR-type effects. Instead of the Julliere result (Eq. (7.43)), one obtains

$$\frac{\Delta R}{R_{\parallel}} = \frac{P^2}{1 - P^2} \frac{R_{FM}^2}{R_{SC}^2} \frac{4}{4(R_{FM}^2/R_{SC}^2) + 2(R_{FM}/R_{SC}) + (1 - P^2)}, \quad (7.51)$$

which is smaller by a factor of $\sim R_{FM}^2/R_{SC}^2$.

There are multiple ways to try to circumvent this difficulty. One approach is to have spin injection and collection take place via tunnel junctions. Rashba has shown [575] that the spin polarization of the current in the semiconductor is determined by the lowest conductance in the total current path. By interposing a (comparatively) high resistance tunnel junction in series with the FM and semiconductor components, it is possible to then achieve efficient spin injection.

A competing approach is to use ferromagnetic semiconductors rather than ferromagnetic metals as the source and drain of spin polarized current. These ferromagnetic materials (see Section 3.3.5) have conductivities comparable to that of the nonmagnetic channel, and therefore do not fall victim to suppressed injection efficiencies.

A third approach to the generation of spin currents touches upon a popular means for their detection, the interaction of charge carriers with light. By using circularly polarized light to excite electron–hole pairs, conservation of quantum mechanical angular momentum ensures that the resulting electrons can have some spin polarization. This may be detected optically as well, using MOKE (in reflection) or Faraday rotation (in transmission). An early cryogenic demonstration of this [576] showed another interesting effect. In the absence of dopants, the spin lifetime of the optically excited electrons is quite short (picoseconds), apparently due to the Bir–Aronov–Pikus exchange between the electrons and the strongly spin-orbit-scattered holes. However, in the presence of an electron gas doped into the system, the lifetime T_2 could be much longer (nanoseconds), as the already-present electrons could recombine with the optically excited holes, eliminating the B-A-P spin relaxation.

Likewise, optical spin injection has been combined with spatially resolved Faraday rotation to observe the drift of spin-polarized carriers through semiconductor structures [577]. In the presence of an electric field to act on the photogenerated electrons, spin polarization has been obseved to traverse hundreds of microns in appropriate materials.

A third optical method, the monitoring of *emitted* light, has been used to detect spin currents. A great example of the use of ferromagnetic semiconductors as spin injectors is the work of Ohno *et al.* [578], in which a layer of GaMnAs is used as an injector of partly spin-polarized holes in a light-emitting diode structure. The emitted light was found to be circularly polarized, consistent with the expected current of angular momentum carried along by the spin-polarized carriers.

7.6.2 Spin Hall effect

There is also great interest in creating spin polarization of mobile charge carriers by spatially separating the spin species, rather than injecting polarized carriers optically or electrically. One recently discovered way to do this is through the *spin Hall effect*. In contrast to the ordinary Hall effect, the spin Hall effect can take place in the absence of an

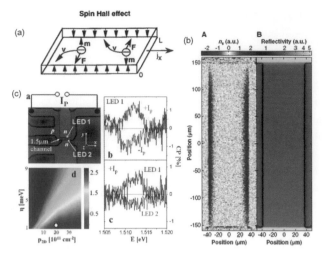

The spin Hall effect. (a) Due to spin–orbit effects, a dc current results in a spin imbalance along sample edges, transverse to the direction of current flow [579]. Shown is the intrinsic spin Hall effect. (b) MOKE has been used to image the spin imbalance of carriers accumulated at the edges of a GaAs bar, likely due to the extrinsic spin Hall effect [580]. (c) Further evidence for spin separation, in the form of circularly polarized light emission of opposite helicity at opposing transverse edges of a nanostructured 2d electron/2d hole nanostructure [581].

applied external magnetic field. Figure 7.40(a) shows the situation. The flow of a dc current longitudinally in a material generates a spin imbalance (but no net charge imbalance!) in the transverse direction. (It should be noted that there is also a quantized spin Hall effect [53], intimately connected with topological properties of surface states; further discussion of this is beyond the scope of this book.)

The essential ingredient is spin–orbit coupling, which can lead to both extrinsic and intrinsic spin Hall phenomena. In the extrinsic case [579], with proper spin–orbit scattering and band structure, there is an inherent asymmetry in the scattering of charge carriers by impurities that depends on the carrier spin direction. Disorder scattering thus leads to a systematic separation of spin species, with carriers of one spin orientation ending up dominantly on one transverse edge of the sample, and carriers of the opposite spin orientation ending up on the opposing edge.

The intrinsic case [582] arises because of the Rashba spin–orbit coupling [583] term in the Hamiltonian for the electrons. Suppose that carriers are moving in a crystal where inversion symmetry is broken along the z-axis. This may be particularly relevant in a 2d electron gas. One version of spin–orbit coupling in this situation (neglecting the explicit lattice potential) is

$$H = \frac{p^2}{2m} - \frac{\lambda_R}{\hbar} \sigma \cdot (\hat{\mathbf{z}} \times \mathbf{p}), \qquad (7.52)$$

where λ_R is the Rashba couping constant, σ is the Pauli spin matrices, and $\hat{\mathbf{z}}$ is the unit vector in the z direction. Because of this coupling, as electrons are accelerated in the x-direction (by the application of a longitudinal voltage to drive a longitudinal current), a torque is effectively exerted on their spins that depends on the y-component of the electron momentum. As a result, carriers moving transversely to the longitudinal current acquire a spin polarization such that there is a spin imbalance between the two transverse edges of the system. If written as a spin Hall conductivity (the 2d transverse spin current density divided by the longitudinal electric field), this intrinsic spin Hall effect is predicted to have a universal value.

The spin Hall effect has been measured recently in semiconductor systems, and it can be challenging to determine if it is extrinsic or intrinsic. One clear example [580] is shown in Fig. 7.40(b), in which MOKE is used to measure the net spin accumulation at the edges of a GaAs bar. This particular experiment is believed to examine the extrinsic spin Hall effect, since the accumulation of spin shows no particular dependence on the crystallographic orientation of the specimen. A second [581] is shown in Fig. 7.40(c), where light emission is produced at the edges of a 2d of a 2d hole gas near a 2d electron gas. The emitted light is circularly polarized by an amount proportional to the current, and reverses polarization when the current direction is flipped, consistent with expectations of the intrinsic spin Hall effect building up spin polarization at the edge of the hole gas.

7.7 Nanomagnetism: other applications

This chapter has largely dealt with magnetoelectronics, but there are other areas of science and engineering where nanomagnetism, or at least the development of nanoscale magnetic materials, is proving to be very fruitful. For completeness, here we briefly describe two.

Magnetic nanoparticles have proven to be a potential boon for medical diagnostics and treatment [584, 585], relying on two particular pieces of physics. First, magnetic nanoparticles or nanomaterials can act as contrast agents in magnetic resonance imaging (MRI). Conventional MRI looks at the signal from the magnetic moment of protons, typically those in the hydrogen nuclei of water molecules. Because different tissue types within the body contain different amounts of water, MRI is a useful imaging technique that can reveal significant soft-tissue contrast (unlike ordinary x-ray imaging). In particular, conventional MRI is based on pulse sequences that rely on the coherent precession of the protons. Any interaction that causes dephasing or decoherence of the protons relative to each other affects the signal. Magnetic nanoparticles produce local inhomogeneous fields and interact magnetically with the precessing protons, significantly shortening both T_2 and T_2^*. Magnetic nanoparticles may be introduced into the body directly into an area of interest, or they may be "targeted" through biochemical surface modification to accumulate in particular tissues, including tumors. Through their dephasing interactions with protons, they can greatly heighten MRI contrast and aid in imaging and diagnostics.

Magnetic nanoparticles have also been demonstrated in treatment approaches, through a technique called *magnetic hyperthermia* [586]. The idea is to use radio frequency electromagnetic fields to heat the nanoparticles, in turn raising the temperature of the surrounding biological tissue by several degrees. Since biological processes can be extremely sensitive to such modest temperature increases, hyperthermia may be used as a directed means to kill tumors. Because of their large effective susceptibility and the characteristic timescale for the reorientation of **M**, superparamagnetic nanoparticles are very effective for this purpose.

A related treatment approach is to use magnetic nanoparticles for drug delivery. Nanoparticles may be conjugated to drugs, and either targeted to the desired site, or potentially guided with external magnetic fields. Ideally, one could use the nanoparticles' MRI contrast functionality to guarantee that they had reached the correct location. Hyperthermia could also be a means of triggering drug release, such as from a thermoactive polymer.

The ability to exert forces on magnetic nanoparticles via magnetic field gradients is central to another proposed use, *magnetic separation*. The nanoparticles may be conjugated to desired analytes and then pulled from solution using strong magnetic field gradients. This is a relatively old idea [587], but has seen new excitement recently with the development of chemically synthesized magnetic nanoparticles. Magnetic separation may be done for the purposes of biological or medical assays [588], but it may also be used for the removal of, e.g., toxins from water [589].

The physics of magnetic separation in the case of nanoparticles is somewhat nontrivial, as can be seen on simple grounds. Consider a single-domain nanoparticle of diameter d, with saturation magnetization M_{sat}. The force experienced by such a particle in a field gradient is

$$\mathbf{F} = \frac{4\pi}{3}(d/2)^3 M_{\text{sat}}(\hat{\mathbf{m}} \cdot \nabla)\mathbf{B}, \tag{7.53}$$

where $\hat{\mathbf{m}}$ is the unit vector in the direction of the magnetization. This scales like d^3. However, as we will see in Chapter 10, the viscous forces from the surrounding liquid scale like d. Furthermore, the typical fluctuating forces on the nanoparticle due to Brownian motion are $\sim k_{\text{B}}T/d$. At first glance, it appears that scaling to smaller particle diameters will make it much more difficult to use field gradients to remove particles from fluid. This argument is correct for individual particles.

If one considers interparticle interactions, the situation improves. An order-of-magnitude estimate for the field gradient in the immediate vicinity of another nanoparticle is $\mu_0 M_{\text{sat}}/d$. For reasonable materials and nanoparticle diameters, this can easily dwarf externally applied field gradients. Interparticle forces can be dominant, and the collective response of an ensemble of nanoparticles can be used to remove them from solution [589, 590]. Similar collective effects on a much larger scale are responsible for the properties of magnetorheological fluids [591, 592], which exhibit drastic changes in their mechanical properties as a function of applied magnetic field.

7.8 Summary and perspective

We examined the basis of magnetism in solids, the interplay of magnetic degrees of freedom with electrical conduction, and the role of magnetic materials in modern data storage technologies, specifically as follows.

- Magnetic materials and responses. Materials can have different magnetic responses, including magnetic ordering brought about by their electronic structure. We defined magnetic units and terminology, with an eye toward ferromagnetism, since ferromagnetic materials have had the greatest technological utility.

- Magnetization and domains. In ferromagnetic or ferrimagnetic materials, $\mathbf{M(r)}$ generally assumes the configuration which minimizes the total energy, including that due to stray fields. The result is the formation of domains in larger structures, with often nanoscale domain walls. There are a number of experimental techniques for assessing the spatial distribution of magnetization at the nanoscale, including optics (MOKE), electron optics (electron holography), and scanned probe microscopy (MFM, MRFM, spin polarized STM).

- Magnetization and electronic transport. The interplay between $\mathbf{M}$ and electronic current leads to a number of magnetoresistive effects, from the intrinsic (AMR, CMR) to those associated with the engineering of artificial nanostructures (GMR, TMR, EMR). These magnetoresistive effects provide the basis for magnetic data storage technologies, and can be used to infer domain wall motion. Conversely, polarized currents may be used to manipulate magnetization through spin transfer torque.

- Magnetic data storage. Magnetic data storage has increased remarkably in capacity and areal density over the last several decades. This growth has entailed challenges in both magnetic disk media (the superparamagnetic limit; the switch to vertical recording) and read-head technologies. GMR- and TMR-based magnetic RAM and related approaches (racetrack memory) are candidates for nonvolatile storage, the nanoscale resurgence of ideas dating back to core memory from the middle of the last century.

- Spin in nonmagnetic materials. Net spin polarization can be created in nonmagnetic materials through spin injection, either electrically (from materials with net spin polarization at the Fermi level) or optically (in semiconductors, via circularly polarized light). While angular momentum is a conserved quantity, there are multiple relaxation mechanisms that act to disrupt net spin polarization of charge carriers in nonmagnetic materials. Spin–orbit interactions can lead to spin Hall effects. Dissipationless spin currents are possible, raising hopes for information processing with low intrinsic heating.

- Magnetic nanoparticles and their uses. Magnetic nanoparticles have found uses in other applications, including assays and manipulation of particles and cells. Their individual magnetic responses allow their use as contrast agents in MRI and as local heating sources when excited via appropriate rf magnetic fields. Collective responses of interacting nanoparticle ensembles can lead to nontrivial effects, including magneto-rheological properties.

In the next chapter, we move from considering magnetic response to the interaction of (nano)materials with electromagnetic radiation, a subject now known by the trendy name of *photonics*.

7.9 Further reading

- *Magnetism and magnetic materials* There are a number of good books that treat magnetism, magnetic materials, and magnetic units. Examples include *Physics of Ferromagnetism* by Chikazumi (Oxford University Press, 2009), *Modern Magnetic Materials: Principles and Applications* by O'Handley (Wiley-Interscience, 1999), and *Introduction to Magnetism and Magnetic Materials* by Jiles (CRC Press, 1998). Gaj and Kossut recently edited a review volume, *Introduction to the Physics of Diluted Magnetic Semiconductors* (Springer, 2011).
- *Mapping magnetism on the nanoscale* Sidles *et al.* have written an excellent review article, "Magnetic resonance force microscopy", *Rev. Mod. Phys.* **67**, 249–265 (1995). More recently, Weisendanger has comprehensively reviewed spin polarized STM in "Spin mapping at the nanoscale", *Rev. Mod. Phys.* **81**, 1495–1550 (2009).
- *Magnetic data storage* An interesting historical look at magnetic recording (though a bit dated now) is *Magnetic Recording: the First 100 Years*, edited by Daniel, Mee, and Clark (Wiley-IEEE, 1998). A recent special issue of the Proceedings of the IEEE (*Proc. IEEE* **96**(11), November, 2008) contained a number of review articles dedicated to current and future developments.
- *Spin injection and spintronics* A recent comparative review of the whole topic is *Spin Electronics* (Springer, 2010), edited by Awschalom *et al.* Two freely available sets of lecture notes on semiconductor spintronics are available via the electronic preprint Arxiv, as article numbers 0711.1461 ("Semiconductor spintronics", Fabian *et al.*, 2007) and 0801.0145 ("Lecture Notes on Semiconductor Spintronics", Dietl, 2008).
- *Other uses of magnetic nanomaterials* Gubin has edited *Magnetic Nanoparticles* (Wiley-VCH, 2009), a volume covering the synthesis, characterization, and properties of magnetic nanoparticles as well as their application to biological systems. Likewise, Varadan, Chen, and Xi have recently written *Nanomedicine: Design and Applications of Magnetic Nanomaterials, Nanosensors, and Nanosystems* (Wiley, 2008). A very highly cited review article on this topic is Gupta and Gupta, "Synthesis and surface engineering of iron oxide nanoparticles for biomedical applications", *Biomaterials* **26**, 3995–4021 (2005).

Exercises

7.1   *Physics of hard drive design* Hard drives today function using a single "head" made of two parts, one that writes and one that reads. Here we examine some of the physics issues that determine the way drive heads are designed and how they operate.

Hard drive media have an extremely smooth substrate coated with a thin polycrystalline film (~ 50 nm thick) made from a cobalt alloy. As we saw in our discussion of demagnetizing factors (Eq. (8.30) and following), generally for thin films it is energetically favorable for the magnetization to lie in the plane of the film. Each bit is stored as a region of disk material that is magnetized either parallel or antiparallel to the direction of the disk's rotation. When two adjacent bits are antialigned, the net vertical component of the magnetic induction is significant.

The write part of the HDD head is a permalloy yoke, as shown in Fig. 7.33. When the current loop around the yoke is energized, the yoke magnetization approaches its saturation value. The yoke has a small gap.

(a) If the saturation magnetization of the yoke material is M_s and the gap in the yoke is small, explain why the magnetic induction in the center of the gap, B_g, is approximately $\mu_0 M_s$. In (b) you'll need to know that the saturation magnetization for this permalloy is 8×10^5 A/m.

(b) The x component of the fringing magnetic field from the gap in the yoke approximately obeys

$$H_x = \frac{B_g}{\mu_0 \pi} \tan^{-1} \left(\frac{zd}{z^2 - (d/2)^2} \right)$$

a distance z directly beneath the center of the gap. Here B_g is the magnetic induction at the center of the gap and d is the width of the gap. If the coercive field of the magnetic disk material is 10^5 A/m, and the gap is 50 nm, how close does the write head need to be to the disk surface for H_x at the surface to be $\sim 3\times$ the coercive field? This is one constraint on the fly height of drive heads.

7.2 *Reading hard drive data* There have been multiple generations of readout techniques for magnetic media. Here we consider two.

(a) Assume that a typical bit is 80 nm by 80 nm patch of the cobalt alloy described above (corresponding to roughly 10^9 bits/in^2). Even though it's not too good an approximation, we can get the right order of magnitude for our calculation by treating two adjacent, antialigned bits as point dipoles located 80 nm apart. At the midpoint between the two bits, at an "altitude" of 25 nm, what is the magnitude of the vertical component of B due to the magnetization of those two bits?

(b) Old-style (pre-1994) hard drive read-heads worked inductively. As the disk rotated past beneath the head, the fields like those calculated in (a) would sweep through the read-head, which (looking something like the yoke from Exercise 1) was composed of some highly permeable ($\mu_r \sim 10,000$) material with a coil of, say, 10 turns (so for a given field ten times as much flux goes through the coils as for a single turn). Assume a turn size something like 100 microns by 100 microns, and a linear speed of the disk medium under the head of ~ 50 m/s, and use the number from (a) for the typical field associated with one bit. Using Faraday's law, estimate the voltage pulse generated in the coil for for each bit (antialigned pair) that it sweeps by.

(c) Modern hard-drive read-heads work magnetoresistively. Typical magnetoresistive heads change their resistance by about 1% in a field of 5 Gauss. Using the basic idea from part (a) of this problem, how close to the surface of the disk would a TMR read-head need to be to experience that sort of resistance change?

7.3   *A scaling problem* Using the material parameters for the cobalt disk medium from above, pretend individual bits are roughly square, with side length L. Again model two adjacent bits as point dipoles separated by that distance L. Estimate how small bits can be made before magnetic interactions between nearest neighbors lead to collective effects (and individual bits are no longer independent). Hint: think about coercive fields.

7.4   *Energetics of magnetism* Iron has a Curie temperature of 1043 K. Using $0.3k_B T_c$ as a rough proxy for the exchange energy, estimate the energy per unit area of an abrupt domain wall at which the magnetization reverses direction from one atomic layer to the next. Assume a typical density of atoms as 10^{14} per cm^2. This is why domain walls are generally not one atom thick.

7.5   *Lorentz microscopy* For 100 kV electrons passing through a 20 nm thick piece of material with $\mu_0 M_{sat} = 1$ Tesla oriented perpendicular to the beam direction, what angular deflection is expected? Suppose that this TEM is capable of atomic-resolution imaging of nonmagnetic materials, using a detector 1 mm behind the sample, with detector pixels 1 micron in extent. (Hint: think about angular deflections that could be detected with such a system.) Roughly how weak a magnetization could be detected, when compared with a nonmagnetic region?

Photonics   8

We begin this chapter with a brief review of electromagnetic radiation and its interactions with materials and interfaces. We consider nonlinear optical effects, and give an overview of lasers. After a quick review of photonics technology in the context of telecommunications, we consider *nanophotonics* – optical phenomena involving nanostructured materials. This includes dielectric mirrors and their three-dimensional generalization, photonic band gap structures, as well as plasmonic nanostructures. Research in plasmonics has become extremely fast-paced lately, in part because of the availability of new experimental and computational tools, and in part because of the promise of surface enhanced spectroscopies and more exotic effects such as "perfect" lenses and invisibility cloaks.

As always, entire books have been written on the various components of nanophotonics. Here we will look specifically at the "nano" aspects of these electromagnetic phenomena, with an emphasis on the underlying physics, the synergy between electromagnetic metamaterials and electronic structure of materials, and how those ideas lead to rich, fascinating, and useful phenomenology.

8.1 Electromagnetic radiation in a nutshell

In the many-photon limit, photonics is better known as the manipulation of classical electromagnetic radiation. All of the interesting effects in nanophotonics originate from underlying equations that govern the interactions of EM radiation and matter, and the electromagnetic response functions of the matter itself. In full generality this can be incredibly messy. Boundary conditions for the $\mathbf{E}$ and $\mathbf{B}$ fields must be satisfied at all times everywhere, including interfaces between materials that can have very different intrinsic properties. The materials themselves can have electromagnetic response functions (e.g., the dielectric function and the magnetic permeability) that can be tensorial, nonlinear, and strongly dependent on frequency.

8.1.1 Maxwell and waves

Let's start with Maxwell's equations:

$$\nabla \cdot \mathbf{D} = \rho, \qquad (8.1)$$

$$\nabla \cdot \mathbf{B} = 0, \qquad (8.2)$$

$$\nabla \times \mathbf{E} = -\frac{\partial}{\partial t} \mu \mathbf{H}, \tag{8.3}$$

$$\nabla \times \mathbf{H} = \frac{\partial}{\partial t} \epsilon \mathbf{E} + \mathbf{J}. \tag{8.4}$$

Equation (8.1), with $\mathbf{D} \equiv \epsilon \mathbf{E} \equiv \kappa \epsilon_0 \mathbf{E}$, and ρ as the volume charge density source term, is nothing more than Gauss' Law. Equation (8.2) is the usual restriction that there are not, as far as we know, an free magnetic monopoles. Equation (8.3) is Faraday's Law, where we have explicitly written out $\mathbf{B} = \mu \mathbf{H}$. Lastly, Eq. (8.4) is Ampere's Law, where we have included both the displacement current and "free" current density source term. In writing the equations this way, we have already simplified things from complete generality, in which the dielectric permittivity, ϵ, and the magnetic permeability, μ, are tensorial.

Start by ignoring the source terms (that is, $\mathbf{J} = 0$, $\rho = 0$), and taking the curl of Eq. (8.3):

$$\nabla \times (\nabla \times \mathbf{E}) = \nabla \times \left(-\frac{\partial}{\partial t} \mu \mathbf{H} \right)$$

$$= -\mu \frac{\partial}{\partial t} (\nabla \times \mathbf{H})$$

$$= -\mu \epsilon \frac{\partial^2 \mathbf{E}}{\partial t^2}. \tag{8.5}$$

Here we've simplified further by assuming that μ is independent of spatial coordinates (clearly not correct when there are interfaces or other inhomogeneities) and not time-varying, and the same restrictions apply to ϵ. Using a standard vector identity, $\nabla \times (\nabla \times \mathbf{A}) = \nabla(\nabla \cdot \mathbf{A}) - \nabla^2 \mathbf{A}$, we arrive at a wave equation for the electric field in a uniform medium:

$$\nabla^2 \mathbf{E} = \mu \epsilon \frac{\partial^2 \mathbf{E}}{\partial t^2}. \tag{8.6}$$

One can do the analogous manipulations starting from Eq. (8.4) and end up with a similar wave equation for $\mathbf{H}$:

$$\nabla^2 \mathbf{E} = \mu \epsilon \frac{\partial^2 \mathbf{E}}{\partial t^2}. \tag{8.7}$$

The derivation of this again implicitly assumes the restrictions that ϵ, μ are uniform in space and not varying in time. Indeed, once gauges are chosen there are corresponding wave equations for the scalar and vector potentials as well. Rather than relaxing those assumptions, it is often sufficient to break up a system of interest into domains over which those restrictions apply, and then worry about matching boundary conditions at the interfaces.

8.1.2 Fourier analysis

In nice, uniform, linear media, solutions to Eqs. (8.6) and (8.7) can be written as plane waves. It is often convenient to write down the solution for a particular Fourier component, knowing that linearity and superposition allow us to construct more complicated wave structures out of the plane waves:

$$\mathbf{E} = \mathbf{E_{0k}}e^{i(\mathbf{k}\cdot\mathbf{r}-\omega t)}, \tag{8.8}$$

$$\mathbf{H} = \mathbf{H_{0k}}e^{i(\mathbf{k}\cdot\mathbf{r}-\omega t)}. \tag{8.9}$$

We discuss the complex notation and its implications below. It's common physics shorthand to describe harmonic time dependences and phase differences. Here $\mathbf{E_{0k}}$ and $\mathbf{H_{0k}}$ are (complex) vectors that describe the *amplitude* (and relative phase) of the time- and space-dependent fields.

These trial solutions effectively convert the differential equations into algebraic equations, with $\partial/\partial t \to -i\omega$, and $\nabla \to i\mathbf{k}$:

$$(-k^2 + \mu\epsilon\omega^2)\mathbf{E_{0k}}e^{i(\mathbf{k}\cdot\mathbf{r}-\omega t)} = 0,$$

$$(-k^2 + \mu\epsilon\omega^2)\mathbf{H_{0k}}e^{i(\mathbf{k}\cdot\mathbf{r}-\omega t)} = 0. \tag{8.10}$$

The requirement that these be true for any chosen amplitude leads to the *dispersion relation* for electromagnetic plane waves:

$$-k^2 + \mu\epsilon\omega^2 = 0. \tag{8.11}$$

This tells us the *phase velocity* of the waves:

$$v_p \equiv \frac{\omega}{k} = \frac{1}{\sqrt{\mu\epsilon}}. \tag{8.12}$$

The phase velocity is the velocity of a surface of constant phase as the EM wave propagates. In vacuum, v_p reduces to $1/\sqrt{\mu_0\epsilon_0} = c$.

The *index of refraction*, n, is defined as the ratio of the speed of light in vacuum to the speed of light (phase velocity) in a medium:

$$n \equiv \sqrt{\frac{\mu\epsilon}{\mu_0\epsilon_0}}, \tag{8.13}$$

so that $v_p = c/n$.

Faraday's Law provides the constraint linking the amplitudes of the electric and magnetic fields:

$$i\mathbf{k} \times \mathbf{E_{0k}}e^{i(\mathbf{k}\cdot\mathbf{r}-\omega t)} = i\mu\omega\mathbf{H_{0k}}e^{i(\mathbf{k}\cdot\mathbf{r}-\omega t)}. \tag{8.14}$$

$$\to \mathbf{H_{0k}} = \frac{1}{\mu\omega}\mathbf{k} \times \mathbf{E_{0k}} = \sqrt{\frac{\epsilon}{\mu}}\frac{\mathbf{k}}{k} \times \mathbf{E_{0k}} = \sqrt{\frac{\epsilon}{\mu}}\hat{\mathbf{e}}_k \times \mathbf{E_{0k}}. \tag{8.15}$$

It is useful to define $Y \equiv \sqrt{\epsilon/\mu}$ as the characteristic *admittance* of a medium.

From Gauss' Law, we know that the electric field for an EM wave in a uniform medium in the absence of free charges is *transverse* to the direction of propagation:

$$\nabla \cdot \epsilon\mathbf{E} = 0 \to \mathbf{k} \cdot \mathbf{E_{0k}} = 0. \tag{8.16}$$

Using the Faraday result again, since $\mathbf{H_{0k}} \sim \mathbf{k} \times \mathbf{E_{0k}}$, we see that the magnetic field is also transverse. An example of such a wave is drawn at one instant in time in Fig. 8.1.

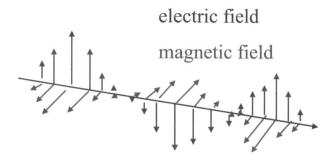

Fig. 8.1 A propagating, linearly polarized electromagnetic wave. The electric and magnetic fields are both transverse to the propagation wavevector.

When we introduced our notation, we mentioned that $\mathbf{E}_{0k}$ and $\mathbf{H}_{0k}$ are complex vectors. This means that each cartesian component is complex. The physical field at some moment in time is the *real* part of the complex vector. For example, consider

$$\mathbf{E}_{0k}e^{-i\omega t} = (E_x\hat{\mathbf{x}} + E_y e^{i(\pi/2)}\hat{\mathbf{y}})e^{-i\omega t}. \tag{8.17}$$

At $\omega t = 0$, the (real part of the) electric field points along x, and at $\omega t = \pi/2$, the field points along y. One could write the real part of this as

$$\mathbf{E}_{0k}(t) = E_x\cos(\omega t)\hat{\mathbf{x}} + E_y\sin(\omega t)\hat{\mathbf{y}}. \tag{8.18}$$

With differing magnitudes of E_x and E_y, this is elliptically polarized light. With identical magnitudes, it is circularly polarized light. Writing things this way makes the notation nice and compact, but the tradeoff is that we have to remember when complex notation is appropriate, and when we need to worry about the real parts of the fields.

As long as we are discussing polarization, consider two waves of the same amplitude, E_0, and orthogonal polarizations that differ in frequency by some small amount, $\Omega \ll \omega$. The total electric field at some fixed position is the superposition of that from the two waves:

$$\mathbf{E} = E_0\hat{\mathbf{y}}e^{-i\omega t} + E_0\hat{\mathbf{z}}e^{-i(\omega+\Omega)t}$$
$$= E_0(\hat{\mathbf{y}} + \hat{\mathbf{z}}e^{-i\Omega t})e^{-i\omega t}. \tag{8.19}$$

When written this way, the polarization at this position varies in time at frequency Ω. A slow detector that averages over times long compared to $1/\Omega$ would see an average of all the polarizations.

Now suppose the amplitudes of the two frequency components also differed slightly. We could write that case as:

$$\mathbf{E} = E_0\frac{1+p}{\sqrt{1+p^2}}\hat{\mathbf{y}}e^{-i\omega t} + E_0\frac{1-p}{\sqrt{1+p^2}}\hat{\mathbf{z}}e^{-i(\omega+\Omega)t}$$
$$= E_0\frac{1-p}{\sqrt{1+p^2}}\hat{\mathbf{y}}e^{-i\omega t} + E_0\frac{2p}{\sqrt{1+p^2}}\hat{\mathbf{y}}e^{-i\omega t} + E_0\frac{1-p}{\sqrt{1+p^2}}\hat{\mathbf{z}}e^{-i(\omega+\Omega)t}. \tag{8.20}$$

Here $p < 1$ characterizes the difference in amplitudes. If we consider what a slow detector of y-polarized light intensity would find, we see that Eq. (8.20) would be equivalent to an *unpolarized* wave containing the fraction $(1 - p)^2/(1 + p^2)$ of the intensity, and a y-polarized wave containing $4p^2/(1 + p^2)$ of the energy. Colloquially, the combined waves have a fractional polarization of $4p^2/(1 + p^2)$ in the y-direction.

8.1.3 Superposition and velocities

One incentive for working with Fourier components $\mathbf{E}_{0\mathbf{k}}$ is that in linear media, we can write a generic field configuration as a superposition of plane waves. Starting from a given electric field configuration at $t = 0$, $\mathbf{E}(\mathbf{r}, t = 0)$, we can use the properties of Fourier series to write

$$\mathbf{E}_{0\mathbf{k}} + \mathbf{E}^*_{0-\mathbf{k}} = \frac{2}{(2\pi)^{3/2}} \int \mathbf{E}(\mathbf{r}, t = 0) e^{-i\mathbf{k}\cdot\mathbf{r}} d^3\mathbf{r}. \tag{8.21}$$

Knowing the direction of propagation, we can pick out the Fourier terms that are of interest, and figure out the configuration of the field at some future time:

$$\mathbf{E}(\mathbf{r}, t) = \frac{1}{(2\pi)^{3/2}} \int \mathbf{E}_{0\mathbf{k}} e^{i(\mathbf{k}\cdot\mathbf{r} - \omega(\mathbf{k})t)} d^3\mathbf{k}. \tag{8.22}$$

This kind of analysis makes it easy to introduce the concepts of *group velocity* and *dispersion*. Consider a generic scalar wave packet built up out of Fourier components this way:

$$\psi(\mathbf{r}, t = 0) = \frac{1}{(2\pi)^{3/2}} \int A(\mathbf{k}) e^{i\mathbf{k}\cdot\mathbf{r}} d^3\mathbf{k}. \tag{8.23}$$

IIere $A(\mathbf{k})$ is proportional to the contribution of the plane wave with wavevector $\mathbf{k}$. To time-evolve this is straightforward; we just time evolve each component:

$$\psi(\mathbf{r}, t) = \frac{1}{(2\pi)^{3/2}} \int A(\mathbf{k}) e^{i(\mathbf{k}\cdot\mathbf{r} - \omega(\mathbf{k})t)} d^3\mathbf{k}. \tag{8.24}$$

If $\omega(\mathbf{k})$ is simple, linear dispersion ($\omega(\mathbf{k}) = ck$, where c is the phase velocity), then $\psi(\mathbf{r}, t)$ defined in Eq. (8.24) has a simple interpretation. Change variables such that $\mathbf{r}' \equiv \mathbf{r} - (\omega t/k)\hat{\mathbf{e}}_\mathbf{k}$. Then Eq. (8.24) shows that $\psi(\mathbf{r}, t) = \psi(\mathbf{r}', t = 0)$. That is, the whole wave packet has been translated in space, undistorted.

If the dispersion $\omega(\mathbf{k})$ is more complicated, then the situation is trickier. Suppose that the wave packet is "localized" in $\mathbf{k}$ around some particular $\mathbf{k}_0$. Then we can Taylor expand:

$$\omega(\mathbf{k}) \approx \omega(\mathbf{k}_0) + (\mathbf{k} - \mathbf{k}_0) \cdot \nabla_\mathbf{k}\omega|_{\mathbf{k}_0}$$
$$= \omega_0 + (\mathbf{k} - \mathbf{k}_0) \cdot \mathbf{v}_\mathrm{g}. \tag{8.25}$$

Here we have defined the group velocity, $\mathbf{v}_\mathrm{g} \equiv \nabla_\mathbf{k}\omega(\mathbf{k})$. Plugging into Eq. (8.24), we get

$$\psi(\mathbf{r}, t) \approx \frac{1}{(2\pi)^{3/2}} \int A(\mathbf{k}) e^{i(\mathbf{k}\cdot\mathbf{r} - (\omega_0 + (\mathbf{k}-\mathbf{k}_0)\cdot\mathbf{v}_\mathrm{g})t)} d^3\mathbf{k}$$
$$= e^{i(\mathbf{v}_\mathrm{g}\cdot\mathbf{k}_0 - \omega_0)t} \frac{1}{(2\pi)^{3/2}} \int A(\mathbf{k}) e^{i\mathbf{k}\cdot(\mathbf{r} - \mathbf{v}_\mathrm{g}t)} d^3\mathbf{k}$$
$$= e^{i(\mathbf{v}_\mathrm{g} - \mathbf{v}_\mathrm{p})\cdot\mathbf{k}_0 t} \psi(\mathbf{r} - \mathbf{v}_\mathrm{g}t, 0). \tag{8.26}$$

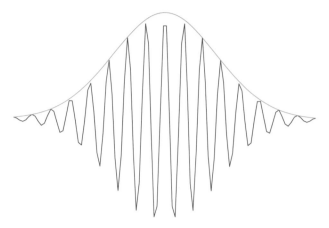

Fig. 8.2 A wavepacket formed by the superposition of plane-waves with differing k centered on k_0. Each plane-wave component travels with phase velocity ω/k, while the envelope of the wave packet as a whole propagates with group velocity v_g.

So, the *envelope* of the wave packet propagates with velocity $\mathbf{v}_g$, while the individual k components propagate with their phase velocity. Figure 8.2 tries to show this, though it is easier to see and understand in animated form.

We note that there are other velocities that are relevant to EM propagation. One of these is the *energy* velocity, at which electromagnetic energy density is propagated through a medium. The magnitude of the *front* velocity is the speed at which an initial step-function-like wave front moves. Most importantly, and most challenging to define, is the *signal* velocity, corresponding with the propagation of information. Only the last of these is required, by causality and special relativity, to be less than c, the speed of light in the vacuum. The constraints obeyed by these velocities, and how those constraints are enforced, can be very subtle, and have prompted discussions by the likes of Sommerfeld [593] and Brillouin [594] that have persisted into the twenty-first century [595]. With the development of novel nanostructured photonic materials, there has been a resurgence in experimental examinations of these issues.

8.1.4 Energy propagation

Let's look at energy propagation, again in uniform, isotropic, time-independent media (for simplicity) by starting from Ampere's Law and dotting with $\mathbf{E}$:

$$\mathbf{E} \cdot (\nabla \times \mathbf{H}) = \mathbf{E} \cdot \mathbf{J} + \mathbf{E} \cdot \frac{\partial}{\partial t} \epsilon \mathbf{E}. \tag{8.27}$$

Recalling another vector identity, $\nabla \cdot (\mathbf{E} \times \mathbf{H}) = \mathbf{H} \cdot (\nabla \times \mathbf{E}) - \mathbf{E} \cdot (\nabla \times \mathbf{H})$, we can see

$$\mathbf{H} \cdot (\nabla \times \mathbf{E}) - \nabla \cdot (\mathbf{E} \times \mathbf{H}) = \mathbf{E} \cdot \mathbf{J} + \frac{1}{2} \frac{\partial}{\partial t} \mathbf{E} \cdot \epsilon \mathbf{E}. \tag{8.28}$$

Plugging in Faraday's Law and further rearranging,

$$\nabla \cdot (\mathbf{E} \times \mathbf{H}) = -\mathbf{E} \cdot \mathbf{J} - \frac{1}{2} \frac{\partial}{\partial t} \mathbf{E} \cdot \epsilon \mathbf{E} - \frac{1}{2} \frac{\partial}{\partial t} \mathbf{H} \cdot \mu \mathbf{H}. \tag{8.29}$$

This has the form of a continuity equation. We define the *Poynting vector*, $\mathbf{S} \equiv \mathbf{E} \times \mathbf{H}$, and recall that the electric energy density is $\mathcal{E}_E \equiv (1/2)\mathbf{E} \cdot \epsilon\mathbf{E}$ and the magnetic energy density is $\mathcal{E}_H \equiv (1/2)\mathbf{H} \cdot \mu\mathbf{H}$. Then we can write

$$\nabla \cdot \mathbf{S} + \mathbf{E} \cdot \mathbf{J} = -\frac{\partial}{\partial t}(\mathcal{E}_E + \mathcal{E}_H). \tag{8.30}$$

Interpretation is straightforward. The first term on the left side is the flux of $\mathbf{S}$ out of a differential volume. The term on the right shows that this flux coincides with a decrease with time of the electromagnetic energy stored in that volume. The second term on the left is a "sink" term, showing that another way of removing electromagnetic energy from the differential volume is to have the electric field do work on the charge carriers in the current density $\mathbf{J}$. This all tells us that $\mathbf{S}$ is an energy current density, and therefore has units of intensity (W/m^2).

Since the energy densities and Poynting vector are quadratic in the electric and magnetic fields, we must use care when considering these quantities in the plane wave language that we've so far used to discuss EM radiation. We must evaluate them using the real parts of the (complex notation) fields. For example, for a time-independent ϵ,

$$\mathcal{E}_E = \frac{1}{2}\mathbf{E} \cdot \epsilon\mathbf{E} = \frac{\epsilon}{2}\left(\frac{1}{2}(\mathbf{E}_{0k}e^{i(\mathbf{k}\cdot\mathbf{r}-\omega t)} + \mathbf{E}_{0k}^*e^{-i(\mathbf{k}\cdot\mathbf{r}-\omega t)})\right)^2. \tag{8.31}$$

This looks messy, but the situation becomes simpler if we care only about the time-average of $\mathcal{E}_E$. The oscillatory parts average to zero, and we are left with

$$\langle \mathcal{E}_E \rangle = \frac{\epsilon}{4}\mathbf{E}_{0k} \cdot \mathbf{E}_{0k}^*, \tag{8.32}$$

and analogously for the magnetic energy density,

$$\langle \mathcal{E}_H \rangle = \frac{\mu}{4}\mathbf{H}_{0k} \cdot \mathbf{H}_{0k}^*. \tag{8.33}$$

Following the same approach for the time-averaged Poynting vector, we find

$$\langle \mathbf{S} \rangle = \frac{1}{2}\left((\mathbf{E}_{0k} \times \mathbf{H}_{0k}^*) + (\mathbf{E}_{0k}^* \times \mathbf{H}_{0k})\right). \tag{8.34}$$

Recalling Eq. (8.15), we find

$$\langle \mathbf{S} \rangle = \hat{\mathbf{e}}_k\sqrt{\frac{\epsilon}{\mu}}|\mathbf{E}_{0k}|^2. \tag{8.35}$$

This is a convenient expression, as it allows us to convert EM intensities into average electric field amplitudes.

What about finding the total energy density or Poynting vector for a generic wave packet built out of a number of $\mathbf{k}$ components? Generally we would need to write down the full expression for the real fields, as in Eq. (8.22), and work with quadratic expressions in those. Fortunately, for linear media we get a bit lucky, and the result for the energy densities ends up as

$$\langle \mathcal{E}_E \rangle = \frac{\epsilon}{4}\int \mathbf{E}_{0k} \cdot \mathbf{E}_{0k}^* d^3k. \tag{8.36}$$

We can find the energy density (or Poynting vector contribution) at each frequency, and add them. Nonlinear dielectric or magnetic material responses make things more complicated.

8.1.5 Restrictions on material response

The previous expressions can become very complicated when real materials are involved. Remember that ϵ originates from the response of the medium's charge distribution to electric fields; similarly, μ comes from the medium's ability to generate either bulk magnetization or induced currents. As a result, both parameters can be frequency dependent (material response depends on the dynamics of internal degrees of freedom), anisotropic, complex (in the sense of having phase lags between driving fields and material response), and spatially varying. You can readily see why nanostructured optical media can be a rich playground.

Let's work with just $\epsilon(\omega)$ for simplicity. A complex $\epsilon(\omega)$ means that the electric polarization of the medium can be out of phase with the driving electric field. However, $\epsilon(\omega)$ can't have an arbitrary functional form, for important physics reasons. The big example of this is causality: the response of a material can only depend on the history of the material, rather than its future. That is,

$$\mathbf{D}(t)/\epsilon_0 = \mathbf{E}(t) + \int_0^\infty f(\tau)\mathbf{E}(t - \tau)d\tau. \qquad (8.37)$$

Fourier transforming, we get

$$\mathbf{D}(\omega) = \epsilon(\omega)\mathbf{E}(\omega), \qquad (8.38)$$

where

$$\epsilon(\omega)/\epsilon_0 = 1 + \int_0^\infty f(\tau)e^{i\omega\tau}d\tau. \qquad (8.39)$$

Writing $\epsilon(\omega) \equiv \epsilon'(\omega) + i\epsilon''(\omega)$, such that ϵ' and ϵ'' are real, we see immediately that $\epsilon(-\omega) = \epsilon^*(\omega)$ implies $\epsilon'(-\omega) = \epsilon'(\omega)$, and $\epsilon''(-\omega) = -\epsilon''(\omega)$. These relationships plus assumptions about analyticity of $\epsilon(\omega)$ in the upper half of the complex plane end up relating $\epsilon'(\omega)$ and $\epsilon''(\omega)$ to each other. The end results are:

$$\epsilon'(\omega)/\epsilon_0 = 1 + \frac{2}{\pi}P\int_0^\infty \frac{\omega'\epsilon''(\omega')}{\omega'^2 - \omega^2}d\omega',$$

$$\epsilon''(\omega)/\epsilon_0 = -\frac{2\omega}{\pi}P\int_0^\infty \frac{\epsilon'(\omega') - 1}{\omega'^2 - \omega^2}d\omega'. \qquad (8.40)$$

These are the *Kramers–Krönig relations*, and can come in very handy in characterizing the optical properties of new media. We shall see shortly that $\epsilon''(\omega)$ is related to electrical conductivity and absorption of EM radiation. To make a long story short, the Kramers–Krönig relations mean that one can measure an absorption spectrum over a large bandwidth and then infer dielectric properties, and vice versa.

The Kramers–Krönig relations also restrict what novel optical materials can do. As we shall see, there is much interest in optical *metamaterials*, engineered systems that are homogeneous on scales of an EM wavelength of interest, but possess unusual optical properties. For example, there is a class of metamaterials called negative index media that can exhibit a variety of surprising properties, including unconventional refraction and superlensing. The causality restriction constrains the frequency range over which such effects can occur, and the magnitude of those effects.

8.1.6 Boundary conditions and interfaces

Now that we have looked at uniform media, it is time to consider the role of interfaces in the propagation of EM radiation. Much like the impact of interfaces on electronic structure, the boundary conditions that must hold for EM fields lead to consequences that are the foundation for much of the excitement in micro- and nanophotonics. The requirements at an interface are as follows.

- The tangential component of $\mathbf{E}$ must be continuous across the interface. This comes from Eq. (8.3).
- The longitudinal component of $\mathbf{B}$ must be continuous across the interface. This is equivalent to the fact that magnetic monopoles don't exist.
- The tangential component of $\mathbf{H}$ must be continuous for equal μ in the two media, and with no current sheets at the interface. This comes from Eq. (8.4).
- The longitudinal component of $\mathbf{D}$ must be continuous, in the absence of free charges at the interface. This follows from Eq. (8.1).

Let's assume that we have the situation shown in Fig. 8.3(a), where an electromagnetic wave of (complex) amplitude $\mathbf{E}_{0i}$ propagating (wavevector $\mathbf{k}_i$) in a medium of dielectric function ϵ_1 is incident on an interface with a medium of dielectric function ϵ_2. For now we assume both media have identical and uninteresting magnetic properties, so $\mu_1 = \mu_2 = \mu$. We know that there can be a reflected wave with amplitude $\mathbf{E}_{0r}$ and wavevector $\mathbf{k}_r$, and a transmitted wave with amplitude $\mathbf{E}_{0t}$ and wavevector $\mathbf{k}_t$. Define $\mathbf{n}$ to be the unit vector normal to the interface (directed into medium 2). Translating the boundary conditions into the language we've been using for propagation of electromagnetic radiation, respectively these become:

$$(\mathbf{E}_{0i} + \mathbf{E}_{0r} - \mathbf{E}_{0t}) \times \mathbf{n} = 0, \tag{8.41}$$

$$(\mathbf{k}_i \times \mathbf{E}_{0i} + \mathbf{k}_r \times \mathbf{E}_{0r} - \mathbf{k}_t \times \mathbf{E}_{0t}) \cdot \mathbf{n} = 0, \tag{8.42}$$

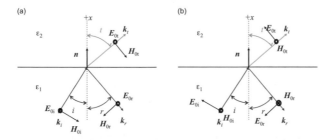

An EM wave incident on an interface between materials of differing dielectric properties. The boundary conditions on the electric and magnetic fields determine the reflected and transmitted component magnitudes, phases, and propagation wavevectors. (a) The s-polarized case; (b) the p-polarized case. Reflected and transmitted components differ for the two polarizations, implying that the polarization of a general wave at non-normal incidence is altered by the interface.

Fig. 8.3

$$(\mathbf{k}_i \times \mathbf{E}_{0i} + \mathbf{k}_r \times \mathbf{E}_{0r} - \mathbf{k}_t \times \mathbf{E}_{0t}) \times \mathbf{n} = 0, \tag{8.43}$$

$$(\epsilon_1 (\mathbf{E}_{0i} + \mathbf{E}_{0r}) - \epsilon_2 \mathbf{E}_{0t}) \cdot \mathbf{n} = 0. \tag{8.44}$$

The *frequencies* of the incident, reflected, and transmitted waves must all be the same (assuming linear optical response), since the reflected and transmitted waves have their physical origins in the response of the material to the incident wave. The fields due to the waves are

$$\mathbf{E}_i = \mathbf{E}_{0i} e^{i(\mathbf{k}_i \cdot \mathbf{r} - \omega t)},$$

$$\mathbf{H}_i = \sqrt{\frac{\epsilon_1}{\mu}} \hat{\mathbf{e}}_{\mathbf{k}_i} \times \mathbf{E}_i, \tag{8.45}$$

$$\mathbf{E}_r = \mathbf{E}_{0r} e^{i(\mathbf{k}_r \cdot \mathbf{r} - \omega t)},$$

$$\mathbf{H}_r = \sqrt{\frac{\epsilon_1}{\mu}} \hat{\mathbf{e}}_{\mathbf{k}_r} \times \mathbf{E}_r. \tag{8.46}$$

$$\mathbf{E}_t = \mathbf{E}_{0t} e^{i(\mathbf{k}_t \cdot \mathbf{r} - \omega t)},$$

$$\mathbf{H}_t = \sqrt{\frac{\epsilon_2}{\mu}} \hat{\mathbf{e}}_{\mathbf{k}_t} \times \mathbf{E}_t. \tag{8.47}$$

The magnitudes of the wavevectors are

$$|\mathbf{k}_i| = |\mathbf{k}_r| = k = \omega \sqrt{\mu \epsilon_1},$$

$$|\mathbf{k}_t| = k_t = \omega \sqrt{\mu \epsilon_2}. \tag{8.48}$$

There are two particular (mutually perpendicular) polarization directions for the radiation that are the simplest to consider; we can write an arbitrary polarization case as a linear combination of these options. The *s*-polarized, or "transverse electric" (TE), case has the electric field in the plane of the interface, perpendicular to $\mathbf{n}$. The *p*-polarized, or "transverse magnetic" (TM), case has the magnetic field in the plane of the interface, perpendicular to $\mathbf{n}$.

Let's consider the *s*-polarized case first, as shown in Fig. 8.3(a). Given the incident wave, we want to find the amplitudes and directions of the reflected and transmitted waves. Applying our interface conditions at $x = 0$, and knowing that they must hold for all y and z, from Eq. (8.41), we find that $E_{0i} + E_{0r} = E_{0t}$. Applying Eq. (8.42), $k_{i,y} = k_{r,y} = k_{t,y}$. Since $k_i = k_r = k$, this immediately implies that the angle of incidence, θ_i, is equal to the angle of reflection, θ_r, with those angles defined as in the figure.

Similarly, we find $k_i \sin \theta_i = k_t \sin \theta_t$. This implies Snell's Law,

$$n_1 \sin \theta_i = n_2 \sin \theta_t, \tag{8.49}$$

since $k/\omega = n/c$. We now know θ_t and θ_r in terms of given quantities.

Appling Eq. (8.43) as well gives

$$E_{0i} n_1 \cos \theta_i - E_{0r} n_1 \cos \theta_r = E_{0t} n_2 \cos \theta_t. \tag{8.50}$$

We now have two equations for our two remaining unknowns, the reflected and transmitted amplitudes. After some algebra we find

$$E_{0r} = \frac{n_1 \cos \theta_i - n_2 \cos \theta_t}{n_1 \cos \theta_i + n_2 \cos \theta_t} E_{0i}$$

$$E_{0t} = \frac{2 n_1 \cos \theta_i}{n_1 \cos \theta_i + n_2 \cos \theta_t} E_{0i}. \tag{8.51}$$

This is a Fresnel formula.

We could perform the exact same analysis for the p-polarized case, shown in Fig. 8.3(b), and we arrive at another Fresnel formula:

$$E_{0r} = \frac{n_2 \cos \theta_i - n_1 \cos \theta_t}{n_2 \cos \theta_i + n_1 \cos \theta_t} E_{0i}$$

$$E_{0t} = \frac{2 n_1 \cos \theta_i}{n_2 \cos \theta_i + n_1 \cos \theta_t} E_{0i}. \tag{8.52}$$

Since Eqs. (8.51) and (8.52) differ, an incident wave of arbitrary polarization can lead to a reflected wave (for example) with quite different polarization.

8.1.7 Evanescent waves

It is clear from Snell's Law that something unusual has to happen when $n_1 \sin \theta_i > n_2$. The result is *total internal reflection*, when $\cos \theta_t$ becomes purely imaginary. The reflected amplitude must be equal to the incident amplitude (conservation of energy). The reflected wave does acquire a polarization-dependent phase shift relative to the incident wave, however. For s-polarized waves,

$$\frac{E_{0r}}{E_{0i}} = \exp \left[i 2 \tan^{-1} \left[\sqrt{\frac{n_1^2 \sin^2 \theta_i - n_2^2}{n_1^2 (1 - \sin^2 \theta_i)}} \right] \right] = \exp(i \phi_s), \tag{8.53}$$

and for p-polarized waves,

$$\frac{E_{0r}}{E_{0i}} = \exp \left[i 2 \tan^{-1} \left[\left(\frac{n_1}{n_2} \right)^2 \sqrt{\frac{n_1^2 \sin^2 \theta_i - n_2^2}{n_1^2 (1 - \sin^2 \theta_i)}} \right] \right] = \exp(i \phi_p). \tag{8.54}$$

Because of these phase shifts, the tangential electric field boundary condition at the interface does require there to be a nonzero E_{0t}. Therefore, there is an electromagnetic wave, an *evanescent wave*, in the second medium, even when the incident light is totally internally reflected in medium 1. How can we reconcile this, particularly since $|E_{0r}| = |E_{0i}|$, and no energy can therefore be transported away into medium 2? Applying the other boundary conditions, we find (choosing the physically reasonable positive root) that

$$k_{tx} = i \frac{\omega}{c} \sqrt{n_1^2 \sin^2 \theta_i - n_2^2} \equiv i \kappa. \tag{8.55}$$

Since κ here is real, for the fields we find

$$\mathbf{E}_t = E_{0t} \hat{\mathbf{z}} e^{i(k_{ty} y - \omega t) - \kappa x}, \tag{8.56}$$

$$\mathbf{H}_t = \sqrt{\frac{\epsilon_2}{\mu}} \hat{\mathbf{e}}_{k_t} \times E_{0t} \hat{\mathbf{z}} e^{i(k_{ty} y - \omega t) - \kappa x} = \sqrt{\frac{\epsilon_2}{\mu}} (k_{ty} \hat{\mathbf{x}} - i \kappa \hat{\mathbf{y}}) E_{0t} e^{i(k_{ty} y - \omega t) - \kappa x}. \tag{8.57}$$

Fig. 8.4   An optical waveguide. In the ray optics picture, total internal reflection confines the light to the interior of the higher index material. In the wave-field picture, Maxwell's equations with appropriate boundary conditions support eigenmodes bound to the interior of the higher index material.

The evanescent fields decay exponentially moving into medium 2. From Eq. (8.55), it is clear that the relevant decay length is comparable to the wavelength of the radiation in the two media. For free space $\lambda = 500$ nm, $n_1 = 1.5$, $n_2 = 1$, and $\theta_i = 60$ degrees, $1/\kappa \approx 100$ nm.

While there is electromagnetic field in medium 2, and therefore there is electromagnetic energy density there, there is no dissipation in that medium. The evanescent state is localized to the interface, and energy does not propagate into medium 2. This is clear from the time-averaged Poynting vector:

$$\langle \mathbf{S}_t \rangle \propto |E_{0t}|^2 \exp(-2\kappa x)\hat{\mathbf{y}}. \tag{8.58}$$

The energy propagates along the plane of the interface.

This kind of total internal reflection is the physics that permits classic optical waveguides like optical fibers, as shown in Fig. 8.4. There are two major approaches for analyzing optical waveguide structures. With *ray optics*, one considers only the direction of $\mathbf{k}$. This throws out intensity, polarization, and phase information, but can at times be revealing and is certainly much simpler mathematically. The other limit is the *wave-field* method, which actually solves Maxwell's equations in and outside the waveguide, keeping track of all boundary conditions.

Propagation in waveguides has certain essential features. As is apparent from the ray optics picture, one can think of propagation as a series of lateral bounces down the wave-guide. This implies that, even independent of other materials considerations, longitudinal propagation of energy, signals, etc. is typically slower than c/n. The wave-field approach involves solving the boundary value problem of radiation in the material cavity. From prior experience, we expect solutions to this boundary value problem to lead to well defined normal (eigen)modes that are naturally orthogonal to each other. This implies that in a linear medium, a waveguide will only pass radiation of particular wavelengths. Moreover, in the ideal case, populating one mode does not affect propagation in the other modes.

Finally, the wave-field approach finds the evanescent fields that result from the internal reflection process. Those evanescent fields are as much a part of the propagating modes as the fields within the waveguide material. This implies that placing some material within the evanescent fields can couple to those modes, and this is a standard means of coupling waves into and out of waveguides.

8.1.8 Real materials: dispersion

A full, quantum mechanical treatment of the optical properties of real dielectrics is beyond our scope. However, it is useful to examine a classical toy model that duplicates some of the essential features seen in real systems. The dielectric response of an insulator is determined by the polarization induced by the external electromagnetic field. To model this, consider a charge $-e$ of mass m bound by a harmonic potential (natural frequency ω_0), with some damping, γ, interacting with an EM wave:

$$m[\ddot{\mathbf{r}} + \gamma\dot{\mathbf{r}} + \omega_0^2\mathbf{r}] = -e\mathbf{E}(\mathbf{r}, t). \tag{8.59}$$

If the electric field has a harmonic time dependence with frequency ω, and therefore the moving charge has a driven response at the same frequency, we can find the polarization, assuming N of these charges per unit volume:

$$\mathbf{P} = \frac{Ne^2}{m}(\omega_0^2 - \omega^2 - i\omega\gamma)^{-1}\mathbf{E}(t). \tag{8.60}$$

This model clearly has the response of a damped, driven harmonic oscillator, possessing a resonance at a frequency given, for light damping ($\gamma \ll \omega_0^2$), by $\sqrt{\omega_0^2 - \gamma^2/4}$. This is a classical model useful for approximating the situation in real solids when there can be an electronic transition or resonance.

From Eq. (8.60) we can find the complex dielectric response:

$$\epsilon \equiv \epsilon_0 + \frac{\mathbf{P}}{\mathbf{E}} = \epsilon_0 + \frac{Ne^2}{m}(\omega_0^2 - \omega^2 - i\omega\gamma)^{-1}. \tag{8.61}$$

Figure 8.5 shows the real and imaginary parts of the dielectric response around the resonance. On the low frequency side of the resonance, the charge displaces nearly in phase with the external field, leading to a real polarization and therefore dielectric function enhanced beyond that of vacuum. Near the resonance, the charge can't respond in phase with the drive anymore, and its motion is 90 degrees out of phase with the electric field; the electric field is doing work on the charge, transferring energy from the field into the mechanical motion of the charge. Above the resonance, the charge's motion has fallen behind the driving field, leading to an inverted dielectric response. While this classical picture is simplistic, these basic features of the dielectric response survive a quantum mechanical treatment in many cases.

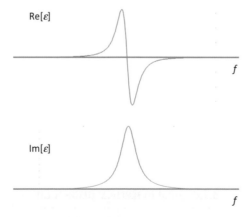

A sketch of the complex dielectric response of a material near a resonance. Depending on the frequency, the polarization of the material can lead or lag the external electric field.

Fig. 8.5

The complex index of refraction is $n(\omega) \equiv \sqrt{\epsilon(\omega)\mu/(\epsilon_0\mu_0)}$. The imaginary part of the propagation wavevector $k = n(\omega)\omega/c$ gives the absorption coefficient, describing the exponential decay of the electromagnetic field with distance propagated into the material.

8.1.9 EM waves and conductors

While the harmonically bound charge model can describe the interactions of light with dielectrics, we must also consider the interaction of electromagnetic radiation with conductors, in which charges are free to move. First, we examine the characteristic timescale for the decay of a nonequilibrium charge distribution within a conductor, in a classical picture. Because charge is conserved the current density obeys a continuity equation:

$$\nabla \cdot \mathbf{J} = \nabla \cdot \sigma \mathbf{E} = -\frac{\partial}{\partial t}\rho, \tag{8.62}$$

where ρ is the volume charge density, and we've used the definition of the conductivity, σ, to relate current density and electric field. Combining with Gauss' Law, $\nabla \cdot (\epsilon \mathbf{E}) = \rho$, we find

$$\frac{\sigma}{\epsilon}\rho = -\frac{\partial}{\partial t}\rho, \tag{8.63}$$

which implies a characteristic timescale $\tau_c \equiv \epsilon/\sigma$. Plugging in typical numbers for a good conductor like copper gives $\tau_c \sim 10^{-19}$ s. As a practical matter, the actual timescale is longer in copper, because the real band structure alters the dielectric function and conductivity as a function of frequency. Still, the point is, even a simple classical argument says that charge relaxation is fast in a good conductor. This makes intuitive sense given our previous results looking at electrostatic screening in conductors. In a good conductor with a high density of mobile charge, screening requires only minute displacements of the charges, and that can happen very quickly when the charges can move readily.

So far we have neglected the possibility of large-scale, collective motion of the electrons. Such motion would dynamically create regions of excess electron density with instantaneous net negative charge, and other regions with an instantaneous deficit of electron density and net positive charge. The resulting electric fields then drive redistribution of the mobile charge. The collective modes of the electronic fluid that arise from these considerations are called *plasmons*, and are discussed further in Section 8.6.

Neglecting such collective modes for the moment, to see how EM waves propagate in a conductor we return to our derivation of the wave equation Eq. (8.6), this time allowing currents to flow. This gives

$$\nabla^2 \mathbf{E} - \mu\frac{\partial}{\partial t}\mathbf{J} - \mu\epsilon\frac{\partial^2 \mathbf{E}}{\partial t^2} = 0. \tag{8.64}$$

Using the definition of conductivity and assuming σ is time-independent changes the second term to $-\mu\sigma(\partial \mathbf{E}/\partial t)$. Plugging in our trial plane wave solution gives

$$(-k^2 + i\omega\mu\sigma + \mu\epsilon\omega^2)\mathbf{E}_{0\mathbf{k}}e^{i(\mathbf{k}\cdot\mathbf{r}-\omega t)} = 0. \tag{8.65}$$

We can solve for the real and imaginary parts of $k = a + ib$, taking positive roots and assuming that all three components of k have the same ratio of real to imaginary parts. Defining $k_0 \equiv \omega\sqrt{\epsilon\mu}$,

$$a = k_0 \frac{1}{\sqrt{2}}\left(1 + \sqrt{1 + 1/(\omega\tau_c)^2}\right)^{1/2},$$

$$b = k_0 \frac{1}{\sqrt{2}}\left(-1 + \sqrt{1 + 1/(\omega\tau_c)^2}\right)^{1/2}. \tag{8.66}$$

The dimensionless product $\omega\tau_c$ sets a natural scale. If $\omega\tau_c \ll 1$, the material is a "good" conductor, so that charge relaxation happens fast relative to the timescale over which the field from the eave varies. Conversely, if $\omega\tau_c \gg 1$, the material is a poor conductor, and $b \to 0$ while $a \to k_0$. The poor conductor limit gives small corrections to what we have already seen by assuming $\sigma = 0$. A small b means a small imaginary part of $\mathbf{k}$ relative to the real part of $\mathbf{k}$. The wave decays exponentially as it propagates, but does so over many wavelengths.

In the good conductor limit, $a \approx b \approx k_0/\sqrt{2\omega\tau_c}$. The wave decays within a conductor on a length scale given by

$$\delta = \sqrt{\frac{2}{\omega\mu\sigma}}. \tag{8.67}$$

The distance δ is known as the *skin depth*. If the correct frequency-dependent conductivity is plugged into this expression, in gold at optical frequencies $\delta \sim 25$ nm. *Nanoscale conductors can be on the same scale as the skin depth, and in general can interact strongly with electromagnetic waves.*

Let's re-examine the derivation of the equivalent of Eq. (8.51) in the case where medium 2 is a good conductor. Applying our boundary conditions on the tangential component of $\mathbf{E}$, we still find $E_{0i} + E_{0r} = E_{0t}$. The good conductor character of medium 2 means $k_{tx} \approx (1 + i)/\delta$. Using our other boundary condition on $\mathbf{H}$, we find

$$E_{0r} = -\left(1 - \frac{2n_1 \cos\theta_i}{n_2}\frac{1-i}{\sqrt{2}}\sqrt{\omega\tau_c}\right)E_{0i},$$

$$E_{0t} = \frac{2n_1 \cos\theta_i}{n_2}\frac{1-i}{\sqrt{2}}\sqrt{\omega\tau_c}E_{0i}. \tag{8.68}$$

For $\omega\tau_c \ll 1$, as we've assumed, then $E_{0r} \to -E_{0t}$. The reflected wave has nearly the same amplitude as the incident wave, but with a π phase shift. Similarly, the amplitude of the transmitted wave is very small. This tells us what anyone who has looked at a polished piece of metal or a silvered mirror knows: good conductors reflect very well. Note that k_{tx} is non-zero, unlike the total internal reflection case. This implies that there is some energy propagating into the conducting medium. Examining the Poynting vectors carefully and self-consistently will reveal that any energy lost to the reflected wave is balanced by energy transmitted into medium 2. Moreover, energy lost by the transmitted wave as it propagates in the $+x$ direction is balanced by the Ohmic ($\mathbf{J} \cdot \mathbf{E}$) losses in the medium. We will return to optical interactions with conductors when we discuss plasmons.

8.1.10 Diffraction

Before we discuss manipulating light at length scales much smaller than its wavelength, we need to understand why this is generally a difficult task. To do so, we consider the general problem of radiation impinging on an opaque screen that has holes of some size and shape. We want to know the resulting intensity pattern at some distance away, as shown from above in schematic form in Fig. 8.6. For simplicity, we use *scalar* diffraction theory. This is a conscious choice to throw away information about polarization and the detailed orientations of the **E** and **H** fields.

As shown, a plane wave is incident upon a screen at $z = 0$. Conceptually, a useful approach is Huygen's Principle, where we can visualize the each hole in the screen as an array of closely spaced point sources of spherical waves. This has a mathematical basis in the treatment of boundary value problems.

We will work with a general scalar field, $\psi(\mathbf{r}, t)$, and assume that this is a reasonable proxy for, say, the E-field amplitude of an electromagnetic wave. We assume that this scalar field is governed in free space by a wave equation. Performing separation of variables (so that the harmonic time dependence is factored out), we are left with the Helmholtz equation, a generic time-independent wave equation for $\psi(\mathbf{r})$:

$$\nabla^2 \psi + k^2 \psi = 0. \tag{8.69}$$

We are trying to solve a boundary value problem, wherein we are given the value of ψ on the screen at $z = 0$, and want to know ψ elsewhere. The Green's function for Eq. (8.69) is

$$\phi(\mathbf{r}) = \frac{e^{ik|\mathbf{r}-\mathbf{r}'|}}{|\mathbf{r} - \mathbf{r}'|}. \tag{8.70}$$

This satisfies $(\nabla^2 + k^2)\phi(\mathbf{r}) = -4\pi \delta(\mathbf{r} - \mathbf{r}')$, the Helmholtz equation with a delta function source term. This is the mathematical basis for Huygen's Principle.

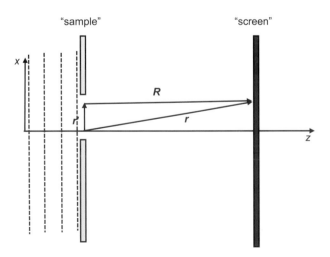

Fig. 8.6 Geometry considered for generic scalar diffraction. A plane wave is incident on an opaque screen at $z = 0$ containing a hole. Of interest is the field at some position $\mathbf{r}$ with $z > 0$. A point on the screen has position $\mathbf{r}'$, and $\mathbf{R} \equiv \mathbf{r} - \mathbf{r}'$.

Using the Green's function, we find the diffracted wave at $z > 0$ in terms of ψ on the screen. Define the origin to be at the center of the screen. The vector $\mathbf{r}$ points to the location at $+z$ where we want to know ψ. The vector $\mathbf{r}'$ points to a particular location on the $z = 0$ surface, which can be the screen or the holes in the screen. The vector $\mathbf{R} \equiv \mathbf{r} - \mathbf{r}'$ points from $\mathbf{r}'$ to $\mathbf{r}$. The solution[1] is called the Kirchoff diffraction integral:

$$\psi_d(\mathbf{r}) = -\frac{1}{2\pi} \int_S da' \, \psi_i(\mathbf{r}') \mathbf{n}' \cdot \left(\frac{e^{ikR}}{R} \left(ik - \frac{1}{R} \right) \hat{\mathbf{e}}_R \right). \tag{8.71}$$

Here $\hat{\mathbf{e}}_R$ is the unit vector $\mathbf{R}/|\mathbf{R}|$, $\psi_d(\mathbf{r})$ is the diffracted wave, and $\psi_i(\mathbf{r}')$ is the incident wave at the screen. The integral is over the screen, da' is a scalar area element at $\mathbf{r}'$, and $\mathbf{n}'$ is a unit normal vector at $\mathbf{r}'$.

Strip away the math, and this is basically adding up, including their relative phases, the contributions of waves originating at each point $\mathbf{r}'$, weighted by the value of $\psi(\mathbf{r}')$. The general case is *Fresnel* diffraction. In our case, *Frauenhofer* diffraction, we are going to assume an incident plane wave for $z < 0$, and further assume that R is very large compared to r'. In that case, $R \approx r - \hat{\mathbf{e}}_r \cdot \mathbf{r}'$. Defining (assuming) $\mathbf{k} \equiv k\hat{\mathbf{e}}_r$, then

$$\psi_d(\mathbf{r}) \approx -\frac{ik}{4\pi} \frac{e^{ikr}}{r} (\cos\theta_i + \cos\theta_d) \int_S da' \, \psi_i(\mathbf{r}') e^{-i\mathbf{k}\cdot\mathbf{r}'}. \tag{8.72}$$

The angle θ_i is the angle between the incident plane wave and the z-axis, while θ_d is the angle between $\mathbf{r}$ and the z-axis.

The important point of Eq. (8.72) becomes clear if we assume normal incidence and an opaque screen, so that $\psi_i = 0$ everywhere except at the holes, where it equals ψ_0. Then the integral on the right hand side is a 2d Fourier transform of the holes in the screen. For example, consider a single hole in the screen that is actually a rectangular slit of with $2a$ in the x direction and height $2b$ in the y direction, centered on the origin. In that case,

$$\psi_d(\mathbf{r}) \approx -\frac{ik}{4\pi} \frac{e^{ikr}}{r} (1 + \cos\theta_d) \psi_0 \int_{-a}^{+a} dx' \int_{-b}^{+b} dy' e^{-i(k_x x' + k_y y')}$$

$$= -\frac{ik}{4\pi} \frac{e^{ikr}}{r} (1 + \cos\theta_d) \psi_0 \frac{4}{k_x k_y} \sin k_x a \sin k_y b. \tag{8.73}$$

The intensity is proportional to the square of the amplitude:

$$I = I_0 \frac{k^2 (1 + \cos\theta_d)^2}{\pi^2 r^2} \frac{\sin^2 k_x a \sin^2 k_y b}{k_x^2 k_y^2}. \tag{8.74}$$

At some fixed position x, y, z ($z > 0$), given the original incident k, we can find r, the components of k as well as θ_d, and compute the intensity. Plotted in Fig. 8.7 for fixed z, $y = 0$, as a function of x, we see the characteristic single slit diffraction pattern. The narrower the initial slit, the wider the resulting diffraction pattern.

One can perform the same analysis for diffraction from a circular hole of radius a in the screen centered at the origin. Working in cylindrical coordinates, for comparatively small

---

[1] This problem is solved in many graduate level texts, including *Classical Electrodynamics* by J. D. Jackson [596].

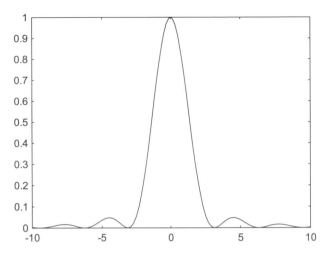

Fig. 8.7 Diffraction pattern in the far field from a single slit at normal incidence.

θ_d, the intensity is

$$I = I_0 \frac{k^2 a^4}{4r^2} \left(\frac{2J_1(ka \sin \theta_d)}{ka \sin \theta_d} \right). \tag{8.75}$$

Here $J_1(u)$ is the first-order Bessel function of the first kind. Like its more familiar trigonometric cousins, it is rather oscillatory as a function of its argument. The first zero of $J_1(u)$ occurs at $u \approx 3.832$. The resulting diffraction pattern is called an *Airy* pattern, with a central spot surrounded by rings. A little algebra shows that the angle of the first dark ring around the bright spot in the center occurs at $\theta_d = 1.22 \frac{\lambda}{2a}$. Here the tradeoff between hole size and diffraction pattern scale is even more obvious. Because of the form of the Frauenhofer integral, diffracted bright spots in the far-field limit ($r \gg \lambda$) always have a minimum size on the order of λ.

8.1.11 The "near field"

In treating far field diffraction, we have assumed that the relevant wavevector of the radiation is well approximated as $\mathbf{k} = k\hat{\mathbf{e}}_{\mathbf{r}}$. However, that is not the case *immediately* behind (z just barely > 0) the opaque screen. The opaque screen, with its sharp-edged holes, imprints short distance scale (features much smaller than λ) information on the amplitude of the field. Suppressing the y-direction for simplicity, we can write the scalar field just beyond the screen in terms of Fourier components of different wavevectors:

$$\psi(k_x) = \int_{-\infty}^{+\infty} \psi(x, z = 0) \exp(ik_x x) dx, \tag{8.76}$$

where $\psi(x, z = 0)$ is ψ_0 where there are holes in the screen, and zero elsewhere. The scalar field beyond the screen is a superposition of components having various x-component wavevectors k_x.

Far field "Airy" diffraction pattern from a circular aperture, as described by Eq. (8.75); image from Indiana University. **Fig. 8.8**

To recover the amplitude at some point (x, z), one should propagate each Fourier component in z, and then take the inverse transform:

$$\psi(x, z) = \int_{-\infty}^{+\infty} dk_x [\psi(k_x, z = 0) \exp(-ik_x x)] \exp(ik_z z). \qquad (8.77)$$

Recall, however, that $k^2 = k_x^2 + k_z^2 \to k_z = \sqrt{k^2 - k_x^2}$. For $k_x > \omega/c$, the quantity under the square root is *negative*. That means that components of the scalar field with $k_x > \omega/c$ decay exponentially in z as one moves away from the screen. As a result, for $z \gg \lambda$,

$$\psi(x, z) \approx \int_{-\omega/c}^{+\omega/c} dk_x \psi(k_x, z = 0) \exp(-ik_x x) \exp\left(i\sqrt{k^2 - k_x^2} z\right). \qquad (8.78)$$

In other words, all of the components that contain high lateral spatial resolution information about the holes in the screen are lost in the far field. These decaying components are the "near field". We shall return to this in Section 8.5, when we discuss using nanoscale physics methods to recover this information in the far field.

8.1.12 Nonlinear effects

In our discussions of EM wave propagation thus far we have taken great advantage of linearity. This is not always legitimate. While not strictly a nanoscale physics issue, nonlinear optical phenomena are quite general, important in optical communications and technologies, and may be relevant in some nanophotonic systems. Therefore we will take a very abbreviated look at some of the basic ideas of nonlinear optics.

In our original definition of $\epsilon(\omega)$ and the Kramers–Krönig relations, Eq. (8.40), we used the idea that the polarization, $\mathbf{P}$, of a medium can only depend on the electric field at earlier times. We assumed the linear approximation of the most general kind of relationship between $\mathbf{E}$ and $\mathbf{P}$. Keeping the tensorial nature of the susceptibility tenosor $\chi_{\alpha\beta}$,

$$P_\alpha(\mathbf{r}, t) = \epsilon_0 \sum_\beta \int dt_1 \int d\mathbf{r}_1 \chi_{\alpha\beta}(\mathbf{r} - \mathbf{r}_1, t - t_1) E_\beta(\mathbf{r}_1, t_1), \qquad (8.79)$$

where P_α is the α component of $\mathbf{P}$, and E_β is the β component of $\mathbf{E}$. Then our definition of $\epsilon(\omega)$ is related to the Fourier transform of the linear susceptibility tensor, $\chi_{\alpha\beta}$.

We could continue to include higher order terms, of course:

$$P_\alpha^{NL}(\mathbf{r}, t) = \epsilon_0 \sum_{\beta\gamma} \int d\mathbf{r}_1 d\mathbf{r}_2 dt_1 dt_2 \chi_{\alpha\beta\gamma}^{(2)}(\mathbf{r} - \mathbf{r}_1, t - t_1; \mathbf{r} - \mathbf{r}_2, t - t_2) E_\beta(\mathbf{r}_1, t_1) E_\gamma(\mathbf{r}_2, t_2)$$

$$+ \epsilon_0 \sum_{\beta\gamma\delta} \int \cdots \int d\mathbf{r}_1 d\mathbf{r}_2 d\mathbf{r}_3 dt_1 dt_2 dt_3 \chi_{\alpha\beta\gamma\delta}^{(3)}(\mathbf{r} - \mathbf{r}_1, t - t_1; \mathbf{r} - \mathbf{r}_2,$$

$$t - t_2; \mathbf{r} - \mathbf{r}_3, t - t_3)$$

$$\times E_\beta(\mathbf{r}_1, t_1) E_\gamma(\mathbf{r}_2, t_2) E_\delta(\mathbf{r}_3, t_3) \cdots . \qquad (8.80)$$

The first term shows how the second-order nonlinear susceptibility tensor, $\chi^{(2)}$, relates the α component of the polarization at $(\mathbf{r}, t)$ to the β component of $\mathbf{E}$ at $(\mathbf{r}_1, t_1)$ and the γ component of $\mathbf{E}$ at $(\mathbf{r}_2, t_2)$, where $t_1, t_2 \le t$. Clearly this gets even uglier if carried to higher orders.

When viewed from the familiar territory of the linear treatment, the consequence of the nonlinearity and dependence on more than one time in the past is to mix frequencies. That is, feeding in two electric field contributions, one oscillating at ω_1 and another at ω_2 can result in a polarization response at other frequencies, thanks to the nonlinear response. We can Fourier transform the time dependences of the nonlinear susceptibilities to arrive at the quantities $\chi_{\alpha\beta\gamma}^{(2)}(\omega_1, \omega_2)$ and $\chi_{\alpha\beta\gamma\delta}^{(3)}(\omega_1, \omega_2, \omega_3)$.

We can at least see in a classical sense where various nonlinear effects originate by considering nonlinear modifications to our harmonic oscillator approach of Eq. (8.59). Working in one dimension for simplicity,

$$\ddot{x} + \Gamma \dot{x} + \omega_0^2 x + a x^2 + b x^3 + \cdots = \frac{-e}{m} E(t), \qquad (8.81)$$

where we have introduced nonlinear terms in the restoring force (anharmonic terms in the potential energy of the charged oscillator), and used $\Gamma \equiv \gamma/m$ from our prior definition of γ as a phenomenological damping term. A cubic term in the potential energy of the charged oscillator gives the ax^2 piece of the force, while a quartic term in the potential likewise gives the cubic contribution to the force. Physically, a cubic term in the potential energy originates in a solid with *broken inversion symmetry*. This makes sense, given that a cubic potential energy would be odd around the origin, in contrast to either the unperturbed harmonic potential or a quartic contribution.

The best plan for small nonlinearities is to try a perturbative solution of the form $x(t) = x^{(1)}(t) + x^{(2)}(t) + x^{(3)}(t) + \cdots$, where $x^{(n)}(t)$ is proportional to the nth power of the electric field. We can plug this in and group like powers of field to find:

$$\ddot{x}^{(1)}(t) + \Gamma \dot{x}^{(1)}(t) + \omega_0^2 x^{(1)}(t) = \frac{-e}{m}E(t),$$

$$\ddot{x}^{(2)}(t) + \Gamma \dot{x}^{(2)}(t) + \omega_0^2 x^{(2)}(t) + a(x^{(1)}(t))^2 = 0,$$

$$\ddot{x}^{(3)}(t) + \Gamma \dot{x}^{(3)}(t) + \omega_0^2 x^{(3)}(t) + 2ax^{(1)}(t)x^{(2)}(t) + b(x^{(1)}(t))^3 = 0. \tag{8.82}$$

This can be solved iteratively. We know the solution to the linear part; it's the usual damped harmonic oscillator.

Things get interesting when we consider driving terms with two different frequencies, $E(t) = E_1 \exp(-i\omega_1 t) + E_2 \exp(-i\omega_2 t)$. Substituting this in to the right side of Eq. (8.82) shows that the presence of *the nonlinear terms in the force lead to a polarization response at frequencies other than just the driving frequencies* ω_1 *and* ω_2.

Second-order nonlinear effects

The squared term in the force equation (related to $\chi^{(2)}$) leads, in the $x^{(2)}$ part of the polarization, to contributions proportional to $E(t)^2$ via the $(x^{(1)}(t))^2$ term. The resulting contributions include:

- **Second harmonic generation** – terms proportional to $\exp(-i(2\omega_1)t)$ and $\exp(-i(2\omega_2)t)$. Classically, since the polarization can oscillate at twice the driving frequencies, it can produce radiation at those doubled frequencies. In a full quantum mechanical treatment, second harmonic generation is a two-photon process, but still originates from the same kind of term in the potential energy of displaced charges.
- **Sum and difference generation** – terms proportional to $\exp(-i(\omega_1 - \omega_2)t)$ and $\exp(-i(\omega_1 + \omega_2)t)$.
- **Optical rectification** – Because of the difference frequency generation, dc terms appear in the polarization (!) that are proportional in magnitude to $a|E_1|^2$ and $a|E_2|^2$.

One additional piece of physics, the **electro-optic effect**, bears special notice because of its technological impact on telecommunications. Consider just the a term, and its effect on a dc electric field applied with the usual ac field of some wave.

$$\ddot{x} + \Gamma \dot{x} + \omega_0^2 x + ax^2 = \frac{-e}{m}\left[\frac{E_0}{2}(e^{i\omega t} + e^{-i\omega t}) + E_{dc}\right]. \tag{8.83}$$

Assume a solution of the form $x(t) = x_0 + (1/2)(x_1 \exp(i\omega t) + x_1^* \exp(-i\omega t))$. Plugging in, the dc terms give us

$$\frac{1}{2}\omega_0^2 x_0 + a\left[x_0^2 + \frac{|x_1|^2}{2}\right] = \frac{-e}{m}E_{dc}, \tag{8.84}$$

while the ac terms give us

$$x_1[\omega_0^2 - \omega^2 - i\Gamma\omega + 2ax_0] = \frac{-e}{m}E_0. \tag{8.85}$$

The dc polarization response and the ac polarization response are coupled. The net result of this is an effective index of refraction that depends upon the dc field. Physically, oscillations of charges around their equilibrium positions give the index of refraction. The dc field statically displaces the charges away from their equilibrium positions, so that they now oscillate about a point where the charge's potential energy's curvature is different because of the cubic contribution. As a result, the frequency response of those oscillations can be shifted by the dc field.

Properly describing electro-optic effects in real materials requires the full tensorial dielectric response. Generally, $\mathbf{D} = \tilde{\epsilon} \cdot \mathbf{E}$, where $\tilde{\epsilon}$ is the dielectric tensor. Inverting, we have

$$\mathbf{E} = \tilde{\eta} \cdot \mathbf{D} = \frac{1}{\epsilon_0} \begin{pmatrix} 1/n_x^2 & 0 & 0 \\ 0 & 1/n_y^2 & 0 \\ 0 & 0 & 1/n_z^2 \end{pmatrix} \cdot \mathbf{D}, \tag{8.86}$$

where we've chosen coordinates along the principal axes of the material's dielectric response, and n_x is the index of refraction for radiation propagating in the x-direction, etc. The tensor $\tilde{\eta}$ is modified by the dc field in an electro-optic material:

$$\tilde{\eta} \rightarrow \frac{1}{\epsilon_0} \begin{pmatrix} 1/n_x^2 & 0 & 0 \\ 0 & 1/n_y^2 & 0 \\ 0 & 0 & 1/n_z^2 \end{pmatrix} + \begin{pmatrix} \delta\eta_1 & \delta\eta_6 & \delta\eta_5 \\ \delta\eta_6 & \delta\eta_2 & \delta\eta_4 \\ \delta\eta_5 & \delta\eta_4 & \delta\eta_3 \end{pmatrix}, \tag{8.87}$$

where

$$\delta\tilde{\eta} = \tilde{r} \cdot \mathbf{E}_{dc} = \begin{pmatrix} r_{1x} & r_{1y} & r_{1z} \\ \cdot & \cdot & \cdot \\ \cdot & \cdot & \cdot \\ \cdot & \cdot & \cdot \\ r_{6x} & r_{6y} & r_{6z} \end{pmatrix} \cdot \mathbf{E}_{dc}. \tag{8.88}$$

The rs are called electro-optic coefficients, and relate the change in $1/(\epsilon_0 n^2)$ in some propagation direction to an applied dc electric field in a different direction. To get a sense of scale, for GaAs, $r_{4x} = r_{5y} = r_{6z} = 5.9 \times 10^{-12}$ m/V near 900 nm incident wavelength. Assume propagation in the z-direction and an applied dc field in the x-direction. The result is a change in index for z-propagating modes with electric field along the y-direction, and the magnitude of the effect is about $\delta n \approx -(1/2)n^3 r E_{dc,x}$. For 900 nm incident wavelength, $n \approx 3.6$, and a dc electric field of 10^7 V/m, the index changes by around 0.0013.

It is difficult to get much larger index changes via the electro-optic effect, both because the electro-optic coefficients are small, and because the breakdown dc electric fields for materials with large rs tend to be poor. One of the favorite industrial electro-optic materials is $LiNbO_3$, which has coefficients 4–5 times larger than those of GaAs. Systems with large electro-optic coefficients have lots of charge arranged in non-centrosymmetric ways within the unit cell. This generally leads to other effects that can be important, such as piezoelectricity and pyroelectricity. Nanostructuring and tuning of material properties to enhance desirable effects without introducing other problems is a very active area of research in nonlinear optical materials.

Third-order nonlinear effects

The cubic term in the force equation has similar consequences in the $x^{(3)}$ contribution to the polarization response. These include **third harmonic generation**, analogous to the second harmonic generation described above, and **intensity-dependent propagation**, which we briefly outline here. Start from the 1d model with only the cubic term:

$$\ddot{x} + \Gamma\dot{x} + \omega_0^2 x + bx^3 = \frac{-e}{2m}E_0[\mathrm{e}^{\mathrm{i}\omega t} + \mathrm{e}^{-\mathrm{i}\omega t}]. \tag{8.89}$$

The solution to the unperturbed ($b = 0$) problem is $x_0(t) = (1/2)[x_0\mathrm{e}^{\mathrm{i}\omega t} + x_0^*\mathrm{e}^{-\mathrm{i}\omega t}]$, where

$$|x_0|^2 = \frac{e^2}{|m(\omega_0^2 - \omega^2 - \mathrm{i}\Gamma\omega)|^2}|E_0|^2. \tag{8.90}$$

Plugging in $x_0(t)$ and looking at the third-order term, we find

$$b(x^{(1)}(t))^3 = \frac{3}{8}b|x_0|^2[x_0\mathrm{e}^{\mathrm{i}\omega t} + x_0^*\mathrm{e}^{-\mathrm{i}\omega t}] + \frac{1}{8}b[x_0^3\mathrm{e}^{\mathrm{i}3\omega t} + (x_0^*)^3\mathrm{e}^{-\mathrm{i}3\omega t}]. \tag{8.91}$$

Note that if we look just at the equation that must be obeyed for the polarization linearly proportional to $E(t)$, we now have

$$\ddot{x}^{(1)}(t) + \Gamma\dot{x}^{(1)}(t) + \left(\omega_0^2 + \frac{3}{8}|x_0|^2\right)x^{(1)}(t) = \frac{-e}{2m}E_0[\mathrm{e}^{\mathrm{i}\omega t} + \mathrm{e}^{-\mathrm{i}\omega t}]. \tag{8.92}$$

The restoring force experienced by a classical charge now has a contribution proportional to the intensity, $|E_0|^2$, which affects the displacement of the charge. The resulting additional contribution to the polarizability may be plugged into the Claussius–Mossotti relation, Eqs. (5.9) and (5.10), to find the index of refraction in the presence of this nonlinearity. The result depends on intensity. If the index in the limit of low intensity is n_{L}, then the third-order nonlinear contribution leads to $n \approx n_{\mathrm{L}} + n_2 I$, where n_2 has units of inverse intensity, such as cm^2/W.

For glass fiber over most frequencies of interest, $n_2 \sim 10^{-16}$ cm^2/W. That means that at 1 $\mathrm{W/cm}^2$ intensity, the index of refraction is only modified by a part in 10^{16}. Nonlinearities in many media are very small like this, which is why they are outside our daily experience. In contrast, near the onset of band absorption in semiconductors, nonlinearities may be considerably larger. For example, InSb at an incident wavelength of 5 μm has $n_2 \sim 10^{-3}$ cm^2/W. The downside from the standpoint of technological leverage is that large nonlinearities are typically coincident with large losses.

The nonlinear enhancement of n can lead to some very interesting effects. While these are rather beside the point of this work, we briefly mention three without going into the mathematical details. The first is *self-focusing*. Imagine a beam propagating in z with a greater intensity in the center of its cross-section (that is, not a plane wave), impinging upon a slab of material with this sort of optical nonlinearity. That greater intensity would lead to a locally higher index of refraction. This would effectively turn the nonlinear medium into a sort of converging lens or a waveguide. Any transversely propagating component would be refracted back toward the beam axis. In some regime of parameters, there are stable propagating solutions with particular functional forms for the transverse shape of the beam. However, if the initial power is too high, there are no stable propagating solutions.

Propagating into the nonlinear medium, the beam becomes more tightly focused, leading to a greater index gradient, leading to further focusing, in a catastrophic runaway. The result will be material damage once the local intensity is too large.

A second feature of third-order nonlinear media is the propagation of *solitons*. Imagine sending an optical pulse with some carrier frequency (and relatively small bandwidth). With ordinary dispersion, the higher frequency components of the pulse will experience a higher index, and the pulse will tend to spread in space along its propagation direction as it moves. However, in a nonlinear medium, the largest amplitude portion of the pulse will experience a higher index, which tends to compress the pulse. If these parameters balance, the result is dispersion-free pulse propagation. The analog of this effect in traveling surface waves on water was first recognized by J. S. Russell in 1834 in the Union Canal in Scotland. For various technical reasons, solitons in optical waveguides have not taken off in deployed telecommunications technology, though that may change, as nanoscale control over materials and their properties makes available more tuning parameters.

A final example of third-order optical nonlinearity is the *optical Kerr effect*, which is relevant when the total intensity affecting the index originates from two counterpropagating waves. In the language of dielectric susceptibility, χ, here, $\mathbf{P} \approx \chi_L \mathbf{E} + \chi^{(3)} |\mathbf{E}|^2 \mathbf{E} = \chi_L \mathbf{E} + \chi^{(3)} \mathbf{E} \mathbf{E}^* \mathbf{E}$. For two waves of amplitude $\mathbf{E}_+$ ($\mathbf{E}_-$) propagating in the $+z$ ($-z$) direction, $\mathbf{E} = [\mathbf{E}_+ \exp(ik_+ z) + \mathbf{E}_- \exp(-ik_- z)] \exp(-i\omega t)$. It turns out that the currect solutions for the propagation constants k_+ and k_- for the two components lead to different effective indices for the different propagation directions:

$$n_+ = n_L + n_2 |\mathbf{E}_+|^2 + 2n_2 |\mathbf{E}_-|^2,$$
$$n_- = n_L + n_2 |\mathbf{E}_-|^2 + 2n_2 |\mathbf{E}_+|^2. \tag{8.93}$$

In interferometers based on counterpropagating waves, such as optical gyroscopes, this can be a major issuc. Phase differences for the two waves can arise because of slight differences in their intensities.

8.1.13 Dielectric multilayers

Returning to linear media, it has been known for decades that layered dielectric structures may be used to engineer the optical response of interfaces. The basic idea is straightforward. Figure 8.9 shows an antireflection coating, of the sort routinely used in optical components. The rays have been drawn with a slight deviation from normal incidence to clarify thc idcntification of different particular rays. We will discuss the normal incidence case, to avoid complications due to polarization.

A wave from a medium with index $n = 1$ is incident on a medium of higher index, $n_a = 1.4$. From the Fresnel formulae (Eqs. (8.51) and (8.52)), we see that the reflected wave, labeled "(1)", produced at that interface picks up a π phase shift relative to the incident wave. The light transmitted at the $n - n_a$ interface accumulates additional phase while propagating through medium a. Some of that light is then reflected of the second interface ($n_a - n_b$, where $n_b > n_a$), accumulates more phase propagating back to to the $n - n_a$ interface, and then (some) of that light (labeled "(2)") re-enters the original medium.

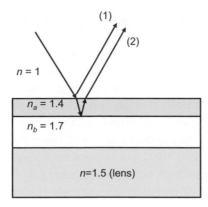

Fig. 8.9

Cartoon of an antireflection coating, with rays slightly tilted from normal incidence for clarity. Layer thicknesses have been chosen so that waves reflected from the various interfaces interfere destructively on the incident side.

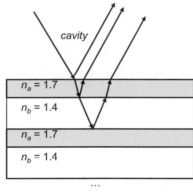

Fig. 8.10

Dielectric mirrors. Each bilayer is constructed so that at a particular frequency reflected waves from the interfaces interfere constructively in the incident medium. A multilayer stack can be designed to continue this interference for every interface. Arbitrarily good reflectance is possible as the number of periods is increased, with no loss for ideal dielectrics.

By choosing the thickness of medium a properly, wave (2) may be chosen to be exactly π out of phase with wave (1), leading to destructive interference. This reduces the reflected intensity relative to what it would have been if the incident wave had impinged directly on the underlying lens material.

Dielectric mirrors are based on the same concept, only this time the interfaces and layer thicknesses are chosen so that all of the reflected components that make it back to the original medium interfere *constructively*. Figure 8.10(a) shows the basic idea. At the $n - n_a$ interface, the reflection coefficient $r \equiv E_{0r}/E_{0t} = (n - n_a)/(n + n_a)$. As before, for $n_a > n$, $r < 0$, so there is a π phase shift for that reflection. For enhanced reflection, we need to choose the thickness t_a such that the light reflected from the $a - b$ interface comes back to the surface with exactly a π phase shift relative to the incident light. Since $k_a = n_a \omega / c$, this means we want $2 \times k_a t_a = \pi$. This gives $t_a = (\pi/2)(c/(2\pi v))(1/n_a) = (1/4)(\lambda_0/n_a)$.

That is, the desired layer thickness t_a is one quarter of the wavelength of the radiation in medium a.

Suppose you continued the stack, and wanted constructive interference between the reflected waves from each succeeding interface. Going from b to a gives $r = (n_b - n_a)/(n_b + n_a)$, and since $n_b < n_a$, there is a π phase shift at that interface. Leave t_a as found above, and choose t_b to be $(1/4)(\lambda_0/n_b)$. The total phase shift experienced by a wave coming in from the original medium and ending back up in the original medium after bouncing off the $b - a$ interface is then $\pi/2 + \pi/2$ (for travel through layer a and layer b) $+\pi$ (for the $b - a$ reflection) $+\pi/2 + \pi/2$ (back up through layers b and a), for a total of $2\pi + \pi$, which is in phase with both waves (1) and (2). Thus, reflection is enhanced by constructive interference of the reflected components that make it back to the original medium.

A series of layers of alternating high and low index, with each layer set to be $1/4$ of the in-layer wavelength in thickness, is sometimes called a "quarter-wave stack". Each bilayer is called a "period". A dielectric mirror of this type is clearly constructed for a particular limited wavelength range. For normal incidence at the design wavelength, dielectric mirrors are often superior to metal mirrors, exhibiting higher reflectance and lower loss.

Knowing the reflection coefficient for each interface, it is possible to sum the series of reflection coefficients for a stack of m periods. For an incident "cavity" index n_c and a final substrate index n_s,

$$r_m = -\frac{1 - \alpha}{1 + \alpha}, \qquad \alpha \equiv \frac{n_s n_c}{n_a^2} \left(\frac{n_b}{n_a}\right)^{2m}. \tag{8.94}$$

The keys to higher reflectivity are to have a smaller α, most readily achieved by having a larger index contrast ($n_b/n_a \ll 1$), or more periods. It is possible to solve for the number of periods required to achieve a given reflectance $R = |r_m|^2$:

$$m \geq 0.5 \ln\left[\frac{n_a^2(1 - \sqrt{R})}{n_s n_c(1 + \sqrt{R})}\right] \Big/ \ln(n_b/n_a). \tag{8.95}$$

Likewise, with a more detailed analysis, one can determine what happens at a free-space wavelength λ a bit away from the design-optimized wavelength, λ_0. The change in λ_0/λ required to suppress the reflectance to half of its maximum value is $2/\pi \sin^{-1}((n_a - n_b)/(n_a + n_b))$.

For a more general treatment of propagation through dielectric multilayers, the appropriate formulation is that of *transfer matrices*. This approach is completely analogous to the matrix approach taken in Section 6.4.3 for quantum transport. In the limit of propagation in the z-direction normal to the interfaces, assuming a transverse electric field (and otherwise ignoring polarization effects), we can factor out the harmonic time dependence of the fields and worry only about the spatial dependence. For a medium with a local admittance $Y \equiv \sqrt{\epsilon/\mu}$,

$$E(z) = E_{0+}e^{-ikz} + E_{0-}e^{+ikz} = E_+(z) + E_-(z),$$

$$H(z) = \frac{1}{Y}[E_{0+}e^{-ikz} + E_{0-}e^{+ikz}] = \frac{1}{Y}[E_+(z) + E_-(z)]. \tag{8.96}$$

We have written the fields at a particular location z as a linear combination of waves propagating in the $\pm z$ directions. Rearranging, we can turn this around:

$$E_+(z) = \frac{1}{2}[E(z) + YH(z)],$$

$$E_-(z) = \frac{1}{2}[E(z) - YH(z)]. \qquad (8.97)$$

In matrix language,

$$\begin{pmatrix} E \\ H \end{pmatrix} = \begin{pmatrix} 1 & 1 \\ Y^{-1} & Y^{-1} \end{pmatrix} \begin{pmatrix} E_+ \\ E_- \end{pmatrix}, \quad \begin{pmatrix} E_+ \\ E_- \end{pmatrix} = \begin{pmatrix} 1 & Y \\ 1 & -Y \end{pmatrix} \begin{pmatrix} E \\ H \end{pmatrix}. \qquad (8.98)$$

For propagation between two locations z_1 and z_2, we can find the relative phases of the fields readily: $E_+(z_2) = E_{0+}e^{-ikz_2} = E_{0+}e^{-ikz_1} \cdot e^{ik(z_1-z_2)} = E_+(z_1)e^{-ikl}$, where $l = z_2 - z_1$. Similarly, $E_-(z_2) = E_-(z_1)e^{+ikl}$. In matrix form,

$$\begin{pmatrix} E_+(z_1) \\ E_-(z_1) \end{pmatrix} = \begin{pmatrix} e^{ikl} & 0 \\ 0 & e^{-ikl} \end{pmatrix} \begin{pmatrix} E_+(z_2) \\ E_-(z_2) \end{pmatrix}. \qquad (8.99)$$

The matrix above is a *propagation matrix*, relating the propagating fields at one location to those at another, within the same material.

At an interface between one type of material and another, boundary conditions hold, in particular the continuity of E and H across the interface. In terms of E_+ and E_-,

$$E_+ + E_- = E'_+ + E'_-,$$

$$\frac{1}{Y}(E_+ - E_-) = \frac{1}{Y'}(E'_+ - E'_-), \qquad (8.100)$$

where the unprimed (primed) quantities refer to the material at lower (higher) z. Introducing the notion of reflection and transmission coefficients from one material to the other, we can rewrite:

$$\begin{pmatrix} E'_+ \\ E'_- \end{pmatrix} = \frac{1}{\tau'} \begin{pmatrix} 1 & \rho' \\ \rho' & 1 \end{pmatrix} \begin{pmatrix} E_+ \\ E_- \end{pmatrix}, \qquad (8.101)$$

where

$$\rho' \equiv \frac{Y - Y'}{Y + Y'} = \frac{n' - n}{n' + n} \qquad \tau' \equiv \frac{2Y}{Y + Y'} = \frac{2n'}{n' + n}. \qquad (8.102)$$

The equalities in Eq. (8.102) hold for fixed μ across the interface. The reflection and transmission coefficients are defined at the interface as $E'_+ = \rho'E'_-$, $E_- = \tau'E'_-$, $E_- = \rho E_+$, and $E'_+ = \tau E_+$. With these, we can write a *scattering matrix*, expressing the magnitudes of the waves heading away from the interface in terms of the magnitudes of the waves heading toward the interface:

$$\begin{pmatrix} E'_+ \\ E_- \end{pmatrix} = \begin{pmatrix} \tau & \rho' \\ \rho & \tau \end{pmatrix} \begin{pmatrix} E_+ \\ E'_- \end{pmatrix}. \qquad (8.103)$$

With this toolkit, one can treat complicated multilayer structures with relative ease. We can represent propagation through each material with an appropriate propagation matrix, and each interface with an appropriate matching matrix. Multiplying all the matrices together can then yield the overall effect of the multilayer. An excellent discussion of this general approach is presented by Prof. S. J. Orfanidis (www.ece.rutgers.edu/orfanidi/ewa/) [597].

8.2 Lasers

A basic overview of lasers and laser physics is needed for discussions of nanophotonics. A number of classic, outstanding books exist for a more detailed discourse on lasers, including *Laser Physics* by Sargent, Scully and Lamb [598], *Lasers* by Siegman [599], and *Principles of Lasers* by Svelto [600].

Colloquially, lasers are bright, monochromatic sources of coherent light, meaning that there is a very small spread in the wavelength of a laser's emission, and that there is a well defined phase of the electromagnetic radiation coming out. The origins of these properties are inherently quantum mechanical, as we shall see.

8.2.1 Laser basics

The word "laser" originated as an acronym, Light Amplification by the Stimulated Emission of Radiation. Therefore it makes sense to begin any discussion of lasers with the interaction of a quantum system with radiation, and the physics of absorption, stimulated emission, and spontaneous emission.

Let's begin initially with a quantum system that starts in some initial state a, and is subjected to a perturbation (Hamiltonian $H' = V_0 \cos \omega t$) that is harmonic in time. For example, one can think about an electronic state with some electric dipole moment, and the interaction with a harmonically varying electric field would provide such a perturbation. We consider transitions to some final state b. The matrix element for the perturbation is then $\langle \psi_b | V_0 | \psi_b \rangle \equiv V_{0,ba}$. Section A.1.5 is a brief derivation of time-dependent perturbation theory in quantum mechanics, and tackles this case. If the unperturbed energies[2] of states a and b are, respectively, E_a^0 and E_b^0, we define $\omega_{ba} \equiv (E_b^0 - E_a^0)/\hbar$. It's useful to define the function $F(t, \Omega) \equiv 2 \sin^2(\Omega t/2)/\Omega^2$. Carrying out the perturbation calculation, the probability at some time t that the system has made a transition from state a to state b is

$$P_{ba}(t) = F(t, \omega_{ba} + \omega)|V_{0,ba}|^2 + F(t, \omega_{ba} - \omega)|V_{0,ba}|^2 + \text{cross terms.} \qquad (8.104)$$

At long times ($t \gg \Omega^{-1}$), $F(t, \Omega) \to \pi t \delta(\Omega)$, where δ is the Dirac delta function, and the cross terms average away to nothing. Equation (8.104) shows that at long times, the only transitions that occur are those that raise or lower the energy of the system by $\hbar \omega$. The first term corresponds to *stimulated emission*, while the second term corresponds to *absorption*. As we saw when we discussed Fermi's Golden Rule in Section 2.1.6, the total transition rate ends up depending on both the matrix element of the perturbation between

---

[2] Not electric fields!

initial and final states (clear from here), and on the density of final states available and the quantum statistics of the particles (not explicitly shown here).

The simplest type of perturbation describing the interaction of matter and radiation is the electric dipole interaction. In that case, $H'(t) = \mathbf{E}_0 \cdot (-e\mathbf{r}) \cos \omega t$, where $\mathbf{r}$ is the position of an electron relative to the origin. If θ is the angle between the radiation electric field $\mathbf{E}_0$ and the dipole moment, we can write the absorption *rate* for going into a particular state b as

$$W_{\text{abs}} = \frac{\pi I(\omega)}{\hbar^2 c \epsilon_0} \cos^2 \theta |\mathbf{p}_{ba}|^2,\tag{8.105}$$

where $I(\omega)$ is the intensity of the radiation, and $\mathbf{p}_{ba} \equiv \langle \psi_b | - e\mathbf{r} | \psi_a \rangle$. Note that $\mathbf{p}_{ba} = \mathbf{p}_{ab}^* \rightarrow |\mathbf{p}_{ba}|^2 = |\mathbf{p}_{ab}|^2$. This last relation implies a key point: **the stimulated emission rate, W_{stim}, is identical to the absorption rate**, under the influence of the same radiation. This is called *detailed balance*. Averaging over all polarization directions, $\langle \cos^2 \theta \rangle \rightarrow 1/3$.

Einstein A and B coefficients

Well before a complete quantum theory of electrodynanmics was formulated, Einstein was able to use thermodynamics to make a brilliant calculation about *spontaneous* emission. Here is the basic idea. In some cavity, there are N_a atoms in the ground state, and N_b atoms in an excited state, and an energy density $\rho(\omega_{ba})$ of radiation at the appropriate frequency ω_{ba} to cause transitions. In this case, the rate of absorption transitions from state a to state b should be

$$\dot{N}_{ba} = B_{ba} N_a \rho(\omega_{ba}).\tag{8.106}$$

From Eq. (8.105) we identify B_{ba} as $\pi |\mathbf{p}_{ba}|^2/(3\hbar^2 \epsilon_0)$, where we've averaged over θ. This makes sense; the total rate of transitions from a to b has to be proportional to the amount of radiation available and the number of atoms in state a.

The emission rate, the rate of transitions from b to a, has two contributions:

$$\dot{N}_{ab} = A N_b + B_{ab} N_b \rho(\omega_{ba}).\tag{8.107}$$

The first term is the spontaneous emission rate, and the second is stimulated emission. As we've already shown, $B_{ba} = B_{ab} \equiv B$. The parameters A and B are known as the Einstein A and B coefficients. As alluded, calculating A_{ab} directly from first principles requires quantum field theory to describe the zero-point electromagnetic field.

Einstein evaded this challenge via extreme cleverness. In steady state, the emission and absorption rates must balance. Also, in thermal equilibrium, the ratio of excited state to ground state population must be given by a Boltzmann factor. Together, these conditions imply

$$\frac{N_b}{N_a} = \exp\left(-\frac{\hbar \omega_{ba}}{k_B T}\right) = \frac{A + B\rho(\omega_{ba})}{B\rho(\omega_{ba})}.\tag{8.108}$$

One could solve this for $\rho(\omega_{ba})$, which we know must be given by Planck's blackbody radiation formula if we are in thermal equilibrium. The result lets us find A in terms of

B (and implicitly the photon density of states, which comes in by our assumption of a 3d cavity for the blackbody formula), and we find the (dipole approximation) spontaneous emission rate without needing quantum field theory:

$$W_{\text{spont}} = \frac{\omega_{ba}^3}{3\pi c^3 \hbar \epsilon_0} |\mathbf{p}_{ab}|^2. \tag{8.109}$$

This treatment highlights three essential points.

- In thermal equilibrium, it is impossible to have more of the atoms in the excited state than in the ground state, because of the Boltzmann factor. This kind of "population inversion" would require negative absolute temperature, not possible in equilibrium. As $T \to \infty$, $N_b \to N_a$ from below.
- In a two-level system like this, even out of equilibrium in the presence of driving radiation at the resonant frequency ω_{ba}, $N_b \leq N_a$. Turning up the intensity of the radiation increases the absorption rate, but also increases the stimulated emission rate equally. Again, no population inversion.
- The derivation of the spontaneous emission rate involves the photon density of states. This means that placing an atom in a cavity designed to have a restricted photon density of states can alter the spontaneous emission rate, a quantity that seems intuitively like an intrinsic property of an atom. This is remarkable, and is the basis for *cavity quantum electrodynamics (QED)*.

We now know enough to consider the LASER idea. Imagine a medium filled with two-level systems like the ones discussed above, with radiation incident upon that medium at the frequency ω that is resonant with the level splitting. The change in radiation energy density due to absorption is $\dot{\rho} = -N_a \hbar \omega W_{\text{abs}}$. The change due to stimulated emission is $(\dot{\rho}) = +N_b \hbar \omega W_{\text{stim}}$. Neglecting spontaneous emission, the net change in radiation energy density is proportional to $N_b - N_a$, the difference between excited and ground state populations. If one could somehow achieve population inversion, it would be possible to get a net gain for the radiation energy density. Hence light amplification by the stimulated emission of radiation. Note that the coherent nature of lasing results from the bosonic character of photons; stimulated emission tends to go into photon states that are already populated.

Figure 8.11 shows the basic concept. Some pumping process is used to establish a population inversion. An optical cavity of some sort establishes particular electromagnetic modes that become populated. Because of the quality factor of the cavity, photons occupying those modes stick around much longer than other photons, and thus have many more opportunities to cause stimulated emission. Light is coupled out of the system through a "leaky" part of the cavity. The many different varieties of laser are distinguished chiefly by their level structure (related to how population inversion is achieved) and cavity type.

Level structures

As we have seen, directly pumping on the transition in a two-level situation can't lead to population inversion. However, it is possible to get lasing with only two levels, provided there is some other means of establishing a very nonthermal population distribution

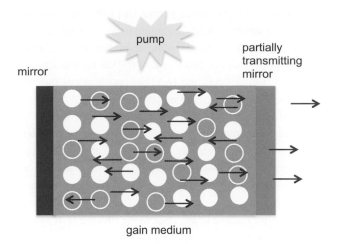

Cartoon of a basic laser. A medium containing emitters is pumped somehow into a population inverted condition. Emitted photons stimulate the emission of more photons at the same frequency and phase. A cavity allows the build-up of a large photon population in a particular mode, with leakage allowing light to escape.

Fig. 8.11

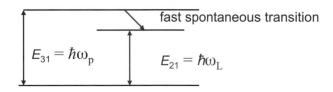

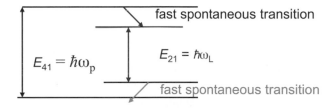

Energy level arrangements conducive to lasing. Top: three-level lasing; bottom: four-level lasing.

Fig. 8.12

between the two levels. For example, the original ammonia maser [601] (microwave amplification by the stimulated emission of radiation) was based on a two-level hyperfine transition in $^{14}NH_3$. Starting from a beam that was a mix of ground and excited state molecules, an inhomogenous electric field was used to select out a beam composed mostly of the excited species, which was then directed into an appropriate microwave cavity. Electrically pumped semiconductor lasers operate on a similar idea, with electrical bias and current used to produce a nonthermal carrier distribution.

Other commonly used lasing systems use either three or four levels, with appropriate kinetics leading to the inversion situation needed for lasing. Figure 8.12 shows the basic idea. In the three-level case, a light source pumps the 1–3 transition. A fast spontaneous transition (radiative or otherwise) leads to population of level 2. If the spontaneous emission rate from level 2 back to ground state level 1 is slow relative to the 3–2 rate, pumping may build up population in level 2, and the lasing transition is from 2–1. The generic four-level case is similar, requiring two fast, spontaneous transitions. The kinetics of these cases may be treated with a master equation approach.

One traditional technique to make a lasing medium is to dope a transparent (at the pump and lasing energies) host material with metal ion impurities, with optical transitions provided by excitation of the ion electronic levels. The classic ruby laser uses chromium impurities in the otherwise transparent Al_2O_3 host as the source of the key optical transitions. A pump excites the Cr^{+3} ions into broad upper levels that decay rapidly through nonradiative processes, populating a doublet (two states split by 3.6 meV) of metastable levels. The transition of those excited levels back to the ground state is the lasing transition, producing light at a free space wavelength of 694.3 nm.

Two other examples of this use rare earth metals rather than transition metals. Because of their f electrons, a larger number of transitions may be available in rare earth dopants. In the case of Nd^{+3} ions in yttrium aluminum garnet ($Y_3Al_5O_{12}$, also known as YAG), at least five lasing level combinations are possible, with emission from 807 nm to 1.8 μm in wavelength. Critical to telecommunications, an effective four-level laser can be made using Er^{+3} ions in silicate glass. The natural emission line that is most useful is at 1.536 μm wavelength. This wavelength corresponds to a minimum in the wavelength-dependent absorption of SiO_2, meaning that erbium-doped lasers are perfect for transmitting information through glass optical fibers.

8.2.2 Semiconductor optical properties

Before discussing semiconductor lasers, we look briefly at optical processes in semi-conductors.[3] The same processes we have been discussing (absorption, spontaneous emission, stimulated emission) remain relevant; we just need to treat the electronic states correctly. For the purposes of this discussion, we will not dwell on exciton effects, the shifting of energy scales due to the binding energy of electron–hole pairs and other Coulomb interactions between carriers. While these effects alter precise energy scales and lineshapes, the underlying physical processes about which we care most are not changed drastically.

Think about optically driven transitions in a generic semiconductor. These transitions must conserve both energy and momentum, and recall that a useful way to think about electrons in periodic solids is as Bloch states, characterized by a band index and a wavevector, **k**. Each band has an effective mass due to the dispersion of that band state's energy as a function of **k**. For an ideal semiconductor at zero temperature, the only transitions that are relevant for absorption or emission involve Bloch states in different bands.

---

[3] This discussion is patterned on lecture notes of Prof. R. Victor Jones of Harvard University.

To calculate transition rates, we need to work with Bloch states as initial and final states for the electrons, and we need an appropriate form of the perturbing Hamiltonian from the radiation, analogous to Eq. (8.105). The general form for the Hamiltonian for charged matter interacting with an electromagnetic field has a term $H = (1/2m)(\mathbf{p} - q\mathbf{A})^2 + \cdots$, where $\mathbf{p}$ here is the momentum, q is the charge, and $\mathbf{A}$ is the vector potential. Expanding this out and looking just at the interaction part gives us

$$H' = -\frac{q}{m}\mathbf{p} \cdot \mathbf{A} = -\frac{q}{2m}A_0\mathbf{p} \cdot \hat{\mathbf{e}}\exp(i\mathbf{k}_{rad} \cdot \mathbf{r} - i\omega t) + \cdots, \qquad (8.110)$$

where we've assumed the usual plane wave form for $\mathbf{A}$, and $\hat{\mathbf{e}}$ is the unit vector in the direction of $\mathbf{A}$.

For absorption, an initial Bloch state for an electron in the valence band is $\psi_{v,\mathbf{k}}(\mathbf{r}) = u_{v,\mathbf{k}}(\mathbf{r})\exp(-i\mathbf{k} \cdot \mathbf{r})$, with $u_{v,\mathbf{k}}(\mathbf{r})$ having the spatial periodicity of the lattice, as with all Bloch functions. Similarly, the final state in the conduction band would be $\psi_{c,\mathbf{k}'}(\mathbf{r}) = u_{c,\mathbf{k}'}(\mathbf{r})\exp(-i\mathbf{k}' \cdot \mathbf{r})$. The matrix element for an electronic transition from the valence to the conduction band is then

$$\langle \psi_{v,\mathbf{k}}|H'|\psi_{c,\mathbf{k}'}\rangle = \frac{e}{2m}A_0\exp(-\omega t)M_{v\mathbf{k}c\mathbf{k}'}, \qquad (8.111)$$

where

$$M_{v\mathbf{k}c\mathbf{k}'} = \int d\mathbf{r}[u_{v,\mathbf{k}}(\mathbf{r})\exp(-i\mathbf{k} \cdot \mathbf{r})(\hat{\mathbf{e}} \cdot \mathbf{p}\exp(i\mathbf{k}_{rad} \cdot \mathbf{r}))u_{c,\mathbf{k}'}^*(\mathbf{r})\exp(i\mathbf{k}' \cdot \mathbf{r})]. \qquad (8.112)$$

The best way to gain further insight here is to expand our Bloch wavefunctions. Remember that, since the u functions have the same periodicity as the lattice they may be written as a series using reciprocal lattice vectors (see Eq. (2.39)): $u_{n,\mathbf{k}}(\mathbf{r}) = \sum_j u_j(n, \mathbf{k})\exp(i\mathbf{G}_j \cdot \mathbf{r})$. Plugging in:

$$M_{v\mathbf{k}c\mathbf{k}'} = \sum_j \sum_l u_j(v, \mathbf{k})u_l^*(c, \mathbf{k})\int d\mathbf{r}[\exp(-i(\mathbf{k} - \mathbf{k}_{rad} + \mathbf{G}_j) \cdot \mathbf{r})(\hat{\mathbf{e}} \cdot \mathbf{p})\exp(i(\mathbf{k}' + \mathbf{G}_l)\cdot\mathbf{r})]. \qquad (8.113)$$

Remember, $\mathbf{p}$ here is the momentum operator, $-i\hbar\nabla$. This gives us

$$M_{v\mathbf{k}c\mathbf{k}'} = \sum_j \sum_l (\hbar\hat{\mathbf{e}} \cdot (\mathbf{k}' + \mathbf{G}_l)u_j(v, \mathbf{k})u_l^*(c, \mathbf{k})\int d\mathbf{r}[\exp(-i(\mathbf{k} - \mathbf{k}_{rad} - \mathbf{k}' + \mathbf{G}_j - \mathbf{G}_l) \cdot \mathbf{r})]. \qquad (8.114)$$

Also recall that we can write $\mathbf{r} = \mathbf{R}_b + \mathbf{r}'$, where $\mathbf{R}_b$ points to unit cell b, and $\mathbf{r}'$ points to a position within that unit cell. This means we can write the matrix element as a product of two parts, one that deals with the wavefunctions within a unit cell, and one that deals with the arrangement of unit cells:

$$M_{v\mathbf{k}c\mathbf{k}'} = \sum_j \sum_l \hbar\hat{\mathbf{e}} \cdot (\mathbf{k}' + \mathbf{G}_l)u_j(v, \mathbf{k})u_l^*(c, \mathbf{k})\int_{cell} d\mathbf{r}'[\exp(-i(\mathbf{k} - \mathbf{k}_{rad} - \mathbf{k}' + \mathbf{G}_j - \mathbf{G}_l)\cdot\mathbf{r}')]$$

$$\times \sum_b \exp(i(\mathbf{k}' - \mathbf{k} + \mathbf{k}_{rad}) \cdot \mathbf{R}_b). \qquad (8.115)$$

Because of this last term, this whole expression only gives a nonzero total result (when summed over a whole sample of many unit cells) if $\mathbf{k}' + \mathbf{k}_{rad} \approx \mathbf{k}$.

This result is basically conservation of momentum. Note that for visible light, $\mathbf{k}_{\text{rad}}$ is usually small compared to the Bloch wavevectors. In the language of band structure, this means that optical transitions are essentially "vertical". This is an illustration of why indirect gap semiconductors are not optically active at the minimum gap energy; radiation cannot provide the necessary change in $\mathbf{k}$ to boost a carrier from the top of the valence band (usually near $\mathbf{k} = 0$) to the bottom of the conduction band (at a nonzero wavevector somewhere else in the Brillouin zone, for an indirect gap system). Note the pivotal role played here by the periodicity of the lattice. While true momentum is always conserved, it is only in the infinite periodic lattice situation that $\mathbf{k}$ is strictly conserved between the carriers and the radiation. Breaking down the lattice symmetry, such as through nanostructuring material, can bend this selection rule, and alter optical transition rates.

Given the band structure for a semiconductor system, we can use this approach to find the matrix elements for interband transitions. As is clear from Eq. (8.115), there is a polarization factor (on the order of one) times the relevant integral. Beyond computing matrix element between states at the band edges, to get the full transition rates via Fermi's Golden Rule, we need to worry about the densities of states, energetic constraints, and the actual distribution functions of the carriers.

Figure 8.13 shows the situation for a generic undoped, direct-gap semiconductor at $T = 0$. With the filled valence band and empty conduction band, one would assume that absorption should only kick in once the radiation has sufficient energy to overcome the band gap. Hence the material should be transparent for $\hbar\omega < E_c - E_v$, and absorption should pick up at higher photon energies.

Emission is slightly more subtle. Suppose we have an electron at some energy E_2 in the conduction band, and a corresponding hole in the valence band at energy E_1. Band structure plays a role here. We know $\hbar\omega = E_2 - E_1$, by energy conservation. We also know that $k_c \approx k_v$, from our discussion above. For parabolic bands, $\hbar k_c = \sqrt{2m_e^*(E_2 - E_c)}$, and

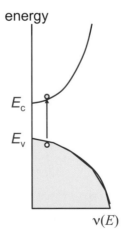

energy

E_c

E_v

$v(E)$

Fig. 8.13   Energy level diagram for absorption in an undoped, direct-gap semiconductor at $T = 0$.

similarly for the valence band. This implies

$$E_2 - E_1 = \frac{m_h^*}{m_e^*}(E_v - E_1). \tag{8.116}$$

The different effective masses of the two bands (corresponding to different densities of states at a given k) mean that the energies of the initial and final states, E_2 and E_1, are not necessarily centered around the middle of the gap.

The result is, when integrating over a certain range of k, different numbers of states in the conduction and valence bands are counted, since

$$\frac{\partial E}{\partial k} = \frac{\hbar^2 k}{m^*} \rightarrow dE_1 = \frac{m_e^*}{m_h^*} dE_2. \tag{8.117}$$

To calculate actual transition rates in the presence of light of some intensity $I(\omega)d\omega$, we need to count all the states compatible with conservation of energy. To do this, we can compute the *joint density of states*:

$$\nu_{cv}(\hbar\omega) = \frac{2}{8\pi^3} \int d\mathbf{k}(\delta(E_c - E_v - \hbar\omega)). \tag{8.118}$$

The delta function expresses the energy constraint, which picks out a surface in **k**-space. The result is a surface integral,

$$\nu_{cv} = \frac{2}{8\pi^3} \int\int \frac{dS}{|\nabla_k(E_c - E_v)|_{E_c - E_v = \hbar\omega}}. \tag{8.119}$$

This is calculable for a given band structure, and conversely this helps explain why optical measurements can tell us much about underlying band structures. For single-valley parabolic bands, the result is

$$\nu_{cv} = \left(\frac{2m_e^* m_h^*}{m_e^* + m_h^*}\right)^{1/2} \sqrt{\hbar\omega - E_g}. \tag{8.120}$$

In the context of semiconductors, pumping is any process that leads to a distribution of electrons and holes that is significantly out of equilibrium, as shown in Fig. 8.14. Here there

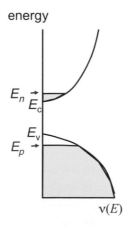

Energy level diagram for a direct-gap semiconductor pumped out of equilibrium.

Fig. 8.14

are different "quasi-Fermi levels" for the electrons (E_n) and holes (E_p). This is most readily established by optical pumping or the application of an electrical bias across a junction of some sort.

Consider what happens when light of intensity per unit bandwidth $I(\omega)$ is incident upon such a material. Neglecting spontaneous emission, we can write down some simple rate equations:

$$W_{1\rightarrow 2} = B_{12}\frac{1}{c}I(\omega)d\omega\nu_{\mathrm{cv}}(\omega)(f_{\mathrm{v}}(E_1)(1 - f_{\mathrm{c}}(E_2))),$$

$$W_{2\rightarrow 1} = B_{21}\frac{1}{c}I(\omega)d\omega\nu_{\mathrm{cv}}(\omega)(f_{\mathrm{c}}(E_2)(1 - f_{\mathrm{v}}(E_1))). \qquad (8.121)$$

Here E_2 is some energy in the conduction band, and E_1 is some energy in the valence band. These equations express the constraints that there must be states available which conserve energy (via ν_{cv}), and that transitions can only take electrons from filled states and place them into empty states. Thus the first equation corresponds to absorption (promoting a valence band electron into the conduction band), and the second, stimulated emission. As usual, $B_{12} = B_{21}$, and these coefficients may be found from the **k**-average of the matrix elements in Eqs. (8.111)–(8.115).

From the difference in these rates, we can find the characteristic (inverse) length scale that describes the change in intensity of the light as it propagates through the semiconductor medium:

$$\gamma(\omega) \equiv \frac{\mathrm{d}I(\omega)/\mathrm{d}z}{I(\omega)} = \frac{\mathrm{power/volume}}{I(\omega)} = \frac{\hbar\omega(W_{2\rightarrow 1} - W_{1\rightarrow 2})}{I(\omega)\mathrm{d}\omega}. \qquad (8.122)$$

This can be rewritten as

$$\gamma(\omega) = B_{21}\frac{1}{c}\nu_{\mathrm{cv}}(\omega)(f_{\mathrm{c}}(E_2) - f_{\mathrm{v}}(E_1)). \qquad (8.123)$$

Really to find the total γ we should integrate over all possible E_1 and E_2.

There are three cases to consider. First, and simplest, if $\hbar\omega < E_{\mathrm{g}}$, then not much happens. While there can be some small amount of absorption as carriers already in the conduction band are promoted even higher, there is no interband absorption or emission because of Pauli exclusion. Here $\gamma \approx 0$, and the material is essentially transparent.

Second, when $E_{\mathrm{g}} < \hbar\omega < E_n - E_p$, there are states available such that $f_{\mathrm{c}}(E_2) \approx 1$ and $f_{\mathrm{v}}(E_1) \approx 0$ due to the pumping. There can then be stimulated emission processes that take electrons from E_2 in the conduction band and deposit them at E_1 in the valence band. The result is $\gamma \propto \sqrt{\hbar\omega - E_{\mathrm{g}}}$, and there is net *gain*. As the beam propagates, its intensity grows with propagation distance like $\exp(\gamma x)$. The frequency range over which there is net gain $((E_n - E_p - E_{\mathrm{g}})/(2\pi\hbar))$ is called the *gain bandwidth*, and is set by the bias we can apply. In GaAs, this can be tens of THz.

Third, when $\hbar\omega > E_n - E_p$, the photons have sufficient energy to take electrons from below E_p ($f_{\mathrm{v}}(E_1) = 1$) and put them in empty states ($f_{\mathrm{c}}(E_2) = 0$) above E_n. This is *absorption*, a loss mechanism that competes with the gain from stimulated emission. As $\hbar\omega$ is increased further, the loss wins out and $\gamma \propto -\sqrt{\hbar\omega - E_{\mathrm{g}}}$. Figure 8.15 shows the qualitative result of all three of these regimes.

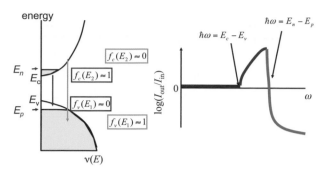

Optical processes in a pumped semiconductor, showing the three relevant regimes (and neglecting excitonic effects). **Fig. 8.15**

Including spontaneous emission in the discussion brings up two other effects, depending on the pumping mechanism. Light generated by spontaneous emission after optical pumping of a semiconductor system is *photoluminescence*. Similarly, if pumping is through electrical means, optical emission is called electroluminescence.[4] Note again that we have ignored excitonic effects and angular momentum concerns, and have been discussing these processes solely with regard to energy conservation and Pauli exclusion.

8.2.3 Light emitting diodes and semiconductor lasers

Now that we know something about the optical processes that take place in semiconductors, we can consider optoelectronic devices, particularly light emitting diodes (LEDs) and semiconductor lasers. With the development of methods to grow high quality semiconductor structures, particularly in direct gap systems like the III–Vs, these devices have become incredibly commonplace.

Light emitting diodes are based on the same ideas described in the previous section. In a biased *pn* junction, carriers are driven out of equilibrium and drift into the depletion region. If spontaneous emission processes are sufficiently rapid compared to nonradiative transitions, light is generated as the carriers recombine. One can imagine actually making a *pn* junction like this from a direct-gap semiconductor like GaAs; this would be a "homojunction" device.

There is a complication, however. In our electrically pumped system the spontaneously emitted photons have a typical energy $E_n - E_p > E_g$. This means that the photons generated in the recombination region can be readily absorbed by the bulk of the semiconductor material. The clever way to mitigate this concern is to produce structures where the source and drain regions are transparent at the frequency of the emitted photons. In a laser, this is equivalent to working in a four-level scheme, where the pump has an energy $E_4 - E_1$, greater than the emission energy $E_3 - E_2$.

Semiconductor growth technology is the key to achieving this, since it permits the creation of heterojunction emitters, like that shown in Fig. 8.16(a). By modulating the

---

[4] For added confusion, sometimes "electroluminescence" refers to a different process, when high energy (tens or more eV) electrons are used to excite phosphors.

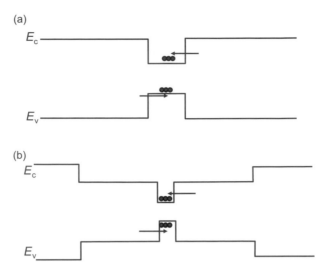

Fig. 8.16 Diagram of a semiconductor heterojunction emitter. (a) A quantum well interaction region surrounded by wider bandgap source and drain. (b) A more complex compositional arrangement to optimize light guiding.

composition during growth, the source and drain portions of the device, appropriately doped p and n, can have larger band gaps than the interaction region where recombination is to take place. If the narrower gap region is so small that quantum confinement effects become relevant, then the device is called a quantum well emitter. More complicated gradients of composition are also possible, as shown schematically in Fig. 8.16(b). Since refractive index is related to band gap, the composition gradient can be used to engineer optical waveguiding effects. Such a "graded index separate confinement heterostructure" is known as a GRINSCH [602].

Cavities

As alluded to previously, going from a spontaneously emitting optical system to a laser generally requires a cavity, so that the photons produced have many opportunities to stimulate emission of other photons into already-occupied modes.

An extremely simple "Fabry–Pérot" cavity may be formed just by cleaving a slab of active (pn junction) material with mirror-like facets. The modes of this cavity are the standing electromagnetic waves that fit inside. In a long 1d cavity, those allowed modes will be spaced in frequency by $c/2nL$, where n is the index of refraction of the material and L is the length of the system. That frequency spacing is just the inverse of the light travel time up and back in the cavity. For GaAs in the near infrared and $L = 250$ μm, the modes are spaced by about 170 GHz. Note that the gain bandwidth can readily exceed the mode spacing in this case. The result, as shown in Fig. 8.17, can be a multimode laser (provided the threshold for lasing is met).

Various cavity types are possible. The goals in semiconductor lasers are usually to have single mode operation with high stability, ease of manufacturing, low parasitic losses, and

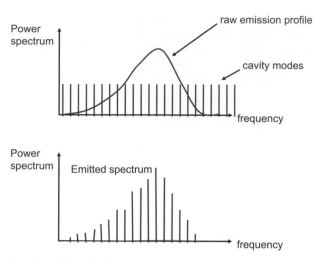

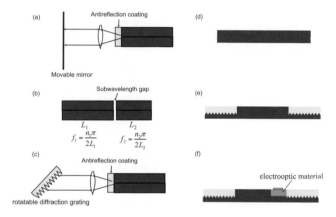

A multimode laser, where relatively broadband emission populates multiple, discrete cavity modes.

Fig. 8.17

A variety of possible cavities for lasing.

Fig. 8.18

ideally some measure of tunability. External cavities are possible, as shown in Fig. 8.18. However, while these often provide some tunability, the requirements of mechanical stability and ease of manufacturing can be difficult to meet.

More common, especially in consumer electronics, are approaches that incorporate the cavity directly in the device structure. In a *distributed feedback* (DFB) laser [603, 604], a waveguide material textured to form a grating is in proximity to the active region, coupled by evanescent fields. The DFB grating elects the mode with intensity commensurate with the periodic index variation. Like most grating-based ways of selecting frequency, the linewidth goes like $1/N$, the number of ridges.

Another approach is the *distributed Bragg reflector* (DBR), with gratings or dielectric mirrors positioned outside the active region [605]. Some measure of tunability is possible in these structures, either through electro-optic effects operating on part of the cavity, or otherwise changing the index of refraction.

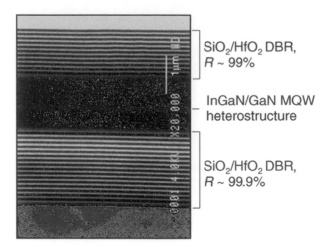

Fig. 8.19 A VCSEL designed to emit in the blue, based on an InGaN/InN multiple quantum well structure. The reflectivity, R, of the dielectric stacks is indicated. From [606].

The vertical version of the DBR uses dielectric mirrors grown from multilayers of semiconductor or dielectric. Such a structure, shown in Fig. 8.19, is known as a vertical cavity surface emitting laser (VCSEL). The emitting face is determined by the "leaky" side of the cavity, which possesses fewer periods of the dielectric multilayer. This particular VCSEL has been engineered so that confinement effects in the multiple quantum well interaction region enhance the desired emission wavelength.

It is important to note that many of the electrically driven solid state lasers in consumer electronics are also not purely diode lasers, but are *diode-pumped solid state* (DPSS) lasers. In these devices, a laser diode based on the emission and cavity techniques outlined here is often used to produce a powerful pump beam in the infrared. This pump is then applied to a crystal such as a rare-earth-doped YAG, which then operates as an optically pumped four-level laser. The emitted light from the optically pumped crystal can then be passed through a nonlinear optical crystal for frequency doubling. This is the basis for many green laser pointers, for example.

Quantum cascade lasers

One additional semiconductor laser technique is sufficiently clever and nano-based that it really demands its own subheading. Recall our discussion back in Section 5.5.5 where we pointed out that multiple quantum wells in superlattices may be grown with sufficient tunneling coupling to produce minibands. In an appropriately designed structure, arrays of quantum wells may be grown in succession, so that when a voltage bias is applied, an electron may tunnel through the structure, periodically making an optical transition from a "conduction" miniband to a "valence" miniband. Such a band structure under bias is shown schematically in Fig. 8.20.

This kind of structure, which relies on intersubband transitions, when coupled with a cavity, is called a *quantum cascade (QC) laser* [299], and there are good reviews on

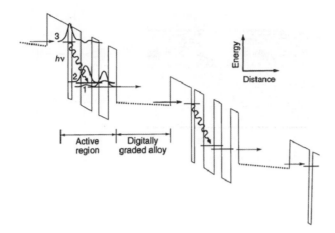

A quantum cascade laser, from [299]. Electrons tunnel sequentially through a series of minibands, periodically making photon-emitting inelastic transitions in active regions.

Fig. 8.20

this topic [607, 608]. QC lasers have two remarkable properties. First, a single electron can generate multiple photons as it tunnels through the structure, meaning that quantum efficiency (conventionally defined as number of photons out / number of electrons in) can exceed one. Second, QC lasers partially free engineers from the tyranny of band structures presented by Nature. While there are no convenient direct-gap semiconductors with energy gaps of 50 meV, "bandgap engineering" in QC lasers enables the construction of systems that emit with that energy. QC lasers have been employed from the mid-infrared to the terahertz regime, with wavelengths exceeding 100 microns [608].

8.3 A brief overview of optical communications

The vast majority of the world's communications traffic runs, at some point in its path from sender to receiver, through an optical link. Optical communications technology has been a driver in the development of optoelectronic devices. An outstanding review of the history of optical communications may be found in the book by Hecht [609], and online (e.g., at www.corning.com). The book by Agrawal [610] is a great, very complete reference to the field. As the demand for higher bandwidth communication between computers and higher speed information traffic within computers both continue to increase, there continues to be pressure to develop new, nano-enabled optical hardware.

Fiber optic communications works in a manner conceptually similar to AM (amplitude modulation) radio. A carrier frequency is selected in a low attenuation band of an optical fiber, such as $\lambda = 1.55$ μm (free space), equivalent to 193.5 THz. To good precision, the power spectrum looks like a function sharply peaked around that fundamental frequency. To send information, we modulate the amplitude of that carrier so that, e.g., large amplitude corresponds to a "1", and small amplitude corresponds to a "0". As a result of the modulation, the power spectrum picks up sidebands in addition to the peak at the carrier

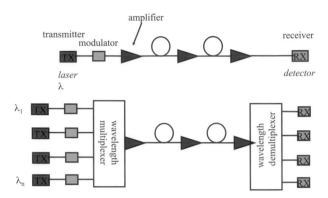

Fig. 8.21 Essential components of a fiber optic communicaton network.

frequency. The more rapid the modulation, the farther away the sidebands lie from the central frequency.

With a single frequency, to send multiple signals requires time sharing, also known as time division multiplexing. The standard approach these days instead uses multiple, distinct carrier frequencies within a low attenuation band of the fiber, a technique called *wavelength division multiplexing* (WDM). The ability to send multiple signals with different "colors" down a fiber simultaneously is only possible because of superposition (and therefore requires very small optical nonlinearities in the medium). Near 1.55 μm, the regime of comparatively low attenuation is around 120 nm wide in wavelength, or 15 THz in frequency. To share this band, the separate carriers have to be spaced sufficiently in frequency so that their sidebands don't run into each other.

Figure 8.21 shows a generic fiber optic communication network in its simplest form. The essential components are sources (lasers), modulators (that encode digital information into the carrier), multiplexers (that place multiple carriers into a single fiber), the fiber itself, amplifiers (that compensate for loss in the fibers by boosting the signal strength), demultiplexers, and detectors (that take the optical signals and re-encode them into electrical signals). Typical attenuation for optical power through a modern fiber is 0.2 dB/km, equivalent to a transmission coefficient of 0.955 through 1 km of glass. This amount of attenuation requires amplifiers to be placed every 80–120 km.

Glass fiber itself is designed to guide light rather well in several bands: the C band (around 1550 nm), the L band (around 1625 nm), and also near 1310 nm with slightly worse attenuation characteristics [611]. Typical core diameters are around 10 μm, with cladding 125 μm thick. Fiber is drawn from large "preforms" made by furnace melting of incredibly pure SiO_2 "soot" (and any deliberate dopants). That soot is prepared by combustive chemistry from raw ultrapure Si.

Optical fibers can be modified to provide other functionality beyond simple signal propagation. For example, it is possible to modify the index of refraction of the fiber core in a periodic pattern in space, so that the fiber core itself can act like a dielectric mirror for some particular wavelength [612]. This is called a *fiber Bragg grating*, and may be used in concert with other elements to selectively add or drop signals. Because of the strong sensitivity of the dielectric mirror to the spatial period, the optical properties of these gratings are extremely sensitive to thermal expansion and other perturbations, making them excellent sensor candidates [613]. By varying the spatial period of the index variation along the length of the fiber, the result is a "chirped" grating, such that various wavelengths propagate differing distances into the grating before reaching a portion meeting the Bragg condition and reflecting. Combining the chirped grating idea and the thermal response, it's possible to incorporate a fiber grating and a cleverly designed heater to compensate actively for dispersion in the non-grating fiber [614].

The optical modulator is an essential element to any communications network; it is the device that encodes digital information onto the carrier wave generated by a solid state laser. Modulating optical output billions of times per second is extremely challenging. Because of transients, thermal stability, and physical limits imposed by standard laser designs, the optical signal is not modulated by actually turning the laser off and on at the desired data rate. Rather, a modulator is placed in the optical path, with electrically tunable optical properties such that the transmission throught the device may be switched at very high speeds.

The main approach to optical modulation in practice is based on the Mach–Zender interferometer [615]. A typical example is shown in Fig. 8.22, where an incoming optical signal is split into two paths, using waveguides implanted or etched into an electro-optic material such as $LiNbO_3$. By applying appropriate voltages, the electro-optic effect alters the optical path length of one arm of the interferometer relative to the other. The interference can then be tuned between constructive and destructive, modulating the output of the interferometer. Typical switching rates with these structures is between 10 and 40 Gb/s. It is important to note that switching the interference at GHz data rates is a significant RF engineering challenge, since at these frequencies one cannot treat the electrodes like a simple DC capacitor. Moreover, because electro-optic coefficients are small, accumulating sufficient phase difference between the two arms requires a macroscopic size for the interferometer, on the order of a few centimeters. Recently approaches involving silicon-based optical modulators have been advanced, wherein the relative phase shift is controlled not by standard electro-optic coefficients, but through dispersion caused by the presence of gate-modulated charge carriers [616]. Research into novel optical modulators continues to be very active.

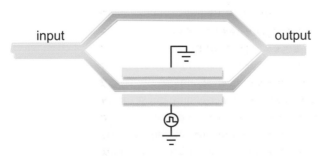

Optical modulator based on a Mach–Zender interferometer.   Fig. 8.22

Often signals propagating at different wavelengths need to be combined into a single optical path, or split into separate streams. These tasks are accomplished, respectively, by multiplexers and demultiplexers. The simplest form of (de)multiplexer can be a prism, where the dispersion in the index of refraction provides a natural means of bending optical signals by a frequency-dependent amount. Alternatively, combined systems of waveguides, lenses, and Bragg reflectors may be integrated to perform these tasks.

During propagation through long fiber runs, it is necessary to boost the strength of the signal. Ideally one would like this to be done without breaking the fiber, detecting the photons electrically, and then having another laser and modulator there to boost the signal. Instead, the standard approach is to use a fiber amplifier [617, 618]. Erbium ions are doped directly into a section of fiber, which is in turn pumped optically. The signal photons act upon the excited erbium ions, inducing stimulated emission and boosting the signals *in situ*, without the need for additional detection/generation/modulation hardware.

Conversion of optical signals back into electronic data does require detection of the photons and a transduction mechanism. A thorough discussion of recent innovations in high-speed optical detection is presented by Bottacchi *et al.* [619]. While charge-coupled devices based on silicon can be very sensitive, approaching single-photon detection, such devices are too slow for the gigabit per second data rates expected in telecommunications systems. Photoconductivity provides a natural coupling mechanism, typically using a direct-gap semiconductor material with a bandgap tuned to the wavelength range of interest. A standard approach is the PIN photodiode, composed of layers of *p*-doped, intrinsic, and *n*-doped material (e.g., InGaAs), respectively, with the undoped layer meant to function as an absorber where electron–hole pairs are created. One key materials requirement is that the exciton binding energy be small, so that the carriers produced are readily dissociated and collected at modest bias conditions.

Further sensitivity may be achieve using *avalanche photodiodes* (APDs). In an APD under bias, charge carriers produced by absorption of a photon accelerate in the presence of the electric field within the device, and can acquire kinetic energy exceeding E_g such that scattering events can lead to the creation of additional electron–hole pairs. This process, called avalanching, can give enhanced sensitivity (more charge per photon) relative to standard photoconduction. However, the recovery time of the device must be fast to allow high data rates.

Rather than relying on doping to provide internal electric fields to dissociate charge carriers, it is also possible to use the built-in electric field that occurs at the Schottky barrier between a metal and an appropriate semiconductor. Alternately, one can create inter-digitated metal electrodes on a semiconductor surface (a *metal–insulator–metal* (MIM) or *metal–semiconductor–metal* (MSM) system), wherein an applied dc bias between the metal fingers provides that needed electric field to sweep up the photo-produced charges.

The bottom line is that optical communications technology is at the core of modern Information Age society. The never-ending quest for higher bandwidth pushes technological innovation, and the call for greater optical capacity and integration does not stop with long-haul telecommunications or internet connectivity. There is increased discussion of the need for optical interconnects integrated with semiconductor fab, either on-chip or for high bandwidth busses. As we will see in subsequent sections, nanoscale physics presents opportunities for improving the performance of photonic systems. The propagation, generation, and detection of light can all be modified when appropriate materials are engineered at the nanoscale, smaller than the wavelength of the radiation of interest.

8.4 Photonic band gap systems

In Section 8.1.13, we saw that it is possible to design periodic dielectric multilayer structures to reflect particular incident wavelengths. In that particular case, the layer thicknesses were chosen so that the reflected rays that made it back to the multilayer surface were all in phase, modulo 2π. The physics at work here is actually much more general. Another way of viewing this is that with that particular periodic arrangement of dielectric interfaces, there are no optical modes that would propagate through the full multilayer while still being compatible with the boundary conditions on the fields that must be met at all the interfaces. It turns out that this is exactly analogous to the physics behind the formation of electronic band gaps in solids as described in Section 2.2.6. This insight and its potential applications were pointed out by Yablonovitch [620, 621].

To see this, start from the wave equation. For simplicity we will work in 1d, but the results generalize to higher dimensionality. We will assume an infinitely extended, spatially uniform μ and a spatially varying $\epsilon = \kappa\epsilon_0$, and normal incidence with propagation in the x direction. The (transverse) electric field must obey

$$\frac{\partial^2 E(x,t)}{\partial x^2} = \frac{\kappa(x)}{c^2}\frac{\partial^2 E(x,t)}{\partial t^2}. \tag{8.124}$$

Assuming the usual harmonic time dependence, $E(x,t) = E(x)\exp(-i\omega t)$, we can do separation of variables and look at the resulting time-independent equation for $E(x)$. With a slight rearrangement,

$$-\frac{\partial^2}{\partial x^2}E(x) - \frac{\kappa(x)-1}{c^2}\omega^2 E(x) = \frac{\omega^2}{c^2}E(x). \tag{8.125}$$

We are interested in the case when $\kappa(x)$ is periodic, so that $\kappa(x+d) = \kappa(x)$ for some spatial period, d. Equation (8.125) for the electric field looks remarkably like the time-independent Schrödinger equation for a particle in a 1d potential, Eq. (2.2). We already know the answer to this – the solutions are in the form of Bloch waves! That is,

$$E(x) = \exp(i\gamma x)u_{b,\gamma}(x), \quad u_{b,\gamma}(x+d) = u_\gamma(x). \tag{8.126}$$

The allowed propagating modes that extend throughout the periodic dielectric structure look like plane waves modulated by a function with the same periodicity as the dielectric. Just as in the electronic problem, we can label the states by a band index b and a wavevector γ. There are only extended mode solutions for certain ranges of ω (the photonic bands), while for other ranges of ω there are no extended solutions to Eq. (8.125) of the form Eq. (8.126).

We can define a Brillouin zone in reciprocal space based on reciprocal lattice vectors relevant to the real-space periodicity of the dielectric. For values of γ that coincide with a reciprocal lattice vector of the periodic dielectric, the Bragg condition is satisfied, and the electromagnetic wave is strongly diffracted. An explicit example of this in 1d, completely analogous to the Krönig–Penney model of Section 2.2.5, is worked out in detail by Yeh and Yariv [605]. What this means is that *any* periodic 1d dielectric structure will act like a dielectric mirror for some wavelength!

This generalizes to higher dimensions, like the electronic structure problem. It is relatively straightforward as a practical matter to create a 2d periodic dielectric structure that can have photonic band gaps for propagation in certain directions in the *xy* plane. The entire semiconductor industry is based on controlled patterning of 2d surfaces at length scales smaller than the wavelength of visible light, so a variety of lithographic definition and etching/deposition modalities are available.

In three dimensions, the fabrication problem is significantly more complicated, but the physics analogy to the electronic case remains valid. There are photonic analogs of the "spaghetti" band diagrams discussed in Chapter 3. The size and frequency range of the photonic band gap depends in a nontrivial way on the direction of γ in reciprocal space. Achieving a full three-dimensional photonic band gap over some frequency range, so that propagation for that frequency range is forbidden in all directions, is challenging but has been achieved in microwave [622] and infrared [623, 624] wavelengths. While progress has been made [625], success in the visible has remained elusive [626].

One reason why the visible (and shorter) wavelengths have been challenging is the constraint on the scale of spatial modulation that is relevant. The dielectric mirror case is illustrative here. To achieve a photonic band gap in some direction at a given free space wavelength λ_0 requires modulation of the dielectric function with a spatial periodicity of around $\lambda_0/4n$, where n is the index of refraction of the medium. Into the visible this becomes increasingly difficult to achieve.

8.4.1 Microcavities and waveguides

In the electronic structure problem, we have discussed how deviations from an infinite periodic potential can lead to electronic states that are not delocalized Bloch waves. Examples include surface states exponentially localized to an interface, and the bound electronic states around donor dopants. Analogous, localized optical bound states can exist at defects in PBG systems.

Figure 8.23 shows two examples of such microcavity modes, in 1d [627] and 2d [628], respectively. In both cases, a PBG system is created by the etching of an array of holes in a III–V semiconductor heterostructure that contains a gain medium. By deliberately omitting a hole, a single, isolated defect is created hosting a localized optical mode. At the defect site, light with an energy in the photonic band gap cannot propagate into the surrounding structure. The surrounding PBG medium acts as like a dielectric mirror, with modes in the gap decaying exponentially within, rather than propagating as in Eq. (8.126). As a result, the defect acts as an extremely small cavity.

The small size, comparable to a wavelength, affects the efficiency of lasing and the inherent timescales associated with optical processes through *cavity QED* (quantum electrodynamics) effects. Our discussion in Section 8.2.1 about Einstein A and B coefficients

Defects in photonic band gap systems act as cavities. Shown here are examples in one and two dimensions, adapted   Fig. 8.23
from [627] and [628], respectively.

made use of Fermi's Golden Rule in finding the transition rates for both stimulated and
spontaneous transitions. Those transition rates depend not just on the density of states
for the electrons undergoing transitions, but also on the density of states for the photons
involved in the optical processes. Throughout our discussion we had been tacitly assuming
the standard, 3d density of states for photons, $g(\omega) = \frac{\omega^3}{\pi^2 c^3}$. That expression gives the
number of photon states per unit volume per unit angular frequency, and is derived by
exactly the same **k**-space counting arguments used to obtain Eq. (2.21) and following.

A cavity, however modifies the density of states [629]. A cavity of quality factor Q
has one state per frequency range $\omega/(2\pi Q)$, and that state occupies an effective mode
volume V. This leads to an effective photon density of states on resonance for the cavity
of $g_c = Q/\omega V$. Carrying this through the calculation for spontaneous emission, assuming
that the transition is resonant with the cavity, we arrive at the *Purcell factor*, the ratio of
emission rate with the cavity relative to the free space emission rate:

$$f_P \equiv \frac{3}{4\pi^2} \frac{Q\lambda_0^3}{V}, \qquad (8.127)$$

where λ_0 is the free-space wavelength of the radiation of the transition, and the vacuum is
assumed to be empty ($\kappa = 1$). This is a wholly remarkable result: a property of a physical
system that at first glance seems intrinsic (say the spontaneous emission rate of an excited
hydrogen atom) can instead be manipulated by placing the system in a tuned cavity. This
cavity QED physics has been verified in many systems, starting with atomic transitions
[630].

Similarly, if the cavity is detuned from the optical transition, so that the effective photon
density of states at the optical transition energy is greatly reduced relative to the free-space
value, then spontaneous emission is dramatically suppressed. This was predicted for free-
space cavities [631] and later specifically for the microcavities comprising defects in PBG
structures [620]. Semiconductor nanocrystals have since been used to confirm these PBG
predictions [632, 633] in 2d photonic crystals.

More than modifying overall spontaneous and stimulated emission rates, PBG systems
are considerably more versatile because of their inherent directionality. By controlling the
directional photonic band structure, it is possible to favor emission and permit propagation
in some directions but not others. For example, consider a typical GaAs/AlGaAs light
emitting diode. Quantum efficiency (number of photons produced per e–h pair) can be

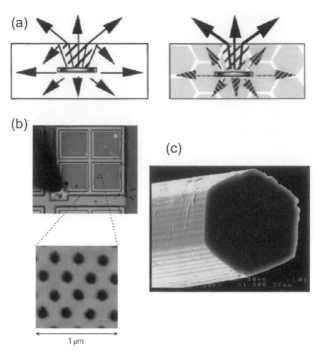

Photonic band gaps to control emission and propagation. (a) By restricting the available states for photons, a PBG-structured LED should be able to avoid total internal reflection for most of the generated photons, improving efficiency. Adapted from [634]. (b) An experimental implementation, in a GaN-based diode. Adapted from [635]. (c) An early photonic band gap optical fiber, constructed to favor enhanced confinement in the central region. Reproduced from [636].

exceedingly high ($\sim 99\%$). Photons are typically emitted isotropically. However, because of the high index of refraction, only those intersecting the wafer surface within 16 degrees of normal are able to be transmitted at all; the rest experience total internal reflection. Only 3–15% of the light generated is therefore useful [634]. With a correctly designed PBG LED, though, one can stack the deck so that the LED active region only couples to a single propagating mode. Figure 8.24(a, b) shows the theoretical concept [634] and an experimental implementation of this idea [635].

This idea of disallowing propagation in particular directions has also been applied to extended systems like optical fibers, as in Fig. 8.24(c) and discussed extensively by Russell [637]. Drawn down from an array of hollow glass tubes, photonic crystal fibers are meant to exhibit a 2d PBG transverse to the propagation direction, greatly reducing losses due to scattering [638].

Extended "defects" in PBG structures, by providing regions not subject to the propagation restrictions of the overall dielectric periodicity, can function as effective waveguides [640, 641]. A particular example in 3d is shown in Fig. 8.25 ([639]). In two-dimensional PBG systems, introducing defects is often a matter of standard lithographic patterning. Success in three dimensions requires greater sophistication, in this case the use of two-photon polymerization and confocal optics. A major appeal of PBG defect waveguides is

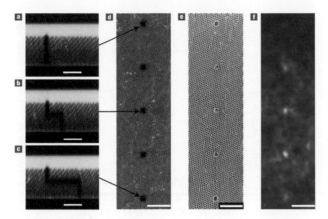

Linear defects in a 3d photonic band gap material as a waveguide. A 3d silica inverse opal designed to have a band gap **Fig. 8.25**
around 1.5 μm free space wavelength, patterned to contain linear channels of voids, imaged in cross-section by
fluorescence (a)–(c) and face-on (d), and face-on via electron microscopy (e). In transmission at 1.48 μm, spots in (f)
indicate that light at that wavelength can propagate through the material via the waveguides, even navigating two
sharp corners. Reproduced from [639].

the ability to "steer" light around corners that are sharp on the scale of the wavelength
[640].

Thanks to the optical confinement afforded by cavities in PBG systems, these defects
may be engineered to support much higher local optical intensities than that supplied
directly by incident radiation. These systems can then be optimized to leverage nonlinear
optical effects. Nonlinear couplings combined with waveguiding have led to attempts at all-
optical switching, an optical transistor with light input to one port "gating" the propagation
of light between two other ports [642].

Other uses of PBG materials for optical switching have been put forward. Many PBG
systems consist of a solid structure containing open volumes that may be infiltrated with
another material. If that infiltrant has a dielectric function that is switchable through some
external stimulus, the PBG may be controlled and modified *in situ*. For example, liquid
crystals, reorientable by externally applied dc electric fields, are an essential component
of many switchable displays thanks to their anisotropic optical properties. A photonic
crystal containing a liquid crystal fluid can have a PBG that may be reversibly switched
by application of an external bias [642]. Similarly, a switchable PBG structure has been
demonstrated based on infiltration of an optically switchable liquid crystal material [644],
as shown in Fig. 8.26. The molecules in the liquid may be driven optically between two
stable conformations, the bent *cis* with isotropic optical properties, and the straight *trans*,
which takes on nematic order[5] and an anisotropic index of refraction. Regions of PBG
response can then be patterned into the material by optical exposure at the wavelength
required to drive the molecular isomerization.

---

[5] In a nematically ordered liquid crystal, the positions of the individual molecules are randomly distributed, as in
a liquid, but there is a shared orientation to the elongated molecules.

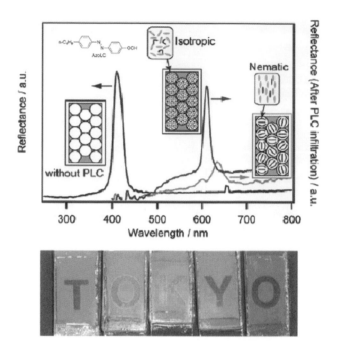

A switchable photonic band gap system incorporating an optically isomerizable liquid crystal. Optical driving between the *cis* and *trans* forms of the liquid crystal molecules drastically changes the metamaterial's optical properties. Top: reflectance data comparing the optical response of a particular inverse opal structure empty and with the infiltrated liquid crystal in two configurations. Inset: the optically switchable molecule. Bottom: a series of PBG inverse opals that have been illuminated through a mask, showing that local transmission can be switched optically. Adapted from [644].

Clearly PBG systems can possess extraordinary optical properties that are a direct consequence of the ability to pattern dielectric structure on scales smaller than the relevant wavelength of light. As discussed, these properties arise directly from the spatially periodic modulation of the dielectric function. Computational capabilities now exist to permit sophisticated modeling and design. Beyond designer optical media, waveguides and cavities based on defects in PBG structures offer opportunities for optical and optoelectronic devices. Manufacturing continues to be a major challenge, particularly for structures intended to operate in the visible or at shorter wavelengths.

Point defects in PBG systems act as cavities because they lead to localized photonic states in the photonic band gap. The trapped optical mode decays exponentially in space away from the defect, since propagation is forbidden at that frequency, thanks to the collective interference of components reflected off the dielectric interfaces in the surrounding structure. This is completely analogous to electronic localization of a defect state, and that mathematical underpinning raises an interesting question. We saw back in Chapter 6 that *disorder* in the potential energy for electrons can also cause localization of electronic states. Can a spatially disordered dielectric function lead to analogous localized optical modes?

The random laser [645, 646] is a clear demonstration that the answer is "yes". Rather than creating a traditional cavity for lasing, in a random laser an effective cavity supporting

8.5 Nanophotonics: near-field optics   407

localized optical modes is created in a disordered aggregate of roughly wavelength-scale dielectric particles. As in Section 6.4.2, interference between possible photon trajectories leads to the localization of light into standing waves that decay exponentially in the disordered medium [647]. This is an effective confinement of the light, and if the disordered medium is also a gain medium, the result can be lasing. Emission takes place preferentially in the directions where the confinement of the light is weakest, as in a regular "leaky" laser cavity.

8.5 Nanophotonics: near-field optics

Near-field optics is the name for a collection of techniques for accessing that short distance scale information stored in the optical near field. Here we give a schematic approach that shows one way this can be done and why it works, patterned after the discussion in Paesler and Moyer [648] and Vigoureux *et al.* [649].

Consider Figure 8.27, where we refer to the original apertures as the "sample", and our quantity of interest is what arrives in the far field at the screen. From our discussion of diffraction, we learned one way to think about the effect of an opaque screen with an aperture. Immediately past the aperture, the short distance scale information about the sharp edges of the aperture is, in Fourier language, encoded in components of the electromagnetic field with large values of k_x, the wavevector component transverse to the edge of the slit and the propagation direction. Because k is of fixed magnitude, very large values of k_x imply imaginary values of k_z; those components of the field decay evanescently moving away from the screen.

In the far field, we found that the amplitude of the diffracted wave (in the language of scalar diffraction) is proportional to the Fourier transform of the (amplitude of the incident wave at the) aperture. For the example shown in Fig. 8.27(a), the relevant quantity evaluated at $z = 0$ would be

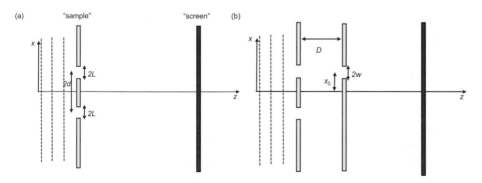

A near-field optical situation. (a) A diagram defining the geometry in question. (b) An aperture is interposed between the "sample" and the far-field "screen".   Fig. 8.27

$$F(k_x, z = 0) = \int_{d-L}^{d+L} E_0 \exp(ik_x x)dx + \int_{-d-L}^{-d+L} E_0 \exp(ik_x x)dx$$

$$= 4E_0 \cos(k_x d)\frac{\sin(k_x L)}{k_x}, \tag{8.128}$$

where we assume normal incidence and a uniform amplitude, E_0, across the transverse direction x. We would find the far-field amplitude at some position z by recalling that $k_z = \sqrt{k^2 - k_x^2}$, and integrating up our Fourier components, as in Eq. (8.78):

$$E(x, z) = \int_{-\infty}^{+\infty} dk_x [F(k_x, z = 0)\exp(-ik_x x)]\exp(ik_z z)$$

$$\approx \int_{-\omega/c}^{\omega/c} dk_x [F(k_x, z = 0)\exp(-ik_x x)\exp(i\sqrt{k^2 - k_x^2}z)], \tag{8.129}$$

where in that second line we assume $z \gg \lambda$ and throw away the evanescently decaying components (high $|k_x|$).

Now we consider the situation in Fig. 8.27(b), where we have interposed an aperture of width $2w$ displaced along the propagation direction by an amount $z = D$ at some transverse position x_0. From the above, the field amplitude at the aperture is given by

$$E(x, z = D) = \int_{-\infty}^{+\infty} dk_x' [F(k_x', z = 0)\exp(-ik_x' x)\exp(i\sqrt{k^2 - k_x'^2}D)]. \tag{8.130}$$

We need to Fourier transform this to find the eventual far-field result:

$$F(k_x, z = D) = \int_{x_0-w}^{x_0+w} E(x', z = D)\exp(ik_x x')dx'. \tag{8.131}$$

Now, the amplitude of the field at the $z = z_0$ screen is given by

$$E(x, z = z_0 \gg \lambda) \approx \int_{-\omega/c}^{+\omega/c} dk_x [F(k_x, z = D)\exp(-ik_x x)\exp(i\sqrt{k^2 - k_x^2}(z_0 - D))]$$

$$= \int_{-\omega/c}^{+\omega/c} dk_x \left[\int_{x_0-w}^{x_0+w} \left[\int_{-\infty}^{+\infty} dk_x' \left[F(k_x', z = 0)\exp(-ik_x' x') \right. \right. \right.$$

$$\left. \left. \times \exp(i\sqrt{k^2 - k_x'^2}D) \right] \exp(ik_x x')dx' \right]$$

$$\left. \times \exp(-ik_x x)\exp(i\sqrt{k^2 - k_x^2}(z_0 - D)) \right]. \tag{8.132}$$

This is not a simple expression, but it contains the essential ingredient of the success of near-field optics: *a tip, aperture, or particle in the near field of a sample can scatter what would ordinarily be evanescent components into propagating components that survive to the far field*. Without the intervening aperture at $z = D$, a high-k_x' component at $z = 0$ would decay evanescently and not survive to the far field. With the aperture, however, that evanescent component ends up getting scattered into the far field by affecting the components of k_x that do survive to the far field. The efficacy with which this takes place depends on D; if $D \gg \lambda$, then we are back to the original situation, since the high-k_x' component will have decayed before ever interacting with the aperture. The width of the aperture is also of critical importance, since that determines the effective bounds on k_x'.

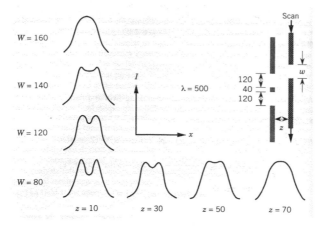

Near-field optical microscopy. Adapted from Fig. 4.3 of Paesler and Moyer [648].   **Fig. 8.28**

Figure 8.28 illustrates this point precisely. The mathematics are evaluated numerically by Vigoureux *et al.* [649] to show the far-field intensity as a function of the transverse position x for various aperture–sample separations, for a sample consisting of two slits. It is unsurprising that with an aperture wider than the slits it is not possible to resolve the two slits. More interesting is the clear demonstration that a small subwavelength aperture positioned in the near field of the sample is successful in transducing the subwavelength size of the sample slits into the far-field intensity.

The basic idea of NSOM was originally suggested in 1928 by E. H. Synge [650] and independently by J. A. O'Keefe [651]. However, the practical demonstration of NSOM had to wait for the development of key technologies, including highly sensitive, stable, and fast photodetectors; effective optical waveguides; stable positioning control at the nanometer level; and material control at the nanometer level for the fabrication of apertures and tips.

The various modes of operation for NSOM have already been mentioned in Section 4.1.6 and figures therein. Interpretation of contrast in NSOM is a persistent challenge. The microscopic configuration of the metal and the aperture are generally not known. Tip–sample interactions can actually modify the system being studied or the properties of the tip itself. For example, fluorescence lifetimes can be altered significantly because the presence of the tip can modify the effective photon density of states or enable nonradiative pathways for energy flow.

8.6 Nanophotonics: plasmonics

In a good metal the conduction electrons, a Fermi liquid as described in Chapter 2, act much like an incompressible fluid that can move around the positively charged background provided by the ion cores of the lattice. The collective normal modes of that electronic fluid, when quantized, are *plasmons*, in analogy with phonons and photons. Since plasmon oscillations involve the displacement of charge, they often couple strongly to electromagnetic radiation. With the advent of nanoscale patterning tools, sophisticated optical techniques, and advanced computational capabilities, it is now possible to design structures that support engineered plasmon responses. The rise of *plasmonics* holds much promise for the control and manipulation of light on deep subwavelength scales, as we discuss below. There are several strong review articles about plasmonics [652, 653, 654, 655, 656].

8.6.1 Simple plasmons

Consider a volume of material uniformly and, instantaneously displacing the conduction electrons from their equilibrium position by some small amount $\mathbf{r}$. On one side of the material there would be an excess of negative charge, while on the opposite side there would be a corresponding excess of positive charge density due to the ion cores left behind. Classically, an electron within the volume will obey Newton's Second Law, $m\ddot{\mathbf{r}} = -e\mathbf{E}$, where $\mathbf{E}$ is the local electric field. In this situation the local electric field inside the volume may be found by applying Gauss' Law (assuming $\kappa = 1$ here for the ion cores) and using the surface charge density caused by the displacement of the charge, $\mathbf{E} = \frac{ne}{\epsilon_0}\mathbf{r}$, where n is the volume density of the conduction electrons. Plugging this in yields a harmonic time dependence with a natural frequency

$$\omega_{\mathrm{p}}^2 = \frac{ne^2}{m\epsilon_0}. \tag{8.133}$$

This is the classical *plasma frequency* for bulk charge density oscillations.

In this idealized case, the electrons would oscillate back and forth about the ion cores forever. In a more realistic treatment, these bulk plasmons are damped for two reasons. First, this approach neglects any scattering of the electrons due to disorder; the motion of the electrons is really the flow of a current, and nonzero resistivity of the material serves to damp the motion. Second, the plasma oscillations here describe a time-varying electric dipole; as such, this oscillation should damp due to the creation of dipole radiation. This coupling to radiation is important and works both ways: plasmons may be excited by radiation as well. The electric field of the electromagnetic wave can couple to mobile charge density, while the changing magnetic field can likewise induce current flow.

We have been discussing plasmons in terms of real-space charge motion because it is easy to visualize and develop physical intuition. In terms of the quantum picture of the electron gas, plasmons are particular coherent superpositions of electron–hole excitations with energy $\hbar\omega_{\mathrm{p}}$. From this point of view, the damping out of plasmon excitations can have two contributions. The first is the relaxation of the electronic distribution back toward a thermal Fermi–Dirac situation. This involves inelastic scattering of the electrons (and holes), such that the "hot" electrons lose energy and fall back toward E_{F}. This is analogous to the T_1 process of Section 7.3.2. The second is the dephasing of the electron–hole excitations. The real-space displacement of charge corresponds to an orchestrated motion of the electrons in $\mathbf{k}$-space. When the various components fall out of step with each other due to elastic (and inelastic) scattering, a dephasing process analogous to T_2 relaxation in magnetic resonance, the plasmon mode decays into (uncoordinated) hot electrons and holes.

8.6.2 Surface plasmons

In addition to the bulk plasmon excitations, one can consider a spatially periodic displacement of charge at a surface, as shown in Fig. 8.29(b). Describing this periodicity by a wavevector $\mathbf{q}$, one can ask what the relationship between $\mathbf{q}$ and the frequency ω of the oscillations of the charge density back and forth is. Classically, the solution is given by solving Maxwell's equations at that interface, where the dielectric function of the metal is given by $\epsilon_m \equiv \epsilon_m' + i\epsilon_m''$, and the dielectric function of the other space, a dielectric, is assumed to be purely real, ϵ_d. We can find the surface mode by assuming propagation in the x-direction in the plane of the interface, and evanescent decay out of the plane into the dielectric or the metal. Further assuming the losses in the metal to be small, so that $|\epsilon_m'| \gg |\epsilon_m''|$, one arrives [657] at

$$q_x' \approx \sqrt{\frac{\epsilon_m' \epsilon_d}{\epsilon_m' + \epsilon_d}} \frac{\omega}{c},$$

$$q_x'' \approx \sqrt{\frac{\epsilon_m' \epsilon_d}{\epsilon_m' + \epsilon_d}} \frac{\epsilon_m'' \epsilon_d}{2\epsilon_m'(\epsilon_m' + \epsilon_d)} \frac{\omega}{c}. \tag{8.134}$$

As expected for a mode confined to the interface propagating in the x direction, the wavevectors in the z directions in the metal and in the dielectric end up both having

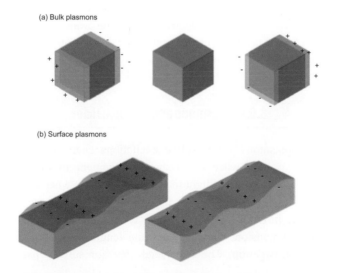

(a) Bulk plasmons

(b) Surface plasmons

Plasmon excitations. (a) Bulk plasmons in a nanoparticle where the electrons oscillate uniformly about the positive ions, over the course of half a period of oscillation. (b) Surface plasmons with some wavevector q at two times separated by half a period of oscillation.

Fig. 8.29

imaginary components:

$$q_{z,m} \approx i\sqrt{\frac{\epsilon_m'^2}{\epsilon_m' + \epsilon_d}} \frac{\omega}{c},$$

$$q_{z,d} \approx i\sqrt{\frac{\epsilon_d^2}{\epsilon_m' + \epsilon_d}} \frac{\omega}{c}. \tag{8.135}$$

For reference, approximate dielectric functions for silver and gold have been tabulated based on experiments [658]. At 820 nm wavelength, for example, $\epsilon_{Ag} \approx -32.8 + 0.458i$, and $\epsilon_{Au} \approx -25.8 + 1.63i$.

Because these propagating excitations are strongly coupled to the evanescent electromagnetic field confined near the interface, they are sometimes called *surface plasmon polaritons*. This idealized case indicates key aspects of plasmons that survive in more sophisticated treatments.

- The plasmon dispersion relation depends sensitively on both the dielectric properties of the metal and of any surrounding medium. This is the underpinning of surface plasmon-based sensing.
- To get long propagation lengths, one wants low loss in the metal.
- The geometry of the problem is very important, since it's the boundary conditions of the electromagnetic field that determine the relationship between the spatial distributions of charge and fields.
- At planar interfaces, the surface plasmons have a different wavelength from free space light of the same frequency. That means that exciting those modes optically requires changing the spacing of oscillating charged regions (using the evanescent field from total internal reflection in a nearby dielectric [659] or the evanescent field that penetrates through the metal from a dielectric coated with a metal film [660]) or breaking the planar symmetry (for example, see Dionne *et al.* [661]).

8.6.3 Plasmons and nanoparticles

Propagating surface plasmons are wavelike excitations characterized by well-defined wavelengths. As we've seen when discussing confinement of electrons in quantum nanostrucures, or the confinement of light in optical structures, when considering metallic nanostructures it makes sense to consider the size of the metal system relative to the wavelength of surface plasmons at that frequency. An extended metal (and dielectric) structure larger in d dimensions than the plasmon length scale is d-dimensional with respect to plasmon propagation. The interfacial case for surface plasmons is the 2d situation.

A metal nanoparticle comparable in all dimensions relative to the plasmon wavelength at the frequency of interest is zero-dimensional, analogous to the way a gate defined quantum dot comparable in size to the Fermi wavelength is zero-dimensional from the point of view of electronic confinement. As in the electronic case,[6] the effect of boundary conditions in the confined system is to limit the allowed frequencies to a discrete spectrum of *local surface plasmon resonances* (LSPRs).

A comparatively simple case is that of a spherical metal particle of radius a in a dielectric. One can solve Maxwell's equations for this system, applying the boundary conditions that the tangential total electric field must be continuous across the metal–dielectric interface, and the normal components of **D** across the interface must be related appropriately by the dielectric functions of the metal and dielectric. This can be solved exactly, as was pointed out by Gustav Mie more than 100 years ago [662], though the solution is cumbersome. It is common to work in the (quasi)static approximation, so that the electric field $\mathbf{E} = -\nabla\Phi$, where Φ is the scalar potential. The appropriate language for describing the Φ is in terms of spherical harmonics, referenced to a particular axis direction that we can take to be z. The natural modes of the system are then described as multipoles. There is a series of LSPR modes, identified by how many nodes the scalar potential has when going from the $+z$-axis to the $-z$-axis (the l number in the spherical harmonics) and how many nodes when going around the z-axis azimuthally (the m number in the spherical harmonics). Higher multipoles are more energetically expensive because they involve charge displacements on shorter distance scales. These higher multipole LSPR modes are only excited if the quasistatic approximation is relaxed [652].

The spherical symmetry of the system as a whole is broken by the propagation direction of incident radiation (the z-axis) and the direction of the radiation's electric field polarization (say along the x-axis). When considering the simplest case of a uniform driving electric field across the particle, the mode that couples directly to the radiation is the dipole mode. What falls out is a resonance condition; the electric fields inside and outside the sphere (as well as the resulting polarizability of the sphere) are proportional to terms that have $\epsilon_m(\omega) + 2\epsilon_d$ in the denominator. The polarizability in particular becomes

$$\alpha = 4\pi a^3 \frac{\epsilon_m - \epsilon_d}{\epsilon_m + 2\epsilon_d}. \tag{8.136}$$

When the metal is not too lossy (ϵ_m mostly real) and the surrounding medium is a good dielectric, something interesting must happen when this denominator vanishes. This is the dipolar plasmon resonance. Keeping track of the details, the incident radiation is scattered (cross section $\sim a^6$) and absorbed (cross section $\sim a^3$); which dominates depends on particle size and the dielectric properties of the metal. The dependence of the plasmon resonance (and hence the scattering and absorption of nanoparticles) on the dielectric response of the surrounding medium is the mechanism used in LSPR sensing [663].

Many other shapes of particle are relevant beyond simple spheres [664]. Exact solutions to the problem exist for, e.g., ellipsoids [665] with principle axis lengths a, b, c, and quasistatic approximate solutions [666] for such geometries can give expressions similar to Eq. (8.136) that are somewhat revealing,

$$\alpha_i = \frac{4}{3}\pi abc \frac{\epsilon_m - \epsilon_d}{\epsilon_d + L_i(\epsilon_m - \epsilon_d)}. \tag{8.137}$$

---

[6] Or, for that matter, in the acoustic case, as in a drumhead or sound board comparable in size to the transverse sound wavelength.

where i labels principal axes and L_i are depolarizing factors that depend on geometry. For a spherical particle, $L_{i=1,2,3} = 1/3$, recovering the prior result. We can pick three principal axes, see that each axis has a dipole resonance determined by the dielectric functions, and that those resonances are degenerate. In general, for $L_i \neq L_j$, the degeneracy is lifted. For the polarization direction along an axis, the longer the particle is in that direction, the more red-shifted the resonance [652]. In a perfect ellipsoid with polarization aligned exactly along a principal axis, it is possible to excite that one particular mode without exciting the others. This is a general feature of LSPR modes in nanoparticles: the overall symmetry of the particle structure establishes the selection rules for exciting particular normal modes.

The particular nanoparticle geometry also strongly affects the local electromagnetic field distribution produced by the plasmon excitations [667]. Nanoparticles with sharp points and tips can lead to local field enhancements significantly larger than what one would find for smooth geometries such as spheres or disks, as shown in Fig. 8.30. This is sometimes called the "lightning rod effect". If a boundary condition must be met for the electric field at the metal surface, and that metal surface is very tightly curved, as at a point or corner, the local electric field can become very large and strongly varying over distance scales comparable to the metal radius of curvature. In calculating the response of realistic structures it is important to remember that "unphysical" boundary conditions (infinitely sharp corners, infinitely sharp interfaces between materials of different dielectric responses) can produce quantitatively misleading answers.

Often we are interested in the plasmon resonances of composite systems. Examples include multiple nanoparticles coupled together within their evanescent near fields; a nanoparticle coupled to a nearby metal surface; and a metal STM tip in close proximity to a planar metal sample. Mathematically, these cases involve the coupling of resonant subsystems, each with normal modes, and the need to find the new normal modes of the coupled system. When coupling two identical resonators together, the new normal modes of the hybrid system are symmetric and antisymmetric combinations of the original modes that end up at lower and higher frequencies, respectively, as in Fig. 8.31(a).

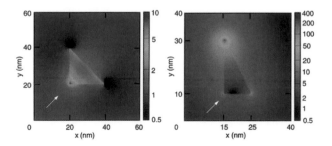

Fig. 8.30 Plasmon resonances in sharp silver nanoparticles, as calculated through finite element methods. The lateral dimensions are in nanometers, and the propagation direction of the incident field is indicated by the arrows, with electric field polarization lying in the plane. Note the logarithmic color scale for the local electric field amplitude. Sharp corners lead to enormous predicted field enhancements through the lightning rod effect. Adapted from [668].

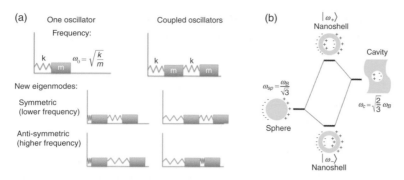

Plasmon hybridization. (a) Coupling two mechanical resonators together gives a combined system with new normal   **Fig. 8.31**
modes, in this case corresponding to symmetric and antisymmetric motion of the masses. (b) Plasmon resonances
may also be coupled via electromagnetic interactions, leading to the formation of new normal modes. In this case, the
plasmon resonances of a dielectric core/metal shell nanoparticle ("nanoshell") result from hybridization of the
plasmon modes of a bulk metal sphere and a spherical void in a metal. Adapted from [669].

A similar procedure, *plasmon hybridization*, may be followed for coupling plasmonic
systems [669, 670].

This is a very useful way to think about plasmon modes of complex systems. For
example, Fig. 8.31(b) shows that the plasmon modes of a nanoshell (a spherical metal shell
around a dielectric core) may be written as the hybridization of the modes of a solid spherical nanoparticle and a spherical dielectric void in an otherwise uniform metal. The coupling
between these modes in the nanoshell arises through the electrostatic interaction between
charges on the inner and outer surfaces of the metal shell. Looking just at the dipolar
modes of each subsystem, the results are a red-shifted bonding mode (with the two dipolar
plasmons in phase with each other) and a blue-shifted antibonding mode (with the two
dipolar plasmons oscillating antisymmetrically, π radians out of phase with each other).

Linear combinations of different plasmon modes and their interference can lead to some
remarkable effects. For example, not all plasmon modes are optically active; those that do
not feature a prominent overall electric dipole tend to be "dark" because they consequently
have a relatively poor coupling to radiation. Interference between a broad bright mode and a
higher Q dark modes in a plasmonic structure can result in a Fano resonance, a sharp dip in
the scattering spectrum of the plasmonic structure [671, 672]. Such a sharp feature is ideal
for use in sensing applications, since its shift with changes in the dielectric environment is
easy to detect.

The hybridization picture also explains the origin of the "tip-induced plasmon", a highly
localized plasmon mode found when a metal tip is separated at the nanoscale from a metal
surface (assuming both metals have good dielectric properties for plasmons) [673, 674].
These local plasmon modes may be excited by sufficiently energetic tunneling electrons
[673, 675]. The tip-induced plasmon is "built" out of linear combinations of the continuum
of surface plasmons on the metal tip and the metal surface. The extremely local coupling
of the two sets of plasmon modes leads to the formation of a new, localized normal mode,

with a basic dipolar sort of excitation (charge of one sign on the tip; charge of the opposite sign on the surface beneath the tip, as in a capacitor, but oscillating back and forth in polarity at optical frequencies).

8.6.4 Surface-enhanced spectroscopies

When optically excited, the amplitude of the local electromagnetic field near a plasmonically active nanostructure can exceed the amplitude of the incident radiation by a significant factor, $g(\omega)$. This field enhancement factor depends on frequency through the plasmonic response of the metal structure. This enhancement of the local field has profound consequences for some optical spectroscopies used for chemical detection and analysis. Since (first-order) scattering processes in general are proportional to the local intensity, it is not surprising that a locally increased intensity $\propto g^2$ results in an enhanced scattering rate. Spectroscopic techniques that leverage plasmonic near-fields are *surface-enhanced spectroscopies*.

The most famous and dramatic surface-enhanced spectroscopy is *surface-enhanced Raman spectroscopy* (SERS). Initially discovered in the 1970s [676, 677], SERS benefits even more than most techniques from enhanced electromagnetic fields. Molecules and solids have polarizability tensors $\tilde{\alpha}$ that relate their induced dipole moment to the magnitude and direction of some incident electric field. Suppose a particular constituent bond's motion affects $\tilde{\alpha}$, and motion has a resonant frequency ω_0. In Raman scattering [678, 679, 680], incident light at frequency ω interacts with the system, and because of the polarizability tensor and the molecular dynamics, inelastic transitions are possible, such that the frequency of the outgoing radiation, ω', can be $\omega - \omega_0$ ("Stokes shifted"), ω ("Rayleigh scattered"), or $\omega + \omega_0$ ("anti-Stokes shifted"). By carefully measuring the energy of scattered light, one can infer the (Raman active components of the) vibrational spectrum of the system of interest, a kind of vibrational fingerprint. If $\hbar\omega$ happens to coincide with the energy of an electronic transition in the system, the result is resonant Raman scattering.

SERS is relevant when molecules of interest interact directly with the evanescent, enhanced fields near the surface of nanostructured metal. First, the local intensity is increased over the free-space incident intensity by a factor of $g(\omega)^2$. When the quantum mechanics of the actual emission process is considered, the effective density of states for the emitted photons is also enhanced over the free-space value by $g(\omega')^2$ [681]. The result is that the total Raman scattering rate is enhanced by $g(\omega)^2 g(\omega')^2$, or roughly g^4 if the field enhancement is roughly frequency independent over the range between ω and ω'. This "electromagnetic enhancement" can be enormous, exceeding 10^{10} in certain nanostructures. With such a huge enhancement, single-molecule Raman spectroscopy becomes possible [682, 683, 684].

For SERS, the relevant polarizability tensor is that of the molecule plus the proximal metal. The symmetries of the free molecule are broken by image charge effects in the metal and any perturbation of the molecular orbitals due to interactions. As a result, some vibrational modes that are Raman-inactive in the gas or solution phases can be seen in SERS, and relative intensities of Raman modes can also be modified. In addition to the plasmonic enhancement, there can also be a "chemical enhancement" contribution, due to changes in the Raman cross-section from bonding, changes in the HOMO–LUMO gap (and hence the resonance Raman condition), and resonances between the photon energy and the energy difference between molecular levels and the metal Fermi level [685]. However, the plasmonic enhancement is generally orders of magnitude larger.

SERS was originally discovered using electrochemically roughened metal films. Understanding of the role of surface plasmons and "hot spots" coincided with advances in nanofabrication methods, so that there are many examples of nanostructured metal surfaces engineered to produce large Raman enhancements [663, 686, 687]. Some of the largest electromagnetic enhancements come from nanoscale gaps between plasmonically active nanoparticles [688], with the field enhancement increasing steeply as the gap distance is reduced. This trend cannot continue without bound; once significant charge transfer across the nanoscale gap (via tunneling, or direct conduction) can take place during an optical cycle, the hybridized plasmon mode active across the gap is "shorted out" [689, 690]. In other words, once the plasmonic systems actually touch, they become a single system, no longer supporting a hybridized gap mode.

A variant of SERS using extended electrodes rather than nanoparticles is *tip-enhanced Raman scattering* (TERS) [691, 692]. TERS specifically employs the tip plasmon mode described above. An advantage of TERS is that the tip may be used for scanned probe microscopy as well as local plasmonic enhancement. This means that particular nanoscale features on a metal surface, including individual molecules of interest, may be imaged via SPM and then characterized via TERS. Tip plasmons are also employed in planar nanogap structures [689].

Other surface-enhanced spectroscopics take advantage of the plasmonic field enhancement afforded by metal nanostructures. In *surface-enhanced infrared absorption* (SEIRA), the locally enhanced intensity ($\propto g(\omega)^2$) leads directly to increased absorption by molecular species in the near field. Absorption lines corresponding to the infrared-active vibrational modes can then be detected on the spectrum of the light scattered by the plasmonic structure [693].

While SEIRA operates by the plasmon-enhanced intensity, *surface-enhanced fluorescence* (SEF) leverages the effectively enhanced photon density of states ($\propto g(\omega')^2$) at the emitted frequency [694]. Fluorescence is the radiative decay, after direct photoexcitation, of a system from an electronically excited state into a lower energy state. Conventional fluorescence is a spontaneous emission process, and as in Eq. (8.127), increasing the photon density of states at the emission energy beyond the ordinary free-space value can modify the rate of spontaneous emission.

The radiative decay process always competes with nonradiative pathways. When a molecule is in close proximity to a metal surface, there is some (strongly distance-dependent) probability that the transition can produce an electron–hole pair in the metal rather than an emitted photon. Likewise, the transition can excite traveling surface plasmons that then decay nonradiatively. Such processes are mechanisms behind "quenching" of fluorescence. The fluorescence yield can only be enhanced relative to the free molecule case if the relative boost to the radiative rate exceeds the increase in quenching.

8.6.5 Plasmonics for optical antennas

In a conventional radio receiver antenna, the electric field from free-space electromagnetic radiation exerts oscillating, radio-frequency forces on mobile charge carriers in a metal structure. At the connection point of the antenna the result is an oscillating voltage that may be amplified and otherwise manipulated. A conventional receiving antenna harvests electromagnetic energy from propagating radiation and deposits that energy into a circuit for further use. Conversely, a broadcast radio antenna uses a time-varying voltage applied to a metal structure to produce oscillating charge distributions that radiate into the far field. In conventional antennas, maximum efficiency of either reception or transmission is achieved when the antenna is designed to be geometrically resonant with a particular wavelength (half-wave or quarter-wave antennas, for example). However, nonresonant operation is still possible, at the cost of reduced efficiency. For example, the wavelength at 1000 kHz, the middle of the AM radio band, is 300 m, yet a one meter automobile antenna can still receive AM radio signals.

Metal nanostructures can function as optical frequency antennas, very much in analogy with their much larger radio frequency counterparts. However, plasmon resonances at the nanoscale, with no analogy in macroscale antennas, modify the appropriate size scaling [695, 696].

Plasmonic structures functioning as receiving antennas are being put forward for photodetector [697, 698] and photovoltaic applications [699]. In these schemes the key is to use the plasmonic response of metal nanostructures to control the near-field. In some antenna approaches a resonant plasmonic structure has a large cross-section for interacting with the incident radiation, while the near-field enhanced intensity is concentrated into a deep subwavelength region [697, 698] for, e.g., absorption by a photoconductive element. For photovoltaic systems, often the issue is increasing net absorption via interactions with the plasmonic near-field. The confinement of intensity to the surface of the resonant metal structure acts in some ways like a cavity, increasing the interaction between photons and the surrounding medium. Losses due to the dielectric properties of the metal can be significant, so that care must be taken in design to ensure that the appropriate figure of merit is actually improved by the inclusion of plasmonic components.

On the broadcast side, plasmonic structures function in two ways, tuning the scattering into the far field, and modifying the spontaneous emission properties of photon-producing elements. In an example of the former, a "bow tie" plasmonic antenna patterned on the output facet of a semiconductor laser can alter the spatial distribution of the emitted radiation [700]. Similarly, semiconductor nanocrystals deposited at the appropriate location on a nanoscale metal Yagi–Uda antenna [701] have strongly directed emission dictated by the antenna geometry.

Plasmonic alteration of photon emission rates has been demonstrated in a number of experiments [702, 703], the most dramatic of which use plasmon resonances as effective cavities for lasing [704, 705]. In these cases the enhanced emission rate due to plasmon-based modifications to the photon density of states is sufficient to outstrip any enhanced nonradiative relaxation mechanisms or plasmonic losses.

By generating very large local field enhancements when driven on resonance, plasmonic antennas are also useful as nonlinear optical systems. Two-photon photoluminescence [706] and supercontinuum generation [695] are only two of the possibilities demonstrated when nanoscale optical antennas are illuminated with intense incident radiation.

8.6.6 Plasmons for metamaterials

We have seen how nanostructured dielectric media can have very rich optical properties such as photonic band gaps. Similarly, we have seen how individual or small numbers of metallic nanostructures can exhibit rich optical responses thanks to plasmonic effects. It should come as no surprise that metamaterials incorporating plasmonic constituents have become extremely active areas of research in recent years.

Extraordinary optical transmission

One of the simplest possible versions of a plasmonic metamaterial is an array of subwavelength holes in a metal membrane that is optically thick enough to be otherwise opaque. Ignoring plasmon response and simply looking at classical transmission, the transmitted power through an aperture of diameter $a < \lambda$ should scale like $(a/\lambda)^4$, as found by presciently nano-oriented Hans Bethe [707]. Instead, if the metal has good plasmonic properties and (properly sized) holes are regularly spaced (by an appropriate amount) in an array (of proper thickness with a large number of holes), the optical transmission through the membrane can exceed classical expectations by orders of magnitude [708]. From the number of qualifiers in that sentence it should be clear that something both plasmonic and cooperative is responsible.

The accepted explanation for this phenomenon, termed *extraordinary optical transmission*, is that the top and bottom surfaces of the metal membrane support surface plasmon resonances [709]. These resonances are hybridized via coupling along the walls of the holes, and both the hole size and hole spacing determine the detailed response as a function of incident wavelength. Further investigations [710, 711] make clear that the specific shape of the holes is very important; the holes themselves are plasmonic objects with shape resonances. The periodic arrangement of the holes is essential, since constructive interference of the propagating surface plasmons boosts the effect. Spacing the holes too far apart allows plasmon losses to take their toll, suppressing the transmission until the holes are effectively independent from one another.

"Left-handed" media

The effective dielectric properties of ensembles of metal structures may be exploited in more dramatic fashion. Veselago considered [712] the possibility of materials with simultaneously negative values of ϵ and μ. While the Kramers–Krönig relations forbid this possibility over too broad a range of wavelengths, through the use of metals it is possible to achieve this condition over a narrow band. At first glance, this might seem boring, since $\epsilon \times \mu$ would still be positive, leading to a positive phase velocity. However, one can interpret the index of refraction in this case as

$$n(\omega) \equiv \sqrt{\frac{\epsilon(\omega)}{\epsilon_0}} \sqrt{\frac{\mu(\omega)}{\mu_0}}. \tag{8.138}$$

Since each quantity under the square root is negative, so is the resulting index of refraction. Such a system is sometimes called a *negative index material*. Because Snell's law in this circumstance would say

$$\sin \theta_i = \pm |n| \sin \theta_r, \tag{8.139}$$

this kind of material can exhibit "negative" refraction, with refracted light bending with the opposite sense relative to the surface normal. At a frequency where $n < 0$, $\mathbf{E} \times \mathbf{H} = -\mathbf{k}$; that is, the electric and magnetic fields obey a left-hand rule, prompting the alternative name for such materials as *left-handed media*.

As in the photonic band gap systems, initial testing of these ideas is simpler at microwave frequencies, when it is possible to construct metamaterials out of essentially macroscopic components. An array of metal rods can take advantage of the dielectric response of the metal to take on a negative $\epsilon(\omega)$ over some frequency range when the electric field is polarized along the rods. Split-ring metal resonators can similarly mimic $\mu(\omega) < 0$ around a chosen frequency. Combining the two into a single periodic array [713] allowed the demonstration of negative refraction of a microwave beam [714].

Note that there are some subtleties at work in these systems, particularly regarding the effects of dispersion and the transport of energy. Exploiting negative index effects in the visible generally requires fabrication methods that work significantly below the scale of a wavelength of interest [715], as in photonic band gap systems.

One potential application of negative index media is in improved lenses, capable of evading the diffraction limit [717]. As initially proposed [718], a slab of material with $n = -1$ at some wavelength can function as a perfect lens, as shown schematically in Fig. 8.32(a). Initially divergent rays become convergent upon entering the negative index medium. Diverging again before exiting the slab, they become converging upon crossing the last interface, focusing far more tightly than the classical diffraction limit. The effects of dispersion and loss limit whether "perfect" lensing could ever be achieved in a practical metamaterial [719]. However, the use of metal plasmonic response (essentially a consequence of $\epsilon < 0$) has been demonstrated to improve optical resolution in the near field [716], as shown in Fig. 8.32(b). Having the metamaterial incorporate a gain medium that may be optically pumped has also been put forward as means of compensating for losses associated with the metal components [720].

Perhaps the most glamorous application suggested for metamaterials is as *invisibility cloaks*. A metamaterial can have, over some bandwidth, an engineered, spatially varying effective index of refraction. In principle, a metamaterial may be structured to steer light of a desired wavelength around an object, so that in the far field it can be impossible to tell that the light interacted with anything at all. A two-dimensional example in microwave frequencies [721] is shown in Fig. 8.33(a, b). Figure 8.33(c) is an electron micrograph

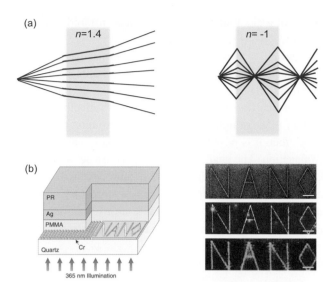

Improved lenses with negative index materials. (a) A slab of negative index material can focus diverging rays (right) while a slab of conventional dielectric does not. (b) A thin silver film leads to improved photolithographic resolution. At right, the top image is an ion micrograph of the metal mask; the middle image is an AFM image of the silver-enhanced-focusing result, while the bottom image is an AFM image of a control result omitting the silver layer. Scale bar is 2 μm. Adapted from [716].

Fig. 8.32

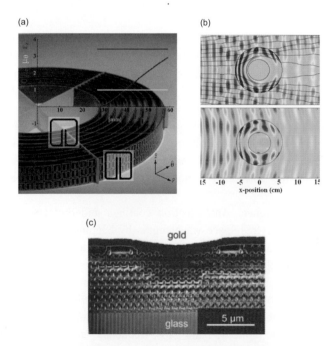

"Invisibility" cloaks. (a) A microwave-frequency (8.5 GHz) metamaterial cloak designed to bend radiation at around the central region. (b) A comparison of simulation (top) and experiment (bottom). Adapted from [721]. (c) An all-dielectric version of a 3d cloak meant to hide the gold "bump" at the top of the image. Adapted from [722].

Fig. 8.33

of an implementation of a cloak that works in the near infrared [722], albeit without any plasmonic components.

The basis for the design of these materials is *transformation optics* [723], an approach based on coordinate mapping [724, 725]. Suppose one wants light to propagate in some complex way, as in guiding the light around an object. If one can come up with a mathematical mapping that transforms the complicated trajectory in the lab coordinates $\mathbf{r}$ to a simple trajectory (e.g., straight line propagation through empty space) in some other coordinate system $\mathbf{r}'$, then it is possible to find the spatially varying $n(\mathbf{r})$ that would give the desired lab propagation. The challenge is then to construct a metamaterial that has the appropriate $\epsilon(\omega, \mathbf{r})$ and $\mu(\omega, \mathbf{r})$. Manipulation on subwavelength scales requires the use of evanescent fields, and plasmon resonances can be necessary. The bottom line is that metamaterials and plasmonics provide a remarkable ability manipulate ϵ and ω beyond the bounds found in natural bulk materials, and that in turn paves the way for manipulation of light in amazing ways.

8.7 Summary

We examined electromagnetic radiation and how the ability to engineer structures on the nanoscale allows us to manipulate and control optical and optoelectronic phenomena. Main points include the following.

- Material response and electromagnetic fields. The response of materials to electromagnetic fields, in the form of $\epsilon(\omega)$ and $\mu(\omega)$, is the basis for how we interact with electromangnetic radiation. Causality, the fact that material response can only depend on past values of the electromagnetic field, leads to restrictions on the real and imaginary parts of the permittivity and the permeability.
- The importance of boundaries and interfaces. The boundary conditions that must apply to the electromagnetic field at material interfaces determine transmission and reflection at interfaces. Even when light is totally reflected at an interface, there is some evanescent field extending into the "forbidden" region, though energy does not propagate away from the interface there.
- Engineering of interference. Antireflection coatings and dielectric mirrors may be constructed for specific free-space wavelengths by keeping track of the phase of the propagating light, and arranging for constructive or destructive interference of forward propagation, respectively. Doing so requires structuring material on scales of a fraction of a wavelength. A transfer matrix approach very similar to that used for electronic transport can be used to find the properties of dielectric superlattices. Spatially periodic dielectric structures are mathematically analogous to spatially periodic potentials for electrons. There can be photonic band gaps, frequency ranges for which there are no propagating waves in particular directions.
- Nonlinear phenomena. At higher intensities and in some materials nonlinear optical phenomena are possible. Examples include nonzero electro-optic coefficients, frequency

doubling, sum and difference frequency generation, optical rectification, and third harmonic generation.

- Diffraction and the near field. The wave nature of light leads to phenomena like diffraction. The far-field intensity pattern from a plane wave propagating through an aperture looks like the square of the spatial Fourier transform of the aperture. This limits the resolution of far field optical imaging to $\sim \lambda$. The short-distance ($< \lambda$) spatial information about the aperture is contained in evanescent components that decay in the far field limit. An object in the near field can scatter these components into the far field, enabling NSOM and the evasion of the diffraction limit.

- Emission, absorption, and all that. Transition rates in quantum systems due to interactions with radiation depend on the matrix connecting initial and final electronic states (giving selection rules), the Pauli principle (electrons can only go from full to empty states), and the density of states for photons. Manipulating the photon density of states with a cavity can change the rate of spontaneous emission. Stimulated emission and absorption rates are generally the same.

- Lasers. If one can achieve population inversion and photons stick around long enough to trigger many stimulated emission transitions, lasing is possible. Lasers can be optically or electrically pumped.

- Optical communications. Fiber optic communication is the backbone of the Information Age. Critical components include laser sources, optical modulators, multiplexers, fiber, and detectors.

- Nanophotonics. Photonic band gap systems can guide and trap light on scales small compared to λ. This can enable devices like ultrasmall lasers and enhanced light-emitting diodes.

- Plasmonics. The dielectric function of metals can lead to collective electronic oscillations. Plasmons can be propagating or confined in metal nanostructures. Confinement leads to a discrete spectrum of modes. As subsystems are allowed to interact, plasmons can hybridize, forming new normal modes of the coupled system. Plasmonic structures on resonance can have near fields that are greatly enhanced, enabling surface-enhanced spectroscopies and optical antenna effects.

- Metamaterials. By creating periodic metal and dielectric nanostructures, metamaterials are possible with designed values of $\epsilon(\omega)$ and $\mu(\omega)$ unavailable in naturally occurring materials. Negative index materials and invisibility cloaks are two examples of possible applications.

8.8 Suggested reading

- *Classical electricity and magnetism* For a refresher on Maxwell's equations and the interactions of light and matter, in order of increasing sophistication, recommendations are *Electricity and Magnetism*, 3rd Edn. by E. O. Purcell and D. J. Morin (Cambridge University Press, 2012); *Introduction to Electrodynamics*, 3rd Edn. by D. J. Griffiths (Benjamin Cummings, 1999); *Classical Electrodynamics*, 3rd Edn. by J. D. Jackson

(Wiley, 1998); and *Electrodynamics of Continuous Media*, 2nd Edn., volume 8 of the famous series by Landau, Pitaevskii, and Lifshitz. For information about wave propagation, I recommend *Electromagnetic Waves and Antennas* by S. J. Orfanidis (2002) [597].

- *Physical optics* A standard reference text on optics is *Optics*, 4th Edn. by Eugene Hecht (Addison Wesley, 2001). A less expensive alternative is *Modern Optics* by G. R. Fowles (Dover, 1989).

- *Lasers* Two standard books on lasers are the extremely comprehensive *Lasers*, by A. E. Siegmann (University Science Books, 1986) and *Principles of Lasers*, 5th Edn., by O. Svelto (Springer, 2009). For quantum cascade lasers, the article [607] by Gmachl *et al.* is very useful.

- *Photonic band gaps* A definite reference text on PBG systems is *Photonic Crystals: Molding the Flow of Light*, 2nd Edn. by Joannopoulos, Johnson, Winn, and Meade (Princeton, 2008). The review article [726] by López highlights some of the fabrication challenges in PBG systems.

- *Nanophotonics* The best references I've found for nanophotonics are the broad look in *Nanophotonics*, by P. N. Prasad (Wiley-Interscience, 2004) and the nicely pedagogical *Principles of Nano-Optics* by Novotny and Hecht (Cambridge University Press, 2006). While there are many books about near-field optics, I found *Near-Field Optics: Theory, Instrumentation, and Applications* by M. A. Paesler and P. J. Moyer (Wiley-Interscience, 1996) very helpful.

- *Optics for telecommunications* There are numerous books about optical telecommunications, though many focus on the detailed ins and outs of networks. For a readable, historical picture, try *City of Light: the Story of Fiber Optics*, by Jeff Hecht (Oxford University Press, 1999). A comprehensive reference text is *Fundamentals of Photonics*, 2nd Edn. by B. E. A. Saleh and M. C. Teich (Wiley-Interscience, 2007), as is *Fiber-optic Communications Systems*, 3rd Edn. by G. P. Agrawal (Wiley-Interscience, 2002).

- *Plasmonics* Stefan Maier's *Plasmonics: Fundamentals and Applications* (Springer, 2007) is a good start. *Reviews of Modern Physics* has recently published articles on extraordinary transmission [727] and plasmonic systems as resonators [728]. The review [652] by Maier and Atwater is also quite comprehensive.

- *Metamaterials* A recent edited volume with useful chapters is *Tutorials in Metamaterials*, edited by M. A. Noginov and V. A. Podolskiy (CRC Press, 2011). Good review articles on negative-index media include a long review by Ramakrishna [729] and a shorter piece by Liu and Zhang [730].

Exercises

8.1 *Dielectric mirrors* It was mentioned in the text that at their design wavelength, dielectric mirrors can have superior performance relative to metal mirrors, with higher reflectance and lower losses. Briefly explain why.

8.2 *Optics for the extreme ultraviolet (EUV)* As we've discussed before, the easiest way to improve the resolution of optical lithography is to switch to a shorter wavelength of light. Unfortunately, silica-based lens materials have an absorption edge in the UV, so that they no longer provide a nice means of manipulating light at wavelengths much smaller than the now-cutting-edge 157 nm. One possible way around this is to use reflective optics – basically bouncing UV light off curved mirrors rather than through lenses to achieve focusing and magnification.

(a) Read the following article:

```
www.intel.com/technology/itj/q31998/pdf/euv.pdf
```

and the description here:

```
xdb.lbl.gov/Section4/Sec_4-1.html
```

of x-ray properties of multilayers. What is a complication of working with multilayers in this frequency range that is generally not relevant when working in the visible? One common combination of materials for multilayer mirrors in the UV/x-ray is silicon and molybdenum. Can you suggest two reasons why this is an attractive materials combination?

(b) Go to

```
www-cxro.lbl.gov/optical_constants/multi2.html
```

and play around with the applet there. Specifically, look at wavelengths from 5 nm to 20 nm. Keep the materials fixed at Si and Mo, and vary the multilayer period around its original default value of 6.9 nm. What happens to the wavelength of maximum reflectance as the multilayer period is varied? What happens to the maximum reflectance itself? Why do you think the maximum reflectance drops when the multilayer period deviates from its optimal value, in both directions?

(c) What do you think happens if the surfaces are rough, so that the layer thicknesses aren't ideal for a given set of indices? Explain. Can you explain why this is a serious issue for working at wavelengths like those in part (b)?

8.3 *Distributed Bragg reflectors* We discussed multilayer dielectric mirrors in Section 8.1.13, and I showed some formulae about their performance. Now go to the website

```
tnweb.tn.utwente.nl/cops/education/5_oc_optlay.html
```

and play around with the java applet there.

(a) Consider your incident wavelength of interest to be 550 nm. Using the indices n_1 (called n_a in Fig. 8.10) = 1.5 and n_2 (called n_b in Fig. 8.10) = 1.33, and assuming $n_c = n_s = 1$, what layer thicknesses should be used to get high reflectivity centered around that wavelength?

(b) Using those parameters and the "table" tab of the applet to get the numbers, plot the reflectivity at that wavelength as a function of m, the number of periods, from $m = 1$ to $m = 20$. Compare with the formula that was presented in Eq. (8.94), and discuss their agreement.

(c) While doing part (b), also keep track of the width in wavelength of the reflectance peak. Compare its behavior with the formula presented in Eq. (8.95), and discuss their agreement.

(d) Play around with the applet. What happens when the layer thicknesses are not chosen to have any particular relationship to each other – do you still see regions of high reflectivity? Also, what happens when the index contrast is very large? Do you agree with the statement that dielectric mirrors are a particular example of a one-dimensional photonic band gap system?

8.4 *Giant birefringent optics (GBO)* Read the paper by Weber *et al., Science* 287, 2451 (2000).

(a) What is the main weakness of dielectric mirrors for polarized light, when compared to traditional mirrors?

(b) Explain briefly what allows so-called GBO materials to avoid this problem.

8.5 *Photon density of states* In Section 8.4.1 we mentioned the idea the effect of a cavity on the photon density of states. Here we discuss this in a little more detail.

(a) Think only about the electric field for a second, and consider a hollow, (3d) box of side length L with perfectly conducting walls. Count up the number of allowed modes within the box up to some frequency ω, remembering that for each mode the electric field must vanish at the walls. So, the lowest frequency (longest wavelength) mode along the x direction, for example, will have a wavelength equal to $2L$, and the next allowed mode will have a wavelength equal to L, and so on. It's easiest to count in k-space. This is completely analogous to the way we figured out the density of states for electrons back in Chapter 2. Including the fact that each spatial mode can have two transverse polarizations, show that the number of modes in the cavity up to frequency ω is $N = (L^3/(3\pi^2 c^3))\omega^3$, and therefore the photon density of states (number of states per volume per frequency) is just what was described above, $\omega^2/(\pi^2 c^3)$.

(b) Quantum mechanics states that each such mode should contain $\frac{1}{2}\hbar\omega$ of zero-point energy. Inside the cavity, our boundary conditions restrict the allowed modes so that wavelengths longer than $2L$ are forbidden. What this implies is that the energy density of the *empty space* inside the cavity is lower than that outside the cavity! Assuming the same cavity as in (a), integrate up the "missing" energy density using the photon density of states above, integrating from $\omega = 0$ up to $\omega = c\pi/L$, the minimum allowed frequency in the cavity. Then multiply by the volume of the cavity to show that the missing energy is given by $\frac{\pi^2}{8}\frac{\hbar c}{L}$. How much energy are we talking about for a cavity of side length 10 nm? In detailed treatments, these kinds of vacuum energy differences are closely related to the van der Waals interactions between the cavity walls, and can result in significant forces at nanometer scales, as we'll see later. This is called the *Casimir effect*.

8.6 *Acousto-optic modulation* As we mentioned above, longitudinal sound waves may be used to modify the index of refraction because the associated strain field changes the density of the medium.

(a) Consider a longitudinal acoustic wave propagating through a crystal in the z direction. If the maximum strain amplitude is $\delta\ell/\ell \sim 10^{-4}$, estimate (order of magnitude) the maximum change in index that results due to the sound wave. Hint: think about the Clausius–Mossotti relation.

(b) One approach to AO modulation is to use *surface acoustic waves*. One approach to this is to set up a standing sound wave between two "launchers" as discussed above. The incoming optical wave can then diffract off the spatially periodic index modulation that results from the antinodes of the acoustic field. Assuming a sound speed of 1000 m/s and a desire to diffract a 900 nm wavelength optical signal by first order diffraction through an angle of 10 degrees, what spacing between antinodes of the acoustic wave is needed? What should the frequency of the sound wave be?

(c) If one wanted to use this SAW device as an optical switch, approximately what would be the switching time, assuming the acoustic launchers are roughly 200 microns apart?

(d) If one wants to launch SAWs using the piezoelectric properties of the acoustic material, one needs to make launchers using interleaved electrodes typically separated by $1/4$ the acoustic wavelength. For GaAs, with a speed of sound of 5.34 km/s, if one wants to work at an acoustic frequency of 40 GHz, how far apart should the launcher "fingers" be?

8.7 *A "random" laser* As we discussed in this chapter, two essential ingredients for a laser are a gain medium (either pumped electrically or optically), and a cavity. Let's consider an unconventional kind of laser that is, in fact, highly *disordered*.

(a) In Chapter 6, we introduced the concept of localization of electrons – the idea that quantum mechanical electrons diffracting off a disordered array of scatterers have an enhanced probability of ending up back where they started due to interference effects. In the strong localization regime, the elastic mean free path for electrons becomes shorter than the Fermi wavelength; that is, $k_F \ell_e < 1$. For electrons, this means that the Bloch wave picture of delocalized states extending over many lattice sites is an inappropriate description of the electron states. Instead, the electronic states are "localized" – basically standing waves confined to a small region. An analogous effect for light exists! What do you think is the analogous condition to the one above to be in this regime? Name one major difference between the light case and the electronic case (apart from the fact that electrons are charged – there are at least two other big differences).

(b) Go and read Cao *et al., Phys. Rev. Lett.* **82**, 2278 (1999). As you can see, the investigators used localized states of light as cavities (!) so that they could get lasing from a highly disordered film of semiconductor nanoparticles. Write a sentence or two describing the main light localization process that they consider. What test do the authors do to demonstrate that their nanoparticle film actually localizes light? What effect does this particular mechanism have on the size dependence of the lasing region?

(c) What is special about the emission directionality from random lasers? So, we've seen in this chapter that you can make a cavity out of: (1) nice, ordered reflecting surfaces; (2) defects in photonic crystals that lead to standing waves;

and (3) even disorder-induced standing waves produced by a highly messy gain medium.

8.8   *Chemically synthesized nanocrystal lasers* Go read Eisler *et al., Appl. Phys. Lett.* **80**, 4614 (2002).

(a) Describe in a few sentences the gain medium and cavity for this system.

(b) According to the paper, different nanocrystal/titania composites were made to lase at 583 nm, 607 nm, and 625 nm, all with a single DFB grating line spacing of 350 nm. How did the authors manage to get this to work? That is, how did they vary the color of the optical transition, and how did they maintain commensurability with a fixed physical grating spacing for these different colors?

Micro- and nanomechanics

From the earliest discussions of nanotechnology, there has been an interest in the development of mechanical devices with moving parts on the nanoscale. Biological systems demonstrate that this is possible, as cells routinely employ molecular machines with mechanical motions and degrees of freedom. Here we give an overview of the principles of the mechanics of solids, also known as *continuum mechanics*, with its concepts of stress, strain, and elastic moduli. As its name implies, continuum mechanics is based on the assumption of matter being continuous, rather than made up of discrete atomic units. In analogy with our previous treatment of electronic structure, it is interesting to explore the limits of this approach, which clearly must fail in the limit that chemistry becomes a more appropriate formalism. We also consider the origins of irreversibility in mechanical systems, including plastic deformation and friction, and see that nanoscale tools have been invaluable in increasing our understanding. After surveying micro- and nanoelectromechanical systems (MEMS and NEMS) as they are employed in current technologies, we will conclude with a discussion of the frontiers of nanomechanics.

9.1 Basics of solid continuum mechanics

We need to define some basic terms so that we can discuss the elastic properties of solids, the relationship between the deformation of a piece of material and the forces acting on that object. Consider a block of material at rest being acted upon by several forces that sum vectorially to zero, so that the block is not accelerating. Now imagine dividing the block into two pieces. Clearly the sum of forces acting on each piece must equal zero, since neither piece is accelerating, as before. That implies that one piece of the material is exerting forces on the other piece, and vice versa. In this example, the material deforms due to the actions of the external forces, developing the necessary internal forces to maintain static equilibrium as a result of that deformation. The elastic properties of the material determine those deformations.

Rather than working with forces, the appropriate quantity to consider is *stress*, defined as force per unit area. Because forces have a directionality, and areas also have a directionality (the orientation of the unit normal vector to an area, $\hat{n}$), stresses are tensorial. The individual components of the stress tensor are scalars. We can write the stress tensor components as σ_{ij}, where the first index indicates the direction of the surface normal and the second index the direction of the force. When the two directions are colinear, the

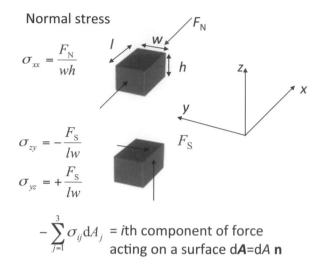

Normal stress

$$\sigma_{xx} = \frac{F_N}{wh}$$

$$\sigma_{zy} = -\frac{F_S}{lw}$$

$$\sigma_{yz} = +\frac{F_S}{lw}$$

$$-\sum_{j=1}^{3} \sigma_{ij} dA_j = i\text{th component of force}$$
acting on a surface $dA = dA\ \mathbf{n}$

Fig. 9.1   Stresses are forces per unit area and are inherently tensorial. The relevant force has both a magnitude and a direction, and the associated surface element has a direction given by a unit normal vector.

component is a *normal stress*. As shown in Fig. 9.1, the normal stress on the surface facing toward us is $\sigma_{xx} = -F_N/wh$, where F_N is the normal force. As shown, σ_{xx} is *compressive*. If the force was oppositely directed, tending to elongate the system of interest, the stress would be *tensile*.

Stresses where the force direction is perpendicular to the surface normal are called *shear stresses*. The response of a continuous medium to shear is what differentiates between solids and liquids, as we shall see.

The response of a material to stresses depends on the deformation of the material. Motion of some location in the material is not sufficient; rather, true deformation requires motion of some point in the material relative to neighboring points, such that the distance between those points is altered. Define $\mathbf{u}(\mathbf{r})$ to be the local displacement of a medium from its unperturbed state. Two points that started out separated by a vector $\mathbf{dr}$ end up separated by

$$\mathbf{dr}' = \mathbf{dr} + \mathbf{dr} \cdot (\nabla \mathbf{u}). \tag{9.1}$$

We can look at one component of the tensor $\nabla \mathbf{u}$:

$$\frac{\partial u_i}{\partial r_j} = \frac{1}{2}\left(\frac{\partial u_i}{\partial r_j} + \frac{\partial u_j}{\partial r_i}\right) + \frac{1}{2}\left(\frac{\partial u_i}{\partial r_j} - \frac{\partial u_j}{\partial r_i}\right). \tag{9.2}$$

That first part defines the strain tensor,

$$\tilde{\epsilon} \equiv \frac{1}{2}\left(\frac{\partial u_i}{\partial r_j} + \frac{\partial u_j}{\partial r_i}\right), \tag{9.3}$$

while the other part is (for $i \neq j$) equivalent to a uniform, rigid rotation about the k axis. From the definition we can see that

$$\text{Tr}\,\tilde{\epsilon} = \nabla \cdot \mathbf{u} = \frac{\delta V}{V},\tag{9.4}$$

where we have taken the trace of the strain tensor, and $\delta V/V$ is the fractional change in the volume of the medium in question.

The strain tensor contains the information we need about the distortion of the material. To relate that back to stresses, the simplest approach is to use a form of Hooke's law, the idea that the restoring force exerted by a spring is linearly proportional to the distance that spring is stretched. In keeping with that approach and our sign conventions, for a linearly elastic medium we write the components of the stress tensor as

$$\sigma_{ij} = -\lambda \delta_{ij} \nabla \cdot \mathbf{u} - \gamma \left(\frac{\partial u_i}{\partial r_j} + \frac{\partial u_j}{\partial r_i} \right),\tag{9.5}$$

where λ and γ here play the roles of spring constants. By making λ and γ simple scalars, we are assuming that the material in question is isotropic in its properties. Rearranging, we define $K \equiv \lambda + (2/3)\gamma$, and get

$$\sigma_{ij} = -K\delta_{ij}\text{Tr}\,\tilde{\epsilon} - 2\gamma \left(\epsilon_{ij} - \frac{1}{3}\delta_{ij}\text{Tr}\,\tilde{\epsilon} \right).\tag{9.6}$$

Often we want to invert this and find the strain as a function of the stress:

$$\epsilon_{ij} = -\frac{1}{9K}\delta_{ij}\text{Tr}\,\tilde{\sigma} - \frac{1}{2\gamma} \left(\sigma_{ij} - \frac{1}{3}\text{Tr}\,\tilde{\sigma} \right).\tag{9.7}$$

Consider three particular cases.

- Uniform pressure, p. In this case $\sigma_{ij} = p\delta_{ij}$. Inspection of Eq. (9.7) shows that the strain tensor also has no off-diagonal terms, and that $\epsilon_{ij} = -(p/3K)\delta_{ij}$. An isotropic cube under uniform pressure can be compressed, but it remains a cube.
- Axial stress, with $\sigma_{zz} = p$, all other components zero. Here $p > 0$ is a tensile stress, tending to extend the object in the z direction. Then

$$\sigma_{zz} = \left(\frac{9K\gamma}{3K + \gamma} \right) \epsilon_{zz} = Y\epsilon_{zz},\tag{9.8}$$

where we have defined Y, the *Young's modulus*. Also,

$$\epsilon_{xx} = \epsilon_{yy} = -\frac{1}{2} \left(\frac{3K - 2\gamma}{3K + \gamma} \right) \epsilon_{zz} = \nu\epsilon_{zz},\tag{9.9}$$

where we define ν, the *Poisson ratio*. A rod pulled on at both ends stretches by an amount linearly proportional to the tensile load, and also gets skinnier as it elongates.
- Uniform shear, with $\sigma_{xy} = \sigma_{yx} = -f$, all other components zero. Then

$$\epsilon_{xy} = \epsilon_{yx} = \frac{f}{2\gamma},\tag{9.10}$$

and the angle (as shown in Fig. 9.2) $\theta = f/\gamma$. We call γ the *shear modulus*. This is the distinction between solids and fluids. Fluids have no shear modulus; they deform continuously under shear.

Uniform pressure p

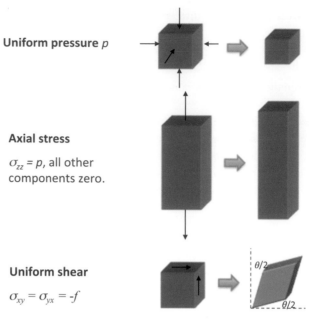

Axial stress

$\sigma_{zz} = p$, all other
components zero.

Uniform shear

$\sigma_{xy} = \sigma_{yx} = -f$

Fig. 9.2   Different common stress geometries.

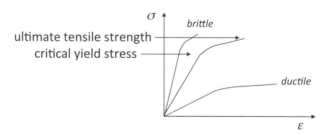

Fig. 9.3   Typical stress–strain curves for common solid materials.

To get a sense of scale, for stainless steel $Y \sim 200$ GPa, $\gamma \sim 76$ GPa. For brittle but strong single crystal silicon, $Y \sim 156$ GPa, $\gamma \sim 65$ GPa. For silver, a soft, ductile metal, $Y \sim 75$ GPa, $\gamma \sim 27$ GPa. Note that the shear modulus is always considerably smaller than the Young's modulus. This makes some physical sense on intuitive grounds (though one must be careful of subtleties). Distorting atomic bonds along their bond direction, as when stretching or compressing a simple cubic material along a principal axis, for example, should be energetically more expensive than tilting bonds perpendicular to their direction.

These moduli are directly applicable only in the regime of linear elasticity. Figure 9.3 shows some relevant regimes. While stress and strain are linearly proportional for the low stress (strain) limit, eventually there is an applied stress beyond which the deformation ceases to be linear and reversible. In the case of tensile loads, this limit is called the *critical yield stress*.

Beyond the critical yield stress many materials exhibit *plastic deformation*, such that upon removing the applied stress the strain curve does not retrace itself. If stress continues to be applied, the material deforms until failure takes place at the *ultimate tensile strength*, σ_{max}. Materials for which the strain at failure is comparatively small (say a few percent or less) are said to undergo *brittle fracture*. Materials for which the strain at failure can approach many tens of percent, and which also tend to exhibit "necking" (plastic deformation to a configuration with a smaller cross-sectional area near the point of failure), fail through *ductile fracture*.

Plastic deformation and failure involve the breaking of bonds; therefore it should not be surprising that understanding these phenomena requires a more microscopic point of view than continuum elasticity, as we discuss in Section 9.1.4. For now, however, having introduced the concepts of elastic moduli, stresses, and strains, we have enough information to understand a fair bit about the mechanics of solids.

9.1.1 Bending

With this minimal introduction to elasticity, we can look at particular types of deformation that are particularly relevant to micro- and nanomechanics: bending and torsion.

Figure 9.4 shows a side view of a piece of beam that extends in the z direction. The piece of beam is of length dz, and at a particular y value, the transverse dimension of the beam in and out of the page is $w(y)$.

Suppose that the beam is being flexed so that the top surface is (slightly) concave up, but nothing is moving. To have this situation, there must be some externally applied torque, and the material will develop internal stresses in such a way that the net torque on the material (external plus from the neighboring material) must be zero. Likewise, the total force on the piece of beam must also sum to zero.

From the drawing and common sense, it is clear that the top surface of the beam ($y = y_2$) is in compression. It is being squeezed, and an elastic response force must be generated (that pushes in the $+z$ direction on the right-hand face at $y = y_2$) to resist that squeezing. Similarly, the bottom surface of the beam ($y = y_1$) is in tension, and the elastic response generates a force (that pulls in the $-z$ direction on the right-hand face at $y = y_1$) that tries to unstretch that bottom surface. This implies that at some intermediate y value, locally the beam is unstrained. The surface defined by the y values where this is true is called the *neutral surface*. For convenience, let's pick our coordinates such that $y = 0$ at the neutral surface.

For small bending deflections, with this coordinate choice we can approximate the local strain along the z direction on the right-hand surface as being proportional to y, and the

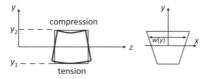

A segment of a bending beam. The beam is extended in the z-direction, while the transverse y direction is up and down on the page, and the x direction is into the page.

Fig. 9.4

local normal stress σ_{zz}, is linearly proportional to the local strain. The total elastic response force in the z direction is then

$$F_z = \int_{y_1}^{y_2} \sigma_{zz}(y) \cdot w(y)\mathrm{d}y = \int_{y_1}^{y_2} (ky)w(y)\mathrm{d}y, \qquad (9.11)$$

where k is some constant of proportionality (more on this below). To within a numerical factor, the right-hand side of this is identical to the expression for the centroid of the cross-section along the y direction. That is,

$$\bar{y} \equiv \frac{\int_{y_1}^{y_2} yw(y)\mathrm{d}y}{\int_{y_1}^{y_2} w(y)\mathrm{d}y}. \qquad (9.12)$$

Since the total force must equal zero, we see that at a given z, $\bar{y} = 0$, meaning that the neutral surface is located at the y position of the centroid of the cross-section.

Now we need to find the total torque due to that elastic response. For convenience, we pick our axis of interest for the torque (or *moment*) to run in and out of the page along $y = 0$ (the *neutral axis*) at the right-hand face. The moment is then

$$M(z) = \int_{y_1}^{y_2} y \cdot \sigma_{zz}w(y)\mathrm{d}y = \int_{y_1}^{y_2} y \cdot (ky)w(y)\mathrm{d}y. \qquad (9.13)$$

Now the right-hand side is proportional to the *areal moment of inertia* about the neutral axis. The areal moment of inertia is

$$I_{\mathrm{a}} \equiv \int_{y_1}^{y_2} (y - (\bar{y}))^2 w(y)\mathrm{d}y. \qquad (9.14)$$

The larger the areal moment of inertia, the larger the restoring moment for a given amount of flexing.

Physically, what does this mean? When a beam or member is flexed about an axis perpendicular to its length, it develops internal stresses to resist that bending. On one side of the neutral axis, those stresses tend to resist squeezing, while on the other side of the neutral axis, they resist stretching. Looking at patches of area $(w(y)\mathrm{d}y)$, one can find the force contributed by each patch. Patches farther away in y from the neutral axis contribute both more force (because the strain and therefore the responding stress are larger in magnitude), and even more torque (because they have an additional moment arm). This means that the cross-sections that best resist bending are those that have most of their area away from the neutral axis. This is the reason an I-beam is the preferred shape for girders; an I-beam is stiffer against (appropriately directed) bending than a simple rectangular beam with the same cross-sectional area. Notice, too, that the largest stresses in the beam at a given section are at the locations farthest from the neutral axis.

We now have enough information to take a mathematical approach at computing the shape of a beam subject to various loads. If our little section of beam bends through an angle $\delta\theta$ along its length, then the compressive strain at some location y away from the neutral axis is $(y\delta\theta)/\delta z$. Then $\sigma_{zz}(y) = Y(y\delta\theta)/\delta z$. Plugging in to our expression for the bending moment,

$$M(z) = \int_{y_1}^{y_2} y \cdot Y\left(y\frac{\partial\theta}{\partial z}\right)w(y)\mathrm{d}y = Y\frac{\partial\theta}{\partial z}I_{\mathrm{a}}. \qquad (9.15)$$

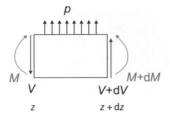

A segment of beam subjected to a variety of loads. Torques about an axis in the x-direction are shown as appropriately **Fig. 9.5** directed bending moments M and $M + dM$; shear forces are also applied at both ends, V and $V + dV$ in magnitude; and finally a transverse load of p per unit length. If the piece of beam is not accelerating or changing angular velocity, then torques and forces must balance.

As expected, a bigger I_a implies that a given bending torque produces a smaller rate of angular deflection. If $u_y(z)$ is the y-displacement of the beam from its unloaded condition, then $\theta(z) \approx \partial u_y / \partial z$. This implies

$$M(z) = YI_a(z) \frac{\partial^2 u_y}{\partial z^2}, \tag{9.16}$$

the Bernoulli–Euler equation.

We can generalize this. Look at the small segment of beam in Fig. 9.5, of length dz. It is subjected to the indicated bending moments at the two ends, as well as shear forces (V and $V + dV$, respectively) and a transverse load, p Newtons per unit length. Balancing forces and torques to first order in all the differential quantities gives

$$V dz = -dM \rightarrow \frac{\partial M}{\partial z} = -V,$$

$$dV = -p dz \rightarrow \frac{\partial V}{\partial z} = -p. \tag{9.17}$$

Differentiating and plugging in, we get a fourth-order differential equation for the deflection of the beam:

$$\frac{\partial^2 M}{\partial z^2} = p \rightarrow \frac{\partial^2}{\partial z^2}\left(YI_a \frac{\partial^2 u_y}{\partial z^2}\right) = p(z). \tag{9.18}$$

The product YI_a shows up all the time in elasticity analyses and is called the *flexural rigidity* of the beam.

To find the actual shape of a loaded beam, we would need to solve Eq. (9.18) with appropriate boundary conditions on u_y at the ends of the beam. For a horizontal clamped end of a beam, $u_y = 0$ as well as the first derivative with respect to z, $u_y' = 0$. For a free end, there are no bending moments ($u_y'' = 0$) and no shearing loads ($u_y''' = 0$). A hinged end has no displacement ($u_y = 0$) and no bending moment ($u_y'' = 0$).

9.1.2 Torsion

We can do an analogous analysis for the case of torsion rather than bending. Figure 9.6 shows a little segment of a beam extended in the z-direction. Consider twisting the beam

Fig. 9.6 A segment of beam under torsion.

about the z-axis, and for now worry only about small rates of twist, so that cross-sections in the pre-twist $x-y$ plane remain planar after the twisting. The material develops shear stresses to resist the twising motion.

Suppose the added twist between the right and left surfaces of the element shown is $\delta\phi$. The shear strain at some position x, y is linearly proportional to that element's radial position $r = \sqrt{(x^2 + y^2)}$. The farther out from the twist axis, the greater the displacement in the circumferential direction. This shear strain results in the generation of a restoring shear stress. Quantitatively, at a radius r, the shear strain is $r\delta\phi/\delta z$. Summing up the contributions to the total restoring *torque* about the z-axis from such patches of area dA,

$$T(z) = \int r \cdot \gamma \frac{r\delta\phi}{\delta z} dA = \gamma \frac{\partial\phi}{\partial z} \int r^2 dA, \tag{9.19}$$

where we have approximated the derivative, inserted the shear modulus appropriately, and the integral is over the cross-section of the beam at constant z. In analogy with the bending case, the largest stresses on a given cross-section occur at the locations farthest radially from the twist axis.

That last integral is the *polar areal moment of inertia*,

$$J_a \equiv \int r^2 dA. \tag{9.20}$$

For a circular rod of diameter d, $J_a = \pi d^4/32$, and for a thin tube of radius a and wall thickness t, $J_a \approx 2\pi a^3 t$.

The usual boundary condition for a torsion member is to have the initial twist be zero. Saying that the beam starts at $z = 0$ and has a length L, if we apply a twisting torque T to the end of the beam,

$$\phi(z = L) = \int (T(z)/\gamma J(z)) dz. \tag{9.21}$$

For a uniform cross-section beam this reduces to $\phi(z = L) = TL/\gamma J$.

For beams with noncircular cross-sections, the situation is more complicated because of the boundary conditions that must be satisfied at, e.g., corners. For rectangular crosssections, conformal mapping has been used to solve this, with the maximum shear stress $\sigma_{xy,\text{max}}$ and maximum twist angle ϕ defined by

$$\sigma_{xy,\text{max}} = \frac{T}{\alpha b c^2}, \tag{9.22}$$

Table 9.1 Parameters for torsional response of a rectangular cross-section beam with sides of length b and c, as described in Eqs. (9.22) and (9.23).

b/c	1.00	1.50	2.00	3.00	6.00	10.0	∞
α	0.208	0.231	0.246	0.267	0.299	0.312	0.333
β	0.141	0.196	0.229	0.263	0.299	0.312	0.333

$$\phi = \frac{TL}{\beta bc^3 \gamma},\qquad(9.23)$$

where the beam side lengths are b and c, and the parameters α and β are defined in Table 9.1.

Common sense still holds. The maximum shear stress still occurs far from the central torsion axis; longer torsion members twist more under a given applied torque, as do skinnier members and those with smaller shear moduli.

9.1.3 Oscillations

Often we are interested in understanding the dynamic response of a system, not just its static deformation or displacement in response to an external load. As when we examine classical dynamics at the undergraduate level, there are two approaches. We can directly apply Newton's second law to each piece of a solid, finding the forces acting on each section of material and solving $F = ma$. Alternatively, a more generalizable method is to use the Lagrangian approach and variational calculus, allowing us to find the equations of motion for the generalized coordinates that describe the solid. That is the approach we take here [731].

The Lagrangian here is the difference between the kinetic and potential energies. In a continuous elastic solid it is simplest to express these quantities in terms of the kinetic and potential energy *densities*. Consider a beam along the z-axis extending from $z = 0$ to $z = L$, with our parameter of interest being the local transverse displacement of the beam, $u_y(z, t)$. Assume that the beam has mass density ρ, elastic modulus Y, cross-sectional area A, and areal moment of inertia I. This kind of flexing beam problem is a common one in micro- and nanomechanics, as we shall see.

The kinetic energy per unit volume of a little piece of beam of length dz is just $(1/2)\rho A dz(\dot{u}_y^2)$. Similarly, the elastic potential energy per unit volume results from a bending deflection, a nonzero curvature $\partial^2 u_y/\partial z^2$. The potential energy for a similar little piece of beam is $(1/2)YI(\partial^2 u_y/\partial z^2)^2 dz$. Putting these together, the full Lagrangian is

$$\mathcal{L}[u_y(z, t)] = \int_0^L \left(\left(\frac{1}{2}\rho A \dot{u}_y^2 \right) - \left(\frac{1}{2}YI\left(\frac{\partial^2 u_y}{\partial z^2} \right)^2 \right) \right) dz.\qquad(9.24)$$

The action $\mathcal{S}$ is the time integral of this Lagrangian from some time t_1 to t_2, and $\mathcal{S}$ is a functional that depends on the position and time dependence of the lateral displacement,

$u_y(z,t)$. The Hamilton–Lagrange approach to dynamics tells us that the classical solution $u_y(z,t)$ is the function of position and time that extremizes S if we consider small variations $\delta u_y(z,t)$ that vanish at the initial and final times. Because we are often interested in driven motion, we further consider the end at $z = L$ to be "free" and subjected to some external driving force $F(t)$, though more general driving terms are possible.[1]

The expression for the variation in the action is

$$\delta S = \delta \int_{t_1}^{t_2} \mathcal{L} dt$$

$$= \int_{t_1}^{t_2} \left(\delta \int_0^L \left(\frac{1}{2}\rho A (\dot{u}_y)^2 \right) - \left(\frac{1}{2} YI \left(\frac{\partial^2 u_y}{\partial z^2} \right)^2 \right) \right) dz \, dt + F(t)\delta u_y(L,t). \quad (9.25)$$

To arrive at equations of motion, we can use standard methods of the calculus of variation, in which we treat $u_y(z,t)$ and $\dot{u}_y(z,t)$ as independent. We can then interchange the order of integration, and assuming that $\delta S = 0$ we can arrive at expressions that must hold true. These turn out to be the equations of motion and boundary conditions for $u_y(z,t)$. In this case, we find:

$$\rho A \frac{\partial^2 u_y}{\partial t^2} + YI \frac{\partial^4 u_y}{\partial z^4} = 0, \quad 0 < z < L, \quad (9.26)$$

$$YI \frac{\partial^3 u_y}{\partial z^3}\big|_{z=L} + F(t) = 0, \quad \frac{\partial^2 u_y}{\partial z^2}\big|_{z=L} = 0. \quad (9.27)$$

We also need to apply the appropriate boundary conditions at $z = 0$, depending on whether the beam is rigidly anchored, hinged, or free.

Equation (9.26) is a fourth-order wave equation that has oscillatory solutions as a function of time. As in other wave equations we have seen, it is the imposition of boundary conditions that forces a particular discrete spectrum of allowed modes. For the undriven case, we can write separate variables and write the solution in the form

$$u_y(z,t) = \exp(-i\Omega t) \times (B_1 \sin(\alpha z \sqrt{\Omega}) + B_2 \cos(\alpha z \sqrt{\Omega})$$
$$+ B_3 \sinh(\alpha z \sqrt{\Omega}) + B_4 \cosh(\alpha z \sqrt{\Omega})), \quad (9.28)$$

where we are using the parameter $\alpha \equiv (\rho A/YI)^{1/4}$. The four coefficients, B_1-B_4, are essentially the integration constants associated with solving a spatially fourth-order partial differential equation. For a particular mode, we can define a spatial periodicity $k_n \equiv \alpha \sqrt{\Omega_n}$, so that the argument of each trig or hyperbolic trig function is then $k_n z$.

One very common scenario in micro- and nanomechanics is the *rigidly cantilevered beam*. The $z = 0$ boundary conditions are simple: $u_y(0,t) = 0$, since the fixed end of the beam can never move; and $\partial u_y/\partial z = 0$ at $z = 0$ for all t, as in the static case. These boundary conditions make clear immediately that $B_2 = -B_4$, and $B_1 = -B_3$.

At $z = L$, there can be no bending moment at the free end ($\to \partial^2 u_y/\partial z^2 = 0$ at $z = L$), and there is no shear force at the free end ($\to \partial^3 u_y/\partial z^3 = 0$ at $z = L$). Together these imply

---

[1] An analogous treatment of longitudinal or torsional deformation is possible, provided that one uses the appropriate expressions for potential and kinetic energy densities.

$$\cos(k_n L)\cosh(k_n L) = -1. \tag{9.29}$$

Equation (9.29) may be solved numerically to find the allowed values, $k_n L = 1.875, 4.694, 7.855, 10.996, 14.137, \ldots$. The actual frequencies are then

$$\Omega_n = \sqrt{\frac{YI}{\rho A}} k_n^2. \tag{9.30}$$

These modes are higher in frequency than what would be found if the $\partial u_y/\partial z$ boundary condition was relaxed at $z = 0$. The derivative boundary condition at the clamped end forces the beam to have more curvature, corresponding to an increased potential energy density relative to the pinned case.

Another very common variant is the *doubly clamped beam*. In that case, the boundary conditions at both ends of the beam are $u_y = 0$ and $\partial u_y/\partial z = 0$ for all t. In this case the boundary condition at $z = L$ now gives

$$\cos(k_n L)\cosh(k_n L) = 1, \tag{9.31}$$

which picks out $k_n L = 4.730, 7.853, 10.996, 14.132, \ldots$. The relationship in Eq. (9.30) still applies. As above, these frequencies are higher than the modes for, e.g., a guitar string, which is pinned rather than clamped at the two ends.

The transverse motion of an anchored elastic beam then supports a discrete set of normal modes. For more complicated material shapes such as membranes, the allowed normal modes depend in detail on the geometry of the elastic medium. If these mechanical structures are excited by an external driving force at some frequency ω, from our undergraduate experience with mechanical resonators we expect a large amplitude response (and a shift in the phase of the motion relative to the drive) when $\omega \to \Omega_n$.

The resonant response is limited by *damping* or *loss*, physics that we have neglected in our treatment thus far. We quantify the amount of damping with the quality factor, Q, of the resonance, defined in the usual way as $Q \equiv 2\pi \times$(total energy in oscillator)/(energy lost per cycle of oscillation). In the limit of weak damping ($Q \gg 1$), the resonant frequency is very close to the ideal normal mode frequency. In a classical harmonic oscillator, damping is usually modeled through a force proportional to the instantaneous velocity of the oscillator.

The detailed expression for the damping in real mechanical resonators depends on the source of the dissipation. The elastic material itself can have intrinsic losses; in the usual complex notation these would appear as an imaginary contribution to Y. In the nanoscale limit there can be other intrinsic contributions to damping, as we shall see in Section 9.4.1. These intrinsic effects are best observed when the oscillator is in a high vacuum environment.

Damping frequently results from interactions between the oscillator and its surroundings. When the oscillator interacts directly with a gas, viscous drag analogous to the "air resistance" modeled in simple undergraduate mechanics can provide a damping force approximately linear in the local velocity of the oscillator relative to the gas. If the gas is very dilute, so that the mean free path of the gas molecules is comparable to the size of the oscillator, then treating the gas as possessing a bulk viscosity is no longer appropriate, and the damping force is modified accordingly. In the other limit, if a surrounding fluid medium is very dense, then true hydrodynamic effects can modify the damping. This latter case is readily experienced by moving one's open hand up and down near the bottom of a sink full of water. As the hand is raised and lowered, hydrodynamic drag forces are apparent when the water rushes in or is squeezed out of the space between the hand and the sink bottom.

9.1.4 Dislocations and crack propagation

Our discussion of continuous elastic media has so far ignored the atomic nature of solids. However, we need a microscopic picture to understand the elastic properties of materials when they experience irreversible "plastic" deformation. Imagine shearing a block of material until it deforms plastically, so that the top half of the block is permanently displaced horizontally relative to the bottom half, without actually sundering the block in two. While this clearly involves the rearrangement of bonds at the atomic scale, it does not proceed by breaking all the bonds, translating the upper portion of the block, and then reattaching some of the bonds. Instead, in a crystalline material such a deformation proceeds by the propagation of *dislocations*, extended defects in the periodic lattice arrangement of the atoms. There are encyclopedic references on the theory of dislocations [732].

Figure 9.7(a) shows a three-dimensional crystal with a plane of atoms terminating inside the bulk to form an *edge dislocation*. To shear the upper portion of the material to the right by one lattice spacing, one can imagine rearranging the bonds such that the terminated plane becomes complete, and the adjacent plane becomes the terminated plane. This is analogous to moving a carpet a few inches at a time by pushing along a lump rather than trying to slide the entire carpet uniformly at once. The terminating line of atoms is the *dislocation line*. Under shear loading sufficient to rearrange the atomic bonding, an edge dislocation propagates along the direction of the shear force, perpendicular to the dislocation line.

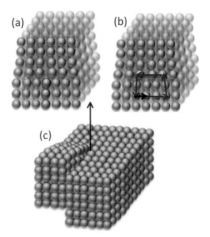

Fig. 9.7   Dislocations. (a) Edge dislocation, highlighting the "extra" atomic plane. (b) Edge dislocation with path enclosing the dislocation line (red) showing the Burgers vector (black). (c) Screw dislocation, with dislocation line indicated by arrow.

In addition to the dislocation line, a dislocation is also parametrized by a *Burgers vector*. Consider a rectangular closed path in some lattice plane in the absence of a dislocation, defined by taking integer steps along the unit lattice vectors in that plane. Now follow the same set of steps, but on a trajectory that encloses a dislocation line passing normal to the plane. Because of the dislocation, the trajectory will no longer close on itself. The vector (in units of lattice basis vectors) that would close the loop is the Burgers vector for that dislocation. As is shown in Fig. 9.7(b), the Burgers vector for an edge dislocation is perpendicular to the dislocation line.

This construction makes it explicit that a dislocation is associated with a strain field. The material just "above" the dislocation line is under compressive strain (lattice parameters slightly smaller than those in the material far from any defects), while just "below" the dislocation line the material is under tensile strain (the lattice parameter is locally increased to accommodate the additional plane of atoms). This dipolar strain field implies a corresponding stress field related through the tensorial elastic response.

Figure 9.7(c) shows a *screw dislocation*. In this case, the Burgers vector is parallel to the dislocation line. When regions on opposite sides of the dislocation line are sheared antiparallel to each other along the dislocation line direction, the dislocation line propagates perpendicular to itself. The classic example of this is the tearing of a phone book. Again, the dislocation line is accompanied by a strain field, though it involves shear strain rather than compressive or tensile strain.

Plastic deformation of crystalline materials involves the propagation of dislocations in response to applied stresses. Since the motion of the dislocations involves the rupture of atomic bonds, this tends to be a dissipative process. Depending on the configuration of dislocations, it is also possible for the number of dislocations within a material to multiply as the material is plastically deformed. One well-known mechanism is the Frank–Read source [733].

The propagation of dislocations under plastic deformation is generally hindered by the interactions between dislocations. A classic example of the consequences of this is *work hardening*, as seen when you bend a paperclip. As-made, a metal paperclip contains some density of dislocations as well as dislocation sources. When plastically deformed by bending, these dislocations propagate and the number of dislocations grows. However, if you try to bend the paperclip back to its original shape, you find that considerably greater stress must be applied. This is the result of interactions between dislocations via their stress fields, and the fact that, depending on the dislocation type, dislocations cannot move "through" each other. The eventual result (seen when a paperclip is bent back and forth repeatedly) is brittle fracture.

Note that when an edge dislocation does propagate all the way to the boundary of the material, the volumes of material above and below the plane in which the dislocation line propagates shift relative to each other by the length of the Burgers vector. Figure 9.8 shows an example of a consequence. As a tensile load is applied to a (polycrystalline) copper rod, the material slips along planes due to the propagation of edge dislocations. The cross-sectional area bearing the load thus shrinks, increasing the local tensile stress, leading to further formation and propagation of dislocations. The local shrinking of the cross-section

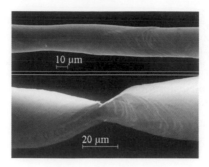

Fig. 9.8   Necking during ductile fracture, with clearly visible signatures of slip planes. Adapted from [734].

is called *necking*, and the time-dependent plastic deformation under constant load is called *creep*.

The increase of local stress due to local reduction of load-bearing area is closely related to the concept of *stress concentrators*. The local increase of stress actually depends on more than just the area over which load is distributed; because of the boundary conditions that must apply at free surfaces of the material, the actual shape of the free surface (e.g., a smooth indentation or a sharp notch) matters. Local stress distributions may be found with finite element techniques (all within the continuum approximation), and engineers have developed a number of analytical and semi-empirical methods for estimating local maximum stresses.

It should not be surprising that these methods collide with active areas of research when considering the propagation of cracks during fracture. At the tip of a propagating crack, elastic energy is sufficiently concentrated to break atomic bonds, and the relevant lengthscales are inherently "nano". An entire discipline (fracture mechanics) is concerned with crack propagation because of its importance in engineering. While much progress has been made (a full discussion of the field is well beyond the scope of this book), nanoscale characterization tools continue to be applied to this fascinating problem.

When material volumes (or grain sizes in polycrystalline systems) approach the lengthscales associated with dislocation formation and propagation, unsurprisingly the plastic deformation properties of materials can deviate from bulk expectations, as shown in Fig. 9.9. One example is superplasticity [735], when fine-grained materials are able to deform plastically to strains much larger than non-nanostructured solids. This phenomenon has been studied for decades, and the advent of nanoscale characterization tools such as TEM and precision x-ray methods are providing valuable new information regarding the competition between different processes (dislocation glide; grain boundary sliding) [736].

A second example is the ability of nanostructured materials such as nanowires to accommodate larger strains before failure than their bulk crystalline counterparts. For example, silicon is of interest for rechargeable battery applications because it can accommodate large amounts of intercalated lithium. When lithiated, the crystal lattice can swell anisotropically with strains exceeding 200%. Bulk silicon will fracture under these conditions, but thanks to their large amount of free surface, nanowires can respond much more reversibly to the distortions associated with lithiation and delithiation, as shown in Fig. 9.10.

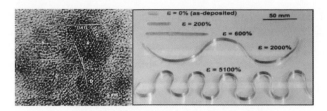

Superplastic deformation in nanocrystalline copper. When grain sizes become small compared to the scale of dislocation propagation and dislocation sources, and a significant fraction of atoms are at grain boundaries, remarkable deformations can be possible. Adapted from [737].

Fig. 9.9

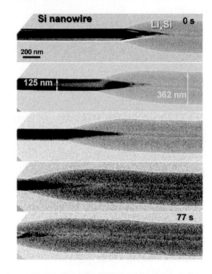

TEM image of a silicon nanowire during electrochemical lithiation. The material can deform dramatically during this process without fracture. Adapted from [738].

Fig. 9.10

9.2 Tribology

Tribology is the study of friction, a topic that is clearly of interest if one wants to construct mechanical devices that involve sliding at solid interfaces. There has been great progress in our understanding of the microscopic origins of friction in recent years, enabled in part by the development of surface science and nanoscale tools. Excellent reviews of this subject are available, ranging from the encyclopedic [739] to concise works including a historical perspective [740, 741, 742].

In first-year undergraduate physics we are taught three basic "laws" that make up a good empirical description of frictional phenomena at the macroscale:

- the frictional force, F_f, acting at the interface between two objects is directly proportional to the normal force, F_N, at that interface, with the *friction coefficient* μ_{stat} (μ_{kin}) defined as the proportionality constant for static (sliding) friction;

- the frictional force is independent of the contact area of the interface;
- in the case of sliding friction, the frictional force is only weakly dependent if at all on the relative velocity of the two surfaces.

In the light of subsequently learned physics knowledge, all three of these laws are rather surprising. For example, from elasticity considerations, it seems like it might make more sense to talk about a frictional shear stress (a force per unit area). The area independence of the frictional force seems very odd, implying that the frictional force acting, e.g., to keep a car on a curved road would be identical regardless of the size of the car's tires. At minimum we would expect some failure of these empirical laws as the contact area approaches the nanometer scale, if not well before.

The limits of these laws are directly related to the microscopic origins of friction. Friction is unique among the forces typically taught at the undergraduate level, in that it appears to violate conservation of (mechanical) energy. When work is done by the friction force, kinetic energy rooted in the motion of macroscopic degrees of freedom is transferred instead to microscopic degrees of freedom, usually the vibrational modes of the large numbers of atoms and molecules that make up the macroscopic bodies. This is the origin of the macroscopic irreversibility that we call heating.

This irreversibility confounded early attempts (e.g., by Coulomb) to treat friction as a consequence of microscopically interlaced surface asperities. As shown in Fig. 9.11, interlocking serrated surfaces subject to a normal force F_N will resist continuous lateral sliding below a threshold horizontal force proportional to F_N. This is qualitatively similar to the case of static friction. However, once sliding begins in this geometry there is no dissipation; any energy that must be put into the system to push the upper object "up hill" comes back out when the object is sliding back down.

While appealing to our macroscopic intuition, the idea that friction originates purely from irregular surfaces faces another serious problem beyond that above. An atomically flat, single-crystal surface brought into contact with a commensurate, mating, atomically flat, single-crystal surface (in ultrahigh vacuum conditions, to avoid contaminating adsorbates) will actually bond very strongly, rather than gliding frictionlessly. This makes physical sense, since the total energy of the combined system would be lowered by forming bonds and eliminating the interface entirely. The lesson to appreciate here is that

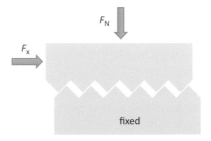

Fig. 9.11   Friction cannot be explained only by a model of interlocking surface asperities. While increasing F_N means a larger F_x is initially required to move the upper block horizontally, there is no dissipation in this system.

surface chemistry considerations play a major role in microscopic frictional properties of interfaces. We return to this idea below.

Surface asperities are important in understanding friction at macroscopic scales. A large group at Cambridge University in the mid-1900s worked on improved modeling of friction to account for this [743]. A central idea, inferred from studies of "hot spots" arising at sliding frictional surfaces, is that the true contact area between macroscopic bodies is usually far smaller than the apparent contact area, because it is difficult to make very flat surfaces. In this picture, the irreversibility and heating of macroscopic friction come about by the shearing away of contact asperities through plastic deformation.

If the true contact area is A, and the shear strength of the material in question is τ_{max}, then the frictional force in this model would be given by $F_f = A\tau_{max}$. However, we still need to understand the dependence on the normal force. Suppose that A is proportional to the normal force, so that tiny contacts squish out into larger contacts as the normal load is increased. The contact pressure, P, would then be constant, leading to a friction coefficient

$$\mu \equiv \frac{F_f}{F_N} = \frac{A\tau_{max}}{F_N} = \frac{\tau_{max}}{P}. \tag{9.32}$$

Various models of elastic and plastic asperities satisfy the assumed relationship between true contact area and normal force [744].

There are problems with this approach, however [740]. A maximum shear stress that can be obtained is $\sim \sigma_{max}/2$, and the maximum pressure that one can expect (well into the plastic regime, to allow for contact deformation) is about $3\sigma_{max}$, implying that the largest friction coefficient should be around 1/6. Larger values of μ are often observed. Moreover, friction can take place without any detectable irreversible damage to the rubbing surfaces. This again brings to the fore the roles of surface chemistry, molecular adhesion, and microscopic degrees of freedom.

9.2.1 Modern techniques: quartz crystal microbalance

The quartz crystal microbalance (QCM) tracks the mechanical resonance properties of a piezoelectric quartz substrate as a function of perturbations, particularly mass loading (see Section 4.2.1). QCM can be prepared with atomically flat and clean surfaces in UHV. QCM methods to examine adsorbed layers and friction are reviewed in detail by Krim [741]. The mechanical interactions between adsorbates and the QCM allow QCM-based studies of friction, typically at low temperatures such that atoms of interest (e.g., noble gases, to avoid strong chemical binding effects) will adsorb readily. Surface scattering methods can be used to determine the coverage of adsorbates. If an adsorbate sticks rigidly to the surface, then the effective mass of the resonator increases and there is a consequential drop in resonant frequency. However, an adsorbate slides frictionlessly on the QCM, there is no increase in the mechanical loading of the resonator. In the intermediate situation, frictional transfer of energy between the QCM and the adsorbate manifests as both a change in frequency and a decreases in Q of the resonator.

One of the most interesting observations [745], looking at solid Xe layers on single-crystal Ag films, gives insight into a mechanism for friction without irreversible,

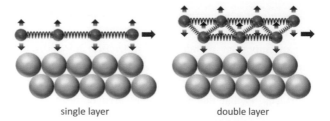

single layer   double layer

Fig. 9.12 Comparison of monolayer vs. bilayer adsorbates. Microscopic frictional interactions are greater in the bilayer case because there are more vibrational degrees of freedom in which to deposit mechanical energy. Adapted from [740].

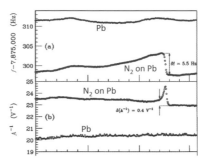

Fig. 9.13 Evidence for an electronic component to friction. Frequency (a) and amplitude (b) of a QCM resonator coated with lead with and without an adsorbed nitrogen film, as a function of temperature. With the nitrogen, the superconducting transition in the lead is clearly visible, indicating that the electrons in the metal contribute to the total friction in the normal state. Adapted from [748].

destructive deformation of materials. In that work, frictional losses are found to be larger for bilayers of Xe than for monolayers, despite nominally identical Xe/Ag interfaces. The explanation [746, 747] is that it is easier to excite phonon modes in the bilayer, with many more microscopic vibrational degrees of freedom, than in the monolayer, as shown schematically in Fig. 9.12. In the microscopic limit, imagine the QCM substrate trying to move relative to the adsorbed overlayer. When the interaction energy due to their relative displacements exceeds some threshold, it can be energetically favorable to excite a phonon in the overlayer (or substrate). In this way, energy can be transferred from the comparatively macroscopic motion of the resonator into the microscopic vibrational degrees of freedom.

When the substrate is an electrical conductor, there is another possible avenue for energy flow. As mentioned in Section 3.2.1, the electrons in a bulk metal are gapless, meaning they can support excitations with energy costs approaching zero. Some very elegant QCM experiments have demonstrated that there can be an electronic contribution to friction as well, with macroscopic mechanical energy transferred into exciting electrons above the Fermi sea. In this work [748], a QCM resonator is coated with a film of a metal that can superconduct, and frictional response of an adsorbed overlayer is examined as the temperature is cycled in and out of the superconducting regime. In the superconducting state (see Section 3.2.3), low energy electronic excitations are "gapped out" by the onset of superconductivity. As a result, there is a reduction in friction when the substrate becomes superconducting, even when the overlayer is electrically insulating. Recent scanned probe measurements [749] have confirmed this phenomenon.

The scientific bottom line here: non-destructive friction can arise, even in the absence of strong chemical interactions between surfaces, when pathways exist to transfer energy from macroscopic motion into low energy excitations. Correspondingly, at the nanoscale, when the number of degrees of freedom is reduced, the prospect exists for reducing these contributions to frictional interactions.

9.2.2 Modern techniques: scanned probe microscopy

Atomic force microscopy relies on the measurement of local interactions between a tip and a surface, while providing nanometer-scale imaging of the topography. Lateral force microscopy makes use of the twist of the tip due to the transverse force from the sample as the probe is scanned horizontally across a sample surface. In the absence of effective frictional interaction, this lateral force should be zero. One can also perform the equivalent of lateral force microscopy using a scanning tunneling microscope tip attached to a tuning-Fork-style mounting. STM can also be used to manipulate objects, sometimes while measuring the forces required for manipulation. Nanomanipulation, imaging, and the preparation of well-characterized, reproducible tips enable scanned probe methods to characterize frictional interactions as a function of relative crystallographic orientation and direction of relative motion.

The importance of crystallographic commensurability is clearly shown in some scanned probe manipulation experiments. Figure 9.14(a) shows the lateral force required to push a multiwalled carbon nanotube into rolling across a clean graphite surface. The force varies in a complicated way depending on which part of the irregular MWNT is in contact with the graphite, but this pattern neatly retraces itself with each revolution. The frictional force is not random, but is a fingerprint of the detailed short-range interactions at the interface,

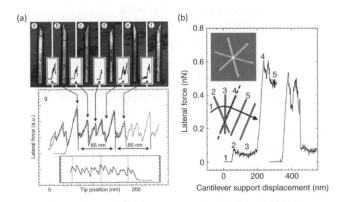

Friction in UHV depends on chemical interactions. (a) A MWNT rolling on graphite has frictional interactions with the substrate that are complicated yet reproducibly determined by the contact interaction between surfaces. Adapted from [750]. (b) Frictional interactions between a smooth MWNT and a graphite surface are enhanced in orientations such that there is commensurability between the tube surface and the underlying graphene lattice. Adapted from [751].

Fig. 9.14

determined by the material structure. Figure 9.14(b) makes this even more explicit. When a smooth MWNT is pivoted on the graphite surface, there are three specific orientations of the MWNT with greatly increased static friction. These orientations, offset from one another by 120°, are those for which the graphene-like outermost layer of the MWNT is in orientational registry as if it was an additional layer of the underlying graphite.

Crystallographic commensurability leads to increased static friction, and vice versa. Scanned probe force measurements have supplied further evidence of this. On quasicrystal surfaces, along particular directions the atoms are arranged with a spatial periodicity as in a conventional crystal. Along other directions, the atomic arrrangement is aperiodic. A specially prepared single-crystal tungsten STM tip has been used [752] to compare frictional interactions for motion along the periodic direction (when there is commensurability between the surface atom spacing and that of the tip) and the aperiodic direction. Much lower friction is found in the aperiodic/incommensurate scan direction. When the surface is allowed to become disordered through oxidation, the difference between scan directions goes away.

In the limit of complete incommensurability and chemical cleanliness, *superlubricity*, the complete absence of frictional interactions, is possible. This has also been demonstrated using single-crystal scanned probe tips in UHV conditions [753].

9.3   Microelectromechanical systems (MEMS)

The motivation for this discussion of solid mechanics and the origins of frictional forces is the rise of micro- and nanoelectromechanical systems (MEMS and NEMS), true mechanical devices intended to operate with moving parts far below the scale of traditional machines. Despite a very limited portfolio of design motifs, MEMS devices now constitute a multibillion dollar annual market, with applications including consumer electronics, automotive systems, telecommunications, and sensing.

9.3.1   Fabrication

The fabrication of MEMS has largely been based on the planar techniques of semiconductor manufacturing, as described in Chapter 4. Advantages of this approach include massively parallel and uniform patterning, leveraging highly developed etch and deposition processscs; and easy interfacing with on-chip control circuitry. The main disadvantages are a loss of design flexibility due to the planar geometry, and a comparatively limited set of working materials. Much of what we think of as machinery (gears, motors) requires rotational freedom and complicated motion in three dimensions, difficult to achieve from top-down, planar processing. As a result, many MEMS/NEMS devices are very simple mechanically, often consisting of cantilevered and clamped beams, with motion resulting from flexure or torsion.

As with semiconductor device manufacturing, MEMS fabrication involves patterning (generally through lithography, as in Section 4.4), additive processes, and subtractive

Polysilicon cantilevers, suspended by wet etching of an underlying oxide layer. Some of the cantilevers remain stuck   Fig. 9.15
down on the substrate due to surface tension effects while drying. Adapted from [754].

processes. The additive steps include physical vapor deposition, sputtering, PECVD, and electrodeposition (for metals). Because most relevant mechanical action requires parts free to move, subtractive processes are extremely important.

Wet etching is commonly used, with chemical selectivity and material interfaces limiting the region of material removal. An example is shown in Fig. 9.15, where a polysilicon beam is cantilevered by the removal (using buffered HF) of underlying SiO_2. In practice, *surface tension* is a major challenge in using wet etches to free mechanically fragile structures. We will discuss surface tension further in Chapter 10; for now, it is enough to know that surface tension results from the mutual attraction of molecules in the liquid state, which tends to make it energetically favorable for liquids to deform to minimize the amount of liquid–vapor interface. When removing the etchant or a rinsing liquid, the passage of the liquid–vapor interface across the structure can exert forces large enough to displace or deform parts. At the interface with air, water has a surface tension of 7.2×10^2 N/m. An example in the exercises shows that the resulting forces in a realistic geometry can lead to very large displacements of parts.

Supercritical drying avoids these surface tension forces. In a controlled volume, the liquid used in the etching/rinsing process is displaced by a liquid with a readily accessible critical point. For example, the liquid–vapor transition line for CO_2 terminates at a critical point at $P_c = 7.4$ MPa and $T_c = 31.1°C$. Once the rinsing liquid is displaced by liquid CO_2 (stabilized near ambient temperature by increased pressure), the pressure is further elevated above P_c. The system is then steered around the critical point by increasing the temperature above T_c, dropping the pressure back to one atmosphere, and then lowering the temperature back to ambient conditions. By traversing the supercritical part of the phase diagram, no liquid–vapor interface exists to exert surface tension forces. This approach is routinely taken in the "release" step of MEMS processing, as well as in the preparation of biological samples for electron microscopy.

Dry etching with RIE is also commonly used in MEMS fabrication, thanks to its chemical selectivity and ability to create nearly vertical sidewalls. A variation of RIE popular in both MEMS and microfluidics fabrication is deep RIE, colloquially known as the Bosch process (see Fig. 4.16), that can leave vertical sidewalls while etching silicon at high rates and thicknesses of hundreds of microns.

500 nm

Fig. 9.16   Supercritical drying. Ultrathin wet-etched Si structures dried by conventional blowing of nitrogen gas (upper panel) and supercritical CO_2 to avoid surface tension (lower panel). Adapted from [755].

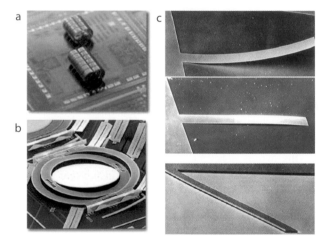

Fig. 9.17   Effects of residual stress in suspended structures. (a) Electroplated Cu inductors designed to curl once freed from the substrate [756]. (b) A gimble mount for a MEMS mirror, lifted by springs [757]. (c) Polysilicon cantilevers made under varying conditions to tune residual stress (courtesy K. Pister, UC Berkeley).

Control of material stresses is also exceptionally important in MEMS fabrication. Everyone who has seen peeling paint knows what can happen when a thin film under tensile stress can become mechanically decoupled from a substrate. Tensile film stress may be used to a design advantage, as shown in Fig. 9.17(a), where electrodeposited copper structures pull away from the substrate to avoid parasitic losses, or Fig. 9.17(b), where previously tensile-stressed members contract, lifting a gimballed Si mirror above the substrate for increased range of motion. By controlling deposition conditions in PECVD, film stresses in some materials may be tailored to allow upward, downward, or zero curvature (Fig. 9.17(c)). This last image reminds us that physical intuition based on macroscopic objects can be misleading at small scales. A macroscopic cantilevered beam with such proportions seems like it should sag under its own weight. However, a quick calculation (see exercises) shows that the stiffness is more than sufficient to avoid visible gravity-induced deflections at the nanoscale.

9.3.2 Driving

Mechanically actuating a MEMS device may be done in a variety of ways. *Inertial drive* involves accelerating the MEMS substrate. In an accelerating reference frame comoving with the substrate, there is an apparent force proportional to the mass of the free-to-move MEMS component and the acceleration. This is the method whereby most AFM cantilevers are excited. A piezo-actuator shakes the cantilever's mount at a frequency near the resonance for the cantilever bending mode. In a sense, this kind of inertial loading is the same mechanism behind MEMS-based accelerometers, as we discuss below. With this approach it is not typically possible to actuate multiple devices independently on a single substrate.

Some kind of electromechanical actuation is more common. While direct integration of piezoelectrics is possible, many schemes avoid using non-CMOS compatible materials because of the inherent fabrication challenges. Direct electrostatic drive is one possibility, using the force between capacitor plates. Consider a resonator with an integrated electrode, capacitively coupled to a fixed electrode, with a total bias $V = V_{DC} + V_{AC} \exp(-i\omega t)$ applied between the plates. At a given bias, for a displacement of the resonator along some coordinate z, the force is going to be $F = -dU/dz$, where U is the energy stored in the capacitor, $1/2 CV^2$. The result is an electrostatic attractive force

$$F \propto \frac{dC}{dz}V^2 \approx \frac{dC}{dz}V_{DC}^2 + 2\frac{dC}{dz}V_{DC}V_{AC}\exp(-i\omega t) + \cdots, \qquad (9.33)$$

where we have omitted terms proportional to V_{AC}^2. Static displacements can be produced by V_{DC}, and the oscillating force at the drive frequency is also tuned by V_{DC}.

For simple vacuum gap parallel plate capacitors, $C = A\epsilon_0/d(z)$, where A is the area of the plates and $d \equiv d_0 - z$, where d is the gap spacing with z the displacement from some initial value d_0. In this case $dC/dz = -A\epsilon_0/d^2$, meaning that the force is strongly dependent on the displacement, diverging as $z \to d_0$.

This nonlinearity can be problematic, even well before $z \to d_0$. The "pull-in" or "snap-down" instability is a challenge of electrostatic drive most easily illustrated with this parallel plate toy model. If the movable plate is anchored by an effective spring constant k, then the equilibrium displacement under a certain bias V is found by balancing the elastic restoring force and the electrostatic attraction between the plates. At a given displacement z,

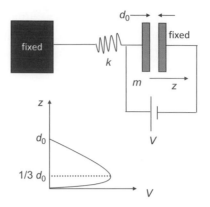

Fig. 9.18 Voltage-driven snapdown in capacitor structures. If z exceeds a critical value $z*$, then reduction of V can result in the mobile capacitor plate snapping down ($z \rightarrow d_0$) instead of the return of z to zero.

$$V = \sqrt{\frac{2kz}{\epsilon_0 A}}(d_0 - z). \qquad (9.34)$$

For a given voltage, Eq. (9.34) may be satisfied by two different displacements. As plotted in Fig. 9.18, $dV/dz = 0$ at a particular displacement, $z^* = d_0/3$, corresponding to a critical voltage $V^* = \sqrt{8kd_0^3/27\epsilon_0 A}$. Consider increasing V slowly from $V = 0$, with the initial $z = 0$. As long as $z < z*$, decreasing V will allow the spring to reduce z. However, once $z = z^*$, upon reduction of V the system can just as well end up on the other branch of solutions, with $z \rightarrow d_0$ unavoidably.

Structures that do snap down can remain stuck, permanently. Short-range forces (van der Waals interactions, adhesion due to adsorbates) can exceed the spring's restoring force even when V is reduced to zero. Note that no relative biasing of the two electrodes will produce a repulsive electrostatic force to push them apart. To mitigate snap-down concerns, movement range may be restricted and active feedback (on-chip, to avoid parasitic capacitances and delays) can adjust V.

To avoid this, many MEMS systems employ a *comb drive* geometry, as shown in Fig. 9.19. Here $C \approx 2N\epsilon_0 Lt/d$, where N is the number of "teeth", L is length of the overlapping region of adjacent teeth, d the spacing between adjacent teeth edges, and t the thickness of the combs. Since displacement is a change in $L = L_0 + z$, the capacitive force is proportional to $dC/dz = 2N\epsilon_0 t/d$, which is conveniently independent of z. The force no longer depends on the displacement over the range of travel of the comb.

Magnetic forces and torques may also be employed in driving MEMS devices. The most common mechanism is magnetomotive, taking advantage of the Ampère's law force on a current-carrying wire in a static magnetic field. For example, a thin-film metal wire may be patterned lithographically on top of a suspended, doubly-clamped beam, with contacts on either side of the anchor points. In a static magnetic field normal to the plane of the device, an AC current in the wire will exert a lateral force on the beam. The need for an external magnetic field and for a continuous current path are restrictive, but not insurmountable.

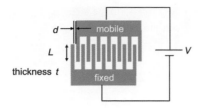

Comb drive approach to electrostatic actuation.

Fig. 9.19

To get a sense of scale, consider an Si beam 50 nm thick, 50 nm wide, and 600 nm in length, anchored at both ends. From Eqs. (9.30) and (9.31), plugging in material properties for Si, the lowest order transverse mode has a frequency $\nu_1 = 1.18$ GHz. The effective spring constant can be found by imagining applying a transverse force localized to the middle of the beam and computing the lateral displacement of that spot. Here $k = Ywt^3/4L^3 = 19.6$ N/m. Suppose a current I is passed through this beam in the presence of a perpendicular magnetic induction $B = 1$ T, comparable to that at the surface of a strong ferromagnet. A reasonable current would be 10 μA, giving a total lateral force acting on the wire of $ILB \sim 6$ pN. If all of this load were acting at the beam midpoint, it would produce a lateral displacement on the order of 300 fm, about 0.1% of an atomic diameter. Detecting small displacements is clearly of interest.

9.3.3 Detection

As we shall see, the displacements involved in MEMS/NEMS devices are often exceedingly tiny, with the average displacement in advanced devices sometimes considerably smaller than an atomic diameter. Fortunately, transducing such infinitesmal motions into detectable signals can be done through multiple methods of varying difficulty and sophistication.

Some transduction methods have been discussed in the context of scanned probe microscopy (see Chapter 4). Tunneling current is exponentially sensitive to relative displacements between conductors at sufficiently small distances. Deflection of a reflected optical beam and piezoresistive response have both been employed to detect cantilever deflections in atomic force microscopy.

Capacitive response is a common means of displacement sensing in MEMS devices. Consider a fixed electrode and a capacitively coupled counterelectrode integrated with the moving part of the MEMS device. If the resulting capacitor is charged through a sufficiently large resistor that the RC time greatly exceeds the period of the interesting mechanical motion, then the charge on the capacitor plates is approximately fixed. Since $Q = CV$, $\delta Q = C\delta V + V\delta C = 0$. For simple parallel plates with spacing d, $\delta V/V = \delta d/d$. In clean, controlled measurement systems, $\delta V \sim 10^{-6}$ V is readily detected, implying that at biases of a few volts it is possible to discern motion on the sub-Ångstrom scale. A disadvantage of this approach for on-chip sensing and control is the need to integrate possibly large resistors. Direct measurement of capacitance is also possible, with AC bridge

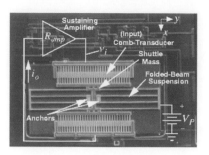

Fig. 9.20   MEMS resonator with integrated drive electronics. Adapted from [758].

methods readily giving $\delta C/C \sim 10^{-6}$ resolution. Again, for reasonable geometries this gives atomic-scale displacement resolution (averaged over an electrode surface).

Alternatively, consider the situation when a fixed DC voltage is applied to the capacitor. Now changing the capacitance through the motion of some coordinate z leads to a current,

$$i = V_{DC}\frac{dC}{dz}\frac{dz}{dt}. \tag{9.35}$$

This is particularly advantageous if the MEMS device is a oscillator and operation involves the system ringing on resonance. For a general driven, damped harmonic oscillator on resonance, the driving force is $\pi/2$ out of phase with the displacement. If $z = z_0 \exp(i\omega t)$, then i in Eq. (9.35) is automatically shifted by $\pi/2$ relative to the displacement. On-chip electronics can produce a drive signal proportional to this current, automatically creating a phase-locked loop for resonant excitation of the oscillator. Figure 9.20 shows such a mechanical resonator with integrated drive electronics.

Magnetomotive detection can also be used. In the earlier example of magnetomotive drive, we estimated a lateral displacement of a clamped, suspended beam on the order of 3×10^{-13} m. However, if that beam was oscillating at its resonant frequency (~ 1 GHz), the maximum lateral *speed* would be around 0.3 mm/s. This rather macroscopic velocity is key to magnetomotive detection. A wire attached to the beam (even the same wire carrying the magnetomotive drive current) moving perpendicular to magnetic induction produces an electromotive voltage along its length as it cuts magnetic flux. That voltage is proportional to the wire length L, the induction B, and the effective speed of the wire, $\dot{z}$: $V_{EMF}(t) \propto LB\dot{z}(t)$.

In either electrical or magnetomotive drive, and with an electrically based detection scheme, it is generally possible to consider an effective drive current I that produces, through the mechanical motion of the device, a displacement sensed through a voltage V. For alternating currents, the displacement and hence the detected voltage can have some phase with respect to the drive that depends on the drive frequency and the mechanical response of the device. This means that an electrically driven MEMS devices can be modeled, electrically, by a complex impedance, $Z(\omega)$. This may be written in terms of an equivalent circuit, extremely helpful for analysis of certain applications, as we shall see below.

9.3.4　Applications

While small compared to the market for now-ubiquitous semiconductor micro/nanoelectronics, the MEMS marketplace is large (tens of billions of dollars per year) and strikingly diverse [759]. The most widespread applications are inertial sensing (accelerations and rotations), pressure sensing, and optical/communications devices, as well as inkjet printer heads.

Inertial sensing

Inertial sensing is one of the best-known applications of MEMS. Have a test mass suspended vertically or held laterally attached to a substrate by micromachined members that act as springs. When the substrate is accelerated, in the accelerating frame of reference the test mass experiences inertial forces and torques, resulting in displacements relative to the substrate that may be sensed. MEMS accelerometers are a mainstay of automotive technology, acting as the triggering sensor for deployment of airbags. These sensors leverage the best aspects of MEMS: high precision sensing, cheap mass production, and good reproducibility. One challenge is that the overall size of devices and the comparatively low density of Si leads to low inertial masses (and hence small displacements).

While the accelerometers associated with airbags must have a very stiff response, MEMS inertial sensors readily able to detect much lower accelerations (fractions of a g) have become nearly ubiquitous in the mobile communications market. Tablet devices and mobile phones that alter their display direction based on the device orientation with respect to gravity are all using MEMS accelerometers. Similar sensors in laptop computers are able to detect freefall and park the read head of the hard drive before the device hits the ground [760]. Likewise, the Nintendo Wii videogame console introduced in 2006 relied on three-axis MEMS accelerometers in its controllers.

MEMS rotation sensors have also been developed [761]. These are colloquially known as gyroscopes, though they do not function as true gyroscopes, which rely a rapidly rotating structure of significant moment of inertia to define an angular moment. MEMS gyroscopes are true inertial sensors because they produce a signal due to the inertial forces associated with being in a rotating (and therefore accelerating) frame of reference.

The Coriolis effect is most relevant here. In a frame of reference that rotates (with respect to "the fixed stars") with an angular velocity Ω, a particle moving at velocity $\mathbf{v}$ experiences an apparent acceleration $\mathbf{a}_{Cor} = 2\mathbf{v} \times \Omega$. This effect originates from a moving object's tendency to continue straight-line motion in a nonaccelerating frame of reference. From the rotating frame, this inertial tendency looks like a transverse acceleration. The saving grace for MEMS devices is that velocities can actually lead to detectable Coriolis accelerations even when displacements are very small. Consider the tuning fork structure shown in Fig. 9.21(a). If the resonant frequency of the tuning fork is 100 MHz and the amplitude of the fork arm motion at the tips is 0.1 nm, that implies a peak velocity of 1 cm/s. In a frame rotating about the z-axis as shown, the Coriolis acceleration would push the tips along the x-axis direction, out of the plane of the tuning fork. This lateral displacement would then be sensed.

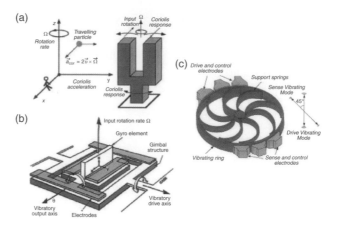

Fig. 9.21 MEMS rotation sensors based on Coriolis forces. (a) A tuning fork design. (b) A coupled torsional resonator approach, trying to gain additional sensitivity due to resonant motion. (c) A flexural ring design, so that Coriolis accelerations excite an out-of-plane mode. Adapted from [761].

Other schemes use resonator designs to increase the magnitude of the Coriolis displacement. Figure 9.21(b) shows a coupled torsional resonator design. Resonance of the outer torsional structure gives periodic motion. In the presence of rotation about the vertical axis shown, this would lead to a periodic Coriolis force acting to drive the inner torsional resonator (designed to be on resonance).

More complex geometries are possible involving flexural rings, as in Fig. 9.21(c). In-plane resonance modes of the ring leads to out-of-plane deformation due to Coriolis forces. With an array of detection electrodes distributed circumferentially, rotation rates about more than one axis may be inferred.

Optics and communications

One communications application for MEMS devices is for filters in radio frequency technologies. Reliable mass fabrication of on-chip inductors and capacitors for integrated RLC filters can be very challenging, with small deviations in fabrication parameters leading directly to changes in the filter's frequency response and quality factor. Mechanical resonators are one alternative to purely electronic rf filters. The resonance frequency of a cantilevered or doubly clamped beam is controlled by the dimensions of the structure and the elastic properties of the material, which tend to be extremely reproducible and uniform. Consider sending a broadband high frequency signal into the drive of a mechanical resonator. Thanks to its Q, the resonator picks out only the component at the resonant frequency, and the displacement sensing mechanism acts as the filter's output. A MEMS resonator filter can have a much smaller footprint than a surface-acoustic wave filter (based on launching, propagation, and detection of sound using interdigitated electrodes and a piezoelectric substrate), and requires no special materials.

More general filters are possible as well, through simple mechanical couplings rather than complex circuit designs. Figure 9.22 shows an example [762]. Once again, as we saw

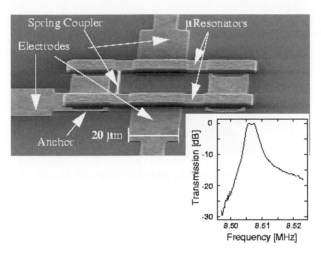

Mechanical bandpass filter made by coupling two narrower-band mechanical resonators. Adapted from [762].   **Fig. 9.22**

in our discussions of atomic hybridization, multiple quantum wells, and plasmonics, we see the power of hybridization. Start with two nominally identical resonators. Couple these resonators together mechanically, in this case with a thin piece of Si that acts as a spring. The frequency response of the coupled system is spread out relative to that of the uncoupled resonators. Rather than having two resonances degenerate in frequency, the coupled system has "bonding" and "antibonding" modes centered about the single-resonator frequency. The net result is a coupled system that acts as a bandpass filter, with the width of the pass band set in part by the strength of the coupling, and the center of the band defined by the individual resonators. Through such methods and the identification of MEMS devices as essentially electronic systems with effective impedances, multipole filter implementations can be comparatively simple and reproducible.

Optical applications tend to rely on MEMS structures for controllable movement of reflective elements. Digital mirror devices are one example, such as the optical cross-connect [763] shown in Fig. 9.23. To route entire data streams from one optical fiber into another, the macroscale approach involves macroscopic lenses and mirrors on electrically driven servos. The MEMS-enabled equivalent uses individually addressable MEMS mirrors, positioned through electrostatic actuation. Mass fabrication and a small footprint make this approach attactive, though high operating voltages and precision fiber alignment are still necessary.

A more broad application of digital mirror technology is the use of mirror arrays in digital projectors. Texas Instruments refers to this approach as "digital light processing" (DLP) [765]. An extremely bright white light source is fed through a color wheel to produce cycles of red, green, and blue. This light is then reflected off a synchronized digital mirror device, with particular mirrors corresponding to pixels in the projected image. The outgoing light then passes through a lens system and is projected for viewing.

Digital mirror devices also have imaging applications. Compressive sensing [766] takes advantage of some of the same mathematics behind image compression. Rather than having

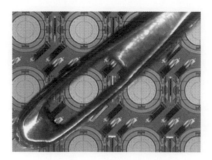

Fig. 9.23   Optical cross-connect, based on MEMS mirrors, made by Lucent Technologies. Needle eye for scale. From *EDN Newsletter*, February 1, 2002 [764].

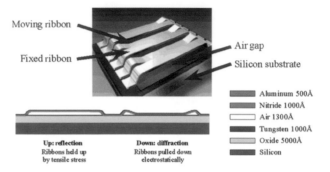

Fig. 9.24   Grating light valve, based on MEMS mirrored grating elements. Polling particular elements changes the effective periodicity of the grating, and hence the scattered wavelength. Adapted from [768].

a 2d array of photodetectors as in a conventional CCD, a single photodetector may be employed [767], with its input light being sampled statistically from a field of view by an array of mirrors. Software keeping track of which mirrors were used to produce particular signals on the single pixel can then statistically reconstruct an image.

An alternative MEMS display and spectroscopy technology is based on switchable diffraction gratings [768]. As shown in Fig. 9.24, a reflective grating structure is constructed such that electrically polling of different combinations of the suspended elements alters the effective spacing, and therefore the scattered wavelength, of the grating.

Other applications

MEMS devices are also relevant to other aspects of modern information technology. MEMS devices have been considered for final-stage positioning actuators for hard disk drive read heads. More famously, MEMS devices based on piezoelectric materials and silicon membranes are at the heart of the ink dispensing process in many inkjet printer heads [769]. As shown in Fig. 9.25, MEMS processing techniques including deep etching, SOI, and wafer bonding go into assembling a print head. When polled the piezo material flexes the upper thin Si surface, applying a pressure pulse to the ink, which passes through

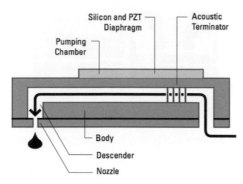

Cross-section of a single jet from an inkjet printer head, with integrated piezo transducer, flexible diaphragm, and Si **Fig. 9.25**
pillars for acoustic damping [769].

an array of Si pillars (to damp acoustic oscillations) and out of an etched orifice. While
there are alternative inkjet print head designs, all are of similar complexity and make use of
MEMS or MEMS-like materials processing. This is the essential reason why inkjet printers
themselves are inexpensive, while the print head cartridges are comparatively pricey: a
tremendous amount of engineering goes into those devices.

The ability to measure very small deformations of thin silicon membranes makes
MEMS platforms ideal for pressure sending. At this writing a number of commercial
vendors sell integrated packages containing a thin MEMS diaphragm and capacitive
measurement electronics. Such devices are now commonplace in the automotive industry
for tire pressure monitoring. These integrated devices have also been widely adopted in
medical applications, particularly blood pressure monitoring.

In short, MEMS devices, most commonly based on planar Si processing technologies,
have become an incredibly diverse global technology. Thus far the devices most broadly
applied tend to be of comparatively simple mechanical design (membranes, clamped
beams, suspended and hinged structures), rather than looking like miniaturized versions
of large-scale machines with many moving parts and complex linkages. This drive toward
simplicity stems largely from manufacturing processes and capabilities, as well as the
challenges of lubrication and adhesion at these small scales.

9.4 Nanoelectromechanical systems

Is there a technological need for even further miniaturization of mechanical devices?
In micro/nanoelectronics, Moore's Laws tending toward ever-smaller device features
have been motivated largely by economic forces, the steadily rising demand for greater
capabilities in a smaller platform. It is not at all clear that such a demand presently exists
in the MEMS arena. Here we will briefly look at one particular technological goal, and
outline the main scientific motivation to look at true nanoelectromechanical systems.

True NEMS would be advantageous for examining the ultimate limits of mass sensing.
Mass loading produces a detectable shift in the frequency f_0 of a mechanical resonator that

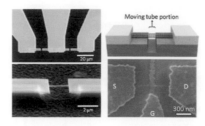

Fig. 9.26 Doubly clamped NEMS beams for amu-level mass sensing. At left, SiC beams [770], and at right, an individual carbon nanotube [771].

scales like $f_0(\delta m/m)$. All other things being equal, a less massive resonator and a higher operating frequency would favor a greater signal. Both of these constraints favor pushing to the nanoscale limit, with the eventual goal of having the sensitivity to detect mass at the limit of a single amu. This would allow, for example, the ability to distinguish between isotopologues of small molecules. Two designs based on doubly clamped NEMS resonators are shown in Fig. 9.26, made from SiC [770] and a carbon nanotube [771]. The operating frequencies of these structures are ~500 MHz and 1.86 GHz, respectively. Steady progress has been made over the years. Actual operation of these devices as mass spectrometers is complicated by sample preparation, the need for exquisite vacuum cleanliness and electronics stability, and interpretation of the data (a molecule landing on an antinode strongly affects the resonator frequency, while one landing on a node has no effect). NEMS fabrication, drive, and detection at their limits are essential for this technology, which could have wide applications.

Scientific goals are some of the strongest motivators in the drive toward NEMS limits. Specific questions include the following.

- *What are the physical mechanisms of and limits to damping in the nanoscale limit?* In the ultimate molecular-scale case, molecular vibrational modes couple to one another via anharmonicity, and energy can be transferred to substrate phonon modes. How do damping mechanisms evolve as resonator structures are reduced in size?
- *Can we measure quantum forces?* There are inherently quantum mechanical interactions that can lead to mechanical forces on the nanoscale. NEMS devices can be useful tools to measure these effects and test our understanding of the relevant physics.
- *Can we create and manipulate quantum harmonic oscillators consisting of large numbers of atoms?* Thinking of a hydrogen atom bound to a benzene molecule as an example of a quantum harmonic oscillator does not seem far-fetched. What about a semiconductor paddle made from millions of atoms?

We very briefly look at each of these in turn.

9.4.1 Physical limits to damping and stability

Many proposed applications of NEMS involve resonators, and usually a higher quality factor, Q, is desirable. Figure 9.27 demonstrates why the question of damping ($\propto Q^{-1}$) in the nanoscale limit is of interest. While macroscopic, single-crystal silicon resonators

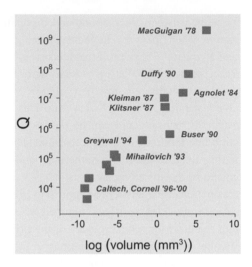

Fig. 9.27

Damping of mechanical resonators as a function of size. The fundamental limits of damping are of much interest in NEMS research [773].

can have Q values exceeding 10^9, there is a general trend toward lower Q as resonator size is decreased. The issue is to what extent is this an "extrinsic" effect from some aspect of device design and quality, and to what degree are there intrinsic physical limitations on Q?

Intrinsic limits on Q would come from the connection between any resonator and the outside world. These are sometimes called *clamping losses*. For example, a doubly clamped beam is by necessity connected to a bulk substrate through the clamping points. Even in a material with no defects or irreversible deformation, as the beam moves, the strain at the clamping points can couple the resonator motion to phonon modes of the substrate at some level. Since the resonator modes are then not exact eigenmodes of the resonator plus substrate system, the resonator modes themselves therefore have some lifetime and hence a finite Q. These effects can be treated in some detail [772]. For reasonable choices of material parameters, resonator dimensions, and operating frequencies (hundreds of MHz), the intrinsic limits to Q are not far from what is seen ($\sim 10^6$ for an Si resonator with vertical displacement and $f_0 = 150$ MHz). Interestingly, that analysis shows that torsional resonators tend to be inherently less damped by these clamping effects.

However, there is much evidence that there are also extrinsic sources of dissipation. Experiments [774] in millimeter-scale oxide-free single-crystal Si resonators prepared in UHV ($Q \sim 8 \times 10^7$) found tens of percent decreases in Q with the deposition of sub-nanometer silicon oxide films. In semiconductor NEMS resonators [773], the internal friction Q^{-1} at low temperatures ($T < 100$ K) was found to be $\sim 10^{-4}$, weakly temperature dependent, and to exhibit very slow relaxational phenomena and history-dependence (warming vs. cooling). This phenomenology is consistent with dissipation from coupling of resonator motion to dynamical defects, the same two-level systems (TLS) discussed back in Section 6.4.2. Recently experimentalists have found that it is possible in some resonators to avoid the presence of active TLS, either through hydrogen passivation [774] or through large, static tensile stress [775].

The fundamental relationship between fluctuations and dissipation further limits the stability of NEMS resonators when they are in thermal contact with the rest of the universe. Johnson [776] and Nyquist [777] showed that a resistor R in thermal equilibrium at temperature T supports a nonzero fluctuating voltage across itself, with the mean square voltage noise per unit bandwidth given by $S_V = 4k_B TR$ V^2/Hz. The short-circuit current noise through such a resistor is $S_I = 4k_B T/R$ A^2/Hz. This Johnson–Nyquist noise is a consequence of thermal fluctuations, and can be derived [777] by considering an electromagnetic transmission line terminated by such resistors. Thermodynamics dictates a certain energy density in the modes of the cavity in thermal equilibrium, and transmitted power can only come from and be dissipated by the resistors.

The same fluctuation–dissipation argument may be applied to MEMS/NEMS resonators. Recall that these devices can be modeled as effective impedances. If such a resonator is in thermal equilibrium with the electromagnetic modes of the drive and sense lines then by necessity there are must be voltage/current fluctuations in those lines, and therefore displacement and force noise. Mean square force noise will be proportional to $m\omega_0 k_B T/Q$, where m is the effective mass of the resonator and ω_0 is the resonant frequency. The constant of proportionality and m will depend on the resonator design in detail. Mean square displacement noise is easier to find, directly from the classical equipartition theorem that $\langle x^2 \rangle = k_B T/m\omega_0^2$.

Finally, even at $T = 0$, the very act of attempting to measure the displacement or force affects the motion of a MEMS/NEMS resonator. This *backaction* is an unavoidable consequence of coupling the resonator to the degrees of freedom used for detection. Backaction can be useful and is essential for some schemes to reduce position noise below classical limits [778]. A thorough discussion of backaction is presented by Poot and van der Zant [779].

9.4.2 Quantum forces

The very small distance scales and extremely high measurement sensitivities possible with MEMS/NEMS devices make it possible to measure forces that are inherently quantum mechanical. The *Casimir force* is one example of this [780, 781]. Consider two perfectly conducting plates separated by a distance d. In the dramatic approach to describing the Casimir effect, remember that the vacuum is filled with zero-point fluctuations of the electromagnetic field at all frequencies.[2] Because of the boundary conditions on the electric field that must be met at the surfaces of the plates, modes of particular polarizations with wavelengths longer than d cannot exist between the plates. In this language the vacuum fluctuations exert an effective pressure that pushes the plates together. Calculating this rigorously gives a force between plates of area A equal to

$$F_C = -\frac{\pi^2 \hbar c}{240} \frac{A}{d^4}. \tag{9.36}$$

---

[2] This touches on some of the same issues as Section 8.2.1. However, note that the Casimir force may equivalently be explained [782] as a relativistically retarded version of the van der Waals interaction, without resorting to vacuum energy densities directly.

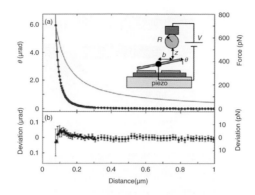

Fig. 9.28

Measurement of Casimir force. Any electrostatic interaction between a MEMS paddle and a Au sphere is nulled away by adjusting V. The remaining attractive interaction produces an angular deflection (black points) that is well described by the Casimir force expectation (red curve, with residuals shown below). Adapted from [784].

Similarly, the force between a perfectly conducting sphere of radius R and a plate is

$$F_C = -\frac{\pi^3 \hbar c}{360} \frac{R}{d^3}.$$ (9.37)

The signs indicate an effective attraction between the bodies. Interestingly, the force survives when the materials are realistic conductors or even dielectrics, though it is possible to choose material properties that can make the force repulsive [783].

The steep dependence of F_C on decreasing d makes the Casimir force both an ideal measurement target for MEMS/NEMS devices, and an effect that must be considered when designing mechanical devices to operate on very small scales. In any experiment designed to measure F_C, precautions must be taken to avoid electrostatic forces, which typically have a $1/d$-like distance dependence.

Based on a macroscopic torsion balance approach [785] to measuring F_C, Fig. 9.28 shows a MEMS scheme to probe for these effects down to $d < 100$ nm. The torsional stiffness of the MEMS paddle is calibrated by using substrate electrodes to apply a known torque. The torsional deflection is determined by the difference in capacitance between the two sides of the paddle and the substrate. Placing the whole device on a piezo-stage, the distance between one side of the paddle and a Au-coated sphere is varied. At comparatively large separations, the voltage difference between the paddle and the sphere is varied to null any electrostatic torque from the sphere. Upon a much closer approach, F_C exerts a readily detectable torque on the paddle, quantitatively consistent with theoretical calculations that account for realistic frequency-dependent dielectric functions of the materials.

9.4.3 True quantum mechanics?

With advances in fabrication and measurement, it has become possible to construct mechanical devices that act as quantum systems governed by a small number of degrees of freedom. Figure 9.29 shows one example, an aluminum nitride cantilevered resonator that may be cooled sufficiently to function as a quantum harmonic oscillator [786].

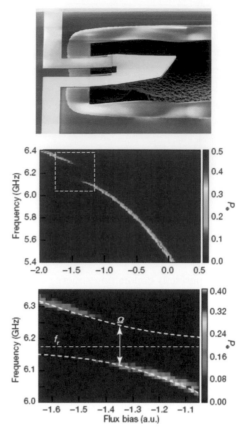

Fig. 9.29   True quantum mechanics. This silicon nitride cantilever (top) resonates at 6.1 GHz. At milliKelvin temperatures, it has
been coupled coherently to a superconducting quantum bit with an energy splitting that may be tuned into resonance
with the cantilever. The avoided crossing shown is one piece of evidence that the cantilever may exist in a coherent
quantum state. Adapted from [786].

It is easier to observe quantum properties of some resonator of natural frequency ω_0 if
the thermal occupation of the resonator modes is small. This implies $k_B T \ll \hbar\omega_0$, so that
milliKelvin temperatures are required when operating frequencies are ~ 1 GHz. In the case
of the device shown, the operating frequency is 6.1 GHz and the operating temperature is
25 mK.

The temperature requirement is necessary but not sufficient, and it is challenging to
formulate an experimental design that really tests for "quantumness". For a true quantum
resonator, it should be possible to place the system in a superposition state, for example, but
direct measurements of the resonator's displacement cannot demonstrate this. Instead, the
investigators capacitively coupled the mechanical resonator to a superconducting quantum
bit that functions as a two-level quantum system with a tunable energy splitting. Spectroscopy on the qubit as the qubit energy splitting is tuned through the mechanical resonator
splitting shows an avoided crossing, a hallmark of coherent coupling. By performing

time-resolved pulsed operations on the qubit, they were able to demonstrate that the mechanical resonator can be placed into a superposition state that Rabi oscillates with nanosecond coherence times.

The ability to take structures composed of many thousands of atoms and place them in quantum superpositions is quite remarkable. The field of quantum optics is based in part on the ability to use atomic resonances as quantum systems to couple to optical fields in cavities. Now that mechanical resonators can have quantum properties, these results may be leveraged in the new subfield of quantum optomechanics. Laser-cooling techniques from atomic physics may be used to remove energy from MEMS/NEMS resonators [787, 788], and it is possible to achieve strong coupling between mechanical resonances and optical cavity modes [789]. These advances may well pave the way for quantum-limited sensing and quantum information processing based on machines with moving parts, a remarkable idea.

9.5 Summary

After introducing the basics of continuum elasticity, we have examined some key issues regarding mechanics in micro- and nanoscale systems, including applications in industry and frontier topics.

- Through continuum elasticity, we can understand the mechanical deformation and normal vibrational modes of solid structures, including cantilevers, clamped beams, and torsional resonators.
- Maximum stresses typically occur far from bending or twisting axes of such structures, implying that surface effects can be very important in, e.g., damping.
- Continuum elasticity works very well down to the nanoscale, with the above caveat about surface effects. True plastic deformation including the motion of dislocations does involve breaking of interatomic bonds.
- Truly understanding frictional interactions requires nanoscale physics and chemistry. Tools developed for nanoscale characterization have been essential in understanding the connections between friction, chemical commensurability, and the excitation of discrete degrees of freedom.
- It is possible to drive mechanical motion by multiple approaches and detect mechanical motion at incredibly small amplitudes. Even though displacements may be fractions of an atomic diameter, velocities can be macroscopic if the frequency of motion is sufficiently high.
- MEMS devices are a multibillion dollar industry, with applications in inertial sensing, optoelectronics, inkjet printers, and other sensing applications. Ideal applications are those that can employ cheap, reproducible, planar fabrication based on semiconductor manufacturing methods.
- Micro- and nanoscale mechanical systems both face some fundamental physics issues (e.g., TLS and the limits of damping) and may be used as tools to address other basic

science questions. For example, such devices have been used to examine Casimir forces, have been demonstrated to function as true quantum harmonic oscillators, and may be coupled nonclassically to light and other quantum systems.

9.6 Suggested reading

- *Elasticity and mechanics of solids* For the physics ideas behind elasticity, it is hard to do better than *Theory of Elasticity*, 3rd edn. by Landau, Pitaevskii, Lifshitz, and Kosevich (Butterworth-Heinemann, 1986). A more accessible discussion of continuum mechanics may be *Introduction to Continuum Mechanics*, 4th edn., by Lai, Rubin, and Krempl (Elsevier, 2009). For the Lagrangian density approach to the dynamics of beams, I found *Dynamics of Mechanical and Electromechanical Systems* by S. H. Crandall (Krieger, 1982) to be helpful.
- *MEMS devices in general* There are several books available that are more engineering focused on MEMS design and fabrication. Good overviews are *Practical MEMS* by Ville Kaajakari (Small Gear, 2009) and *MEMS/NEMS: Handbook Techniques and Applications* by C. T. Leondes (Springer, 2006). In terms of review articles, for inertial sensing an older but solid article is that by Yazdi *et al.* [761]. For NEMS devices, the article by Ekinci and Roukes [790] is nice.
- *Friction* There are a large number of books with engineering approaches to tribology and lubrication, such as Bhushan's *Introduction to Tribology* (Wiley, 2013). Good review articles including an historical perspective are Braun and Naumovets [742] and an excellent work by Krim [741].
- *Quantum limits of mechanics* A good overview of nanomechanical devices and the fundamental issues facing them is Cleland, *Foundations of Nanomechanics: from Solid-State Theory to Device Applications* (Springer, 2003). A review article about nanomechanical systems in the quantum regime is that by Poot and van der Zant [779]. Bordag *et al.* have also written an overview of the Casimir force [781].

Exercises

9.1 *Silicon, its mechanical properties, and MEMS/NEMS sensors* Assume that the Young's modulus E for single crystal Si is 169 GPa, the shear modulus is 79.9 GPa, and the density of Si is 2.33 g/cm^3.

(a) Consider a torsion paddle supported by two torsion members of square cross-section (width a, thickness a, each with length L). What is the torsional spring constant for this system (with both torsion members)? That is, give an expression for K in the formula $\tau = K\phi$, where τ is the torque (directed along the torsional axis) applied to the paddle, and ϕ is the twist angle of the paddle about the torsional axis.

(b) Suppose that $a = 0.5$ m, and $L = 5$ μm, and let's think about thermal noise in this system. The classical equipartition theorem says that this resonator should have an average potential energy of $1/2k_B T$, due to thermal fluctuations. What is the root-mean-square angular twist of the paddle at 300 K? If the distance from the axis of rotation to the outermost radius of the paddle is 5 μm, how much vertical deflection of the paddle edge would this correspond to?

(c) It is possible to use the spin of electrons to apply a torque using electrical currents. Consider sending in (using the torsion members as wires) a current (I) of electrons all spin-polarized in the same direction along the torsional axis, and allow those spins to relax (back to a 50/50 distribution of aligned/antialigned with the torsional axis) on the paddle. This corresponds to a net rate of flow of angular momentum into the paddle given by $(\hbar/2)(I/e)$ that must be balanced by a restoring torque from the torsion members in steady state. How much spin-polarized current would need to be flowing to produce a spin-induced angular deflection twice as big as the rms thermal twist angle at room temperature? The answer you get should be an unreasonably large current.

(d) Nonetheless, this experiment has actually been done! The experimenters worked at very low temperatures and also took advantage of the resonant Q of the torsional paddle by applying the spin-polarized current as an AC current resonant with the torsional mode. Read the paper (Zolfagharkani *et al.*, *Nature Nanotech.* **3**, 720 (2008)) here (dx.doi.org/10.1038/nnano.2008.311). What was the source of polarized electrons, and where were they supposed to relax? Describe in a sentence or two the main control experiment performed by the investigators.

9.2 *MEMS-based optical cavities* Effects you never worry about in large systems. Consider a MEMS-based optical device that is a 1d optical cavity (called a Fabry–Perot etalon) defined by a fixed mirror and a MEMS mirror (mounted on an Si membrane with a position tunable by electrostatics). Suppose the mass of the movable mirror and its membrane is 3×10^{-12} kg, and the effective spring constant for the movable mirror is $k = 2.32$ N/m.

(a) First we'll worry about thermal noise. Consider a cavity of length $L = 5$ microns. This will have allowed modes spaced in frequency by $c/2L = 30$ THz. Each mode will have some bandwidth in frequency related to the reflectivity of the mirrors. Now, thermal vibrations of the movable mirror will shift the frequency position of the center frequency of the etalon by an amount $\sqrt{(\Delta v^2)} = \frac{2(c/2L)}{\lambda}\sqrt{k_B T/k}$. In a sentence or two, explain why this expression makes sense.

(b) Suppose the etalon is designed to operate around 1.5 microns wavelength. Suppose further that the bandwidth of a typical mode (how far in frequency one must go away from the center frequency for the amplitude to be attenuated by 3 dB, for fixed mirror positions) is 2 GHz. How does the room temperature thermal jitter in center frequency from (a) compare to this bandwidth? What does this imply about the performance of this gadget?

(c) Now we'll worry about radiation pressure (!). The force on the movable mirror due to the radiation pressure of the optical energy inside the cavity is given by $2P_c/c$, where P_c is the optical power in the cavity. This can be rewritten in terms of the reflectance of the mirror and the power coupled out of the cavity as $2P_{\text{out}}/(1 - R)c$. For a reflectance of 0.9998, estimate the out-coupled power required for the radiation-pressure induced mirror displacement to be comparable to the thermal noise in the mirror displacement. This tells you when radiation pressure can cause optical nonlinearity in this system. These radiation pressure effects can be used to do optical cooling of microscale mirrors. See dx.doi.org/10.1038/nature05231.

9.3   *Optically driving nanomechanical devices* Read over Li *et al., Nature* **456**, 480 (2008) (dx.doi.org/10.1038/nature07545).

(a) Instead of radiation pressure, this optical driving scheme is different. Briefly describe the physical origins of the driving force in this setup, acting on the suspended, doubly-clamped Si beam. As described in this paper, do you think this force can ever be repulsive? Why or why not?

(b) How do the authors detect the motion of the suspended beam?

(c) What are some of the advantages of this mode of operation of nanomechanical devices?

Micro- and nanofluidics 10

I am an old man now, and when I die and go to Heaven there are two matters on which I hope for enlightenment. One is quantum electrodynamics and the other is the turbulent motion of fluids. And about the former I am rather more optimistic.

Sir Horace Lamb

We have examined the mechanical response of solids as they approach the nanoscale, paying particular attention to the adequacy of the continuum approach to elasticity and the origins of friction and dissipation, as macroscale motions excite microscopic degrees of freedom. We now turn to fluids with similar goals in mind. After a brief overview of some concepts of fluid mechanics and a discussion of dimensional analysis, we discuss fluid flows of particular interest in the micro- and nanoscale. Microfluidic applications are discussed in brief, and the chapter concludes with a look at nanofluidic frontiers.

10.1 Basic fluid mechanics

Fluids are materials that are unable to resist shear and take on the shapes of their containers. In the case of a gas, the typical separation between constituent particles is considerably larger than the size of a particle; in contrast, in a liquid the molecular constituents are essentially "cheek by jowl". One additional length scale is the characteristic size of a fluid container, which we will call L. Another is the mean free path, ℓ, for collisions between the fluid molecules. We can define a dimensionless quantity called the *Knudsen number*, $Kn \equiv \ell/L$. When $Kn \ll 1$, the statistical description of the fluid as an effective continuum is valid. For liquids, this implies that we should not run into problems with the continuum description until the liquid is confined on a scale comparable to molecular dimensions.

While it deforms continuously under shear, a fluid does exert shear stresses on an adjacent solid (or fluid) interface. In Fig. 10.1, we consider translating the upper fluid surface by an amount δl while holding the bottom fluid surface fixed. If this takes place in a time δt, the rate of change of the angle α is

$$\frac{\partial \alpha}{\partial t} = \frac{\partial(\partial l/\partial t)}{\partial y} = \frac{\partial u}{\partial y},\qquad(10.1)$$

where we have implicitly defined u as the x-component of the (local) fluid velocity. In analogy to the solid case, when we defined the shear modulus as the ratio between the

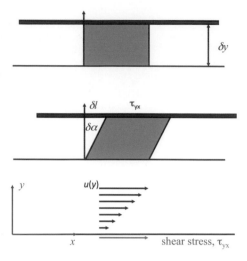

Fig. 10.1 Definition of viscosity from shearing rate of a fluid. Consider a region of fluid (top panel). The bottom of the fluid region is considered immobile, while the upper upper face of the fluid region is sheared. The shear rate, $\partial\alpha/\partial t$, is equal to the gradient of the x-component of the velocity in the y direction, $\partial u/\partial y$, and in a Newtonian fluid this is linearly proportional to the shear stress, τ_{yx}.

shear strain and the shear stress (Eq. (9.10)), here we define the *viscosity* μ as the ratio[1] between the *rate* of shear strain and the shear stress:

$$\tau_{yx} = \mu\frac{\partial u}{\partial y}. \tag{10.2}$$

Fluids for which μ is independent of the velocity itself are called *Newtonian fluids*. Another useful quantity is the *kinematic viscosity*, $\nu \equiv \mu/\rho$, where ρ is the density of the fluid. As Eq. (10.2) makes clear, velocity gradients within the fluid are responsible for viscous shear stresses. The microscopic origin of the viscosity is the transport of momentum between adjacent layers of fluid via scattering events between constituent molecules.

The viscosity may also be considered a diffusion constant for momentum. In ordinary diffusion of particles familiar from statistical mechanics, Fick's Law says that a gradient in the number density of particles, n, results in a number current density, $\mathbf{J}_n = -D\nabla n$. Here $|\mathbf{J}_n|$ is the number of particles per area per time crossing a plane normal to $\mathbf{J}_n$. In the example considered in Fig. 10.1, where $u(x, y, z)$ describes the x-component of the velocity field, the fluid at higher y has a larger density of x-momentum, given by $\rho u(y)$, than the fluid at smaller y. Assuming uniform ρ, the gradient in x-momentum density is just proportional to the velocity gradient, $\rho\partial u/\partial y$. The result is a momentum current density, with a flow of x-momentum in the $-y$ direction; this is the shear stress delivered to the lower surface (microscopically due to collisions between the fluid molecules and the interface). The units of momentum current density (momentum per unit area per time) are the same units as stress.

---

[1] Yes, we once again use μ as a physical parameter. Truly, physicists and engineers were sometimes woefully uncreative. Since the context here is very different from the discussions of charge mobility, magnetic permeability, magnetic moments, and coefficients of friction, hopefully no confusion will result.

While many common fluids of interest in micro/nanofluidics are reasonably Newtonian, complex fluids, particularly suspensions of micro- or nanoparticles, often have viscosities with strong dependences on shear rate. The most common types of non-Newtonian fluids are:

- *pseudoplastic* – viscosity drops with increasing shear rate. This is also called "shear-thinning", and examples include mayonnaise and toothpaste.
- *dilatant* – viscosity increases with increasing shear rate. Examples of "shear-thickening" fluids are corn starch in water and silly putty.
- *thixotropic* – viscosity drops in time after a shear rate is turned on. An example is yogurt, which becomes less viscous the longer a shear rate is applied.
- *rheopeptic* – viscosity increases with time after application of a shear rate.

For the most part, we will limit our discussion to Newtonian fluids.

We will also approximate most of our fluids of interest as incompressible, so that ρ is constant in space and time. Clearly this may be an unwise approximation for gas-based microfluidic devices such as microscale turbines or refrigerators. By assuming incompressibility, we are also ignoring possibly interesting physics such as longitudinal sound waves.

10.2 A digression: dimensional analysis

Fluid mechanics is often where students first encounter named dimensionless numbers, such as the Mach number (ratio of an object's speed to the speed of sound in the surrounding medium) or the Reynolds number (a more complex quantity). We will take a moment and see why these dimensionless products (of dimension-ful quantities) are actually rather profound.

The central ideas are concepts called *scaling* and *similarity*. It is a central feature of physics, though one often neglected in everyday experience, that Nature does not care about our choice of units. That is, actual physical phenomena in real systems are unaffected by whether we choose to measure length in meters or inches or furlongs. This seems obvious, but it has profound consequences, because many of the parameters that we use when constructing physics models of phenomena use interrelated units. For a concrete example, consider the motion of an ideal pendulum, with a bob of mass m attached to a massless rod of length ℓ, sitting on the Earth's surface and thus experiencing a gravitational acceleration g. We know from experience that a pendulum executes motion with some period τ.

If τ only depends on m, ℓ, and g, then on dimensional grounds alone, one can argue that τ is independent of m: no other parameter contains mass as a unit to cancel out any contribution from $[m] = M$. The only combination of ℓ (units of $[\ell] = L$) and g (units $[g] = L/T^2$) that has units of time is $\sqrt{\ell/g}$. Thus, τ must be some (dimensionless) constant times $\sqrt{\ell/g}$. The correct constant in this case is 2π, which we find by actually solving the mechanics problem or (more appropriate for complex systems) by performing careful experiments and analyzing the data.

In this context, we would say that the period *scales* like $1/\ell$, meaning that fixing the other parameters in the problem we know how the period would vary with ℓ, even if we don't know the detailed motion. Alternatively, we could imagine constructing pendula to function on different planets. Two pendula that had identical ratios g/ℓ (or ℓ/g) would be *similar*, in the same sense of usage as the expression "similar triangles".

10.2.1 Buckingham Pi Theorem

The dimensional argument above is a particular example of the Buckingham Pi Theorem. Discovered by Buckingham in 1914 [791], we state it here without proof, in plain language. Suppose the physical response of a system can be determined by n dimensioned variables, $x_1, ..., x_n$. (In our example above, τ is some function of g and ℓ, so $n = 3$.) If these variables involve r independent units or dimentions (e.g., M, L, T), then one can define $n - r$ independent dimensionless products of the form $\Pi_i = x_1^a x_2^b ...$, where a, b, etc. are rational numbers. The physical evolution of the system can then be written in the form

$$\Pi_1 = \Phi(\Pi_2, \Pi_3, ..., \Pi_{n-r}), \tag{10.3}$$

where Φ is some function with a form that depends on the underlying physics. The proof of this theorem rests on precisely the idea that true underlying physics cannot depend on our choice of units.

Looking at our pendulum example in this context, we find that we have three variables of interest, τ, g, and ℓ, that use two independent physical units, L and T. This implies the existence of one dimensionless product. By inspection, $\Pi_1 = \tau\sqrt{g/\ell}$, though $\Pi_1 = \sqrt{\ell/g}/\tau$ would work just as well. Since there are no other independent dimensionless products that can be made, the function Φ can only be some dimensionless constant, C, so that Eq. (10.3) reduces to $\tau\sqrt{g/\ell} = C$.

Let's do a more complicated example, this time involving fluid flow. Suppose we are interested in how much pressure gradient ∇P (units = $ML^{-2}T^{-2}$) we will have to apply to get a fluid of mass density ρ (units ML^{-3}) and viscosity μ (units $ML^{-1}T^{-1}$) to flow through some pipe of diameter d (units L) and wall roughness e (units L) at some speed v (units L/T). This kind of problem is clearly relevant to the design of fluidic networks, and we know from everyday experience that the details of this kind of fluid flow can be very complicated; think about the patterns made by a little drop of ink placed into water flowing through a pipe. We have $n = 6$ variables of interest here. This time the total number of dimensional units is $r = 3$, and the units are M, L, and T. We now need to find $n - r = 3$ dimensionless products, and then we can describe the response of the physical system by writing

$$\Pi_1 = f(\Pi_2, \Pi_3), \tag{10.4}$$

where f is some presumably complicated function.

For this example, there is consensus on which dimensionless parameters to choose. The combination of variables ρv^2 has the same units as pressure, and is a measure of the dynamical or inertial pressure that would be exerted by bringing fluid moving at velocity v to rest. Therefore, $\rho v^2/d$ would have the same units as a pressure gradient, and one could then consider $\Pi_1 = \nabla P/(\rho v^2/d)$ as one of our dimensionless products. We pick this to be Π_1 to play the role of our *dependent* variable in Eq. (10.4), since the pressure gradient is what we wish to find.

For the second product, an obvious choice would be $\Pi_2 = \epsilon^a d^b$. Since each variable has units of length, this implies that $a = -b$, and the simplest (though not unique) situation would be to pick $\Pi_2 = e/d \equiv \epsilon$. That ratio has a sensible physical interpretation as the pipe's relative wall roughness compared to its diameter.

Lastly, we can notice that, as ρv^2 has units of pressure or stress, so does $\mu(v/d)$, where v/d acts as a velocity gradient. We can compare inertial pressure with shear stress in our third product,

$$\Pi_3 = \frac{\rho v^2}{\mu v/d} = \frac{\rho v d}{\mu} \equiv \mathrm{Re}. \tag{10.5}$$

This combination of variables is known as the Reynolds number.

When $\mathrm{Re} \gg 1$, a fluid flow is dominated by the inertial dynamics of the fluid. One consequence of this is the ability of fluid elements to barrel through surrounding fluid, creating whorls and eddies, with a resulting cascade of fluid kinetic energy down to smaller and smaller lengthscales. This is called *turbulence*. Conversely, when Re is small, viscous forces control the motion of the fluid, smoothing out velocity gradients. The result is *laminar flow*.

The Buckingham Pi Theorem then says that we should be able to summarize the rich physics of this complicated fluid mechanics problem with

$$\frac{d\nabla P}{\rho v^2} = f(\mathrm{Re}, \epsilon). \tag{10.6}$$

In keeping with historical convention, the (unknown) function f is usually defined slightly differently, so that

$$\nabla P = \frac{2\rho v^2}{d} f(\mathrm{Re}, \epsilon). \tag{10.7}$$

Here f is called Fanning's *friction factor*. If the 2 is in the denominator in Eq. (10.7), f is the Darcy–Weisbach friction factor.

This example demonstrates the power of dimensional analysis. Because of the complexity of the physics, we generally cannot determine f from first principles.[2] However, we can go into the laboratory and perform a systematic series of measurements, empirically mapping out f as a function of its dependent variables, Re and ϵ. This was done by Moody in the early 1940s [792], and the result is shown in Fig. 10.2, known as a Moody chart. Thanks to the phenomenological determination of f as a function of dimensionless variables, engineers may now calculate the pressure drop in a pipe to reasonable precision, given the fluid and pipe specifications, regardless of whether that fluid is water, liquid methane, or gasoline.

To see this power of similitude in action, consider the test of the first atomic bomb in Alamagordo, NM in June of 1945. Immediately after World War II, the technical details of the bomb and its testing were classified, including the energy yield of the explosion.

---

[2] The laminar flow regime for certain pipe shapes is an exception.

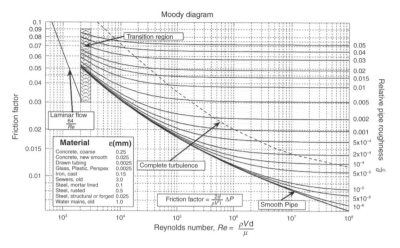

Fig. 10.2 A Moody chart showing the Darby–Weisbach friction factor for fluid flow through a pipe. Image from Wikimedia commons, created by S. Beck and R. Collins, University of Sheffield.

However, *Life* magazine published photographs of the Trinity test with time indices and scale bars, showing the growth of the fireball as a function of time. This is a problem amenable to dimensional analysis. Dump a large amount of energy E into a point, such that a spherical shockwave has expanded into surround air of density ρ to a radius r by a time t. If these are the only important parameters, we have $n = 4$, $r = 3$, implying that there is only a single dimensionless product: $\Pi = (Et^2)/(\rho r^5)$ that should remain *constant* during this process. To find the numerical value of that constant, one needs to do an experiment, such as detonating a large amount of explosives on a test tower and filming the resulting shockwave propagation with high-speed cameras. The British and the Soviets did analyses very similar to this, and with that empirical constant were able to use the *Life* photographs to estimate the yield of the first atomic bomb [793].

10.2.2 Common dimensionless quantities

Finding dimensionless products can be tricky; in general, one must consider products of rational powers of variables of interest, with consideration of the units leading to algebraic equations for the exponents. The choice of dimensionless parameters for a given problem is generally not unique. For example, rather than Π_1 and Π_3, one could just as well choose $\Pi_1' \equiv \Pi_1/\Pi_3$ and $\Pi_3' \equiv \Pi_1 \times \Pi_3$. A brute-force algorithmic approach to finding dimensionless combinations of variables is certainly possible. There is also something of an art to such analysis. If one starts with too few dimensioned variables in trying to describe some phenomenon, then one will arrive at too few dimensionless parameters, meaning that attempts to analyze data from real systems in a scaling analysis will fail. That is, plotting many experimental measurements of Π_3 a function of Π_1 and Π_2 will not give a collapse of the data onto a well-defined manifold. Alternatively, starting with too many dimensioned

Table 10.1 Some common dimensionless products and their interpretation. Here R, L, and d are length scales; c_s is the speed of sound; C_p is the specific heat at constant pressure; κ_T is the thermal conductivity; and H is the convective heat transfer coefficient.

Name	Expression	Interpretation
Reynolds number, Re	$\frac{\rho v d}{\mu}$	inertial forces/viscous forces
Mach number, M	v/c_s	speed/speed of sound
Froude number, Fr	$\frac{v}{\sqrt{Rg}}$	speed/wave speed
Prandtl number, Pr	$\mu C_p/\kappa_T$	viscous diffusion/thermal diffusion
Nusselt number, Nu	Hd/κ_T	convective/conductive heat transfer
Knudsen number, Kn	ℓ/L	mean free path/system size

parameters can lead to an excess of dimensionless products, obscuring potentially simple universality of physical response.

Fortunately, over time conventions have developed for a number of dimensionless products. Table 10.1 lists some common dimensionless parameters and their physical interpretations. More exhaustive tables are available online.

10.3 Laminar flow

Now that we have the preliminaries out of the way, we can actually begin a discussion of the physics of fluid flow relevant for micro- and nanoscale systems. As stated previously, we are typically concerned with *laminar* flow, when Re is small and viscous forces dominate. Often we are interested in *steady* flow, when all parameters such as the velocity field $\mathbf{u}(\mathbf{r}, t)$ have reached time-independent steady-state values. Later, when we discuss electrokinetic effects, we will have occasion to look at unsteady, time-dependent fluid response. Many microfluidic systems work with "confined" flows, where there is no liquid/vapor interface. However, there are times ("open" flows, for example, or when considering wetting) that it is necessary to worry about surface tension and interfacial chemistry effects. Often we are interested in understanding the steady-state $\mathbf{u}(\mathbf{r})$, given the fluid properties and applied boundary conditions.

Empirically, the transition from laminar to turbulent flow is largely governed by the Reynolds number. Micro/nanofluidic systems tend to be in the laminar regime because a characteristic length scale appears in the numerator of Re. For viscosities of common fluids like water ($\mu \approx 0.001$ Pa-s near room temperature), length scales in the micro regime, and laboratory-scale flow speeds (say hundreds of microns per second), the result is Re $< \sim 1$. This is fortuitous, in the sense that the steady-state, vicosity-dominated, smooth velocity profiles can often be found with good accuracy, given a small number of simplifying assumptions.

One critical supposition is the *no-slip condition*, the boundary condition that specifies that there is no relative motion of the fluid and adjacent solid in steady flow. This simplifying approximation is generally very good on the macroscale, and is justified by the idea of very strong viscous interactions at the solid interface. However, it can be challenging to apply this idea under all circumstances without running into some unphysical situations. Reminiscent of the discussion of friction at solid interfaces, we shall find that in the nano regime the no-slip condition is an idealization, and that detailed surface interactions can be extremely important.

10.3.1 Control volume and Eulerian analysis

There are two major approaches to deriving the governing equations for particular fluid mechanics problems, with the problem type determining which procedure is simpler. One method is *control volume analysis*. A region in space is demarked, and basic conservation laws are applied.

- Total mass of fluid must be conserved. The rate of fluid flow into the volume must equal the rate of flow out of the volume.
- Total momentum must be conserved. Any net momentum outflow from the volume must be supplied by external forces acting on the volume.
- Total energy must be conserved. We generally will not concern ourselves with this criterion, which requires tracking the temperature of the fluid as well as its velocity.

We will first demonstrate this approach by considering flow along *streamlines* and deriving a famous result about the evolution of pressure and velocity along a steady flow in the absence of viscosity. Streamlines are the trajectories traced out by particles in the flowing fluid. By definition, streamlines cannot cross each other; at an intersection of two streamlines, the future trajectory of a particle would not be uniquely defined. This property makes it handy to define the sides of a control volume by a locus of streamlines, as in Fig. 10.3. This ensures that no fluid can flow out the sides of the control volume, allowing a quasi-1d description of the flow worrying only about motion along the streamline at the center of the control volume. Suppose this flow is directed in the s direction, and we think about a small control volume extending from s to $s + ds$. The velocity of the fluid along the flow direction is then $u(s)$ at the "entrance" to the control volume, and $u(s + ds) \equiv u + du$ at the "exit" of the control volume. Likewise, the area of the entrance is A and the area of

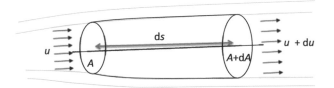

Control volume set up along a streamline, with sidewalls defined by streamlines. Considering mass and momentum fluxes for this situation (incompressible, inviscid flow) lets us derive the famous Bernoulli equation.

the exit is $A + \mathrm{d}A$. Note that the $+s$ direction could be inclined at an angle $\sin\theta = \mathrm{d}z/\mathrm{d}s$ relative to the horizontal x axis.

Applying conservation of mass, we find

$$\rho A u = \rho(A + \mathrm{d}A)(u + \mathrm{d}u). \tag{10.8}$$

More generally, for an incompressible fluid,

$$\int_{\text{surf}} \rho \mathbf{u} \cdot \mathrm{d}\mathbf{A} = 0, \tag{10.9}$$

where $\mathrm{d}\mathbf{A}$ is an area element and the integral is over the surface of the control volume. In differential form, $\nabla \cdot \mathbf{u} = 0$. These are all forms of the fluid continuity equation.

For conservation of momentum for steady, incompressible flow, in integral form

$$\mathbf{F}_s + \mathbf{F}_b = \int_{\text{surf}} (\rho\mathbf{u})\mathbf{u} \cdot \mathrm{d}\mathbf{A}, \tag{10.10}$$

where $\mathbf{F}_s$ and $\mathbf{F}_b$ are forces acting on the surface and bulk, respectively, of the control volume. A relevant surface force is the pressure of the surrounding fluid, which is p at the entrance to the control volume, and $p + \mathrm{d}p$ at the exit of the control volume. Viscous forces are also surface forces, though we neglect those in this example. Here the net pressure force in the $+s$ direction is

$$F_s = pA - (p + \mathrm{d}p)(A + \mathrm{d}A) + (p + \mathrm{d}p/2)(\mathrm{d}A), \tag{10.11}$$

where the last term accounts for the average pressure force on the sidewalls of the control volume. Keeping only terms first-order in the differentials, $F_s = -A\mathrm{d}p$.

A relevant body force could be gravity, which acts in the $-z$ direction. The gravitational force proportional to the total mass in the control volume and the gravitational acceleration, g. The component of that force that points along the $+s$ direction would be given by

$$F_{b,s} = -\rho g[(A + \mathrm{d}A/2)\mathrm{d}s]\frac{\mathrm{d}z}{\mathrm{d}s} = -\rho g(A + \mathrm{d}A/2)\mathrm{d}z \approx -\rho g A \mathrm{d}z, \tag{10.12}$$

where the product in square brackets gives the volume of the control volume, and we have taken the component of gravity along the streamline.

Looking at momentum conservation along the $+s$ direction, using Eq. (10.10),

$$F_s + F_b = u(-\rho u A) + (u + \mathrm{d}u)(\rho(u + \mathrm{d}u)(A + \mathrm{d}A)). \tag{10.13}$$

Plugging in and keeping only terms first-order in the differentials, we are left with

$$\mathrm{d}p + \rho g\mathrm{d}z + u\mathrm{d}u = 0, \tag{10.14}$$

where the differentials are taken along a streamline. This implies that along each streamline,

$$p + \rho gz + \left(\frac{1}{2}u^2\right) = \text{const.} \tag{10.15}$$

This is the famed Bernoulli equation for flow along streamlines, often cited in undergrad physics discussions of airplane flight. Remember, this comes from simultaneously solving the continuity and momentum equations for a differential control volume along a streamline, for steady, inviscid, incompressible flow.

In general, with a control volume, continuity provides the conditions on the velocities and flow areas that are needed to conserve mass. The net rate of change of momentum of the fluid going through the volume must equal the total *vector* sum of the surface and body forces on the fluid.

Instead of working with a fixed volume, the *Eulerian* approach follows an individual fluid blob as it passes through the velocity field that is assumed to exist everywhere. The blob must obey Newton's Laws, and in general may be rotated and deformed (sheared, stretched) as it flows along. At time t, the particle is at position x, y, z with velocity $\mathbf{u}(x, y, z, t)$. An instant dt later, the particle's velocity should then be $\mathbf{u}(x + dx, y + dy, z + dz, t + dt)$, where we have accounted for the blob's motion. By properly using the chain rule, the change in velocity experienced by the blob in that instant is

$$d\mathbf{u}_{\text{blob}} = \frac{\partial \mathbf{u}}{\partial x} dx + \frac{\partial \mathbf{u}}{\partial y} dy + \frac{\partial \mathbf{u}}{\partial z} dz + \frac{\partial \mathbf{u}}{\partial t} dt. \tag{10.16}$$

The acceleration of the blob is then

$$\mathbf{a}_{\text{blob}} = \frac{\partial \mathbf{u}}{\partial x} \frac{\partial x}{\partial t} + \frac{\partial \mathbf{u}}{\partial y} \frac{\partial y}{\partial t} + \frac{\partial \mathbf{u}}{\partial z} \frac{\partial z}{\partial t} + \frac{\partial \mathbf{u}}{\partial t}$$
$$= \mathbf{u} \cdot \nabla \mathbf{u} + \frac{\partial \mathbf{u}}{\partial t}. \tag{10.17}$$

This is the "total" derivative of the blob's velocity with respect to time. The blob accelerates both because of the explicit time dependence of the velocity field, and because it gets carried into regions with different velocities.

The Eulerian approach is then to write the equivalent of $\mathbf{F} = m\mathbf{a}$ for a blob of fluid. Considering a blob of fluid with sides dx, dy, dz and density ρ, it is possible to look at the surface and body forces on the blob. Clearly pressure can exert forces normal to the blob's faces. The trickier analysis involves keeping track of all the shear stresses, which related to velocity gradients through the viscosity. If the three cartesian components of $\mathbf{u}$ are (u, v, w), then for example the shear stress

$$\tau_{xy} = \tau_{yx} = \mu \left(\frac{\partial v}{\partial x} + \frac{\partial u}{\partial y} \right). \tag{10.18}$$

The other shear stresses may be obtained by cyclically permuting the x, y, z components. The viscosity can also add a component to the normal stresses, so that

$$\sigma_{xx} = -p - \frac{2}{3} \mu \nabla \cdot \mathbf{u} + 2\mu \frac{\partial u}{\partial x}. \tag{10.19}$$

Again, the other components may be found by cycling $x \to y \to z$, $u \to v \to w$. Keeping track of all the stress-related forces acting on the faces of our blob (and remembering that $u(x, y, z) \neq u(x + dx, y, z)$, for instance), we can write Newton's Second Law for the blob. For the case of steady ($\partial \mathbf{u}/\partial t = 0$) and incompressible ($\nabla \cdot \mathbf{u} = 0$) flow, and adding in gravity, the result is

$$\rho \frac{d\mathbf{u}_{\text{blob}}}{dt} = \rho \mathbf{g} - \nabla p + \mu \nabla^2 \mathbf{u}. \tag{10.20}$$

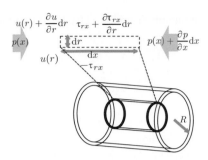

Control volume for the problem of laminar flow in a circular pipe. Close-up shows a side-section of part of the annular   Fig. 10.4
control volume. Conserving mass and momentum leads to the derivation of the velocity profile $u(r)$ associated with
Poiseuille flow.

This is the *Navier–Stokes equation* for viscous, incompressible flow. Solving for the
velocity field in this formalism is then a boundary value problem.

10.3.2 Laminar flow in a round pipe and beyond

A useful, general result is the velocity profile for incompressible, viscous fluid flow in a
round pipe, in the "fully developed" limit (meaning that we are far from the pipe inlet). By
symmetry, the velocity has to point along the pipe direction, and (ignoring gravity) it can
depend only on r, the distance from the center of the pipe. We will use an annular control
volume, with an inner radius at r and an outer radius at $r + dr$, extending from x to $x + dx$.
   The pressure force in the $+x$ direction on the left end of the control volume is
$F_x = p2\pi r dr$; conversely, on the right end of the control volume, we find $F_x = -(p +
(\partial p/\partial x)dx)2\pi r dr$. We also need to find the forces in the x direction due to shear stresses.
Remember that τ_{rx} is the force per unit area in the $+x$ direction acting on a surface with a
normal pointing in the $+r$ direction. Looking at the shear stress contributions,

$$F_{x,\text{inner}} = (-\tau_{rx})2\pi r dx, \qquad (10.21)$$

$$F_{x,\text{outer}} = \left(\tau_{rx} + \frac{\partial \tau_{rx}}{\partial r}dr\right)2\pi(r + dr)dx, \qquad (10.22)$$

where Eqs. (10.21) and (10.22) are the forces acting on the inner and outer surfaces of the
annulus, respectively.
   Adding up all of these contributions, and realizing that, in steady state, the total force on
the control volume must be zero, after some algebra we find

$$\frac{1}{r}\frac{\partial(r\tau_{rx})}{\partial r} = \frac{\partial p}{\partial x}. \qquad (10.23)$$

Integrating both sides with respect to r gives

$$r\tau_{rx} = \frac{1}{2}r^2\frac{\partial p}{\partial x} + c_1, \qquad (10.24)$$

where c_1 is a constant of integration. From Eq. (10.2) defining the viscosity, we then find

$$\frac{\partial u}{\partial r} = \frac{1}{2\mu} r \frac{\partial p}{\partial x} + \frac{c_1}{r\mu}. \tag{10.25}$$

Integrating both sides again with respect to r gives

$$u(r) = \frac{1}{4\mu} r^2 \frac{\partial p}{\partial x} + \frac{c_1}{\mu} \ln r + c_2, \tag{10.26}$$

where c_2 is another integration constant.

Clearly c_1 must be zero; otherwise the velocity would unphysically diverge at $r = 0$. To find c_2, we apply the no-slip condition at $r = R$, the wall of the pipe, leading to

$$u(r) = \frac{R^2}{4\mu} \left(\frac{\partial p}{\partial x} \right) \left[\left(\frac{r}{R} \right)^2 - 1 \right]. \tag{10.27}$$

The velocity profile for fully developed laminar flow is *parabolic*, a regime called *Poiseuille flow*.

We can also compute the total volume flow rate for this case,

$$\text{vol. rate} = \int_0^R u(r) 2\pi r \, dr = -\frac{\pi}{8} \frac{R^4}{\mu} \left(\frac{\partial p}{\partial x} \right). \tag{10.28}$$

A similar approach may be taken for open flows, where the top surface at $y = d$ is a liquid–vapor interface rather than a hard wall, as in Fig. 10.5. For a flow that is very wide (we can ignore the side-walls) and shallow (we can ignore the pressure difference between the top and bottom of the liquid), driven by a longitudinal pressure gradient $\partial p / \partial x$, we have a velocity profile $u(y)$. Doing a control volume analysis for a rectangular control volume leads to

$$\frac{\partial \tau_{yx}}{\partial y} = \frac{\partial p}{\partial x} \rightarrow \mu \frac{\partial^2 u}{\partial y^2} = \frac{\partial p}{\partial x}. \tag{10.29}$$

This time the boundary conditions are no-slip at the bottom of the liquid, $u(y = 0) = 0$, and no stress at the free surface, $\partial u / \partial y = 0$ at $y = d$.

For a general geometry of channel, the best (and mathematically equivalent) approach in solving laminar flow is to solve the Navier–Stokes equations (often numerically) with appropriate boundary conditions.

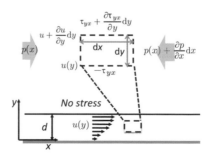

Fig. 10.5   Control volume for the problem of laminar flow in a wide, shallow (p independent of y), open channel. Flow velocity is assumed to be zero at $y = 0$, and shear stress must vanish at the free surface ($y = d$).

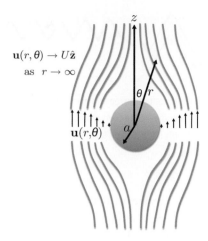

Situation relevant for the Stokes drag law. The velocity field is uniform far from the sphere.

Fig. 10.6

10.3.3 Flow around a small particle

Another well-known problem highly relevant for micro- and nanofluidic applications is trying to compute the net viscous drag force on a spherical particle of radius a moving in an incompressible medium with uniform flow speed U far from the particle. This problem, shown schematically in Fig. 10.6, was solved by Stokes in 1851 as a boundary value problem in spherical coordinates. The detailed solution is available online and is beyond our scope; suffice it to say that the approach is to find the velocity field that satisfies the Navier–Stokes (momentum) equation, continuity of the incompressible fluid, and guarantees that the flow is always tangential to the surface of the sphere. That is, $u_r(r = a) = 0$, while far from the sphere $u_r(r \to \infty) = U \cos\theta$ and $u_\theta(r \to \infty) = -U \sin\theta$. Once the velocity field has been found, the shear forces at the particle surface may be computed and integrated. The final result for the drag force is

$$F_{\mathrm{d}} = 6\pi \mu U a. \qquad (10.30)$$

At low Re, it is quite easy to have such drag forces play a major role in the motion of small particles in fluids (colloidal particles, cells, etc.).

10.3.4 Low Reynolds number

Note that our physical intuition is built around our daily experiences, which generally happen to be in the limit of large Re and turbulence. Some phenomena that take place when Re is very low therefore seem surprising. Here we point out a couple of examples. A thorough, engaging discussion of "Life at low Reynolds number" was written by E. M. Purcell [794], specifically looking at the fluid environment experienced by single-celled life such as bacteria.

First, consider the problem of trying to mix two fluids. Turbulence is exceedingly helpful with this task, as is seen when stirring milk into coffee. The dominance of inertial forces

Fig. 10.7 Lack of mixing in the laminar regime. Water containing red and black dye is injected into a microfluidic channel at flow speeds as high as 0.50 m/s. For water under these conditions in this geometry, Re $\ll$ 2000, with the system remaining firmly in the laminar regime. Only interdiffusion takes place, rather than turbulent mixing. Adapted from [795], reprinted with permission of AAAS.

in turbulent systems leads to the breakup of fluid streams into smaller and smaller eddies. These eddies lead to very high concentration gradients near their edges, so that a thin milk-rich eddy is surrounded by coffee-rich fluid. These eddies rapidly approach a transverse size such that the typical diffusion time for a molecule of fluid to cross an eddy becomes usefully small. In this way concentration gradients are smoothed out, and milk and coffee may be thoroughly mixed in seconds.

An example of the contrasting situation at very low Re is shown in Fig. 10.7. A stream of red-dyed water and a stream of black-dyed water are introduced side by side into a microfluidic channel, with dimensions and flow rates such that Re is small [795]. On the time- and lengthscales of the experiment, there is no transition to turbulence. That means that the only mixing between the fluids that takes place is through interdiffusion at their interface. Notice that the comparatively sharp interface between the colors survives flow patterns that turn sharp corners.

A second example concerns swimming. When you swim in a pool, one consequence of the large Re is that you coast if you stop moving your arms and legs. Your inertia and the momentum of your motion is more than enough to overcome the comparatively weak viscous drag forces that your body experiences, so that you can readily drift distances comparable to or larger than your own size.

The situation encountered by a bacterium is the antithesis of this. For a bacterium 10^{-6} m long, swimming at 30 microns per second in water, Re$\sim 3 \times 10^{-5}$. Assuming that the bacterium is roughly spherical, we can use the Stokes drag law, Eq. (10.30), to estimate the coasting distance. If the bacterium has a radius R and a density ρ (comparable to that of water) and an instantaneous velocity v, then

$$\rho \left(\frac{4\pi R^3}{3} \right) \frac{dv}{dt} \approx -6\pi \mu R v. \tag{10.31}$$

Solving this first-order differential equation for $v(t)$ assuming an initial velocity v_0, we find

$$v = v_0 \exp \left(-\frac{9\mu}{2\rho R^2} t \right). \tag{10.32}$$

A typical stopping time is then around 150 microseconds. This may be integrated further to find the stopping distance, which is around 1 nm!

The consequences of this lack of mixing and the lack of inertial coasting are profound for living things, as discussed by Purcell. The implications for micro- and nanoscale machines or systems designed to operate in fluid environments are similarly enormous.

10.4 Surface interactions

Details of the interactions at fluid interfaces (liquid–solid, liquid–gas, and even liquid–liquid) can be of crucial importance in micro- and nanofluidic environments. For small volumes of material, the surface-to-volume ratio can become large, so that surface forces and surface energies can swamp out bulk contributions. The surface energy is the energy per unit area associated with the presence of an interface between two materials. For the particular case of a liquid–vapor interface, the surface energy is often called a *surface tension*. When there is a net attractive interaction between liquid molecules, in the absence of other physics the liquid can lower its overall energy by minimizing the area of the liquid–vapor interface; this minimizes the number of liquid molecules that are not "fully satisfied" in the sense of being in contact with as many other liquid molecules as possible. The units of surface tension are energy per area, or force per unit length, and the symbol used is typically σ.[3] As a thermodynamic quantity, surface tension is analogous to pressure. The work done by a system to change its volume by dV is $-pdV$; similarly, the work done by a system to expand an interface with surface tension σ by an area change dA is $+\sigma dA$. It can be difficult to measure absolute surface energies directly; rather, differences between different surface energies are what determine experimental observables.

The rising energetic contribution of surfaces relative to bulk energies at small system sizes has clear consequences. All other things being equal, two small drops of a liquid will merge when brought together, to minimize the surface area by forming a larger drop. Similarly, the creation of a tiny bubble of vapor within a liquid is energetically expensive. Such bubbles will tend to collapse if the pressure forces from the vapor within the bubble cannot balance the surface tension forces shrinking the liquid–vapor interface. This kind of bubble collapse can be very intense in terms of energy density, leading to phenomena like cavitation and sonoluminescence [796, 797]. For a rough sense of scale, the room temperature liquid–vapor surface tensions for water, octane, and mercury are about 0.072, 0.022, and 0.436 J/m^2, respectively.

A common way to characterize relative surface energies is to measure *contact angle*, θ_c, for a drop of a liquid of interest on a solid (Fig. 10.8) [798]. Qualitatively, spreading the drop creates both more liquid–solid and liquid–vapor interface. Therefore the contact angle reflects the relative affinity of the liquid for itself and for the solid surface and vapor phase. A contact angle less than 90 degrees implies a tendency for the liquid to wet the solid, while a contact angle much greater than 90 degrees (such as mercury on glass or water on a waxed car) indicates the opposite. Looking at the forces acting on the liquid–solid contact

---

[3] Hopefully the appropriate subscripts will avoid possible confusion with stresses, and the context will avoid confusion with the electrical conductivity.

(a)

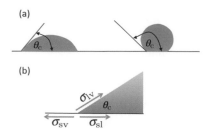

(b)

Fig. 10.8 Contact angles. (a) Contact angles for comparatively wetting (left) and non-wetting (right) liquid–solid interactions. (b) Thinking of the surface energies as forces per unit length of interface, the contact angle reveals a relationship between surface energies (Eq. (10.33)) that must hold in equilibrium.

line, in static equilibrium

$$\sigma_{sv} - \sigma_{sl} = \sigma_{lv} \cos \theta_c. \tag{10.33}$$

This relationship goes back a long way [799].

A *spreading coefficient*, $S \equiv \sigma_{sv} - \sigma_{sl} - \sigma_{lv}$, helps describe the tendency of a liquid drop to spread on a surface. A large positive S implies good wetting; conversely, "high energy" surfaces have large values of σ_{sl}, so that spreading is energetically expensive.

The surface energies between liquids and solids, σ_{sl}, come about because of microscopic interactions at the interface, including what we call chemical interactions. The sensitivity to surface details is similar to the case of sliding friction (Section 9.2.2). This explains why contact angle results can vary, for example, depending on humidity (and whether there is an adsorbed water layer on the solid). At the same time, this sensitivity presents opportunities.

If we define the local attractive interaction per unit area between liquid and the underlying solid to be V_{sl}, then we would find $\sigma_{sl} = \sigma_{sv} + \sigma_{lv} - V_{sl}$. If instead we worried about an interface between the liquid and itself (which should have no net surface energy), then $0 = \sigma_{lv} + \sigma_{lv} - V_{ll}$, where V_{ll} is the (positive if attractive) interaction per unit area between the liquid and itself. Combining these, we find that $S = V_{sl} - V_{ll}$. If the liquid–liquid attraction is stronger than the liquid–solid attraction, $S < 0$, and the liquid tends not to wet the solid surface.

This analysis is at the root of why fluoropolymers like Teflon and waxed surfaces are comparatively "waterproof". The attractive hydrogen bonding interaction between water molecules is much stronger than the van der Waals interaction between water molecules and the saturated bonds of the fluoropolymer or wax. In the van der Waals limit, the attractive interaction between two species is related to their dielectric polarizabilities, α (see Section 5.3.1), so that $V_{ll} \sim \alpha_l^2$, $V_{sl} \sim \alpha_s \alpha_l$. Since the polarizability of most hard solids is larger than that of most liquids, wetting is often favored. Of course, stronger interactions than van der Waals can be relevant, from hydrogen bonding all the way to covalent attachment. Moreover, because of strong surface interactions, the structure of the liquid surface (parameters like average molecular spacing and orientation) can be quite different from those in the bulk.

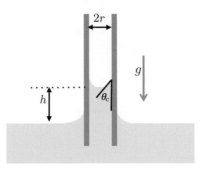

Capillary forces. For a wetting interface, the liquid-vapor surface tension force can be enough to hold up a significant column of liquid in a small tube.

Fig. 10.9

Surface tension forces can be strong enough to overcome gravity; this is called *capillary action*, and provides an additional experimental means of assessing relative surface energies. Assume that the liquid–vapor surface tension is the largest surface energy in the problem. For a circular tube of radius r inserted in a liquid of density ρ, with a meniscus making a contact angle θ_c as shown in Fig. 10.9, the surface tension force must support the weight of the column of liquid:

$$\rho g(\pi r^2 h) = 2\pi r \sigma_{lv} \cos \theta_c \rightarrow h = \frac{2\sigma_{lv} \cos \theta_c}{\rho g r}. \qquad (10.34)$$

Note that this argument works for contact angles greater than 90°, too. For such non-wetting interactions, the capillary actually pushes the liquid down, and the curvature of the meniscus is of the opposite sign, with maximum liquid height in the center of the tube. This is the physics that allows water striders, insects with rather hydrophobic exoskeletons, to "walk" on water surfaces while dimpling the liquid–vapor interface downward.

Capillary action can be very powerful in nanoscale systems. For a 10 nm interior diameter tube, the formula suggests that water could be drawn up kilometer (!) distances. This capillary physics is relevant to the ability of trees to draw water up to their tops, though several other effects are also important [800].

Because of the prominent role of interfacial chemistry, surface interactions can be tuned and engineered. The most common case is the use of surfactants dissolved in the bulk to modify the surface tension of liquids. This (the lowering of the liquid–vapor interfacial energy cost) is the reason why far larger bubbles may be blown in soapy water than in pure water. Similarly, soapy water can have such strong wetting tendencies on some surfaces that it can displace oily substances, as shown in commercials for dishwashing soap. Alternatively, one can modify just the liquid–vapor interface, for example by putting a drop of oil on a water surface, making it no longer possible for water striders to support themselves.

Chemical functionalization of solid surfaces is the natural way to engineer the energetics of the solid–liquid interface. Self-assembled monolayers are one approach. Figure 10.10 shows an example of how this can work [801]. An SAM is created from molecules containing hydrophobic endgroups, but with a photoreactive segment in the center of the

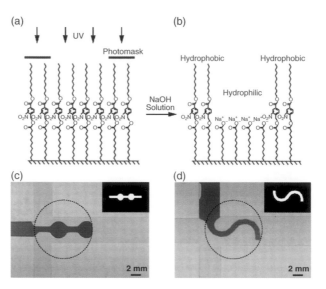

Fig. 10.10 Manipulating wetting via self-assembled monolayers. The solid–liquid surface energy is altered by a SAM that presents either hydrophobic or hydrophilic end groups, depending on whether photochemistry has taken place. From [801], reprinted with permission of AAAS.

molecule. Ultraviolet light can cleave the exposed molecules, leaving behind hydrophilic groups. By patterning the UV exposure, it is possible to guide aqueous solutions using complex patterns of wettability. Similarly, SAMs incorporating reversibly photoreactive molecules allow photoswitchable wetting [802], as shown in Figure 10.11. Depending on the illumination history, the molecules in question present ends that are more or less oleophilic. A gradient in the illumination leads to a gradient in the solid–liquid surface energy, which acts to push an oil droplet along the interface. We shall revisit solid–liquid interfacial energy manipulation in Section 10.7.1, when we discuss the fate of the no-slip condition.

10.5 Electrolytes

In addition to worrying about simple hydrodynamic effects, a great deal of microfluidic work concerns exerting forces on fluids and on particles within fluids via electrical means. These electrokinetic effects are possible in electrolytes, fluids that contain dissolved ionic species. To understand electrokinetic effects we must first consider how the charged species in the fluid respond to electric fields, what happens at liquid–solid interfaces, both at the boundaries of the fluid volume and at the surface of any suspended particles.

A typical electrolyte is a buffer solution containing equal amounts of positive and negatively charged ionic species (cations and anions, respectively), though the solution

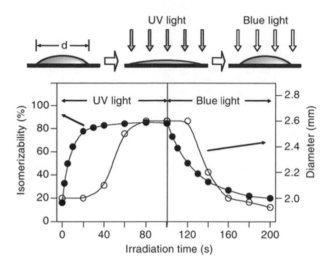

More manipulation of wetting via self-assembled monolayers. A reversibly photoswitchable SAM can be made more (less) oleophilic upon exposure to UV (blue) light. From [802], reprinted with permission of AAAS.

Fig. 10.11

is overall charge neutral.[4] Labeling the charged species by the index i, the valence of each ion of that species in solution is z_i, and the average concentration of each species is n_i per unit volume. The species can be atomic ions (e.g., Na^+, Cl^-), molecular ions, and even solvated protons (H^+, sometimes written as H_3O^+) and solvated electrons [805]. The local charge density is then

$$\rho_e = \sum_i z_i e n_i(\mathbf{r}), \tag{10.35}$$

which must obey Poisson's equation. If the local electrostatic potential is $\phi(\mathbf{r})$, then

$$\nabla^2 \phi = \frac{\rho_e}{\kappa \epsilon_0}, \tag{10.36}$$

where κ is the relative dielectric function of the solvent. In equilibrium, charge neutrality implies an equilibrium potential distribution ϕ_0 that satisfies $\nabla^2 \phi_0 = 0$. We can then consider local deviations from this equilibrium situation, $\hat{\phi} \equiv \phi - \phi_0$. Local excursions from equilibrium are permitted due to thermal fluctuations, with a local charge density given by

$$\rho_e = \sum_i z_i e n_i \exp\left(-\frac{z_i e \hat{\phi}}{k_B T}\right), \tag{10.37}$$

where a Boltzmann factor accounts for the local energy cost. Plugging into Eq. (10.36),

$$\nabla^2 \hat{\phi} = -\frac{1}{\kappa \epsilon_0} \sum_i z_i e n_i \exp\left(-\frac{z_i e \hat{\phi}}{k_B T}\right). \tag{10.38}$$

---

[4] A limiting case is an *ionic liquid*, essentially an organic salt consisting of small molecule cations and anions, that happens to be a liquid at room temperature [803, 804].

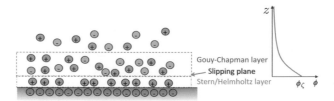

Fig. 10.12 Charge double-layer. A solid interface immersed in an electrolyte can acquire a net surface charge (here negative). Ions in the electrolyte adsorb strongly (the Stern/Helmholtz layer) and partially screen this charge. Above a slipping plane, the remaining surface charge is screened on the scale of λ_D, with the potential at the slipping plane designated the zeta potential (assuming $\phi = 0$ in the bulk, neutral solution).

Expanding the exponential for small perturbations/high temperatures, and using the charge neutrality constraint that $\sum_i z_i e n_i = 0$,

$$\nabla^2 \hat{\phi} \approx \frac{e^2}{\kappa \epsilon_0 k_B T} \sum_i z_i^2 n_i \hat{\phi} \equiv \frac{1}{\lambda_D^2} \hat{\phi}. \qquad (10.39)$$

Here we define the *Debye length*,

$$\lambda_D \equiv \left(\frac{e^2}{\kappa \epsilon_0 k_B T} \sum_i z_i^2 n_i \right)^{-1/2}, \qquad (10.40)$$

the characteristic length scale over which charge perturbations are screened by the ionic charge in an electrolyte, rather analogous to the Thomas–Fermi screening length Eq. (5.20) in metals. In this case, the higher the temperature, the longer the distance scale; conversely, the larger the ionic charge concentrations, the shorter the screening length.

One situation of recurring interest is the quasi-1d geometry, where a solid/electrolyte interface is at some wall potential ϕ_w that differs from the bulk solution potential, which we define to be zero. For a symmetric electrolyte (two ionic species with equal valences z) with average concentration n_0,

$$\frac{d^2 \hat{\phi}}{dx^2} = \frac{2 e z n_0}{\kappa \epsilon_0} \sinh \left(\frac{z e \hat{\phi}}{k_B T} \right). \qquad (10.41)$$

This is the Poisson–Boltzmann equation, and for the boundary condition above the solution is

$$\hat{\phi} = \phi_w \exp \left(-\frac{x}{\lambda_D} \right). \qquad (10.42)$$

The Debye length is precisely the screening length.

This implies that the perturbing potential of walls (or charged particles in a solution) is almost complete screened over a distance of $\sim 3\lambda_D$. In pure water with a pH of 7, the screening results from the H^+ and OH^- ions that exist in equilibrium; at room temperature in this case $\lambda_D \sim 1$ micron. In contrast, in 1 M KCl solution in water, $\lambda_D \sim 0.3$ nm. Here we have used the dc value for κ in water, approximately 80. Depending on the ionic concentration, the Debye length can readily span from the nano to the micro scale.

A general view of solid–electrolyte interfaces in these situations has arisen, called the *double-layer* picture [806]. When a solid material is placed into the solution, that solid surface has some particular affinity for charge determined by its chemical termination. Some ions from the solution, with charges opposite to that of any surface charge of the solid, experience such strong electrostatic interactions with the wall that they become effectively immobilized at the interface. This is the *Stern* or *Helmholtz* layer [807, 808]. Beyond this is a more diffuse layer, where the net ionic charge is the same sign as that of the Stern layer, but the ions are increasingly mobile. This is often called the *Gouy–Chapman* layer [809, 810]. It is the charges in this Gouy–Chapman layer that produce the Debye screening described above. The electrostatic potential at the "slipping plane", beyond which the fluid is considered mobile, is often called ϕ_ζ, the *zeta potential*.

Suppose, after the formation of the Stern layer, that the net surface charge density of the interfacial "wall" is σ_w. At some distance x away from the interface the local charge density is approximately $\rho_e \approx -(\kappa\epsilon_0/\lambda_D^2)\hat{\phi}$, using Gauss' Law and our knowledge of Debye screening. We can integrate up this charge density moving away from the interface, knowing that by $x = \infty$ the charge on the wall should be fully screened:

$$\sigma_{GC} = \int_0^\infty \rho_e(x)\mathrm{d}x = -\sigma_w. \tag{10.43}$$

In this case,

$$\hat{\phi}(x) \approx \frac{\sigma_w \lambda_D}{\kappa\epsilon_0} \exp\left(-\frac{x}{\lambda_D}\right). \tag{10.44}$$

Ideally, the surface charge density at the solid electrolyte interface should be a fixed property of the particular material system. That is, one would expect a clean fused silica wall touching a 1 mM KCl solution in water to always develop the same σ_w.

We began the discussion above by declaring that the overall solution is charge neutral. However, we then found that the local ion concentration in the solution arranges itself to screen charged perturbations such as interfacial charges. Suppose a solid wall has a net positive σ_w, leading to a net negative σ_{GC} that extends out around $3\lambda_D$ into the electrolyte. Tacit in the discussion is the idea that the fluid system is sufficiently large that the positive charges necessary to achieve overall charge neutrality of the solution are in there, just "far away". Clearly, as fluid volumes approach the micro- and nanoscale, this assumption cannot hold, and the double-layer screening of interfacial charges can deviate from the large size limit.

10.5.1 Electro-osmosis

One strong piece of experimental evidence for this charged double-layer and Debye screening picture is the kinetic response of such a liquid to an externally applied electric field **F**. Consider the geometry in Fig. 10.13, with laminar flow assumed and channel dimensions much larger than λ_D. When an electric field in the x direction is imposed by an external voltage supply, only the regions of fluid containing net ionic charge density should experience a net electric force. In steady state, this force should be balanced by

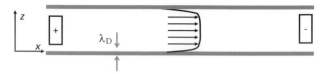

Electro-osmosis in the steady flow limit. An electric field is applied along the flow direction. The bulk of the fluid is neutral, while the double-layer regions near the walls have net charge (in this example, +). The steady flow result is Poiseuille flow or plug flow, with no velocity gradient in the channel more than a Debye length from the walls.

viscous forces so that there is no net acceleration of that fluid. Ordinarily we expect the no-slip condition to apply at the channel walls. Far from the channel walls, there is no electric force; therefore, in steady state there should be no viscous force, implying that the velocity field far from the walls should be uniform. This electrolyte fluid motion driven by coupling to the net charge near interfaces is rather inappropriately named *electro-osmosis*.

The Navier–Stokes momentum equation in this situation is

$$\rho \frac{du}{dt} = 0 = \eta \nabla^2 u(y) + \rho_e(y) F, \qquad (10.45)$$

where we assume that the fluid velocity is purely in the x-direction and depends only on y, the transverse distance from the lower (in the figure) wall. Near that lower interface,

$$\eta \frac{\partial^2 u}{\partial y^2} \approx \frac{\sigma_w F}{\lambda_D} \exp\left(-\frac{y}{\lambda_D}\right). \qquad (10.46)$$

Using the no-slip condition, near the bottom wall

$$u(y) \approx \frac{\sigma_w \lambda_D F}{\eta} \left(1 - \exp\left(-\frac{y}{\lambda_D}\right)\right). \qquad (10.47)$$

Near the far wall of the channel the velocity should look like this with y replaced by the distance from the channel wall, assuming that solid surface is identical to the one at $y = 0$. Far from the double layer, this reduces to

$$u(y \to \infty) \to u_0 \approx \frac{\sigma_w \lambda_D F}{\eta}. \qquad (10.48)$$

This kind of velocity profile is called *plug flow*. In the limit of uniform wall surface charge, a Debye length small compared to the channel geometry, and no applied pressure gradient, this can be generalized beyond just the 1d case:

$$\mathbf{u}_0(\mathbf{r}) \approx \frac{\sigma_w \lambda_D}{\eta} \mathbf{F}(\mathbf{r}). \qquad (10.49)$$

We can then define an electro-osmotic mobility, $\mu_{eo} \equiv u_0/F$.

Electro-osmosis gives us one experimental way to infer quantities like σ_w, or equivalently ϕ_ζ. Bear in mind that one cannot in general simply apply a local probe technique like electrostatic force microscopy, because an AFM tip would carry with it its own charge double-layer.

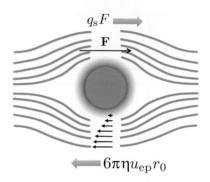

$$q_s F \longrightarrow$$
$$F \longrightarrow$$

$$\longleftarrow 6\pi\eta u_{ep} r_0$$

Electrophoresis. Effective net charge on a particle due to the double-layer results in an electrophoretic force in the presence of an electric field. In steady state this is balanced by viscous drag, setting the electrophoretic velocity.

Fig. 10.14

10.5.2 Electrophoresis

Now assume that there are small particles of interest suspended in the electrolyte, and that the concentration of particles is sufficiently low that the basic double layer physics still works. In this case each particle will have, around itself, its own double layer. The result will be an effective surface charge on each particle, where we are assuming that the particles are dilute enough to be considered independent, and again the compensating charges that ensure overall neutrality of the solution are somewhere "far away".

Because of the net charge per particle, an electric field will exert a corresponding force on each particle that in the steady state ends up balanced by viscous drag. The motion of the particles is *electrophoresis*. The particles move at at electrophoretic velocity u_{ep} relative to the fluid, which moves relative to the fixed channel walls at the electro-osmotic velocity.

Assuming a spherical particle of effective radius r_0 (which could differ from the strict "dry" radius of the particle due to adsorption of ions), the Stokes drag law implies

$$q_s F = 6\pi \eta u_{ep} r_0, \tag{10.50}$$

where q_s is the effective charge of the particle. For a spherical particle, and defining the radius r_0 as the same location where the electrostatic potential is ϕ_ζ,

$$q_s = 4\pi r_0^2 \left(-\kappa\epsilon_0 \left(\frac{\partial\phi}{\partial r} \right)_{r=r_0} \right) \approx 4\pi\kappa\epsilon_0 r_0 \phi_\zeta \left(1 + \frac{r_0}{\lambda_D} \right). \tag{10.51}$$

If $\lambda_D \gg r_0$, then

$$u_{ep} = \frac{2}{3} \frac{\kappa\epsilon_0\phi_\zeta F}{\eta}. \tag{10.52}$$

In the opposite limit of $\lambda_D \ll r_0$, the surface charge density is similar to that in the flat interface situation, and we end up finding

$$u_{ep} = \frac{\kappa\epsilon_0\phi_\zeta F}{\eta}. \tag{10.53}$$

The similarity of these limiting cases explains why electrophoresis is applicable over a broad range of particle sizes, from several microns (cells) down to a few nanometers

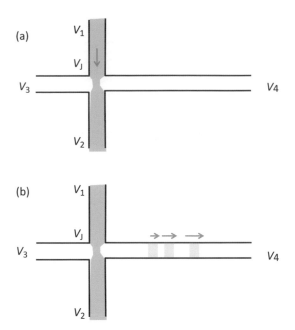

Fig. 10.15   Microfluidic capillary electrophoresis. By cycling voltages properly, "plugs" of particle-containing fluid (gray) may be extracted into the side channel. (a) Initially, voltages are set to drive EO and EP motion of the fluid/particles down, as indicated. (b) The same system after cycling voltages and allowing sideways flow of fluid plugs. The differences in electrophoretic velocities of the particles leads to the breakup of the plug into three distinct bands, corresponding to three different particle types. The width of the plugs is limited by diffusion.

(semiconductor nanocrystals). Electrophoresis is an experimental means of estimating the zeta potential for materials under particular conditions. For systems like coiled DNA, ϕ_ζ depends strongly on the molecular weight and valence state of the molecule that consitutes the particle, giving electrophoresis further analytical utility.

Figure 10.15 shows schematically how electro-osmosis and electrophoresis may be implemented in a microfluidic channel or capillary. A junction may be constructed between two channels filled with electrolyte, with the y-oriented channel containing particles for analysis. Electro-osmotic flow is directed by the application of voltages to the channel ends. Suppose the electrolyte/wall interactions are such that the electro-osmotic flow is directed from high to low voltage. Treating the electrolyte fluid in the channels as a simple effective resistor, the potential at the junction V_j may be found in terms of the other voltages. Setting $V_4 = 0$ as ground, and applying $+V_1$ as shown, one can set $V_3 = V_4$ to be slightly greater than $V_j \approx V_1/2$. The result is electro-osmotic flow between ports 1 and 2, with some confinement of the flow at the intersection, as shown.

Now briefly swap the direction of flow, by setting $V_4 = 0$, $+V_3$, and $V_1 = V_2 \sim V_3/2$. This pushes a "plug" of fluid toward port 4. By cycling voltages appropriately, multiple plugs can be sent downstream toward that port, separated by particle-free buffer electrolyte.

Any particles in each of those plugs will then separate out into bands sorted by ϕ_ζ as they undergo electrophoresis. Typically the particles are then imaged, either by direct microscopy or through other techniques such as fluorescence. The longer they are allowed to flow toward port 4, the larger the spatial separation between bands consisting of different particle types. The bands do get smeared over time due to diffusion of the particles in the electrolyte, with Brownian diffusion superposed on top of the electrophoretic and electro-osmotic motion. For a diffusion constant D, the bands get smeared over a lengthscale $\sim \sqrt{DL/u_0}$ as the fluid is driven electro-osmotically over a distance L. Being able to perform these experiments on small scales is clearly advantageous, as this minimizes blurring of any bands.

10.5.3 Dielectrophoresis

One can also use the dielectric response of particles and their ionic screening clouds to move particles in an electrolyte. This approach, *dielectrophoresis* [811], is rich and complex, as it depends on the frequency-dependent response of the particle, the screening cloud, and the fluid itself. Good reviews include those by Gascoyne and Vykoukal [812] and Pethig [813].

Electrodes are used to produce an electric field $\mathbf{F}(\mathbf{r}, t)$ within the fluid. If the particle is small compared to the distance scale over which the electric field varies, then one can treat the particle like a point dipole. The dielectric response of the particle (and its screening ions) leads to an induced dipole moment, $\mathbf{p}_{ind} \propto \mathbf{F}$. This induced dipole then experiences a force $\propto \mathbf{p}_{ind} \cdot \nabla\mathbf{F}$. Since the dipole is induced, the time-averaged force in this limit is really proportional to ∇F^2.

Note that in general a time-dependent $\mathbf{F}$ implies that one must be concerned about the frequency dependence of the dielectric response of the fluid, the particle material, and the screening cloud itself. For a harmonic time dependence, phase lags imply that the dipole moment could be instantaneously oriented parallel or antiparallel to $\mathbf{F}$ depending on the phase. That is, the dielectrophoretic force can either attract particles to regions of strong field, or repel particles from regions of strong field, depending on the material system and the frequency applied.

Nanostructures are ideal for dielectrophoresis, thanks to the ability to fabricate electrodes with features at the nanoscale, including points and corners with tight radii of curvature. Modest voltages (for example, below the threshold for the onset of electrochemistry) can produce both large electric fields and large field gradients. Dielectrophoresis has been employed for diverse applications, including cell sorting in biological contexts [814] and the assembly of nanoparticles into larger structures [815].

10.6 Microfluidic devices

Extensive reviews have been written about microfluidics, particularly since the advent of "soft lithography" (Section 4.4.4), including articles by Whitesides [816], by Squires and Quake [817], and by Stone, Stroock, and Ajdari [818]. A number of strong textbooks now also exist, discussing some of the topics described in this chapter in considerably greater depth and detail, as described in the Suggested Reading at the end of the chapter.

Microfluidics has not yet achieved the technological deployment level and market share of the other technologies that we have discussed (semiconductor micro/nanoelectronics; magnetic data storage; photonics; microelectromechanical systems). This is largely a question of the comparatively recent perfection of relatively inexpensive and simple fabrication methods. A broadened definition of microfluidics can encompass inkjet printers, which are able to dispense micron-scale droplets of ink through a microfabricated array of orifices on demand. However, if "true" microfluidics requires circulation and switching of fluid flow through integrated channels, pumps, and valves, then the subdiscipline has only existed at an industrial level for less than two decades.

The main envisioned area of application for microfluidics is in "lab-on-a-chip" technologies. By enabling the circulation, routing, mixing, and storage of small quantities of fluids on demand, microfluidic systems can perform complex chemical and biochemical procedures using much more reduced amounts of analytes or reactants than traditional, "full size" bench approaches. If the required fluid networks can be created inexpensively, reliably, and in large quantities, the potential exists for revolutionary changes in, for example, point-of-service medical diagnostics.

10.6.1 Essential components

The components necessary to create microfluidic (or nanofluidic) networks are rather analogous to those needed to control the flow of electrical current. Rather than wires, switches, and batteries to carry, direct, and source electrical current, fluidic networks require flow channels, valves, and pumps. Moreover, measuring the rate of fluid flow is often necessary, and for chemistry purposes it can be desirable to mix precise amounts of distinct fluids, a process for which there is no simple electronic analog.

Similar in spirit to MEMS as discussed in Chapter 9, the fabrication methods applied to microfluidics are often related to those used in planar semiconductor fabrication as in Chapter 4. Some microfluidic systems are "hard", based on the lithographic patterning and etching of rigid substrates such as silicon wafers or glass slides, the latter of particular interest for optical access to the fluid. Mechanical robustness is a virtue, as is the relative chemical inertness of glass when exposed to many fluids. However, properly sealing channel networks in rigid structures can be difficult, involving adhesives or wafer bonding. Likewise, moving parts such as valves can be challenging to implement with hard materials. An extensive discussion of fabrication of hard microfluidic systems is in Chapter 3 of Nguyen and Wereley [819].

Soft lithographic techniques involving PDMS have been revolutionary for microfluidics [820, 821, 822]. Recall that PDMS is an inexpensive silicone polymer that cures near room temperature, is biocompatible, and is optically transparent in the visible. Smooth, flat PDMS can form water-tight, intimate seals with clean, flat solid surfaces such as Si wafers or glass slides. Depending on surface treatment, PDMS can be peeled easily from other materials. Rigid materials may be patterned with typical lithographic and etching

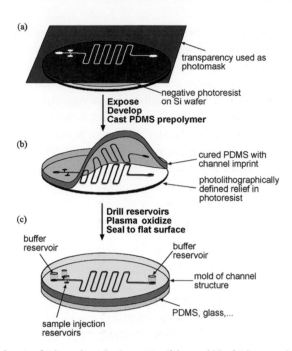

(a)

transparency used as
photomask

negative photoresist
on Si wafer

Expose
Develop
Cast PDMS prepolymer

(b)

cured PDMS with
channel imprint

photolithographically
defined relief in
photoresist

Drill reservoirs
Plasma oxidize
Seal to flat surface

(c)

buffer
reservoir

buffer
reservoir

mold of channel
structure

PDMS, glass,...

sample injection
reservoirs

PDMS replica molding for microfluidic applications. A negative of the would-be fluidic network is created using photolithographic patterning of a thick photoresist. A PDMS mold of the developed resist pattern is then made and attached to a flat substrate. Image from bdml.stanford.edu/twiki/bin/view/Rise/PDMSProceSS.html.

Fig. 10.16

techniques and then serve as a "master". PDMS inverted replicas of surface features may then be created, as shown in Fig. 10.16.

Upon exposure to an oxygen plasma, the surface of PDMS becomes hydrophilic, but more importantly the treated surface will then tend to bond covalently with either SiO_2 or other treated PDMS. In this way, PDMS slabs can act as permanent sealing surfaces for etched rigid structures, and complex networks may be constructed through layer-by-layer assembly of PDMS components. The mechanical flexibility of PDMS and its ability to form reversible or irreversible seals provide functionality for active components (valves, pumps) that would otherwise be challenging to achieve, as described below.

For mass production, particularly with an eye toward inexpensive, disposable fluidic networks for biomedical applications, there has been considerable work on replica molding or embossing of more rigid polymer materials, such as PMMA [823].

Flow channels

In rigid substrates, flow channels can be created through lithography coupled with either wet or dry etching methods. The variability of wet etch rates with chemical composition of the substrate and crystallographic direction can enable some clever structures, such as the pits shown in Fig. 4.15. Deep RIE has proven useful for creating flow channel networks, with the capability of etching tens or hundreds of microns into Si substrates.

Alternately, some techniques employ polymer resists as structural materials. For example, SU-8 photoresist is expressly designed to produce extremely thick polymer layers that can support high aspect ratio structures after exposure and development. Planar channel networks may be created directly by lithography, leaving the unexposed resist in place. Such resists can also be etched anisotropically using oxygen RIE. The resulting channels can be coated or further modified should the sidewall chemistry be important.

Instead of patterning channels, hard materials may be patterned to form ridges (inverse channels), and then used as templates for replica molding of PDMS. This has become the standard approach for creating large, complex microfluidic networks.

Fluid interconnects

Just as integrated circuit layout often requires interconnects or vias for electrical linkage of different layers of devices, in microfluidic networks it is often desirable to connect multiple layers of flow channels. With rigid materials this can be very challenging because of the need to achieve leak-free seals, ideally with a minimum of incorporation of additional materials like gaskets. The same sealing challenges exist in trying to connect rigid microfluidic networks to external, macroscale reservoirs such as syringe pumps.

In PDMS, the interconnect problem becomes extremely simple, as shown in Fig. 10.17(a). PDMS layers with appropriate grooves are fabricated individually, and an interconnect layer with vertical ports is similarly molded from a rigid master. After O_2 plasma treatment, the layers are stacked and joined. Dramatic, complex results are possible, as shown in Fig. 10.17(b).

Connecting PDMS structures with external fluid supplies can also be comparatively easy, thanks to the tendency of PDMS to form leak-tight seals against smooth surfaces.

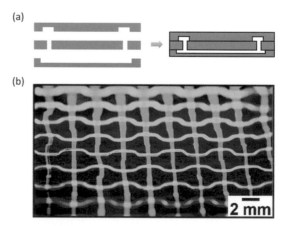

(a)

(b)

2 mm

Fig. 10.17   Fluidic interconnects using PDMS. (a) Separate layers of PDMS are created with flow channels running in desired directions parallel to the plane of interest; a third PDMS layer consisting of vertical interconnect penetrations is also constructed. (b) Following oxygen plasma treatment, these layers may be laminated into a stack that can have complicated fluidic topology. Image adapted from [822].

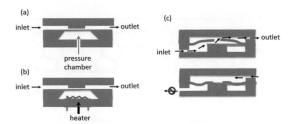

Examples of microfluidic valves. (a) Normally open pneumatic valve, where pressurizing the pressure chamber flexes a **Fig. 10.18** membrane that pinches off flow between the inlet and outlet. (b) A similar design, but actuated via heating of the sealed fluid volume. (c) A simple one-way valve leveraging the motion of a thin PDMS membrane.

A syringe needle, for example, may be inserted directly through PDMS into a reservoir volume formed by replica molding, with the PDMS self-sealing to the needle surface.

Valves

Valves are essential to controlling fluid flow, and face two critical challenges: reversible formation of a good seal to block fluid flow when required, and on-demand actuation. In rigid microfluidic platforms, the sealing challenge again can require interfacing with other, more flexible materials as gaskets and sealing surfaces. In PDMS microfluidics the polymer material itself comprises both sealing surfaces.

Valve designs are tied to actuation schemes. The most common actuation mechanism is pneumatic, where a pressure difference is exploited to move a sealing surface, either closing a normally-open path, or opening a normally-closed path. The pressure can be applied to some controlled volume (Fig. 10.18(a)), to switch flow between an inlet and an outlet. The pressure in the sealed chamber can be increased via heating (Fig. 10.18(b)), an approach called thermopneumatics. Alternatively, with properly designed flexible seals, the pressure difference that switches the flow can be applied directly from the inlet to the outlet (Fig. 10.18(c)) as in a one-way or check valve. As in MEMS/NEMS, electrostatic or electromagnetic forces may be used to actuate a valve, though this requires integration of electronic circuitry with fluidic components.

Surface tension may also be used to manipulate and switch flow, in the form of bubbles. As shown in Fig. 10.19, two-phase flow (in this case liquid and vapor) can lead to rich phenomenology, including complicated pattern formation. Geometry coupled with surface tension can energetically favor passage of bubbles in one direction, for example.

Pumps

In the absence of electrokinetic effects, moving liquids through flow networks requires establishing controlled pressure gradients. In many applications this is done through connections to macro-scale syringes that act as pressurized reservoirs. Ideally it would be preferable to have integrated pumping "on chip".

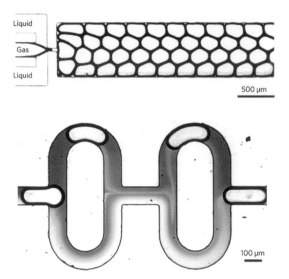

Fig. 10.19 Bubbles in microfluidic systems. Adding surface tension effects into the microfluidic toolkit can lead to complex pattern formation and additional tools for manipulation and control of fluids at small scales. Adapted from [816].

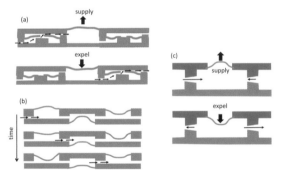

Fig. 10.20 Three approaches to microfluidic pumping. (a) Pneumatic actuation with two one-way valves. (b) Peristaltic pumping, where the directionality is provided by the cyclic pneumatic actuation. (c) Hydrodynamic pumping. While there is back-flow in this scheme, the broken symmetry of the inlet/outlet combined with viscosity leads to a net directionality of flow as the chamber volume is pneumatically cycled.

To achieve pumping requires some cyclic actuation, and some break in symmetry (geometry or time) that distinguishes a flow direction. Common approaches are shown in Fig. 10.20. One-way valves by their nature break the symmetry of flow direction. A chamber connected to an inlet and an outlet via appropriately oriented one-way valves can be manipulated by cycling the chamber geometry pneumatically. As the chamber is expanded, fluid fills the volume through the inlet; conversely, as the chamber is squeezed, fluid leaves through the outlet valve, so that one cycle results in a net transfer of fluid through the pump.

Figure 10.20(b) shows *peristaltic pumping*. The flow direction symmetry is broken by the order in which several sequential chambers are compressed and expanded. This approach can use valves to prevent backflow, or rely on hydrodynamics and pressure gradients to minimize it. Pneumatic manipulation of the chambers requires additional valving and pressure control [824].

The ultimate limit of the valveless hydrodynamic approach is shown in Fig. 10.20(c). An inlet nozzle and an outlet diffuser break the flow direction symmetry through their geometry and the resultant coupling to the viscosity of the fluid. While there is some backflow on every chamber compression, the viscous fluid is more readily pumped in the forward direction with each compression [825].

Flow sensors

Quantifying the rate of fluid flow in microfluidic networks can also be challenging due to the often tiny volumes of fluid being pumped [826, 827]. For example, a flow channel 10 μm wide, 10 μm deep, with an averaged (across the channel face) flow speed of 100 μm/s corresponds to a volume flow rate of 0.6 nL/min, and the situation only becomes more challenging as device scale is reduced.

Apart from direct optical imaging of entrained particles, additional methods for flow speed assessment include differential pressure measurements, drag force measurements, and time-of-flight studies of the fluid using, for example, the transport of heat.

The friction factor analysis of Eq. (10.7) combined with exact results for laminar, pressure-driven flow give a direct connection between pressure gradient and flow speed. For laminar flow,

$$\frac{\Delta P}{L} = u \frac{\mu}{2D^2} \, \mathrm{Re} \, f, \tag{10.54}$$

where $D \equiv$ the "hydraulic diameter" of the channel (four times the cross-sectional area divided by the perimeter). For circular channels, Eq. (10.27) can be manipulated to show that $\mathrm{Re}\, f = 64$. For rectangular channels, depending on the aspect ratio, $\mathrm{Re}\, f = 50 - 60$.

For the 100 μm/s flow discussed above, with $D = 10$ μm, $\mu = 10^{-3}$ kg/m-s for water, and $\mathrm{Re}\, f \approx 50$, we find $\Delta P/L \approx 500$ Pa/m, implying a pressure drop along a 100 μm channel of 0.05 Pa. While it is not easy, such small pressure differences are measurable.

One pressure measurement technique is to use capacitive sensing to detect the deformation of a thin diaphragm or membrane, as shown in Fig. 10.21. Using the continuum elasticity methods of Chapter 9, the deflection as a function of radius, $y(r)$, of a circular membrane of radius R and thickness t anchored at its perimeter and subjected to a pressure differential p is given by

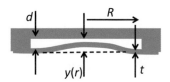

Capacitive pressure transducer, based on a flexible membrane with an integrated electrode.

Fig. 10.21

$$y(r) \approx \frac{3(1 - \nu^2)}{16Yt^3}(R^2 - r^2)^2 p, \tag{10.55}$$

where Y is the Young's modulus and ν is the Poisson ratio of the membrane material. The capacitance between this membrane and a fixed, flat electrode initially (when $p = 0$) a distance d away is sensitive to the deflection. Integrating up contributions to the capacitance, the total change is

$$\frac{\Delta C}{C} \approx \frac{(1 - \nu^2)}{16Yt^3 d}R^4 p. \tag{10.56}$$

For reasonable materials and dimensions, the sensitivity to pressure changes can easily be well below 0.01 Pa.

Alternatively, the deflections of such membranes may be detected by piezoresistive measurements of the membrane material or a coating on it. Optical techniques can also be applied to detect membrane deflections, and these approaches may be calibrated by the application of known pressure differentials.

Viscous drag on some flow obstruction is another means of accessing flow speed. The Stokes drag law is a particular realization of a more general expression. The drag force on an object (say a cantilever) inserted into a flow has two contributions, one linear in the flow velocity (analogous to the viscous drag of the Stokes drag law), and one quadratic in flow rate ($\sim \rho u^2$, originating from the momentum transfer between the fluid and the object). In the laminar (low Re) regime, the first term dominates. The deflection of a cantilever, paddle, or perforated membrane (measured optically or piezoresistively) then transduces the flow speed.

Thermal techniques are also used. For a system where the flow of the fluid itself takes place more quickly than thermal conduction through the fluid or the channel walls, time-of-flight measurements are possible. A heater integrated into the wall of the flow channel may be pulsed, locally elevating the temperature of a volume of fluid. Downstream, a temperature sensor (resistive thermometer or thermocouple) integrated into the channel wall can respond to the arrival of the "slug" of warm fluid, with the time delay and known heater/thermometer distance giving the flow speed.

Another thermal approach is to use a single resistive element as both a heater and temperature sensor. The temperature of the element is elevated above that of the fluid via Joule heating. The flow continually brings cool fluid into contact with the heater while carrying warm fluid away. Using the temperature dependence of the heater's resistance as an effective thermometer, the current applied to the heater can be varied to maintain a constant heater resistance (and hence average temperature). The feedback current can then be monitored to infer the changes in the flow rate. Such a thermal feedback system can be calibrated with known flow rates, or empirical dimensionless expressions relating heat transfer to parameters like Re may be used to try to infer quantitative information about flow speeds.

Mixing

We rely on diffusion to perform the intimate mixing of fluids, and at the macroscale turbulence effectively stirs mixing fluids into eddies with nanoscale transverse dimensions,

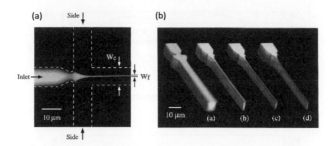

Fig. 10.22

Flow focusing. (a) Inflowing side streams of fluid may be used to hydrodynamically compress the width of a central stream, to shorten the timescale for diffusive mixing across the transverse direction. (b) Confocal microscopy shows squeezing down to a transverse width below 100 nm. Adapted from [828].

so that the timescale for diffusion across such eddies is conveniently short. As mentioned in Section 10.3.4, in the absence of turbulence it can be challenging to mix fluids efficiently in microfluidic systems. Given the interest in using microfluidics to study chemical processes, this is problematic.

There are several approaches to enhancing mixing in microfluidic environments. Figure 10.22(a) shows an example of *flow focusing*. Using inflowing side channels with well-chosen flow rates and pressures, a liquid of interest may be "squeezed" down hydrodynamically to small (tens of nanometers) transverse dimensions. This small flow width leads to diffusive mixing times across the flow down to the microsecond regime [828]. This can be generalized and parallelized, with multiple parallel streams being introduced. This allows the production of controlled concentration gradients flowing parallel to one another within a single channel. A related approach is *laminate* mixing, where two (or more) input streams are split into many parts, and then recombined in an interleaved manner, again seeking to create comparatively narrow adjacent regions of contrasting fluid types.

A second approach that is directly physics-based is the use of *chaotic* mixing. Surface relief or other flow channel geometry can be structured to introduce chaotic motion of fluid elements in the flow. In Fig. 10.23(a), periodic surface relief along the floor of a flow channel couples to the viscosity to lead to vortical motion during flow down the channel. Breaking the symmetry of that surface relief on length scales smaller than the period of the vortical motion, as shown in Fig. 10.23(b), can lead to the break up of the streamlines into whorls and eddies on a transverse scale similar to that seen in turbulent systems, enabling more rapid diffusive mixing [829].

10.6.2 Integration

Where soft lithography has truly been remarkably enabling is in the integration of fluidic devices into complex networks, often inspired in operation and topology by microelectronic circuits [824, 830]. Thanks to pneumatically actuated valves based on PDMS, multiplexed fluid flow is possible, with the idea shown in Fig. 10.24(a). The horizontal lines shown reside in a pneumatic control layer, with the wider regions indicating regions with flexible "floor" membranes that can be deflected downward with the application of pressure,

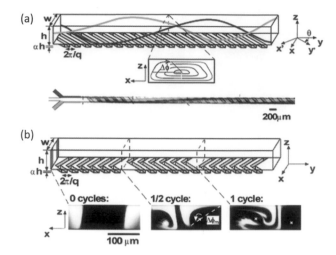

Fig. 10.23   Chaotic mixing. (a) Spatially periodic surface relief can couple to viscosity to instill vorticity into laminar flow, as shown by the helical motion of a stream containing two dyes. (b) Breaking up the helical motion by altering the symmetry of the surface relief on a length scale comparable to the fluid's motion can break the streams up into narrow eddies, facilitating mixing. Adapted from [829], with permission of the AAAS.

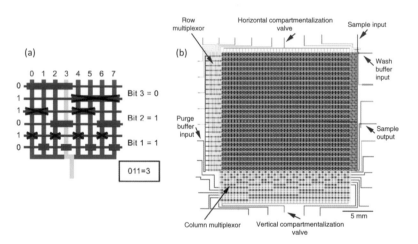

Fig. 10.24   Multiplexed microfluidics. (a) Layered pneumatically actuated valves configured so that each horizontal line clamps off half of the underlying vertical channels. This arrangement allows binary multiplexing. (b) A practical implementation with two multiplexing networks, resulting in 1000 individually addressable chambers that may be filled or flushed as desired. Adapted from [830], with permission of the AAAS.

pinching off flow paths in the underlying vertical layer. Note that the placement of these valves means that each horizontal line pinches off half of the available flow paths, allowing binary specification of a particular open channel or channels.

This is taken to its logical limit in Fig. 10.24(b), where two such binary multiplexing systems have been arranged to allow fluid addressing of specific chambers out of an array

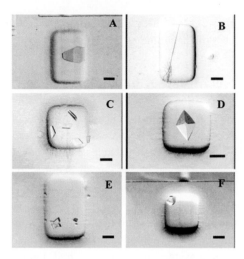

Microfluidics for protein crystallization. The quiescent and well-controlled environment possible in PDMS microfluidic chambers is amenable to the crystallization of proteins for structural analysis via x-ray diffraction. Adapted from [831]. **Fig. 10.25**

of 1000, each chamber with a volume of about 250 picoliters [830]. With this approach, if desired an operator can purge or fill individual chambers one at a time. This capability can enable massively parallel chemical assays, or combinatorial chemistry approaches.

One particular application is shown in Fig. 10.25 [831]. A difficult but standard approach to obtaining the microscopic structure of complicated proteins is to crystallize the macromolecules for x-ray crystallography. Protein crystallization can be very finicky, requiring precisely controlled reaction conditions and a quiescent environment. Integrated microfluidics enables the exploration of a huge parameter space of crystallization conditions while ideally using only a minuscule amount of protein. The ability to vary nearly buffer volumes, pumping rates, reactor geometries, etc. with comparative ease increases the appeal of the microfluidic approach [832]. Similar ideas have been applied to drug discovery [833]. These successes and the manifest versatility of microfluidic tools imply that these techniques will have a long, exciting, innovative future.

10.7 Nanofluidics

The utility and potential technological impact of microfluidics should now be clear. Analogous to our approach of previous chapters, we now consider the nanoscale frontier of fluids, looking at the new physics that becomes relevant at the nanoscale, and considering these changes bring to light new opportunities. Two natural areas of concern come to mind.

- Do hydrodynamic or traditional fluid mechanics/dynamics assumptions fail in the nanoscale limit? If so, what are the consequences?
- Does the continuum approximation underlying the standard treatment of fluid mechanics break down? If so, what are the consequences?

When answering these questions we discuss how fluids are confined and studied on the nanoscale, and possible uses of "nanofluidics".

10.7.1 Fate of the no-slip condition

The no-slip boundary condition can, in fact, fail badly at the nanoscale. It is an idealization that assumes sufficiently strong attractive interactions between the fluid constituent particles and the solid wall, so that fluid particles are brought to a relative halt upon collisions with the interface. However, as we saw in Section 10.4, it is possible to tune fluid–solid interactions by using surface chemistry or other means to alter the interfacial energies.

There must be limits to how badly the no-slip condition can fail at the macroscale, in the sense that it has functioned as a good description of macroscopic fluid flow even when considering, for example, water flowing through hydrophobic fluoropolymer tubing. That implies that velocity profiles on the macroscale still show a steady decrease as a sidewall boundary is approached, as in Fig. 10.26. Slip may be characterized by a *slip length*, L_{slip}, defined as the distance into the wall material that one would have to traverse before the fluid velocity extrapolates to zero. In flow systems with channel dimensions on the micro/nanoscale, it is not surprising that L_{slip} can be noticeable for poorly wetting interfaces.

An example of an experiment to probe this effect is shown in Fig. 10.27(a). Two atomically smooth (mica-coated) cylinders of radius R are immersed in fluid of viscosity μ, one fixed and one attached to an actuator that oscillates their vertical separation, d,

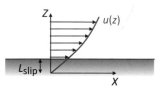

Fig. 10.26 Definition of slip length. The flow velocity decreases near boundaries but does not reach zero unless the velocity field's trend is extrapolated into the solid region by a distance L_{slip}.

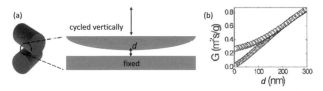

Fig. 10.27 Demonstration of slip. (a) Cylinders coated with atomically smooth mica are crossed and immersed in a fluid, with a separation between them d that is varied in time by oscillating the upper cylinder. (b) The parameter G, defined in the text, is expected to vary linearly with d if the no-slip condition is obeyed. Circles are data for tetradecane on bare mica, a wetting interaction. Diamonds are for tetradecane on mica coated with a SAM that discourages wetting, leading to a failure of the no-slip condition, with slip lengths on the 1 micron scale. Adapted from [834].

at a frequency ν such that the surrounding fluid remains in the laminar regime. Solving the Navier–Stokes equations for this geometry assuming no slip leads to a prediction that the parameter $G \equiv 6\pi R^2 \nu / F_{peak} = d/\mu$, where F_{peak} is the peak force applied in the oscillatory cycle. One can perform a series of measurements of the peak force as a function of d for a variety of fluid combinations or surface chemistry treatments of the cylinders. The no-slip expectation would be that a plot of G vs. d should extrapolate linearly to zero as $d \to 0$.

A representative result [834] is shown in Fig. 10.27(b). When the fluid (in this case tetradecane) wets the cylinders (bare mica) well, the expected no-slip result is observed (circle symbols). However, when the cylinder surfaces are functionalized with a methyl-terminated self-assembled monolayer, leading to a weaker wetting interaction, there is a pronounced deviation. Further analysis implies that poor wetting in this geometry can lead to hydrodynamic forces $10^{-2} - 10^{-4}$ times what would be expected in the absence of slip. Slip lengths as large as the micron scale are inferred at high frequencies and small cylinder separations.

For a more extreme example of this, consider the situation of water flowing through carbon nanotubes. Recall that Poiseuille regime flow rates like the radius of the flow channel to the fourth power, assuming no slip, suggesting that it should be difficult to flow water through truly nanoscale "pipes" like carbon nanotubes. However, the graphene surface is comparatively hydrophobic (contact angle near 95 °) [835], suggesting that this might be a good system in which to observe slippage. Experiments [836, 837] have shown flow rates of water orders of magnitude higher than the no-slip expectation.

By controlling the interactions between the fluid and the adjacent solid walls, it is clearly possible to achieve fluid slip that is readily measurable in micro/nanoscale systems. On a complementary note, it is also possible to use nanoscale engineering of material structure and composition to achieve giant *effective* liquid–solid surface energies on the macroscale, and hence remarkable slip. Consider creating a surface with nano- and microscale relief such that the true surface area greatly exceeds the apparent macroscale surface area. This effectively boosts the energy cost per macroscale liquid–solid interface area, leading to large contact angles. Combining appropriate heirarchal structural motifs and chemical termination, it is possible to create *superhydrophobic* surfaces, so that contact angles for water (and other polar liquids) can approach 180 °, as has been done with forests of carbon nanotubes [838], leading to slip lengths of several microns [839]. Related approaches can create *superoleophobic* surfaces that "repel" nonpolar liquids like oils and alkanes [840], and clever combinations can produce *superomniphobic* surfaces that are resistant to wetting by both polar and non-polar liquids [841]. The resulting macroscale effects can be very striking, and the potential for industrial applications seems significant.

10.7.2 Fate of the continuum approximation

The Knudsen number tells us that we should run into problems with the continuum description of fluids when the fluid is so confined that constituent molecules collide with the boundaries of their environment at rate comparable to or exceeding their collisions with each other. Likewise, common sense dictates that confining fluid molecules on scales comparable to the molecular volume should lead to deviations from the properties of the bulk, unhindered fluid.

What should happen, particularly for liquids, under these circumstances? When highly confined or within molecular scales of a solid interface, the relative positions of the molecules can become more constrained. The density–density correlation function (Eq. (3.1)) begins to resemble that of a solid, with peaks developing at particular interparticle separations, rather than just at the nearest-neighbor distance. On the molecular scale, the concept of a continuum, bulk density fails.

Remarkably, this can be seen in experiments. For example, Figure 10.28(a) extends the atomically smooth mica cylinder approach by silvering the back sides of the mica sheets [842]. The result is an optical cavity partly filled by the fluid between the cylinders. Using a micron diameter spot of light and a nanoscale separation between the cylinders, the experiment can interrogate attoliter volumes of fluid. The transmittance of the silver/mica/fluid/mica/silver stack is sensitive to the index of refraction of the fluid, in this case cyclohexane. Recall that the fluid index of refraction, n, is directly related to the spatial density of the fluid molecules (Eqs. (5.9, 5.10)). By using white light, the investigators can cross-correlate between different frequencies, since genuine fluctuations in the molecular arrangement (as opposed to spurious noise) should affect the whole spectral response. The results in Fig. 10.28(b) are striking in two ways. First, it is clear that when the cylinder separation approaches the molecular size, there are discrete steps in that separation, implying that there are preferred molecular arrangements that satisfy the constrained geometry. Second, in the very confined limit, the effective index of refraction fluctuates enormously, consistent with tremendous variations in the instantaneous (number) density of the 2d confined fluid.

The fluid's mechanical properties may be strongly affected. An essential difference between a fluid and a solid is that a fluid has no zero-frequency shear modulus, and instead has a viscosity. Under strong confinement, one might expect some intermediate properties, and this has also been seen experimentally. For example, again using the crossed mica cylinders geometry, Zhu and Granick [843] look at the mechanical response when an oscillatory force at some frequency ω is applied to slide one cylinder across the other with

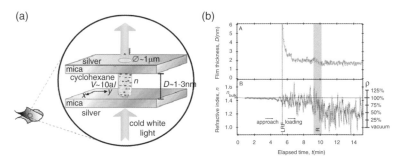

Fig. 10.28 Density fluctuations of confined fluids. (a) A version of the mica cylinder approach is used to create a nanoscale optical cavity. Cross-correlation of optical fluctuations across a broad wavelength band distinguishes genuine density fluctuations. (b) Cylinder spacing exhibits plateaus corresponding to molecular layering as cylinders are forced together; this coincides with very large effective density fluctuations. Adapted from [842], with permission from AAAS.

water as the nanoconfined fluid between the mica surfaces. At nanoscale separations, they observe large, frequency-dependent enhancements of the inferred (complex) viscosity, and these depend on the relative crystallographic orientation of the mica sheets. The implication is that the fluid molecules take on certain ordered arrangements at the mica surfaces. (Note that this experiment also raises another issue mentioned in Section 10.5. The effective ionic concentration in the nanoconfined water is much larger than the bulk ion concentration, and in the region of interest charge neutrality is not preserved, because the double layer regions of the two surfaces overlap.)

The lack of a shear modulus also implies that transverse sound waves do not propagate in fluids. As a fluid is nanoconfined and its viscous response looks increasingly like a zero frequency shear modulus, new vibrational modes related to transverse sound should arise. Low frequency vibrational modes have been observed in nanoconfined water [845, 844] using terahertz optical spectroscopy. As shown in Fig. 10.29(a), reverse micelles, in which a surfactant holds water molecules in a nanoscale volume, are a convenient tool for examining truly nanoscale fluid confinement. The absorption in the low terahertz regime (1 THz $= 33$ cm^{-1}) increases dramatically (Fig. 10.29(b)) as reverse micelle radius is decreased to the nanometer limit, and the peak absorption moves to higher frequencies. New vibrational modes are emerging as confinement is made more strict, and those modes are becoming progressively stiffer.

Note that these confinement effects really become noticeable only in the limit that the confining volume or dimension approaches the scale of the fluid constituents. This implies that, absent interactions with solid boundaries, liquids comprising ensembles of identical, single atoms should be well described by the continuum approximation down to nearly the atomic scale. Consider the pinch-off of a drop of mercury [846], where the connection between the drop and a larger liquid reservoir is a truncated cone of opening angle θ, with a truncation diameter d. A scaling analysis (Section 10.2) shows that if the time remaining before pinch-off is t, then $d(t) \propto \left(\frac{\sigma_{lv} t^2}{\rho}\right)^{1/3}$ where ρ is the (continuum) fluid density, and σ_{lv} is the (continuum) liquid–vapor surface tension that governs the pinch-off process.

Mercury has an electrical conductivity σ_e, and for the truncated cone geometry, the electrical resistance of the constriction between the drop and the bulk fluid should be $R =$

(a)   (b)

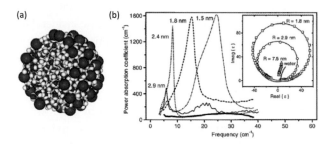

Evolution of vibrational response of nanoconfined fluids. (a) A reverse micelle structure may be used to confine water within a small volume. Adapted from [844], with permission from AAAS. (b) Water confined in reverse micelles exhibits dramatic changes in its far-infrared optical response due to changes in its vibrational properties. Unconfined water response is shown in by the heavy black curve; labeled curves correspond to inverse micelles of various calculated internal radii. Adapted from [845].

Fig. 10.29

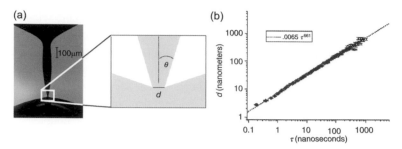

Fig. 10.30 Pinch-off of a fluid drop and the fate of the continuum approximation. (a) A mercury drop pinching off from a larger reservoir has a constriction geometry that approaches a universal shape. (b) The diameter of the constriction (inferred from the electrical resistance) as a function of time remaining to pinch-off obeys the expectations of a continuum scaling treatment down to constriction diameters of a few atoms. Adapted from [846].

$2 \cot(\theta)/(\pi \sigma_e d)$. For the inviscid case (not obviously valid here), $\theta \to \approx 18°$ as $d \to 0$. This means that an electrical measurement of the resistance as a function of time should allow the inference of $d(t)$. The results of such a measurement are shown in Fig. 10.30(b), where it is found that $d(t)$ does, indeed, vary approximately like $t^{2/3}$, as predicted by a continuum-approximation scaling analysis, down to constriction diameters of few atoms.

10.7.3 Manipulating nanoscale fluids and nanoscale objects within fluids

Applications for nanofluidics are emerging and tend to take advantage of the control of fluid velocity gradients afforded by the nanoscale engineering of structures. The sorting of macromolecules and nanoparticles gives two examples of this idea.

Macromolecules such as DNA can be complicated in solution, tending to coil up due to basic entropic reasons – energetics aside, there are far more bunched-up conformations of a macromolecule than there are extended conformations. Consider a desire to stretch out macromolecules, perhaps to sequence individual strands of DNA. This might entail threading individual macromolecules through channels too narrow to allow multiple molecules to pass. Coiled macromolecules impinging on such nanofluidic channels simply cannot fit. However, if the fluid containing the molecules is first incident on an arrangement of micro/nanoscale obstacles [847], the situation changes considerably. Because of the no-slip condition and the strong effects of viscosity in this low Re regime, the obstacles establish strong velocity gradients around themselves. These gradients exert shear, tending to uncoil and extend the molecules, allowing them to entrain into the nanoscale channels, as shown in Fig. 10.31.

A similar clever idea may be used to sort nanoparticles by size. An array of mesoscale obstacles staggered appropriately can force the laminar flow to divide asymmetrically as it passes each row of obstacles. Depending on the relative sizes of the nanoparticles (carried along in the flow via hydrodynamic drag) and the widths of the flow branches, particles of a chosen size can be shunted progressively in one lateral direction. Figure 10.32 shows the results, with an obstacle array designed to pull off various sizes of fluorescent particles sequentially.

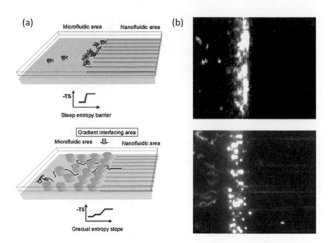

Entropic sieve. (a) Principle behind the entropic sieve. Viscosity-driven velocity gradients around mesoscale obstacles shear long-chain DNA molecules so that they uncoil and can be entrained into nanofluidic channels. (b) Fluorescence microscopy images demonstrating this idea, corresponding respectively to the diagrams in (a). Adapted from [847].

Fig. 10.31

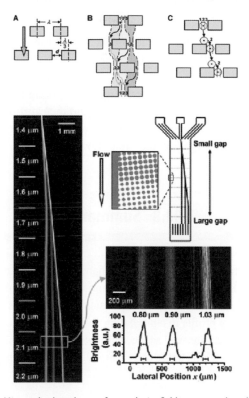

Microfluidic particle sorting. Microscale obstacles can force velocity fields to vary on lengthscales comparable to the sizes of entrained particles. This can be used to sort micro/nano objects by size. Adapted from [848], with permission of AAAS.

Fig. 10.32

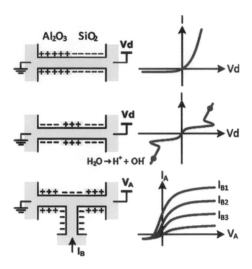

Fig. 10.33 Nanofluidic devices leveraging ionic gradients. Combining nanoscale flow channels with varying wall materials produces sharp gradients in ionic charge density, particularly if transverse dimensions are small compared to λ_D. The resulting ionic current-voltage characteristics can be rich. Adapted from [849].

The nanofluidic size regime also enables more sophisticated manipulation of electrokinetic effects, particularly if flow channel materials may be engineered on the nanoscale. Figure 10.33 shows some of the possibilities. Flow channels may be designed so that the sign of the Stern/Gouy–Chapman layer charge varies along the solid–liquid interface. Depending on the concentration of ions in solution and their diffusion rates, it is possible to create nanofluidic analogs of diodes and transistors for fluid flow [849]. The ability to rectify or amplify ion-charge-specific fluid flow could be useful in lab-on-a-chip applications.

10.8 Summary

We introduced the general concepts of fluid mechanics, emphasizing the laminar, incompressible regime as that most relevant to micro/nanofluidics. After presenting dimensional analysis and scaling, we solved a couple of classic fluid problems of interest (fluid flow along a streamline; velocity profile in a pipe; drag force on a spherical object). We touched on surface and interface effects as well as electrokinetics. After a discussion of microfluidic devices, we concluded with a look at the truly nano regime. Key points to take away include the following.

- Continuum fluid properties and viscosity. Ordinary fluids are approximated as continuous media described by a density ρ and a velocity field $\mathbf{u}(\mathbf{r})$. Fluids deform continuously when sheared. Viscosity, originating from the microscopic dissipative interactions between fluid particles, acts as a diffusion constant for momentum. In a

Newtonian fluid, viscosity produces a shear stress proportional to the gradient of the flow velocity.

- Dimensional analysis. The constraint that the behavior of the physical universe is independent of our choice of units, and the requirement that our units systems must be internally consistent, leads to the Buckingham Pi Theorem. A system parametrized by n dimensional variables using r fundamental units can be described using $n - r$ dimensionless products. The Reynolds number is an example, and is interpreted as the ratio between inertial and viscous forces.

- Control volumes and Navier–Stokes. Either we can apply conservation of mass and conservation of momentum to a region (control volume), or we can apply Newton's Laws to a blob of fluid (leading to the Navier–Stokes equations). In either case, we can then find equations describing $\mathbf{u}(\mathbf{r})$. The no-slip condition ($\mathbf{u} \to 0$ at a solid wall) is commonly assumed.

- Low Reynolds number. In micro/nanofluidics we generally operate in the low Re/laminar flow regime, implying that viscous forces dominate, turbulence is absent, and velocity fields are often rather simple.

- Poiseuille flow and Stokes drag law. Laminar flow in a pipe leads to parabolic velocity profiles and volume flow rates that scale linearly with the average pressure gradient. A spherical particle feels a viscous drag force linear in its radius, the viscosity, and its velocity relative to the fluid.

- Surface effects. Wetting and contact angles depend on the relative magnitudes of surface energies. Surface energies originate from the competition between intermolecular interactions (liquid–liquid; liquid–vapor; liquid–solid), and these can be tuned via surface chemistry. The no-slip approximation can fail with appropriate surface interactions when examined on the nanoscale.

- Electrokinetics. Electrolytes are liquids containing dissolved ions. Solid interfaces with electrolytes are generally charged and are screened on the Debye screening length scale via electric double layers. Screening improves with increasing ionic concentration and decreasing temperature. Applied electric fields can exert forces on the net-charged double layers, leading to electro-osmotic flow. Particles in solution can be net-charged, and electric fields lead to electrophoretic flow relative to the fluid. Dielectric response in inhomogeneous electric fields can also produce dielectrophoretic forces.

- Nanofluidics. Fluids confined on lengthscales comparable to the molecular regime can act more like solids, with hints of an emerging shear modulus; likewise, molecules can take on specific conformations when confined near surfaces. The continuum approximations of well-defined density, surface tension, etc. break down at the molecular scale.

10.9 Suggested reading

- *Basic fluid mechanics* There are many good introductory fluid mechanics texts. The all-time classic is probably *Fluid Mechanics*, 2nd Edn. by Landau and Lifshit (Butterworth-Heinemann, 2003). A similarly physicsy take is given in G. Falkovich's *Fluid Mechanics:*

a Short Course for Physicists (Cambridge University Press, 2011). My discussion of the basics is patterned after that in *Introduction to Fluid Mechanics*, 7th Edn. by R. W. Fox, A. T. McDonald, and P. J. Pritchard (Wiley, 2009).

- *Dimensional analysis and scaling* A very complete take on dimensionless products and the like is J. C. Gibbings' *Dimensional Analysis* (Springer, 2011). Sanjoy Mahajan has a broader, fun approach in *Streetfighting Mathematics* (MIT, 2012).

- *Microfluidics* The landscape of microfluidics books has been evolving rapidly. For a very complete description of fabrication, particularly regarding hard materials, try *Fundamentals and Applications of Microfluidics*, 2nd Edn. by N.-T. Nguyen and S. T. Wereley (Artech, 2006). Another complete look is given by Patrick Tabeling's *Introduction to Microfluidics* (Oxford University Press, 2010). Henrik Bruus' *Theoretical Microfluidics* (Oxford University Press, 2008) presents a more rigorous look at the fluid mechanics, including electrokinetic effects, than this chapter. As mentioned previously, there are also excellent review articles by Whitesides [816] and by Squires and Quake [817] that focus on soft microfluidics.

- *Nanofluidics* At present there are few pedagogical books that touch on the nanoscale limits of fluid mechanics; most nanofluidics books are collections of topical reviews. Brian Kirby's *Micro- and Nanoscale Fluid Mechanics: Transport in Microfluidic Devices* (Cambridge University Press, 2013) has a very complete treatment of the physics of microfluidics and touches on the nano limit. *Nanofluidics* by P. Abgrall and N.-T. Nguyen (Artech, 2010) is comparatively brief but good. A strong review article with an emphasis on electrokinetics in nanoscale channels is R. B. Schoch, J. Han, and P. Renaud, "Transport phenomena in nanofluidics" [850].

Exercises

10.1 *Melting point suppression for nanoparticles* It is well known that nanoparticles melt at a temperature, T_m, that is below the melting point of the bulk system, T_b. Consider a nanoparticle of diameter d made out of atoms of a diameter a. Further, let's define the energy cost of having at atom in the bulk of the material to be α, and the energy cost of having an atom at the surface of the material to be γ. (Note that deep down these two parameters aren't independent! They both reflect some underlying energy per bond that holds an atom in the material. It can be OK to use non-independent variables like this as long as we're careful with our counting.) Treat the temperatures as having the units of energy (this is equivalent to working in a units system where $k_B = 1$, which is fine; alternatively, you can write $k_B T_m$ and $k_B T_b$ if that makes you more comfortable).

(a) How many independent variables do we have, and how many different fundamental units do we have? How many independent dimensionless products can we therefore expect to have?

(b) We can use physical reasoning to pick which combinations of dimensionless ratios make the most sense and are easiest to interpret, and we can make more progress. Explain why it makes physical sense for one dimensionless parameter to be $[(d^2/a^2)\gamma]/[(d^3/a^3)\alpha] = a\gamma/d\alpha$. The relatively simple interpretation of this ratio is the reason that we chose to use the non-independent variables α and γ.

(c) Like the pendulum example, we also have additional knowledge about our physical system. When the particle size becomes very large, the melting point has to end up at the bulk value. Please write a simple expression for (T_m/T_b) that obeys this constraint and should apply reasonably well when the melting point suppression (which is basically a surface effect) is relatively small. Two-word hint: linear approximation. A research paper relevant to this result is Nanda *et al.*, *Phys. Rev. A* **66**, 013208 (2002).

10.2 *Capacitively actuated fluid deformation* Suppose one wants to use an electric field F to alter the shape of some blob of incompressible fluid with surface tension σ. This is possible in principle because the fluid has some dielectric response, given by $\epsilon \equiv \kappa\epsilon_0$, so it has the same units as ϵ_0. Consider an initially spherical blob of material with a radius r, and ignore gravity! Work in SI units.

(a) In the presence of an electric field, the blob will deform. Since the fluid is incompressible, its volume will remain constant, but its asperity (how elongated it is: bigger asperity = more elongated) will change. Let's define the asperity as a dimensionless number α. How many variables do we have, and how many different fundamental units do we have? How many independent dimensionless products can we therefore expect to have? (Remember the example with the pendulum: we still counted the dimensionless variable θ as a variable! Recall, too, that a dimensionless independent variable, when raised to some power, remains dimensionless.)

(b) Assuming that the asperity is expected to be linearly proportional to the electric field, give an expression for α in terms of the other variables. Do you think capacitive manipulation of a liquid blob's shape is likely to be easy or difficult at the nanoscale?

(c) Writing the answer to (b) as a dimensionless quantity that needs to remain constant, try to give a one or two sentence interpretation of the significance of this quantity.

10.3 *Control volume analysis of a hydraulic jump* A hydraulic jump is shown in side-view in Fig. 10.34. The dotted line represents the control volume to use. Assume the channel here extends a distance b into the page. Assume the only force acting on the fluid in the control volume is pressure due to the fluid at either end of the volume.

(a) Given that the hydrostatic pressure at a depth d is given by $\rho g d$, show that conservation of momentum implies

$$\frac{u_1^2}{g}y_1 + \frac{y_1^2}{2} = \frac{u_2^2}{g}y_2 + \frac{y_2^2}{2},$$

independent of atmospheric pressure.

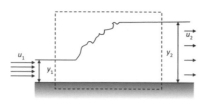

Fig. 10.34   Hydraulic jump.

(b) Combining the result from part (a) with conservation of mass, show that the ratio of final depth to initial depth is given by

$$\frac{y_2}{y_1} = \frac{1}{2}\left(\sqrt{1 + 8\mathrm{Fr}_1^2} - 1\right),$$

where Fr is the Froude number, $u/\sqrt{gy}$. Notice that you can only have a hydraulic jump from a region where Fr > 1 to a region where Fr < 1. The Froude number here is the ratio of flow velocity to the speed of surface waves on the fluid. A similar, analogous relationship holds for speeds on either side of a shock wave in compressible flow, where there the relevant parameter is the Mach number.

10.4 *Viscous, laminar, incompressible flow between two parallel plates* Assume a fluid of viscosity μ is flowing steady state in the x−direction between two plates, one at $y = 0$ and one at $y = a$. The x-velocity $u(y)$ is a function of y, and we will assume the no-slip condition at the walls, so that $u(y = 0) = u(y = a) = 0$.

(a) We will ignore gravity. Consider a control volume of differential size, $dxdydz$. Assume that the pressure in the center of the control volume (CV) is $p(x)$, and that the shear stress in the center of the control volume is $\tau_{yx}(y)$. There is no net change in the momentum of the fluid as it passes through the control volume, so the pressure forces on the left and right walls of the cv must be balanced by the shear forces on the top and bottom surfaces of the cv. Use this information to show that $-\partial p/\partial x + \partial \tau_{yx}/\partial y = 0$.

(b) Integrate with respect to y, and plug in the definition of shear stress in terms of viscosity to find an equation relating u and y in terms of the pressure gradient. Remember to use your boundary conditions on u to find the constants of integration. Show that the velocity profile ends up being parabolic:

$$u = \frac{a^2}{2\mu}\frac{\partial p}{\partial x}\left(\left(\frac{y}{a}\right)^2 - \left(\frac{y}{a}\right)\right).$$

Sketch this profile.

10.5 *The no-slip condition* This problem refers to Joseph et al., "Slippage of water past superhydrophobic carbon nanotubes forests in microchannels", *Phys. Rev. Lett.* **97**, 156104 (2006).

(a) Explain very briefly what the superhydrophobic surface is in this experiment and how it is made. Also, take a look at Figures 1 and 2 from *Phys. Rev. Lett.* **96**, 066001 (2006), which show the Cassie state of flow. Give a one or two sentence qualitative explanation for why you might expect to see large slip lengths for water flowing over such a surface.

(b) Explain how Joseph *et al.* were able to measure velocity profiles and therefore infer slip lengths. What is the Wenzel state? Physically, why does the slip length differ so much between the Wenzel and Cassie states?

10.6 *Nanofluidic membranes* Look at the following paper: Holt *et al.*, "Fast mass transport through sub-2-nanometer carbon nanotubes", *Science* **312**, 1034 (2006).

(a) Read the description of how the membranes are prepared. How is the pore size estimated (two ways)?

(b) For non-hydrocarbon gases, how does relative flow rate scale with molecular weight of the gas? Why is this expected? What is the suggested explanation for the enhancement of gas flow rate above that expected for the Knudsen regime?

(c) Conceptually, why might it be difficult to interpret the enhanced flow of water through these very small pores in terms of a slip length? Think about size scales.

10.7 *The no-slip condition in nanoscale channels* This problem refers to the paper by Cheng and Giordano, "Fluid flow through nanometer-scale channels", *Phys Rev. E* **65**, 031206 (2002). These investigators examine pressure-driven flow through microfabricated channels.

(a) Read the description of the fabrication technique for the flow channels involved, and the diagnostic methods used to determine the actual height of the flow channels. Why is measuring the channel height so important? Can you think of two other methods that could have been used to measure the height? From what we've discussed about flows, in the nanoscale limit do you think the electrolyte conductivity method of estimating channel height would over- or under-estimate the true height of the channel, and why?

(b) Consider the sample associated with Fig. 3 of Cheng and Giordaw. Let's analyze this channel as if its the case of laminar flow between two parallel plates in Exercise 10.4. For a typical pressure difference across the channel of 1 atm = 101 kPa, and water with a viscosity of 10^{-3} kg/m-s, estimate the expected peak flow velocity using the given channel dimensions. Also, calculate the expected volume flow rate (integral of velocity over the cross-section of the channel) in m^3/s (assuming again that the channel is much wider than it is tall, so that this is effectively a 1d integration). How does this compare with the value given in the figure?

11 Bionanotechnology: a very brief overview

Biology has clear, direct relevance to nanoscale science and technology. The organelles within our cells are exquisite nanoscale machines, with the capability to fabricate complex structures with molecular precision in a fluctuating electrolytic environment. Individual protein molecules can function as motors and pumps, transducing chemical energy into useful mechanical or electrochemical work. Biological systems can build complex structures from the nano to the macroscale incorporating inorganic as well as organic constituents. Moreover, there is a tremendous societal drive toward greater understanding of this biological apparatus, motivated by the quest for basic knowledge, the desire to leverage biological mechanisms to accomplish useful tasks, and the obvious ramifications for clinical treatment of disease.

Because of the vast diversity of biological systems,[1] this chapter does not *remotely* attempt to survey all of bionanotechnology. Rather, I will emphasize a handful of key concepts relevant to molecular and cell biology and highlight major research directions. View this as a very simple primer more than as a textbook-depth explication. For a book-length discussion of many of these topics, I recommend D. S. Goodsell's *Bionanotechnology: Lessons from Nature* (Wiley-Liss, 2004) as a good place to start.

11.1 Basic elements and tools of bionano

11.1.1 Intermolecular interactions relevant to bionano

Biological activity has evolved to take place in a complex, fluctuating chemical environment near room temperature (though extremophile bacteria push the limits of this generalization). So that biological systems can respond to their environments, metabolize nutrients, and generally carry on the business of living, they support many mechanical and chemical processes that operate on energy scales comparable to $k_B T$. As a result, while covalent interactions at the eV energy scale remain very important, many interesting biological materials and processes depend on *noncovalent* interactions closer to the thermal energy available from the surroundings.

Figure 11.1 schematically highlights four relevant noncovalent interactions. Electrostatic interactions with energies ~ 100 meV can result from the net charge of molecular (or structural) species. As we have discussed in Section 10.5, these interactions are screened

---

[1] The fact that I am not trained in biology or bioengineering also weighed heavily on this choice.

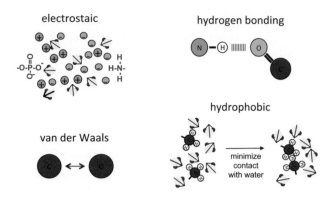

Noncovalent interactions relevant in biological systems.

Fig. 11.1

on distances comparable to λ_D, which depends on ionic concentration. Molecular species can be charged in solution, and large molecules can have net valences that are considerably larger than one (making them *polyelectrolytes*). The net electrostatic interactions reflect the ionic charge and its compensation by charge double-layer effects.

Van der Waals interactions are generally attractive and can take place between any two atoms, and result from fluctuating dipole–dipole interactions. As a result, the distance dependence for the atom–atom interactions is a potential that scales like $1/r^6$, making this a short range (< 1 nm) effect. For atoms with typical polarizabilities, the energy scale of this interaction is ~ 10 meV.

Hydrogen bonding interactions are somewhere between a simple attractive interaction between permanent dipoles and covalent bonding [851]. The classic example is the interaction between water molecules. In water, the covalent OH bond is polar, with the oxygen having a slight excess of negative charge, while the hydrogen has a slight excess of positive charge. The directionality of the interaction is a bit more complicated that simple dipole–dipole, because the precise distribution of charge in the oxygen is sensitive to the covalent bonding of that atom (in chemist language, the location of the "lone pairs" matters). The general hydrogen bonding result is a directional, attractive interaction between a terminal hydrogen and an oxygen or nitrogen atom. The "bond length" of a hydrogen bond is usually around 0.18 nm, while a "fully" covalent OH bond is more like 0.1 nm. The energy of the interaction depends on the electronegativity of the atom covalently bound to H, the other atom (O, N) participating in the bond, and typical distances. For the case of hydrogen bonding in liquid water, the attractive interaction (net enthalpy gain) is an energy of around 240 meV per hydrogen bond [852]. For reference, the enthalpy of formation of a covalent OH bond is closer to 5 eV.

The strength of hydrogen bonding is indirectly responsible for *hydrophobic interactions*. It can be overall energetically favorable for saturated hydrocarbon groups (such as the -CH_3 groups shown) to come together, minimizing their area of contact with the surrounding water. This effective attractive interaction between nonpolar/hydrophobic groups is really a complementary way of describing the attractive interactions between the water molecules; hence hydrophobic interactions are on an energy scale similar to that of hydrogen bonds. This assignment of a hydrophobic effective interaction is somewhat analogous to defining a buoyant force supporting a boat, when really the operative principle is the net gravitational attraction on the surrounding water.

A particularly strong example of noncovalent binding is that between *biotin*, a small molecule ligand, and *streptavidin*, a protein with a molecular weight of around 53,000 Daltons. There are multiple hydrogen bonding interactions between the ligand and the protein, as well as hydrophobic and van der Waals interactions that favor the binding [853]. The result is a binding energy of nearly 0.9 eV per biotin/streptavidin pair. Because of the strength and specificity of this interaction, the small molecular size of biotin, and the expertise that has developed in functionalizing surfaces with both molecules, the biotin/streptavidin combination has become a standard motif for biotechnology and biosensing applications [854, 855, 856].

11.1.2 Building blocks

Figure 11.2 shows the four most common building blocks of biological systems. *Sugars* (Fig. 11.2(a)) are small carbohydrates, readily water soluble. For our purposes, *lipids* (Fig. 11.2(b)) are moderately small molecules generally containing some saturated hydrocarbon component (nonpolar), and terminated with a polar head group. The saturated hydrocarbons are often described as oily, fatty, or waxy. Often the head group can donate a proton to the surrounding aqueous environment, as is the case for the lipid in the figure. These molecules are therefore sometimes called fatty acids.

Nucleotides (Fig. 11.2(c), Fig. 11.3) are more complicated, consisting of a phosphate block (one, two, or three phosphate groups bound together), a five-membered sugar (ribose or deoxyribose, depending on whether one of the carbons is attached to -H or -OH), and a

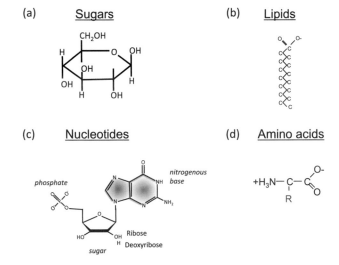

Fig. 11.2   Building blocks for biological (nano)materials. Most biological systems are built up from these basic, organic ingredients.

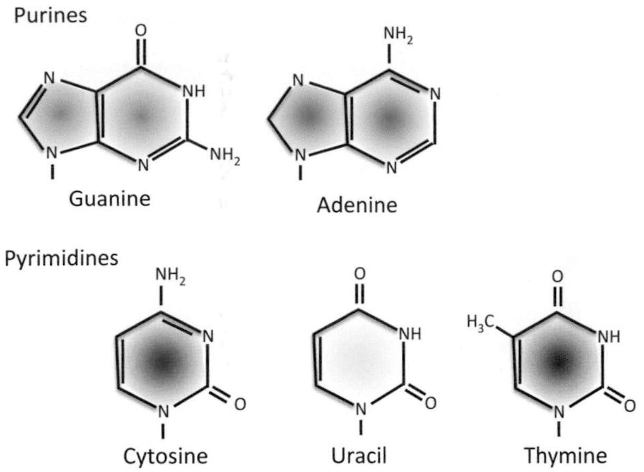

The most important nitrogenous bases. Guanine, adenine, cytosine, and thymine occur in DNA. Uracil takes the place of thymine in RNA.

Fig. 11.3

"base" made from a carbon/nitrogen ring compound. There are five nitrogenous bases that occur naturally. The six-membered rings (thymine, uracil, and cytosine) are pyrimidines, while the 6–5 systems (guanine and adenine) are purines.

Nucleosides are structurally similar molecules that lack the phosphate groups. Nucleoside di- and triphosphates play critical roles in chemical energy transport and metabolism within cells. The most famous examples are *ATP* (adenosine triphosphate) and *ADP* (adenosine diphosphate). Fully accounting for energetics, including interactions between water molecules, it is energetically favorable (by roughly 0.3 eV per ATP) for ATP to decompose into ADP and a phosphate group in water. Cells metabolize food (sugars) to produce excess ATP (above what would be expected for thermodynamic equilibrium), and many molecular machines convert ATP into ADP, taking advantage of the net release of chemical energy.

Peptides (Fig. 11.2(d)) are short oligomers of *amino acids*. Amino acids may have some side chains that distinguish the type of amino acid, attached to a small backbone terminated at one end by an amine group (-NH$_2$) and at the other end by a carboxylic acid (-COOH). In eukaryotes (living things with cells that have nuclei), there are 21 naturally occurring amino acids. The carboxylic acid group of one amino acid can react with the amine group of another, forming a covalent C-N "peptide" bond.

Each of these various components may be used to build larger structures, as shown in Fig. 11.4. Sugars may be linked together covalently to form polysaccharides such as starch. Lipids tend to organize into larger structures through noncovalent interactions, as we discuss below.

Nucleotides can polymerize in particularly rich ways. In single-stranded *DNA* (deoxyribonucleic acid) and *RNA* (ribonucleic acid), the (single) phospates bond covalently, and the bases tend to stack so that the planar structures are lying flat upon one another (though

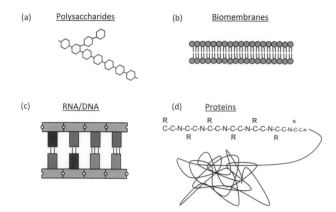

(a) Polysaccharides   (b) Biomembranes

(c) RNA/DNA   (d) Proteins

Fig. 11.4 Collective forms of biological building blocks. Sugars polymerize to form polysaccharides. Lipids self-organize into bilayers in the presence of water, forming the basis for many biological membranes. Nucleotides link up via their sugar and phosphate "backbone" and interactions between the bases. Amino acids are polymerized into proteins (polypeptides).

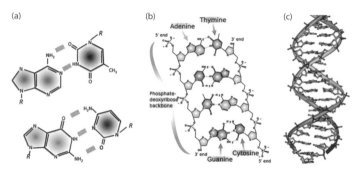

(a)   (b)   (c)

Fig. 11.5 Structure of DNA. (a) Hydrogen bonding of complementary bases. (b) Covalent bonding along the phosphate–sugar backbone and hydrogen bonding between the bases holds together double-stranded DNA. Image credit: Wikimedia commons (Madeline Price Ball). (b) DNA double helix structure. Image: public domain, wikimedia commons (Jerome Walker).

not perfectly vertically aligned). In naturally occurring DNA, the bases are adenine (A), thymine (T), cytosine (C), and guanine (G). In RNA, uracil (U) appears instead of thymine.

Thanks to hydrogen bonding, other morphologies of DNA and RNA are possible. The particular configurations of hydrogen, oxygen, and nitrogen atoms in the nucleotide bases leads to strong, specific interactions between thymine (or uracil) and adenine, and between cytosine and guanine. In DNA, these interactions can lead to specific binding between *complementary strands* of DNA, leading to the formation of the famed double helix configuration, as shown in Fig. 11.5(b). Likewise, during the process of transcription, complementary RNA strands may bind specifically to single strands of DNA. The specificity and selectivity of complementary nucleotide binding is essential to the robustness of life and a key aspect of the utility of these molecules as tools for use at the nanoscale.

Other DNA and RNA morphologies are possible. A long, appropriately sequenced single strand of DNA or RNA can fold over on itself, stabilized by hydrogen bonding. Branched structures joining multiple single strands into complex geometries can be designed and engineered [857], now that the technology exists for obtaining DNA oligomers with predefined sequences. These possible structures have been reviewed at the technical [858, 859, 860] and more popular level [861], and examples are shown in Section 11.2.2.

The hydrogen bonds of complementarity that build up the larger DNA and RNA structures may be broken by elevated temperature or through modifying the solution environment, such as changes in the pH. This process is called "melting" or "denaturing". Return the conditions (temperature, pH) to normal, and the double-stranded structures will rehybridize.

One additional point of terminology concerns the labeling of ends of single RNA and DNA strands. The carbon atoms of the pentagonal-ring sugar attached to the last base are numbered clockwise, with the $1'$ position linking the sugar to the base. A strand end with a dangling phosphate group (from the carbon in the $5'$ position) is called $5'$, or 5-prime. The other end of the strand, with an intact -OH group at the carbon in the $3'$ position, is called $3'$. This asymmetry in the sugar backbone gives single-stranded RNA/DNA a directionality, determining the order in which genes are transcribed, among other things.

Double-stranded DNA can also be terminated in different ways, as shown in Fig. 11.6. With *blunt ends*, each strand has the same number of bases, with all bases paired complementarily. With *sticky ends*, at least one strand overhangs the other, leaving some uncomplemented bases available for further interactions. With a *frayed end*, the complementarity of the two strands is broken for some number of final bases.

Proteins are polymers of amino acids. Proteins play many vital roles within living things, including acting as enzymes, ion pumps, molecular motors, structural elements, etc. The precise sequence of amino acids in a protein is its *primary structure*. Because proteins can be very large, having molecular weights of tens of thousands of Daltons, it is often more useful to describe them in terms of structural motifs. Roughly speaking, the two most common larger-scale structures that arise are α-helices and β-sheets. Both of these *secondary structures* are stabilized by hydrogen bonding between different portions of the amino acid residues along the peptide backbone. *Tertiary structure* then describes the largest-scale arrangement of helices and sheets for large proteins. The arrangement of multiple proteins into larger superstructures is quaternary structure.

The famous "protein folding problem" is often summarized as the challenge of predicting the final, usually quite complicated, tertiary structure of a protein given the original amino acid sequence [862]. Not all proteins necessarily have well-defined tertiary structure in biological environments, further complicating this issue [863].

Proteins often perform tasks more sophisticated than simply acting as structural members and supports. Two examples are shown in Fig. 11.8. In panel (a) is an enzyme known as F1-ATP synthase [864]. This enzyme is inside the mitochondria in your cells, attached to mitochondrial membrane. It binds ADP and a phosphate group, and in the presence of a proton concentration gradient (higher outside the membrane than inside), it catalyzes the formation of ATP, which is then released into the mitochondria. Interestingly, this protein is also a molecular motor, with an inner stalk performing a 120° rotation with

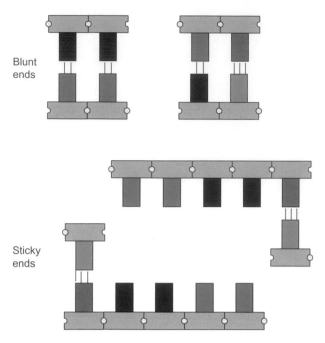

Blunt
ends

Sticky
ends

Fig. 11.6   DNA ends may be either blunt (top) or sticky (bottom).

each ATP synthesis [865]. Panel (b) shows a calcium pump, a transmembrane protein complex that consumes one ATP molecule (producing ADP and phosphate) while pumping two Ca^{2+} ions across the lipid bilayer that composes an intracellular membrane, the sarcoplasmic reticulum [866]. These two proteins demonstrate an essential capability: they can selectively bind very specific molecules (ADP and ATP, respectively) while being immersed in a biochemical soup containing many other compounds.

The specificity of protein binding to smaller molecules (ligands) is key to *many* biological functions. Two major contributors to this binding specificity are steric fit (a geometrical lock-and-key interaction between the protein and the ligand) and noncovalent interactions (hydrogen bonding, hydrophobicity). Sophisticated models of these effects to understand binding kinetics have been constructed [867]. For our purposes, the specificity of binding is something that can be a tool to achieve other ends, such as the specific detection of some analyte or the manipulation of specific molecules or materials.

Finally, individual lipid molecules tend to organize collectively into larger-scale structures through their noncovalent interactions. A common motif is the *Lipid bilayer* membrane, as shown in Fig. 4.24. Cell membranes, and within cells organelle membranes and nuclear membranes, are all composed of lipid bilayers, with various proteins residing on, within, or threaded across the bilayers. Given that these membranes must respond dynamically to many circumstances (e.g., bifurcating during cell division; opening up vacuoles to take in materials), it is no surprise that there is great interest in understanding their properties and the chemical/physical mechanisms at work. Within the bilayer, the lipid molecules (and proteins carried along for the ride) are able to diffuse. Lipid bilayers can

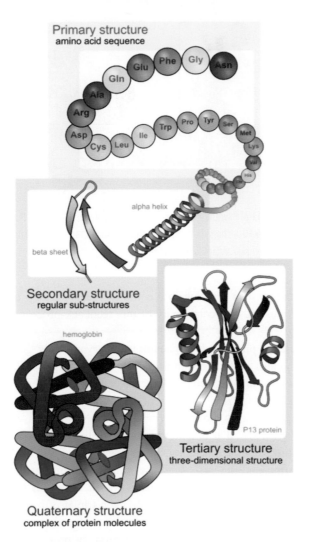

Fig. 11.7

Protein structure. Amino acid sequence (primary structure) leads to hydrogen bonding and steric hindrance building up alpha helices and beta sheets (secondary structure). These helices and sheets then fold and organize into tertiary structure, while multiple folded proteins can interact and form complexes (quaternary structure). Image: wikimedia commons, public domain (Mariana Ruiz Villarreal).

be supported on planar substrates for greater ease of study [868, 869] of the membranes themselves and the molecular interactions of the proteins within them.

11.1.3 From DNA to proteins

The so-called "standard dogma" of cell molecular biology is that DNA encodes genes; specific segments of that DNA are transcribed into RNA; the RNA is then used as a template by ribosomes within the cell to build proteins out of amino acids. These proteins

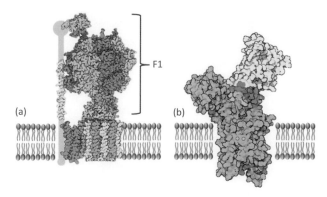

Fig. 11.8 Proteins playing active roles. (a) F1-ATP synthase, an enzyme consisting of several proteins that also acts as a rotory motor. Its related protein complex F0 anchors the complex to the lipid membrane. Adapted from [864]. (b) A calcium pump transmembrane protein complex. Adapted from [866].

and other circumstances then function in a complex feedback loop to alter and control which genes are expressed. The complexity of this feedback process is at the heart of a large portion of "systems biology".

In greater (though still physicist-level) detail: in eukaryotic cells, genetic information is stored in the cell nucleus in double-stranded DNA that is further coiled around proteins called histones, and these are grouped into larger structures called chromosomes. Enzymes are able to unwind and "unzip" portions of the DNA for transcription. An enzyme called *RNA polymerase* crawls along the (unzipped) DNA strands, assembling a piece of *messenger RNA* (mRNA) that is complementary to the DNA sequence. The mRNA (possibly with further modifications) makes its way out of the nucleus and into the cell cytoplasm. A ribosome interacts with the mRNA, starting at one end, facilitating interactions with pieces of *transfer RNA* (tRNA), each of which has three bases that complement three adjacent bases on the mRNA. Each triplet of bases is called a *codon*, and each tRNA carries with it a particular amino acid. The ribosome and its associated apparatus links together the amino acids into the growing protein and frees them from the now unburdened tRNA, which is released. When a codon is reached that translates as "stop", the protein is complete and released.

11.1.4 Viruses

Viruses are not alive, but they are an important ingredient in the bionanotechnology tool kit. A virus consists primarily of a protein coat or capsid, tens to hundreds of nanometers in size, enclosing a small amount of genetic material (RNA or DNA) that codes for the proteins that make up the virus.

Viruses do not metabolize and do not reproduce by themselves. Rather, they reproduce by hijacking the genetic transcription and protein building capabilities of host cells. The protein or proteins that make up the capsid can stimulate a host cell to take in the virus. The capsid either falls apart or is broken down in the cell's cytoplasm, freeing

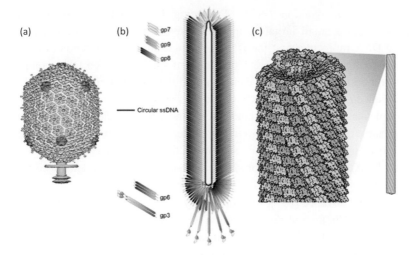

Fig. 11.9

Various viruses. (a) The head capsid of a T4 bacteriophage, from a cryoTEM reconstruction [870]. Top to bottom, the length of this structure is about 120 nm. (b) Diagram of a M13 filamentous bacteriophage. The actual virus is about 9 nm in diameter and 900 nm long. Adapted from [871]. (c) Tobacco mosaic virus, about 18 nm in diameter and 300 nm long. From [872].

the genetic payload. The cell's own machinery both replicates the virus' genetic material and transcribes it to produce the proteins that constitute the capsid. The details of these processes depend greatly on the type of virus. New virus particles (virions) then self-assemble within the cellular environment. The new virions are eventually released, either through cell death or other means.

Figure 11.9 shows examples of several viruses. These show the rich structural motifs possible from the self-assembly of even a small number of protein components. The comparative simplicity of the viral structure and its inherently nanoscale size make viruses tempting tools for use in nanotechnological applications. Viruses may be used as scaffolds for the ordered assembly of other materials [873, 874, 875], or as templates for the growth of inorganic materials [876], as discussed in Section 11.2.3.

11.1.5 Biochemistry tools

Biochemists and molecular biologists have developed an impressive array of tools and techniques that have enabled the explosive growth in our ability to manipulate DNA, RNA, and to a lesser extent proteins. Among these are the following.

- *Restriction enzymes*. Restriction enzymes are proteins that can cleave double-stranded DNA at specific sites determined by nucleotide sequences [877]. There are thousands of naturally occurring restriction enzymes, many of which are available commercially. Depending on the enzyme, DNA may be cut to produce blunt ends or sticky ends. Restriction enzymes may also be engineered to perform cuts at specific sites by splicing synthesized nucleotide sequences into the enzyme framework [878].

- *Polymerase chain reaction* (PCR). Producing large quantities of identical DNA molecules from a small initial sample has proven invaluable. To replicate a double-stranded DNA molecule, first the DNA is melted into single strands through heating. At a lower temperature, short bits ("primers") of single-stranded DNA complementary to part of a desired sequence are then bound to the single strands near the $3'$ ends. An enzyme called *DNA polymerase* is bound to the primed strands. This enzyme then builds the complement to the single strand out of individual nucleotides (deoxynucleotide triphosphates (dNTPs)) in the surrounding buffer solution, working toward the $5'$ end of the original single strand. At the conclusion of this process, there are now two double-stranded DNA molecules, each produced by building the complement to the original single strands.

  The essential advance [879], for which Kary Mullis was awarded a share of the 1993 Nobel Prize in Chemistry, was in using a thermally stable form of DNA polymerase that could survive the elevated temperature required to denature the double-stranded DNA. With such an enzyme, it became possible to carry out repeated DNA doubling by careful cycling of the temperature. The resulting exponential growth in the amount of reproduced DNA as a function of the number of cycles is the origin of the "chain reaction" moniker.

- *Antibodies and immune response*. The immune system in animals is truly remarkable and far too complex to explain in detail here. At its heart, the immune system uses certain types of white blood cells to produce proteins called *antibodies*. These antibodies have tip structures that can take on millions of variations in shape (and hydrogen bonding/hydrophobic interaction arrangement). Through these tip structures, antibodies can bind with great specificity to *antigens*, which can be biological (e.g., proteins associated with infectious bacteria or viruses) or other "foreign" substances. Through a complicated feedback mechanism, the immune system is able to determine which antibodies are effectively binding a would-be invader, and produce more of those antibodies so that the foreign agent may be targeted for removal. The upshot of this is that the immune system of animals is a ready-made factory for the production of proteins with highly specific binding. Such proteins can be grown deliberately and then incorporated into various devices, such as sensors to detect cancer markers.

11.2 Leveraging biology

One strong research direction of bionanotechnology is the goal of leveraging biological capabilities and structures to accomplish desired nanoscience or nanoengineering tasks. While biological systems can't do everything, they can do some things very well, including selective binding of molecules and materials, and self-assembly into structural motifs. Here are instances of biological components being adapted for interesting applications, or biological systems used as the inspiration for nanoscale control of materials.

11.2.1 Molecular motors

There are a number of biological molecules, particularly proteins or complexes of proteins, that function as motors, consuming fuel (typically through the conversion of ATP to ADP) and transducing the liberated chemical energy into mechanical motion. Examples include F1-ATP synthase; the rotary motor transmembrane protein complex that drives bacterial flagella; and kinesin, which "walks" along protein microtubules in cells, pulling cargo.

The detailed functioning of motor proteins can be extremely rich and complex, requiring highly sophisticated experiments to determine. Similarly, at present we are comparatively far from being able to design, *ab initio*, wholly new molecular motors that perform similar conversion of chemical energy into cyclical, controlled mechanical motion of three-dimensional nanoscale structures. However, in recent years many investigators have taken to adapting biological molecular motors for custom purposes [880].

An example [881] is shown in Fig. 11.10. Using the biotin/streptavidin binding scheme, the investigators attached nickel nanowire propellers (fabricated lithographically) to the rotating subunit of F1-ATPase. The investigators could then use optical microscopy to track the orientation of the propellers *in situ* while ATP was supplied to the motors via

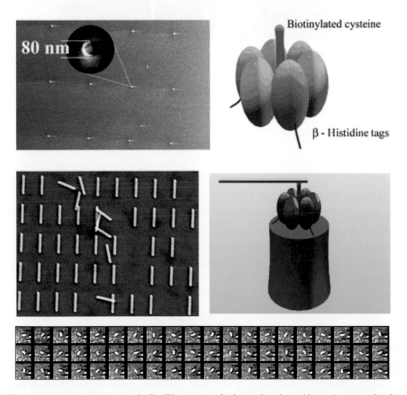

Using F1-ATPase as a biomotor. Upper panels: F1-ATPase is attached to pedestals, and biotin/streptavidin chemistry is used to attach Ni nanowires to the rotary shaft of the enzyme. Bottom: in the presence of ATP, the motor unit rotates, 120° per molecule metabolized, as seen in frame captures of an optical microscopy movie. Adapted from [881].

Fig. 11.10

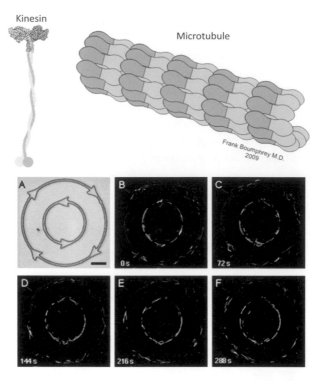

Kinesin

Microtubule

Frank Boumphrey M.D.
2009

Fig. 11.11   Using kinesin to push microtubules. Upper left: kinesin, which metabolizes ATP to "walk" along microtubules. Adapted from [864]. Upper right: microtubule protein complex, showing the directionality of the structure. Adapted from Wikimedia commons (Frank Boumphrey, MD). Bottom: with kinesin tethered to the channel floors, fluorescently tagged microtubules are pushed around a microfluidic network, with net pushing direction set by microfluidic environment. Adapted from [882].

the buffer solution. Imaging clearly showed rotation of the propellers with 120°-separated preferred angular orientations, and a rotation rate related to the ATP concentration. These measurements were proof-of-concept that biological motors could be interfaced with top-down fabricated structures. At the same time, the experiments allowed the investigators to estimate quantitatively the efficiency of the motor, based on the inferred viscous drag torque on the propellers, the rate of ATP consumption, and the free energy release during ATP hydrolysis.

A second example is shown in Fig. 11.11. This is a proof of concept experiment using biological motors to shuttle cargo along preferred paths. *Microtubules* are high aspect ratio structures present in eukaryotic cells, with lengths of microns and outer diameters of around 24 nm. The microtubules are composed of dimers of two varieties of the creatively named protein tubulin. Kinesin is a molecular motor protein that binds to a microtubule and consumes ATP to "walk" down the microtubule, with the direction of motion determined by the orientation of the tubulin dimers. In the experiment of Fig. 11.11 [882], the motor protein was bound to the substrate at the bottom of lithographically defined microfluidic channels. A solution containing fluorescently tagged microtubules was then placed in the channels along with ATP. The microtubules are pushed along by the kinesin, with the preferred direction of the flow set not by orienting the microtubules, but by controlling the channel geometry to favor fluid flow in one direction (see Section 10.6.1). More sophisticated manipulations are possible, permitting this approach to direct cargo transport through relatively complex fluidic networks [883].

11.2.2 DNA as a structural material

Thanks to the tools of molecular biology, it is now possible to synthesize custom DNA sequences readily and cheaply, particularly for relatively short strands. That capability combined with creativity and the realization that DNA can take on morphologies beyond the double helix has led to the use of DNA as a building material for designer nanostructures.

Starting with the creation of nets and lattices from DNA oligomers [857], the field has progressed remarkably far in recent years. Figure 11.12 shows the basic idea behind *DNA origami*. The original basic idea was to use an algorithm to design structures consisting of relatively short DNA segments (16- and 32-base snippets) that can fit together with a longer "backbone" to tile the inside of a polygon [884, 885]. A delicate aspect of this is optimizing the twists and turns of the DNA strands, particularly where to place seams between larger blocks and where to put cross-overs and hairpin turns. Figure 11.13(a) shows the impressive efficacy of this approach in creating (DNA) structures that assemble themselves into complex shapes. By allowing for folded structures and other motifs, DNA origami has been extended into three dimensions (Fig. 11.13(b), (c)). Modifying the DNA constituents to include components responsive to chemical stimuli allows box-like structures that can release cargo upon chemical command [886]. Truss-like organizational motifs have also been realized [887].

The three-dimensional design precision enabled by these DNA tools is remarkable, but not without its limits. These structures must be assembled in a wet chemical environment, and the building materials are intrinsically floppy compared to the common inorganic building blocks of electronics and photonics. To mitigate these issues, functionalized

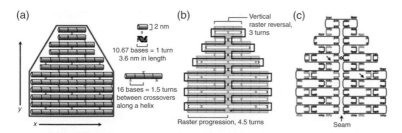

DNA origami, schematically. A desired 2d polygon may be filled in by folded and cleverly cross-linked DNA. This design approach may be automated algorithmically, to produce a DNA sequence likely to self-assemble spontaneously into such a shape. Adapted from [884].

Fig. 11.12

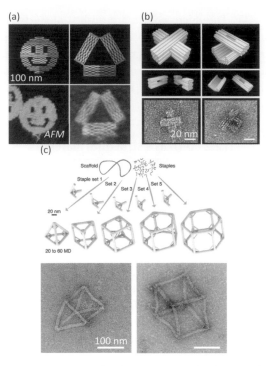

Fig. 11.13 DNA origami examples. (a) Two-dimensional complex shapes, as conceived and as imaged via AFM. Adapted from [884]. (b) Three-dimensional complex shapes, as designed and as imaged via TEM. Adapted from [886]. (c) Truss-like 3d DNA origami structures, as designed and as imaged via TEM. Adapted from [887].

constituents that conjugate to inorganic materials allow DNA origami structures to template the organization and assembly of metal [889, 890] and semiconducting nanoparticles [891].

11.2.3 Biological manipulation of inorganic materials

The DNA origami examples used very specific chemical functionalization to attach and incorporate inorganic particles into DNA structures. We know that biological systems are able to use proteins to great effect in manipulating inorganic materials, including the growth and patterning of bone, seashells, etc. Rather than preparing inorganic materials and bio-based scaffolds with complementary chemical functionalizations, an alternative approach is to use the machinery of biology to find, within the existing biological toolkit, affinities for inorganic materials that may be leveraged.

A combinatorial version of this approach is based on *phage display* [892]. Bacteriophages are viruses that prey upon bacteria; bacteriophage M13, known to infect *E. coli* bacteria, is shown in Fig. 11.9(b). M13 is approximately 900 nm long and 9 nm in diameter, and contains single-stranded DNA with 6407 bases which encodes the proteins that constitute the various parts of the phage's protein coat. The main capsid is a crystal comprising over 2700 copies of one protein (P8), while the ends of the coat involve a few other minor proteins. The P3 protein at one end is essential to the binding of M13 to an *E. coli* host.

Thanks to years of study, scientists know which segments of the M13 genome code for the coat proteins, and can prepare versions of the M13 phage with designer alterations to those proteins. The challenge is then to find an efficient means of screening for proteins that have affinities for desired materials. Figure 11.14 shows the basic idea. Start from a population of M13 with broadly varying compositions of P3, for example. Expose test surfaces of the desired material to this population, washing away those phages that do not have comparatively strong (noncovalent) binding to the surface. The successfully bound phages may then be eluted separately and introduced to *E. coli* for reproduction. This process may be repeated for several generations, selectively "breeding" a population of M13 phages with P3 proteins that have preferential affinity for the display surface. These phages can then be sequenced and analyzed to determine which amino acid combinations function best and to try to gain insight into the binding mechanisms or motifs.

This approach has been employed to develop a library of peptides that bind selectively to semiconductor materials [876]. The investigators started from an initial library of 10^9 different 12-amino acid combination versions of P3 and exposed phages to nominally clean semiconducting surfaces. Each generation of the process (binding, washing, elution, and amplification of populations by a factor of 10^6) selected for good adhesion, and after several generations the genomes of the best binders were analyzed to identify trends. The authors found that peptides functioning as Lewis bases (meaning that they tend to donate an electron pair in chemical interactions) have the strongest binding affinity for GaAs(100), occurring in 55% of the binding genomes, while only appearing 34% of the time in the initial random M13 library. A dramatic display of the binding specificity is shown in Fig. 11.15, where the phages adhering to embedded GaAs lines were fluorescently tagged.

A similar process may be employed to modify the P8 sidewall protein, to arrive at variants that have affinities for particular inorganic materials. For example, populations of M13 have been developed with a P8 binding affinity for Co_3O_4 and Au nanoparticles [894]. Viral templating these nanomaterials (Fig. 11.16(a)) was pursued to create self-organized, large surface area electrodes for lithium ion batteries. A similar strategy, binding

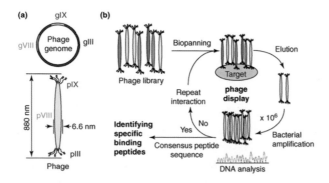

Phage display protocol. Adapted from [893].

Fig. 11.14

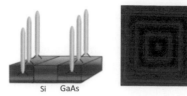

Fig. 11.15   Phage display resulting in binding specificity. Fluorescently tagged phages with terminal proteins selected for binding affinity to GaAs are self-assembled on a planarized substrate consisting of GaAs and Si regions. Adapted from [876].

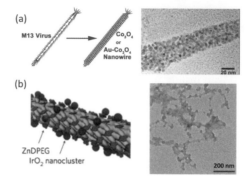

Fig. 11.16   Viral plasmid as templates for inorganic materials. (a) M13 phages designed to favor interactions with Co_3O_4 and Au nanoparticles. Adapted from [894]. (b) Phages instead designed for affinity with IrO_2 for catalytic water splitting. Adapted from [895].

zinc porphyrin and IrO_2 nanoparticles (Fig. 11.16(b)), has been used to develop materials for photocatalytic splitting of water [895].

This kind of leveraging of biological tools and building blocks to manipulate inorganic materials at the nanoscale is just in its infancy. As our ability to manipulate genomes and protein expression continues to improve, the sophistication of materials and architectures is sure to progress.

11.2.4   Nucleation and growth of inorganic crystals

We can turn these ideas around and instead of using biological components themselves to manipulate inorganic systems, we can use the mechanisms exploited by the biological systems to do so. The growth of crystalline calcium carbonate in mollusk shells is the example discussed here.

Calcium carbonate, $CaCO_3$, can form multiple crystal structures. At ambient temperature and pressure, the calcite structure is the most thermodynamically stable, while the similar aragonite structure is favored at higher pressures [896]. However, some mollusks are readily able to biomineralize aragonite, with crystal growth being templated or seeded on proteins in the extracellular matrix of the mollusks. Investigations have shown that while solution conditions (ion concentrations, impurities) are important, the particular protein macromolecules (hydrophobicity/hydrophilicity, charge distributions) play an essential role in determining which structure grows [897, 898].

Testing the role of spatially patterned charge density in controlling mineral nucleation, investigators made use of self-assembled monolayers (SAMs) and soft lithography [899, 900]. SAMs of alkane chains anchored to Au surfaces by thiol linkers can be highly ordered, with known molecular tilt angles depending primarily on the chain length. The terminal groups of the SAMs can be hydrophobic and comparatively inert ($-CH_3$ groups), or acidic/charged ($-CO_2H$ carboxylic acid groups, for example). PDMS-based microcontact printing was used to pattern regions of acid-terminated alkanes, with methyl-terminated alkanes used to fill in the surrounding substrate territory. Depending on the SAM tilt angle, the spacing of the terminal charges was varied to favor nucleation of particular crystallographic faces of $CaCO_3$ [901] (Fig. 11.17).

By tuning the size of the nucleating patches of SAM and the concentration of ions in the solution, the researchers could control the kinetics of crystal nucleation and growth, such that only individual single crystals grew, with a well-defined crystallographic orientation and without extraneous crystals, as shown in Fig. 11.18. Each nucleation site consumed the ions from the surrounding supersaturated solution, with diffusion limiting the inflow of reactants. By spacing the nucleation sites appropriately, the concentration everywhere except those sites is kept below the threshold required to start other crystals. The result is a very homogeneous population of crystals. Note that this is very similar in spirit to the approach in Section 4.2.6 for solution-phase growth of size-homogeneous nanocrystals.

11.2.5 Gecko feet

Geckos are able to walk up walls and across ceilings made from a variety of materials without the need for sticky adhesive secretions [902]. Rather, their feet are coated with hair-like structures called setae that are tens of microns long and a few microns in diameter, each further augmented with nanoscale projections called spatulae [903]. These structures are

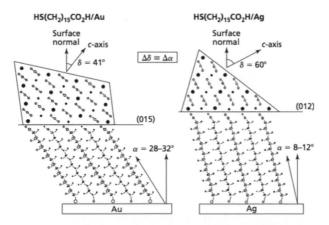

Self-assembled monolayers for templating inorganic crystallization. Varying the length of the alkane chain alters the tilt angle and hence the spacing of charges and the favored nucleation plane of calcium carbonate. Adapted from [901].   **Fig. 11.17**

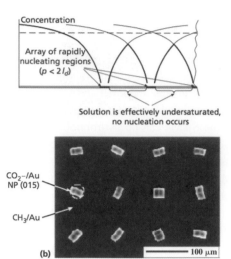

Fig. 11.18   Controlled single-crystal growth. Control over the density of nucleation sites and reactant concentrations can favor templated single-crystal growth of inorganic materials. Adapted from [901].

shown in Fig. 11.19(a). The remarkable adhesive properties of the gecko foot are ascribed to van der Waals coupling between the foot and the surface being climbed, with the large surface area, hierarchal structure, and compliant nature of these nanostructures ensuring very large effective contact areas [904].

These biological structures have inspired a number of engineered, nanostructured material immitations. Examples shown in Fig. 11.19(b–d), include molded polyimide nanohairs [908, 906], multiwalled carbon nanotube patches [907], and polymer nanorods fabricated using large electric fields [905]. These systems have been very effective at duplicating many of the sought-after properties of gecko feet, permitting impressive demonstrations of reversible yet very strong adhesion [909]. This may be the clearest example of using nanofabrication capabilities to mimic biological structures and succeeding in reaping the expected benefits.

11.3   Nanotechnology and biosensing

Detecting biological information and monitoring biological processes using nanoscale structures are enormously powerful motivations for the development of a large variety of sensing techniques. Sensing requires transduction of some desired chemical or physical effect (e.g., the presence of a cancer marker protein) into some readily detectable signal (e.g., a change in electrical conduction, a change in optical properties). Sensing of biologically relevant information can take place on a lab bench using prepared samples of biological material, using living cells in a cell culture (*in vitro*), or within a living multicellular organism (*in vivo*). A recurring challenge for sensing is the specificity of

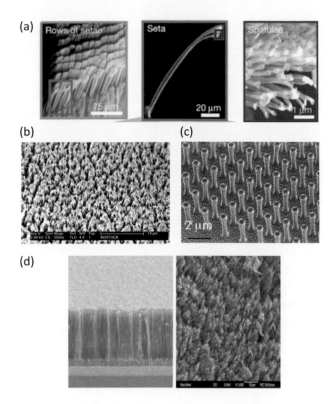

Mimicking gecko feet. (a) Natural gecko feet have a hierarchal structure that maximizes van der Waals interactions with substrates and favors self-cleaning. Adapted from [904]. (b)–(d) Mimetic structures. Polyimide nanohairs (b), (c) adapted from [905, 906]; and multiwalled carbon nanotube patches (d), adapted from [907].

Fig. 11.19

the sensor. It is not enough to have some nanodevice that responds to the presence of some protein; one must be able to distinguish that response from what would take place in the presence of other proteins, other molecules, or changes in pH. The leveraging of the specificity of biological binding motifs is a key ingredient in nano-enabled biosensing.

Entire books have been written on the subject of biosensing enabled through nano-techniques or materials [910, 911, 912], so this discussion will be very brief. Here are a handful of representative examples of nanotechnology-enabled biosensing.

11.3.1 Nanophotonics

In addition to the surface-enhanced spectroscopies discussed in Section 8.6.4, there are a variety of nanophotonics sensing approaches. While it predates the nano sobriquet by decades, *Förster resonant energy transfer (FRET)* is an inherently nano phenomenon employed extensively for sensing in biological systems. In FRET, a fluorophore is excited by an external light source. The radiative lifetime and emission intensity of the fluorophore can be greatly reduced due to energy transfer (via the optical near-field, traditionally through dipole–dipole interactions) to a nearby acceptor molecule. As a near-field effect, the modification of the fluorescence depends very steeply on distance [913]. Therefore, with appropriately chosen and biochemically linked fluorophores and acceptors, FRET has become an extremely powerful tool for monitoring interactions between biomolecules at the single molecule level [914].

Semiconductor nanocrystal quantum dots combined with the development of a number of bioconjugation strategies for these materials [915] have made nanotechnology even more relevant to FRET. The nanocrystals may be conjugated to molecules that bind specifically to desired biological structures, permitting direct optical imaging through fluorescence [916]. These nanoscale inorganic fluorophores have proven very useful as participants in FRET [917].

The other common biosensing approach based on nanophotonics starts from some nanostructured material, surface, or nanoparticle that has a well defined optical response. This system is functionalized or conjugated to encourage specific binding to an analyte of interest, with the idea that the binding will perturb the local dielectric environment of the photonic system, producing a detectable change in optical response. Examples of this basic idea include photonic crystal-based sensors [918], whispering gallery resonator sensors [919], and surface plasmon-based biosensing [663, 920].

11.3.2 Nanoelectronics

Rather than optical transduction, many biosensors are based on the idea of local (bio)chemical modulation of electronic properties of some device. The field effect transistor (Section 6.5) is a natural starting point, given the strong dependence of the source–drain conductance on the electrostatic potential of the gate (relative to the source and drain). Imagine a FET, but instead of (or in addition to) a gate electrode, the top of the gate dielectric is functionalized to bind with a bioanalyte of choice. Bias the device near the threshold for turn-on (or pinch-off) of the channel. Even if the binding is not covalent, the resulting redistribution of charge (either from polarization or due to the charge double layer accompanying a larger analyte in solution) can change the effective gate potential, modulating the conductance. Arrays of silicon FETs functionalized with appropriate single-stranded DNA have proven effective at identifying particular base sequences [921]. Organic semiconductors raise the possibility of directly functionalizing the semiconductor channel for selective binding [922].

The standard two-dimensional FET layout is not necessarily optimal for high sensitivity. By working with effectively one-dimensional channels, the sensitivity to individual binding events can be greatly enhanced, since the trajectory of every charge carrier moving from source to drain would likely be affected by every binding site. The bottom-up growth of 1D materials like semiconductor nanowires (Section 4.2.5) and carbon nanotubes (Section 4.2.5) has enabled many investigations along these lines.

Figure 11.20 shows an example of such an implementation. A silicon nanowire is configured as the channel of a MOSFET, with the wire surface functionalized with biotin. The specific binding of streptavidin produces a clear change in the nanowire conductance by changing the local charge environment of the channel. Null results in control experiments involving clean buffer solution, unfunctionalized wires, and streptavidin already bound

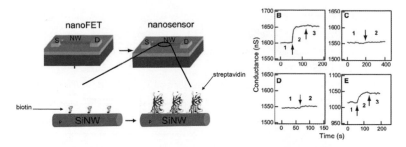

Nanowire-based electronic sensing. An Si nanowire can be a sensor, with analyte binding acting as an effective gate, as shown here with biotin/streptavidin binding. (B) Adding (1) buffer, (2) 250 nM streptavidin, (3) buffer. (C) Same conditions, but on a nanowire without the biotin. (D) Biotinylated nanowire, with (1) buffer, (2) 250 nM streptavidin already bound to biotin, (3) buffer. (E) Biotinylated nanowire, with (1) buffer, (2) 25 pM streptavidin, (3) buffer. Adapted from [855].

Fig. 11.20

with biotin are consistent with the specific binding of this system. Combining multiple active sensors with similar or complementary detectable analytes has produced selective detection thresholds for cancer markers comparable to the best commercial schemes [923]. Semiconductor nanowires produced by top-down processing have also achieved very high levels of performance, with the added benefits of improved uniformity and wafer-scale fabrication [924].

Carbon nanotubes have been similarly employed to examine biomolecules. In addition to electronically transducing the binding of biotin/streptavidin [925], nanotubes have been conjugated to enzymes. As the enzymes perform their catalytic function, the proteins change their conformation, and through their own charge and accompanying double-layer, they modulate the nanotube source–drain conductance [926]. For some complex enzymes, it has proven possible to monitor the time variation of the conductance and deduce information about the steps of the enzyme's operation [927].

11.3.3 Nanomechanical sensors

Mechanical transduction (detected either optically or piezoresistively) of biomolecular binding is another pathway toward nanoenabled biosensing. Working within solutions usually precludes the mechanical resonance techniques associated with AFM and other micro/nanomechanical methods because of viscous damping. However, changes in static cantilever deflection thanks to analyte binding can work very well [928]; an example of this is shown in Fig. 11.21(a), where hybridization with a 12-base sequence of single-stranded DNA produces detectable deflections of functionalized cantilevers even at nanomolar concentrations of analyte [929].

A clever inverted approach creates cantilevers that include an internal, integrated microfluidic channel [930]. In this way, analyte binding can take place on the *interior* of the sealed cantilever, while the cantilever itself can be operated in a mechanically resonant mode in gas or vacuum surroundings. Viscous damping is comparatively unimportant in this scheme, though fabrication with proper internal functionalization of the channel is

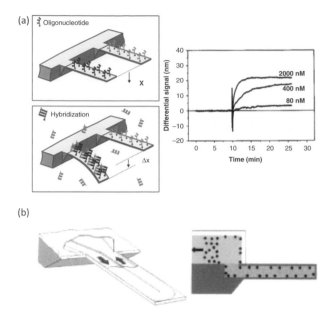

Cantilever-based electronic sensing. (a) Binding to functionalized cantilevers changes the surface stress, creating flexure deformation that can be detected optically or piezoresistively. As shown, detection of 12-mer single-stranded DNA at the nM level. Adapted from [929]. (b) Mechanical resonances may still be used for detection of binding, provided the fluid is enclosed within a sealed channel integrated into the cantilever. Adapted from [930]

a challenge. This technique has been employed to detect specific binding of proteins at sub-nanomolar concentrations of analyte [931].

11.4   Frontiers

The biochemical toolkits now available open up remarkable possibilities for bionanotechnology. Specific DNA sequences may be produced on demand in an automated way. Relatively short oligopeptides are similarly available. Nanoparticles can be bioconjugated with molecules tailored to target specific proteins such as cancer markers, with the idea of using nanoparticles not just for diagnosis *in vitro*, but for imaging and therapeutic applications *in vivo*.

Nanoscale materials such as nanocrystal quantum dots are advancing systems biology, improving our understanding of the complex interactions that govern gene expression and protein manufacture in living creatures. Semiconductor nanowires are being used as nanoscale voltage probes and molecular delivery systems in the study of cultured cells. The ability to create designer genetic sequences raises the possibility of tailoring or *ab initio* engineering cellular nanomachinery to synthesize desired products. This is one facet of *synthetic biology*.

Top-down nanoscience approaches, long associated with the quest for atomic-resolution perfection in patterning or manufacturing, can take a page from the biological world's ability to function in the presence of disorder and a complex, fluctuating environment. Between mimicking biological approaches to nanostructured materials, leveraging biological nanomachinery and materials directly, the use of nanofabricated tools to examine biological systems, and the rapidly increasing capability to produce custom biological systems, it is certain that the bio and the nano will have a long, intertwined future.

11.5 Summary

In our whirlwind survey of the bionano landscape, we began by reviewing the basic building blocks of biological materials and their interactions. We considered how biological approaches can be leveraged, either through immitation or through the direct appropriation of biological nanosystems. We concluded by examining the sensing domain of bionanotechnology. The highlight points of this overview are summarized here.

- Noncovalent interactions are tremendously important in biological systems. These include van der Waals, electrostatic, hydrogen bonding, and so-called hydrophobic interactions.
- The main building blocks of biological systems are sugars, nucleic acids, amino acids, and lipids. These organize into larger structures, respectively polysaccharides, DNA/RNA, proteins, and lipid bilayers.
- The central dogma of molecular biology is that genes coded in DNA get transcribed into RNA, and RNA is used by ribosomes to template the construction of proteins. The interaction of such proteins is one component of the feedback network that regulates the expression of genes.
- Viruses are not alive, but can leverage cellular machinery to reproduce.
- Biological components can be put to custom use. Molecular motors can be isolated incorporated into top-down structures; DNA can be a structural material of great sophistication; viral evolution may be employed to find proteins with affinities for particular inorganic materials.
- Biological nanosystems can be imitated. Examples include SAM patterning of surface charge to favor nucleation of particular oriented inorganic crystals, and branched nanoscale fibers to act as dry adhesives in the manner of gecko feet.
- Nanoscience devices may be used to detect biomolecules with high sensitivity. Common transduction methods involve photonic structures, electronic transport, and changes in mechanical properties upon molecular binding.
- There is enormous potential for bionanotechnology that is only beginning to be realized, particularly thanks to the synchronous maturation of biochemical tools and nanomaterials.

11.6 Suggested reading

- *Cell and molecular biology* There are several encyclopedic tomes that are standard undergraduate textbooks for this topic. Two books from a physics perspective are Sneppen and Zocchi's *Physics in Molecular Biology* (Cambridge University Press, 2005) and the more-than-complete *Physical Biology of the Cell*, 2nd edn. by R. Phillips, J. Kondev, J. Theriot, and H. Garcia (Garland, 2012). For a very readable (though not uncontroversial) narrative account without great rigor, I recommend *Life Itself* by Robert Rosen (Columbia University, 2005). A gorgeous book at an intermediate level is *The Machinery of Life* by David Goodsell (Springer, 2009) [932].
- *Biomaterials* A nice recent review of DNA origami is that by Kuzuya and Komiya, *Nanoscale* **2**, 309-321 (2010) [933].
- *Biomimetics* Two good review articles that survey using biological systems as inspiration for nanoscale materials design and creation are C. Sanchez *et al.*, *Nat. Mater.* **4**, 277–288 (2005) [934] and B. Bhushan, *Phil. Trans. R. Soc. A* **367**, 1445–1486 (2009) [935].
- *Photonics and nanobio* The review article by X. Michalet *et al.*, *Science* **307**, 538–544 (2005)[916] is a standard when discussing the application of semiconductor nanocrystals to biological imaging and sensing. Similarly, the revew by Medintz *et al.*, *Nat. Mater.* **4**, 435-446 (2005) [915] is a great reference for bioconjugation strategies and quantum dots for FRET.
- *Biosensing in general* One book that provides a very broad introduction to the challenges of biosensing is *Chemical Sensors and Biosensors: Fundamentals and Applications*, by F.-G. Banica (Wiley, 2012) [936].
- *Frontiers in synthetic biology* A landmark paper in the pursuit of "synthetic" cells is D. G. Gibson *et al.*, *Science* **329**, 52–56 (2010) [937], where the authors report a bacterial cell with a chemically synthesized (but not created *de novo*) genome. An example of using molecular biology tools to recreate a virus (polio) was reported earlier by Cello *et al.*, *Science* **297**, 1016–1018 (2002) [938].

Exercises

11.1 *Van der Waals interactions* Argue that the potential energy of the van der Waals interaction should scale like $1/r^6$, assuming that the interaction results from a fluctuating electric dipole $|\mathbf{p}|$ from one atom inducing an instantaneous dipole on an atom a distance d away. Explain why this interaction should be attractive.

11.2 *Energy consumption of biological motors* A typical molecular motor draws energy from the metabolism of ATP to ADP. An example discussed in this chapter is F1-ATPase. Through very clever experiments such as those of Toyabe *et al.*, *Phys. Rev. Lett.* **104**, 198103 (2010) [939], it has been possible to examine the energetics at work here in detail.

(a) In a few sentences, how do the authors of that paper apply controlled torques to the rotary shaft part of individual F1-ATPase molecules?

(b) Under certain conditions (plenty of ATP, ADP, and phosphate in solution), the authors find that in the absence of an applied torque, the F1-ATPase is able to rotate at 5.08 revolutions per second. Under an applied torque of 17.7 pN-nm/rad, that rotation rate slowed to 2.58 rev/sec. What is the power dissipated working against that external torque? Extrapolating linearly, what is the total power being dissipated without any applied torque, and what must be the viscous torque acting on the shaft just from the setup used?

(c) What do the authors say is the efficiency of the conversion of free energy into mechanical motion? Assuming that about 0.3 eV (net) is released per ATP molecule that is metabolized, at what rate is ATP being consumed under these conditions?

11.3 *Biosensing based on a 2d FET* Suppose you want to build a biosensor based in a field-effect transistor geometry. Consider an FET with a gate dielectric with $\kappa = 3.9$ and thickness 2 nm, based on an Si channel with a mobility of 1000 cm^2/Vs, somewhat miraculously configured so that the system sits exactly at threshold when the gate dielectric surface is exposed to the solution under analysis. Suppose that each binding event of a particular analyte molecule carries with it an effective surface charge of $3e$. For a channel of length $L = 100$ nm, how many binding sites per unit area would be needed to produce a change in conductance (per unit channel width) of 10^{-6} S/cm?

11.4 *Lipids and diffusion* One demonstration that lipids can diffuse around within bilayers is shown in Fig. 11.22. Two lipid species are mixed in a patch of supported bilayer, constrained within a rectangular patch [940]. The species have different molecular charges (one negative and fluorescently tagged, one approximately neutral) within the wet bilayer. Detailed modeling of this system is a bit complicated, but we can get a sense of some of the physics at work.

(a) Read the paper by J. T. Groves *et al.* [940]. The authors are able to come up with a mathematical model of the concentration profile by finding expressions for the chemical potential for each of the two lipid species, and then saying that in equilibrium the chemical potential gradients are zero. What are the three main contributions to the chemical potentials?

(b) Why aren't the two species of lipids perfectly segregated? How would you expect the equilibrium profile of the relative concentrations of the two species to change if this experiment were performed at higher temperatures?

(c) In an applied electric field of 10–15 V/cm, the authors estimate a typical net force acting on the charged component of the bilayer of 8.9×10^{-17} N/molecule. What effective net charge per molecule does this imply? Is this reasonable? The authors also say it takes around an hour to reach a steady state profile for patches around 350 μm across. Assuming that this time is about 3τ, where τ is a characteristic diffusion timescale for the molecules within the membrane, roughly estimate D, the diffusion constant for the molecules.

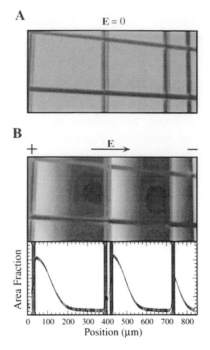

Fig. 11.22   Lipids can diffuse within bilayers. Supported lipid bilayers containing two components (one fluorescently tagged), with different molecular charges when in solution conditions, constrained into rectangular patches. Top panel: no external electric field. Bottom panel: applied electric field results in a concentration gradient of the charged species. Adapted from [940]

11.5   *Sensing based on lipid bilayers* Lipid bilayers can enable extremely sensitive detection of bioanalytes under the right circumstances. Colloidal nanoparticles suspended in solution can undergo a phase transition between aggregated (crystallized) and fluid regimes, depending on the effective interactions between the nanoparticles. Look at the paper by M. M. Baksh *et al.*, *Nature* **427**, 139–141 (2004) [941] and answer the following questions.

(a) What is the colloid system in question made out of? What physical interaction is the basis for the pair-wise interaction between colloidal components?

(b) Describe briefly how this system is able to function as a sensor to detect the presence of a particular protein? What observable is calculated from optical micrographs to assess the state of the colloid ensemble?

(c) What do you think limits the ultimate sensitivity of this technique?

Nanotechnology and the future   12

Be a scientist: Save the world!

Richard E. Smalley

As previous chapters have shown, new technologies now exist to manipulate matter down to the atomic scale. Material properties at these scales differ from those of bulk systems as different physical processes become relevant. Nanotechnology is based on engineering these new properties into useful devices. These capabilities are already having a significant technological impact on many disciplines, from consumer electronics and information technology to optical communications to biology.

Where is this all going? Is reality going to live up to the early hype surrounding nanotechnology? Will nanoscale engineering lead to a new industrial revolution, enabling solutions to many of the major problems facing humanity? Alternatively, will practitioners of this new science and engineering metadiscipline bring forth "a nightmarish, dystopic future too small to see" (*The Onion* [942])? In this chapter we look briefly at two particular areas: nanotechnology's potential impact on humanity's energy crisis, and the potential dangers of these new capabilities. Disclaimer: more so than the rest of this book, this chapter is a bit of an opinion piece, and should be viewed as such.

12.1 Nanotechnology and energy

Humanity faces, in very real terms, an energy crisis, though there are political disagreements on the extent of the problem. The *per capita* demand for energy, both for transportation and electricity, of the "first world" nations is vast and growing. Moreover, as economic development spreads over the planet, increasing portions of the global populace are experiencing higher standards of living (generally agreed to be a good thing), and a corresponding increase in energy demands. There are very serious questions about the sustainability of this trajectory, in terms of economic and environmental impact. There are also certainly negative consequences for the species if we are unable to continue raising the global standard of living. Below we analyze the situation in more detail, and discuss the promise that nanotechnology may contribute significantly to the solution of these problems.

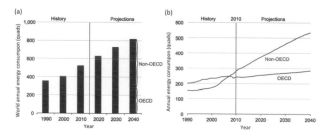

Fig. 12.1 Global energy demand, historical and projected. (a) Total demand in quadrillions of BTUs per year. OECD = Organization for Economic Cooperation and Development; largely developed economies. (b) Specific breakdown showing rapid growth of energy consumption in the developing world. Adapted from [943].

12.1.1 Energy demand

Figure 12.1(a) illustrates one component of the energy problem. Global energy demand has nearly tripled over the last forty years. First, a word about units. A common historical measure of energy in the power generation community is the BTU (British Thermal Unit), approximately 1.06 kJ.[1] A "quad" is short for one quadrillion BTUs, or 1.06×10^{18} J, or 33 GW-years. Projected annual global energy demand is certain to exceed 1000 quads by the middle of the century. To put this in perspective, a typical large electrical generating plant produces around 1 GW of power. The world would therefore need something like 33,000 such power plants to accomodate this anticipated demand.

Large numbers like this immediately raise several questions. Do sufficient fuel sources exist to supply this demand, even if the economic capacity is found to build these power plants? If so, for how long can such a demand be met? How severe are the environmental consequences of meeting such a demand? These questions can lead to heated political disagreements.

For example, there are large but finite supplies of "fossil fuels", consumable materials such as oil, coal, and natural gas (methane), that exist as end products of geologic processes tens to hundreds of millions of years in duration. The finiteness of the supply of readily accessible and refinable petroleum is well known, and some argue that we have already entered the age of "peak oil" [944], meaning that global oil production and refining have passed their supply-determined maximum and will now decline. Considerably larger supplies (in terms of latent energy content) of coal and natural gas exist.

However, the central issues remain. First, these materials were built up over geologic timescales, and as a species we are consuming them at a rate that, if unmodified, will deplete these supplies in a *much* shorter period, hundreds of years. Second, exploiting these fuels produces large quantities of pollutants, particularly greenhouse gases such as carbon dioxide. Significant evidence exists for global climate change [945]. It is absolutely certain that anthropogenic use of fossil fuels has significantly changed the atmospheric content of carbon dioxide over the last century. The global climate is a complicated, nonlinear system with many feedback and feedforward mechanisms and is not completely

---

[1] The BTU is the amount of energy required to raise the temperature of one gallon of water one degree Fahrenheit.

understood. For this reason, some still argue the extent of humanity's impact on the climate. However, it is difficult to disagree with the following. When concerned about the response of such a complex system, it is unwise to drastically alter a known forcing term (CO_2 concentration, for example). Moreover, there are significant environmental costs (including energy consumption!) of the extraction and transportation of the fuels themselves.

Supply problems exist for traditional nuclear power as well. While the energy content of nuclear fuels such as uranium is considerable, there are still finite amounts of readily accessible ore. Extracting, refining, and enriching the fissile material and preparing fuel rods all take energy. Nuclear power generation does produce some radioactive waste material, in the form of residual "spent" fuel and structural components of the reactor that have been "activated" via neutron exposure. The most efficient mitigations of the waste issue are the reprocessing of spent fuel and the use of breeder reactors and fuel cycles that run off more than standard enriched uranium. However, fuel reprocessing is expensive (so much so that the United States long considered it economically unviable at market prices for electricity) and technically complex. Moreover, the other fuel cycles such as those involving plutonium carry added security risks due to the threat of nuclear weapons proliferation. To put matters succinctly, it is not necessarily a good idea to have plutonium-based nuclear reactors in politically unstable parts of the world.

This issue comes to the fore in Figure 12.1(b), which shows another component of the global energy challenge. The developing world makes up an appreciable fraction of the rising energy demand. The developed world used cheap fossil fuels and a lack of environmental concern to achieve comparatively early industrialization. Having realized some of the negative consequences of this, the developed world is trying to convince the developing world not to use the same approach, despite the short-term appeal of some fossil fuels (particularly cheap coal). Moreover, people living in comparatively undeveloped economies (e.g., subsistence farmers in sub-Saharan Africa or rural Asia) currently have extremely modest energy demands because of their relatively impoverished status. Raising the standard of living of the world's poor up to that of those living in developed economies will require an enormous commitment of infrastructure, and will further increase rapidly growing energy demands.

Can we achieve our implied common goals (a high standard of living and its prerequisite, access to adequate energy supplies, for all of humanity, without environmental destruction or major unrest)? Is some fraction of humanity consigned to a low standard of living? These are major questions facing our species.

12.1.2 How nano may help

There are a number of realistic ways that nanotechnology and nanomaterials may help address these issues.

- Nano-based materials can improve the energy efficiency of technology, to decrease the energy demands of the developed world.
- Nano-enabled technologies can improve power generation, distribution, and energy storage.

- Nano-based materials have the potential to aid in environmental remediation, helping to remove waste products produced by meeting our energy demands.

Here we look at each of these, spotlighting a sample of the possible roles and impacts for nanotechnology. This is by no means exhaustive, and is meant as an illustration of potential.

Efficiency

Technologies based on nanomaterials have the potential to reduce sharply the energy consumption of high end economies through more efficient use of energy. Examples include the following.

- *Light, strong, multifunctional materials* Much energy consumption is associated with transportation of people and goods. Light, strong composite materials such as those based on carbon nanotubes, graphene, and other nanomaterials can reduce the mass of both goods and vehicles while maintaining necessary structural strength and performance. The deployment of (microstructured) carbon fiber composites in airplanes and cars has highlighted both the potential to reap energy savings and the challenge inherent in economically scaling up manufacturing of new materials. Recent work to create multifunctional systems incorporating nanomaterials (e.g., windows that incorporate solar cells; structural materials that can act as batteries) is likely to be increasingly important.
- *Electrical transmission* A significant portion of the world's generated electrical energy is lost to heat as it is transmitted through power lines. Copper is an extremely good electrical conductor, but it is worth considering whether nanotechnology can improve the overall efficiency of electrical distribution. One possibility is the use of nanomaterials to enable *local* generation and storage of electrical energy, to be discussed in the next section. Another direction would be the development of improved electrical conductors. Conducting cables based on carbon nanotubes have some promise of electrical conduction comparable to that of copper at considerably less weight. High temperature superconducting wires by their nature incorporate nanostructured materials. Should the ultimate atomic-scale "materials by design" challenge of room temperature superconductivity be achieved and widely deployed,[2] the world could be reshaped in terms of low loss energy distribution.
- *Solid state lighting* Another significant percentage of global electricity consumption (by some counting, as much as 20% !) is in the form of electric lighting. Traditional incandescent bulbs are dreadfully inefficient, providing only around 20 lumens/W. Fluorescent lighting, based on electrical discharge as an excitation source for phosphors, is considerably more efficient, providing $\sim$60–100 lm/W. Even greater efficiencies ($\sim$ 200 lm/W) are possible through solid state lighting based on light emitting diodes. LED bulbs are available and becoming considerably more affordable. The standard LED approaches employ either co-packaged red, green, and blue LEDs, or UV LEDs

---

[2] Note that to be useful such a material would have to have large critical current density, large critical fields, and ideally ease of manufacturing and handling.

that illuminate phosphor material. As discussed in Chapter 8, improvements such as nanostructuring the LED semiconductor material as a photonic band gap system have the potential to enable even greater efficiency of light extraction.

Power generation

Nanostructures and nanotechnology can also contribute to the generation of electricity and chemical fuels, illustrated here through the example of solar energy. The flux of energy from the Sun in the form of light at the Earth's orbit is approximately 1.3 kW/m^2. Thanks to atmospheric scattering and absorption, this is reduced to approximately 340 W/m^2 at the Earth's surface (in the daytime in direct sunlight).

To get a sense of scale, the projected annual energy demand of the world in 2030 is around 700 quads, corresponding to a continuous generation rate of about 24 TW. If solar energy could be converted into the desired forms of energy (electricity and liquid fuels) at a generous 10% efficiency, the world's energy needs would require about 7×10^{10} m^2 of solar cells, corresponding to a square some 263 km on a side. This is large, but not inconceivable; for comparison, this is comparable to the area printed per month by one of the world's large newspapers.

Most of the world's solar power is generated through silicon photovoltaic cells, as shown schematically in Fig. 12.2. A photon incident on a doped *pn* junction is absorbed, producing photogenerated charge carriers. Thanks to the comparatively large relative dielectric constant κ and the correspondingly small exciton binding energy (see Section 5.4), the built-in electric field within the *pn* junction is sufficient to separate the electrons and holes, leading to the build-up of a photovoltage between the *p* and *n* sides. Defects

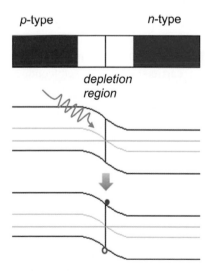

Homojunction photovoltaic system. A photon is absorbed in the depletion region of a *pn* junction. The built-in electric field is sufficient to separate the electron and the hole. This leads to either an open-circuit photovoltage between the electrodes, or a short-circuit photocurrent.

Fig. 12.2

and disorder can be detrimental, since charge trapping and electron–hole recombination reduce the useful photovoltaic effects. While silicon is an indirect gap semiconductor, it has achieved preeminence in photovoltaic manufacture for several reasons, including outstanding material quality, depth of understanding of its processing and doping, and comparative inexpense.

The *Schockley–Queisser limit* on efficiency in *pn* junction cells [946] assumes ordinary sunlight illuminating a single junction with each photon producing one electron–hole pair. Taking into account blackbody radiation, thermalization of extra energy, and the spectrum of sunlight, such a single junction cell is limited to a maximum conversion efficiency of around 34%.

Nanostructured materials play roles in several avenues being explored for advanced photovoltaics. Some of these attempt to evade the constraints of the Schockley–Queisser limit, while others focus on ease or expense of manufacture.

- The most straightforward idea, intended to increase efficiency, is to replace or augment silicon (or other single semiconductors) with more complex multilayer, *multijunction cells* consisting of several semiconductors, with bandgaps chosen to better match the solar spectrum. III–V semiconductor multilayers are particularly appealing because of the proven ability to grow high quality material [947]. Nanoscale layer thicknesses, atomically smooth interfaces, and precise control of dopant concentrations are essential ingredients.

- Using selective etching (either of SOI [948] or III-V semiconductor [949] multilayer substrates) makes it possible to prepare *ultrathin single crystal materials*, and these can be integrated into flexible, printable photovoltaics [950]. Wiring remains a challenge, but the benefits of this approach include consuming much less semiconductor per unit area than traditional rigid cells and adaptability to nonplanar geometries.

- As mentioned in Section 8.6.5, *plasmon-enhanced photovoltaics* are being intensely investigated [951, 952, 699]. Plasmonic structures may be used to concentrate light, effectively enhancing the absorption of photons by the semiconductor and allowing greater production of carriers with less semiconductor material. However, the engineering of these systems is challenging: plasmonic materials are lossy; plasmonic field enhancements are greatest in immediate proximity to the conductor, but metal/semiconductor interfaces exhibit band bending, trap formation, and recombination. With proper materials engineering, however, this approach has much promise.

- There has been much interest in combining chemically synthesized semiconductor nanocrystals with organic semiconducting polymers to produce *hybrid organic/inorganic photovoltaics* [953, 954]. The optical absorption of the nanocrystals may be tuned through quantum confinement even at fixed chemical composition. Charge separation can take place at the nanocrystal/polymer interface. The inorganic semiconductor nanocrystals can be comparatively good transporters of electron-like carriers, while the organic semiconductor is often a hole-transporting system (meaning that the mobility is higher and the trap density is lower for holes). One appeal of this approach is the idea of a solution-processable photovoltaic material, such that roll-to-roll or inkjet-like printing processes could be used to produce large areas of photovoltaic surface.

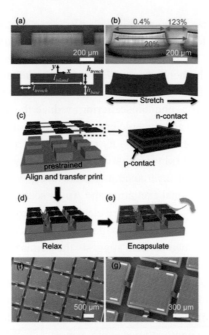

Flexible photovoltaic cells. GaAs photovoltaic stacks 3 μm thick are laminated onto a PDMS backing and linked by mechanically tolerant interconnects, allowing non-planar and flexible geometries. Reproduced from [950].

Fig. 12.3

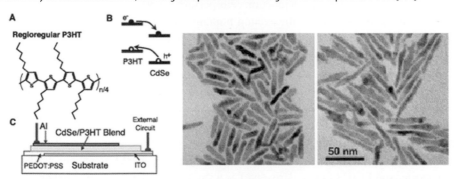

Hybrid photovoltaic cells. Semiconducting polymers are blended with semiconductor nanocrystals (in this case CdSe rods). Absorption can take place in the nanocrystals, with charge separation at the interfaces and charge transport in both the nanocrystals and the polymer. Adapted from [954].

Fig. 12.4

These hybrid materials present many challenges, however. Transport of electrons throughout the composite material to a collecting electrode would require a continuous, percolative network of nanocrystals with well-controlled interfacial properties (few charge traps or recombination centers). Chemical synthesis of the nanocrystals commonly requires ligand functionalization to prevent flocculation; these ligands must either be removed or tailored to permit charge transfer across the nanocrystal/organic semiconductor interface. The organic semiconductor must be chemically stable, highly pure, comparatively high mobility for charge transport, and support favorable chemical interactions with the nanocrystals to allow their dispersal and network formation.

- Thanks to their small size, nanocrystals have also been touted as having the potential to greatly increase photogenerated carrier yield through *multiexciton generation and carrier multiplication*. The idea [955] is intriguing. In an ordinary photovoltaic semiconductor, a photon with energy $\hbar\omega \gg E_g$ typically generates a single electron–hole excitation that is "hot", because that is what is favored by energy and $\mathbf{k}$ conservation. Inelastic processes (electron–phonon scattering) reduce the electron energy to the bottom of the conduction band and the hole energy to the top of the valence band relatively quickly, so that the excess energy $\hbar\omega - E_g$ is lost to heat. However, if the requirement of $\mathbf{k}$ conservation is relaxed (physics that happens naturally in nanoparticles, thanks to the breaking of translational invariance at the surface), two other processes can take place at a higher rate, collisional excitation and Auger scattering. In principle, one could have a photon of energy $3E_g$ produce three electron–hole pairs of energy E_g (neglecting exciton binding). Working with a narrow-gap semiconductor and having each photon able to produce more than two carriers suggests the demise of the Schockley–Queisser limit. Carrier multiplication does take place in demonstration devices [956, 957, 958, 959], though the process is highly sensitive to shape and surface termination of the nanocrystals and has proven challenging to quantify and leverage [960].
- From the standpoint of solution processing and cost, there has been a serious effort in *purely organic photovoltaics*, with nanostructured morphology to improve both separation and transport of charge. There are numerous reviews of this topic [961, 962, 963, 964]. Simple *pn* junction cells perform poorly in organic semiconductors. As mentioned in Section 5.4, the exciton binding energy depends strongly on effective masses m^* and on the relative dielectric function κ. In organic semiconductors, m^* is typically comparable to the free electron mass, and $\kappa \sim 3$, so that electrons and holes are much more deeply bound than in silicon or GaAs. As a result, the built-in electric field within a *pn* junction is typically insufficient to separate the photogenerated charges. Instead, the driving design principle is to have charge separation at the interface between electron- and hole-transporting organic semiconductors, thanks to local mismatches in molecular levels. The organic semiconductors may be polymers or small molecules, with one particularly popular combination being poly(3-hexylthiophene) as the hole transporter and PCBM (a C_{60} derivative) as a small molecule electron transporter.

  As in the hybrid approaches described above, this topic is inherently focused on nanoscale materials issues. The nanoscale morphology of the semiconductors is critically important, with a drive toward maximizing the interfacial area between the chemical species, yet the need to optimize material ordering and paths for carrier transport.
- Another hybrid approach involves *photoelectrochemical cells*, as reviewed extensively by Grätzel [965]. Rather than deal entirely with electron and hole transport within a semiconductor, photoelectrochemical cells use redox chemistry to transport charge, as shown in Fig. 12.5(a). A photon is absorbed in the semiconductor, generating an electron–hole pair. Band bending within the semiconductor due in part to contact with an electrolyte leads to a built-in field that separates the charges. As shown, the hole oxidizes a species within the electrolyte; that species then diffuses to the counterelectrode where it grabs an electron and is reduced back to its original state.

  A variant on this approach is the *dye-sensitized* solar cell, as shown in Fig. 12.5(b) and reviewed by, among others, Hagfeldt *et al.* [966]. A dye molecule at the semiconductor

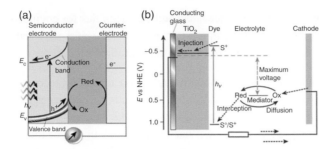

Photoelectrochemical cells. (a) Absorption takes place in the semiconductor, while a redox couple is used to transport charge, rather than traditional conduction. (b) A dye-sensitized cell, with a small molecule dye acting as the absorption site. Adapted from [965].

Fig. 12.5

surface acts as the source of photoexcited charges [967, 966]. This allows the selection of semiconductor to be somewhat decoupled from the electrochemical properties of the redox species and the dye, and the dye may be chosen to support particular chemical pathways or match specific parts of the incident spectrum. Solid-state hole-transporting materials may also be used rather than electrochemical redox couples.

As usual there is nanoscience associated with control of the interfacial chemistry at the semiconductor surface. Nanostructured semiconductor materials are also directly relevant to this approach; the large specific surface area of nanocrystalline semiconductors implies that cells with a small macroscale footprint can contain far larger interfacial surface sites for the electrochemistry. Moreover, rough surface sites (defects, step edges) can favor electrochemical processes relative to smooth, macroscopic single crystals. The great appeal of dye-sensitized cells, which have been demonstrated with power conversion efficiencies exceeding 10%, is the hope that they can be manufactured cheaply, without the need for high purity materials.

- Finally, another approach to energy capture is *photocatalysis* or *artificial photosynthesis*, where in analogy to photosynthesis sunlight drives chemical reactions to store its energy in chemical form. The prototypical system for this is the photoelectrochemical splitting of $2H_2O$ into $2H_2 + O_2$. There have been many review articles written on this topic [968, 969, 970, 971, 972]. A central theme is that storing energy in the form of chemical fuels can have real advantages over direct conversion to electricity, depending on the application. For example, the volume energy density of a lithium-ion battery (see the next section) is around 2 MJ/L, while that for gasoline is around 36 MJ/L.

The basic principle of photosynthesis of chemical fuel is essentially that of the photoelectrochemical cells described above. A photon incident on some engineered structure (often a nanostructured semiconductor, the example I will use here) is absorbed, producing photogenerated charge carriers. These carriers are spatially separated (e.g., by band bending in a semiconductor). The electron at the bottom of the conduction band (and hole at the top of the valence band) ideally end up at some surface site on the semiconductor or a directly coupled metal patch, where an incident chemical species can be reduced (or oxidized), without making irreversible chemical changes to the surface. The physics and chemistry behind heterogeneous catalysis is generally quite complicated [973, 974]. Nanostructured materials provide high specific surface area, many types of surface sites, and opportunities for engineering and control [975]. There is a long-running interest in producing useful photocatalytic materials based on transition metal oxides (like TiO_2 [976]) rather than expensive metals.

This only scratches the surface of how nanoscale science and technology can have an impact on solar power. The ability to control interfaces and engineer material structures down to the nanometer scale has ramifications for other energy-generating technologies, from the small scale (piezoelectric power harvesting [977]; improved thermoelectrics [978, 979]) to new materials for macroscale systems (turbines [980], corrosion-resistant coatings for piping [981], radiation-tolerant structures for nuclear power [982]).

Energy storage

In addition to energy production, nanostructured materials are already beginning to play a critical role in electrical energy storage [983, 984]. The two most relevant technologies are batteries [985] and supercapacitors [986].

In batteries energy is stored electrochemically, through reactions performed at the two electrodes mediated by an electrolyte. Figure 12.6 shows a diagram of a lithium ion battery of the type commonly in use. During discharge, positive lithium ions (cations) are de-intercalated from the anode and transported via the electrolyte to the cathode, where they are reduced. At the same time, electrons flow from the anode through the load to the

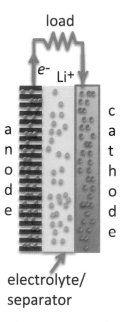

Fig. 12.6 Lithium-ion battery. There is enormous interest in new nanostructured materials for the anode and cathode, though meeting the competing requirements of stability, useful lithiation/delithiation potentials, electrical conductivity, and high specific surface area is challenging.

cathode, meaning that current comes from the cathode (labeled as the positive terminal on the battery). When being recharged, a positive voltage is applied to the cathode relative to the anode, driving current the other direction. While charging, lithium ions are oxidized at the cathode, transported by the electrolyte, and inserted into the anode. The open circuit voltage of the battery depends on the difference in lithiation potentials between the anode and cathode materials, typically around 3.5 V in commercial Li-ion cells. An ion-permeable separator physically isolates the anode and cathode while permitting Li^+ motion. In typical Li-ion cells, as of this writing, the anode is graphitic carbon, while the cathode is $LiCoO_2$, though other transition metal oxides are also employed.

One figure of merit for batteries is the mass-specific capacity, often quoted in units of mA-h/g. The typical capacity of graphite as an anode is around 370 mA-h/g. Clearly desirable features in any improved battery technology would be higher overall capacity (and specific capacity), the ability to accommodate very rapid charge and discharge rates, and longevity without degradation through many charge/discharge cycles. The electrodes also need to be at least moderately electrically conductive. Nanomaterial-based approaches to the electrode materials have been proposed to address one or more of these perceived needs.

- Silicon anodes can offer a theoretical capacity [987] of over 4000 mA-h/g, thanks to the remarkable ability of Si to stabilize alloyed stoichiometries as lopsided as $Li_{4.4}Si$. However, accommodation of large lithium fillings results in enormous structural distortions (a 300% volume change! [988]). Bulk Si cannot sustain this strain upon repeated charge/discharge cycles and fractures into powder. However, nanostructured silicon, with an abundance of free surface to permit strain relaxation, can be considerably more robust. As shown in Fig. 12.7, silicon nanowires [987] are platform for capitalizing on this nanoscale elasticity. Porous silicon films etched from the bulk is another way of achieving the requisite surface to volume ratio [989]. Other methods seek to create composites comprising nanostructured Si particles supported with carbon-based binders.
- When Li-ion batteries are cycled, a chemically complex and difficult to access layer of material (the "solid-electrolyte interphase") tens of nm thick builds up on the electrode surfaces. This layer is critical to maintaining the stability of the electrode materials

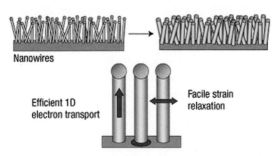

Nanowire battery. Silicon nanowires have been demonstrated as high capacity Li-ion battery anodes, with their large free surface allowing them to accommodate the strain associated with lithiation. From [987].   **Fig. 12.7**

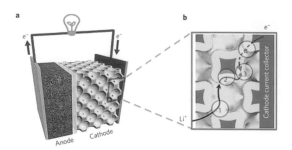

Fig. 12.8 Inverse opal battery design. The large open volume favors rapid kinetics for charge and discharge. From [990].

upon repeated cycling. Nanoscale characterization techniques, such as *in situ* TEM through an electrochemical nanocell [738], are enabling new studies of this layer and the lithiation/delithiation processes, as shown in Fig. 9.10.

- The rate at which Li-ion batteries may be charged and discharged is also of critical interest. From simple kinetics, high speed operations would be most encouraged with high surface area electrodes readily accessible by electrolyte. Inverse opal structures of the type used for photonic band gap applications (Fig. 4.23) can be used to create open frameworks with these properties, and then coated with active electrode materials, as shown in Fig. 12.8. The result [990] is a battery that can charge in seconds, provided sufficient power is available.[3]

Another approach for comparatively high discharge powers is to use capacitors rather than batteries. In capacitors, energy is stored electrostatically through the arrangement of charge on two non-reacting electrodes and the polarization of a dielectric medium. There are no reaction kinetics to act as a rate limiting process. A *supercapacitor* consists of very high surface area electrodes, with typical specific capacitances as large as hundreds of Farads per gram (!).

Nanostructured electrode materials are essential to achieve the required specific surface areas of tens of m^2/g. The natural timescale associated with charging or discharging is RC, where R is some total resistance of the circuit, including any effective internal resistance of the capacitor itself. The desire for highest speed operations implies that the electrodes must be good conductors with low internal resistance, despite their high surface area structure. The electrolyte, typically an ionic liquid, and the electrode materials must be electrochemically inert at the desired operating voltages.

Figure 12.9 shows an example of a supercapacitor material, made from a nanoporous, graphitic carbon structure [991]. The capacitance of the resulting device is determined not merely by the simple geometry of the electrode, but also by the properties of the ionic liquid when in a nanoconfined space (Section 10.5). In some structures there can be a significant contribution of the "quantum capacitance" as well, corrections to the classical capacitance due to the true density of electronic states in the nanostructured electrodes.

---

[3] Of course, rapid charging tolerance of the battery is not necessarily the key to electric cars that charge fully in minutes at a service station. The *power* required to deliver a full charge to an automotive range electric vehicle in the time it takes to fill up a tank with a liquid fuel is on the order of megawatts.

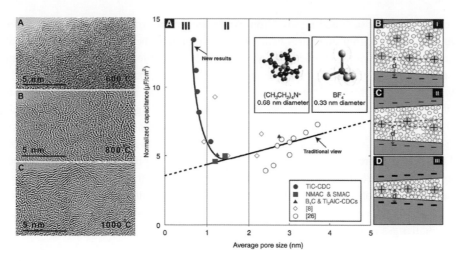

Supercapacitor electrode material. By treating titanium carbide material at very high temperatures, it is possible to create graphitic carbon with nm-scale pores. Using this as a supercapacitor electrode with an ionic liquid electrolyte, the pore size and electrical double layer physics couple to enable very high capacitances. Adapted from [991].

Fig. 12.9

In addition to energy storage with the capability of rapid (high power) discharge, supercapacitors are desired for the conversion from ac to dc power. An ac voltage at 50 or 60 Hz, as from a generator, is rectified using diodes, and the remaining ripple is smoothed away using an RC filter. To succeed at this task, the supercapacitor must have a sufficiently low internal resistance (and ion kinetics) that it remains an effective capacitor at frequencies higher than around 120 Hz. Supercapacitors have been demonstrated that meet this criterion through careful engineering of the nanostructured electrode material, such as growth of carbon nanotube forests integrated directly on graphene synthesized on Ni electrode bases [992].

Environmental remediation

Beyond energy generation and storage, nanostructured materials are already helping to mitigate the environmental impact of energy consumption. The most well-known example of this is the supported precious metal nanoparticle catalyst in the catalytic converters on automobile exhaust systems. These catalysts promote the complete combustion of hydrocarbon fuels, reducing the amount and nastiness of the resulting pollutants.

Nanomaterial catalysts have also shown great promise for general environmental cleanup. Nanoscale titania particles have great demonstrated efficacy in photocatalytic decomposition of hydrocarbon pollutants [974] in aqueous solutions. Incident photons can lead to redox chemistry at the particle surfaces that favor the formation of reactive oxygen species such as ozone, which in turn can oxidize organic contaminants.

Beyond catalysis, properly functionalized nanoparticles of some materials have the potential to serve as scavengers of contaminants, as mentioned in Section 7.7. The nanoparticles' enormous specific surface area allows gram-scale quantities of magnetic nanoparticles to adsorb and remove from water an amount of contamination that would have required kilogram quantities of larger-scale magnetic beads. This is made possible by leveraging the unique magnetic properties (single magnetic domain; large interparticle forces due to field gradients) inherent in nanoscale magnetic particles [589].

Wide-spread adoption of nanomaterials for environmental remediation faces the same challenges as any scale-up of nanotechnology. Producing large amounts of high quality, well-characterized nanomaterials remains challenging in general. These materials must then interact with the environmental hazard or contaminant in an efficient, controlled way, and the resulting product materials must be handled and disposed appropriately. Properly assessing the environmental impact of nanomaterials, even those meant to clean up the environment, is not a trivial challenge. Still, as environmental concerns loom large in general and in the energy industry in particular, the promise of nanotechnology to address these challenges is encouraging.

12.2   Dangers of nanotechnology?

Nanotechnology is frequently portrayed in science fiction as having a significant "dark side". Of particular (fictional) concern are conjectured self-reproducing nanomachines, able to break down materials in their immediate environment and build additional copies of themselves. On a less fantastical level, there are more grounded concerns about the environmental health and safety of nanomaterials, rooted in past human experience with the unintended consequences of new materials and manufacturing processes. Let's look at these in turn.

12.2.1   Environmental hazards

Some new technologies have had unanticipated, negative environmental consequences. Examples include: contamination of ground water with organic solvents in something as innocuous as dry cleaning; lead in paint and gasoline; pesticide interactions with non-target species such as birds; pulmonary disease associated with powdered asbestos; algal blooms due to agricultural fertilizer runoff; plastic microbeads from cleaning products ending up in fish; etc. Given this history, it is not surprising that some are quite concerned about the possible environmental impacts of "nanotechnology" in general, and nanomaterials in particular.[4]

There are both technical and regulatory questions surrounding the wide-spread application of nanomaterials. As we have discussed, the electronic, magnetic, optical, mechanical, and chemical interaction properties of nanoscale materials can be very different from their bulk counterparts. Moreover, some nanomaterials are of a size that can be suspended readily in air or water, making possible their unintentional dispersal into the environment

---

[4] In the early twenty-first century some groups went so far as to call for a moratorium on all nanoscience research until safety concerns could be properly assessed!

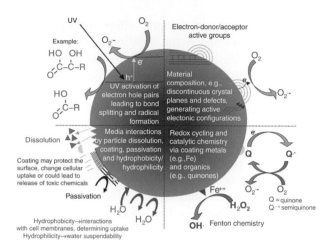

Nanostructured materials and toxicology. Using a nanoparticle as an example, this schematic summarizes a number of possible biological interactions when nanomaterials are taken into living systems. From [993].   **Fig. 12.10**

and thereafter into living systems. Before sensible environmental health and safety procedures and policies can be enacted regarding the handling of nanomaterials, it is now widely recognized that it is necessary to perform careful studies of their dispersal and biological consequences. This has led to the creation of new subdisciplines.

Nanotoxicology focuses on the interactions of nanomaterials with living organisms, with a particular interest in possible deleterious effects on health [993]. Clearly we (and the rest of the world's flora and fauna) are exposed to some distribution of naturally occurring nanoscale materials all the time, and therefore have adapted to surviving in such an environment. However, it is reasonable to worry about possible health consequences of the large-scale industrial deployment of nanomaterials such as semiconductor nanocrystals, carbon nanotubes, or ultraviolet-absorbing nanoparticles such as those employed already in sunscreens. The people most likely to face the largest exposures to these materials are the workers associated with their mass production and their disposal, though the propagation of nanomaterials into the general environment will lead to some general increase in exposure.

Nanotoxicologists perform careful, controlled studies of the interactions of nano-materials with living things, with the intent of mapping out possible pathologies and the exposure levels associated with illness. Ordinary toxicology studies involving chemical compounds, like many experiments involving biological systems, can be challenging to design, extremely complex, and difficult to interpret. Nanotoxicology requires the same level of care and expertise, combined with an understanding of the nanomaterials themselves. As with other potential toxins, it is important to understand rates of uptake depending on type of exposure, what biological processes may be modified by the uptake of nanomaterials, and whether nanomaterials will accumulate in particular systems (e.g., the liver, fat cells, etc.) or be excreted. The number of nanotoxicology papers has grown exponentially since 2000, now numbering over 1500.

Along with assessing the potential harmful effects of various concentrations of nanomaterials, it is important to understand how such materials would behave in the environment. *Environmental nanoscience* examines the fate of nanomaterials outside the laboratory or factory – how do these materials propagate? Do they undergo chemical modifications as they are exposed to environmental conditions such as sunlight or moisture? Do nanomaterials aggregate in living systems? This fledgling discipline is also off to a rapid start and is likely to have a long future.

On a hopeful note, it would appear that industry is applying lessons learned through the course of the problems mentioned at the beginning of this section. There has been a recognition that environmental health and safety studies are important, as is the establishment of manufacturing and handling standards. Moreover, industrial companies seem to have been engaging environmental organizations and governments about these issues at a comparatively early stage, rather than dismissing concerns as unreasonable or fear-mongering. This is not necessarily altruistic; companies recognize that unintended negative consequences could be bad for business or lead to government regulations that they perceive as restrictive.

My best guess is that nanomaterials will prove to be as rich and mixed a blessing as "chemicals". Some nanomaterials will undoubtedly be found to be harmful, just as we now know that benzene raises the likelihood of cancer, thalidomide causes birth defects, and chlorofluorocarbons are harmful to the ozone layer. That being said, we are far more conscious of such potential dangers than at any time in the past, and we have extensive experience in developing protocols for the proper handling and disposal of hazardous materials. Our ability to anticipate and mitigate such problems should be neither over nor underestimated. At the same time, some nanomaterials may well prove to be an incredible boon, just as some small molecule drugs can function as antibiotics, synthetic fertilizers have helped make it possible to feed billions of people, and some polymers have become incredible, ubiquitous building materials. Vigilence regarding the health and environmental consequences of nanoscale technologies is necessary to make sure that we reap the benefits and minimize the costs of our growing ability to manipulate matter down to the atomic scale.

12.2.2 Molecular manufacturing and "grey goo"

Leaning more toward the fantastic, since the 1980s some have advanced the idea of *molecular manufacturing* or *atomically precise manufacturing* as the ultimate endpoint of nanofabrication. The idea is that through ever greater control over both top-down and bottom-up fabrication techniques, it should be possible to build up macroscale systems with something approaching atomic level control over composition and morphology. Combined with advanced calculational techniques with predictive power, this could permit a true revolution in materials-by-design and the creation of structures with incredible properties, including ultradense information processing capabilities.

Aspirational though those goals are, it should be clear from the discussions throughout this book that it is extremely difficult to see how such control could ever be possible in general, let alone practical. Richard Smalley and Eric Drexler engaged in a debate about this in the pages of *C & E News* in 2003, and the arguments for (Drexler) and against (Smalley) atomically precise manufacturing are worth reading in their words [994, 995]. In a brief, atomic-scale positioning of component atoms from the top down is extremely difficult, requiring tremendous control of chemical environments that is only possible for some systems, usually in planar geometries and in ultrahigh vacuum environments. Even in those circumstances the atoms are still subject to the limitations of chemistry, reflecting the remarkably robust properties of atomic orbitals in determining the energetically favorable configurations in which atoms prefer to bind. Even if a particular arrangement of atoms appears calculationally to be an energetic minimum, there is no guarantee that such an endpoint can be reached by simple means, as any synthetic chemist or materials grower can attest. Entropy is also an unavoidable consideration; forming defects, even those that cost energy, is generally favored if there are many ways to do so (Section 5.1).

The rate of any top-down process is the driver toward consideration of bottom-up methods, including self-reproducing "assemblers". If a top-down machine could properly position and bind one atom every 0.1 ns, constructing a sugar-cube-sized piece of material would still require hundreds of thousands of years. Clearly this would never be a practical manufacturing method. Therefore, runs the argument, the only hope of creating macroscopic amounts of such material would be for many (billions) of assembling machines to work in parallel, and the most practical way to achieve such volumes would be for such assemblers to construct, among other things, copies of themselves. For the sake of robustness, it would make sense for the design of the assemblers to be fault-tolerant, so that the occasional defect is not a complete show-stopper. The "grey goo problem" is the concern that such assemblers could get out of control, disassembling matter around themselves and reorganizing it into more assemblers in a runaway process, turning the entire world or large parts of it into grey goo.

The grey goo scenario should sound a bit familiar. One can make an argument that the runaway reproduction of some kinds of assemblers has already taken place at least once, with the propagation of living creatures on the Earth. The "green goo" that has taken over the planet has taken fault tolerance to spectacular heights.

However, just as life shows what can happen to runaway, self-reproducing systems over long periods, it also highlights the challenges that, in my view, make any inorganic, human-made assembler program unlikely to get off the ground. Living systems work with a comparatively small subset of building materials, and rarely require macroscale structural perfection. While these systems are fault tolerant in a sense, they tend to be fragile, with most of their mechanisms failing to function under comparatively modest changes in environmental conditions of temperature, pressure, pH, etc. Constructing from scratch an inorganic assembler able to take advantage of a broad material palette (including metals and semiconductors), with atomically precise functionality and some kind of ability for self-propagation, seems more difficult and demanding than the creation of entirely synthetic life *ab initio*.

Therefore, the achievement of large-scale, atomically precise manufacturing does not seem possible to me in the foreseeable future. Similarly, grey goo appears far less likely an environmental concern than more banal but far more likely hazards such as climate change or pandemic disease. When considering self-propagating systems that have the potential to get out of control, it is sobering to realize that the capability *already exists* to synthesize deadly pathogens. Grey goo, because it is so technologically far fetched, is not remotely worth that level of concern.

12.3 Prognosis

Quo vadis, nanostructures and nanotechnology?

Our understanding of the physics and chemistry of nanoscale systems has grown enormously in recent years. This progress shows little signs of slowing, as fabrication and characterization capabilities continue to advance and become more widespread and accessible. Hopefully this book has given you an appreciation for the physics at work at the nanoscale, the exciting possibilities that this presents for future technologies, and some sense of the challenges facing the transition of interesting nanoscale physics from the laboratory, through engineering development, and into a functional industrial role.

Way back in the introduction, I asked whether nanotechnology was a "disruptive technology", a specific advance in capabilities with ramifications so large that society itself is reshaped. Depending on where one sets the bar, examples of true disruptive technologies include controlled production and use of fire, the wheel, writing, the printing press, the steam engine, the electric motor, the automobile, the airplane, antibiotics, the transistor, the digital computer, the integrated circuit, the laser, DNA sequencing, etc.

Does "nanotechnology" fit the bill? Perhaps. I do think that our ability to structure materials at the nanometer scale, combined with the resulting changes in properties that take place, will have an enormous impact on many facets of our lives. However, the breadth of what is labeled nanotechnology, and the eventual ubiquity of many technologies involving some nano capability, will make the situation much more diffuse than the examples listed above. Continued technological advancement will undoubtedly bring with it profound societal changes, and nanoscale science will be connected to many facets of that advancement, but it's not clear to me that people will look back and say that there was a single nano revolution. By analogy, synthetic plastics are everywhere, but most people don't think we live in a "plastic age".

On the other hand, many people would say that we live in an "Information Age". Nearly all facets of life in the economically developed world have been touched by the ready availability of large-scale computing power and communications networks, even though each of these has required a multitude of inventions and innovations. The Information Age is an *emergent* product of these building blocks, in analogy to the properties of solids and liquids emerging from the collective response of individual atomic entities. It is conceivable that the reach of broadly defined nanotechnology will one day be so encompassing that we could speak of a "Nano Age", even though it is an aggregate result of many nano-based materials, processes, and inventions. If this does happen, you are living through what may be the dawn of this period, and it will be exciting to see what emerges!

12.4 Summary

We considered the "energy problem" as an example of how our ability to manipulate and characterize materials on the nanometer scale can have impact on many facets of a complex, societally important issue. I also briefly editorialized about possible hazards of nanotechnology and nanomaterials, concluding that some concerns are more realistic than others. Nano capabilities will be relevant to many technologies, and potentially profound changes in society may emerge from the confluence of many applications of nanomaterials, nanodevices, and nanocharacterization.

- Energy production and consumption is a serious societal problem. There is no "magic bullet" that will take the place of fossil fuels. However, rising nanocapabilities can contribute to improved generation, storage, utilization, and transmission capabilities.
- Nanostructures can improve efficiencies significantly. This is true for light/strong nanomaterials in structural applications and improved solid state lighting, for example.
- Solar generation of useful energy is an example of the many potential nanocontributions. These include enhanced photovoltaics, photoelectrochemical cells, and photocatalytic generation of chemical fuels.
- Nanomaterials can be applied to environmental cleanup, a related societal concern. Approaches include photocatalytic decomposition of wastes and nanoparticle methods for scavenging pollutants.
- It is important to understand potential environmental risks of nanomaterials. This requires studies of both toxicology and the fate of nanomaterials in the environment.
- Atomically precise manufacturing on large scales is, in my opinion, exceedingly unlikely for the foreseeable future. The obstacles in terms of positioning, material flexibility, environmental stability, etc. do not appear surmountable in generality.
- Biological creations run amok are a more realistic concern than "grey goo". We know that rapidly spread biological constructs with negative impacts can be made now, while grey goo requires nanoassemblers that are unlikely to exist for decades, if ever.

12.5 Suggested reading

- *The energy problem* There are a large number of books about the general predicament of energy production and usage within environmental constraints. A well-written survey bearing in mind climate impact is *Energy Demand and Climate Change: Issues and Resolutions* by F. H. Cocks (Wiley, 2011) [996]. Another is *The Energy Problem* by R. S. Stein and J. Powers (World Scientific, 2011) [997].
- *Nano and energy* A great overview that dovetails with many of the ideas presented here is *Nanotechnology for the Energy Challenge*, second edn., edited by J. García-Martínez (Wiley, 2013) [998]. A particularly nice review article about the materials challenges for

Li-ion batteries is that by Goodenough and Kim, in *Chem. Mater.* **3**, 587–693 (2009) [999].

- *Nanotoxicology, environmental health and safety* These areas are so rapidly evolving, it is difficult to come up with established overview texts. The nanotoxicology review articles by Oberdörster *et al. Env. Health Persp.* **113**, 823–839 (2005) [1000] and Krug & Wick, *Angew. Chemie Int. Edn.* **50**, 1260–1278 (2011) [1001] are helpful at providing context. For the fate of nanomaterials in the environment, I suggest S. J. Klaine *et al.*, *Env. Tox. Chem.* **27**, 1825–1851 (2008) [1002].

- *Atomically precise manufacturing and grey goo* The classic *inspirational* work on this topic is K. E. Drexler's *The Engines of Creation* (Anchor, 1987) [1003], with remarkable vision (though outcomes that are unphysically optimistic in my view). A now-classic science fiction work along these lines is Neal Stephenson's *The Diamond Age* (Random House, 2003) [1004].

Appendix: Common quantum mechanics and statistical mechanics results

A.1 A review of perturbation theory

Most quantum mechanics problems are not solvable in closed form with analytical techniques. To extend our repertoire beyond just particle-in-a-box, a number of approximation techniques have been developed. A large class of these fall under the heading of "perturbation theory", in which we consider our system to obey Hamiltonian H that may be written as

$$H = H^0 + \lambda H^1 + \lambda^2 H^2 + \cdots , \tag{A.1}$$

where H^0 is an exactly solvable Hamiltonian, λ is a small parameter, and the other terms may therefore be taken as small corrections.

These notes are a quick review of how to deal with systems that obey such Hamiltonians. We'll be using Dirac notation and the Schrödinger formalism, in which the states are time-dependent. We begin with time-independent perturbation theory, and will then move on to consider time-dependent problems. For now, we'll only deal with single-particle problems, too.

A.1.1 Time-independent

We're going to use the time-independent Schrödinger equation, and look for the energy eigenvalues and eigenstates of H. First, suppose that our *unperturbed* problem,

$$H^0|\psi^0\rangle = E^0|\psi^0\rangle \tag{A.2}$$

may be solved exactly, giving an energy eigenvalue spectrum E_j^0 with corresponding eigenstates $|\psi_j^0\rangle$. Remember, because H^0 is a Hermitian operator, its eigenvalues are real, and its eigenvectors $|\psi_j^0\rangle$ form a complete set.

The strategy we're going to take is straightforward. We're going to assume that our perturbative corrections to the Hamiltonian, $\lambda H^1 + \lambda^2 H^2 + \cdots$, lead to corresponding perturbative corrections to the eigenvalues and eigenstates. That is,

$$\begin{aligned} E_j &= E_j^0 + \lambda E_j^1 + \lambda^2 E_j^2 + \cdots , \\ |\psi_j\rangle &= |\psi_j^0\rangle + \lambda|\psi_j^1\rangle + \lambda^2|\psi_j^2\rangle + \cdots . \end{aligned} \tag{A.3}$$

For now, we'll plug into the time-independent Schrödinger equation, and use some linear algebra to solve for terms of interest.

First, we need to check normalization. Taking the inner product of the proposed eigenstate $|\psi_j\rangle$ with itself and setting equal to 1, and grouping terms by orders in λ, we see

$$\langle\psi_j^0|\psi_j^0\rangle = 1$$
$$\langle\psi_j^1|\psi_j^0\rangle + \langle\psi_j^0|\psi_j^1\rangle = 0, \text{ etc.} \tag{A.4}$$

So, we can take $\langle\psi_j^1|\psi_j^0\rangle = 0$ without any problems.

A.1.2 Nondegenerate, 1st and 2nd order

Plugging into the Schrödinger equation up to second order,

$$(H^0+\lambda H^1+\lambda^2 H^2)(|\psi^0\rangle+\lambda|\psi^1\rangle+\lambda^2|\psi^2\rangle) = (E_j^0+\lambda E_j^1+\lambda^2 E_j^2)(|\psi^0\rangle+\lambda|\psi^1\rangle+\lambda^2|\psi^2\rangle). \tag{A.5}$$

For simpicity's sake, let's assume initially that our original, unperturbed eigenvalue spectrum is nondegenerate. That is,

$$E_j^0 \neq E_k^0 \tag{A.6}$$

for all $j \neq k$.

Now let's take the inner product of Eq. (A.5) with $\langle\psi_j^0|$. Again, let's look order by order in λ. To zeroth order, we get

$$\langle\psi_j^0|H^0|\psi_j^0\rangle = \langle\psi_j^0|E^0|\psi_j^0\rangle, \tag{A.7}$$

which we know is true because $|\psi_j^0\rangle$ is the solution to the unperturbed problem with eigenvalue E_j^0.

To first order, we find

$$\langle\psi_j^0|H^1|\psi_j^0\rangle = E_j^1. \tag{A.8}$$

So, the first-order correction to the energy eigenvalue of state j is the matrix element of the first-order perturbation Hamiltonian between the unperturbed states.

Now we want to find $|\psi_j^1\rangle$, the first-order correction to the state $|\psi_j^0\rangle$. Take the inner product of Eq. (A.5) with $\langle\psi_k^0|$ this time, with $k \neq j$. The zeroth order term vanishes. The first-order term gives:

$$\langle\psi_k^0|H^1|\psi_j^0\rangle = (E_k^0 - E_j^0)\langle\psi_k^0|\psi_j^1\rangle. \tag{A.9}$$

Rearranging,

$$\langle\psi_k^0|\psi_j^1\rangle = \frac{\langle\psi_k^0|H^1|\psi_j^0\rangle}{(E_k^0 - E_j^0)}. \tag{A.10}$$

Since the unperturbed eigenfunctions form a complete set, we can write the correction to the eigenfunction as

$$|\psi_j^1\rangle = \sum_k |\psi_k^0\rangle\langle\psi_k^0|\psi_j^1\rangle$$

$$= \sum_{k\neq j} |\psi_k^0\rangle \frac{\langle\psi_k^0|H^1|\psi_j^0\rangle}{(E_k^0 - E_j^0)}. \qquad (A.11)$$

Now that we have an expression for $|\psi_j^1\rangle$, we can plug that into Eq. (A.5), take the inner product of both sides with $\langle\psi_j^0|$ as before, and look at the second-order terms.

$$E_j^2 = \langle\psi_j^0|H^1|\psi_j^1\rangle$$

$$= \sum_{k\neq j} \langle\psi_j^0|H^1|\psi_k^0\rangle \frac{\langle\psi_k^0|H^1|\psi_j^0\rangle}{(E_k^0 - E_j^0)}$$

$$= \sum_{k\neq j} \frac{|\langle\psi_k^0|H^1|\psi_j^0\rangle|^2}{(E_k^0 - E_j^0)}. \qquad (A.12)$$

We've now found the first- and second-order corrections to the unperturbed energy spectrum, and the first-order correction to the unperturbed states.

A.1.3 Degeneracies

If the unpertrubed system has degenerate levels, things can get messy. The perturbation can "lift the degeneracy", meaning that two unperturbed states with *identical* unperturbed energics can end up getting *different* perturbation corrections to their energies and the states themselves. A classical example: a compass in zero magnetic field has all orientations degenerate. However, an external magnetic field in the plane of the compass needle breaks the degeneracy, and picks out a unique ground state of the perturbed system.

Suppose we start out with two unperturbed degenerate states, $|\xi_f^0\rangle$ and $|\xi_g^0\rangle$. Rather than just using $|\xi_f^0\rangle$ and $|\xi_g^0\rangle$ as the basis for our perturbation calculation, we want to find linear combinations of the two to use instead. Ideally we want the perturbation to not mix the originally degenerate states. That is, we want to find a combination of coefficients c_f, c_g and use $c_f|\xi_f^0\rangle + c_g|\xi_g^0\rangle$ (and the combination orthogonal to that) as the starting points for our perturbation calculation. We determine the values of c_f, c_g by requiring that the perturbation be diagonal in the new basis.

A.1.4 Summary

We've seen that perturbation theory may be used to solve problems that are *close* to exactly solvable unperturbed problems. The perturbation changes the energy eigenvalues from their unperturbed values. It also alters a particular unperturbed state by mixing it with the other unperturbed states by an amount related to the matrix element of the perturbation; see Eq. (A.8). That change to the states leads to a higher order correction to the energy, Eq. (A.12).

There are systematic approaches to carry out perturbation theory to higher orders. The most well-known example is the diagrammatic technique developed in quantum field theory by Richard Feynman.

A.1.5 Time-dependent

What happens if we have an explicitly time-dependent Hamiltonian? The classic example of this is behavior of a system under the influence of electromagnetic radiation. Here the time-dependence of the electric and magnetic fields is harmonic, and the results can be profound (Stokes shifts in Raman spectroscopy; lasing; etc.).

Here we briefly introduce a few ways of dealing with time-dependent problems. We discuss the sudden and adiabatic approximations, and time-dependent perturbation theory. We end with a derivation of Fermi's Golden Rule, an expression used commonly for determining rates of processes in solid state systems.

A.1.6 Sudden and adiabatic approximations

Suppose we consider turning on some perturbing Hamiltonian, H', in addition to our unperturbed Hamiltonian, H^0. Suppose further that the system was initially in a state $|\psi(0)\rangle$.

Consider the case where the perturbation is turned on *instantaneously* at $t = 0$ and is left on for all future times. This is a time when we can apply the sudden approximation. If the eigenfunctions of the new Hamitonian $H^0 + H'$ are labeled $|\phi'_j\rangle$ with energies E'_j, then we can write down the state for all future times:

$$\phi(t > 0) = \sum_j |\phi'_j\rangle\langle\phi'_j|\psi(0)\rangle e^{-iE'_j t/\hbar}. \tag{A.13}$$

We project the original state onto the new eigenstates, and time-evolve those new eigenstates according to the new Hamiltonian. This approximation is valid provided the turn-on time of H' is short compared to the relevant timescales of $\sim \hbar/E'_j$.

Similarly, we consider the adiabatic limit, in which H' is turned on infinitesmally slowly (or at least slowly compared to all relevant energy scales, given by $\hbar/\Delta E_{jk}$, where ΔE_{jk} is the energy spacing between any two adjacent levels). If the adiabatic approximation is a good one, then the turn-on of the perturbation may be sufficiently gentle that each old eigenstate $|\phi^0_j\rangle$ smoothly evolves into a new eigenstate $|\phi'_j\rangle$. If the perturbation would have caused old levels to cross, instead one generally finds an "avoided crossing". The detailed analysis of the probability of ending up in the "other" level after an avoided crossing as a function of perturbation rate was done by Landau and Zener back in the 1930s. See Landau and Lifschitz [1006], or Rubbermark *et al.*, *Phys. Rev.* A **23**, 3107 (1981).

A.1.7 Transition probabilities

What if the time dependence is more subtle than just turning on a perturbation and leaving it there? Then we need time-dependent perturbation theory. Begin with our unperturbed

Hamiltonian, H^0, and again assume a nondegenerate eigenvalue spectrum E_j^0. (As above, the degenerate case is a reasonably straightforward development from the nondegenerate case.) At time $t = 0$ we consider beginning to perturb our system with some $H'(t)$.

At any time t we can expand the true, unknown $|\Psi(t)\rangle$ in the time-evolved versions of the unperturbed eigenstates, $|\psi_j^0\rangle$. That is:

$$|\Psi(t)\rangle = \sum_j c_j(t) e^{-iE_j^0 t/\hbar} |\psi_j^0\rangle, \tag{A.14}$$

where the $c_j(t)$ are the time-dependent coefficients of the expansion. We can interpret this as saying that, if we turned the perturbation H' off instantly at time t, the probability of finding the system in state $|\psi_j^0\rangle$ is given by $|c_j(t)|^2$. Note (without proof) that this expansion automatically preserves normalizalization as long as the perturbation $H'(t)$ is real.

Plug the state given in Eq. (A.14) into the time-dependent Schrödinger equation, using the total Hamiltonian, $H^0 + H'$:

$$i\hbar \sum_j \dot{c}_j(t) e^{-iE_j^0 t/\hbar} |\psi_j^0\rangle + i\hbar \sum_j \frac{-i}{\hbar} E_j^0 c_j(t) e^{-iE_j^0 t/\hbar} |\psi_j^0\rangle = \sum_j (H^0 + H'(t)) c_j(t) e^{-iE_j^0 t/\hbar} |\psi_j^0\rangle. \tag{A.15}$$

Drop identical terms from both sides after using the H^0 operator on the right hand side. Then take the inner product of both sides with $\langle\psi_k^0| e^{iE_k^0 t/\hbar}$, and get

$$i\hbar \dot{c}_k(t) = \sum_j c_j(t) \langle\psi_k^0|H'(t)|\psi_j^0\rangle e^{i(E_k^0 - E_j^0)t/\hbar}. \tag{A.16}$$

This is exact, so far. Define

$$\omega_{kj} \equiv \frac{E_k^0 - E_j^0}{\hbar},$$
$$H_{kj}' \equiv \langle\psi_k^0|H'(t)|\psi_j^0\rangle. \tag{A.17}$$

One can now try writing $H'(t)$ as a power series in some small parameter λ, and parallel our earlier analysis. However, a more common and useful approach is an iterative one, described below. Let's assume that at $t = 0$ the system was in a particular unperturbed eigenstate, $|\psi_a^0\rangle$. Therefore, $c_a(t = 0) = 1$, and $c_{j\neq a}(t = 0) = 0$. That is, $c_a(t = 0) = \delta_{ja}$. Plug this into Eq. (A.16), and we get

$$c_a(t) = \frac{1}{i\hbar} \int_0^t H_{aa}'(\tau) d\tau + 1, \quad j = a,$$
$$c_j(t) = \frac{1}{i\hbar} \int_0^t H_{ja}'(\tau) e^{i\omega_{ja}\tau} d\tau, \quad j \neq a. \tag{A.18}$$

We can define a transition probability, the probability of finding the system in (approximately) state j at time t after starting out in state a, as

$$P_{ja}(t) = |c_j(t)|^2 = \frac{1}{\hbar^2} \left| \int_0^t H_{ja}'(\tau) e^{i\omega_{ja}\tau} d\tau \right|^2. \tag{A.19}$$

From this expression it is already possible to see where selection rules come from: if the matrix element of the perturbation between the initial and (approximate) final states, H'_{ja}, is zero, then to first order the transition is forbidden.

Consider something like our sudden approximation case from before, where H' is turned on at $t = 0$. In particular, let's assume

$$H'(t) = V_0 \cos(\omega_r t) = V_0 \left[\frac{e^{i\omega_r t} + e^{-i\omega_r t}}{2} \right] \tag{A.20}$$

for $t > 0$. Now we're representing the true new eigenstates approximately by the old unperturbed states.

If we actually plug this into our probability expression, we find, first for $\omega_r = 0$,

$$P_{ja} = \frac{|V_{0,ja}|^2}{\hbar^2} \frac{\sin^2(\omega_{ja} t/2)}{(\omega_{ja}/2)^2}. \tag{A.21}$$

Remember, $\sin^2(x)/x^2$ is strongly peaked around $x = 0$. That implies that *transitions between initial and final states are favored if energy is conserved*, that is, if $\omega_{ja} \approx 0$. Let's define

$$F(t, \omega_{ja}) \equiv \frac{2 \sin^2 \omega_{ja} t/2}{\omega_{ja}^2}. \tag{A.22}$$

This quantity will be useful in a while.

If we do the $\omega_r \neq 0$ case, we find

$$c_j(t) = -i \frac{|V_{0,ja}|^2}{\hbar^2} \left[\frac{e^{i(\omega_{ja}+\omega_r)t} - 1}{i(\omega_{ja} + \omega_r)} + \frac{e^{i(\omega_{ja}-\omega_r)t} - 1}{i(\omega_{ja} - \omega_r)} \right]. \tag{A.23}$$

If we take the magnitude-squared of this to find P_{ja}, we get

$$P_{ja}(t) = F(t, \omega_{ja} + \omega_r)|V_{0,ja}|^2 + F(t, \omega_{ja} - \omega_r)|V_{0,ja}|^2 + \quad \text{crossterms}. \tag{A.24}$$

The function $F(t, \omega)$ has some interesting properties. For fixed $\omega \neq 0$, its t-dependence is just $\sim \sin^2$. However, at fixed t, its peak at $\omega = 0$ is $t^2/2$, and its half-width is $2\pi/t$. So, as t increases, $F(t, \omega)$ becomes more and more strongly peaked around $\omega = 0$. In fact,

$$\int_{-\infty}^{+\infty} F(t, \omega) d\omega = \pi t, \tag{A.25}$$

and one can show that $F(t, \omega) \to \pi t \delta(\omega)$ as $t \to \infty$.

For us, this means that at long times (but not times so long that the initial state gets depleted!) transitions only occur to states that conserve energy. For our $\omega_r \neq 0$ case above, we find transitions are probable between states j and a only if they differ in energy by an amount $\hbar \omega_r$. These two processes correspond to absorption and stimulated emission.

A.1.8 Fermi's Golden Rule

Consider the $\omega_r \neq 0$ case from above, and suppose that there are a number of states $\nu(E)$ that satisfy the conservation of energy condition between initial and final states. Also, suppose those final states also share the same matrix element with the initial state, $V_{0,ja}$. In that case, the *rate* at which transitions occur is given by

$$\frac{\mathrm{d}P_{ja}(t)}{\mathrm{d}t} = \frac{2\pi}{\hbar}|V_{0,ja}|^2 \nu(E). \qquad (A.26)$$

This relation is called *Fermi's Golden Rule*, and is commonly used to determine the rates of many quantum mechanical processes.

Remember that we're doing perturbation theory here, and asking the question, if a system starts out in (unperturbed) state a, what is the probability of finding it later in (approximately) unperturbed state j. The key to our approximations is that the time-dependent perturbation is small, that we wait long enough for our F functions to act like energy-conserving delta functions, but not so long that the initial state's population is changed much. These are conditions for the validity of Fermi's Golden Rule, which is fundamentally a first-order time dependent perturbation theory result.

Note that:

- If the initial and final states in question are not connected by the perturbing potential (that is, the matrix element in question is zero), the rate of the process is also zero. This is an example of the origin of selection rules.
- If the number of available final states is zero, the rate of the process is also zero. The importance of this point can't be overestimated. As we'll see, it's the reason that semiconductors are semiconductors, that band insulators are band insulators, etc.

A.1.9 Conclusion

This concludes our review of perturbation theory. The main results to take away from this are the first- and second-order corrections to the energies in time-independent problems, and Fermi's Golden Rule. For a more complete discussion of perturbation methods, most quantum mechanics textbooks will be useful references. Particularly valuable are Cohen-Tannoudji's [1007] and Sakurai's [1008] texts.

A.2 The vector potential

For a thorough, in-depth discussion of the vector potential, I recommend the textbooks by Griffiths [1005] and Jackson [596].

Just as it is often helpful to think about the scalar potential ϕ, defined such that $-\nabla\phi = \mathbf{E}$, it is very useful to think about the vector potential, defined such that $\nabla \times \mathbf{A} = \mathbf{B}$. Thanks to vector calculus identities, this definition ensures $\nabla \cdot \mathbf{B} = 0$.

Remember that the fields are the physically measurable quantities, through the forces exerted on test charges. That implies that ϕ is only defined up to an arbitrary constant ϕ_0, since $-\nabla\phi = -\nabla(\phi + \phi_0) = \mathbf{E}$. Similarly, $\mathbf{A}$ is only defined up to the gradient of a scalar function, since $\nabla \times \mathbf{A} = \nabla \times (\mathbf{A} + \nabla f)$ because $\nabla \times \nabla f = 0$. The ability to add scalar constants to ϕ and gradients of scalar functions to $\mathbf{A}$ without affecting the physical fields is called the freedom to choose a *gauge*.

From the point of view of quantum mechanics for a charged particle, ϕ and $\mathbf{A}$ are very important. The mechanical momentum of a charged particle (charge q) is not conserved in the presence of an electric field, because a force $-q\mathbf{E}$ is being exerted on the particle. Similarly, the mechanical momentum of a charged particle moving in the presence of magnetic induction $\mathbf{B}$ is not conserved because of the Lorentz force, $q\mathbf{v} \times \mathbf{B}$. However, adding $q\mathbf{A}$ to the mechanical momentum does result in a quantity that is conserved, the "canonical momentum", in the absence of electric fields. If we call the canonical momentum $\mathbf{p}$, then the *mechanical* part of the momentum is given by $\mathbf{p} - q\mathbf{A}$. In quantum mechanics, the canonical momentum is replaced by the momentum operator, which in the real-space representation is $\mathbf{p} \equiv -i\hbar\nabla$. The bottom line is, the presence of a vector potential implies a gradient in the phase of the wavefunction of a charged particle, as given in Eq. (6.24).

A.3 The Einstein relation

A common result from statistical physics is the Einstein relation that connects the diffusion constant and mobility. Consider some gas of charged particles (charge q) in the presence of an electric field. There are two contributions to the current: the *drift* part due to the carrier response to the electric field, and the *diffusion* part due to the tendency of particles to diffuse away from regions of high concentration. In equilibrium, the total current density must be zero:

$$\mathbf{J} = \sigma\mathbf{E} - qD\nabla n = 0, \tag{A.27}$$

where n is the spatial density of particles and σ is the electrical conductivity. Assuming isotropic material response and rearranging,

$$\sigma = \frac{qD\nabla n}{|\mathbf{E}|}. \tag{A.28}$$

More progress can then be made when considering specific systems. Consider electrons in semiconductors at high temperature, for example, for a uniform chemical potential (Fermi level) as required in equilibrium, the local density $n(\mathbf{r})$ depends on the effective density of states in the conduction band, N_C, and the local conduction band energy, $E_C(\mathbf{r})$, as described in Eq. (3.10):

$$n(\mathbf{r}) = N_C \exp\left(-\frac{E_C(\mathbf{r}) - E_F}{k_B T}\right). \tag{A.29}$$

We can differentiate this and find

$$\frac{\nabla n}{n} = -\frac{\nabla E_C(\mathbf{r})}{k_B T}. \tag{A.30}$$

The gradient in the conduction band level is tied to the local electric field, $\nabla E_C = -q\mathbf{E}$, since work must be done to push a charge q against an electric field.

For semiconductors it is more common to consider the mobility rather than the conductivity, and write $\sigma = nq\mu$. In that case, we can simplify Eq. (A.28) to

$$\frac{\mu}{D} = \frac{q}{k_B T}.$$
(A.31)

This (or its reciprocal) is the classical Einstein relation.

For the situation in metals, when the electron gas is degenerate, the local density of electrons is related to the density of states at the Fermi level, $\nu_{3d}(E_F)$:

$$\nabla n = q\mathbf{E}\nu_{3d}(E_F).$$
(A.32)

Combining with Eq. (A.28),

$$\sigma = q^2 D \nu_{3d}(E_F).$$
(A.33)

This form of the Einstein relation allows one to extract a physically meaningful value for D based on experimental parameters. Heat capacity measurements on a metal can be used to infer $\nu_{3d}(E_F)$, and conductivity can be measured readily.

References

[1] "The Lycurgus Cup", www.britishmuseum.org/research/collection_online/collection_object_details/collection_image_gallery.aspx?partid=1&assetid=1066673&objectid=61219 (2014).

[2] O. D. Sherby, "Ultrahigh carbon steels, Damascus steels and ancient blacksmiths", *ISIJ international* **39**, 637–648 (1999).

[3] K. Kinoshita, "Electrochemical uses of carbon", electrochem.cwru.edu/encycl.art-c01-carbon.htm (2001).

[4] H. Ohnishi, Y. Kondo, and K. Takayanagi, "Quantized conductance through individual rows of suspended gold atoms", *Nature* **395**, 780–783 (1998).

[5] C. Kittel, *Introduction to Solid State Physics*, Eighth ed. (New York, John Wiley & Sons, 2004).

[6] N. W. Ashcroft and N. D. Mermin, *Solid State Physics* (New York, Brooks Cole, 1976).

[7] W. A. Harrison, *Solid State Theory* (New York, Dover, 1980).

[8] W. A. Harrison, *Electronic Structure and the Properties of Solids* (New York, Dover, 1989).

[9] R. P. Feynman, R. B. Leighton, and M. Sands, *The Feynman Lectures on Physics*, volume III (New York, Addison Wesley, 1965).

[10] D. C. Ralph, C. T. Black, and M. Tinkham, "Spectroscopic Measurements of Discrete Electronic States in Single Metal Particles", *Phys. Rev. Lett.* **74**, 3241–3244 (1995).

[11] N. Nilius, T. M. Wallis, and W. Ho, "Development of one-dimensional band structure in artificial gold chains", *Science* **297**, 1853–1856 (2002).

[12] D. K. Ferry, private communication (2009).

[13] M. Topinka, B. LeRoy, S. Shaw, *et al.*, "Imaging coherent electron flow from a quantum point contact", *Science* **289**, 2323–2326 (2000).

[14] S. Frank, P. Poncharal, Z. Wang, and W. A. de Heer, "Carbon nanotube quantum resistors", *Science* **280**, 1744–1746 (1998).

[15] M. P. Marder, *Condensed Matter Physics* (New York, Wiley, 2000).

[16] L. C. Pauling, *The Nature of the Chemical Bond and the Structure of Molecules and Crystals* (Ithaca, NY, Cornell University, 1960).

[17] J. C. Phillips, *Bonds and Bands in Semiconductors* (Academic, New York, 1973).

[18] L. C. Pauling, "The nature of the chemical bond IV. The energy of single bonds and the relative electronegativity of atoms", *J. Am. Chem. Soc.* **54**, 3570–3582 (1932).

[19] E. Hückel, "Quantum-theoretical contributions to the benzene problem. I. The electron configuration of benzene and related compounds", *Z. Physik* **71**, 204–286 (1931).

[20] J. C. Slater, "Wave functions in a periodic potential", *Phys. Rev.* **51**, 846–851 (1937).

[21] C. Herring, "A new method for calculating wave functions in crystals", *Phys. Rev.* **57**, 1169–1177 (1940).

[22] E. Antončík, "Approximate formulation of the orthogonalized plane-wave method", *J. Phys. Chem. Solids* **10**, 314–320 (1959).

[23] J. C. Phillips and L. Kleinman, "New method for calculating wave functions in crystals and molecules", *Phys. Rev.* **116**, 287–294 (1959).

[24] P. Hohenberg and W. Kohn, "Inhomogeneous electron gas", *Phys. Rev.* **136**, B864–B871 (1964).

[25] W. Kohn and L. J. Sham, "Self-consistent equations including exchange and correlation effects", *Phys. Rev.* **140**, A1133–A1138 (1965).

[26] N. F. Mott, "The basis of the electron theory of metals, with special reference to the transition metals", *Proc. Phys. Soc. London Series A* **62**, 416–421 (1949).

[27] R. de Picciotto, M. Reznikov, M. Heiblum, V. Umansky, G. Bunin, and D. Mahalu, "Direct observation of a fractional charge", *Nature* **389**, 162–164 (1997).

[28] L. Saminadayar, D. C. Glattli, Y. Jin, and B. Etienne, "Observation of the $e/3$ fractionally charged Laughlin quasiparticle", *Phys. Rev. Lett.* **79**, 2526–2529 (1997).

[29] L. Landau, "Theory of the Fermi liquid", *Sov. Phys. JETP* **3**, 920–928 (1956).

[30] D. Pines and P. Nozières, *The Theory of Quantum Liquids*, Vol. I (New York, Addison Wesley, 1989).

[31] G. Baym and C. Pethick, *Landau Fermi-Liquid Theory: Concepts and Applications* (New York, Wiley, 1992).

[32] J. Bardeen, L. N. Cooper, and J. R. Schrieffer, "Microscopic theory of superconductivity", *Phys. Rev.* **106**, 162–164 (1957).

[33] J. Bardeen, L. N. Cooper, and J. R. Schrieffer, "Theory of superconductivity", *Phys. Rev.* **108**, 1175–1204 (1957).

[34] J. R. Schrieffer, *Theory of Superconductivity*, Advanced Book Classics (New York, Westview, 2001).

[35] M. Tinkham, *Introduction to Superconductivity*, 2nd edn. (New York, Dover, 2004).

[36] C. P. Poole, H. A. Farach, and R. J. Creswick, *Superconductivity* (New York, Academic, 1996).

[37] W. Meissner and R. Oschenfeld, "Ein neuer Effekt bei Eintritt der Supraleitfähigkeit", *Naturwiss.* **21**, 787–788 (1933).

[38] R. C. Jaeger, *Introduction to Microelectronic Fabrication*, volume 5 of *Modular Series on Solid State Devices*, 2nd edn. (New York, Prentice Hall, 2002).

[39] H. Ohno, "Making nonmagnetic semiconductors ferromagnetic", *Science* **281**, 951–956 (1998).

[40] H. Ohno, "Properties of ferromagnetic III-V semiconductors", *J. Mag. Mag. Mater.* **200**, 110–129 (1999).

[41] T. Dietl, "Ferromagnetic semiconductors", *Semicond. Sci. Technol.* **17**, 377–392 (2002).

[42] Y. D. Park, A. T. Hanbicki, S. C. Erwin, *et al.*, "A group-IV ferromagnetic semiconductor: Mn_xGe_{1-x}", *Science* **295**, 651–654 (2002).

[43] A. H. MacDonald, P. Schiffer, and N. Samarth, "Ferromagnetic semiconductors: Moving beyond (Ga,Mn)As", *Nate. Mater.* **4**, 195–202 (2005).

[44] T. Schallenberg and H. Munekata, "Preparation of ferromagnetic (In,Mn)As with a high Curie temperature of 90 K", *Appl. Phys. Lett.* **89**, 042507 (2006).

[45] H. Ohno, D. Chiba, F. Matsukura, *et al.*, "Electric-field control of ferromagnetism", *Nature* **208**, 944–946 (2000).

[46] S. Koshihara, A. Oiwa, M. Hirasawa, *et al.*, "Ferromagnetic order induced by photogenerated carriers in magnetic III-V semiconductor heterostructures of (In,Mn)As/GaSb", *Phys. Rev. Lett.* **78**, 4617–4620 (1997).

[47] S. M. Sze and K. K. Ng, *Physics of Semiconductor Devices*, 3rd edn. (New York, Wiley-Interscience, 2006).

[48] B. E. Deal and C. R. Helms, eds., *The Physics and Chemistry of SiO_2 and the Si-SiO_2 interface* (Berlin, Springer, 1993).

[49] D. A. Muller, T. Sorsch, S. Moccio, F. H. Baumann, K. Evans-Lutterodt, and G. Timp, "The electronic structure at the atomic scale of ultrathin gate oxides", *Nature* **399**, 758–762 (1999).

[50] A. A. Demkov, L. R. C. Fonseca, E. Verret, J. Tomfohr, and O. F. Sankey, "Complex band structure and the band alignment problem at the Si-high-k dielectric interface", *Phys. Rev. B* **71**, 195306 (2005).

[51] J. P. O'Sullivan and G. C. Wood, "The morphology and mechanisms of formation of porous anodic films on aluminum", *Proc. R. Soc. London A* **317**, 511–543 (1970).

[52] H. Masuda, H. Yamada, M. Satoh, H. Asoh, M. Nakao, and T. Tamamura, "Highly ordered nanochannel-array architecture in anodic alumina", *Appl. Phys. Lett.* **71**, 2770–2772 (1997).

[53] C. L. Kane and E. J. Mele, "Quantum spin Hall effect in graphene", *Phys. Rev. Lett.* **95**, 226801 (2005).

[54] C. L. Kane and E. J. Mele, "Z 2 topological order and the quantum spin Hall effect", *Phys. Rev. Lett.* **95**, 146802 (2005).

[55] L. Fu, C. L. Kane, and E. J. Mele, "Topological insulators in three dimensions", *Phys. Rev. Lett.* **98**, 106803 (2007).

[56] H. Zhang, C.-X. Liu, X.-L. Qi, X. Dai, Z. Fang, and S.-C. Zhang, "Topological insulators in Bi_2Se_3, Bi_2Te_3 and Sb_2Te_3 with a single Dirac cone on the surface", *Nat. Phys.* **5**, 438–442 (2009).

[57] Y. Chen, J. Analytis, J.-H. Chu, *et al.*, "Experimental realization of a three-dimensional topological insulator, Bi_2Te_3", *Science* **325**, 178–181 (2009).

[58] Y. Xia, D. Qian, D. Hsieh, *et al.*, "Observation of a large-gap topological-insulator class with a single Dirac cone on the surface", *Nat. Phys.* **5**, 398–402 (2009).

[59] J. E. Moore, "The birth of topological insulators", *Nature* **464**, 194–198 (2010).

[60] X.-L. Qi and S.-C. Zhang, "The quantum spin Hall effect and topological insulators", *Physics Today* **63**, 33–38 (2010).

[61] M. Z. Hasan and C. L. Kane, "Colloquium: topological insulators", *Rev. Mod. Phys.* **82**, 3045 (2010).

[62] X.-L. Qi and S.-C. Zhang, "Topological insulators and superconductors", *Rev. Mod. Phys.* **83**, 1057 (2011).

[63] M. M. Qazilbash, K. S. Burch, D. Whisler, *et al.* "Correlated metallic state of vanadium dioxide", *Phys. Rev. B* **74**, 205118 (2006).

[64] F. Walz, "The Verwey transition – a topical review", *J. Phys.: Condens. Matter* **14**, R285–R340 (2002).

[65] J. García and G. Subías, "The Verwey transition – a new perspective", *J. Phys.: Condens. Matter* **16**, R145–R178 (2004).

[66] P. W. Anderson, "Ordering and antiferromagnetism in ferrites", *Phys. Rev.* **102**, 1008–1013 (1956).

[67] J. Q. Wu, Q. Gu, B. S. Guiton, N. P. deLeon, O. Y. Lian, and H. Park, "Strain-induced self organization of metal-insulator domains in single-crystalline VO_2 nanobeams", *Nano Lett.* **6**, 2313–2317 (2006).

[68] H. Zeng, C. T. Black, R. L. Sandstrom, P. M. Rice, C. B. Murray, and S. Sun, "Magnetotransport of magnetite nanoparticle arrays", *Phys. Rev. B* **73**, 020402(R) (2006).

[69] P. R. Wallace, "The band theory of graphite", *Phys. Rev.* **71**, 622–634 (1947).

[70] J. C. Slonczewski and P. R. Weiss, "Band structure of graphite", *Phys. Rev.* **109**, 272–279 (1958).

[71] R. Saito, M. Fujita, G. Dresselhaus, and M. S. Dresselhaus, "Electronic structure of chiral graphene tubules", *Appl. Phys. Lett.* **60**, 2201–2203 (1992).

[72] M. I. Katsnelson, "Graphene: carbon in two dimensions", *Materials Today* **10**, 20–27 (2007).

[73] A. K. Geim and N. V. Novoselov, "The rise of graphene", *Nat. Mater.* **6**, 183–191 (2007).

[74] J. S. Bunch, Y. Yaish, M. Brink, K. Bolotin, and P. L. McEuen, "Coulomb oscillations and Hall effect in quasi-2d graphite quantum dots", *Nano Lett.* **5**, 287–290 (2005).

[75] Y. B. Zhang, J. P. Small, W. V. Pontius, and P. Kim, "Fabrication and electric-field dependent transport measurements of mesoscopic graphite devices", *Appl. Phys. Lett.* **86**, 073104 (2005).

[76] K. S. Novoselov, A. K. Geim, S. V. Morozov, *et al.*, "Electric field effect in atomically thin carbon films", *Science* **306**, 666–669 (2004).

[77] K. S. Novoselov, A. K. Geim, S. V. Morozov, *et al.*, "Two-dimensional gas of massless Dirac fermions in graphene", *Nature* **438**, 197–200 (2005).

[78] Y. B. Zhang, Y. W. Tan, H. L. Stormer, and P. Kim, "Experimental observation of the quantum Hall effect and Berry's phase in graphene", *Nature* **438**, 201–204 (2005).

[79] C. Berger, Z. M. Song, X. B. Li, *et al.*, "Electronic confinement and coherence in patterned epitaxial graphene", *Science* **312**, 1191–1196 (2006).

[80] C. Mattevi, H. Kim, and M. Chhowalla, "A review of chemical vapour deposition of graphene on copper", *J. Mat. Chem.* **21**, 3324–3334 (2011).

[81] S. Das Sarma, S. Adam, E. H. Hwang, and E. Rossi, "Electronic transport in two-dimensional graphene", *Rev. Mod. Phys.* **83**, 407–470 (2011).

[82] J. Wilson and A. Yoffe, "The transition metal dichalcogenides discussion and interpretation of the observed optical, electrical and structural properties", *Adv. Phys.* **18**, 193–335 (1969).

[83] R. F. Frindt, "Superconductivity in ultrathin NbSe$_2$ layers", *Phys. Rev. Lett.* **28**, 299–301 (1972).

[84] J. N. Coleman, M. Lotya, A. ONeill, *et al.*, "Two-dimensional nanosheets produced by liquid exfoliation of layered materials", *Science* **331**, 568–571 (2011).

[85] R. Friend and A. Yoffe, "Electronic properties of intercalation complexes of the transition metal dichalcogenides", *Adv. Phys.* **36**, 1–94 (1987).

[86] M. Monthioux and V. Kuznetsov, "Who should be given credit for the discovery of carbon nanotubes?", *Carbon* **44**, 1621–1623 (2006).

[87] S. Iijima, "Helical microtubules of graphitic carbon", *Nature* **354**, 56–58 (1991).

[88] S. Iijima and T. Ichihashi, "Single-shell carbon nanotubes of 1-nm diameter", *Nature* **363**, 603–605 (1993).

[89] D. S. Bethune, C. H. Klang, M. S. de Vries, *et al.*, "Cobalt-catalysed growth of carbon nanotubes with single-atomic-layer walls", *Nature* **363**, 605–607 (1993).

[90] M. S. Dresselhaus, G. Dresselhaus, P. Eklund, and R. Saito, "Carbon nanotubes", *Physics World* **11**, 33–38 (1998).

[91] M. S. Dresselhaus, G. Dresselhaus, J. C. Charlier, and E. Hernández, "Electronic, thermal and mechanical properties of carbon nanotubes", *Phil. Trans. R. Soc. Lond. A* **362**, 2065–2098 (2004).

[92] R. S. Lee, H. J. Kim, J. E. Fischer, A. Thess, and R. E. Smalley, "Conductivity enhancement in single-walled carbon nanotube bundles doped with K and Br", *Nature* **388**, 255–257 (1997).

[93] V. Derycke, R. Martel, J. Appenzeller, and P. Avouris, "Carbon nanotube inter- and intramolecular logic gates", *Nano Lett.* **1**, 453–456 (2001).

[94] V. Derycke, R. Martel, J. Appenzeller, and P. Avouris, "Controlled doping and carrier injection in carbon nanotube transistors", *Appl. Phys. Lett.* **80**, 2773–2775 (2002).

[95] J. Cao, Q. Wang, and H. J. Dai, "Electromechanical properties of metallic, quasimetallic, and semiconducting carbon nanotubes under stretching", *Phys. Rev. Lett.* **90**, 157601 (2003).

[96] H. Maki, T. Sato, and K. Ishibashi, "Direct observation of the deformation and the band gap change from an individual single-walled carbon nanotube under uniaxial strain", *Nano Lett.* **7**, 890–895 (2007).

[97] E. D. Minot, Y. Yaish, V. Sazonova, J. Y. Park, M. Brink, and P. L. McEuen, "Tuning carbon nanotube band gaps with strain", *Phys. Rev. Lett.* **90**, 156401 (2003).

[98] S. Zaric, G. N. Ostojic, J. Kono, *et al.*, "Optical signatures of the Aharonov-Bohm phase in single-walled carbon nanotubes", *Science* **304**, 1129–1131 (2004).

[99] G. Fedorov, A. Tselev, D. Jimenez, *et al.*, "Magnetically induced field effect in carbon nanotube devices", *Nano Lett.* **7**, 960–964 (2007).

[100] A. Bachtold, C. Strunk, J.-P. Salvetat, *et al.*, "Aharonov-Bohm oscillations in carbon nanotubes", *Nature* **397**, 673–676 (1999).

[101] J. Cumings, P. G. Collins, and A. Zettl, "Peeling and sharpening multiwall nanotubes", *Nature* **406**, 586 (2000).

[102] A. Kis, K. Jensen, S. Aloni, W. Mickelson, and A. Zettl, "Interlayer forces and ultralow sliding friction in multiwalled carbon nanotubes", *Phys. Rev. Lett.* **97**, 025501 (2006).

[103] L. D. Landau, "Über die Bewegung der Elektronen in Kristalgitter", *Phys. Z. Sowjetunion* **3**, 644–645 (1933).

[104] S. I. Pekar, *Untersuchangen Über die Electronentheorie der Kristalle* (Akademie-Verlag, 1954).

[105] H. Frölich, "Electrons in lattice fields", *Adv. Phys.* **3**, 325–361 (1954).

[106] T. Holstein, "Studies of polaron motion, Part I. The molecular-crystal model", *Ann. Phys.* **8**, 325–342 (1959).

[107] T. Holstein, "Studies of polaron motion, Part II. The 'small' polaron", *Ann. Phys.* **8**, 343–389 (1959).

[108] V. Coropceanu, J. Cornil, D. A. da Silva Filho, Y. Olivier, R. Silbey, and J.-L. Brédas, "Charge transport in organic semiconductors", *Chem. Rev.* **107**, 926–952 (2007).

[109] R. G. Kepler, "Charge carrier production and mobility in anthracene crystals", *Phys. Rev.* **119**, 1226–1229 (1960).

[110] J. O. H. LeBlanc, "Hole and electron drift mobilities in anthracenc", *J. Chem. Phys.* **33**, 626 (1960).

[111] W. Warta and N. Karl, "Hot holes in naphthalene: high, electric-field-dependent mobilities", *Phys. Rev. B* **32**, 1172–1182 (1985).

[112] M. E. Gershenson, V. Podzorov, and A. F. Morpurgo, "Colloquium: Electronic transport in single-crystal organic transistors", *Rev. Mod. Phys.* **78**, 973–989 (2006).

[113] H. E. Katz, "Recent advances in semiconductor performance and printing processes for organic transistor-based electronics", *Chem. Mater.* **16**, 4748–4756 (2004).

[114] M. C. J. M. Vissenberg and M. Matters, "Theory of the field-effect mobility in amorphous organic transistors", *Phys. Rev. B* **57**, 12964–12967 (1998).

[115] W. F. Pasveer, J. Cottaar, C. Tanase, *et al.*, "Unified description of charge-carrier mobilities in disordered semiconducting polymers", *Phys. Rev. Lett.* **94**, 206601 (2005).

[116] Z. G. Soos, "Theory of π-molecular charge-transfer crystals", *Ann. Rev. Phys. Chem.* **25**, 121–153 (1974).

[117] J. B. Torrance, "The difference between metallic and insulating salts of tetracyanoquinodimethane (TCNQ): how to design an organic metal", *Acc. Chem. Res.* **12**, 79–86 (1979).

[118] A. E. Underhill, "Molecular metals and superconductors", *J. Mater. Chem.* **2**, 1–11 (1992).

[119] M. R. Bryce, "Recent progress in conducting organic charge-transfer salts", *Chem. Soc. Rev.* **20**, 355–390 (1991).

[120] D. Jerome, A. Mazaud, M. Ribault, and K. Bechgaard, "Superconductivity in a synthetic organic conductor (TMTSF)$_2$PF$_6$", *J. de Phys. Lett.* **41**, L95–L98 (1980).

[121] S. T. Hannahs, J. S. Brooks, W. Kang, L. Y. Chiang, and P. M. Chaikin, "Quantum Hall effect in a bulk crystal", *Phys. Rev. Lett.* **63**, 1988–1991 (1989).

[122] E. Coronado, J. R. Galan-Mascaros, C. J. Gomez-Garcia, and V. Laukhin, "Coexistence of ferromagnetism and metallic conductivity in a molecule-based layered compound", *Nature* **408**, 447–449 (2000).

[123] A. J. Heeger, S. Kivelson, J. R. Schrieffer, and W. P. Su, "Solitons in conducting polymers", *Rev. Mod. Phys.* **60**, 781–850 (1988).

[124] R. E. Peierls, *Quantum Theory of Solids*, Oxford Classic Texts in the Physical Sciences (Oxford, Oxford University Press, 1955).

[125] A. J. Heeger, "Nobel lecture: Semiconducting and metallic polymers: the fourth generation of polymeric materials", *Rev. Mod. Phys.* **73**, 681–700 (2001).

[126] W. P. Su, J. R. Schrieffer, and A. J. Heeger, "Solitons in Polyacetylene", *Phys. Rev. Lett.* **42**, 1698–1701 (1979).

[127] W. P. Su, J. R. Schrieffer, and A. J. Heeger, "Soliton excitations in polyacetylene", *Phys. Rev. B* **22**, 2099–2111 (1980).

[128] R. C. Haddon, A. F. Hebard, M. J. Rosseinsky, *et al.*, "Conducting films of C$_{60}$ and C$_{70}$ by alkali-metal doping", *Nature* **350**, 320–322 (1991).

[129] A. F. Hebard, M. J. Rosseinsky, R. C. Haddon, *et al.*, "Superconductivity at 18 K in potassium-doped C$_{60}$", *Nature* **350**, 600–601 (1991).

[130] L. Forró and L. Mihály, "Electronic properties of doped fullerenes", *Rep. Prog. Phys.* **64**, 649–699 (2001).

[131] O. Gunnarsson, M. Calandra, and J. E. Han, "Colloquium: Saturation of electrical resistivity", *Rev. Mod. Phys.* **75**, 1085–1099 (2003).

[132] P. W. Anderson, "Infrared catastrophe in fermi gases with local scattering potentials", *Phys. Rev. Lett.* **18**, 1049–1051 (1967).

[133] G. R. Stewart, "Heavy-fermion systems", *Rev. Mod. Phys.* **56**, 755–787 (1984).

[134] M. Gibertini, A. Singha, V. Pellegrini, *et al.*, "Engineering artificial graphene in a two-dimensional electron gas", *Phys. Rev. B* **79**, 241406 (2009).

[135] D. K. G. DeBoer, A. J. G. Leenaers, and W. W. Vandenhoogenhof, "Glancing-incidence X-ray-analysis of thin-layered materials - a review", *X-ray spectrometry* **24**, 91–102 (1995).

[136] K. N. Stoev and K. Sakurai, "Review on grazing incidence X-ray spectrometry and reflectometry", *Spectrochimica Acta Part B: Atomic Spectroscopy* **54**, 41–82 (1999).

[137] M. Krumrey, M. Hoffmann, G. Ulm, K. Hasche, and P. Thomsen-Schmidt, "Thickness determination for SiO2 films on Si by X-ray reflectometry at the Si K edge", *Thin Solid Films* **459**, 241–244 (2004), Proceedings of the 8th European Vacuum Congress Berlin 2003, 23–26 June 2003, featuring the 8th European Vacuum Conference and 2nd Annual Conference of the German Vacuum Society.

[138] S. K. Andersen, J. A. Golovchenko, and G. Mair, "New applications of X-ray standing-wave fields to solid state physics", *Phys. Rev. Lett.* **37**, 1141–1145 (1976).

[139] D. P. Woodruff, "Normal incidence X-ray standing wave determination of adsorbate structures", *Prog. Surf. Sci.* **57**, 1–60 (1998).

[140] K. Siegbahn, "Electron spectroscopy for atoms, molecules, and condensed matter", *Rev. Mod. Phys.* **54**, 709–728 (1982).

[141] L. Yarris, www2.lbl.gov/Science-Articles/Archive/sabl/2006/Jul/04.html (2006).

[142] www.vcbio.science.ru.nl/en/fesem/info/fesemfaq/ (2004).

[143] J. Joo, H. Na, T. Yu, *et al.*, "Generalized and facile synthesis of semiconducting metal sulfide nanocrystals", *J. Am. Chem. Soc.* **125**, 11100–11105 (2003).

[144] G. Binnig, C. F. Quate, and C. Gerber, "Atomic force microscope", *Phys. Rev. Lett.* **56**, 930–933 (1986).

[145] Y. Martin, D. W. Abraham, and H. K. Wickramasinghe, "High-resolution capacitance measurement and potentiometry by force microscopy", *Appl. Phys. Lett.* **52**, 1103–1105 (1988).

[146] J. E. Stern, B. D. Terris, H. J. Mamin, and D. Rugar, "Deposition and imaging of localized charge on insulator surfaces using a force microscope", *Appl. Phys. Lett.* **53**, 2717–2719 (1988).

[147] Y. Martin and H. K. Wickramasinghe, "Magnetic imaging by "force microscopy" with 1000 [A-ring] resolution", *Appl. Phys. Lett.* **50**, 1455–1457 (1987).

[148] M. Fujihira, "Kelvin probe force microscopy of molecular surfaces", *Ann. Rev. Mat. Sci.* **29**, 353–380 (1999).

[149] C. C. Williams, "Two-dimensional dopant profiling by scanning capacitance microscopy", *Ann. Rev. Mat. Sci.* **29**, 471–504 (1999).

[150] C. D. Frisbie, L. F. Rozsnyai, A. Noy, M. S. Wrighton, and C. M. Lieber, "Functional-group imaging by chemical force microscopy", *Science* **265**, 2071–2074 (1994).

[151] A. Noy, D. V. Vezenov, and C. M. Lieber, "Chemical force microscopy", *Ann. Rev. Mat. Sci.* **27**, 381–421 (1997).

[152] M. Rief, F. Oesterhelt, B. Heymann, and H. E. Gaub, "Single molecule force spectroscopy on polysaccharides by atomic force microscopy", *Science* **275**, 1295–1297 (1997).

[153] F. Oesterhelt, D. Oesterhelt, M. Pfeiffer, A. Engel, H. E. Gaub, and D. J. Muller, "Unfolding pathways of individual bacteriorhodopsins", *Science* **288**, 143–146 (2000).

[154] J. H. Hafner, C. L. Cheung, A. T. Woolley, and C. M. Lieber, "Structural and functional imaging with carbon nanotube AFM probes", *Prog. Biophys. Mole. Bio.* **77**, 73–110 (2001).

[155] G. Binnig, H. Rohrer, C. Gerber, and E. Weibel, "Surface studies by scanning tunneling microscopy", *Phys. Rev. Lett.* **49**, 57–61 (1982).

[156] J. Tersoff and D. R. Hamann, "Theory of the scanning tunneling microscope", *Phys. Rev. B* **31**, 805–813 (1985).

[157] P. Sautet and C. Joachim, "Calculation of the benzene on rhodium STM images", *Chem. Phys. Lett.* **185**, 23–30 (1991).

[158] H. L. Edwards, J. T. Markert, and A. L. de Lozanne, "Energy gap and surface structure of $YBa_2Cu_3-xO_7$ probed by scanning tunneling microscopy", *Phys. Rev. Lett.* **69**, 2967–2970 (1992).

[159] K. McElroy, R. Simmonds, J. Hoffman, *et al.*, "Relating atomic-scale electronic phenomena to wave-like quasi-particle states in superconducting $Bi_2Sr_2CaCu_2O_{8+\delta}$", *Nature* **422**, 592–596 (2003).

[160] B. C. Stipe, M. A. Rezaei, and W. Ho, "Single-molecule vibrational spectroscopy and microscopy", *Science* **280**, 1732–1735 (1998).

[161] A. J. Heinrich, J. A. Gupta, C. P. Lutz, and D. M. Eigler, "Single-atom spin-flip spectroscopy", *Science* **306**, 466–469 (2004).

[162] R. Wiesendanger, "Spin mapping at the nanoscale and atomic scale", *Rev. Mod. Phys.* **81**, 1495–1550 (2009).

[163] J. A. Stroscio and D. M. Eigler, "Atomic and molecular manipulation with the scanning tunneling microscope", *Science* **254**, 1319–1326 (1991).

[164] W. Ho, "Single-molecule chemistry", *J. Chem. Phys.* **117**, 11033–11061 (2002).

[165] R. C. Dunn, "Near-field scanning optical microscopy", *Chem. Rev.* **99**, 2891–2928 (1999).

[166] J. W. Hsu, "Near-field scanning optical microscopy studies of electronic and photonic materials and devices", *Mat. Sci. Eng.: R: Reports* **33**, 1–50 (2001).

[167] E. A. Ash and G. Nicholls, "Super-resolution aperture scanning microscope", *Nature* **237**, 510–512 (1972).

[168] D. W. Pohl, W. Denk, and M. Lanz, "Optical stethoscopy: image recording with resolution $\lambda/20$", *Appl. Phys. Lett.* **44**, 651–653 (1984).

[169] U. Dürig, D. W. Pohl, and F. Rohner, "Near-field optical-scanning microscopy", *J. Appl. Phys.* **59**, 3318–3327 (1986).

[170] E. Betzig, J. K. Trautman, T. D. Harris, J. S. Weiner, and R. L. Kostelak, "Breaking the diffraction barrier: optical microscopy on a nanometric scale", *Science* **251**, 1468–1470 (1991).

[171] F. Zenhausern, Y. Martin, and H. K. Wickramasinghe, "Scanning interferometric apertureless microscopy: optical imaging at 10 Angstrom resolution", *Science* **269**, 1083–1085 (1995).

[172] D. L. Smith, *Thin Film Deposition: Principles & Practice* (New York, McGraw-Hill Professional, 1995).

[173] M. Ohring, *Materials Science of Thin Films* 2nd edn. (New York, Academic, 2001).

[174] K. Seshan, *Handbook of Thin Film Deposition Processes and Techniques* 2nd edn. (Norwich, NY, William Andrew, 2002).

[175] A. R. Barron, cnx.org/contents/0b2614e4-aa98-44cL-bf50-idb3ee3c5ecd@2.

[176] A. Ohtomo and H. Y. Hwang, "A high-mobility electron gas at the $LaAlO_3/SrTiO_3$ heterointerface", *Nature* **427**, 423–426 (2004).

[177] Z. Ren, Z. Huang, J. Xu, *et al.*, "Synthesis of large arrays of well-aligned carbon nanotubes on glass", *Science* **282**, 1105–1107 (1998).

[178] M. Meyyappan, L. Delzeit, A. Cassell, and D. Hash, "Carbon nanotube growth by PECVD: a review", *Plasma Sources Sci. & Tech.* **12**, 205–216 (2003).

[179] S. Fan, M. Chapline, N. Franklin, T. Tombler, A. Cassell, and H. Dai, "Self-oriented regular arrays of carbon nanotubes and their field emission properties", *Science* **283**, 512–514 (1999).

[180] A. Cassell, J. Raymakers, J. Kong, and H. Dai, "Large scale CVD synthesis of single-walled carbon nanotubes", *J. Phys. Chem. B* **103**, 6484–6492 (1999).

[181] M. Marchand, C. Journet, D. Guillot, J.-M. Benoit, B. I. Yakobson, and S. T. Purcell, "Growing a carbon nanotube atom by atom: "and yet it does turn"", *Nano Lett.* **9**, 2961–2966 (2009).

[182] X. Wang, Q. Li, J. Xie, *et al.*, "Fabrication of ultralong and electrically uniform single-walled carbon nanotubes on clean substrates", *Nano Letters* **9**, 3137–3141 (2009/08/03/).

[183] T. Yamada, T. Namai, K. Hata, *et al.*, "Size-selective growth of double-walled carbon nanotube forests from engineered iron catalysts", *Nature Nano.* **1**, 131–136 (2006).

[184] S. Han, X. Liu, and C. Zhou, "Template-free directional growth of single-walled carbon nanotubes on a- and r-plane sapphire", *J. Am. Chem. Soc.* **127**, 5294–5295 (2005).

[185] C. Kocabas, S. Hur, A. Gaur, M. Meitl, M. Shim, and J. Rogers, "Guided growth of large-scale, horizontally aligned arrays of single-walled carbon nanotubes and their use in thin-film transistors", *Small* **1**, 1110–1116 (2005).

[186] P. Nikolaev, M. J. Bronikowski, R. K. Bradley, *et al.*, "Gas-phase catalytic growth of single-walled carbon nanotubes from carbon monoxide", *Chem. Phys. Lett.* **313**, 91–97 (1999).

[187] K. S. Kim, Y. Zhao, H. Jang, *et al.*, "Large-scale pattern growth of graphene films for stretchable transparent electrodes", *Nature* **457**, 706–710 (2009).

[188] X. Li, W. Cai, J. An, *et al.*, "Large-area synthesis of high-quality and uniform graphene films on copper foils", *Science* **324**, 1312–1314 (2009).

[189] C. Berger, Z. M. Song, T. B. Li, *et al.*, "Ultrathin epitaxial graphite: 2d electron gas properties and a route toward graphene-based nanoelectronics", *J. Phys. Chem. B* **108**, 19912–19916 (2004).

[190] R. S. Wagner and W. C. Ellis, "Vapor-liquid-solid mechanism of single crystal growth", *Appl. Phys. Lett.* **4**, 89–90 (1964).

[191] A. Morales and C. Lieber, "A laser ablation method for the synthesis of crystalline semiconductor nanowires", *Science* **279**, 208–211 (1998).

[192] X. Duan and C. Lieber, "General synthesis of compound semiconductor nanowires", *Adv. Mater.* **12**, 298–302 (2000).

[193] A. Persson, M. Larsson, S. Stenstrom, B. Ohlsson, L. Samuelson, and L. Wallenberg, "Solid-phase diffusion mechanism for GaAs nanowire growth", *Nat. Mater.* **3**, 677–681 (2004).

[194] K. Dick, K. Deppert, T. Martensson, B. Mandl, L. Samuelson, and W. Seifert, "Failure of the vapor-liquid-solid mechanism in Au-assisted MOVPE growth of InAs nanowires", *Nano Lett.* **5**, 761–764 (2005).

[195] Z. H. Wu, X. Y. Mei, D. Kim, M. Blumin, and H. E. Ruda, "Growth of Au-catalyzed ordered GaAs nanowire arrays by molecular-beam epitaxy", *Appl. Phys. Lett.* **81**, 5177–5179 (2002).

[196] T. Mårtensson, M. Borgström, W. Seifert, B. J. Ohlsson, and L. Samuelson, "Fabrication of individually seeded nanowire arrays by vapour-liquid-solid growth", *Nanotechnology* **14**, 1255 (2003).

[197] M. S. Gudiksen, L. J. Lauhon, J. Wang, D. C. Smith, and C. M. Lieber, "Growth of nanowire superlattice structures for nanoscale photonics and electronics", *Nature* **415**, 617–620 (2002).

[198] L. Lauhon, M. Gudiksen, C. Wang, and C. Lieber, "Epitaxial core-shell and core-multishell nanowire heterostructures", *Nature* **420**, 57–61 (2002).

[199] C. Thelander, T. Mårtensson, M. T. Björk, *et al.*, "Single-electron transistors in heterostructure nanowires", *Appl. Phys. Lett.* **83**, 2052–2054 (2003).

[200] W. Lu, J. Xiang, B. Timko, Y. Wu, and C. Lieber, "One-dimensional hole gas in germanium/silicon nanowire heterostructures", *Proc. Nat. Acad. Sci. US* **102**, 10046–10051 (2005).

[201] C. B. Murray, D. J. Norris, and M. G. Bawendi, "Synthesis and characterization of nearly monodisperse CdE (E = S, Se, Te) semiconductor nanocrystallites", *J. Am. Chem. Soc.* **115**, 8706–8715 (1993).

[202] A. P. Alivisatos, "Semiconductor clusters, nanocrystals, and quantum dots", *Science* **271**, 933–937 (1996).

[203] C. B. Murray, C. R. Kagan, and M. G. Bawendi, "Synthesis and characterization of monodisperse nanocrystals and close-packed nanocrystal assemblies", *Ann. Rev. Mater. Sci.* **30**, 545–610 (2000).

[204] G. Cao, *Nanostructures & Nanomaterials: Synthesis, Properties & Applications* (London, Imperial College, 2004).

[205] Y. Yin and A. P. Alivisatos, "Colloidal nanocrystal synthesis and the organic-inorganic interface", *Nature* **437**, 664–670 (2005).

[206] EPFL, cmi.epfl.ch/etch/PladeKOH.php (2001).

[207] L. N. Pfeiffer, K. W. West, R. L. Willett, H. Akiyama, and L. P. Rokhinson, "Nanostructures in GaAs fabricated by molecular beam epitaxy", *Bell Labs Tech. Journ.* **10**, 151–159 (2005).

[208] www.oxfordplasma.de/images/sem_ti_3.jpg (2004).

[209] L. R. Harriott, "Scattering with angular limitation projection electron beam lithography for suboptical lithography", in *Papers from the 41st International Conference on Electron, Ion, and Photon Beam Technology and Nanofabrication*, volume 15, 2130–2135 (AVS, 1997).

[210] S. Y. Chou, P. R. Krauss, and P. J. Renstrom, "Imprint of sub-25 nm vias and trenches in polymers", *Appl. Phys. Lett.* **67**, 3114–3116 (1995).

[211] S. Y. Chou, P. R. Krauss, and P. J. Renstrom, "Imprint lithography with 25-nanometer resolution", *Science* **272**, 85–87 (1996).

[212] M. D. Austin, H. Ge, W. Wu, *et al.*, "Fabrication of 5 nm linewidth and 14 nm pitch features by nanoimprint lithography", *Appl. Phys. Lett.* **84**, 5299–5301 (2004).

[213] R. W. Jaszewski, H. Schift, B. Schnyder, A. Schneuwly, and P. Groning, "The deposition of anti-adhesive ultra-thin teflon-like films and their interaction with polymers during hot embossing", *Appl. Surf. Sci.* **143**, 301–308 (1999).

[214] W. Zhang and S. Y. Chou, "Multilevel nanoimprint lithography with submicron alignment over 4 in. Si wafers", *Appl. Phys. Lett.* **79**, 845–847 (2001).

[215] Y. N. Xia and G. M. Whitesides, "Soft lithography", *Ann. Rev. Mat. Sci.* **28**, 153–184 (1998).

[216] S. R. Quake and A. Scherer, "From micro- to nanofabrication with soft materials", *Science* **290**, 1536–1540 (2000).

[217] J. A. Rogers, R. J. Jackman, G. M. Whitesides, J. L. Wagener, and A. M. Vengsarkar, "Using microcontact printing to generate amplitude photomasks on the surfaces of optical fibers: a method for producing in-fiber gratings", *Appl. Phys. Lett.* **70**, 7–9 (1997).

[218] Y. Xia, J. A. Rogers, K. E. Paul, and G. M. Whitesides, "Unconventional methods for fabricating and patterning nanostructures", *Chem. Rev.* **99**, 1823–1848 (1999).

[219] Y.-L. Loo, R. L. Willett, K. W. Baldwin, and J. A. Rogers, "Additive, nanoscale patterning of metal films with a stamp and a surface chemistry mediated transfer process: applications in plastic electronics", *Appl. Phys. Lett.* **81**, 562–564 (2002).

[220] S.-H. Hur, D.-Y. Khang, C. Kocabas, and J. A. Rogers, "Nanotransfer printing by use of noncovalent surface forces: applications to thin-film transistors that use single-walled carbon nanotube networks and semiconducting polymers", *Appl. Phys. Lett.* **85**, 5730–5732 (2004).

[221] M. A. Meitl, Z.-T. Zhu, V. Kumar, *et al.* "Transfer printing by kinetic control of adhesion to an elastomeric stamp", *Nat. Mater.* **5**, 33–38 (2006).

[222] X.-M. Zhao, Y. Xia, and G. M. Whitesides, "Soft lithographic methods for nano-fabrication", *J. Mater. Chem.* **7**, 1069–1074 (1997).

[223] F. Hua, A. Gaur, Y. Sun, *et al.*, "Processing dependent behavior of soft imprint lithography on the 1–10-nm scale", *IEEE Trans. Nanotechnology*, **5**, 301–308 (2006).

[224] P. Vettiger, G. Cross, M. Despont, *et al.*, "The "millipede" – nanotechnology entering data storage", *IEEE Trans. Nanotech.*, **1**, 39–55 (2002).

[225] R. D. Piner, J. Zhu, F. Xu, S. Hong, and C. A. Mirkin, "Dip-pen nanolithography", *Science* **283**, 661–663 (1999).

[226] L. L. Sohn and R. L. Willett, "Fabrication of nanostructures using atomic-force-microscope-based lithography", *Appl. Phys. Lett.* **67**, 1552–1554 (1995).

[227] V. Bouchiat and D. Esteve, "Lift-off lithography using an atomic force microscope", *Appl. Phys. Lett.* **69**, 3098–3100 (1996).

[228] E. S. Snow and P. M. Campbell, "AFM fabrication of sub-10-nanometer metal-oxide devices with in situ control of electrical properties", *Science* **270**, 1639–1641 (1995).

[229] R. Held, T. Vancura, T. Heinzel, K. Ensslin, M. Holland, and W. Wegscheider, "In-plane gates and nanostructures fabricated by direct oxidation of semiconductor

heterostructures with an atomic force microscope", *Appl. Phys. Lett.* **73**, 262–264 (1998).

[230] H. Dai, N. Franklin, and J. Han, "Exploiting the properties of carbon nanotubes for nanolithography", *Appl. Phys. Lett.* **73**, 1508–1510 (1998).

[231] K. Salaita, Y. Wang, and C. A. Mirkin, "Applications of dip-pen nanolithography", *Nat. Nano.* **2**, 145–155 (2007).

[232] S. Gustavsson, R. Leturcq, T. Ihn, K. Ensslin, and A. C. Gossard, "Electrons in quantum dots: one by one", *J. Appl. Phys.* **105**, 122401 (2009).

[233] A. Majumdar, P. I. Oden, J. P. Carrejo, L. A. Nagahara, J. J. Graham, and J. Alexander, "Nanometer-scale lithography using the atomic force microscope", *Appl. Phys. Lett.* **61**, 2293–2295 (1992).

[234] I. I. Smolyaninov, D. L. Mazzoni, and C. C. Davis, "Near-field direct-write ultraviolet lithography and shear force microscopic studies of the lithographic process", *Appl. Phys. Lett.* **67**, 3859–3861 (1995).

[235] S. Sun and G. J. Leggett, "Matching the resolution of electron beam lithography by scanning near-field photolithography", *Nano Lett.* **2004**, 1381–1384 (2004).

[236] J. W. Lyding, T.-C. Shen, J. S. Hubacek, J. R. Tucker, and G. C. Abeln, "Nanoscale patterning and oxidation of H-passivated Si(100)-2 x 1 surfaces with an ultrahigh vacuum scanning tunneling microscope", *Appl. Phys. Lett.* **64**, 2010–2012 (1994).

[237] T. C. Shen, C. Wang, G. C. Abeln, *et al.*, "Atomic-scale desorption through electronic and vibrational excitation mechanisms", *Science* **268**, 1590–1592 (1995).

[238] G.-Y. Liu, S. Xu, and Y. Qian, "Nanofabrication of self-assembled monolayers using scanning probe lithography", *Acc. Chem. Res.* **33**, 457–466 (2000).

[239] R. K. Smith, P. A. Lewis, and P. S. Weiss, "Patterning self-assembled monolayers", *Prog. Surf. Sci.* **75**, 1–68 (2004).

[240] J. Chen, M. A. Reed, C. L. Asplund, *et al.*, "Placement of conjugated oligomers in an alkanethiol matrix by scanned probe microscope lithography", *Appl. Phys. Lett.* **75**, 624–626 (1999).

[241] C. L. Haynes and R. P. van Duyne, "Nanosphere lithography: a versatile nanofabrication tool for studies of size-dependent nanoparticle optics", *J. Phys. Chem. B* **105**, 5599–5611 (2001).

[242] J. C. Hulteen and R. P. van Duyne, "Nanosphere lithography: a materials general fabrication process for periodic particle array surfaces", *J. Vac. Sci. Tech. A* **13**, 1553–1558 (1995).

[243] A. Bezryadin, C. N. Lau, and M. Tinkham, "Quantum suppression of superconductivity in ultrathin nanowires", *Nature* **404**, 971–974 (2000).

[244] Y. Zhang, N. W. Franklin, R. J. Chen, and H. Dai, "Metal coating on suspended carbon nanotubes and its implication to metal-tube interaction", *Chem. Phy. Lett.* **331**, 35–41 (2000).

[245] D. S. Hopkins, D. Pekker, P. M. Goldbart, and A. Bezryadin, "Quantum interference device made by DNA templating of superconducting nanowires", *Science* **308**, 1762–1765 (2005).

[246] W. S. Yun, J. Kim, K.-H. Park, *et al.*, "Fabrication of metal nanowire using carbon nanotube as a mask", in *The 46th International Symposium of the American Vacuum Society*, volume 18, 1329–1332 (AVS, 2000).

[247] R. Šordan, M. Burghard, and K. Kern, "Removable template route to metallic nanowires and nanogaps", *Appl. Phys. Lett.* **79**, 2073–2075 (2001).

[248] D. Natelson, R. L. Willett, K. W. West, and L. N. Pfeiffer, "Fabrication of extremely narrow metal wires", *Appl. Phys. Lett.* **77**, 1991–1993 (2000).

[249] F. Altomare, A. M. Chang, M. R. Melloch, Y. Hong, and C. W. Tu, "Ultranarrow AuPd and Al wires", *Appl. Phys. Lett.* **86**, 172501 (2005).

[250] N. A. Melosh, A. Boukai, F. Diana, *et al.*, "Ultrahigh-density nanowire lattices and circuits", *Science* **300**, 112–115 (2003).

[251] G. E. Possin, "A method for forming very small diameter wires", *Rev. Sci. Instr.* **41**, 772–774 (1970).

[252] W. D. Williams and N. Giordano, "Fabrication of 80 Å metal wires", *Rev. Sci. Instr.* **55**, 410–412 (1984).

[253] G. E. Thompson, R. C. Furneaux, G. C. Wood, J. A. Richardson, and J. S. Goode, "Nucleation and growth of porous anodic films on aluminium", *Nature* **272**, 433–435 (1978).

[254] B. T. Holland, C. F. Blanford, and A. Stein, "Synthesis of macroporous minerals with highly ordered three-dimensional arrays of spheroidal voids", *Science* **281**, 538–540 (1998).

[255] G. Whitesides, J. Mathias, and C. Seto, "Molecular self-assembly and nanochemistry: a chemical strategy for the synthesis of nanostructures", *Science* **254**, 1312–1319 (1991).

[256] J. S. Lindsay, "Self-assembly in synthetic routes to molecular devices – biological principles and chemical perspectives – a review", *New J. Chem.* **15**, 153–180 (1991).

[257] G. M. Whitesides and B. Grzybowski, "Self-assembly at all scales", *Science* **295**, 2418–2421 (2002).

[258] F. Schreiber, "Structure and growth of self-assembling monolayers", *Progr. Surf. Sci.* **65**, 151–257 (2000).

[259] K. Ariga, J. P. Hill, M. V. Lee, A. Vinu, R. Charvet, and S. Acharya, "Challenges and breakthroughs in recent research on self-assembly", *Sci. Tech. Adv. Mater.* **9**, 014109 (2008).

[260] M. Park, C. Harrison, P. M. Chaikin, R. A. Register, and D. H. Adamson, "Block copolymer lithography: periodic arrays of 10^{11} Holes in 1 square centimeter", *Science* **276**, 1401–1404 (1997).

[261] J. K. W. Yang, Y. S. Jung, J.-B. Chang, *et al.*, "Complex self-assembled patterns using spares commensurate templates with locally varying motifs", *Nat. Nano.* **5**, 256–260 (2010).

[262] F. S. Bates and G. H. Fredrickson, "Block copolymer thermodynamics: theory and experiment", *Ann. Rev. Phys. Chem.* **41**, 525–557 (1990).

[263] F. S. Bates, "Polymer-polymer phase behavior", *Science* **251**, 898–905 (1991).

[264] S. A. Stauth and B. A. Parviz, "Self-assembled single-crystal silicon circuits on plastic", *Proc. Nat. Acad. Sci.* **103**, 13922–13927 (2006).

[265] A. Guiner, *X-Ray Diffraction: in Crystals, Imperfect Crystals, and Amorphous Bodies* (New York, Dover, 1994).

[266] J. Als-Nielsen and D. McMorrow, *Elements of Modern X-Ray Physics* (New York, Wiley, 2001).

[267] J. Goldstein, D. E. Newbury, D. C. Joy, *et al.*, *Scanning Electron Microscopy and X-ray Microanalysis*, 3rd edn. (New York, Springer, 2003).

[268] D. B. Williams and C. B. Carter, *Transmission Electron Microscopy: a Textbook for Materials Science*, 2nd edn. (New York, Springer, 2009).

[269] C. J. Chen, *Introduction to Scanning Tunneling Microscopy*, Monographs on the Physics and Chemistry of Materials, 2nd edn. (New York, Oxford, 2007).

[270] R. Wiesendanger, *Scanning Probe Microscopy and Spectroscopy: Methods and Applications* (Cambridge, Cambridge University Press, 1994).

[271] F. J. Giessibl, "Advances in atomic force microscopy", *Rev. Mod. Phys.* **75**, 949–983 (2003).

[272] W. A. Hofer, A. S. Foster, and A. L. Shluger, "Theories of scanning probe microscopes at the atomic scale", *Rev. Mod. Phys.* **75**, 1287–1331 (2003).

[273] J. E. Mahan, *Physical Vapor Deposition of Thin Films* (New York, Wiley-Interscience, 2000).

[274] G. A. Ozin and A. C. Arsenault, *Nanochemistry: a Chemical Approach to Nanomaterials* (London, RSC Publishing, 2005).

[275] M. Law, J. Goldberger, and P. D. Yang, "Semiconductor nanowires and nanotubes", *Ann. Rev. Mater. Res.* **34**, 83–122 (2004).

[276] K. Suzuki and B. W. Smith, eds., *Microlithography: Science and Technology*, Optical Science and Engineering, 2nd edn. (Boca Raton, CRC Press, 2007).

[277] C. Mack, *Fundamental Principles of Optical Lithography: The Science of Microfabrication* (New York, Wiley, 2008).

[278] I. Brodie and J. J. Muray, *The Physics of Micro/Nano-Fabrication* (New York, Springer, 2010).

[279] Z. Cui, *Nanofabrication: Principles, Capabilities, and Limits* (New York, Springer, 2008).

[280] A. Busnaina, ed., *Nanomanufacturing Handbook* (Boca Raton, CRC, 2007).

[281] Y. S. Lee, *Self-Assembly and Nanotechnology: a Force Balance Approach* (New York, Wiley-Interscience, 2008).

[282] J. A. Pelesko, *Self Assembly: the Science of Things that Put Themselves Together* (Boca Raton, Chapman and Hall/CRC, 2007).

[283] L. Gross, F. Mohn, N. Moll, P. Liljeroth, and G. Meyer, "The chemical structure of a molecule resolved by atomic force microscopy", *Science* **325**, 1110–1114 (2009).

[284] R. Temirov, S. Soubatch, O. Neucheva, A. Lassise, and F. Tautz, "A novel method achieving ultra-high geometrical resolution in scanning tunnelling microscopy", *New J. Phys.* **10**, 053012 (2008).

[285] C.-l. Chiang, C. Xu, Z. Han, and W. Ho, "Real-space imaging of molecular structure and chemical bonding by single-molecule inelastic tunneling probe", *Science* **344**, 885–888 (2014).

[286] N. S. Rasor and C. Warner, "Correlation of emission processes for adsorbed alkali films on metal surfaces", *J. Appl. Phys.* **35**, 2589–2600 (1964).

[287] I. H. Campbell, S. Rubin, T. A. Zawodzinski, *et al.*, "Controlling Schottky energy barriers in organic electronic devices using self-assembled monolayers", *Phys. Rev. B* **54**, R14321–R14324 (1996).

[288] M. Nonnenmacher, M. P. O'Boyle, and H. K. Wickramasinghe, "Kelvin probe force microscopy", *Appl. Phys. Lett.* **58**, 2921–2923 (1991).

[289] I. Tamm, "A possible kind of electron binding on crystal surfaces", *Phys. Z. Sowjetunion* **1**, 733–746 (1932).

[290] I. Tamm, "A possible kind of electron binding on crystal surfaces", *Z. Phys.* **76**, 849–850 (1932).

[291] K. S. Ralls, W. J. Skocpol, L. D. Jackel, *et al.*, "Discrete resistance switching in submicrometer silicon inversion layers: individual interface traps and low-frequency ($1/f$?) noise", *Phys. Rev. Lett.* **52**, 228–231 (1984).

[292] K. Kanisawa, M. J. Butcher, H. Yamaguchi, and Y. Hirayama, "Imaging of Friedel oscillation patterns of two-dimensionally accumulated electrons at epitaxially grown InAs(111) A surfaces", *Phys. Rev. Lett.* **86**, 3384–3387 (2001).

[293] B. Tanatar and D. M. Ceperley, "Ground state of the two-dimensional electron gas", *Phys. Rev. B* **39**, 5005–5016 (1989).

[294] D. M. Ceperley and B. J. Alder, "Ground state of the electron gas by a stochastic method", *Phys. Rev. Lett.* **45**, 566–569 (1980).

[295] G. Ortiz, M. Harris, and P. Ballone, "Zero temperature phases of the electron Gas", *Phys. Rev. Lett.* **82**, 5317–5320 (1999).

[296] R. L. Anderson, "Germanium-gallium arscnide heterojunction", *IBM J. Res. Dev.* **4**, 283–287 (1960).

[297] J. H. Davies, *The Physics of Low-Dimensional Semiconductors: an Introduction* (Cambridge, Cambridge University Press, 1997).

[298] S. Gwo, K.-J. Chao, C. K. Shih, K. Sadra, and B. G. Streetman, "Direct mapping of electronic structure across $Al_{0.3}Ga_{0.7}As$/GaAs heterojunctions: Band offsets, asymmetrical transition widths, and multiple-valley band structures", *Phys. Rev. Lett.* **71**, 1883–1886 (1993).

[299] J. Faist, F. Capasso, D. L. Sivco, C. Sirtori, A. L. Hutchinson, and A. Y. Cho, "Quantum cascade laser", *Science* **264**, 553–556 (1994).

[300] F. Capasso, C. Gmachl, D. L. Sivco, and A. Y. Cho, "Quantum cascade lasers", *Phys. Today* **55**, 34–40 (2002).

[301] A. I. Yanson, G. R. Bollinger, H. E. van den Brom, N. Agrait, and J. M. van Ruitenbeek, "Formation and manipulation of a metallic wire of single gold atoms", *Nature* **395**, 783–785 (1998).

[302] N. Agrait, A. L. Yeyati, and J. M. van Ruitenbeek, "Quantum properties of atomic-sized conductors", *Phys. Rep.* **377**, 81–279 (2003).

[303] F. J. Himpsel, A. Kirakosian, J. N. Crain, J.-L. Lin, and D. Petrovykh, "Self-assembly of one-dimensional nanostructures at silicon surfaces", *Sol. State Comm.* **117**, 149–157 (2000).

[304] Y. Chen, D. A. A. Ohlberg, and R. S. Williams, "Nanowires of four epitaxial hexagonal silicides grown on Si(001)", *J. Appl. Phys.* **91**, 3213–3218 (2002).

[305] P. J. Pearah, A. C. Chen, K. C. Hsieh, and K. Y. Cheng, "AlGaInP multiple-quantum wire heterostructure lasers prepared by the strain-induced lateral-layer ordering process", *IEEE J. Quant. Electron.* **30**, 608–618 (1994).

[306] S. Tsukamoto, Y. Nagamune, M. Nishioka, and Y. Arakawa, "Fabrication of GaAs quantum wires on epitaxially grown v-grooves by metal-organic chemical-vapor deposition", *J. Appl. Phys.* **71**, 533–535 (1992).

[307] A. R. Goni, L. N. Pfeiffer, K. W. West, A. Pinczuk, H. U. Baranger, and H. L. Stormer, "Observation of quantum wire formation at intersecting quantum-wells", *Appl. Phys. Lett.* **61**, 1956–1958 (1992).

[308] T. Heinzel, R. Held, S. Luscher, K. Ensslin, W. Wegscheider, and M. Bichler, "Electronic properties of nanostructures defined in Ga[Al]As heterostructures by local oxidation", *Physica E* **9**, 84–93 (2001).

[309] L. Samuelson, C. Thelander, M. T. Bjork, *et al.*, "Semiconductor nanowires for 0 d and 1d physics and applications", *Physica E* **25**, 313–318 (2004).

[310] W. Lu and C. M. Lieber, "Semiconductor nanowires", *J. Phys. D: Appl. Phys.* **39**, R387–R406 (2006).

[311] Y. Y. Wu, R. Fan, and P. D. Yang, "Block-by-block growth of single-crystalline Si/SiGe superlattice nanowires", *Nano Lett.* **2**, 83–86 (2002).

[312] G. Sek, K. Ryczko, M. Motyka, *et al.*, "Wetting layer states of InAs/GaAs self-assembled quantum dot structures: effect of intermixing and capping layer", *J. Appl. Phys.* **101**, 063539 (2007).

[313] L. P. Kouwenhoven, D. G. Austing, and S. Tarucha, "Few-electron quantum dots", *Rep. Prog. Phys.* **64**, 701–736 (2001).

[314] X. Peng, L. Manna, W. Yang, *et al.*, "Shape control of CdSe nanocrystals", *Nature* **404**, 56–61 (2000).

[315] D. Goldhaber-Gordon, H. Shtrikman, D. Mahalu, D. Abusch-Magder, U. Meirav, and M. A. Kastner, "Kondo effect in a single-electron transistor", *Nature* **391**, 156–159 (1998).

[316] A. M. Smith and S. Nie, "Semiconductor nanocrystals: structure, properties, and band gap engineering", *Acc. Chem. Res.* **43**, 190–200 (2009).

[317] S. Gaponenko, *Optical Properties of Semiconductor Nanocrystals*, Cambridge Studies in Modern Optics (Cambridge, Cambridge University Press, 1998).

[318] V. Klimov, *Nanocrystal Quantum Dots* (Boca Raton, CRC Press, 2010).

[319] G. Konstantatos, I. Howard, A. Fischer, *et al.*, "Ultrasensitive solution-cast quantum dot photodetectors", *Nature* **442**, 180–183 (2006).

[320] D. V. Talapin and C. B. Murray, "PbSe nanocrystal solids for n- and p-channel thin film field-effect transistors", *Science* **310**, 86–89 (2005).

[321] D. L. Klein, R. Roth, A. K. L. Lim, A. P. Alivisatos, and P. L. McEuen, "A single-electron transistor made from a cadmium selenide nanocrystal", *Nature* **389**, 699–701 (1997).

[322] A. Dorn, H. Huang, and M. G. Bawendi, "Electroluminescence from nanocrystals in an electromigrated gap composed of two different metals", *Nano Lett.* **8**, 1347–1351 (2008).

[323] L. Onsager, "Reciprocal relations in irreversible processes II", *Phys. Rev.* **38**, 2265–2279 (1931).

[324] A. D. Benoit, S. Washburn, C. P. Umbach, R. B. Laibowitz, and R. A. Webb, "Asymmetry in the magnetoconductance of metal wires and loops", *Phys. Rev. Lett.* **57**, 1765–1768 (1986).

[325] P. Drude, "Zue Elektronentheorie der Metalle", *Ann. Physik* **1**, 566–613 (1900).

[326] K. von Klitzing, G. Dorda, and M. Pepper, "New method for high-accuracy determination of the fine-structure constant based on quantized hall resistance", *Phys. Rev. Lett.* **45**, 494–497 (1980).

[327] S. M. Girvin, "The quantum Hall effect: novel excitations and broken symmetries", in *Aspects Topologiques de la Physique en Basse Dimension. Topological Aspects of Low Dimensional Systems*, 53–175 (Lesllies, France, Springer, 1999).

[328] C. Weisbuch and B. Vinter, *Quantum Semiconductor Structures: Fundamentals and Applications* (New York, Academic Press, 1991).

[329] D. C. Tsui, H. L. Stormer, and A. C. Gossard, "Two-dimensional magnetotransport in the extreme quantum limit", *Phys. Rev. Lett.* **48**, 1559–1562 (1982).

[330] R. B. Laughlin, "Anomalous quantum Hall effect: an incompressible quantum fluid with fractionally charged excitations", *Phys. Rev. Lett.* **50**, 1395–1398 (1983).

[331] H. L. Stormer, "Nobel lecture: The fractional quantum Hall effect", *Rev. Mod. Phys.* **71**, 875–889 (1999).

[332] Y. Imry, *Introduction to Mesoscopic Physics* (Oxford, Oxford University Press, 1997).

[333] S. Lee, A. Trionfi, and D. Natelson, unpublished (2005).

[334] A. G. Aronov and Y. V. Sharvin, "Magnetic flux effects in disordered conductors", *Rev. Mod. Phys.* **59**, 755–779 (1987).

[335] F. Pierre and N. O. Birge, "Dephasing by extremely dilute magnetic impurities revealed by Aharonov–Bohm oscillations", *Phys. Rev. Lett.* **89**, 206804 (2002).

[336] C. P. Umbach, C. van Haesendonck, R. B. Laibowitz, S. Washburn, and R. A. Webb, "Direct observation of ensemble averaging of the Aharonov–Bohm effect in normal-metal loops", *Phys. Rev. Lett.* **56**, 386–389 (1986).

[337] B. L. Althsuler, A. G. Aronov, and B. Z. Spivak, "The Aharonov–Bohm effect in disordered conductors", *JETP* **33**, 94 (1981).

[338] D. J. Bishop, R. C. Dynes, and D. C. Tsui, "Magnetoresistance in Si metal-oxide-semiconductor field-effect transitors: evidence of weak localization and correlation", *Phys. Rev. B* **26**, 773–779 (1982).

[339] G. Bergmann, "Weak localization in thin films: a time-of-flight experiment with conduction electrons", *Phys. Rep.* **107**, 1–58 (1984).

[340] P. E. Lindelof, J. Nørrgaard, and J. B. Hansen, "Magnetoresistance in two-dimensional magnesium films", *Z. Phys. B* **59**, 423–428 (1985).

[341] P. W. Anderson, B. I. Halperin, and C. M. Varma, "Anomalous low-temperature thermal properties of glasses and spin glasses", *Phil. Mag.* **25**, 1–9 (1972).

[342] R. C. Zeller and R. O. Pohl, "Thermal conductivity and specific heat of noncrystalline solids", *Phys. Rev. B* **4**, 2029–2041 (1971).

[343] N. O. Birge, B. Golding, and W. H. Haemmerle, "Electron quantum interference and $1/f$ noise in bismuth", *Phys. Rev. Lett.* **62**, 195–198 (1989).

[344] P. Dutta and P. M. Horn, "Low-frequency fluctuations in solids: $1/f$ noise", *Rev. Mod. Phys.* **53**, 497–516 (1981).

[345] M. B. Weissman, "$1/f$ noise and other slow, nonexponential kinetics in condensed matter", *Rev. Mod. Phys.* **60**, 537–571 (1988).

[346] N. O. Birge, B. Golding, and W. H. Haemmerle, "Conductance fluctuations and 1/f noise in Bi", *Phys. Rev. B* **42**, 2735–2743 (1990).

[347] B. L. Altshuler and D. E. Khmelnitskii, "Fluctuation properties of small conductors", *JETP Lett.* **42**, 359–363 (1985).

[348] R. A. Webb, S. Washburn, and C. P. Umbach, "Experimental study of nonlinear conductance in small metallic samples", *Phys. Rev. B* **37**, 8455–8458 (1988).

[349] B. Ludolph and J. M. van Ruitenbeek, "Conductance fluctuations as a tool for investigating the quantum modes in atomic-size metallic contacts", *Phys. Rev. B* **61**, 2273–2285 (2000).

[350] L. P. Lévy, G. Dolan, J. Dunsmuir, and H. Bouchiat, "Magnetization of mesoscopic copper rings: Evidence for persistent currents", *Phys. Rev. Lett.* **64**, 2074–2077 (1990).

[351] V. Chandrasekhar, R. A. Webb, M. J. Brady, M. B. Ketchen, W. J. Gallagher, and A. Kleinsasser, "Magnetic response of a single, isolated gold loop", *Phys. Rev. Lett.* **67**, 3578–3581 (1991).

[352] D. K. Ferry and S. M. Goodnick, *Transport in Nanostructures* (Cambridge, Cambridge University Press, 1997).

[353] J. G. Simmons, "Generalized formula for the electric tunnel effect between similar electrodes separated by a thin insulating film", *J. Appl. Phys.* **34**, 1793–1803 (1963).

[354] E. H. Hauge and J. A. Støvneng, "Tunneling times: a critical review", *Rev. Mod. Phys.* **61**, 917–936 (1989).

[355] R. Landauer and T. Martin, "Barrier interaction time in tunneling", *Rev. Mod. Phys.* **66**, 217–228 (1994).

[356] P. C. W. Davies, "Quantum tunneling time", *Amer. J. Phys.* **73**, 23–27 (2005).

[357] U. Peskin, M. Galperin, and A. Nitzan, "Traversal times for resonant tunneling", *J. Phys. Chem. B* **106**, 8306–8312 (2002).

[358] R. Tsu and L. Esaki, "Tunneling in a finite superlattice", *Appl. Phys. Lett.* **22**, 562–564 (1973).

[359] L. L. Chang, L. Esaki, and R. Tsu, "Resonant tunneling in semiconductor double barriers", *Appl. Phys. Lett.* **24**, 593–595 (1974).

[360] S. Datta, *Electronic Transport in Mesoscopic Systems* (Cambridge, Cambridge University, 1995).

[361] R. Landauer, "Spatial variation of currents and fields due to localized scatterers in metallic conduction", *IBM J. Res. Develop.* **1**, 223–231 (1957).

[362] B. J. van Wees, H. van Houten, C. W. J. Beenakker, *et al.*, "Quantized conductance of point contacts in a two-dimensional electron gas", *Phys. Rev. Lett.* **60**, 848–850 (1988).

[363] D. A. Wharam, T. J. Thornton, R. Newbury, *et al.*, "One-dimensional transport and the quantization of the ballistic resistance", *J. Phys. C - Sol. State Phys.* **8**, L209–L214 (1988).

[364] J. L. Costa-Krämer, N. García, P. García-Mochales, P. A. Serena, M. I. Marqués, and A. Correia, "Conductance quantization in nanowires formed between micro and macroscopic metallic electrodes", *Phys. Rev. B* **55**, 5416–5424 (1997).

[365] E. Scheer, N. Agraït, J. C. Cuevas, *et al.*, "The signature of chemical valence in the electrical conduction through a single-atom contact", *Nature* **394**, 154–157 (1998).

[366] R. de Picciotto, H. L. Stormer, L. N. Pfeiffer, K. W. Baldwin, and K. W. West, "Four-terminal resistance of a ballistic quantum wire", *Nature* **411**, 51–54 (2001).

[367] N. D. Lang and P. Avouris, "Understanding the variation of the electrostatic potential along a biased molecular wire", *Nano Lett.* **3**, 737–740 (2003).

[368] Y. Meir and N. S. Wingreen, "Landauer formula for the current through an interacting electron region", *Phys. Rev. Lett.* **68**, 5612–5615 (1992).

[369] K. L. Shepard, M. L. Roukes, and B. P. van der Gaag, "Experimental measurement of scattering coefficients in mesoscopic conductors", *Phys. Rev. B* **46**, 9648–9666 (1992).

[370] K. L. Shepard, M. L. Roukes, and B. P. van der Gaag, "Direct measurement of the transmission matrix of a mesoscopic conductor", *Phys. Rev. Lett.* **68**, 2660–2663 (1992).

[371] N. van der Post, E. T. Peters, I. K. Yanson, and J. M. van Ruitenbeek, "Subgap structure as a function of the barrier in atom-size superconducting tunnel junctions", *Phys. Rev. Lett.* **73**, 2611–2614 (1994).

[372] M. Cahay, M. McLennan, and S. Datta, "Conductance of an array of elastic scatterers: a scattering-matrix approach", *Phys. Rev. B* **37**, 10125–10136 (1988).

[373] W. Liang, M. Bockrath, D. Bozovic, J. H. Hafner, M. Tinkham, and H. Park, "Fabry-Perot interference in a nanotube electron waveguide", *Nature* **411**, 665–669 (2001).

[374] M. Büttiker, "Coherent and sequential tunneling in series barriers", *IBM J. Res. Dev.* **32**, 63–75 (1988).

[375] S. Datta, "Electrical resistance: an atomistic view", *Nanotech.* **15**, S433–S451 (2004).

[376] S. Datta, *Quantum Transport: Atom to Transistor* (Cambridge, Cambridge University Press, 2005).

[377] M. Riordan and L. Hoddeson, *Crystal Fire: the Birth of the Information Age* (New York, W. W. Norton, 1997).

[378] J. W. Orton, *The Story of Semiconductors* (New York, Oxford University Press, 2009).

[379] C. Jacoboni, C. Canali, G. Ottaviani, and A. A. Quaranta, "Review of some charge transport properties of silicon", *Sol. State Electronics* **20**, 77–89 (1977).

[380] P. M. Smith, M. Inoue, and J. Frey, "Electron velocity in Si and GaAs at very high electric fields", *Appl. Phys. Lett.* **37**, 797–798 (1980).

[381] Y.-F. Chen and M. S. Fuhrer, "Electric-field-dependent charge-carrier velocity in semiconducting carbon nanotubes", *Phys. Rev. Lett.* **95**, 236803 (2005).

[382] I. Meric, M. Y. Han, A. F. Young, B. Ozyilmaz, P. Kim, and K. L. Shephard, "Current saturation in zero-bandgap, topgated graphene field-effect transistors", *Nat, Nano.* **3**, 654–659 (2008).

[383] G. A. Sai-Halasz, M. R. Wordeman, D. P. Kern, S. Rishton, and E. Ganin, "High transconductance and velocity overshoot in NMOS devices at the 0.1 μm gate-length level", *IEEE Electron Dev. Lett.* **9**, 464–466 (1988).

[384] F. Assaderaghi, P. K. Ko, and C. Hu, "Observation of velocity overshoot in silicon inversion layers", *IEEE Electron Dev. Lett.* **14**, 484–486 (1993).

[385] J. H. Stathis, "Reliability limits for the gate insulator in CMOS technology", *IBM J. Res. Dev.* **46**, 265–286 (2002).

[386] J. Robertson, "High dielectric constant gate oxides for metal oxide Si transistors", *Rep. Prog. Phys.* **69**, 327–396 (2006).

[387] P. Packan, S. Akbar, M. Armstrong, *et al.*, "High performance 32nm logic technology featuring 2nd generation high-k+ metal gate transistors", in *Electron Devices Meeting (IEDM), 2009 IEEE International*, 1–4 (IEEE, 2009).

[388] B. van Zeghbroek, *Principles of Semiconductor Devices*, ecee.colorado.edu/~bart/book/book/title.htm

[389] C. A. Mead, "Scaling of MOS technology to submicrometer feature sizes", *J. VLSI Signal Proc.* **8**, 9–25 (1994).

[390] D. J. Frank, R. H. Dennard, E. Nowak, P. M. Solomon, Y. Taur, and H.-S. P. Wong, "Device scaling limits of Si MOSFETs and their application dependencies", *Proc. IEEE* **89**, 259–288 (2001).

[391] H.-S. P. Wong, "Beyond the conventional transistor", *IBM J. Res. Dev.* **46**, 133–168 (2002).

[392] H.-S. P. Wong, K. Chan, and Y. Taur, "Self-aligned (top and bottom) double-gate MOSFET with a 25 nm thick silicon channel", *IEDM Tech. Digest* **427** (1997).

[393] J. M. Hergenrother, S.-H. Oh, T. Nigam, D. Monroe, F. P. Klemens, and A. Kornblit, "The vertical replacement-gate (VRG) MOSFET", *Solid-State Electronics* **46**, 939–950 (2002).

[394] V. Schmidt, H. Reil, S. Senz, S. Karg, W. Riess, and U. Gösele, "Realization of a silicon nanowire vertical surround-gate field-effect transistor", *Small* **2**, 85–88 (2005).

[395] T. Bryllert, L.-E. Wernersson, T. Löwgren, and L. Samuelson, "Vertical wrap-gated nanowire transistors", *Nanotech.* **17**, S227–S230 (2006).

[396] M. Lundstrom and Z. Ren, "Essential physics of carrier transport in nanoscale MOSFETs", *IEEE Trans. Elect. Dev.* **49**, 133–141 (2002).

[397] F. Assad, Z. Ren, D. Vasileska, S. Datta, and M. Lundstrom, "On the performance limits for Si MOSFETs: a theoretical study", *IEEE Trans. Electron Devices,* **47**, 232–240 (2000).

[398] Z. Chen, J. Appenzeller, Y.-M. Lin, *et al.*, "An integrated logic circuit assembled on a single carbon nanotube", *Science* **311**, 1735 (2006).

[399] H. Grabert and M. H. Devoret, eds., *Single Charge Tunneling* (New York, Plenum, 1992).

[400] P. Delsing, T. Claeson, K. K. Likharev, and L. S. Kuzmin, "Observation of single-electron-tunneling oscillations", *Phys. Rev. B* **42**, 7439–7449 (1990).

[401] G.-L. Ingold and Y. V. Nazarov, "Charge tunneling rates in ultrasmall junctions", in H. Grabert and M. H. Devoret, eds., *Single Charge Tunneling*, (New York, Plenum, 1992).

[402] E. Bonet, M. M. Deshmukh, and D. C. Ralph, "Solving rate equations for electron tunneling via discrete quantum states", *Phys. Rev. B* **65**, 045317 (2002).

[403] K. Likharev, "Single-electron transistors: electrostatic analogs of the DC SQUIDS", *IEEE Trans. Magnetics,* **23**, 1142–1145 (1987).

[404] D. Berman, N. B. Zhitenev, R. C. Ashoori, H. I. Smith, and M. R. Melloch, "Single-electron transistor as a charge sensor for semiconductor applications", *Papers from the 41st International Conference on Electron, Ion, and Photon Beam Technology and Nanofabrication* **15**, 2844–2847 (1997).

[405] R. G. Knobel and A. N. Cleland, "Nanometre-scale displacement sensing using a single electron transistor", *Nature* **424**, 291–293 (2003).

[406] M. H. Devoret and R. J. Schoelkopf, "Amplifying quantum signals with the single-electron transistor", *Nature* **406**, 1039–1046 (2000).

[407] S. M. Cronenwett, T. H. Oosterkamp, and L. P. Kouwenhoven, "A tunable Kondo effect in quantum dots", *Science* **281**, 540–544 (1998).

[408] P. Wahl, L. Diekhöner, G. Wittich, L. Vitali, M. A. Schneider, and K. Kern, "Kondo effect of molecular complexes at surfaces: ligand control of the local spin coupling", *Phys. Rev. Lett.* **95**, 166601 (2005).

[409] A. Zhao, Q. Li, L. Chen, *et al.*, "Controlling the Kondo effect of an absorbed magnetic ion through its chemical bonding", *Science* **309**, 1542–1544 (2005).

[410] T. A. Fulton and G. J. Dolan, "Observation of single-electron charging effects in small tunnel junctions", *Phys. Rev. Lett.* **59**, 109–112 (1987).

[411] A. Bezryadin, C. Dekker, and G. Schmid, "Electrostatic trapping of single conducting nanoparticles between nanoelectrodes", *Appl. Phys. Lett.* **71**, 1273–1275 (1997).

[412] K. I. Bolotin, F. Kuemmeth, A. N. Pasupathy, and D. C. Ralph, "Metal-nanoparticle single-electron transistors fabricated using electromigration", *Appl. Phys. Lett.* **84**, 3154–3156 (2004).

[413] U. Meirav, M. A. Kastner, M. Heiblum, and S. J. Wind, "One-dimensional electron gas in GaAs: periodic conductance oscillations as a function of density", *Phys. Rev. B* **40**, 5871–5874 (1989).

[414] D. G. Austing, T. Honda, Y. Tokura, and S. Tarucha, "Sub-micron vertical AlGaAs/GaAs resonant tunneling single electron transistor", *Japanese J. Appl. Phys.* **34**, 1320–1325 (1995).

[415] S. Lüscher, A. Fuhrer, R. Held, T. Heinzel, K. Ensslin, and W. Wegscheider, "In-plane gate single-electron transistor in Ga[Al]As fabricated by scanning probe lithography", *Appl. Phys. Lett.* **75**, 2452–2454 (1999).

[416] Z. Yao, H. Postma, L. Balents, and C. Dekker, "Carbon nanotube intramolecular junctions", *Nature* **402**, 273–276 (1999).

[417] A. Tilke, R. H. Blick, H. Lorenz, and J. P. Kotthaus, "Single-electron tunneling in highly doped silicon nanowires in a dual-gate configuration", *J. Appl. Phys.* **89**, 8159–8162 (2001).

[418] Z. Zhong, Y. Fang, W. Lu, and C. Lieber, "Coherent single charge transport in molecular-scale silicon nanowires", *Nano Lett.* **5**, 1143–1146 (2005).

[419] H. Park, J. Park, A. K. L. Lim, E. H. Anderson, A. P. Alivisatos, and P. L. McEuen, "Nanomechanical oscillations in a single-C_{60} transistor", *Nature* **407**, 57–60 (2000).

[420] J. Park, A. N. Pasupathy, J. I. Goldsmith, *et al.*, "Coulomb blockade and the Kondo effect in single-atom transistors", *Nature* **417**, 722–725 (2002).

[421] W. Liang, M. P. Shores, M. Bockrath, J. R. Long, and H. Park, "Kondo resonance in a single-molecule transistor", *Nature* **417**, 725–729 (2002).

[422] S. Kubatkin, A. Danilov, M. Hjort, *et al.*, "Single-electron transistor of a single organic molecule with access to several redox states", *Nature* **425**, 698–701 (2003).

[423] L. H. Yu and D. Natelson, "The Kondo effect in C_{60} single-molecule transistors", *Nano Lett.* **4**, 79–83 (2004).

[424] H. Sellier, G. P. Lansbergen, J. Caro, *et al.*, "Transport spectroscopy of a single dopant in a gated silicon nanowire", *Phys. Rev. Lett.* **97**, 206805 (2006).

[425] K. Y. Tan, K. W. Chan, M. Möttönen, *et al.*, "Transport spectroscopy of single phosphorus donors in a silicon nanoscale transistor", *Nano Lett.* **10** (2010).

[426] M. W. Keller, A. L. Eichenberger, J. M. Martinis, and N. M. Zimmerman, "A capacitance standard based on counting electrons", *Science* **285**, 1706–1709 (1999).

[427] P. J. Mohr and B. N. Taylor, "CODATA recommended values of the fundamental physical constants: 2002", *Rev. Mod. Phys.* **77**, 1–107 (2005).

[428] P. J. Mohr, B. N. Taylor, and D. B. Newell, "CODATA recommended values of the fundamental physical constants: 2006", *Rev. Mod. Phys.* **80**, 633 (2008).

[429] J. Bylander, T. Duty, and P. Delsing, "Current measurement by real-time counting of electrons", *Nature* **434**, 361–364 (2005).

[430] R. J. Schoelkopf, P. Wahlgren, A. A. Kozhevnikov, P. Delsing, and D. E. Prober, "The radio-frequency single-electron transistor (RF-SET): a fast and ultrasensitive electrometer", *Science* **280**, 1238 (1998).

[431] W. Lu, Z. Ji, L. Pfeiffer, K. W. West, and A. J. Rimberg, "Real-time detection of electron tunneling in a quantum dot", *Nature* **423**, 422–425 (2003).

[432] M. D. LaHaye, O. Buu, B. Camarota, and K. C. Schwab, "Approaching the quantum limit of a nanomechanical resonator", *Science* **304**, 74–77 (2004).

[433] D. M. Pozar, *Microwave Engineering*, 3rd edn. (New York, Wiley, 2004)

[434] R. H. Chen, A. N. Korotkov, and K. K. Likharev, "Single-electron transistor logic", *Appl. Phys. Lett.* **68**, 1954–1956 (1996).

[435] K. Likharev, "Single-electron devices and their applications", *Proc. IEEE* **87**, 606–632 (1999).

[436] G. Zimmerli, R. L. Kautz, and J. M. Martinis, "Voltage gain in the single-electron transistor", *Appl. Phys. Lett.* **61**, 2616–2618 (1992).

[437] P. D. Tougaw and C. S. Lent, "Logical devices implemented using quantum cellular automata", *J. Appl. Phys.* **75**, 1818–1825 (1994).

[438] I. Amlani, A. O. Orlov, G. Toth, G. H. Bernstein, C. S. Lent, and G. L. Snider, "Digital logic gate using quantum-dot cellular automata", *Science* **284**, 289–291 (1999).

[439] A. R. von Hippel, ed., *Molecular Science and Molecular Engineering* (New York, Wiley and MIT Press, 1959).

[440] A. Aviram and M. A. Ratner, "Molecular rectifiers", *Chem. Phys. Lett.* **29**, 277–283 (1974).

[441] A. Nitzan and M. A. Ratner, "Electron transport in molecular wire junctions", *Science* **300**, 1384–1389 (2003).

[442] R. L. McCreery, "Molecular electronic junctions", *Chem. Mater.* **16**, 4477–4496 (2004).

[443] Y. Selzer and D. L. Allara, "Single-molecule electrical junctions", *Ann. Rev. Phys. Chem.* **57**, 693–623 (2006).

[444] D. Natelson, "Single-molecule transistors", in H. S. Nalwa, ed., *Handbook of Organic Electronics and Photonics*, (Stephenson Ranch, CA, American Scientific, 2007).

[445] D. Natelson, "Towards the ultimate transistor", *Physics World* **22**, 27–31 (2009).

[446] J. Chen, M. A. Reed, A. M. Rawlett, and J. M. Tour, "Large on-off ratios and negative differential resistance in a molecular electronic device", *Science* **286**, 1550–1552 (1999).

[447] J. G. Kushmerick, D. B. Holt, J. C. Yang, J. Naciri, M. H. Moore, and R. Shashidhar, "Metal-molecule contacts and charge transport across monomolecular layers: measurement and theory", *Phys. Rev. Lett.* **89**, 086802 (2002).

[448] H. B. Akkerman, P. W. M. Blom, D. M. de Leeuw, and B. de Boer, "Towards molecular electronics with large-area molecular junctions", *Nature* **441**, 69–72 (2006).

[449] H. Park, A. K. L. Lim, A. P. Alivisatos, J. Park, and P. L. McEuen, "Fabrication of metallic electrodes with nanometer separation by electromigration", *Appl. Phys. Lett.* **75**, 301–303 (1999).

[450] D. Natelson, L. H. Yu, J. W. Ciszek, Z. K. Keane, and J. M. Tour, "Single-molecule transistors: electron transfer in the solid state", *Chem. Phys.* **324**, 267–275 (2006).

[451] M. A. Ratner, "Bridge-assisted electron transfer - effective electronic coupling", *J. Phys. Chem.* **94**, 4877–4883 (1990).

[452] H. B. Akkerman, R. C. G. Naber, B. Jongbloed, *et al.*, "Electron tunneling through alkanedithiol self-assembled monolayers in large-area molecular junctions", *Proc. Nat. Acad. Sci. US* **104**, 11161–11166 (2007).

[453] S. H. Ke, H. U. Baranger, and W. T. Yang, "Contact atomic structure and electron transport through molecules", *J. Chem. Phys.* **122**, 074704 (2005).

[454] V. B. Engelkes, J. M. Beebe, and C. D. Frisbie, "Length-dependent transport in molecular junctions based on SAMs of alkanethiols and alkanedithiols: effect of metal work function and applied bias on tunneling efficiency and contact resistance", *J. Am. Chem. Soc.* **126**, 14287–14296 (2004).

[455] L. Venkataraman, J. E. Klare, I. W. Tam, C. Nuckolls, M. S. Hybertsen, and M. L. Steigerwald, "Single-molecule circuits with well-defined molecular conductance", *Nano Lett.* **6**, 458–462 (2006).

[456] S. H. Choi, B. Kim, and C. D. Frisbie, "Electrical resistance of long conjugated molecular wires", *Science* **320**, 1482–1486 (2008).

[457] J. G. Kushmerick, J. Lazorcik, C. H. Patterson, and R. Shashidhar, "Vibronic contributions to charge transport across molecular junctions", *Nano Lett.* **4**, 639–642 (2004).

[458] R. M. Metzger, "Electrical rectification by a molecule: the advent of unimolecular electronic devices", *Acc. Chem. Res.* **32**, 950–957 (1999).

[459] G. J. Ashwell, B. Urasinska, and W. D. Tyrrell, "Molecules that mimic Schottky diodes", *Phys. Chem. Chem. Phys.* **8**, 3314–3319 (2006).

[460] L. H. Yu, Z. Keane, J. Ciszek, *et al.*, "Strong Kondo physics and anomalous gate dependence in single-molecule transistors", *Phys. Rev. Lett.* **95**, 256803 (2005).

[461] C. P. Collier, E. W. Wong, M. Belohradsky, *et al.*, "Electronically configurable molecular-based logic gates", *Science* **285**, 391–394 (1999).

[462] V. Balzani, M. Gomez-Lopez, and J. F. Stoddart, "Molecular machines", *Acc. Chem. Res.* **31**, 405–414 (1998).

[463] C. N. Lau, D. R. Stewart, R. S. Williams, and M. Bockrath, "Direct observation of nanoscale switching centers in metal/molecule/metal structures", *Nano Lett.* **4**, 569–572 (2004).

[464] J. E. Green, J. W. Choi, A. Boukai, *et al.*, "A 160-kilobit molecular electronic memory patterned at 10^{11} bits per square centimetre", *Nature* **445**, 414–417 (2007).

[465] C. Zhang, M.-H. Du, H.-P. Cheng, X.-G. Zhang, A. E. Roitberg, and J. L. Krause, "Coherent electron transport through an azobenzene molecule: a light-driven molecular switch", *Phys. Rev. Lett.* **92**, 158301 (2004).

[466] M. Irie, "Diarylethenes for memories and switches", *Chem. Rev.* **100**, 1685–1716 (2000).

[467] A. Salomon, T. Boecking, C. K. Chan, *et al.*, "How do electronic carriers cross Si-bound alkyl monolayers?" *Phys. Rev. Lett.* **95**, 266807 (2005).

[468] T. He, J. L. He, M. Lu, *et al.*, "Controlled modulation of conductance in silicon devices by molecular monolayers", *J. Am. Chem. Soc.* **128**, 14537–14541 (2006).

[469] S. A. Scott, W. N. Peng, A. M. Kiefer, *et al.*, "Influence of surface chemical modification on charge transport properties in ultrathin silicon membranes", *ACS Nano* **3**, 1683–1692 (2009).

[470] J. R. Heath, P. J. Kuekes, G. S. Snider, and R. S. Williams, "A defect-tolerant computer architecture: opportunities for nanotechnology", *Science* **280**, 1716–1721 (1998).

[471] P. J. Kuekes, D. R. Stewart, and R. S. Williams, "The crossbar latch: logic value storage, restoration, and inversion in crossbar circuits", *J. Appl. Phys.* **97**, 034301 (2005).

[472] M. A. Kastner, "The single-electron transistor", *Rev. Mod. Phys.* **64**, 849–858 (1992).

[473] C. W. J. Beenakker, "Random-matrix theory of quantum transport", *Rev. Mod. Phys.* **69**, 731–808 (1997).

[474] Y. Imry and R. Landauer, "Conductance viewed as transmission", *Rev. Mod. Phys.* **71**, S306–S312 (1999).

[475] B. Ludoph, M. H. Devoret, D. Esteve, C. Urbina, and J. M. van Ruitenbeek, "Evidence for saturation of channel transmission from conductance fluctuations in atomic-size point contacts", *Phys. Rev. Lett.* **82**, 1530–1533 (1999).

[476] C. Untiedt, G. Rubio Bollinger, S. Vieira, and N. Agraït, "Quantum interference in atomic-sized point contacts", *Phys. Rev. B* **62**, 9962–9965 (2000).

[477] D. C. Jiles, *Introduction to Magnetism and Magnetic Materials*, 2nd edn. (New York, CRC, 1998).

[478] R. C. O'Handley, *Modern Magnetic Materials: Principles and Applications* (New York, Wiley-Interscience, 1999).

[479] S. Chikazumi, *Physics of Ferromagnetism*, 3rd edn. (New York, Oxford, 2009).

[480] J. M. D. Coey, "Materials for spin electronics", in M. Ziese and M. J. Thornton, eds., *Spin Electronics*, Lecture Notes in Physics, 277–297 (New York, Springer Berlin/Heidelberg, 2001).

[481] J. M. D. Coey and C. L. Chien, "Half-metallic ferromagnetic oxides", *MRS Bulletin* **28**, 720–724 (2003).

[482] J. Van Kranendonk and J. H. Van Vleck, "Spin waves", *Rev. Mod. Phys.* **30**, 1–23 (1958).

[483] H. Barkhausen, "Zwei mit Hilfe der neuen verstarker entdeckte Erscheinungen", *Z. Phys.* **20**, 401–403 (1919).

[484] C. Kittel, "On the theory of ferromagnetic resonance absorption", *Phys. Rev.* **73**, 155–161 (1948).

[485] L. D. Landau and E. Lifshitz, "On the theory of the dispersion of magnetic permeability in ferromagnetic bodies", *Phys. Z. Sowjetunion* **8**, 153–169 (1935).

[486] T. L. Gilbert, "A Lagrangian formulation of the gyromagnetic equation of the magnetization field", *Phys. Rev.* **100**, 1243 (1955).

[487] T. Gilbert, "A phenomenological theory of damping in ferromagnetic materials", *IEEE Trans. Magnetics,* **40**, 3443–3449 (2004).

[488] F. Bloch, "Nuclear induction", *Phys. Rev.* **70**, 460–474 (1946).

[489] H. Y. Carr and E. M. Purcell, "Effects of diffusion on free precession in nuclear magnetic resonance experiments", *Phys. Rev.* **94**, 630–638 (1954).

[490] F. Bitter, "Experiments on the nature of ferromagnetism", *Phys. Rev.* **41**, 507–515 (1932).

[491] A. Oral, S. J. Bending, and M. Henini, "Real-time scanning Hall probe microscopy", *Appl. Phys. Lett.* **69**, 1324–1326 (1996).

[492] J. G. S. Lok, A. K. Geim, U. Wyder, J. C. Maan, and S. V. Dubonos, "Thermally activated annihilation of an individual domain in submicrometer nickel particles", *J. Mag. Mag. Mater.* **204**, 159–164 (1999).

[493] A. D. Kent, S. von Molnár, S. Gider, and D. D. Awschalom, "Properties and measurement of scanning tunneling microscope fabricated ferromagnetic particle arrays", in *The 6th Joint Magnetism and Magnetic Materials (MMM)-Intermag Conference*, volume 76, 6656–6660 (Melville, NY, AIP, 1994).

[494] A. K. Geim, S. V. Dubonos, J. G. S. Lok, *et al.*, "Ballistic Hall micro-magnetometry", *Appl. Phys. Lett.* **71**, 2379–2381 (1997).

[495] D. Cohen, "Magnetoencephalography: detection of the brain's electrical activity with a superconducting magnetometer", *Science* **175**, 664–666 (1972).

[496] Y. S. Greenberg, "Application of superconducting quantum interference devices to nuclear magnetic resonance", *Rev. Mod. Phys.* **70**, 175–222 (1998).

[497] J. R. Kirtley, C. C. Tsuei, M. Rupp, *et al.*, "Direct imaging of integer and half-integer Josephson vortices in high- *Tc* grain boundaries", *Phys. Rev. Lett.* **76**, 1336–1339 (1996).

[498] W. Wernsdorfer, "From micro- to nano-SQUIDs: applications to nanomagnetism", *Superconductor Science and Technology* **22**, 064013 (2009).

[499] T. J. Silva, S. Schultz, and D. Weller, "Scanning near-field optical microscope for the imaging of magnetic domains in optically opaque materials", *Appl. Phys. Lett.* **65**, 658–660 (1994).

[500] K. Koike and K. Hayakawa, "Domain observation with spin-polarized secondary electrons", *J. Appl. Phy.* **57**, 4244–4248 (1985).

[501] M. R. Scheinfein, J. Unguris, M. H. Kelly, D. T. Pierce, and R. J. Celotta, "Scanning electron-microscopy with polarization analysis (SEMPA)", *Rev. Sci. Instr.* **61**, 2501–2526 (1990).

[502] T. J. Gay and F. B. Dunning, "Mott electron polarimetry", *Review of Scientific Instruments* **63**, 1635–1651 (1992).

[503] J. Unguris, D. T. Pierce, A. Galejs, and R. J. Celotta, "Spin and energy analyzed secondary electron emission from a ferromagnet", *Phys. Rev. Lett.* **49**, 72–76 (1982).

[504] M. E. Hale, H. W. Fuller, and H. Rubinstein, "Magnetic domain observations by electron microscopy", *J. Appl. Phys.* **30**, 789–791 (1959).

[505] P. J. Grundy and R. S. Tebble, "Lorentz electron microscopy", *Adv. Phys.* **17**, 153–242 (1968).

[506] M. S. Cohen, "Wave-optical aspects of Lorentz microscopy", *J. Appl. Phys.* **38**, 4966–4976 (1967).

[507] A. Tonomura, T. Matsuda, J. Endo, T. Arii, and K. Mihama, "Direct observation of fine structure of magnetic domain walls by electron holography", *Phys. Rev. Lett.* **44**, 1430–1433 (1980).

[508] D. Rugar, H. J. Mamin, P. Guethner, *et al.*, "Magnetic force microscopy: general principles and application to longitudinal recording media", *J. Appl. Phys.* **68**, 1169–1183 (1990).

[509] J. A. Sidles, J. L. Garbini, K. J. Bruland, *et al.*, "Magnetic resonance force microscopy", *Rev. Mod. Phys.* **67**, 249–265 (1995).

[510] D. Rugar, R. Budakian, H. J. Mamin, and B. W. Chui, "Single spin detection by magnetic resonance force microscopy", *Nature* **430**, 329–332 (2004).

[511] C. L. Degen, M. Poggio, H. J. Mamin, C. T. Rettner, and D. Rugar, "Nanoscale magnetic resonance imaging", *Proc. Nat. Acad. Sci. US.* **106**, 1313–1317 (2009).

[512] M. Bode, A. Kubetzka, O. Pietzsch, and R. Wiesendanger, "Spin-resolved spectro-microscopy of magnetic nanowire arrays", *Sur. Sci.* **514**, 135–144 (2002).

[513] L. Piraux, S. Dubois, E. Ferain, *et al.*, "Anisotropic transport and magnetic properties of arrays of sub-micron wires", *J. Mag. Mag. Mat.* **165**, 352–355 (1997), symposium E: "Magnetic ultrathin films, multilayers and surfaces".

[514] M. N. Baibich, J. M. Broto, A. Fert, *et al.*, "Giant magnetoresistance of (001)Fe/(001)Cr magnetic superlattices", *Phys. Rev. Lett.* **61**, 2472–2475 (1988).

[515] G. Binasch, P. Grünberg, F. Saurenbach, and W. Zinn, "Enhanced magneto-resistance in layered magnetic structures with antiferromagnetic interlayer exchange", *Phys. Rev. B* **39**, 4828–4830 (1989).

[516] S. S. P. Parkin, N. More, and K. P. Roche, "Oscillations in exchange coupling and magnetoresistance in metallic superlattice structures: Co/Ru, Co/Cr, and Fe/Cr", *Phys. Rev. Lett.* **64**, 2304–2307 (1990).

[517] M. Johnson and R. H. Silsbee, "Interfacial charge-spin coupling: injection and detection of spin magnetization in metals", *Phys. Rev. Lett.* **55**, 1790–1793 (1985).

[518] F. J. Jedema, A. T. Filip, and B. J. van Wees, "Electrical spin injection and accumulation at room temperature in an all-metal mesoscopic spin valve", *Nature* **410**, 345–349 (2001).

[519] J. S. Moodera, L. R. Kinder, T. M. Wong, and R. Meservey, "Large magnetoresistance at room temperature in ferromagnetic thin film tunnel junctions", *Phys. Rev. Lett.* **74**, 3273–3276 (1995).

[520] M. Julliere, "Tunneling between ferromagnetic films", *Phys. Lett. A* **54**, 225–226 (1975).

[521] R. Meservey and P. M. Tedrow, "Spin-polarized electron tunneling", *Phys. Rep.* **238**, 173–243 (1994).

[522] E. Tsymbal, K. Belashchenko, J. Velev, *et al.*, "Interface effects in spin-dependent tunneling", *Prog. Mat. Sci.* **52**, 401–420 (2007),

[523] A. P. Ramirez, "Colossal magnetoresistance", *J. Phys.: Cond. Mat.* **9**, 8171 (1997).

[524] J. M. D. Coey, M. Viret, and S. V. Molnar, "Mixed-valence manganites." *Adv. Phys.* **48**, 167–293 (1999).

[525] N. García, M. Muñoz, and Y.-W. Zhao, "Magnetoresistance in excess of 200% in ballistic Ni nanocontacts at room temperature and 100 Oe", *Phys. Rev. Lett.* **82**, 2923–2926 (1999).

[526] Z. K. Keane, L. H. Yu, and D. Natelson, "Magnetoresistance of atomic-scale electromigrated nickel nanocontacts", *Appl. Phys. Lett.* **88**, 062514 (2006).

[527] K. I. Bolotin, F. Kuemmeth, A. N. Pasupathy, and D. C. Ralph, "From ballistic transport to tunneling in electromigrated ferromagnetic breakjunctions", *Nano Lett.* **6**, 123–1237 (2006).

[528] B. Doudin and M. Viret, "Ballistic magnetoresistance?", *J. Phys.: Cond. Mat.* **20**, 083201 (2008).

[529] S. A. Solin, T. Thio, D. R. Hines, and J. J. Heremans, "Enhanced room-temperature geometric magnetoresistance in inhomogeneous narrow-gap semiconductors", *Science* **289**, 1530–1532 (2000).

[530] S. A. Solin, D. R. Hines, A. C. H. Rowe, *et al.*, "Nonmagnetic semiconductors as read-head sensors for ultra-high-density magnetic recording", *Appl. Phys. Lett.* **80**, 4012–4014 (2002).

[531] J. Nogués and I. K. Schuller, "Exchange bias", *J. Mag. Mag. Mat.* **192**, 203–232 (1999).

[532] A. Hubert and R. Schäfer, *Magnetic Domains: the Analysis of Magnetic Microstructures* (New York, Springer, 1998).

[533] K. Hong and N. Giordano, "Evidence for domain wall tunnelling in a quasi-one dimensional ferromagnet", *J. Phys.: Cond. Mat.* **8**, L301 (1996).

[534] P. M. Levy and S. Zhang, "Resistivity due to domain wall scattering", *Phys. Rev. Lett.* **79**, 5110–5113 (1997).

[535] E. M. Chudnovsky and L. Gunther, "Quantum tunneling of magnetization in small ferromagnetic particles", *Phys. Rev. Lett.* **60**, 661–664 (1988).

[536] H.-B. Braun, J. Kyriakidis, and D. Loss, "Macroscopic quantum tunneling of ferromagnetic domain walls", *Phys. Rev. B* **56**, 8129–8137 (1997).

[537] T. Ono, H. Miyajima, K. Shigeto, K. Mibu, N. Hosoito, and T. Shinjo, "Propagation of a magnetic domain wall in a submicrometer magnetic wire", *Science* **284**, 468–470 (1999).

[538] R. Ferré, K. Ounadjela, J. M. George, L. Piraux, and S. Dubois, "Magnetization processes in nickel and cobalt electrodeposited nanowires", *Phys. Rev. B* **56**, 14066–14075 (1997).

[539] T. Shinjo, T. Okuno, R. Hassdorf, K. Shigeto, and T. Ono, "Magnetic vortex core observation in circular dots of permalloy", *Science* **289**, 930–932 (2000).

[540] C. P. Bean and J. D. Livingston, "Superparamagnetism", *J. Appl. Phys.* **30**, S120–S129 (1959).

[541] F. C. Fonseca, G. F. Goya, R. F. Jardim, *et al.*, "Superparamagnetism and magnetic properties of Ni nanoparticles embedded in SiO_2", *Phys. Rev. B* **66**, 104406 (2002).

[542] W. F. Brown, "Thermal fluctuations of a single-domain particle", *Phys. Rev.* **130**, 1677–1686 (1963).

[543] J. I. Gittleman, B. Abeles, and S. Bozowski, "Superparamagnetism and relaxation effects in granular $Ni\text{-}SiO_2$ and $Ni\text{-}Al_2O_3$ films", *Phys. Rev. B* **9**, 3891–3897 (1974).

[544] P. C. E. Stamp, E. M. Chudnovsky, and B. Barbara, "Quantum tunneling of magnetization in solids", *Int. J. Mod. Phys. B* **6**, 1355–1473 (1992).

[545] W. Wernsdorfer, E. Bonet Orozco, K. Hasselbach, *et al.*, "Macroscopic quantum tunneling of magnetization of single ferrimagnetic nanoparticles of barium ferrite", *Phys. Rev. Lett.* **79**, 4014–4017 (1997).

[546] J. R. Friedman, M. P. Sarachik, J. Tejada, and R. Ziolo, "Macroscopic measurement of resonant magnetization tunneling in high-spin molecules", *Phys. Rev. Lett.* **76**, 3830–3833 (1996).

[547] D. Gatteschi and R. Sessoli, "Quantum tunneling of magnetization and related phenomena in molecular materials", *Angew. Chemie, Int. Ed.* **42**, 268–297 (2003).

[548] G. Christou, D. Gatteschi, D. N. Hendrickson, and R. Sessoli, "Single-molecule magnets", *MRS Bulletin* **25**, 66–71 (2000).

[549] R. J. Elliott, "Theory of the effect of spin-orbit coupling on magnetic resonance in some semiconductors", *Phys. Rev.* **96**, 266–279 (1954).

[550] Y. Yafet, "g factors and spin-lattice relaxation of conduction electrons", *Sol. State Phys.* **14**, 1–98 (1963).

[551] M. I. D'Yakonov and V. I. Perel, "Possibility of orienting electron spins with current", *JETP Letters-USSR* **13**, 467–469 (1971).

[552] M. I. D'Yakonov and V. I. Perel, "Current-induced spin orientation of electrons in semiconductors", *Phys. Lett. A* **A 35**, 459–460 (1971).

[553] G. L. Bir, A. G. Aronov, and G. E. Pikus, "Spin relaxation of electrons scattered by holes", *Zhurnal Eksperimentalnoi I Teoreticheskoi Fiziki* **69**, 1382–1397 (1975).

[554] A. Y. Elezzabi, M. R. Freeman, and M. Johnson, "Direct measurement of the conduction electron spin-lattice relaxation time T_1 in gold", *Phys. Rev. Lett.* **77**, 3220–3223 (1996).

[555] G. Zolfagharkani, A. Gaidarzhy, P. Degiovanni, S. Kettemann, P. Fulde, and P. Mohanty, "Nanomechanical detection of itinerant electron spin flip", *Nature Nano.* **3**, 720–723 (2008).

[556] E. B. Myers, D. C. Ralph, J. A. Katine, R. N. Louie, and R. A. Buhrman, "Current-induced switching of domains in magnetic multilayer devices", *Science* **285**, pp. 867–870 (1999).

[557] M. Tsoi, A. G. M. Jansen, J. Bass, *et al.*, "Excitation of a Magnetic Multilayer by an Electric Current", *Phys. Rev. Lett.* **80**, 4281–4284 (1998).

[558] J. A. Katine, F. J. Albert, R. A. Buhrman, E. B. Myers, and D. C. Ralph, "Current-driven magnetization reversal and spin-wave excitations in Co /Cu /Co pillars", *Phys. Rev. Lett.* **84**, 3149–3152 (2000).

[559] I. N. Krivorotov, N. C. Emley, J. C. Sankey, S. I. Kiselev, D. C. Ralph, and R. A. Buhrman, "Time-domain measurements of nanomagnet dynamics driven by spin-transfer torques", *Science* **307**, 228–231 (2005).

[560] S. Kiselev, J. Sankey, I. Krivorotov, *et al.*, "Microwave oscillations of a nanomagnet driven by a spin-polarized current", *Nature* **425**, 380–383 (2003).

[561] E. D. Daniel, C. D. Mee, and M. H. Clark, eds., *Magnetic Recording: the First 100 Years* (New York, Wiley-IEEE, 1998).

[562] E. Grobowski, "Hitachi global storage technologies HDD roadmap", www1.hitachigst.com/hdd/technolo/overview/storagetechchart.html (2011).

[563] Y. Tanaka, "Perpendicular recording technology: from research to commercialization", *Proc. IEEE* **96**, 1754 –1760 (2008).

[564] R. Wood, Y. Hsu, and M. Schulz, "Perpendicular magnetic recording technology", Technical report, Hitachi Global Storage Technology (2007).

[565] "CPP read-head technology enables smaller form factor storage", Technical report, Fujitsu Computer Products of America (2006).

[566] M. Kryder, E. Gage, T. McDaniel, *et al.*, "Heat assisted magnetic recording", *Proc. IEEE* **96**, 1810 –1835 (2008).

[567] E. Dobisz, Z. Bandic, T.-W. Wu, and T. Albrecht, "Patterned media: nanofabrication challenges of future disk drives", *Proc. IEEE* **96**, 1836–1846 (2008).

[568] J.-G. Zhu, "Magnetoresistive random access memory: the path to competitiveness and scalability", *Proc. IEEE* **96**, 1786–1798 (2008).

[569] S. S. P. Parkin, M. Hayashi, and L. Thomas, "Magnetic domain-wall racetrack memory", *Science* **320**, 190–194 (2008).

[570] S. A. Wolf, D. D. Awschalom, R. A. Buhrman, *et al.*, "Spintronics: A spin-based electronics vision for the future", *Science* **294**, 1488–1495 (2001).

[571] S. A. Wolf, A. Y. Chtchelkanova, and D. M. Treger, "Spintronics – a retrospective and perspective", *IBM J. Res. Dev.* **50**, 101–110 (2006).

[572] S. Datta and B. Das, "Electronic analog of the electro-optic modulator", *Appl. Phys. Lett.* **56**, 665–667 (1990).

[573] G. Schmidt, D. Ferrand, L. W. Molenkamp, A. T. Filip, and B. J. van Wees, "Fundamental obstacle for electrical spin injection from a ferromagnetic metal into a diffusive semiconductor", *Phys. Rev. B* **62**, R4790–R4793 (2000).

[574] G. Schmidt and L. W. Molenkamp, "Spin injection into semiconductors, physics and experiments", *Semiconductor Science and Technology* **17**, 310 (2002).

[575] E. I. Rashba, "Theory of electrical spin injection: tunnel contacts as a solution of the conductivity mismatch problem", *Phys. Rev. B* **62**, R16267–R16270 (2000).

[576] J. M. Kikkawa, I. P. Smorchkova, N. Samarth, and D. D. Awschalom, "Room-temperature spin memory in two-dimensional electron gases", *Science* **277**, 1284–1287 (1997).

[577] J. Kikkawa and D. D. Awschalom, "Lateral drag of spin coherence in gallium arsenide", *Nature* **397**, 139–141 (1999).

[578] Y. Ohno, D. K. Young, B. Beschoten, F. Matsukura, H. Ohno, and D. D. Awschalom, "Electrical spin injection in a ferromagnetic semiconductor heterostructure", *Nature* **402**, 790–792 (1999).

[579] J. E. Hirsch, "Spin Hall effect", *Phys. Rev. Lett.* **83**, 1834–1837 (1999).

[580] Y. K. Kato, R. C. Myers, A. C. Gossard, and D. D. Awschalom, "Observation of the spin Hall effect in semiconductors", *Science* **306**, 1910–1913 (2004).

[581] J. Wunderlich, B. Kaestner, J. Sinova, and T. Jungwirth, "Experimental observation of the spin-Hall effect in a two-dimensional spin-orbit coupled semiconductor system", *Phys. Rev. Lett.* **94**, 047204 (2005).

[582] J. Sinova, D. Culcer, Q. Niu, N. A. Sinitsyn, T. Jungwirth, and A. H. MacDonald, "Universal intrinsic spin Hall effect", *Phys. Rev. Lett.* **92**, 126603 (2004).

[583] Y. A. Bychkov and E. I. Rashba, "Oscillatory effects and the magnetic susceptibility of carriers in inversion layers", *J. Phys. C: Solid State Phys.* **17**, 6039 (1984).

[584] Q. A. Pankhurst, J. Connolly, S. K. Jones, and J. Dobson, "Applications of magnetic nanoparticles in biomedicine", *J. Phys. D: Appl. Phys.* **36**, R167 (2003).

[585] S. Mornet, S. Vasseur, F. Grasset, and E. Duguet, "Magnetic nanoparticle design for medical diagnosis and therapy", *J. Mater. Chem.* **14**, 2161–2175 (2004).

[586] A. Jordan, R. Scholz, P. Wust, H. Fähling, and R. Felix, "Magnetic fluid hyperthermia (MFH): cancer treatment with AC magnetic field induced excitation of biocompatible superparamagnetic nanoparticles", *J. Mag. Mag. Mat.* **201**, 413–419 (1999).

[587] J. Oberteuffer, "Magnetic separation: a review of principles, devices, and applications", *IEEE Trans. Magnetics,* **10**, 223–238 (1974).

[588] O. Olsvik, T. Popovic, E. Skjerve, *et al.*, "Magnetic separation techniques in diagnostic microbiology." *Clinical Microbiology Reviews* **7**, 43–54 (1994).

[589] C. T. Yavuz, J. T. Mayo, W. W. Yu, *et al.*, "Low-field magnetic separation of monodisperse Fe_3O_4 Nanocrystals", *Science* **314**, 964–967 (2006).

[590] G. De Las Cuevas, J. Faraudo, and J. Camacho, "Low-gradient magnetophoresis through field-induced reversible aggregation", *J. Phys. Chem. C* **112**, 945–950 (2008).

[591] J. M. Ginder and L. C. Davis, "Shear stresses in magnetorheological fluids: Role of magnetic saturation", *Appl. Phys. Lett.* **65**, 3410–3412 (1994).

[592] M. R. Jolly, J. D. Carlson, and B. C. Muñoz, "A model of the behaviour of magnetorheological materials", *Smart Mat. Struct.* **5**, 607 (1996).

[593] A. Sommerfeld, "The propagation of light in dispersing media", *Ann. Physik* **44**, 177 (1914).

[594] L. Brillouin, "On the propagation of light in dispersing media", *Ann. Physik* **44**, 203 (1914).

[595] A. Kuzmich, A. Dogariu, L. J. Wang, P. W. Milonni, and R. Y. Chiao, "Signal velocity, causality, and quantum noise in superluminal light pulse propagation", *Phys. Rev. Lett.* **86**, 3925–3929 (2001).

[596] J. Jackson, *Classical Electrodynamics* (New York, Wiley, 1999).

[597] S. J. Orfanidis, "Electromagnetic waves and antennas", www.ece.rutgers.edu~orfanidi/ewa/ (2002).

[598] M. Sargent III, M. O. Scully, and J. W. E. Lamb, *Laser Physics* (New York, Westview, 1978).

[599] A. E. Siegman, *Lasers* (Sausalito, University Science Books, 1986).

[600] O. Svelto, *Principles of Lasers*, 5th edn. (New York, Springer, 2009).

[601] J. P. Gordon, H. J. Zeiger, and C. H. Townes, "The Maser–new type of microwave amplifier, frequency standard, and spectrometer", *Phys. Rev.* **99**, 1264–1274 (1955).

[602] W. T. Tsang, "A graded-index waveguide separate-confinement laser with very low threshold and a narrow Gaussian beam", *Appl. Phys. Lett.* **39**, 134–137 (1981).

[603] H. Kogelnik and C. V. Shank, "Stimulated emission in a periodic structure", *Appl. Phys. Lett.* **18**, 152–154 (1971).

[604] H. Yen, M. Nakamura, E. Garmire, S. Somekh, A. Yariv, and H. Garvin, "Optically pumped GaAs waveguide lasers with a fundamental 0.11 corrugation feedback", *Optics Commu.* **9**, 35–37 (1973).

[605] P. Yeh and A. Yariv, "Bragg reflection waveguides", *Optics Commu.* **19**, 427–430 (1976).

[606] A. Nurmikko and J. Han, "Blue and near-ultraviolet vertical-cavity surface-emitting lasers", *MRS Bulletin* **27**, 502–506 (2002).

[607] C. Gmachl, F. Capasso, D. L. Sivco, and A. Y. Cho, "Recent progress in quantum cascade lasers and applications", *Rep. Progr. Phys.* **64**, 1533 (2001).

[608] B. S. Williams, "Terahertz quantum-cascade lasers", *Nat. Phot.* **1**, 517–525 (2007).

[609] J. Hecht, *City of Light: the Story of Fiber Optics* (New York, Oxford University, 1999).

[610] G. Agrawal, *Fiber-optic Communication Systems*, Number 1 in Series in Microwave and Optical Engineering (New York, Wiley-Interscience, 2002).

[611] "Introduction to optical fibers, dB, attenuation and measurements", Technical report, Cisco Systems, Inc. (2010).

[612] K. O. Hill, Y. Fujii, D. C. Johnson, and B. S. Kawasaki, "Photosensitivity in optical fiber waveguides: application to reflection filter fabrication", *Appl. Phys. Lett.* **32**, 647–649 (1978).

[613] A. Kersey, M. Davis, H. Patrick, *et al.*, "Fiber grating sensors", *J. Lightwave Technol.*, **15**, 1442–1463 (1997).

[614] B. J. Eggleton, A. Ahuja, P. S. Westbrook, *et al.*, "Integrated tunable fiber gratings for dispersion management in high-bit rate systems", *J. Lightwave Technol.* **18**, 1418 (2000).

[615] E. Wooten, K. Kissa, A. Yi-Yan, *et al.*, "A review of lithium niobate modulators for fiber-optic communications systems", *IEEE J. Selected Topics in Quantum Electronics,* **6**, 69–82 (2000).

[616] A. Liu, R. Jones, L. Liao, *et al.*, "A high-speed silicon optical modulator based on a metal oxide semiconductor capacitor", *Nature* **427**, 615–618 (2004).

[617] R. Mears, L. Reekie, I. Jauncey, and D. Payne, "Low-noise erbium-doped fibre amplifier operating at 1.54 μm", *Electron. Lett.* **23**, 1026–1028 (1987).

[618] W. Miniscalco, "Erbium-doped glasses for fiber amplifiers at 1500 nm", *J. Lightwave Technol.,* **9**, 234–250 (1991).

[619] S. Bottacchi, A. Beling, A. Matiss, *et al.*, "Advanced photoreceivers for high-speed optical fiber transmission systems", *IEEE J. Selected Topics in Quantum Electronics,* **16**, 1099–1112 (2010).

[620] E. Yablonovitch, "Inhibited spontaneous emission in solid-state physics and electronics", *Phys. Rev. Lett.* **58**, 2059–2062 (1987).

[621] E. Yablonovitch, "Photonic band-gap structures", *J. Opt. Soc. Am. B* **10**, 283–295 (1993).

[622] E. Özbay, A. Abeyta, G. Tuttle, *et al.*, "Measurement of a three-dimensional photonic band gap in a crystal structure made of dielectric rods", *Phys. Rev. B* **50**, 1945–1948 (1994).

[623] A. Blanco, E. Chomski, S. Grabtchak, *et al.*, "Large-scale synthesis of a silicon photonic crystal with a complete three-dimensional bandgap near 1.5 micrometres", *Nature* **405**, 437–440 (2000).

[624] S. Noda, K. Tomoda, N. Yamamoto, and A. Chutinan, "Full three-dimensional photonic bandgap crystals at near-infrared wavelengths", *Science* **289**, 604–606 (2000).

[625] M. Campbell, D. Sharp, M. Harrison, R. Denning, and A. Turberfield, "Fabrication of photonic crystals for the visible spectrum by holographic lithography", Nature **404**, 53–56 (2000).

[626] D. J. Norris, "Photonic crystals: a view of the future", *Nat. Mater.* **6**, 177–178 (2007).

[627] A. A. Erchak, D. J. Ripin, M. M. P. Rakich, *et al.*, "Photonic band gap microcavity laser embedded in a strip waveguide", Technical report, Massachusetts Institute of Technology (2004).

[628] H. Altug, D. Englund, and J. Vučković, "Ultrafast photonic crystal nanocavity laser", *Nat. Phys.* **2**, 484–488 (2006).

[629] E. M. Purcell, "Spontaneous emission probabilities at radio frequencies", *Phys. Rev.* **69**, 681–681 (1946).

[630] P. Goy, J. M. Raimond, M. Gross, and S. Haroche, "Observation of cavity-enhanced single-atom spontaneous emission", *Phys. Rev. Lett.* **50**, 1903–1906 (1983).

[631] D. Kleppner, "Inhibited spontaneous emission", *Phys. Rev. Lett.* **47**, 233–236 (1981).

[632] D. Englund, D. Fattal, E. Waks, *et al.*, "Controlling the spontaneous emission rate of single quantum dots in a two-dimensional photonic crystal", *Phys. Rev. Lett.* **95**, 013904 (2005).

[633] A. Kress, F. Hofbauer, N. Reinelt, *et al.*, "Manipulation of the spontaneous emission dynamics of quantum dots in two-dimensional photonic crystals", *Phys. Rev. B* **71**, 241304 (2005).

[634] T. F. Krauss and R. M. D. L. Rue, "Photonic crystals in the optical regime: past, present and future", *Prog. Quant. Electr.* **23**, 51–96 (1999).

[635] J. J. Wierer, Jr., A. David, and M. M. Megens, "III-nitride photonic-crystal light-emitting diodes with high extraction efficiency", *Nat. Phot.* **3**, 163–169 (2009).

[636] J. C. Knight, T. A. Birks, P. S. J. Russell, and D. M. Atkin, "All-silica single-mode optical fiber with photonic crystal cladding", *Opt. Lett.* **21**, 1547–1549 (1996).

[637] P. S. Russell, "Photonic-crystal fibers", *J. Lightwave Technol.*, **24**, 4729–4749 (2006).

[638] J. C. Knight, J. Broeng, T. A. Birks, and P. S. J. Russell, "Photonic band gap guidance in optical fibers", *Science* **282**, 1476–1478 (1998).

[639] S. A. Rinne, F. Garcia-Santamaria, and P. V. Braun, "Embedded cavities and waveguides in three-dimensional silicon photonic crystals", *Nat. Phot.* **2**, 52–56 (2008).

[640] A. Mekis, J. C. Chen, I. Kurland, S. Fan, P. R. Villeneuve, and J. D. Joannopoulos, "High transmission through sharp bends in photonic crystal waveguides", *Phys. Rev. Lett.* **77**, 3787–3790 (1996).

[641] S. G. Johnson, S. Fan, P. R. Villeneuve, J. D. Joannopoulos, and L. A. Kolodziejski, "Guided modes in photonic crystal slabs", *Phys. Rev. B* **60**, 5751–5758 (1999).

[642] M. F. Yanik, S. Fan, M. Soljačić, and J. D. Joannopoulos, "All-optical transistor action with bistable switching in a photonic crystal cross-waveguide geometry", *Opt. Lett.* **28**, 2506–2508 (2003).

[643] M. F. Yanik, S. Fan, M. Soljačić, and J. D. Joannopoulos, "All-optical transistor action with bistable switching in a photonic crystal cross-waveguide geometry", *Opt. Lett.* **28**, 2506–2508 (2003).

[644] S. Kubo, Z.-Z. Gu, K. Takahashi, A. Fujishima, H. Segawa, and O. Sato, "Control of the optical properties of liquid crystal-infiltrated inverse opal structures using photo irradiation and/or an electric field", *Chem. Mater.* **17**, 2298–2309 (2005).

[645] D. S. Wiersma and A. Lagendijk, "Light diffusion with gain and random lasers", *Phys. Rev. E* **54**, 4256–4265 (1996).

[646] H. Cao, Y. G. Zhao, S. T. Ho, E. W. Seelig, Q. H. Wang, and R. P. H. Chang, "Random laser action in semiconductor powder", *Phys. Rev. Lett.* **82**, 2278–2281 (1999).

[647] D. S. Wiersma, P. Bartolini, A. Lagendijk, and R. Righini, "Localization of light in a disordered medium", *Nature* **390**, 671–673 (1997).

[648] M. A. Paesler and P. J. Moyer, *Near-Field Optics: Theory, Instrumentation, and Applications* (New York, Wiley Interscience, 1996).

[649] J. M. Vigoureux, F. Depasse, and C. Girard, "Superresolution of near-field optical microscopy defined from properties of confined electromagnetic waves", *Appl. Opt.* **31**, 3036–3045 (1992).

[650] E. Synge, "XXXVIII. A suggested method for extending microscopic resolution into the ultra-microscopic region", *Phil. Mag. Ser.* **76**, 356–362 (1928).

[651] J. A. O'Keefe, "Resolving power of visible light", *J. Opt. Soc. Am.* **46**, 359–359 (1956).

[652] S. A. Maier and H. A. Atwater, "Plasmonics: localization and guiding of electromagnetic energy in metal/dielectric structures", *J. Appl. Phys.* **98**, 011101 (2005).

[653] E. Ozbay, "Plasmonics: merging photonics and electronics at nanoscale dimensions", *Science* **311**, 189–193 (2006).

[654] W. Murray and W. Barnes, "Plasmonic materials", *Adv. Mater.* **19**, 3771–3782 (2007).

[655] D. K. Gramotnev and S. I. Bozhevolnyi, "Plasmonics beyond the diffraction limit", *Nat. Phot.* **4**, 83–91 (2010).

[656] J. A. Schuller, E. S. Barnard, W. Cai, Y. C. Jun, J. S. White, and M. L. Brongersma, "Plasmonics for extreme light concentration and manipulation", *Nat. Mater.* **9**, 193–204 (2010).

[657] L. Novotny and B. Hecht, *Principles of Nano-Optics* (New York, Cambridge University Press, 2006).

[658] P. B. Johnson and R. W. Christy, "Optical constants of the noble metals", *Phys. Rev. B* **6**, 4370–4379 (1972).

[659] A. Otto, "Excitation of nonradiative surface plasma waves in silver by the method of frustrated total reflection", *Zeitschrift für Physik A Hadrons and Nuclei* **216**, 398–410 (1968).

[660] E. Kretschmann, "Die Bestimmung optischer Konstanten von Metallen durch Anregung von Oberflächenplasmaschwingungen", *Zeitschrift für Physik A Hadrons and Nuclei* **241**, 313–324 (1971).

[661] J. A. Dionne, H. J. Lezec, and H. A. Atwater, "Highly confined photon transport in subwavelength metallic slot waveguides", *Nano Letters* **6**, 1928–1932 (2006).

[662] G. Mie, "Beiträge zur Optik trüber Medien, speziell kolloidaler Metallösungen", *Ann. Physik* **330**, 377–445 (1908).

[663] K. A. Willets and R. P. Van Duyne, "Localized surface plasmon resonance spectroscopy and sensing", *Ann. Rev. Phys. Chem.* **58**, 267–297 (2007).

[664] M. Pelton, J. Aizpurua, and G. Bryant, "Metal-nanoparticle plasmonics", *Laser Phot. Rev.* **2**, 136–159 (2008).

[665] N. Calander and M. Willander, "Theory of surface-plasmon resonance optical-field enhancement at prolate spheroids", *J. Appl. Phys.* **92**, 4878–4884 (2002).

[666] C. F. Bohren and D. R. Huffman, *Absorption and Scattering of Light by Small Particles* (New York, John Wiley & Sons, 2012).

[667] K. L. Kelly, E. Coronado, L. L. Zhao, and G. C. Schatz, "The optical properties of metal nanoparticles: the influence of size, shape, and dielectric environment", *J. Phys. Chem. B* **107**, 668–677 (2003).

[668] J. P. Kottmann, O. J. Martin, D. R. Smith, and S. Schultz, "Dramatic localized electromagnetic enhancement in plasmon resonant nanowires", *Chem. Phys. Lett.* **341**, 1–6 (2001).

[669] E. Prodan, C. Radloff, N. J. Halas, and P. Nordlander, "A hybridization model for the plasmon response of complex nanostructures", *Science* **302**, 419–422 (2003).

[670] P. Nordlander, C. Oubre, E. Prodan, K. Li, and M. I. Stockman, "Plasmon hybridization in nanoparticle dimers", *Nano Lett.* **4**, 899–903 (2004).

[671] J. B. Lassiter, H. Sobhani, J. A. Fan, *et al.*, "Fano resonances in plasmonic nanoclusters: geometrical and chemical tunability", *Nano Lett.* **10**, 3184–3189 (2010).

[672] B. Luk'yanchuk, N. I. Zheludev, S. A. Maier, *et al.*, "The Fano resonance in plasmonic nanostructures and metamaterials", *Nat. Mat.* **9**, 707–715 (2010).

[673] R. Berndt, J. K. Gimzewski, and P. Johansson, "Inelastic tunneling excitation of tip-induced plasmon modes on noble-metal surfaces", *Phys. Rev. Lett.* **67**, 3796–3799 (1991).

[674] A. L. Demming, F. Festy, and D. Richards, "Plasmon resonances on metal tips: understanding tip-enhanced Raman scattering", *J. Chem. Phys.* **122**, 184716 (2005).

[675] R. Berndt, J. K. Gimzewski, and P. Johansson, "Electromagnetic interactions of metallic objects in nanometer proximity", *Phys. Rev. Lett.* **71**, 3493–3496 (1993).

[676] M. Fleischmann, P. Hendra, and A. McQuillan, "Raman spectra of pyridine adsorbed at a silver electrode", *Chem. Phys. Lett.* **26**, 163–166 (1974).

[677] D. L. Jeanmaire and R. P. V. Duyne, "Surface Raman spectroelectrochemistry. Part I. Heterocyclic, aromatic, and aliphatic amines adsorbed on the anodized silver electrode", *J. Electroanalytical Chemistry and Interfacial Electrochemistry* **84**, 1–20 (1977).

[678] C. Raman and K. Krishnan, "A new type of secondary radiation", *Nature* **121**, 501–502 (1928).

[679] P. W. Atkins and R. S. Friedman, *Molecular Quantum Mechanics*, Third edn. (New York, Oxford University Press, 1997).

[680] M. A. Ratner and G. C. Schatz, *Quantum Mechanics in Chemistry* (New York, Dover, Mineola, 2002).

[681] H. Wei, F. Hao, Y. Huang, W. Wang, P. Nordlander, and H. Xu, "Polarization dependence of surface-enhanced raman scattering in gold nanoparticle nanowire systems", *Nano Lett.* **8**, 2497–2502 (2008).

[682] K. Kneipp, Y. Wang, H. Kneipp, *et al.*, "Single molecule detection using surface-enhanced raman scattering (SERS)", *Phys. Rev. Lett.* **78**, 1667–1670 (1997).

[683] S. Nie and S. R. Emory, "Probing single molecules and single nanoparticles by surface-enhanced raman scattering", *Science* **275**, 1102–1106 (1997).

[684] P. G. Etchegoin and E. C. Le Ru, "A perspective on single molecule SERS: current status and future challenges", *Phys. Chem. Chem. Phys.* **10**, 6079–6089 (2008).

[685] L. Jensen, C. M. Aikens, and G. C. Schatz, "Electronic structure methods for studying surface-enhanced Raman scattering", *Chem. Soc. Rev.* **37**, 1061–1073 (2008).

[686] M. J. Banholzer, J. E. Millstone, L. Qin, and C. A. Mirkin, "Rationally designed nanostructures for surface-enhanced Raman spectroscopy", *Chem. Soc. Rev.* **37**, 885–897 (2008).

[687] G. V. P. Kumar, "Plasmonic nano-architectures for surface enhanced Raman scattering: a review", *J. Nanophotonics* **6**, 064503 (2012).

[688] J. Jiang, K. Bosnick, M. Maillard, and L. Brus, "Single molecule Raman spectroscopy at the junctions of large Ag nanocrystals", *J. Phys. Chem. B* **107**, 9964–9972 (2003).

[689] D. R. Ward, N. J. Halas, J. W. Ciszek, *et al.*, "Simultaneous measurements of electronic conduction and Raman response in molecular junctions", *Nano Lett.* **8**, 919–924 (2008).

[690] J. Zuloaga, E. Prodan, and P. Nordlander, "Quantum description of the plasmon resonances of a nanoparticle dimer", *Nano Lett.* **9**, 887–891 (2009).

[691] R. M. Stöckle, Y. D. Suh, V. Deckert, and R. Zenobi, "Nanoscale chemical analysis by tip-enhanced Raman spectroscopy", *Chem. Phys. Lett.* **318**, 131–136 (2000).

[692] B. Pettinger, P. Schambach, C. J. Villagmez, and N. Scott, "Tip-enhanced Raman spectroscopy: near-fields acting on a few molecules", *Ann. Rev. Phys. Chem.* **63**, 379–399 (2012).

[693] F. Neubrech, A. Pucci, T. W. Cornelius, S. Karim, A. García-Etxarri, and J. Aizpurua, "Resonant plasmonic and vibrational coupling in a tailored nanoantenna for infrared detection", *Phys. Rev. Lett.* **101**, 157403 (2008).

[694] E. Fort and S. Grésillon, "Surface enhanced fluorescence", *J. Phys. D: Appl. Phys.* **41**, 013001 (2008).

[695] P. Mühlschlegel, H.-J. Eisler, O. J. F. Martin, B. Hecht, and D. W. Pohl, "Resonant optical antennas", *Science* **308**, 1607–1609 (2005).

[696] L. Novotny, "Effective wavelength scaling for optical antennas", *Phys. Rev. Lett.* **98**, 266802 (2007).

[697] R. D. Bhat, N. C. Panoiu, S. R. Brueck, and R. M. Osgood, "Enhancing the signal-to-noise ratio of an infrared photodetector with a circular metal grating", *Opt. Express* **16**, 4588–4596 (2008).

[698] L. Tang, S. E. Kocabas, S. Latif, *et al.*, "Nanometre-scale germanium photodetector enhanced by a near-infrared dipole antenna", *Nat. Phot.* **2**, 226–229 (2008).

[699] H. A. Atwater and A. Polman, "Plasmonics for improved photovoltaic devices", *Nature Mater.* **9**, 205–213 (2010).

[700] N. Yu, E. Cubukcu, L. Diehl, *et al.*, "Plasmonic quantum cascade laser antenna", *Appl. Phys. Lett.* **91**, 173113 (2007).

[701] T. H. Taminiau, F. D. Stefani, and N. F. van Hulst, "Enhanced directional excitation and emission of single emitters by a nano-optical Yagi-Uda antenna", *Opt. Express* **16**, 10858–10866 (2008).

[702] J. N. Farahani, D. W. Pohl, H.-J. Eisler, and B. Hecht, "Single quantum dot coupled to a scanning optical antenna: a tunable superemitter", *Phys. Rev. Lett.* **95**, 017402 (2005).

[703] A. Kinkhabwala, Z. Yu, S. Fan, Y. Avlasevich, K. Muellen, and W. E. Moerner, "Large single-molecule fluorescence enhancements produced by a bowtie nanoantenna", *Nature Phot.* **3**, 654–657 (2009).

[704] M. A. Noginov, G. Zhu, A. M. Belgrave, *et al.*, "Demonstration of a spaser-based nanolaser", *Nature* **460**, 1110–U68 (2009).

[705] R. F. Oulton, V. J. Sorger, T. Zentgraf, *et al.*, "Plasmon lasers at deep subwavelength scale", *Nature* **461**, 629–632 (2009).

[706] P. J. Schuck, D. P. Fromm, A. Sundaramurthy, G. S. Kino, and W. E. Moerner, "Improving the mismatch between light and nanoscale objects with gold bowtie nanoantennas", *Phys. Rev. Lett.* **94**, 017402 (2005).

[707] H. A. Bethe, "Theory of diffraction by small holes", *Phys. Rev.* **66**, 163–182 (1944).

[708] T. Ebbesen, H. Lezec, H. Ghaemi, T. Thio, and P. Wolff, "Extraordinary optical transmission through sub-wavelength hole arrays", *Nature* **391**, 667–669 (1998).

[709] L. Martín-Moreno, F. J. García-Vidal, H. J. Lezec, *et al.*, "Theory of extraordinary optical transmission through subwavelength hole arrays", *Phys. Rev. Lett.* **86**, 1114–1117 (2001).

[710] K. J. K. Koerkamp, S. Enoch, F. B. Segerink, N. F. van Hulst, and L. Kuipers, "Strong influence of hole shape on extraordinary transmission through periodic arrays of subwavelength holes", *Phys. Rev. Lett.* **92**, 183901 (2004).

[711] Z. Ruan and M. Qiu, "Enhanced transmission through periodic arrays of subwavelength holes: the role of localized waveguide resonances", *Phys. Rev. Lett.* **96**, 233901 (2006).

[712] V. G. Veselago, "The electrodynamics of substances with simultaneously negative values of ϵ and μ", *Sov. Phys. Uspekhi* **10**, 509 (1968).

[713] D. R. Smith, W. J. Padilla, D. C. Vier, S. C. Nemat-Nasser, and S. Schultz, "Composite medium with simultaneously negative permeability and permittivity", *Phys. Rev. Lett.* **84**, 4184–4187 (2000).

[714] R. A. Shelby, D. R. Smith, and S. Schultz, "Experimental verification of a negative index of refraction", *Science* **292**, 77–79 (2001).

[715] V. M. Shalaev, "Optical negative-index metamaterials", *Nat. Phot.* **1**, 41–48 (2007).

[716] N. Fang, H. Lee, C. Sun, and X. Zhang, "Sub-diffraction-limited optical imaging with a silver superlens", *Science* **308**, 534–537 (2005).

[717] S. Kawata, Y. Inouye, and P. Verma, "Plasmonics for near-field nano-imaging and superlensing", *Nat. Phot.* **3**, 388–394 (2009).

[718] J. B. Pendry, "Negative refraction makes a perfect lens", *Phys. Rev. Lett.* **85**, 3966–3969 (2000).

[719] R. W. Ziolkowski and E. Heyman, "Wave propagation in media having negative permittivity and permeability", *Phys. Rev. E* **64**, 056625 (2001).

[720] S. Xiao, V. P. Drachev, A. V. Kildishev, *et al.*, "Loss-free and active optical negative-index metamaterials", *Nature* **466**, 735–738 (2010).

[721] D. Schurig, J. J. Mock, B. J. Justice, *et al.*, "Metamaterial electromagnetic cloak at microwave frequencies", *Science* **314**, 977–980 (2006).

[722] T. Ergin, N. Stenger, P. Brenner, J. B. Pendry, and M. Wegener, "Three-dimensional invisibility cloak at optical wavelengths", *Science* **328**, 337–339 (2010).

[723] V. M. Shalaev, "Transforming light", *Science* **322**, 384–386 (2008).

[724] U. Leonhardt, "Optical conformal mapping", *Science* **312**, 1777–1780 (2006).

[725] J. B. Pendry, D. Schurig, and D. R. Smith, "Controlling electromagnetic fields", *Science* **312**, 1780–1782 (2006).

[726] C. López, "Materials aspects of photonic crystals", *Adv. Mater.* **15**, 1679–1704 (2003).

[727] F. J. García de Abajo, "Colloquium: Light scattering by particle and hole arrays", *Rev. Mod. Phys.* **79**, 1267–1290 (2007).

[728] K. Y. Bliokh, Y. P. Bliokh, V. Freilikher, S. Savelév, and F. Nori, "Colloquium: Unusual resonators: plasmonics, metamaterials, and random media", *Rev. Mod. Phys.* **80**, 1201–1213 (2008).

[729] S. A. Ramakrishna, "Physics of negative refractive index materials", *Rep. Progr. Phys.* **68**, 449 (2005).

[730] Y. Liu and X. Zhang, "Metamaterials: a new frontier of science and technology", *Chem. Soc. Rev.* **40**, 2494–2507 (2011).

[731] S. H. Crandall, D. C. Karnopp, E. F. Kunter Jr., and D. C. Pridmore-Brown, *Dynamics of Mechanical and Electromechanical Systems* (Malabar, Florida, Kriegar Publishing, 1982).

[732] J. Hirth and J. Lothe, *Theory of Dislocations* (Wiley, 1982).

[733] F. C. Frank and W. T. Read, "Multiplication processes for slow moving dislocations", *Phys. Rev.* **79**, 722–723 (1950).

[734] B. Yang, C. Motz, W. Grosinger, and G. Dehm, "Stress-controlled fatigue behaviour of micro-sized polycrystalline copper wires", *Mater. Sci. Eng. A* **515**, 71–78 (2009).

[735] T. Nieh, J. Wadsworth, and O. Sherby, *Superplasticity in Metals and Ceramics*, Cambridge Solid State Science Series (Cambridge, Cambridge University Press, 2005).

[736] Z. Budrovic, H. Van Swygenhoven, P. M. Derlet, S. Van Petegem, and B. Schmitt, "Plastic deformation with reversible peak broadening in nanocrystalline nickel", *Science* **304**, 273–276 (2004).

[737] L. Lu, M. Sui, and K. Lu, "Superplastic extensibility of nanocrystalline copper at room temperature", *Science* **287**, 1463–1466 (2000).

[738] X. H. Liu, L. Q. Zhang, L. Zhong, *et al.*, "Ultrafast electrochcmical lithiation of individual Si nanowire anodes", *Nano Lett.* **11**, 2251–2258 (2011).

[739] B. Bhushan, *Introduction to Tribology*, Tribology in Practice Series (New York, Wiley, 2013).

[740] J. Krim, "Resource letter: FMMLS-1: Friction at macroscopic and microscopic length scales", *Amer. J. Phys.* **70**, 890–897 (2002).

[741] J. Krim, "Friction and energy dissipation mechanisms in adsorbed molecules and molecularly thin films", *Adv. Phys.* **61**, 155–323 (2012).

[742] O. Braun and A. Naumovets, "Nanotribology: microscopic mechanisms of friction", *Surf. Sci. Rep.* **60**, 79–158 (2006).

[743] F. Bowden and D. Tabor, *The Friction and Lubrication of Solids*, number 1 in Oxford classic texts in the physical sciences (Oxford, Clarendon Press, 2001).

[744] J. A. Greenwood and J. B. P. Williamson, "Contact of nominally flat surfaces", *Proc. R. Soc. London. Series A. Math. Phys. Sci.* **295**, 300–319 (1966).

[745] C. Daly and J. Krim, "Sliding friction of solid xenon monolayers and bilayers on Ag(111)", *Phys. Rev. Lett.* **76**, 803–806 (1996).

[746] M. Cieplak, E. D. Smith, and M. O. Robbins, "Molecular origins of friction: the force on adsorbed layers", *Science* **265**, 1209–1212 (1994).

[747] M. S. Tomassone, J. B. Sokoloff, A. Widom, and J. Krim, "Dominance of phonon friction for a xenon film on a silver (111) surface", *Phys. Rev. Lett.* **79**, 4798–4801 (1997).

[748] A. Dayo, W. Alnasrallah, and J. Krim, "Superconductivity-dependent sliding friction", *Phys. Rev. Lett.* **80**, 1690–1693 (1998).

[749] M. Kisiel, E. Gnecco, U. Gysin, L. Marot, S. Rast, and E. Meyer, "Suppression of electronic friction on Nb films in the superconducting state", *Nat. Mater.* **10**, 119–122 (2011).

[750] M. Falvo, R. Taylor I, A. Helser, *et al.*, "Nanometre-scale rolling and sliding of carbon nanotubes", *Nature* **397**, 236–238 (1999).

[751] M. R. Falvo, J. Steele, R. M. Taylor, and R. Superfine, "Gearlike rolling motion mediated by commensurate contact: carbon nanotubes on HOPG", *Phys. Rev. B* **62**, R10665–R10667 (2000).

[752] J. Y. Park, D. Ogletree, M. Salmeron, *et al.*, "High frictional anisotropy of periodic and aperiodic directions on a quasicrystal surface", *Science* **309**, 1354–1356 (2005).

[753] M. Hirano, K. Shinjo, R. Kaneko, and Y. Murata, "Observation of superlubricity by scanning tunneling microscopy", *Phys. Rev. Lett.* **78**, 1448–1451 (1997).

[754] P. W. Mertens, T. Bearda, M. Wada, and A. Pacco, "Drying of high aspect ratio structures: a comparison of drying techniques via electrical stiction analysis", *Solid State Phenomena* **145**, 87–90 (2009).

[755] H. Namatsu, K. Yamazaki, and K. Kurihara, "Supercritical drying for nanostructure fabrication without pattern collapse", *Microelectronic Engineering* **46**, 129–132 (1999).

[756] K. Van Schuylenbergh, C. L. Chua, D. K. Fork, J.-P. Lu, and B. Griffiths, "On-chip out-of-plane high-Q inductors", in *Proceedings of IEEE Lester Eastman Conference on High Performance Devices, 2002*, 364–373 (IEEE, 2002).

[757] D. J. Bishop, C. R. Giles, and G. P. Austin, "The Lucent LambdaRouter: MEMS technology of the future here today", IEEE *Comm. Mag.*, **40**, 75–79 (2002).

[758] C.-C. Nguyen, "High-Q micromechanical oscillators and filters for communications", in *Proceedings of 1997 IEEE International Symposium on Circuits and Systems, 1997. ISCAS'97.*, volume 4, 2825–2828 (IEEE, 1997).

[759] E. Mounier and L. Robin, "Status of the MEMS industry", Technical report, Yole Développement (2012).

[760] C. Leondes, *MEMS/NEMS: Handbook Techniques and Applications* (New York, Springer Science+Business Media, Incorporated, 2006).

[761] N. Yazdi, F. Ayazi, and K. Najafi, "Micromachined inertial sensors", *Proc. IEEE* **86**, 1640–1659 (1998).

[762] K. Wang and C.-C. Nguyen, "High-order micromechanical electronic filters", in *Proceedings of Tenth Annual International Workshop on Micro Electro Mechanical Systems, 1997. MEMS'97*, 25–30 (IEEE, 1997).

[763] D. Neilson, V. Aksyuk, S. Arney, *et al.*, "Fully provisioned 112×112 micromechanical optical crossconnect with 35.8 Tb/s demonstrated capacity", in *Optical Fiber Communication Conference*, (Washington, DC, Optical Society of America, 2000).

[764] R. Arensman, "At long last MEMS", *EDN Newsletter* (2002).

[765] L. J. Hornbeck, "Digital light processing and MEMS: reflecting the digital display needs of the networked society", in *Lasers, Optics, and Vision for Productivity in Manufacturing I*, 2–13 (Belliagham, WA, International Society for Optics and Photonics, 1996).

[766] R. G. Baraniuk, "Compressive sensing", *IEEE Signal Processing Magazine*, **24**, 118–121 (2007).

[767] M. F. Duarte, M. A. Davenport, D. Takhar, *et al.*, "Single-pixel imaging via compressive sampling", *IEEE Signal Processing Magazine*, **25**, 83–91 (2008).

[768] D. M. Bloom, "Grating light valve: revolutionizing display technology", in *Electronic Imaging '97*, 165–171 (Belliangham, WA, International Society for Optics and Photonics, 1997).

[769] C. Menzel, A. Bibl, and P. Hoisington, "MEMS solutions for precision micro-fluidic dispensing application", Technical report, Spectra, Inc. (2012).

[770] A. Naik, M. Hanay, W. Hiebert, X. Feng, and M. Roukes, "Towards single-molecule nanomechanical mass spectrometry", *Nature Nanotech.* **4**, 445–450 (2009).

[771] J. Chaste, A. Eichler, J. Moser, G. Ceballos, R. Rurali, and A. Bachtold, "A nanomechanical mass sensor with yoctogram resolution", *Nature Nanotech.* **7**, 301–304 (2012).

[772] I. Wilson-Rae, "Intrinsic dissipation in nanomechanical resonators due to phonon tunneling", *Phys. Rev. B* **77**, 245418 (2008).

[773] P. Mohanty, D. Harrington, K. Ekinci, Y. Yang, M. Murphy, and M. Roukes, "Intrinsic dissipation in high-frequency micromechanical resonators", *Phys. Rev. B* **66**, 085416 (2002).

[774] X. Liu, B. E. White, Jr., R. O. Pohl, *et al.*, "Amorphous solid without low energy excitations", *Phys. Rev. Lett.* **78**, 4418–4421 (1997).

[775] D. R. Southworth, R. A. Barton, S. S. Verbridge, *et al.*, "Stress and silicon nitride: a crack in the universal dissipation of glasses", *Phys. Rev. Lett.* **102**, 225503 (2009).

[776] J. B. Johnson, "Thermal agitation of electricity in conductors", *Nature* **119**, 50–51 (1927).

[777] H. Nyquist, "Thermal agitation of electric charge in conductors", *Phys. Rev.* **32**, 110–113 (1928).

[778] A. Naik, O. Buu, M. LaHaye, *et al.*, "Cooling a nanomechanical resonator with quantum back-action", *Nature* **443**, 193–196 (2006).

[779] M. Poot and H. S. van der Zant, "Mechanical systems in the quantum regime", *Phys. Rep.* **511**, 273–335 (2012).

[780] H. B. Casimir, "On the attraction between two perfectly conducting plates", *Proc. K. Ned. Akad. Wet*, **51**, 150 (1948).

[781] M. Bordag, U. Mohideen, and V. Mostepanenko, "New developments in the Casimir effect", *Phys. Rep.* **353**, 1–205 (2001).

[782] R. L. Jaffe, "Casimir effect and the quantum vacuum", *Phys. Rev. D* **72**, 021301 (2005).

[783] J. N. Munday, F. Capasso, and V. A. Parsegian, "Measured long-range repulsive Casimir–Lifshitz forces", *Nature* **457**, 170–173 (2009).

[784] H. Chan, V. Aksyuk, R. Kleiman, D. Bishop, and F. Capasso, "Quantum mechanical actuation of microelectromechanical systems by the Casimir force", *Science* **291**, 1941–1944 (2001).

[785] S. K. Lamoreaux, "Demonstration of the Casimir force in the 0.6 to 6µm range", *Phys. Rev. Lett.* **78**, 5–8 (1997).

[786] A. D. OConnell, M. Hofheinz, M. Ansmann, *et al.*, "Quantum ground state and single-phonon control of a mechanical resonator", *Nature* **464**, 697–703 (2010).

[787] J. Teufel, T. Donner, D. Li, *et al.*, "Sideband cooling of micromechanical motion to the quantum ground state", *Nature* **475**, 359–363 (2011).

[788] J. Chan, T. M. Alegre, A. H. Safavi-Naeini, J. T. Hill, A. Krause, S. Gröblacher, M. Aspelmeyer, and O. Painter, "Laser cooling of a nanomechanical oscillator into its quantum ground state", *Nature* **478**, 89–92 (2011).

[789] E. Verhagen, S. Deléglise, S. Weis, A. Schliesser, and T. J. Kippenberg, "Quantum-coherent coupling of a mechanical oscillator to an optical cavity mode", *Nature* **482**, 63–67 (2012).

[790] K. L. Ekinci and M. L. Roukes, "Nanoelectromechanical systems", *Rev. Sci. Instr.* **76**, 061101 (2005).

[791] E. Buckingham, "On physically similar systems; illustrations of the use of dimensional equations", *Phys. Rev.* **4**, 345–376 (1914).

[792] L. F. Moody, "Friction factors for pipe flow", *Trans. Asme* **66**, 671–684 (1944).

[793] M. A. Deakin, "GI Taylor and the Trinity test", *International Journal of Mathematical Education in Science and Technology* **42**, 1069–1079 (2011).

[794] E. M. Purcell, "Life at low Reynolds number", *Amer. J. Phys.* **45**, 3–11 (1977).

[795] P. J. Kenis, R. F. Ismagilov, and G. M. Whitesides, "Microfabrication inside capillaries using multiphase laminar flow patterning", *Science* **285**, 83–85 (1999).

[796] E. B. Flint and K. S. Suslick, "The temperature of cavitation", *Science* **253**, 1397–1399 (1991).

[797] B. P. Barber, R. A. Hiller, R. Löfstedt, S. J. Putterman, and K. R. Weninger, "Defining the unknowns of sonoluminescence", *Phys. Rep.* **281**, 65–143 (1997).

[798] W. A. Zisman, "Relation of the equilibrium contact angle to liquid and solid constitution" in F. M. Fowker, ed., *Contact Angle, Wettability, and Adhesion* (New York, American Chemical Society, 1964).

[799] T. Young, "An essay on the cohesion of fluids", *Phil. Trans. R. Soc. Lond.* **95**, 65–87 (1805).

[800] N. M. Holbrook and M. A. Zwieniecki, "Transporting water to the tops of trees", *Physics Today* **61**, 76 (2008).

[801] B. Zhao, J. S. Moore, and D. J. Beebe, "Surface-directed liquid flow inside microchannels", *Science* **291**, 1023–1026 (2001).

[802] K. Ichimura, S.-K. Oh, and M. Nakagawa, "Light-driven motion of liquids on a photoresponsive surface", *Science* **288**, 1624–1626 (2000).

[803] T. Welton, "Room-temperature ionic liquids. Solvents for synthesis and catalysis", *Chem. Rev.* **99**, 2071–2084 (1999).

[804] M. Armand, F. Endres, D. R. MacFarlane, H. Ohno, and B. Scrosati, "Ionic-liquid materials for the electrochemical challenges of the future", *Nat. mater.* **8**, 621–629 (2009).

[805] E. Hart and M. Anbar, *The Hydrated Electron* (Wiley-Interscience, 1970).

[806] D. C. Grahame, "The electrical double layer and the theory of electrocapillarity", *Chem. Rev.* **41**, 441–501 (1947).

[807] H. Von Helmholtz, "Studies of electric boundary layers", *Wied. Ann.* **7**, 337–382 (1879).

[808] O. Stern, "The theory of the electrolytic double-layer", *Zeit. Elektrochem.* **30**, 508–516 (1924).

[809] G. Gouy, "Constitution of the electric charge at the surface of an electrolyte", *J. phys.* **9**, 457–467 (1910).

[810] D. L. Chapman, "LI. A contribution to the theory of electrocapillarity", *The London, Edinburgh and Dublin Philosophical Magazine and Journal of Science* **25**, 475–481 (1913).

[811] H. A. Pohl, "The motion and precipitation of suspensoids in divergent electric fields", *J. Appl. Phys.* **22** (1951).

[812] P. R. Gascoyne and J. Vykoukal, "Particle separation by dielectrophoresis", *Electrophoresis* **23**, 1973 (2002).

[813] R. Pethig, "Review article: Dielectrophoresis: status of the theory, technology, and applications", *Biomicrofluidics* **4**, 022811 (2010).

[814] S. Fiedler, S. G. Shirley, T. Schnelle, and G. Fuhr, "Dielectrophoretic sorting of particles and cells in a microsystem", *Anal. Chem.* **70**, 1909–1915 (1998).

[815] K. D. Hermanson, S. O. Lumsdon, J. P. Williams, E. W. Kaler, and O. D. Velev, "Dielectrophoretic assembly of electrically functional microwires from nanoparticle suspensions", *Science* **294**, 1082–1086 (2001).

[816] G. M. Whitesides, "The origins and the future of microfluidics", *Nature* **442**, 368–373 (2006).

[817] T. M. Squires and S. R. Quake, "Microfluidics: fluid physics at the nanoliter scale", *Rev. Mod. Phys.* **77**, 977 (2005).

[818] H. A. Stone, A. D. Stroock, and A. Ajdari, "Engineering flows in small devices: microfluidics toward a lab-on-a-chip", *Ann. Rev. Fluid Mech.* **36**, 381–411 (2004).

[819] N. Nguyen and S. Wereley, *Fundamentals and Applications of Microfluidics*, Artech House MEMS library (Norwood, MA, Artech House, 2002).

[820] J. R. Anderson, D. T. Chiu, R. J. Jackman, *et al.*, "Fabrication of topologically complex three-dimensional microfluidic systems in PDMS by rapid prototyping", *Anal. Chemi.* **72**, 3158–3164 (2000).

[821] J. M. K. Ng, I. Gitlin, A. D. Stroock, and G. M. Whitesides, "Components for integrated poly(dimethylsiloxane) microfluidic systems", *Electrophoresis* **23**, 3461–3473 (2002).

[822] H. Wu, T. W. Odom, D. T. Chiu, and G. M. Whitesides, "Fabrication of complex three-dimensional microchannel systems in PDMS", *J. Amer. Chem. Soc.* **125**, 554–559 (2003).

[823] J. Narasimhan and I. Papautsky, "Polymer embossing tools for rapid prototyping of plastic microfluidic devices", *J. Micromech. Microeng.* **14**, 96 (2004).

[824] M. A. Unger, H.-P. Chou, T. Thorsen, A. Scherer, and S. R. Quake, "Monolithic microfabricated valves and pumps by multilayer soft lithography", *Science* **288**, 113–116 (2000).

[825] E. Stemme and G. Stemme, "A valveless diffuser/nozzle-based fluid pump", *Sensors and Actuators A: Physical* **39**, 159–167 (1993).

[826] N. Nguyen, "Micromachined flow sensors, a review", *Flow Measurement and Instrumentation* **8**, 7–16 (1997).

[827] S. Wu, Q. Lin, Y. Yuen, and Y.-C. Tai, "MEMS flow sensors for nano-fluidic applications", *Sensors and Actuators A: Physical* **89**, 152–158 (2001), Special Issue: *Micromechanics Section of Sensors and Actuators*, based on contributions revised from the *Technical Digest of the Thirteenth {IEEE} International Workshop on Micro Electro Mechanical Systems (MEMS-2000)*.

[828] J. B. Knight, A. Vishwanath, J. P. Brody, and R. H. Austin, "Hydrodynamic focusing on a silicon chip: mixing nanoliters in microseconds", *Phys. Rev. Lett.* **80**, 3863 (1998).

[829] A. D. Stroock, S. K. Dertinger, A. Ajdari, I. Mezić, H. A. Stone, and G. M. Whitesides, "Chaotic mixer for microchannels", *Science* **295**, 647–651 (2002).

[830] T. Thorsen, S. J. Maerkl, and S. R. Quake, "Microfluidic large-scale integration", *Science* **298**, 580–584 (2002).

[831] C. L. Hansen, E. Skordalakes, J. M. Berger, and S. R. Quake, "A robust and scalable microfluidic metering method that allows protein crystal growth by free interface diffusion", *Proc. Nat. Acad. Sci.* **99**, 16531–16536 (2002).

[832] C. L. Hansen, S. Classen, J. M. Berger, and S. R. Quake, "A microfluidic device for kinetic optimization of protein crystallization and in situ structure determination", *J. Amer. Chem. Soc.* **128**, 3142–3143 (2006).

[833] P. S. Dittrich and A. Manz, "Lab-on-a-chip: microfluidics in drug discovery", *Nat. Rev. Drug Discovery* **5**, 210–218 (2006).

[834] Y. Zhu and S. Granick, "Rate-dependent slip of newtonian liquid at smooth surfaces", *Phys. Rev. Lett.* **87**, 096105 (2001).

[835] F. Taherian, V. Marcon, N. F. van der Vegt, and F. Leroy, "What is the contact angle of water on graphene?", *Langmuir* **29**, 1457–1465 (2013).

[836] M. Majumder, N. Chopra, R. Andrews, and B. J. Hinds, "Nanoscale hydrodynamics: enhanced flow in carbon nanotubes", *Nature* **438**, 44–44 (2005).

[837] J. K. Holt, H. G. Park, Y. Wang, *et al.*, "Fast mass transport through sub-2-nanometer carbon nanotubes", *Science* **312**, 1034–1037 (2006).

[838] K. K. Lau, J. Bico, K. B. Teo, *et al.*, "Superhydrophobic carbon nanotube forests", *Nano Lett.* **3**, 1701–1705 (2003).

[839] P. Joseph, C. Cottin-Bizonne, J.-M. Benoit, *et al.*, "Slippage of water past superhydrophobic carbon nanotube forests in microchannels", *Phys. Rev. Lett.* **97**, 156104 (2006).

[840] A. Tuteja, W. Choi, M. Ma, *et al.*, "Designing superoleophobic surfaces", *Science* **318**, 1618–1622 (2007).

[841] S. Pan, A. K. Kota, J. M. Mabry, and A. Tuteja, "Superomniphobic surfaces for effective chemical shielding", *J. Amer. Chem. Soc.* **135**, 578–581 (2012).

[842] M. Heuberger, M. Zäch, and N. Spencer, "Density fluctuations under confinement: when is a fluid not a fluid?", *Science* **292**, 905–908 (2001).

[843] Y. Zhu and S. Granick, "Viscosity of interfacial water", *Phys. Rev. Lett.* **87**, 096104 (2001).

[844] N. E. Levinger, "Water in confinement", *Science* **298**, 1722–1723 (2002).

[845] J. E. Boyd, A. Briskman, V. L. Colvin, and D. M. Mittleman, "Direct observation of terahertz surface modes in nanometer-sized liquid water pools", *Phys. Rev. Lett.* **87**, 147401 (2001).

[846] J. C. Burton, J. E. Rutledge, and P. Taborek, "Fluid pinch-off dynamics at nanometer length scales", *Phys. Rev. Lett.* **92**, 244505 (2004).

[847] H. Cao, J. O. Tegenfeldt, R. H. Austin, and S. Y. Chou, "Gradient nanostructures for interfacing microfluidics and nanofluidics", *Appl. Phys. Lett.* **81**, 3058–3060 (2002).

[848] L. R. Huang, E. C. Cox, R. H. Austin, and J. C. Sturm, "Continuous particle separation through deterministic lateral displacement", *Science* **304**, 987–990 (2004).

[849] L.-J. Cheng and L. J. Guo, "Ionic current rectification, breakdown, and switching in heterogeneous oxide nanofluidic devices", *ACS Nano* **3**, 575–584 (2009).

[850] R. B. Schoch, J. Han, and P. Renaud, "Transport phenomena in nanofluidics", *Rev. Mod. Phys.* **80**, 839–883 (2008).

[851] E. Arunan, G. R. Desiraju, R. A. Klein, *et al.*, "Defining the hydrogen bond: An account (IUPAC Technical Report)", *Pure Appl. Chem.* **83** (2011).

[852] S. Suresh and V. Naik, "Hydrogen bond thermodynamic properties of water from dielectric constant data", *J. Chem. Phys.* **113**, 9727–9732 (2000).

[853] P. C. Weber, D. Ohlendorf, J. Wendoloski, and F. Salemme, "Structural origins of high-affinity biotin binding to streptavidin", *Science* **243**, 85–88 (1989).

[854] E. P. Diamandis and T. K. Christopoulos, "The biotin-(strept) avidin system: principles and applications in biotechnology." *Clinical Chemistry* **37**, 625–636 (1991).

[855] Y. Cui, Q. Wei, H. Park, and C. M. Lieber, "Nanowire nanosensors for highly sensitive and selective detection of biological and chemical species", *Science* **293**, 1289–1292 (2001).

[856] K. Caswell, J. N. Wilson, U. H. Bunz, and C. J. Murphy, "Preferential end-to-end assembly of gold nanorods by biotin-streptavidin connectors", *J. Amer. Chem. Soc.* **125**, 13914–13915 (2003).

[857] N. C. Seeman, "Nucleic acid junctions and lattices", *J. Theor. Biol.* **99**, 237–247 (1982).

[858] N. C. Seeman, "DNA nanotechnology: novel DNA constructions", *Ann. Rev. Biophy. Biomole. Struc.* **27**, 225–248 (1998).

[859] N. C. Seeman, "Nanomaterials based on DNA". *Ann. Rev. Biochem.* **79**, 65–87 (2009).

[860] A. V. Pinheiro, D. Han, W. M. Shih, and H. Yan, "Challenges and opportunities for structural DNA nanotechnology", *Nature Nanotech.* **6**, 763–772 (2011).

[861] N. C. Seeman, "Nanotechnology and the double helix", *Scientific American* **290**, 64 (2004).

[862] K. A. Dill, S. B. Ozkan, M. S. Shell, and T. R. Weikl, "The protein folding problem", *Ann. Rev. Biophys.* **37**, 289 (2008).

[863] V. N. Uversky, "Natively unfolded proteins: a point where biology waits for physics", *Protein Science* **11**, 739–756 (2002).

[864] D. Goodsell, "ATP synthase: December 2005 molecule of the month", *RCSB Protein Data Bank* (2005).

[865] R. K. Nakamoto, J. A. B. Scanlon, and M. K. Al-Shawi, "The rotary mechanism of the ATP synthase", *Archives of Biochemistry and Biophysics* **476**, 43–50 (2008), Special Issue: *Transport ATPases*.

[866] D. Goodsell, "Calcium pump: March 2004 molecule of the month", *RCSB Protein Data Bank* (2004).

[867] D. W. Miller and K. A. Dill, "Ligand binding to proteins: the binding landscape model", *Protein Science* **6**, 2166–2179 (1997).

[868] E. Sackmann, "Supported membranes: scientific and practical applications", *Science* **271**, 43–48 (1996).

[869] S. G. Boxer, "Molecular transport and organization in supported lipid membranes", *Current Opinion in Chemical Biology* **4**, 704–709 (2000).

[870] M. G. Rossmann, F. Arisaka, A. J. Battisti, *et al.*, "From structure of the complex to understanding of the biology", *Acta Crystallographica Section D* **63**, 9–16 (2007).

[871] M. Ploss and A. Kuhn, "Kinetics of filamentous phage assembly", *Phys. Biol.* **7**, 045002 (2010).

[872] D. Goodsell, "Tobacco mosaic virus: January 2009 molecule of the month", *RCSB Protein Data Bank* (2009).

[873] A. S. Blum, C. M. Soto, C. D. Wilson, *et al.*, "Cowpea mosaic virus as a scaffold for 3-D patterning of gold nanoparticles", *Nano Lett.* **4**, 867–870 (2004).

[874] J. C. Falkner, M. E. Turner, J. K. Bosworth, *et al.*, "Virus crystals as nanocomposite scaffolds", *J. Amer. Chem. Soc.* **127**, 5274–5275 (2005).

[875] Y. Huang, C.-Y. Chiang, S. K. Lee, *et al.*, "Programmable assembly of nanoarchitectures using genetically engineered viruses", *Nano Lett.* **5**, 1429–1434 (2005).

[876] S. R. Whaley, D. English, E. L. Hu, P. F. Barbara, and A. M. Belcher, "Selection of peptides with semiconductor binding specificity for directed nanocrystal assembly", *Nature* **405**, 665–668 (2000).

[877] D. Nathans and H. O. Smith, "Restriction endonucleases in the analysis and restructuring of DNA molecules", *Ann. Rev. Biochem.* **44**, 273–293 (1975).

[878] H. Katada and M. Komiyama, "Artificial restriction DNA cutters as new tools for gene manipulation", *Chem. Bio. Chem.* **10**, 1279–1288 (2009).

[879] R. K. Saiki, D. H. Gelfand, S. Stoffel, *et al.*, "Primer-directed enzymatic amplification of DNA with a thermostable DNA polymerase", *Science* **239**, 487–491 (1988).

[880] M. G. Van den Heuvel and C. Dekker, "Motor proteins at work for nanotechnology", *Science* **317**, 333–336 (2007).

[881] R. K. Soong, G. D. Bachand, H. P. Neves, A. G. Olkhovets, H. G. Craighead, and C. D. Montemagno, "Powering an inorganic nanodevice with a biomolecular motor", *Science* **290**, 1555–1558 (2000).

[882] Y. Hiratsuka, T. Tada, K. Oiwa, T. Kanayama, and T. Q. Uyeda, "Controlling the direction of kinesin-driven microtubule movements along microlithographic tracks", *Biophys. J.* **81**, 1555–1561 (2001).

[883] J. Clemmens, H. Hess, R. Doot, C. M. Matzke, G. D. Bachand, and V. Vogel, "Motor-protein roundabouts: microtubules moving on kinesin-coated tracks through engineered networks", *Lab on a Chip* **4**, 83–86 (2004).

[884] P. W. Rothemund, "Folding DNA to create nanoscale shapes and patterns", *Nature* **440**, 297–302 (2006).

[885] T. Tørring, N. V. Voigt, J. Nangreave, H. Yan, and K. V. Gothelf, "DNA origami: a quantum leap for self-assembly of complex structures", *Chem. Soc. Rev.* **40**, 5636–5646 (2011).

[886] S. M. Douglas, I. Bachelet, and G. M. Church, "A logic-gated nanorobot for targeted transport of molecular payloads", *Science* **335**, 831–834 (2012).

[887] R. Iinuma, Y. Ke, R. Jungmann, T. Schlichthaerle, J. B. Woehrstein, and P. Yin, "Polyhedra self-assembled from DNA tripods and characterized with 3D DNA-PAINT", *Science* **344**, 65–69 (2014).

[888] S. M. Douglas, H. Dietz, T. Liedl, B. Högberg, F. Graf, and W. M. Shin, "Self-assembly of DNA into nanoscale three-dimensional shapes", *Nature.* **459**, 414–418 (2009).

[889] B. Ding, Z. Deng, H. Yan, S. Cabrini, R. N. Zuckermann, and J. Bokor, "Gold nanoparticle self-similar chain structure organized by DNA origami", *J. Amer. Chem. Soc.* **132**, 3248–3249 (2010).

[890] A. M. Hung, C. M. Micheel, L. D. Bozano, L. W. Osterbur, G. M. Wallraff, and J. N. Cha, "Large-area spatially ordered arrays of gold nanoparticles directed by lithographically confined DNA origami", *Nat. Nanotechn.* **5**, 121–126 (2010).

[891] H. Bui, C. Onodera, C. Kidwell, *et al.*, "Programmable periodicity of quantum dot arrays with DNA origami nanotubes", *Nano Lett.* **10**, 3367–3372 (2010).

[892] G. P. Smith, "Filamentous fusion phage: novel expression vectors that display cloned antigens on the virion surface", *Science* **228**, 1315–1317 (1985).

[893] A. Merzlyak and S.-W. Lee, "Phage as templates for hybrid materials and mediators for nanomaterial synthesis", *Current Opinion in Chemical Biology* **10**, 246–252 (2006).

[894] K. T. Nam, D.-W. Kim, P. J. Yoo, *et al.*, "Virus-enabled synthesis and assembly of nanowires for lithium ion battery electrodes", *Science* **312**, 885–888 (2006).

[895] Y. S. Nam, A. P. Magyar, *et al.*, "Biologically templated photocatalytic nanostructures for sustained light-driven water oxidation", *Nat. Nanotech.* **5**, 340–344 (2010).

[896] J. C. Jamieson, "Phase equilibrium in the system calcite-aragonite", *J. Chem. Phys.* **21**, 1385–1390 (1953).

[897] L. Addadi and S. Weiner, "Control and design principles in biological mineralization", *Angewandte Chemie*, International edition in English **31**, 153–169 (1992).

[898] G. Falini, S. Albeck, S. Weiner, and L. Addadi, "Control of aragonite or calcite polymorphism by mollusk shell macromolecules", *Science* **271**, 67–69 (1996).

[899] J. Aizenberg, A. J. Black, and G. M. Whitesides, "Control of crystal nucleation by patterned self-assembled monolayers", *Nature* **398**, 495–498 (1999).

[900] J. Aizenberg, "Crystallization in patterns: a bio-inspired approach", *Adv. Mater.* **16**, 1295–1302 (2004).

[901] J. Aizenberg, "A bio-inspired approach to controlled crystallization at the nanoscale", *Bell Labs Technical Journal* **10**, 129–141 (2005).

[902] P. Maderson, "Keratinized epidermal derivatives as an aid to climbing in gekkonid lizards", *Nature* **203**, 780–781 (1964).

[903] R. Ruibal and V. Ernst, "The structure of the digital setae of lizards", *J. Morphology* **117**, 271–293 (1965).

[904] K. Autumn, Y. A. Liang, S. T. Hsieh, *et al.*, "Adhesive force of a single gecko foot-hair", *Nature* **405**, 681–685 (2000).

[905] M. T. Northen and K. L. Turner, "A batch fabricated biomimetic dry adhesive", *Nanotech.* **16**, 1159 (2005).

[906] A. Geim, S. Dubonos, I. Grigorieva, K. Novoselov, A. Zhukov, and S. Y. Shapoval, "Microfabricated adhesive mimicking gecko foot-hair", *Nat. Mater.* **2**, 461–463 (2003).

[907] B. Yurdumakan, N. R. Raravikar, P. M. Ajayan, and A. Dhinojwala, "Synthetic gecko foot-hairs from multiwalled carbon nanotubes", *Chem. Commu.* 3799–3801 (2005).

[908] M. Sitti and R. S. Fearing, "Synthetic gecko foot-hair micro/nano-structures as dry adhesives", *J. Adhesion Science and Technology* **17**, 1055–1073 (2003).

[909] L. Ge, S. Sethi, L. Ci, P. M. Ajayan, and A. Dhinojwala, "Carbon nanotube-based synthetic gecko tapes", *Proc. Nat. Acad. Sci.* **104**, 10792–10795 (2007).

[910] A. Merkoci, *Biosensing Using Nanomaterials*, Wiley Nanoscience and Nanotechnology Series (New York, Wiley, 2009).

[911] S. Li, J. Singh, H. Li, and I. Banerjee, *Biosensor Nanomaterials* (New York, Wiley, 2011).

[912] J. Li and N. Wu, *Biosensors Based on Nanomaterials and Nanodevices*, Nanomaterials and their Applications (New York, Taylor & Francis, 2013).

[913] T. Förster, "10th Spiers Memorial Lecture. Transfer mechanisms of electronic excitation", *Discussions of the Faraday Society* **27**, 7–17 (1959).

[914] P. R. Selvin, "The renaissance of fluorescence resonance energy transfer." *Nat. Struct. Biol.* **7**, 730–734 (2000).

[915] I. L. Medintz, H. T. Uyeda, E. R. Goldman, and H. Mattoussi, "Quantum dot bioconjugates for imaging, labelling and sensing", *Nat. Mater.* **4**, 435–446 (2005).

[916] X. Michalet, F. F. Pinaud, L. A. Bentolila, *et al.*, "Quantum dots for live cells, in vivo imaging, and diagnostics", *Science* **307**, 538–544 (2005).

[917] I. L. Medintz, A. R. Clapp, H. Mattoussi, E. R. Goldman, B. Fisher, and J. M. Mauro, "Self-assembled nanoscale biosensors based on quantum dot FRET donors", *Nat. Mater.* **2**, 630–638 (2003).

[918] M. R. Lee and P. M. Fauchet, "Two-dimensional silicon photonic crystal based biosensing platform for protein detection", *Optics Express* **15**, 4530–4535 (2007).

[919] S. Arnold, M. Khoshsima, I. Teraoka, S. Holler, and F. Vollmer, "Shift of whispering-gallery modes in microspheres by protein adsorption", *Optics Lett.* **28**, 272–274 (2003).

[920] A. J. Haes, S. Zou, G. C. Schatz, and R. P. Van Duyne, "Nanoscale optical biosensor: short range distance dependence of the localized surface plasmon resonance of noble metal nanoparticles", *J. Phys. Chem. B* **108**, 6961–6968 (2004).

[921] J. Fritz, E. B. Cooper, S. Gaudet, P. K. Sorger, and S. R. Manalis, "Electronic detection of DNA by its intrinsic molecular charge", *Proc. Nat. Acad. Sci.* **99**, 14142–14146 (2002).

[922] B. Crone, A. Dodabalapur, A. Gelperin, *et al.*, "Electronic sensing of vapors with organic transistors", *Appl. Phys. Lett.* **78**, 2229–2231 (2001).

[923] G. Zheng, F. Patolsky, Y. Cui, W. U. Wang, and C. M. Lieber, "Multiplexed electrical detection of cancer markers with nanowire sensor arrays", *Nat. Biotech.* **23**, 1294–1301 (2005).

[924] E. Stern, J. F. Klemic, D. A. Routenberg, *et al.*, "Label-free immunodetection with CMOS-compatible semiconducting nanowires", *Nature* **445**, 519–522 (2007).

[925] A. Star, J.-C. P. Gabriel, K. Bradley, and G. Grüner, "Electronic detection of specific protein binding using nanotube FET devices", *Nano Lett.* **3**, 459–463 (2003).

[926] K. Besteman, J.-O. Lee, F. G. Wiertz, H. A. Heering, and C. Dekker, "Enzyme-coated carbon nanotubes as single-molecule biosensors", *Nano Lett.* **3**, 727–730 (2003).

[927] Y. Choi, I. S. Moody, P. C. Sims, *et al.*, "Single-molecule lysozyme dynamics monitored by an electronic circuit", *Science* **335**, 319–324 (2012).

[928] R. Raiteri, M. Grattarola, H.-J. Butt, and P. Skládal, "Micromechanical cantilever-based biosensors", *Sensors and Actuators B: Chemical* **79**, 115–126 (2001).

[929] J. Fritz, M. Baller, H. Lang, *et al.*, "Translating biomolecular recognition into nanomechanics", *Science* **288**, 316–318 (2000).

[930] T. P. Burg and S. R. Manalis, "Suspended microchannel resonators for biomolecular detection", *Appl. Phys. Lett.* **83**, 2698–2700 (2003).

[931] T. P. Burg, M. Godin, S. M. Knudsen, *et al.*, "Weighing of biomolecules, single cells and single nanoparticles in fluid", *Nature* **446**, 1066–1069 (2007).

[932] D. Goodsell, *The Machinery of Life*, Biomedical and Life Sciences (New York, Copernicus Books, 2009).

[933] A. Kuzuya and M. Komiyama, "DNA origami: fold, stick, and beyond", *Nanoscale* **2**, 309–321 (2010).

[934] C. Sanchez, H. Arribart, and M. M. G. Guille, "Biomimetism and bioinspiration as tools for the design of innovative materials and systems", *Nat. Mater.* **4**, 277–288 (2005).

[935] B. Bhushan, "Biomimetics: lessons from nature – an overview", *Phil. Trans. R. Soc. A: Math., Phys. Eng. Sci.* **367**, 1445–1486 (2009).

[936] F. Banica, *Chemical Sensors and Biosensors: Fundamentals and Applications* (New York, Wiley, 2012).

[937] D. G. Gibson, J. I. Glass, C. Lartigue, *et al.*, "Creation of a bacterial cell controlled by a chemically synthesized genome", *Science* **329**, 52–56 (2010).

[938] J. Cello, A. V. Paul, and E. Wimmer, "Chemical synthesis of poliovirus cDNA: generation of infectious virus in the absence of natural template", *Science* **297**, 1016–1018 (2002).

[939] S. Toyabe, T. Okamoto, T. Watanabe-Nakayama, H. Taketani, S. Kudo, and E. Muneyuki, "Nonequilibrium energetics of a single F_1-ATPase molecule", *Phys. Rev. Lett.* **104**, 198103 (2010).

[940] J. T. Groves, S. G. Boxer, and H. M. McConnell, "Electric field-induced reorganization of two-component supported bilayer membranes", *Proc. Nat. Acad. Sci.* **94**, 13390–13395 (1997).

[941] M. M. Baksh, M. Jaros, and J. T. Groves, "Detection of molecular interactions at membrane surfaces through colloid phase transitions", *Nature* **427**, 139–141 (2004).

[942] "The Onion: nanotechnology infographic", www.theonion.com/articles/nanotechnology,7971/ (2004).

[943] U. E. I. Administration, "International Energy Outlook 2013", Technical Report DOE/EIA-0484(2013), US Department of Energy (2013).

[944] D. Goodstein, *Out of Gas: the End of the Age of Oil*, Norton Paperback (New York, Norton, 2005).

[945] T. F. Stocker, D. Qin, G.-K. Plattner, *et al.*, "Climate change 2013: the physical science basis", Intergovernmental Panel on Climate Change, Working Group I Contribution to the *IPCC Fifth Assessment Report* (AR5)(New York, Cambridge University Press) (2013).

[946] W. Shockley and H. J. Queisser, "Detailed balance limit of efficiency of pn junction solar cells", *J. Appl. Phys.* **32** (1961).

[947] F. Dimroth and S. Kurtz, "High-efficiency multijunction solar cells", *MRS Bull.* **32**, 230–235 (2007).

[948] D.-H. Kim, J.-H. Ahn, W. M. Choi, *et al.*, "Stretchable and foldable silicon integrated circuits", *Science* **320**, 507–511 (2008).

[949] J. Yoon, S. Jo, I. S. Chun, *et al.*, "GaAs photovoltaics and optoelectronics using releasable multilayer epitaxial assemblies", *Nature* **465**, 329–333 (2010).

[950] J. Lee, J. Wu, M. Shi, *et al.*, "Stretchable GaAs photovoltaics with designs that enable high areal coverage", *Adv. Mater.* **23**, 986–991 (2011).

[951] S. Pillai, K. Catchpole, T. Trupke, and M. Green, "Surface plasmon enhanced silicon solar cells", *J. Appl. Phys.* **101**, 093105 (2007).

[952] A. J. Morfa, K. L. Rowlen, T. H. Reilly III, M. J. Romero, and J. van de Lagemaat, "Plasmon-enhanced solar energy conversion in organic bulk heterojunction photovoltaics", *Appl. Phys. Lett.* **92**, 013504 (2008).

[953] W. U. Huynh, X. Peng, and A. P. Alivisatos, "CdSe nanocrystal rods/poly (3-hexylthiophene) composite photovoltaic devices", *Adv. Mater.* **11**, 923–927 (1999).

[954] W. U. Huynh, J. J. Dittmer, and A. P. Alivisatos, "Hybrid nanorod-polymer solar cells", *Science* **295**, 2425–2427 (2002).

[955] A. Nozik, "Quantum dot solar cells", *Physica E: Low-dimensional Systems and Nanostructures* **14**, 115–120 (2002).

[956] R. D. Schaller and V. I. Klimov, "High efficiency carrier multiplication in PbSe nanocrystals: implications for solar energy conversion", *Phys. Rev. Lett.* **92**, 186601 (2004).

[957] R. J. Ellingson, M. C. Beard, J. C. Johnson, *et al.*, "Highly efficient multiple exciton generation in colloidal PbSe and PbS quantum dots", *Nano Lett.* **5**, 865–871 (2005).

[958] R. D. Schaller, M. Sykora, J. M. Pietryga, and V. I. Klimov, "Seven excitons at a cost of one: redefining the limits for conversion efficiency of photons into charge carriers", *Nano Lett.* **6**, 424–429 (2006).

[959] A. J. Nozik, "Nanoscience and nanostructures for photovoltaics and solar fuels", *Nano Lett.* **10**, 2735–2741 (2010).

[960] M. T. Trinh, A. J. Houtepen, J. M. Schins, *et al.*, "In spite of recent doubts carrier multiplication does occur in PbSe nanocrystals", *Nano Lett.* **8**, 1713–1718 (2008).

[961] C. J. Brabec, N. S. Sariciftci, J. C. Hummelen, *et al.*, "Plastic solar cells", *Adv. Func. Mater.* **11**, 15–26 (2001).

[962] H. Spanggaard and F. C. Krebs, "A brief history of the development of organic and polymeric photovoltaics", *Solar Energy Materials and Solar Cells* **83**, 125–146 (2004).

[963] S. Günes, H. Neugebauer, and N. S. Sariciftci, "Conjugated polymer-based organic solar cells", *Chem. Rev.* **107**, 1324–1338 (2007).

[964] B. Kippelen and J.-L. Brédas, "Organic photovoltaics", *Energy & Environmental Science* **2**, 251–261 (2009).

[965] M. Grätzel, "Photoelectrochemical cells", *Nature* **414**, 338–344 (2001).

[966] A. Hagfeldt, G. Boschloo, L. Sun, L. Kloo, and H. Pettersson, "Dye-sensitized solar cells", *Chem. Rev.* **110**, 6595–6663 (2010).

[967] B. Oregan and M. Grfitzeli, "A low-cost, high-efficiency solar cell based on dye-sensitized", *Nature* **353**, 737–740 (1991).

[968] T. J. Meyer, "Chemical approaches to artificial photosynthesis", *Accounts of Chemical Research* **22**, 163–170 (1989).

[969] M. R. Wasielewski, "Photoinduced electron transfer in supramolecular systems for artificial photosynthesis", *Chem. Rev.* **92**, 435–461 (1992).

[970] A. J. Bard and M. A. Fox, "Artificial photosynthesis: solar splitting of water to hydrogen and oxygen", *Accounts of Chemical Research* **28**, 141–145 (1995).

[971] N. S. Lewis and D. G. Nocera, "Powering the planet: Chemical challenges in solar energy utilization", *Proc. Nat. Acad. Sci.* **103**, 15729–15735 (2006).

[972] D. Gust, T. A. Moore, and A. L. Moore, "Solar fuels via artificial photosynthesis", *Accounts of Chemical Research* **42**, 1890–1898 (2009).

[973] C. H. Christensen and J. K. Nørskov, "A molecular view of heterogeneous catalysis", *J. Chem. Phys.* **128**, 182503 (2008).

[974] R. A. Van Santen and M. Neurock, *Molecular Heterogeneous Catalysis: a Conceptual and Computational Approach* (New York, John Wiley & Sons, 2009).

[975] A. T. Bell, "The impact of nanoscience on heterogeneous catalysis", *Science* **299**, 1688–1691 (2003).

[976] X. Chen and S. S. Mao, "Titanium dioxide nanomaterials: synthesis, properties, modifications, and applications", *Chem. Rev.* **107**, 2891–2959 (2007).

[977] S. R. Anton and H. A. Sodano, "A review of power harvesting using piezoelectric materials (2003–2006)", *Smart Materials and Structures* **16**, R1 (2007).

[978] A. I. Boukai, Y. Bunimovich, J. Tahir-Kheli, J.-K. Yu, W. A. Goddard III, and J. R. Heath, "Silicon nanowires as efficient thermoelectric materials", *Nature* **451**, 168–171 (2008).

[979] A. I. Hochbaum, R. Chen, R. D. Delgado, *et al.*, "Enhanced thermoelectric performance of rough silicon nanowires", *Nature* **451**, 163–167 (2008).

[980] J. Nicholls, "Advances in coating design for high-performance gas turbines", *MRS Bulletin* **28**, 659–670 (2003).

[981] P. A. Sørensen, S. Kiil, K. Dam-Johansen, and C. Weinell, "Anticorrosive coatings: a review", *J. Coatings Technology and Research* **6**, 135–176 (2009).

[982] I. Beyerlein, A. Caro, M. Demkowicz, N. Mara, A. Misra, and B. Uberuaga, "Radiation damage tolerant nanomaterials", *Materials Today* **16**, 443–449 (2013).

[983] A. S. Aricò, P. Bruce, B. Scrosati, J.-M. Tarascon, and W. Van Schalkwijk, "Nanostructured materials for advanced energy conversion and storage devices", *Nature Materials* **4**, 366–377 (2005).

[984] Y.-G. Guo, J.-S. Hu, and L.-J. Wan, "Nanostructured materials for electrochemical energy conversion and storage devices", *Advanced Materials* **20**, 2878–2887 (2008).

[985] B. Kang and G. Ceder, "Battery materials for ultrafast charging and discharging", *Nature* **458**, 190–193 (2009).

[986] P. Simon and Y. Gogotsi, "Materials for electrochemical capacitors", *Nat. Mater.* **7**, 845–854 (2008).

[987] C. K. Chan, H. Peng, G. Liu, *et al.*, "High-performance lithium battery anodes using silicon nanowires", *Nat. Nanotech.* **3**, 31–35 (2008).

[988] M.-H. Park, M. G. Kim, J. Joo, *et al.*, "Silicon nanotube battery anodes", *Nano Lett.* **9**, 3844–3847 (2009).

[989] M. Thakur, M. Isaacson, S. L. Sinsabaugh, M. S. Wong, and S. L. Biswal, "Gold-coated porous silicon films as anodes for lithium ion batteries", *J. Power Sources* **205**, 426–432 (2012).

[990] H. Zhang, X. Yu, and P. V. Braun, "Three-dimensional bicontinuous ultrafast-charge and-discharge bulk battery electrodes", *Nature Nanotech.* **6**, 277–281 (2011).

[991] J. Chmiola, G. Yushin, Y. Gogotsi, C. Portet, P. Simon, and P.-L. Taberna, "Anomalous increase in carbon capacitance at pore sizes less than 1 nanometer", *Science* **313**, 1760–1763 (2006).

[992] J. Lin, C. Zhang, Z. Yan, *et al.*, "3-dimensional graphene carbon nanotube carpet-based microsupercapacitors with high electrochemical performance", *Nano Lett.* **13**, 72–78 (2012).

[993] A. Nel, T. Xia, L. Mädler, and N. Li, "Toxic potential of materials at the nanolevel", *Science* **311**, 622–627 (2006).

[994] R. E. Smalley, "Of chemistry, love, and nanobots", *Scientific American* **285**, 76–77 (2001).

[995] K. E. Drexler, "Drexler and Smalley make the case for and against molecular assemblers", *Chemical & Engineering News* **81**, 1 (2003).

[996] F. Cocks, *Energy Demand and Climate Change: Issues and Resolutions* (New York, Wiley, 2011).

[997] R. Stein and J. Powers, *The Energy Problem* (Hackensack, NJ, World Scientific, 2011).

[998] J. García-Martínez, ed., *Nanotechnology for the Energy Challenge*, 2nd edn. (New York, Wiley, 2013).

[999] J. B. Goodenough and Y. Kim, "Challenges for rechargeable Li batteries", *Chem. Mater.* **22**, 587–603 (2009).

[1000] G. Oberdörster, E. Oberdörster, and J. Oberdörster, "Nanotoxicology: an emerging discipline evolving from studies of ultrafine particles", *Env. Health Persp.* **113**, 823–839 (2005).

[1001] H. F. Krug and P. Wick, "Nanotoxicology: an interdisciplinary challenge", *Ange. Chemie Int. Edn.* **50**, 1260–1278 (2011).

[1002] S. J. Klaine, P. J. Alvarez, G. E. Batley, *et al.*, "Nanomaterials in the environment: behavior, fate, bioavailability, and effects", *Env. Tox. Chem.* **27**, 1825–1851 (2008).

[1003] K. Drexler, *Engines of Creation* (New York, Anchor, 1987).

[1004] N. Stephenson, *The Diamond Age*, A Bantam spectra book (New York, Random House Publishing Group, 2003).

[1005] D. Griffiths, *Introduction to Electrodynamics*, Pearson International Edition (Upper Saddle River, NJ, Pearson Education, Limited, 2012).

[1006] L. D. Landau and E. M. Lifshitz, *Quantum Mechanics: Non-Relativistic Theory*, 3rd. edn. (New York, Elsevier, 2013).

[1007] C. Cohen–Tannoudji, B. Diu, and F. Laloë, *Quantum Mechanics*, Vol. 1 (New York, Wiley, 1991).

[1008] J. J. Sakurai and J. J. Napolitano, *Modern Quantum Mechanics*, 2nd edn. (New York, Addison - Wesley, 2010).

Index